# STUDENT'S SOLUTIONS MANUAL

### DAVID ATWOOD
*Rochester Community and Technical College*

# BEGINNING AND INTERMEDIATE ALGEBRA WITH APPLICATIONS AND VISUALIZATION
## THIRD EDITION

## Gary Rockswold
*Minnesota State University, Mankato*

## Terry Krieger
*Rochester Community and Technical College*

**PEARSON**

Boston   Columbus   Indianapolis   New York   San Francisco   Upper Saddle River
Amsterdam   Cape Town   Dubai   London   Madrid   Milan   Munich   Paris   Montreal   Toronto
Delhi   Mexico City   Sao Paulo   Sydney   Hong Kong   Seoul   Singapore   Taipei   Tokyo

The author and publisher of this book have used their best efforts in preparing this book. These efforts include the development, research, and testing of the theories and programs to determine their effectiveness. The author and publisher make no warranty of any kind, expressed or implied, with regard to these programs or the documentation contained in this book. The author and publisher shall not be liable in any event for incidental or consequential damages in connection with, or arising out of, the furnishing, performance, or use of these programs.

Reproduced by Pearson from electronic files supplied by the author.

ISBN-13: 978-0-321-75655-8
ISBN-10: 0-321-75655-X

www.pearsonhighered.com

# Contents

Chapter 1      1

Chapter 2      35

Chapter 3      71

Chapter 4      127

Chapter 5      163

Chapter 6      199

Chapter 7      233

Chapter 8      283

Chapter 9      327

Chapter 10      353

Chapter 11      391

Chapter 12      441

Chapter 13      479

Chapter 14      509

# Chapter 1: Introduction to Algebra

## Section 1.1: Numbers, Variables, and Expressions

1. natural

3. 1

5. factors

7. variable

9. sum

11. quotient

13. The number 4 is a composite number because it has factors other than itself and 1; $4 = 2 \cdot 2$.

15. The number 1 is neither a prime nor a composite number.

17. The number 29 is a prime number because its only factors are itself and 1.

19. The number 92 is a composite number because it has factors other than itself and 1; $92 = 2 \cdot 2 \cdot 23$.

21. The number 225 is a composite number because it has factors other than itself and 1; $225 = 3 \cdot 3 \cdot 5 \cdot 5$.

23. The number 149 is a prime number because its only factors are itself and 1.

25. $6 = 2 \cdot 3$

27. $12 = 2 \cdot 2 \cdot 3$

29. $32 = 2 \cdot 2 \cdot 2 \cdot 2 \cdot 2$

31. $39 = 3 \cdot 13$

33. $294 = 2 \cdot 3 \cdot 7 \cdot 7$

35. $300 = 2 \cdot 2 \cdot 3 \cdot 5 \cdot 5$

37. Yes, the population of a country could be described by the whole numbers because we cannot have a fraction of a person.

39. No, a student's grade point average could not be described by the whole numbers because a grade point average usually contains a decimal point.

41. Yes, the number of apps stored on an iPad could be described by the whole numbers because the number of apps does not contain a fraction or a decimal point.

43. No, the winning time in a 100-meter sprint, could not be described by the whole numbers because the winning time would be expressed in the number of seconds plus a fraction of an seconds.

45. The value of the expression $3x$, when $x = 5$, is $3x = 3(5) = 15$.

47. The value of the expression $9 - x$, when $x = 4$, is $9 - x = 9 - 4 = 5$.

49. The value of the expression $\dfrac{x}{8}$, when $x = 32$, is $\dfrac{x}{8} = \dfrac{32}{8} = 4$.

51. The value of the expression $3(x+1)$, when $x = 5$, is $3(x+1) = 3(5+1) = (3)(6) = 18$.

53. The value of the expression $\frac{x}{2}+1$, when $x=6$, is $\frac{x}{2}+1=\frac{6}{2}+1=3+1=4$.

55. When $x=8$ and $y=14$, $x+y=8+14=22$.

57. When $x=8$ and $y=4$, $6\cdot\frac{x}{y}=6\cdot\frac{8}{4}=6\cdot2=12$.

59. When $x=5$ and $y=3$, $y(x-2)=3(5-2)=(3)(3)=9$.

61. When $x=0$, $y=x+5=0+5=5$.

63. When $x=7$, $y=4x=4\cdot7=28$.

65. When $z=12$, $F=z-5=12-5=7$.

67. When $z=6$, $F=\frac{30}{z}=\frac{30}{6}=5$.

69. When $x=2$ and $z=0$, $y=3xz=3\cdot2\cdot0=0$.

71. When $x=9$ and $z=3$, $y=\frac{x}{z}=\frac{9}{3}=3$.

73. If $x$ is the number, $x+5$ is five more than the number.

75. If $s$ is the cost of a soda, then three times this cost is $3s$.

77. If $n$ is the number, $n+5$ is the sum of the number and 5.

79. If $p$ is the population of a town, $p-200$ is two hundred less than the population.

81. If $z$ is the number, $\frac{z}{6}$ is the number divided by 6.

83. If $s$ is the speed and $t$ is the time, then $st$ is the product of the speed and the time.

85. If $x$ is one number and y is another number, $\frac{x+7}{y}$ is a number plus 7, all divided by another number.

87. $P=100D$ because there are one hundred pennies in one dollar.

89. See Figure 89.  $F=3y$ because there are three feet in one yard.

| Yards (y) | 1 | 2 | 3 | 4 | 5 | 6 | 7 |
|---|---|---|---|---|---|---|---|
| Feet (F) | 3 | 6 | 9 | 12 | 15 | 18 | 21 |

Figure 89

91. $M=3x$; $M$ is miles and $x$ is minutes. If $x=36$, then $M=3x=3(36)=108$ mi.

93. $B=6D$; for each drog there are 6 blims so there are 6 times as many blims as there are drogs.

95. Each album costs \$12. Thus, $C=12x$ where $C$ is cost and $x$ is number of albums.

97. Since the area of a rectangle equals its length times its width, $22\text{ ft}\times9\text{ ft}=198$ square feet.

## Section 1.2 Fractions

1. The person ate $\dfrac{3}{4}$ of the pie. $\dfrac{1}{4}$ of the pie remains.

3. The variable $b$ cannot equal 0.

5. True

7. $\dfrac{ac}{bc} = \dfrac{a}{b}$

9. The reciprocal of $a$ with $a \neq 0$ is $\dfrac{1}{a}$.

11. $\dfrac{a}{b} \cdot \dfrac{c}{d} = \dfrac{ac}{bd}$

13. $\dfrac{a}{b} + \dfrac{c}{b} = \dfrac{a+c}{b}$

15. The largest number that divides evenly into 4 and 12 is 4, so the GCF is 4.

17. The largest number that divides evenly into 50 and 75 is 25, so the GCF is 25.

19. The largest number that divides evenly into 100, 60 and 70 is 10, so the GCF is 10.

21. $\dfrac{3 \cdot 4}{5 \cdot 4} = \dfrac{3}{5} \cdot \dfrac{4}{4} = \dfrac{3}{5} \cdot 1 = \dfrac{3}{5}$

23. $\dfrac{3 \cdot 8}{8 \cdot 5} = \dfrac{3 \cdot 8}{5 \cdot 8} = \dfrac{3}{5} \cdot \dfrac{8}{8} = \dfrac{3}{5} \cdot 1 = \dfrac{3}{5}$

25. $\dfrac{4}{8} = \dfrac{1 \cdot 4}{2 \cdot 4} = \dfrac{1}{2} \cdot \dfrac{4}{4} = \dfrac{1}{2} \cdot 1 = \dfrac{1}{2}$

27. $\dfrac{10}{25} = \dfrac{2 \cdot 5}{5 \cdot 5} = \dfrac{2}{5} \cdot \dfrac{5}{5} = \dfrac{2}{5} \cdot 1 = \dfrac{2}{5}$

29. $\dfrac{12}{36} = \dfrac{1 \cdot 12}{3 \cdot 12} = \dfrac{1}{3} \cdot \dfrac{12}{12} = \dfrac{1}{3} \cdot 1 = \dfrac{1}{3}$

31. $\dfrac{12}{30} = \dfrac{2 \cdot 6}{5 \cdot 6} = \dfrac{2}{5} \cdot \dfrac{6}{6} = \dfrac{2}{5} \cdot 1 = \dfrac{2}{5}$

33. $\dfrac{19}{76} = \dfrac{1 \cdot 19}{4 \cdot 19} = \dfrac{1}{4} \cdot \dfrac{19}{19} = \dfrac{1}{4} \cdot 1 = \dfrac{1}{4}$

35. $\dfrac{3}{4} \cdot \dfrac{1}{5} = \dfrac{3 \cdot 1}{4 \cdot 5} = \dfrac{3}{20}$

37. $\dfrac{5}{3} \cdot \dfrac{3}{5} = \dfrac{5 \cdot 3}{3 \cdot 5} = \dfrac{15}{15} = 1$

39. $\dfrac{5}{6} \cdot \dfrac{18}{25} = \dfrac{5 \cdot 18}{6 \cdot 25} = \dfrac{90}{150} = \dfrac{3 \cdot 30}{5 \cdot 30} = \dfrac{3}{5} \cdot \dfrac{30}{30} = \dfrac{3}{5} \cdot 1 = \dfrac{3}{5}$

41. $\dfrac{4}{1} \cdot \dfrac{3}{5} = \dfrac{4 \cdot 3}{1 \cdot 5} = \dfrac{12}{5}$

43. $\dfrac{2}{1} \cdot \dfrac{3}{8} = \dfrac{2 \cdot 3}{1 \cdot 8} = \dfrac{6}{8} = \dfrac{3 \cdot 2}{4 \cdot 2} = \dfrac{3}{4} \cdot \dfrac{2}{2} = \dfrac{3}{4} \cdot 1 = \dfrac{3}{4}$

45. $\dfrac{x}{y} \cdot \dfrac{y}{x} = \dfrac{xy}{yx} = \dfrac{xy}{xy} = \dfrac{x}{x} \cdot \dfrac{y}{y} = \dfrac{x}{x} \cdot 1 = \dfrac{x}{x} = 1$

47. $\dfrac{a}{b} \cdot \dfrac{3}{2} = \dfrac{a \cdot 3}{b \cdot 2} = \dfrac{3a}{2b}$

49. $\dfrac{1}{4} \cdot \dfrac{3}{4} = \dfrac{1 \cdot 3}{4 \cdot 4} = \dfrac{3}{16}$

51. $\dfrac{2}{3} \cdot \dfrac{6}{1} = \dfrac{2 \cdot 6}{3 \cdot 1} = \dfrac{12}{3} = \dfrac{4 \cdot 3}{1 \cdot 3} = \dfrac{4}{1} \cdot \dfrac{3}{3} = \dfrac{4}{1} \cdot 1 = 4$

53. $\dfrac{1}{2} \cdot \dfrac{2}{3} = \dfrac{1 \cdot 2}{2 \cdot 3} = \dfrac{2}{6} = \dfrac{1 \cdot 2}{3 \cdot 2} = \dfrac{1}{3} \cdot \dfrac{2}{2} = \dfrac{1}{3} \cdot 1 = \dfrac{1}{3}$

55. (a) $\dfrac{1}{5}$       (b) $\dfrac{1}{7}$       (c) $\dfrac{7}{4}$       (d) $\dfrac{8}{9}$

57. (a) $\dfrac{2}{1} = 2$   (b) $\dfrac{9}{1} = 9$   (c) $\dfrac{101}{12}$   (d) $\dfrac{17}{31}$

59. $\dfrac{1}{2} \div \dfrac{1}{3} = \dfrac{1}{2} \cdot \dfrac{3}{1} = \dfrac{1 \cdot 3}{2 \cdot 1} = \dfrac{3}{2}$

61. $\dfrac{3}{4} \div \dfrac{1}{8} = \dfrac{3}{4} \cdot \dfrac{8}{1} = \dfrac{3 \cdot 8}{4 \cdot 1} = \dfrac{24}{4} = 6$

63. $\dfrac{4}{3} \div \dfrac{4}{3} = \dfrac{4}{3} \cdot \dfrac{3}{4} = \dfrac{4 \cdot 3}{3 \cdot 4} = \dfrac{12}{12} = 1$

65. $\dfrac{32}{27} \div \dfrac{8}{9} = \dfrac{32}{27} \cdot \dfrac{9}{8} = \dfrac{32 \cdot 9}{27 \cdot 8} = \dfrac{288}{216} = \dfrac{4 \cdot 72}{3 \cdot 72} = \dfrac{4}{3} \cdot \dfrac{72}{72} = \dfrac{4}{3} \cdot 1 = \dfrac{4}{3}$

67. $\dfrac{10}{1} \div \dfrac{5}{6} = \dfrac{10}{1} \cdot \dfrac{6}{5} = \dfrac{10 \cdot 6}{1 \cdot 5} = \dfrac{60}{5} = \dfrac{12 \cdot 5}{1 \cdot 5} = \dfrac{12}{1} \cdot \dfrac{5}{5} = \dfrac{12}{1} \cdot 1 = 12$

69. $\dfrac{9}{10} \div \dfrac{3}{1} = \dfrac{9}{10} \cdot \dfrac{1}{3} = \dfrac{9 \cdot 1}{10 \cdot 3} = \dfrac{9}{30} = \dfrac{3 \cdot 3}{10 \cdot 3} = \dfrac{3}{10} \cdot \dfrac{3}{3} = \dfrac{3}{10} \cdot 1 = \dfrac{3}{10}$

71. $\dfrac{a}{b} \div \dfrac{2}{b} = \dfrac{a}{b} \cdot \dfrac{b}{2} = \dfrac{ab}{2b} = \dfrac{a}{2} \cdot \dfrac{b}{b} = \dfrac{a}{2} \cdot 1 = \dfrac{a}{2}$

73. $\dfrac{x}{y} \div \dfrac{x}{y} = \dfrac{x}{y} \cdot \dfrac{y}{x} = \dfrac{xy}{xy} = 1$

75. (a) $\dfrac{5}{12} + \dfrac{1}{12} = \dfrac{5+1}{12} = \dfrac{6}{12} = \dfrac{1 \cdot 6}{2 \cdot 6} = \dfrac{1}{2} \cdot \dfrac{6}{6} = \dfrac{1}{2} \cdot 1 = \dfrac{1}{2}$

(b) $\dfrac{5}{12}-\dfrac{1}{12}=\dfrac{5-1}{12}=\dfrac{4}{12}=\dfrac{1\cdot 4}{3\cdot 4}=\dfrac{1}{3}\cdot\dfrac{4}{4}=\dfrac{1}{3}\cdot 1=\dfrac{1}{3}$

77. (a) $\dfrac{18}{29}+\dfrac{7}{29}=\dfrac{18+7}{29}=\dfrac{25}{29}$

(b) $\dfrac{18}{29}-\dfrac{7}{29}=\dfrac{18-7}{29}=\dfrac{11}{29}$

79. Prime factorizations are $9=3\cdot 3$ and $15=3\cdot 5$. The LCD is $3\cdot 3\cdot 5=45$.

81. Prime factorizations are $5$ and $15=5\cdot 3$. The LCD is $5\cdot 3=15$.

83. Prime factorizations are $6=2\cdot 3$ and $8=2\cdot 2\cdot 2$. The LCD is $2\cdot 2\cdot 2\cdot 3=24$.

85. Prime factorizations are $2$, $3$ and $4=2\cdot 2$. The LCD is $2\cdot 2\cdot 3=12$.

87. Prime factorizations are $4=2\cdot 2$, $8=2\cdot 2\cdot 2$ and $12=2\cdot 2\cdot 3$. The LCD is $2\cdot 2\cdot 2\cdot 3=24$.

89. The LCD is 6. $\dfrac{1}{2}=\dfrac{1}{2}\cdot\dfrac{3}{3}=\dfrac{3}{6};\dfrac{2}{3}=\dfrac{2}{3}\cdot\dfrac{2}{2}=\dfrac{4}{6}$

91. The LCD is 36. $\dfrac{7}{9}=\dfrac{7}{9}\cdot\dfrac{4}{4}=\dfrac{28}{36};\dfrac{5}{12}=\dfrac{5}{12}\cdot\dfrac{3}{3}=\dfrac{15}{36}$

93. The LCD is 48. $\dfrac{1}{16}=\dfrac{1}{16}\cdot\dfrac{3}{3}=\dfrac{3}{48};\dfrac{7}{12}=\dfrac{7}{12}\cdot\dfrac{4}{4}=\dfrac{28}{48}$

95. The LCD is 12. $\dfrac{1}{3}=\dfrac{1}{3}\cdot\dfrac{4}{4}=\dfrac{4}{12};\dfrac{3}{4}=\dfrac{3}{4}\cdot\dfrac{3}{3}=\dfrac{9}{12};\dfrac{5}{6}=\dfrac{5}{6}\cdot\dfrac{2}{2}=\dfrac{10}{12}$

97. $\dfrac{5}{8}+\dfrac{3}{16}=\dfrac{5}{8}\cdot\dfrac{2}{2}+\dfrac{3}{16}=\dfrac{5\cdot 2}{8\cdot 2}+\dfrac{3}{16}=\dfrac{10}{16}+\dfrac{3}{16}=\dfrac{10+3}{16}=\dfrac{13}{16}$

99. $\dfrac{25}{24}-\dfrac{7}{8}=\dfrac{25}{24}-\dfrac{7}{8}\cdot\dfrac{3}{3}=\dfrac{25}{24}-\dfrac{7\cdot 3}{8\cdot 3}=\dfrac{25}{24}-\dfrac{21}{24}=\dfrac{4}{24}=\dfrac{1\cdot 4}{6\cdot 4}=\dfrac{1}{6}\cdot\dfrac{4}{4}=\dfrac{1}{6}\cdot 1=\dfrac{1}{6}$

101. $\dfrac{11}{14}+\dfrac{2}{35}=\dfrac{11}{14}\cdot\dfrac{5}{5}+\dfrac{2}{35}\cdot\dfrac{2}{2}=\dfrac{11\cdot 5}{14\cdot 5}+\dfrac{2\cdot 2}{35\cdot 2}=\dfrac{55}{70}+\dfrac{4}{70}=\dfrac{55+4}{70}=\dfrac{59}{70}$

103. $\dfrac{5}{12}-\dfrac{1}{18}=\dfrac{5}{12}\cdot\dfrac{3}{3}-\dfrac{1}{18}\cdot\dfrac{2}{2}=\dfrac{5\cdot 3}{12\cdot 3}-\dfrac{1\cdot 2}{18\cdot 2}=\dfrac{15}{36}-\dfrac{2}{36}=\dfrac{15-2}{36}=\dfrac{13}{36}$

105. $\dfrac{3}{100}+\dfrac{1}{300}-\dfrac{1}{200}=\dfrac{3}{100}\cdot\dfrac{6}{6}+\dfrac{1}{300}\cdot\dfrac{2}{2}-\dfrac{1}{200}\cdot\dfrac{3}{3}=\dfrac{3\cdot 6}{100\cdot 6}+\dfrac{1\cdot 2}{300\cdot 2}-\dfrac{1\cdot 3}{200\cdot 3}=$

$\dfrac{18}{600}+\dfrac{2}{600}-\dfrac{3}{600}=\dfrac{18+2-3}{600}=\dfrac{17}{600}$

107. $\dfrac{7}{8}-\dfrac{1}{6}+\dfrac{5}{12}=\dfrac{7}{8}\cdot\dfrac{3}{3}-\dfrac{1}{6}\cdot\dfrac{4}{4}+\dfrac{5}{12}\cdot\dfrac{2}{2}=\dfrac{7\cdot 3}{8\cdot 3}-\dfrac{1\cdot 4}{6\cdot 4}+\dfrac{5\cdot 2}{12\cdot 2}=$

$\dfrac{21}{24}-\dfrac{4}{24}+\dfrac{10}{24}=\dfrac{21-4+10}{24}=\dfrac{27}{24}=\dfrac{9\cdot 3}{8\cdot 3}=\dfrac{9}{8}\cdot\dfrac{3}{3}=\dfrac{9}{8}\cdot 1=\dfrac{9}{8}$

109. Find the product of $2\frac{1}{2}$ and $1\frac{9}{10}$. First convert $2\frac{1}{2}$ to $\frac{5}{2}$ and $1\frac{9}{10}$ to $\frac{19}{10}$.

$$\frac{5}{2}\cdot\frac{19}{10}=\frac{95}{20}=\frac{19\cdot5}{4\cdot5}=\frac{19}{4}=4\frac{3}{4}\text{ ft.}$$

111. Find the value of one-half of $64\frac{5}{8}$. First convert $64\frac{5}{8}$ to $\frac{517}{8}$.  $\frac{517}{8}\div2=\frac{517}{8}\cdot\frac{1}{2}=\frac{517}{16}=32\frac{5}{16}$ in.

113. Convert the base of the triangle from $1\frac{2}{3}$ to $\frac{5}{3}$.

$$\frac{1}{2}\cdot\frac{5}{3}\cdot\frac{3}{4}=\frac{1\cdot5\cdot3}{2\cdot3\cdot4}=\frac{15}{24}=\frac{5\cdot3}{8\cdot3}=\frac{5}{8}\cdot\frac{3}{3}=\frac{5}{8}\cdot1=\frac{5}{8}\text{ square yards.}$$

115. Add the distance from Smalltown to Middletown and Middletown to Bigtown.

$$3\frac{1}{2}+4\frac{3}{4}=\frac{7}{2}+\frac{19}{4}=\frac{7}{2}\cdot\frac{2}{2}+\frac{19}{4}=\frac{7\cdot2}{2\cdot2}+\frac{19}{4}=\frac{14}{4}+\frac{19}{4}=\frac{14+19}{4}=\frac{33}{4}=8\frac{1}{4}\text{ miles.}$$

117. Multiply the fraction of U.S. vegetarian adults by the fraction of U.S. vegan adults.

$$\frac{4}{125}\cdot\frac{1}{200}=\frac{1\cdot4}{4\cdot50\cdot125}=\frac{1}{6250}$$

119. Find the sum of the motor vehicle deaths and the firearms deaths as a fraction of all accidental deaths. $\frac{31}{42}+\frac{31}{1260}=\frac{31}{42}\cdot\frac{30}{30}+\frac{31}{1260}$.  It follows that

$$\frac{31\cdot30}{42\cdot30}+\frac{31}{1260}=\frac{930}{1260}+\frac{31}{1260}=\frac{930+31}{1260}=\frac{961}{1260}.$$

## Checking Basic Concepts Sections 1.1 and 1.2

1. (a)  Prime

   (b)  Composite;  $28=2\cdot2\cdot7$

   (c)  Neither

   (d)  Composite;  $180=2\cdot2\cdot3\cdot3\cdot5$

2. $\dfrac{10}{3+2}=\dfrac{10}{5}=2$

3. $y=6\cdot5=30$

4. $x+5$

5. $I=12F$,  because there are 12 inches in 1 foot.

6. (a)  The largest number that divides evenly into 3 and 18 is 3, so the GCF is 3.

   (b)  The largest number that divides evenly into 40 and 72 is 8, so the GCF is 8.

7. (a)  $\dfrac{25}{35}=\dfrac{5\cdot5}{7\cdot5}=\dfrac{5}{7}\cdot\dfrac{5}{5}=\dfrac{5}{7}\cdot1=\dfrac{5}{7}$

(b) $\dfrac{26}{39} = \dfrac{2 \cdot 13}{3 \cdot 13} = \dfrac{2}{3} \cdot \dfrac{13}{13} = \dfrac{2}{3} \cdot 1 = \dfrac{2}{3}$

8. $\dfrac{3}{4}$

9. (a) $\dfrac{2}{3} \cdot \dfrac{3}{4} = \dfrac{2 \cdot 3}{3 \cdot 4} = \dfrac{6}{12} = \dfrac{1 \cdot 6}{2 \cdot 6} = \dfrac{1}{2} \cdot \dfrac{6}{6} = \dfrac{1}{2} \cdot 1 = \dfrac{1}{2}$

(b) $\dfrac{5}{6} \div \dfrac{10}{3} = \dfrac{5}{6} \cdot \dfrac{3}{10} = \dfrac{5 \cdot 3}{6 \cdot 10} = \dfrac{15}{60} = \dfrac{1 \cdot 15}{4 \cdot 15} = \dfrac{1}{4} \cdot \dfrac{15}{15} = \dfrac{1}{4} \cdot 1 = \dfrac{1}{4}$

(c) $\dfrac{3}{10} + \dfrac{1}{10} = \dfrac{3+1}{10} = \dfrac{4}{10} = \dfrac{2 \cdot 2}{5 \cdot 2} = \dfrac{2}{5} \cdot \dfrac{2}{2} = \dfrac{2}{5} \cdot 1 = \dfrac{2}{5}$

(d) $\dfrac{3}{4} - \dfrac{1}{6} = \dfrac{3}{4} \cdot \dfrac{3}{3} - \dfrac{1}{6} \cdot \dfrac{2}{2} = \dfrac{3 \cdot 3}{4 \cdot 3} - \dfrac{1 \cdot 2}{6 \cdot 2} = \dfrac{9}{12} - \dfrac{2}{12} = \dfrac{9-2}{12} = \dfrac{7}{12}$

10. Multiply $1\dfrac{2}{3}$ by 2. $\dfrac{5}{3} \cdot \dfrac{2}{1} = \dfrac{5 \cdot 2}{3 \cdot 1} = \dfrac{10}{3} = 3\dfrac{1}{3}$ cups.

## Section 1.3 Exponents and Order of Operations

1. multiply

3. base; exponent

5. $8^3$

7. 2; exponents; subtraction

9. False

11. $3^4$

13. $2^5$

15. $\left(\dfrac{1}{2}\right)^4$

17. $a^5$

19. $(x+3)^2$

21. (a) $2^4 = 2 \cdot 2 \cdot 2 \cdot 2 = 16$

(b) $4^2 = 4 \cdot 4 = 16$

23. (a) $6^1 = 6$

(b) $1^6 = 1$

25. (a) $2^5 = 2 \cdot 2 \cdot 2 \cdot 2 \cdot 2 = 32$

(b) $10^3 = 10 \cdot 10 \cdot 10 = 1000$

27. (a) $\left(\dfrac{2}{3}\right)^2 = \dfrac{2}{3} \cdot \dfrac{2}{3} = \dfrac{2 \cdot 2}{3 \cdot 3} = \dfrac{4}{9}$

(b) $\left(\dfrac{1}{2}\right)^5 = \dfrac{1}{2} \cdot \dfrac{1}{2} \cdot \dfrac{1}{2} \cdot \dfrac{1}{2} \cdot \dfrac{1}{2} = \dfrac{1 \cdot 1 \cdot 1 \cdot 1 \cdot 1}{2 \cdot 2 \cdot 2 \cdot 2 \cdot 2} = \dfrac{1}{32}$

29. $8 = 2 \cdot 2 \cdot 2 = 2^3$

31. $25 = 5 \cdot 5 = 5^2$

33. $49 = 7 \cdot 7 = 7^2$

35. $1000 = 10 \cdot 10 \cdot 10 = 10^3$

37. $\dfrac{1}{16} = \dfrac{1}{2} \cdot \dfrac{1}{2} \cdot \dfrac{1}{2} \cdot \dfrac{1}{2} = \left(\dfrac{1}{2}\right)^4$

39. $\dfrac{32}{243} = \dfrac{2}{3} \cdot \dfrac{2}{3} \cdot \dfrac{2}{3} \cdot \dfrac{2}{3} \cdot \dfrac{2}{3} = \left(\dfrac{2}{3}\right)^5$

41. Perform multiplication before addition: $5 + 4 \cdot 6 = 5 + 24 = 29$

43. Perform division before addition: $6 \div 3 + 2 = 2 + 2 = 4$

45. Perform division before subtraction: $100 - \dfrac{50}{5} = 100 - 10 = 90$

47. $10 - 6 - 1 = 3$

49. $20 \div 5 \div 2 = 4 \div 2 = 2$

51. $3 + 2^4 = 3 + 2 \cdot 2 \cdot 2 \cdot 2 = 3 + 16 = 19$

53. $4 \cdot 2^3 = 4 \cdot 2 \cdot 2 \cdot 2 = 4 \cdot 8 = 32$

55. $(3 + 2)^3 = 5^3 = 5 \cdot 5 \cdot 5 = 125$

57. $\dfrac{4 + 8}{1 + 3} = \dfrac{12}{4} = \dfrac{3 \cdot 4}{1 \cdot 4} = \dfrac{3}{1} \cdot \dfrac{4}{4} = \dfrac{3}{1} \cdot 1 = 3$

59. $\dfrac{2^3}{4 - 2} = \dfrac{2 \cdot 2 \cdot 2}{2} = \dfrac{8}{2} = 4$

61. $10^2 - (30 - 2 \cdot 5) = 10^2 - (30 - 10) = 10^2 - 20 = 10 \cdot 10 - 20 = 100 - 20 = 80$

63. $\left(\dfrac{1}{2}\right)^4 + \dfrac{5 + 4}{3} = \dfrac{1}{2} \cdot \dfrac{1}{2} \cdot \dfrac{1}{2} \cdot \dfrac{1}{2} + \dfrac{9}{3} = \dfrac{1}{16} + 3 = \dfrac{1}{16} + \dfrac{3}{1} \cdot \dfrac{16}{16} = \dfrac{1}{16} + \dfrac{48}{16} = \dfrac{1 + 48}{16} = \dfrac{49}{16}$

65. $2^3 - 8 = 2 \cdot 2 \cdot 2 - 8 = 8 - 8 = 0$

67. $30 - 4 \cdot 3 = 30 - 12 = 18$

69. $\dfrac{4^2}{2^3} = \dfrac{4 \cdot 4}{2 \cdot 2 \cdot 2} = \dfrac{16}{8} = \dfrac{2 \cdot 8}{1 \cdot 8} = \dfrac{2}{1} \cdot \dfrac{8}{8} = \dfrac{2}{1} \cdot 1 = 2$

71. $\dfrac{40}{10}+2=\dfrac{4\cdot 10}{1\cdot 10}+2=\dfrac{4}{1}\cdot\dfrac{10}{10}+2=\dfrac{4}{1}\cdot 1+2=4+2=6$

73. $100(2+3)=100\cdot 5=500$

75. $512\text{ MB}=512\cdot 2^{20}\text{ bytes}=536,870,912\text{ bytes}$

77. (a)   Because $2^7=128,\ k=7.$

(b)   $\dfrac{128}{100}=\dfrac{32\cdot 4}{25\cdot 4}=\dfrac{32}{25}.$   There were 32 males for every 25 females.

79. (a)   $72\div 9=8$ years

(b)   The investment doubles every $72\div 12=6$ years, so in 18 years the amount will double 3 times.

The investment will increase by $2^3=8$ times $10,000 and will equal $80,000.

## Section 1.4  Real Numbers and the Number Line

1. $-b$

3. rational

5. irrational

7. True

9. 1; 4

11. principal

13. approximately equal

15. left

17. (a)   The opposite of 9 is $-9$.

(b)   The opposite of $-9$ is $-(-9)=9.$

19. (a)   The opposite of $\dfrac{2}{3}$ is $-\dfrac{2}{3}.$

(b)   The opposite of $-\dfrac{2}{3}$ is $-\left(-\dfrac{2}{3}\right)=\dfrac{2}{3}.$

21. (a)   $-(-8)=8,$ so the opposite of 8 is $-8.$

(b)   $-(-(-8))=-8,$ so the opposite of $-8$ is $-(-8)=8.$

23. (a)   The opposite of $a$ is $-a.$

(b)   The opposite of $-a$ is $-(-a)=a.$

25. The additive inverse of $t$ is $-t.-t=6$

27. The additive inverse of $-b$ is $-(-b)=b.\,b=\dfrac{1}{2}$

29. $\dfrac{1}{4} = 0.25$

31. $\dfrac{7}{8} = 0.875$

33. $\dfrac{3}{2} = 1.5$

35. $\dfrac{1}{20} = 0.05$

37. $\dfrac{2}{3} = 0.\overline{6}$

39. $\dfrac{7}{9} = 0.\overline{7}$

41. 8 is a natural, whole and rational number, and is an integer.

43. $\dfrac{16}{4} = 4$ is a natural, whole and rational number, and is an integer.

45. 0 is a whole and rational number, and is an integer.

47. $-\dfrac{7}{6}$ is a rational number.

49. $\sqrt{25} = 5$ because $5 \cdot 5 = 25$ and 5 is positive.

51. $\sqrt{49} = 7$ because $7 \cdot 7 = 49$ and 7 is positive.

53. We can estimate the value of $\sqrt{7}$ with a calculator. $\sqrt{7} \approx 2.646$

55. $-4.5 = -4\dfrac{1}{2} = -\dfrac{9}{2}$ is a rational number.

57. $\dfrac{3}{7}$ is a rational number.

59. $\sqrt{11}$ is an irrational number.

61. $\dfrac{8}{4} = 2$ is a natural and rational number, and is an integer.

63. $\sqrt{49} = 7$ is a natural and rational number, and is an integer.

65. $1.\overline{8}$ is a rational number because its decimal repeats the same number.

67.

69.

71.

73.

75. $|5.23| = 5.23$

77. $|-8| = 8$

79. $|2-2| = |0| = 0$

81. $|\pi - 3| = \pi - 3$

83. $|b|$, if $b$ is negative, $= -b$

85. $5 < 7$

87. $-5 > -7$

89. $-\dfrac{1}{3} > -\dfrac{2}{3}$

91. $-1.9 < -1.3$

93. $|-8| = 8$ and $8 > 3$ so $|-8| > 3$.

95. $|-2| = 2$ and $|-7| = 7$ and $2 < 7$ so $|-2| < |-7|$.

97. $>$

99. $=$

101. $-9, -2^3, -3, 0, 1$

103. $-2, -\dfrac{3}{2}, \dfrac{1}{3}, \sqrt{5}, \pi$

105. (a)    The age in 2005 was 25.5.

  (b)    *Answers may vary.*

  (c)    The average age was $\dfrac{25.1 + 25.3 + 25.5 + 25.9}{4} = \dfrac{101.8}{4} = 25.45$.

## Checking Basic Concepts Sections 1.3 and 1.4

1. (a)    $5 \cdot 5 \cdot 5 \cdot 5 = 5^4$

  (b)    $7 \cdot 7 \cdot 7 \cdot 7 \cdot 7 = 7^5$

2. (a)    $2^3 = 2 \cdot 2 \cdot 2 = 8$

  (b)    $10^4 = 10 \cdot 10 \cdot 10 \cdot 10 = 10,000$

  (c)    $\left(\dfrac{2}{3}\right)^3 = \dfrac{2}{3} \cdot \dfrac{2}{3} \cdot \dfrac{2}{3} = \dfrac{2 \cdot 2 \cdot 2}{3 \cdot 3 \cdot 3} = \dfrac{8}{27}$

3. (a)    $64 = 4 \cdot 4 \cdot 4 = 4^3$

  (b)    $64 = 2 \cdot 2 \cdot 2 \cdot 2 \cdot 2 \cdot 2 = 2^6$

4. (a) $6 + 5 \cdot 4 = 6 + 20 = 26$

   (b) $6 + 6 \div 2 = 6 + 3 = 9$

   (c) $5 - 2 - 1 = 2$

   (d) $\dfrac{6-3}{2+4} = \dfrac{3}{6} = \dfrac{1 \cdot 3}{2 \cdot 3} = \dfrac{1}{2} \cdot \dfrac{3}{3} = \dfrac{1}{2} \cdot 1 = \dfrac{1}{2}$

   (e) $12 \div (6 \div 2) = 12 \div 3 = 4$

   (f) $2^3 - 2\left(2 + \dfrac{4}{2}\right) = 8 - 2(2+2) = 8 - 2(4) = 8 - 8 = 0$

5. $5^3 \div 3$, or $\dfrac{5^3}{3}$

6. (a) The opposite of $-17$ is $-(-17) = 17$.

   (b) The opposite of $a$ is $-a$.

7. (a) $\dfrac{3}{20} = \dfrac{3 \cdot 5}{20 \cdot 5} = \dfrac{15}{100} = 0.15$

   (b) $\dfrac{5}{8} = \dfrac{5 \cdot 125}{8 \cdot 125} = \dfrac{625}{1000} = 0.625$

8. (a) $\dfrac{10}{2} = 5$ is a natural and rational number, and is an integer.

   (b) $-5$ is a rational number and is an integer.

   (c) $\sqrt{5}$ is an irrational number.

   (d) $-\dfrac{5}{6}$ is a rational number.

9.

10. (a) $|-12| = -(-12) = 12$

    (b) $|6 - 6| = |0| = 0$

11. (a) $4 < 9$

    (b) $-1.3 < -0.5$

    (c) $|-3| = 3$ and $|-5| = 5$ and $3 < 5$ so $|-3| < |-5|$.

12. $-7, -1.6, 0, \dfrac{1}{3}, \sqrt{3}, 3^2$

## Section 1.5　Addition and Subtraction of Real Numbers

1. sum

3. True

5. absolute value

7. addition

9. addition

11. The opposite of 25 is $-25.\ 25+(-25)=0$

13. The opposite of $-\sqrt{21}$ is $\sqrt{21}$ . $-\sqrt{21}+\sqrt{21}=0$

15. The opposite of 5.63 is $-5.63.\ 5.63+(-5.63)=0$

17. ; $1+3=4$

19. ; $4+(-2)=2$

21. ; $-1+(-2)=-3$

23. ; $-3+7=4$

25. ; $-1+3=2$

27. ; $4+(-5)=-1$

29. ; $-10+20=10$

31. ; $-50+(-100)=-150$

33. $5+(-4)=1$

35. $-1+(-6)=-7$

37. $\dfrac{3}{4}+\left(-\dfrac{1}{2}\right)=\dfrac{3}{4}-\dfrac{1}{2}\cdot\dfrac{2}{2}=\dfrac{3}{4}-\dfrac{2}{4}=\dfrac{1}{4}$

39. $-\dfrac{6}{7}+\dfrac{3}{14}=\dfrac{3}{14}-\dfrac{6}{7}=\dfrac{3}{14}-\dfrac{6}{7}\cdot\dfrac{2}{2}=\dfrac{3}{14}-\dfrac{12}{14}=-\dfrac{9}{14}$

41. $0.6+(-1.7)=0.6-1.7=-1.1$

43. $-52+86=86-52=34$

45. $5-8=5+(-8)=-3$

47. $-2-(-9)=-2+9=7$

49. $\dfrac{6}{7}-\dfrac{13}{14}=\dfrac{12}{14}-\dfrac{13}{14}=\dfrac{12}{14}+\left(-\dfrac{13}{14}\right)=-\dfrac{1}{14}$

51. $-\dfrac{1}{10}-\left(-\dfrac{3}{5}\right)=-\dfrac{1}{10}-\left(-\dfrac{6}{10}\right)=-\dfrac{1}{10}+\dfrac{6}{10}=\dfrac{5}{10}=\dfrac{1}{2}$

53. $0.8-(-2.1)=0.8+2.1=2.9$

55. $-73 - 91 = -73 + (-91) = -164$

57. $10 - 19 = 10 + (-19) = -9$

59. $19 - (-22) + 1 = 19 + 22 + 1 = 42$

61. $-3 + 4 - 6 = -3 + 4 + (-6) = 4 - 3 + (-6) = 1 + (-6) = -5$

63. $100 - 200 + 100 - (-50) = 100 + 100 - 200 + 50 = 200 - 200 + 50 = 50$

65. $1.5 - 2.3 + 9.6 = 1.5 + (-2.3) + 9.6 = -0.8 + 9.6 = 8.8$

67. $-\dfrac{1}{2} + \dfrac{1}{4} - \left(-\dfrac{3}{4}\right) = -\dfrac{2}{4} + \dfrac{1}{4} + \dfrac{3}{4} = \dfrac{2}{4} = \dfrac{1}{2}$

69. $|4 - 9| - |1 - 7| = |-5| - |-6| = 5 - 6 = -1$

71. $2 + (-5) = -3$

73. $-5 + 7 = 2$

75. $-2^3 = -(2 \cdot 2 \cdot 2) = -8.$

77. $-6 - 7 = -6 + (-7) = -13$

79. $6 + (-10) - 5 = 6 + (-10) + (-5) = -4 + (-5) = -9$

81. The highest point in the continental United States is Mt. Whitney at 14,497 feet.

    The lowest point in the continental United States is Death Valley at 282 feet below sea level.

    The difference is $14,497 - (-282) = 14,779$ feet.

83. Take the initial balance and then add to it the deposits and subtract from it the withdrawals.

    $358 - 45 + 37 + 120 - 240 = 358 + 37 + 120 + (-45) + (-240) = \$230$

85. Add the yardage gained or lost in each play. $9 + (-2) + (-1) + 14 + 5 = 28 + (-3) = 25$ yd.

87. The word height is in reference to sea level, the height of Mount Everest is 29,029 feet and the height of the Mariana Trench is $(-35,839)$.

    To find the difference take $29,029 - (-35,839) = 29,029 + 35,839 = 64,868$ feet.


## Section 1.6 Multiplication and Division of Real Numbers

1. product

3. positive

5. $\dfrac{1}{a}$

7. reciprocal or multiplicative inverse

9. $\dfrac{1}{b}$

11. negative

13. $-3 \cdot 4 = -12$

15. $6 \cdot (-3) = -18$

17. $0 \cdot (-2.13) = 0$

19. $-6 \cdot (-10) = 60$

21. $-\dfrac{1}{2} \cdot \left(-\dfrac{2}{4}\right) = \dfrac{2}{8} = \dfrac{1}{4}$

23. $-\dfrac{3}{7} \cdot \dfrac{7}{3} = -\dfrac{21}{21} = -1$

25. $-10 \cdot (-20) = 200$

27. $-0.5 \cdot 100 = -50$

29. $-2 \cdot 3 \cdot (-4) \cdot 5 = -6 \cdot (-20) = 120$

31. $-6 \cdot \dfrac{1}{6} \cdot \dfrac{7}{9} \cdot \left(-\dfrac{9}{7}\right) \cdot \left(-\dfrac{3}{2}\right) = -\dfrac{6}{6} \cdot \left(-\dfrac{63}{63}\right) \cdot \left(-\dfrac{3}{2}\right) = -1 \cdot (-1) \cdot \left(-\dfrac{3}{2}\right) = 1 \cdot \left(-\dfrac{3}{2}\right) = -\dfrac{3}{2}$

33. Negative, because there is an odd number of negative factors

35. $(-5)^2 = (-5)(-5) = 25$

37. $(-1)^3 = (-1)(-1)(-1) = -1$

39. $-2^4 = -2 \cdot 2 \cdot 2 \cdot 2 = -16$

41. $-(-2)^3 = -(-2)(-2)(-2) = -(-8) = 8$

43. $5 \cdot (-2)^3 = 5 \cdot [(-2)(-2)(-2)] = 5 \cdot (-8) = -40$

45. $-10 \div 5 = -\dfrac{10}{1} \cdot \dfrac{1}{5} = -\dfrac{10}{5} = -2$

47. $-20 \div (-2) = -\dfrac{20}{1} \cdot \left(-\dfrac{1}{2}\right) = \dfrac{20}{2} = 10$

49. $-\dfrac{12}{3} = -4$

51. $-16 \div \dfrac{1}{2} = -\dfrac{16}{1} \cdot \dfrac{2}{1} = -\dfrac{32}{1} = -32$

53. $0 \div 3 = 0$

55. $\dfrac{-1}{0} =$ undefined, because division by 0 is not allowed.

57. $\dfrac{1}{2} \div (-11) = \dfrac{1}{2} \cdot -\dfrac{1}{11} = -\dfrac{1}{22}$

59. $-\dfrac{4}{5} \div (-3) = -\dfrac{4}{5} \cdot \left(-\dfrac{1}{3}\right) = \dfrac{4}{15}$

61. $\dfrac{5}{6} \div \left(-\dfrac{8}{9}\right) = \dfrac{5}{6} \cdot \left(-\dfrac{9}{8}\right) = -\dfrac{45}{48} = -\dfrac{15}{16}$

63. $-\dfrac{1}{2} \div 0 =$ undefined, because division by 0 is not allowed.

65. $-0.5 \div \dfrac{1}{2} = -\dfrac{5}{10} \div \dfrac{1}{2} = -\dfrac{5}{10} \cdot \dfrac{2}{1} = -\dfrac{10}{10} = -1$

67. $-\dfrac{2}{3} \div 0.5 = -\dfrac{2}{3} \div \dfrac{5}{10} = -\dfrac{2}{3} \cdot \dfrac{10}{5} = -\dfrac{20}{15} = -\dfrac{4}{3}$

69. $\dfrac{1}{2} = 0.5$

71. $\dfrac{3}{16} = 0.1875$

73. Because $1 \div 2 = 0.5, \ 3\dfrac{1}{2} = 3.5$

75. Because $2 \div 3 = 0.\overline{6}, \ 5\dfrac{2}{3} = 5.\overline{6}$

77. Because $7 \div 16 = 0.4375, \ 1\dfrac{7}{16} = 1.4375$

79. $\dfrac{7}{8} = 7 \div 8 = 0.875$

81. $0.25 = \dfrac{25}{100} = \dfrac{1 \cdot 25}{4 \cdot 25} = \dfrac{1}{4}$

83. $0.16 = \dfrac{16}{100} = \dfrac{4 \cdot 4}{25 \cdot 4} = \dfrac{4}{25}$

85. $0.625 = \dfrac{625}{1000} = \dfrac{5 \cdot 125}{8 \cdot 125} = \dfrac{5}{8}$

87. $0.6875 = \dfrac{6875}{10,000} = \dfrac{11 \cdot 625}{16 \cdot 625} = \dfrac{11}{16}$

89. $\left(\dfrac{1}{3} + \dfrac{5}{6}\right) \div \dfrac{1}{2} = 2.\overline{3}$ or $\dfrac{7}{3}$

91. $\dfrac{4}{5} \div \dfrac{2}{3} \cdot \dfrac{7}{4} = 2.1$ or $\dfrac{21}{10}$

93. $\dfrac{15}{2} - 4 \cdot \dfrac{7}{3} = -1.8\overline{3}$ or $-\dfrac{11}{6}$

95. $\dfrac{17}{40} + 3 \div 8 = 0.8$ or $\dfrac{4}{5}$

97. Multiply the real numbers 202 and $\dfrac{13}{20}$ to obtain $202 \cdot \dfrac{13}{20} = 131.3$. Total admissions for *The Ten Commandments* were about 131 million.

99. $\dfrac{21}{125} = 21 \div 125 = 0.168$

## Checking Basic Concepts Sections 1.5 and 1.6

1. (a) $-4 + 4 = 0$

   (b) $-10 + (-12) + 3 = -22 + 3 = -19$

2. (a) $\dfrac{2}{3} - \left(-\dfrac{2}{9}\right) = \dfrac{2}{3} + \dfrac{2}{9} = \dfrac{6}{9} + \dfrac{2}{9} = \dfrac{8}{9}$

   (b) $-1.2 - 5.1 + 3.1 = -1.2 + (-5.1) + 3.1 = -6.3 + 3.1 = -3.2$

3. (a) $-1 + 5 = 4$

   (b) $4 - (-3) = 4 + 3 = 7$

4. $99 - (-46) = 99 + 46 = 145°F$ is the difference between these two temperatures.

5. (a) $-5 \cdot (-7) = 35$

   (b) $-\dfrac{1}{2} \cdot \dfrac{2}{3} \cdot \left(-\dfrac{4}{5}\right) = -\dfrac{2}{6} \cdot \left(-\dfrac{4}{5}\right) = \dfrac{8}{30} = \dfrac{4}{15}$

6. (a) $-3^2 = -3 \cdot 3 = -9$

   (b) $4 \cdot (-2)^3 = 4 \cdot \left[(-2)(-2)(-2)\right] = 4 \cdot (-8) = -32$

   (c) $(-5)^2 = (-5)(-5) = 25$

7. (a) $-5 \div \dfrac{2}{3} = -\dfrac{5}{1} \cdot \dfrac{3}{2} = -\dfrac{15}{2}$

   (b) $-\dfrac{5}{8} \div \left(-\dfrac{4}{3}\right) = -\dfrac{5}{8} \cdot \left(-\dfrac{3}{4}\right) = \dfrac{15}{32}$

8. The reciprocal of $-\dfrac{7}{6}$ is $-\dfrac{6}{7}$.

9. (a) $\dfrac{-10}{2} = \dfrac{-5 \cdot 2}{1 \cdot 2} = \dfrac{-5}{1} = -5$

   (b) $\dfrac{10}{-2} = \dfrac{5 \cdot 2}{-1 \cdot 2} = \dfrac{5}{-1} = -5$

(c)   $-\dfrac{10}{2} = -\dfrac{5 \cdot 2}{1 \cdot 2} = -\dfrac{5}{1} = -5$

(d)   $\dfrac{-10}{-2} = \dfrac{-5 \cdot 2}{-1 \cdot 2} = \dfrac{-5}{-1} = 5$

10. (a)   $\dfrac{3}{5} = 0.6$

(b)   $3\dfrac{7}{8} = 3.875$ because $\dfrac{7}{8} = 0.875$

## Section 1.7  Properties of Real Numbers

1. commutative; addition

3. associative; addition

5. True

7. distributive

9. identity; addition

11. $-a$

13. $-6 + 10 = 10 + (-6)$

15. $-5 \cdot 6 = 6 \cdot (-5)$

17. $a + 10 = 10 + a$

19. $b \cdot 7 = 7 \cdot b$

21. $(1 + 2) + 3 = 1 + (2 + 3)$

23. $2 \cdot (3 \cdot 4) = (2 \cdot 3) \cdot 4$

25. $(a + 5) + c = a + (5 + c)$

27. $(x \cdot 3) \cdot 4 = x \cdot (3 \cdot 4)$

29. $a + b + c = (a + b) + c = c + (a + b) = c + (b + a) = c + b + a$

31. $4(3 + 2) = (4 \cdot 3) + (4 \cdot 2) = 12 + 8 = 20$

33. $a(b - 8) = ab - 8a$

35. $-4(t - z) = -4t - 4(-z) = -4t + 4z$

37. $-(5 - a) = -(1)(5) - 1(-a) = -5 + a$

39. $(a + 5)3 = 3a + (3)(5) = 3a + 15$

41. $12 - (a - 5) = 12 - a - (-5) = 12 - a + 5 = 17 - a$

43. $a \cdot (b + c + d) = a \cdot ((b + c) + d) = a \cdot (b + c) + ad = ab + ac + ad$

45. $6x + 5x = (6+5)x = 11x$

47. $-4b + 3b = (-4+3)b = (-1)b = -b$

49. $3a - a = (3-1)a = 2a$

51. $13w - 27w = (13-27)w = (13+(-27))w = -14w$

53. Commutative (multiplication)

55. Associative (addition)

57. Distributive

59. Distributive, Commutative (multiplication)

61. Distributive

63. Associative (multiplication)

65. Distributive

67. Identity (addition)

69. Identity (multiplication)

71. Identity (multiplication)

73. Inverse (multiplication)

75. Inverse (addition)

77. $(4+2)+(9+8)+(1+6) = (4+6)+(2+8)+(9+1) = 30$

79. $(45+43)+(5+7) = (45+5)+(43+7) = 100$

81. $129 + 50 - 1 = 179 - 1 = 178$

83. $178 - 100 + 1 = 78 + 1 = 79$

85. $6 \cdot 10 + 6 \cdot 5 = 60 + 30 = 90$

87. $8 \cdot 100 + 8 \cdot 2 = 800 + 16 = 816$

89. $\left(\dfrac{1}{2} \cdot \dfrac{1}{2} \cdot \dfrac{1}{2}\right) \cdot 2 \cdot 2 \cdot 2 = \left(\dfrac{1}{8}\right) \cdot 8 = 1$

91. $\dfrac{7}{6} \cdot \left(\dfrac{1}{2} \cdot \dfrac{1}{2} \cdot \dfrac{1}{2}\right) \cdot \dfrac{8}{7} = \dfrac{7}{6} \cdot \left(\dfrac{1}{8} \cdot \dfrac{8}{7}\right) = \dfrac{7}{6} \cdot \dfrac{1}{7} = \dfrac{1}{6}$

93. (a) $10 \cdot 41 = 410$

    (b) $10 \cdot 997 = 9970$

    (c) $-630 \cdot 10 = -6300$

    (d) $-14,000 \cdot 10 = -140,000$

95. (a) $1000 \cdot 19 = 19,000$

    (b) $100 \cdot (-451) = -45,100$

    (c) $10,000 \cdot 6 = 60,000$

    (d) $-79 \cdot 100,000 = -7,900,000$

97. (a)    $12.56 \div 10 = 1.256$

    (b)    $9.6 \div 10 = 0.96$

    (c)    $0.987 \div 10 = 0.0987$

    (d)    $-0.056 \div 10 = -0.0056$

    (e)    $1200 \div 10 = 120$

    (f)    $4578 \div 10 = 457.8$

99. Because $100 + 75 = 75 + 100,$ this shows the commutative property of addition.

101. Because 1 gallon is $10 \div 10$ gallons, divide 198 by 10 to get 19.8 miles.

103. (a)    Because multiplying by 10 is easy, multiply $13 \cdot (5 \cdot 2) = 13 \cdot 10 = 130$ ft$^3$.

    (b)    The associative property of multiplication.


## Section 1.8  Simplifying and Writing Algebraic Expressions

1. term

3. factors; terms

5. like

7. The expression 91 is a term; its coefficient is 91.

9. The expression $-6b$ is a term; its coefficient is $-6$.

11. The expression $x + 10$ is not a term because it is the sum of two terms.

13. The expression $x^2$ is a term; its coefficient is 1.

15. The expression $4x - 5$ is not a term because it is the difference of two terms.

17. The terms 6 and $-8$ are like because neither term contains a variable.

19. The terms $5x$ and $-22x$ are like because each term contains the same variable raised to the same power.

21. The terms 14 and $14a$ are unlike because one contains a variable while the other does not.

23. The terms $18x$ and $18y$ are unlike because the first term contains a different variable than the second term.

25. The terms $x^2$ and $-15x^2$ are like because each term contains the same variable raised to the same power.

27. The terms $3x^2$ and $\frac{1}{5}x$ are unlike because the variables are the same but they are raised to different powers.

29. The terms $4ab$ and $-3ba$ are like because each term contains the same variables raised to the same powers.

31. $-4x + 7x = (-4 + 7)x = 3x$

33. $19y - 5y = (19 - 5)y = 14y$

35. $28a + 13a = (28 + 13)a = 41a$

37. $11z - 11z = (11 - 11)z = 0z = 0$

39. It is not possible to combine terms because the first term contains a different variable than the second term.

41. It is not possible to combine terms because the first term does not contain a variable and the second term does.

43. $5x^2 - 2x^2 = (5 - 2)x^2 = 3x^2$

45. $8y - 10y + y = (8 - 10 + 1)y = -1y = -y$

47. $5 + x - 3 + 2x = 5 - 3 + x + 2x = 5 - 3 + (1 + 2)x = 2 + 3x = 3x + 2$

49. $-\dfrac{3}{4} + z - 3z + \dfrac{5}{4} = z - 3z + \dfrac{5}{4} - \dfrac{3}{4} = (1 - 3)z + \dfrac{5}{4} - \dfrac{3}{4} = -2z + \dfrac{2}{4} = -2z + \dfrac{1}{2}$

51. $4y - y + 8y = (4 - 1 + 8)y = 11y$

53. $-3 + 6z + 2 - 2z = 6z - 2z - 3 + 2 = (6 - 2)z - 3 + 2 = 4z - 1$

55. $-2(3z - 6y) - z = -6z + 12y - z = -6z - z + 12y = (-6 - 1)z + 12y = -7z + 12y = 12y - 7z$

57. $2 - \dfrac{3}{4}(4x + 8) = 2 - \dfrac{3 \cdot 4}{4}x - \dfrac{3 \cdot 8}{4} = 2 - \dfrac{12}{4}x - \dfrac{24}{4} = 2 - 3x - 6 = 2 - 6 - 3x = -4 - 3x = -3x - 4$

59. $-x - (5x + 1) = -x - 5x - 1 = (-1 - 5)x - 1 = -6x - 1$

61. $1 - \dfrac{1}{3}(x + 1) = 1 - \dfrac{1}{3}x - \dfrac{1}{3} = -\dfrac{1}{3}x + 1 - \dfrac{1}{3} = -\dfrac{1}{3}x + \dfrac{3}{3} - \dfrac{1}{3} = -\dfrac{1}{3}x + \dfrac{2}{3}$

63. $\dfrac{3}{5}(x + y) - \dfrac{1}{5}(x - 1) = \dfrac{3}{5}x + \dfrac{3}{5}y - \dfrac{1}{5}x + \dfrac{1}{5} = \dfrac{3}{5}x - \dfrac{1}{5}x + \dfrac{3}{5}y + \dfrac{1}{5} = \left(\dfrac{3}{5} - \dfrac{1}{5}\right)x + \dfrac{3}{5}y + \dfrac{1}{5} = \dfrac{2}{5}x + \dfrac{3}{5}y + \dfrac{1}{5}$

65. $0.2x^2 + 0.3x^2 - 0.1x^2 = (0.2 + 0.3 - 0.1)x^2 = 0.4x^2$

67. $2x^2 - 3x + 5x^2 - 4x = 2x^2 + 5x^2 - 3x - 4x = (2 + 5)x^2 - (3 + 4)x = 7x^2 - 7x$

69. $a + 3b - a - b = a - a + 3b - b = (1 - 1)a + (3 - 1)b = 0a + 2b = 2b$

71. $8x^3 + 7y - x^3 - 5y = 8x^3 - x^3 + 7y - 5y = (8 - 1)x^3 + (7 - 5)y = 7x^3 + 2y$

73. $\dfrac{8x}{8} = \dfrac{8}{8} \cdot \dfrac{x}{1} = 1 \cdot \dfrac{x}{1} = 1 \cdot x = x$

75. $\dfrac{-3y}{-y} = \dfrac{-3}{1} \cdot \dfrac{y}{-y} = -3 \cdot (-1) = 3$

77. $\dfrac{-108z}{-108} = \dfrac{-108}{-108} \cdot \dfrac{z}{1} = 1 \cdot \dfrac{z}{1} = 1 \cdot z = z$

79. $\dfrac{9x-6}{3} = \dfrac{9x}{3} - \dfrac{6}{3} = \dfrac{9}{3} \cdot \dfrac{x}{1} - \dfrac{6}{3} = 3x - 2$

81. $\dfrac{14z+21}{7} = \dfrac{14z}{7} + \dfrac{21}{7} = \dfrac{14}{7} \cdot \dfrac{z}{1} + \dfrac{21}{7} = 2z + 3$

83. $5x + 6x = (5+6)x = 11x$

85. $x^2 + 2x^2 = (1+2)x^2 = 3x^2$

87. $6x - 4x = (6-4)x = 2x$

89. (a)  Let $w$ be the constant width of the street in feet. The area of each street section equals its

    length times its width. The total area of the street is

    $400w + 350w + 220w + 600w = (400 + 350 + 220 + 600)w = 1570w.$

(b)  If the width = 42 feet, $w = 42.$ Then, $1570w = 1570 \cdot 42 = 65{,}940 \text{ ft}^2.$

91. (a)  Let $x$ be the number of minutes. Then, $20x + 30x = 50x =$ the number of cubic feet of snow

    removed in $x$ minutes.

(b)  Let $x = 48.$ Then, the total number of cubic feet of snow removed in 48 minutes is

    $50x = 50 \cdot 48 = 2400 \text{ ft}^3.$

(c)  First, calculate how many cubic feet of snow the driveway contains. Volume equals length

    times width times height; thus, the driveway contains $30 \cdot 20 \cdot 2 = 1200 \text{ ft}^3$ of snow. If 50

    cubic feet of snow are removed in 1 minute, then 1200 cubic feet of snow are removed in 24

    minutes.

## Checking Basic Concepts Sections 1.7 and 1.8

1. (a)   $y \cdot 18 = 18y$

(b)   $10 + x = x + 10$

2. $5 \cdot (y \cdot 4) = 5 \cdot (4y) = 20y$

3. (a)   $10 - (5+x) = 10 - 5 - x = 5 - x$

(b)   $5(x-7) = 5x - 35$

4. Because $5x + 3x = (5+3)x = 8x,$ this equation illustrates the distributive property.

5. $-4xy + 4xy = (-4+4)xy = 0xy = 0$

6. (a)   $32 + 17 + 8 + 3 = (32+8) + (17+3) = 60$

(b)   $\left(\dfrac{5}{6} \cdot \dfrac{6}{5}\right) \cdot \left(\dfrac{7}{8} \cdot \dfrac{8}{1}\right) = 1 \cdot 7 = 7$

(c)   $567 - 200 + 1 = 367 + 1 = 368$

7. (a) The terms $-3x$ and $-3z$ are unlike because the first term contains a different variable than the second term.

 (b) The terms $4x^2$ and $-2x^2$ are like because each term contains the same variable raised to the same power.

8. (a) $5z+9z=(5+9)z=14z$

 (b) $5y-4-8y+7=5y-8y-4+7=(5-8)y-4+7=-3y+3$

9. (a) $2y-(5y+3)=2y-5y-3=(2-5)y-3=-3y-3$

 (b) $-4(x+3y)+2(2x-y)=-4x-12y+4x-2y=-4x+4x-12y-2y=$

  $(-4+4)x-(12+2)y=0x-14y=-14y$

 (c) $\dfrac{20x}{20}=\dfrac{20}{20}\cdot x=1\cdot x=x$

 (d) $\dfrac{35x^2}{x^2}=35\cdot\dfrac{x^2}{x^2}=35\cdot 1=35$

10. $3x+5x=(3+5)x=8x$

## Chapter 1 Review

1. The number 29 is prime because its only factors are itself and 1.

2. The number 27 is a composite number because it has factors other than itself and 1; $27=3\cdot 3\cdot 3$.

3. The number 108 is a composite number because it has factors other than itself and 1;
 $108=2\cdot 2\cdot 3\cdot 3\cdot 3$.

4. The number 91 is a composite number because it has factors other than itself and 1; $91=7\cdot 13$.

5. The number 0 is neither a prime nor a composite number.

6. The number 1 is neither a prime nor a composite number.

7. $2x-5$, when $x=4$, is $2\cdot 4-5=8-5=3$

8. $7-\dfrac{10}{x}$, when $x=5$, is $7-\dfrac{10}{5}=7-\dfrac{2\cdot 5}{1\cdot 5}=7-\dfrac{2}{1}\cdot\dfrac{5}{5}=7-2\cdot 1=7-2=5$

9. $9x-2y$, when $x=2$ and $y=3$, is $9\cdot 2-2\cdot 3=18-6=12$

10. $\dfrac{2x}{x-y}$, when $x=6$ and $y=4$, is $\dfrac{2\cdot 6}{6-4}=\dfrac{12}{2}=\dfrac{6\cdot 2}{1\cdot 2}=\dfrac{6}{1}\cdot\dfrac{2}{2}=6\cdot 1=6$

11. $y=x-5$, when $x=12$, is $y=12-5\Rightarrow y=7$

12. $y=xz+1$, when $x=2$ and $z=3$, is $y=2\cdot 3+1\Rightarrow y=6+1\Rightarrow y=7$

13. $y=4(x-z)$, when $x=7$ and $z=5$, is $y=4(7-5)\Rightarrow y=4(2)\Rightarrow y=8$

14. $y = \dfrac{x+z}{4}$, when $x = 14$ and $z = 10$, is $y = \dfrac{14+10}{4} \Rightarrow y = \dfrac{24}{4} \Rightarrow y = 6$

15. Three squared increased by five is $3^2 + 5$.

16. Two cubed divided by the quantity three plus one is $2^3 \div (3+1)$.

17. Let $x$ be the number. Then, the product of three and the number is $3x$.

18. Let $x$ be the number. Then, the difference between the number and four is $x - 4$.

19. Since $15 = 3 \cdot 5$ and $35 = 5 \cdot 7$ then the GCF of 15 and 35 is 5.

20. Since $12 = 2 \cdot 2 \cdot 3$, $30 = 2 \cdot 3 \cdot 5$ and $42 = 2 \cdot 3 \cdot 7$ then the GCF of 12, 30 and 42 is $2 \cdot 3 = 6$

21. (a) $\dfrac{5 \cdot 7}{8 \cdot 7} = \dfrac{5}{8} \cdot \dfrac{7}{7} = \dfrac{5}{8} \cdot 1 = \dfrac{5}{8}$

    (b) $\dfrac{3a}{4a} = \dfrac{3}{4} \cdot \dfrac{a}{a} = \dfrac{3}{4} \cdot 1 = \dfrac{3}{4}$

22. (a) $\dfrac{9}{12} = \dfrac{3 \cdot 3}{4 \cdot 3} = \dfrac{3}{4} \cdot \dfrac{3}{3} = \dfrac{3}{4} \cdot 1 = \dfrac{3}{4}$

    (b) $\dfrac{36}{60} = \dfrac{3 \cdot 12}{5 \cdot 12} = \dfrac{3}{5} \cdot \dfrac{12}{12} = \dfrac{3}{5} \cdot 1 = \dfrac{3}{5}$

23. $\dfrac{3}{4} \cdot \dfrac{5}{6} = \dfrac{3 \cdot 5}{4 \cdot 6} = \dfrac{15}{24} = \dfrac{5 \cdot 3}{8 \cdot 3} = \dfrac{5}{8} \cdot \dfrac{3}{3} = \dfrac{5}{8} \cdot 1 = \dfrac{5}{8}$

24. $\dfrac{1}{2} \cdot \dfrac{4}{9} = \dfrac{1 \cdot 4}{2 \cdot 9} = \dfrac{4}{18} = \dfrac{2 \cdot 2}{9 \cdot 2} = \dfrac{2}{9} \cdot \dfrac{2}{2} = \dfrac{2}{9} \cdot 1 = \dfrac{2}{9}$

25. $\dfrac{2}{3} \cdot \dfrac{9}{10} = \dfrac{2 \cdot 9}{3 \cdot 10} = \dfrac{18}{30} = \dfrac{6 \cdot 3}{6 \cdot 5} = \dfrac{3}{5}$

26. $\dfrac{12}{11} \cdot \dfrac{22}{23} = \dfrac{12 \cdot 22}{11 \cdot 23} = \dfrac{264}{253} = \dfrac{11 \cdot 24}{11 \cdot 23} = \dfrac{24}{23}$

27. $4 \cdot \dfrac{5}{8} = \dfrac{4 \cdot 5}{1 \cdot 8} = \dfrac{20}{8} = \dfrac{4 \cdot 5}{4 \cdot 2} = \dfrac{5}{2}$

28. $\dfrac{2}{3} \cdot 9 = \dfrac{2 \cdot 9}{3 \cdot 1} = \dfrac{18}{3} = 6$

29. $\dfrac{x}{3} \cdot \dfrac{6}{x} = \dfrac{x \cdot 6}{3 \cdot x} = \dfrac{6}{3} \cdot \dfrac{x}{x} = 2 \cdot 1 = 2$

30. $\dfrac{2}{3} \cdot \dfrac{9x}{4y} = \dfrac{2 \cdot 9x}{3 \cdot 4y} = \dfrac{18x}{12y} = \dfrac{3x \cdot 6}{2y \cdot 6} = \dfrac{3x}{2y} \cdot 1 = \dfrac{3x}{2y}$

31. One-fifth of three-sevenths equals $\dfrac{1}{5} \cdot \dfrac{3}{7} = \dfrac{1 \cdot 3}{5 \cdot 7} = \dfrac{3}{35}$.

32. (a)  The reciprocal of 8 equals the reciprocal of $\dfrac{8}{1}$, which is $\dfrac{1}{8}$.

    (b)  The reciprocal of 1 equals the reciprocal of $\dfrac{1}{1}$, which is $\dfrac{1}{1}$.

    (c)  The reciprocal of $\dfrac{5}{19}$ which is $\dfrac{19}{5}$.

    (d)  The reciprocal of $\dfrac{3}{2}$ which is $\dfrac{2}{3}$.

33. $\dfrac{3}{2} \div \dfrac{1}{6} = \dfrac{3}{2} \cdot \dfrac{6}{1} = \dfrac{3 \cdot 6}{2 \cdot 1} = \dfrac{18}{2} = \dfrac{9 \cdot 2}{1 \cdot 2} = \dfrac{9}{1} \cdot \dfrac{2}{2} = 9 \cdot 1 = 9$

34. $\dfrac{9}{10} \div \dfrac{7}{5} = \dfrac{9}{10} \cdot \dfrac{5}{7} = \dfrac{9 \cdot 5}{10 \cdot 7} = \dfrac{45}{70} = \dfrac{9 \cdot 5}{14 \cdot 5} = \dfrac{9}{14} \cdot \dfrac{5}{5} = \dfrac{9}{14} \cdot 1 = \dfrac{9}{14}$

35. $8 \div \dfrac{2}{3} = \dfrac{8}{1} \cdot \dfrac{3}{2} = \dfrac{8 \cdot 3}{1 \cdot 2} = \dfrac{24}{2} = \dfrac{12 \cdot 2}{1 \cdot 2} = \dfrac{12}{1} \cdot \dfrac{2}{2} = 12 \cdot 1 = 12$

36. $\dfrac{3}{4} \div 6 = \dfrac{3}{4} \cdot \dfrac{1}{6} = \dfrac{3 \cdot 1}{4 \cdot 6} = \dfrac{3}{24} = \dfrac{1 \cdot 3}{8 \cdot 3} = \dfrac{1}{8} \cdot \dfrac{3}{3} = \dfrac{1}{8} \cdot 1 = \dfrac{1}{8}$

37. $\dfrac{x}{y} \div \dfrac{3}{y} = \dfrac{x}{y} \cdot \dfrac{y}{3} = \dfrac{x \cdot y}{y \cdot 3} = \dfrac{xy}{3y} = \dfrac{x}{3}$

38. $\dfrac{4x}{3y} \div \dfrac{9x}{5} = \dfrac{4x}{3y} \cdot \dfrac{5}{9x} = \dfrac{4x \cdot 5}{3y \cdot 9x} = \dfrac{20x}{27xy} = \dfrac{20}{27y}$

39. The least common denominator for the fractions $\dfrac{1}{8}$ and $\dfrac{5}{12}$ is 24, because 24 is the smallest number that both 8 and 12 divide into evenly.

40. The least common denominator for the fractions $\dfrac{3}{14}$ and $\dfrac{1}{21}$ is 42, because 42 is the smallest number that both 14 and 21 divide into evenly.

41. $\dfrac{2}{15} + \dfrac{3}{15} = \dfrac{2+3}{15} = \dfrac{5}{15} = \dfrac{1 \cdot 5}{3 \cdot 5} = \dfrac{1}{3}$

42. $\dfrac{5}{4} - \dfrac{3}{4} = \dfrac{5-3}{4} = \dfrac{2}{4} = \dfrac{1 \cdot 2}{2 \cdot 2} = \dfrac{1}{2}$

43. $\dfrac{11}{12} - \dfrac{1}{8} = \dfrac{11 \cdot 2}{12 \cdot 2} - \dfrac{1 \cdot 3}{8 \cdot 3} = \dfrac{22}{24} - \dfrac{3}{24} = \dfrac{22-3}{24} = \dfrac{19}{24}$

44. $\dfrac{6}{11} - \dfrac{3}{22} = \dfrac{6 \cdot 2}{11 \cdot 2} - \dfrac{3}{22} = \dfrac{12}{22} - \dfrac{3}{22} = \dfrac{12-3}{22} = \dfrac{9}{22}$

45. $\dfrac{2}{3} - \dfrac{1}{2} + \dfrac{1}{4} = \dfrac{2 \cdot 4}{3 \cdot 4} - \dfrac{1 \cdot 6}{2 \cdot 6} + \dfrac{1 \cdot 3}{4 \cdot 3} = \dfrac{8}{12} - \dfrac{6}{12} + \dfrac{3}{12} = \dfrac{8-6+3}{12} = \dfrac{5}{12}$

46. $\dfrac{1}{6}+\dfrac{2}{3}-\dfrac{1}{9}=\dfrac{1\cdot 3}{6\cdot 3}+\dfrac{2\cdot 6}{3\cdot 6}-\dfrac{1\cdot 2}{9\cdot 2}=\dfrac{3}{18}+\dfrac{12}{18}-\dfrac{2}{18}=\dfrac{3+12-2}{18}=\dfrac{13}{18}$

47. $5\cdot 5\cdot 5\cdot 5\cdot 5\cdot 5=5^6$

48. $\dfrac{7}{6}\cdot\dfrac{7}{6}\cdot\dfrac{7}{6}=\left(\dfrac{7}{6}\right)^3$

49. $x\cdot x\cdot x\cdot x\cdot x=x^5$

50. $3\cdot 3\cdot 3\cdot 3=3^4$

51. $(x+1)\cdot(x+1)=(x+1)^2$

52. $(a-5)\cdot(a-5)\cdot(a-5)=(a-5)^3$

53. (a)    $4^3=4\cdot 4\cdot 4=64$

    (b)    $7^2=7\cdot 7=49$

    (c)    $8^1=8$

54. $2^n=32;\ 2\cdot 2\cdot 2\cdot 2\cdot 2=32$, thus $2^5=32$ and $n=5$

55. $7+3\cdot 6=7+18=25$

56. $15-5-3=(15-5)-3=10-3=7$

57. $24\div 4\div 2=(24\div 4)\div 2=6\div 2=3$

58. $30-15\div 3=30-(15\div 3)=30-5=25$

59. $18\div 6-2=(18\div 6)-2=3-2=1$

60. $\dfrac{18}{4+5}=\dfrac{18}{9}=\dfrac{2\cdot 9}{1\cdot 9}=\dfrac{2}{1}\cdot\dfrac{9}{9}=2\cdot 1=2$

61. $9-3^2=9-9=0$

62. $2^3-8=8-8=0$

63. $2^4-8+\dfrac{4}{2}=(2\cdot 2\cdot 2\cdot 2)-8+2=16-8+2=(16-8)+2=8+2=10$

64. $3^2-4(5-3)=3^2-4(2)=9-8=1$

65. $7-\dfrac{4+6}{2+3}=7-\dfrac{10}{5}=7-2=5$

66. $3^3-2^3=(3\cdot 3\cdot 3)-(2\cdot 2\cdot 2)=27-8=19$

67. (a)    The opposite of $-8$ is $-(-8)=8$.

    (b)    The opposite of $-(-(-3))$ is $-(-(-(-3)))=3$.

68. (a)   The opposite of $-\left(\dfrac{-3}{7}\right)$ is $-\left(-\left(\dfrac{-3}{7}\right)\right)=-\dfrac{3}{7}$.

   (b)   The opposite of $\dfrac{-2}{-5}$ is $-\left(\dfrac{-2}{-5}\right)=-\dfrac{2}{5}$.

69. (a)   $\dfrac{4}{5}=\dfrac{4}{5}\cdot\dfrac{2}{2}=\dfrac{8}{10}=0.8$

   (b)   $\dfrac{3}{20}=\dfrac{3}{20}\cdot\dfrac{5}{5}=\dfrac{15}{100}=0.15$

70. (a)   $\dfrac{5}{9}=0.\overline{5}$

   (b)   $\dfrac{7}{11}=0.\overline{63}$

71. The number 0 is a whole number and a rational number, and is an integer.

72. The number $-\dfrac{5}{6}$ is a rational number.

73. The number $-7$ is a rational number and is an integer.

74. The number $\sqrt{17}$ is an irrational number.

75. The number $\pi$ is an irrational number.

76. The number 3.4 is a rational number.

77.

78. (a)   $\left|-5\right|=5$

   (b)   $\left|-\pi\right|=\pi$

   (c)   $\left|4-4\right|=\left|0\right|=0$

79. (a)   $-5<4$

   (b)   $-\dfrac{1}{2}>-\dfrac{5}{2}$

   (c)   $\left|-9\right|=9$ and $-3<9$ so $-3<\left|-9\right|$.

   (d)   $\left|-8\right|=8$ and $\left|-1\right|=1$ and $8>1$ so $\left|-8\right|>\left|-1\right|$.

80.  $-3,\ -\dfrac{2}{3},\ \sqrt{3},\ \pi-1,\ 3$

81. ○○○○○⌒⌒⌒⌒   $-5+9=4$

82. ○○○○⌣⌣⌣   $4+(-7)=-3$

83.
$-1+2=1$

84. $-2+(-3)=-5$

85. $5+(-4)=1$

86. $-9-(-7)=-9+7=-2$

87. $11\cdot(-4)=-44$

88. $-8\cdot(-5)=40$

89. $11\div(-4)=-\dfrac{11}{4}$

90. $-4\div\dfrac{4}{7}=-\dfrac{4}{1}\cdot\dfrac{7}{4}=-\dfrac{4\cdot7}{4}=-\dfrac{7\cdot4}{1\cdot4}=-\dfrac{7}{1}\cdot\dfrac{4}{4}=-7\cdot1=-7$

91. $-\dfrac{5}{9}-\left(-\dfrac{1}{3}\right)=-\dfrac{5}{9}+\dfrac{1}{3}=-\dfrac{5}{9}+\dfrac{1\cdot3}{3\cdot3}=-\dfrac{5}{9}+\dfrac{3}{9}=-\dfrac{2}{9}$

92. $-\dfrac{1}{2}+\left(-\dfrac{3}{4}\right)=-\dfrac{1\cdot2}{2\cdot2}+\left(-\dfrac{3}{4}\right)=-\dfrac{2}{4}+\left(-\dfrac{3}{4}\right)=-\dfrac{5}{4}$

93. $-\dfrac{1}{3}\cdot\left(-\dfrac{6}{7}\right)=\dfrac{1\cdot6}{3\cdot7}=\dfrac{2\cdot3}{7\cdot3}=\dfrac{2}{7}\cdot\dfrac{3}{3}=\dfrac{2}{7}\cdot1=\dfrac{2}{7}$

94. $\dfrac{\frac{4}{5}}{-7}=\dfrac{4}{5}\div(-7)=\dfrac{4}{5}\cdot\left(-\dfrac{1}{7}\right)=-\dfrac{4}{35}$

95. $-\dfrac{3}{2}\div\left(-\dfrac{3}{8}\right)=-\dfrac{3}{2}\cdot\left(-\dfrac{8}{3}\right)=\dfrac{24}{6}=4$

96. $\dfrac{3}{8}\div(-0.5)=\dfrac{3}{8}\div\left(-\dfrac{1}{2}\right)=\dfrac{3}{8}\cdot\left(-\dfrac{2}{1}\right)=-\dfrac{6}{8}=-\dfrac{3\cdot2}{4\cdot2}=-\dfrac{3}{4}\cdot\dfrac{2}{2}=-\dfrac{3}{4}\cdot1=-\dfrac{3}{4}$

97. $3+(-5)=-2$

98. $2-(-4)=2+4=6$

99. $\dfrac{7}{9}=0.\overline{7}$

100. $2\dfrac{1}{5}=2.2$

101. $0.6=\dfrac{6}{10}=\dfrac{3\cdot2}{5\cdot2}=\dfrac{3}{5}\cdot\dfrac{2}{2}=\dfrac{3}{5}\cdot1=\dfrac{3}{5}$

102. $0.375=\dfrac{375}{1000}=\dfrac{3\cdot125}{8\cdot125}=\dfrac{3}{8}\cdot\dfrac{125}{125}=\dfrac{3}{8}\cdot1=\dfrac{3}{8}$

103. $3+16=16+3$

104. $14\cdot(-x)=-x\cdot14$

105. $-4+(1+3)=(-4+1)+3$

106. $(x \cdot y) \cdot 5 = x \cdot (y \cdot 5)$

107. $5(x+12)=5 \cdot x+5 \cdot 12=5x+60$

108. $-(a-3)=-1 \cdot a-1 \cdot (-3)=-a+3$

109. The equation $y+0=y$ is illustrating the identity property of addition.

110. The equation $b \cdot 1 = b$ is illustrating the identity property of multiplication.

111. The equation $\dfrac{1}{4} \cdot 4 = 1$ is illustrating the inverse property of multiplication.

112. The equation $-3a+3a=0$ is illustrating the inverse property of addition.

113. Commutative (multiplication)

114. Associative (addition)

115. Distributive

116. Commutative (addition)

117. Identity (multiplication)

118. Associative (multiplication)

119. Distributive

120. Identity (addition)

121. Inverse (addition)

122. Inverse (multiplication)

123. $7+9+12+8+1+3=(7+3)+(9+1)+(12+8)=10+10+20=40$

124. $500-199=500-200+1=300+1=301$

125. $25 \cdot 99=25(100-1)=25(100)-25(1)=2500-25=2475$

126. $4581+2000-1=6581-1=6580$

127. $54.98 \times 10 = 549.8$ because we move the decimal point to the right one place.

128. $4356 \div 100 = 43.56$ because we move the decimal point to the left two places.

129. $55x$ is a term; its coefficient is 55.

130. $-xy$ is a term; its coefficient is $-1$.

131. $9xy+2z$ is not a term because it is the sum of two terms.

132. $x-7$ is not a term because it is the difference of two terms.

133. $-10x+4x=(-10+4)x=-6x$

134. $19z-4z=(19-4)z=15z$

135. $3x^2+x^2=(3+1)x^2=4x^2$

136. $7+2x-6+x=7-6+2x+x=7-6+(2+1)x=1+3x=3x+1$

137. $-\dfrac{1}{2}+\dfrac{3}{2}z-z+\dfrac{5}{2}=-\dfrac{1}{2}+\dfrac{5}{2}+\dfrac{3}{2}z-z=\dfrac{-1+5}{2}+\left(\dfrac{3}{2}-1\right)z=\dfrac{4}{2}+\left(\dfrac{3}{2}-\dfrac{2}{2}\right)z=2+\dfrac{1}{2}z=\dfrac{1}{2}z+2$

138. $5(x-3)-(4x+3)=5x-15-4x-3=5x-4x-15-3=(5-4)x-15-3=x-18$

139. $4x^2-3+5x^2-3=4x^2+5x^2-3-3=(4+5)x^2-(3+3)=9x^2-6$

140. $3x^2+4x^2-7x^2=(3+4-7)x^2=0x^2=0$

141. $\dfrac{35a}{7a}=\dfrac{35}{7}\cdot\dfrac{a}{a}=\dfrac{35}{7}\cdot1=5\cdot1=5$

142. $\dfrac{0.5c}{0.5}=\dfrac{0.5}{0.5}\cdot c=1\cdot c=c$

143. $\dfrac{15y+10}{5}=\dfrac{15y}{5}+\dfrac{10}{5}=\dfrac{15}{5}\cdot\dfrac{y}{1}+\dfrac{10}{5}=3y+2$

144. $\dfrac{24x-60}{12}=\dfrac{24x}{12}-\dfrac{60}{12}=\dfrac{24}{12}\cdot\dfrac{x}{1}-\dfrac{60}{12}=2x-5$

145. (a)  Let $x$ be the number of minutes. The first person paints $3\cdot x$ or $3x$, square feet in $x$ minutes. The second person paints $4\cdot x$ or $4x$, square feet in $x$ minutes. Working together the two people paint $3x+4x=(3+4)x=7x$ square feet in $x$ minutes.

   (b)  First convert hours to minutes; 1 hour = 60 minutes.  Replace $x$ with 60 to get
   $7x=7\cdot60=420\text{ ft}^2$.

   (c)  First find the number of square feet the wall contains. The area of the wall is width times height. The area $=8\cdot21=168\text{ ft}^2$. Because the two people paint 7 square feet in 1 minute, it takes them $168\div7=24$ minutes to paint the wall.

146. The area of a triangle is $\dfrac{1}{2}\cdot$ base $\cdot$ height. The base of the triangle is 8 and the height is 4. Thus,

   $\dfrac{1}{2}\cdot8\cdot4=16\text{ ft}^2$.

147. See Figure 147. Since there are 8 pints in 1 gallon, $P=8G$, where $P=$ pints and $G=$ gallons.

| Gallons (G) | 1 | 2 | 3 | 4 | 5 | 6 |
|---|---|---|---|---|---|---|
| Pints (P) | 8 | 16 | 24 | 32 | 40 | 48 |

   Figure 147

148. Because each text message costs 5 cents, $C=0.05x$, where $C$ represents cost and $x$ represents the number of text messages

149. Find the fraction of the population that will be over the age of 65 but under the age of 85. Thus,

   $\dfrac{1}{5}-\dfrac{1}{20}=\dfrac{1\cdot4}{5\cdot4}-\dfrac{1}{20}=\dfrac{4}{20}-\dfrac{1}{20}=\dfrac{3}{20}$.

150. The investment doubles every $72 \div 9 = 8$ years, so in 24 years the investment will double 3 times.

    The investment will increase by $2^3 = 8$ times $25,000 and will equal $200,000.

151. Find the length of each piece when the board is cut into five equal lengths. Because the board

    measures $5\frac{3}{4} = \frac{23}{4}, \frac{23}{4} \div 5 = \frac{23}{4} \cdot \frac{1}{5} = \frac{23}{20} = 1\frac{3}{20}$ feet per piece.

152. Add the individual distances to find the total distance. Thus,

    $3\frac{1}{8} + 4\frac{3}{8} + 6\frac{1}{4} + 1\frac{5}{8} = \frac{25}{8} + \frac{35}{8} + \frac{25}{4} + \frac{13}{8} = \frac{25}{8} + \frac{35}{8} + \frac{50}{8} + \frac{13}{8} = \frac{25+35+50+13}{8} = \frac{123}{8} = 15\frac{3}{8}$ miles.

153. Subtract the withdrawals and add the deposits to the initial balance.

    Thus, $1652 - 78 - 91 + 256 - 638 = \$1101$.

154. The difference between the two temperatures is $108 - (-16) = 108 + 16 = 124°F$.

155. Multiply the real numbers 202 and $\frac{16}{25}$ to obtain $202 \cdot \frac{16}{25} = 129.28$. Total admissions for *Titanic*

    were about 129 million people.

## Chapter 1 Test

1. (a)  The number 29 is a prime number because its only factors are itself and 1.

    (b)  The number 56 is a composite number because it has factors other than itself and 1;
    $56 = 2 \cdot 2 \cdot 2 \cdot 7$.

2. $\frac{5x}{2x-1}$, when $x = -3$, is $\frac{5 \cdot (-3)}{2 \cdot (-3) - 1} = \frac{-15}{-6-1} = \frac{-15}{-7} = \frac{15}{7}$

3. $4^2 - 3 = 16 - 3 = 13$

4. $\frac{24}{32} = \frac{3 \cdot 8}{4 \cdot 8} = \frac{3}{4}$

5. $\frac{2}{10} \cdot \frac{1}{6} = \frac{2 \cdot 1}{10 \cdot 6} = \frac{2}{60} = \frac{2 \cdot 1}{2 \cdot 30} = \frac{1}{30}$

6. The least common denominator for the fractions $\frac{3}{8}$ and $\frac{5}{12}$ is 24, because 24 is the smallest number

    that both 8 and 12 divide into evenly.

7. (a)  $\frac{5}{8} + \frac{1}{8} = \frac{5+1}{8} = \frac{6}{8} = \frac{3 \cdot 2}{4 \cdot 2} = \frac{3}{4}$

    (b)  $\frac{5}{9} - \frac{3}{15} = \frac{5 \cdot 5}{9 \cdot 5} - \frac{3 \cdot 3}{15 \cdot 3} = \frac{25}{45} - \frac{9}{45} = \frac{25-9}{45} = \frac{16}{45}$

    (c)  $\frac{3}{5} \cdot \frac{10}{21} = \frac{3 \cdot 10}{5 \cdot 21} = \frac{30}{105} = \frac{2 \cdot 15}{7 \cdot 15} = \frac{2}{7}$

(d)   $6 \div \dfrac{8}{5} = \dfrac{6}{1} \cdot \dfrac{5}{8} = \dfrac{6 \cdot 5}{1 \cdot 8} = \dfrac{30}{8} = \dfrac{15 \cdot 2}{4 \cdot 2} = \dfrac{15}{4}$

(e)   $\dfrac{5}{12} + \dfrac{4}{9} = \dfrac{5 \cdot 3}{12 \cdot 3} + \dfrac{4 \cdot 4}{9 \cdot 4} = \dfrac{15}{36} + \dfrac{16}{36} = \dfrac{15 + 16}{36} = \dfrac{31}{36}$

(f)   $\dfrac{10}{13} \div 5 = \dfrac{10}{13} \cdot \dfrac{1}{5} = \dfrac{10 \cdot 1}{13 \cdot 5} = \dfrac{10}{65} = \dfrac{2 \cdot 5}{13 \cdot 5} = \dfrac{2}{13}$

8.  $y \cdot y \cdot y \cdot y = y^4$

9.  (a)   $6 + 10 \div 5 = 6 + 2 = 8$

(b)   $4^3 - (3 - 5 \cdot 2) = 64 - (3 - 10) = 64 - (-7) = 64 + 7 = 71$

(c)   $-6^2 - 6 + \dfrac{4}{2} = -36 - 6 + 2 = -40$

(d)   $11 - \dfrac{1 + 3}{6 - 4} = 11 - \dfrac{4}{2} = 11 - 2 = 9$

10.  $\dfrac{7}{20} = 7 \div 20 = 0.35$

11.  (a)   $-1$ is an integer and a rational number.

(b)   $\sqrt{5}$ is an irrational number.

12.

13.  (a)   $2 < |-5|$ because $|-5| = 5$ and $2 < 5$.

(b)   $|-1| > |0|$ because $|-1| = 1$, $|0| = 0$, and $1 > 0$.

14.  (a)   $-5 \div \dfrac{5}{6} = \dfrac{-5}{1} \cdot \dfrac{6}{5} = \dfrac{-5 \cdot 6}{1 \cdot 5} = \dfrac{-6}{1} = -6$

(b)   $-7 \cdot (-3) = 21$

15.  $0.75 = \dfrac{75}{100} = \dfrac{25 \cdot 3}{25 \cdot 4} = \dfrac{3}{4}$

16.  (a)   Distributive

(b)   Associative (multiplication)

(c)   Commutative (addition)

17.  $17 \cdot 102 = 17(100 + 2) = 17(100) + 17(2) = 1700 + 34 = 1734$

18.  (a)   $5 - 5z + 7 + z = 5 + 7 - 5z + z = 12 + (-5 + 1)z = 12 + (-4)z = 12 - 4z$

(b)   $12x - (6 - 3x) = 12x - 6 + 3x = 12x + 3x - 6 = (12 + 3)x - 6 = 15x - 6$

(c)   $5 - 4(x + 6) + \dfrac{15x}{3} = 5 - 4x - 24 + \dfrac{15}{3}x = 5 - 4x - 24 + 5x = 5x - 4x + 5 - 24 =$

$(5 - 4)x - 19 = 1x - 19 = x - 19$

19. (a) Let $x$ be the number of hours. Then, the first person can mow $\frac{4}{3} \cdot x$ acres in $x$ hours, and the

second person can mow $\frac{1}{4} \cdot x$ acres in $x$ hours. Thus,

$$\frac{4}{3}x + \frac{1}{4}x = \frac{16}{12}x + \frac{3}{12}x = \left(\frac{16}{12} + \frac{3}{12}\right) = \frac{19}{12}x.$$

(b) Let $x = 8$. Then, $\frac{19}{12} \cdot 8 = \frac{19 \cdot 8}{12} = \frac{152}{12} = \frac{38 \cdot 4}{12} = \frac{38}{3 \cdot 4} = \frac{38}{3} = 12\frac{2}{3}$ acres.

20. Find $7\frac{4}{5} \div 3$ to find the length of three equal parts.

Because $7\frac{4}{5} = \frac{39}{5}, \frac{39}{5} \div 3 = \frac{39}{5} \cdot \frac{1}{3} = \frac{39}{15} = \frac{13 \cdot 3}{5 \cdot 3} = \frac{13}{5} = 2\frac{3}{5}$ feet.

21. (a) Because $39 \div 3 = 13$, each ticket costs \$13.

Then, let $C = $ cost and $x$ be the number of tickets to obtain $C = 13x$.

(b) Let $x = 17$. Thus, $13 \cdot 17 = \$221$.

22. Subtract the withdrawals from and add the deposits to the initial balance.

Thus, $892 - 57 + 150 - 345 = \$640$.

## Chapter 1 Extended and Discovery Exercises

1. $2 + 2 - 2 - 2 = 4 - 4 = 0; \quad 3 \times 3 + 3 \div 3 = 9 + 1 = 10; \quad 4 \div 4 + 4 - 4 = 1 + 0 = 1;$

$6 \times 6 + 6 - 6 = 36 + 0 = 36; \quad 7 \times 7 + 7 + 7 = 49 + 14 = 63$  *Answers may vary.*

2.

| 16 | 2 | 3 | 13 |
|----|----|----|----|
| 5 | 11 | 10 | 8 |
| 9 | 7 | 6 | 12 |
| 4 | 14 | 15 | 1 |

# Chapter 2: Linear Equations and Inequalities

## Section 2.1 Introduction to Equations

1. solution

3. false

5. solutions

7. $b+c$

9. $bc$

11. equivalent

13. 22

15. 3

17. $x+5=0 \Rightarrow x+5-5=0-5 \Rightarrow x+0=-5 \Rightarrow x=-5;$ To check your answer, substitute $-5$ for $x$ in
the original equation, $-5+5=0.$ This statement is true. Thus, the solution $x=-5$ is correct.

19. $a-12=-3 \Rightarrow a-12+12=-3+12 \Rightarrow a+0=9 \Rightarrow a=9$

21. $9=y-8 \Rightarrow 9+8=y-8+8 \Rightarrow 17=y+0 \Rightarrow 17=y \Rightarrow y=17$

23. $\dfrac{1}{5}=z-\dfrac{3}{2} \Rightarrow \dfrac{1}{5}+\dfrac{3}{2}=z-\dfrac{3}{2}+\dfrac{3}{2} \Rightarrow \dfrac{17}{10}=z+0 \Rightarrow \dfrac{17}{10}=z \Rightarrow z=\dfrac{17}{10}$

25. $t-0.8=4.3 \Rightarrow t-0.8+0.8=4.3+0.8 \Rightarrow t+0=5.1 \Rightarrow t=5.1$

27. $4+x=1 \Rightarrow 4-4+x=1-4 \Rightarrow 0+x=-3 \Rightarrow x=-3$

29. $1=\dfrac{1}{3}+y \Rightarrow 1-\dfrac{1}{3}=\dfrac{1}{3}-\dfrac{1}{3}+y \Rightarrow \dfrac{2}{3}=0+y \Rightarrow \dfrac{2}{3}=y \Rightarrow y=\dfrac{2}{3}$

31. $a$

33. $\dfrac{1}{5}$

35. 6

37. $5x=15 \Rightarrow \dfrac{5x}{5}=\dfrac{15}{5} \Rightarrow x=3$

39. $-7x=0 \Rightarrow \dfrac{-7x}{-7}=\dfrac{0}{-7} \Rightarrow x=0$

41. $-35=-5a \Rightarrow \dfrac{-35}{-5}=\dfrac{-5a}{-5} \Rightarrow 7=a \Rightarrow a=7$

43. $-18=3a \Rightarrow \dfrac{-18}{3}=\dfrac{3a}{3} \Rightarrow -6=a \Rightarrow a=-6$

45. $\dfrac{1}{2}x=\dfrac{3}{2} \Rightarrow \dfrac{2}{1}\cdot\dfrac{1}{2}x=\dfrac{3}{2}\cdot\dfrac{2}{1} \Rightarrow x=3$

47. $\dfrac{1}{2} = \dfrac{2}{5}z \Rightarrow \dfrac{5}{2} \cdot \dfrac{1}{2} = \dfrac{5}{2} \cdot \dfrac{2}{5}z \Rightarrow \dfrac{5}{4} = z \Rightarrow z = \dfrac{5}{4}$

49. $0.5t = 3.5 \Rightarrow \dfrac{0.5t}{0.5} = \dfrac{3.5}{0.5} \Rightarrow t = 7$

51. $-1.7 = 0.2x \Rightarrow \dfrac{-1.7}{0.2} = \dfrac{0.2x}{0.2} \Rightarrow -8.5 = x \Rightarrow x = -8.5$

53. *a*

55. (a)   See Figure 55.

(b)   Let $R$ represent total rainfall and let $x$ represent the number of hours past noon. Start with 3

   inches of rain and then add $\dfrac{1}{2}$, or 0.5, inches per hour after noon,

   $3 + 0.5x = R$, or equivalently $R = 0.5x + 3$.

(c)   At 3 pm, $x = 3$. Substituting $x$ with 3 in the formula, $R = 0.5 \cdot 3 + 3 \Rightarrow R = 4.5$ inches. This

   answer agrees with the table from part (a).

(d)   At 2:15 pm, $x = 2.25$. Substituting $x$ with 2.25 in the formula,

   $R = 0.5 \cdot 2.25 + 3 \Rightarrow R = 4.125$ inches.

| Hours $(x)$ | 0 | 1 | 2 | 3 | 4 | 5 | 6 |
|---|---|---|---|---|---|---|---|
| Rainfall $(R)$ | 3 | 3.5 | 4 | 4.5 | 5 | 5.5 | 6 |

Figure 55

57. (a)   Let $L$ be the length of the football fields and $x$ be the number of fields. Because each field $x$

   contains 300 feet, $L = 300x$.

(b)   Substitute $L$ with 870. Then $870 = 300x$.

(c)   $870 = 300x \Rightarrow \dfrac{870}{300} = \dfrac{300x}{300} \Rightarrow x = \dfrac{870}{300} \Rightarrow x = 2.9$

59. The formula for this scenario is $T = 0.3x$ where $x$ is in days and $T$ is in millions. To find the number

   of days needed for Twitter to add 15 million new accounts, replace the variable $T$ in the formula

   with 15. $T = 0.3x \Rightarrow 15 = 0.3x \Rightarrow \dfrac{15}{0.3} = \dfrac{0.3x}{0.3} \Rightarrow 50 = x$   At this rate it takes 50 days to add 15

   million accounts.

61. a)   The latitude of Winnipeg to the nearest degree is $50° \, N$

   b)   The sun at the equator will be approximately $\dfrac{325}{57} \approx 5.7$ times as intense as they are in

      Winnipeg on March 21$^{\text{st}}$.

63. Let $x$ represent the cost of the car to obtain the equation $0.07x = 1750$. Then the solution is

   $\dfrac{0.07x}{0.07} = \dfrac{1750}{0.07} \Rightarrow x = \dfrac{1750}{0.07} \Rightarrow x = 25,000$. Thus, the cost of the car is \$25,000.

## Section 2.2  Linear Equations

1. constant

3. Exactly one

5. addition, multiplication

7. None

9. $3x - 7 = 0$ is a linear equation. $a = 3$ and $b = -7$.

11. $\frac{1}{2}x = 0$ is a linear equation. $a = \frac{1}{2}$ and $b = 0$.

13. $4x^2 - 6 = 11$ is not a linear equation because it cannot be written in the form $ax + b = 0$. It has a non-zero term containing $x^2$.

15. $\frac{6}{x} - 4 = 2$ is not a linear equation because it cannot be written in the form $ax + b = 0$. It has the variable $x$ in the denominator.

17. $1.1x + 0.9 = 1.8$ is a linear equation.
    $1.1x + 0.9 = 1.8 \Rightarrow 1.1x + 0.9 - 1.8 = 1.8 - 1.8 \Rightarrow 1.1x - 0.9 = 0.$  $a = 1.1$ and $b = -0.9$.

19. $2(x - 3) = 0$ is a linear equation. Use the distributive property to obtain
    $2x - 6 = 0.$ $a = 2$ and $b = -6$.

21. $|3x| + 2 = 1$ is not a linear equation because it cannot be written in the form $ax + b = 0$. It has a non-zero term containing $|x|$.

23. For $x = -1$, substitute $-1$ for $x$ and solve : $-1 - 3 = -4$.

    For $x = 0$, substitute $0$ for $x$ and solve: $0 - 3 = -3$.

    For $x = 1$, substitute $1$ for $x$ and solve : $1 - 3 = -2$.

    For $x = 2$, substitute $2$ for $x$ and solve : $2 - 3 = -1$.

    For $x = 3$, substitute $3$ for $x$ and solve : $3 - 3 = 0$.

    See Figure 23. From the table, we see that the equation $x - 3 = -1$ is true when $x = 2$. Therefore, the solution to the equation $x - 3 = -1$ is $x = 2$.

    | $x$ | $-1$ | $0$ | $1$ | $2$ | $3$ |
    |-----|------|-----|-----|-----|-----|
    | $x - 3$ | $-4$ | $-3$ | $-2$ | $-1$ | $0$ |

    Figure 23

25. For $x = 0$, substitute $0$ for $x$ and solve : $-3(0) + 7 = 0 + 7 = 7$.

    For $x = 1$, substitute $1$ for $x$ and solve: $-3(1) + 7 = -3 + 7 = 4$.

    For $x = 2$, substitute $2$ for $x$ and solve : $-3(2) + 7 = -6 + 7 = 1$.

    For $x = 3$, substitute $3$ for $x$ and solve : $-3(3) + 7 = -9 + 7 = -2$.

For $x = 4$, substitute 4 for $x$ and solve : $-3(4) + 7 = -12 + 7 = -5$.

See Figure 25. From the table, we see that the equation $-3x + 7 = 1$ is true when $x = 2$. Therefore, the solution to the equation $-3x + 7 = 1$ is $x = 2$.

| $x$ | 0 | 1 | 2 | 3 | 4 |
|---|---|---|---|---|---|
| $-3x + 7$ | 7 | 4 | 1 | -2 | -5 |

| $x$ | -2 | -1 | 0 | 1 | 2 |
|---|---|---|---|---|---|
| $4 - 2x$ | 8 | 6 | 4 | 2 | 0 |

Figure 25                                       Figure 27

27. For $x = -2$, substitute $-2$ for $x$ and solve : $4 - 2(-2) = 4 + 4 = 8$.

For $x = -1$, substitute $-1$ for $x$ and solve : $4 - 2(-1) = 4 + 2 = 6$.

For $x = 0$, substitute 0 for $x$ and solve : $4 - 2(0) = 4 - 0 = 4$.

For $x = 1$, substitute 1 for $x$ and solve : $4 - 2(1) = 4 - 2 = 2$.

For $x = 2$, substitute 2 for $x$ and solve : $4 - 2(2) = 4 - 4 = 0$.

See Figure 27. From the table, we see that the equation $4 - 2x = 6$ is true when $x = -1$. Therefore, the solution to the equation $4 - 2x = 6$ is $x = -1$.

29. $11x = 3 \Rightarrow \dfrac{11x}{11} = \dfrac{3}{11} \Rightarrow x = \dfrac{3}{11}$

31. $x - 18 = 5 \Rightarrow x - 18 + 18 = 5 + 18 \Rightarrow x = 23$

33. $\dfrac{1}{2}x - 1 = 13 \Rightarrow \dfrac{1}{2}x - 1 + 1 = 13 + 1 \Rightarrow \dfrac{1}{2}x = 14 \Rightarrow 2 \cdot \dfrac{1x}{2} = 2 \cdot 14 \Rightarrow x = 28$

35. $-6 = 5x + 5 \Rightarrow -6 - 5 = 5x + 5 - 5 \Rightarrow -11 = 5x \Rightarrow \dfrac{-11}{5} = \dfrac{5x}{5} \Rightarrow x = -\dfrac{11}{5}$

37. $3z + 2 = z - 5 \Rightarrow 3z + 2 - 2 = z - 5 - 2 \Rightarrow 3z = z - 7 \Rightarrow 3z - z = z - z - 7 \Rightarrow 2z = -7 \Rightarrow$

$\dfrac{2z}{2} = \dfrac{-7}{2} \Rightarrow z = -\dfrac{7}{2}$

39. $12y - 6 = 33 - y \Rightarrow 12y - 6 + 6 = 33 + 6 - y \Rightarrow 12y = 39 - y \Rightarrow 12y + y = 39 - y + y \Rightarrow$

$13y = 39 \Rightarrow \dfrac{13y}{13} = \dfrac{39}{13} \Rightarrow y = 3$

41. $4(x - 1) = 5 \Rightarrow 4x - 4 = 5 \Rightarrow 4x - 4 + 4 = 5 + 4 \Rightarrow 4x = 9 \Rightarrow \dfrac{4x}{4} = \dfrac{9}{4} \Rightarrow x = \dfrac{9}{4}$

43. $1 - (3x + 1) = 5 - x \Rightarrow 1 - 3x - 1 = 5 - x \Rightarrow -3x = 5 - x \Rightarrow -3x + x = 5 - x + x \Rightarrow$

$-2x = 5 \Rightarrow \dfrac{-2x}{-2} = \dfrac{5}{-2} \Rightarrow x = -\dfrac{5}{2}$

45. $(5t - 6) = 2(t + 1) + 2 \Rightarrow 5t - 6 = 2t + 2 + 2 \Rightarrow 5t - 6 = 2t + 4 \Rightarrow 5t - 2t - 6 + 6 = 2t - 2t + 6 + 4 \Rightarrow$

$3t = 10 \Rightarrow \dfrac{3t}{3} = \dfrac{10}{3} \Rightarrow t = \dfrac{10}{3}$

47. $3(4z-1)-2(z+2)=2(z+1) \Rightarrow 12z-3-2z-4=2z+2 \Rightarrow 10z-7=2z+2 \Rightarrow$

$10z-7+7=2z+2+7 \Rightarrow 10z=2z+9 \Rightarrow 10z-2z=2z-2z+9 \Rightarrow 8z=9 \Rightarrow \dfrac{8z}{8}=\dfrac{9}{8} \Rightarrow z=\dfrac{9}{8}$

49. $7.3x-1.7=5.6 \Rightarrow 7.3x-1.7+1.7=5.6+1.7 \Rightarrow 7.3x=7.3 \Rightarrow \dfrac{7.3x}{7.3}=\dfrac{7.3}{7.3} \Rightarrow x=1$

51. $-9.5x-0.05=10.5x+1.05 \Rightarrow -9.5x-10.5x-0.05=10.5x-10.5x+1.05 \Rightarrow$

$-20x-0.05=1.05 \Rightarrow -20x-0.05+0.05=1.05+0.05 \Rightarrow -20x=1.1 \Rightarrow \dfrac{-20x}{-20}=\dfrac{1.1}{-20} \Rightarrow x=-0.055$

53. $\dfrac{1}{2}x-\dfrac{3}{2}=\dfrac{5}{2} \Rightarrow \dfrac{1}{2}x-\dfrac{3}{2}+\dfrac{3}{2}=\dfrac{5}{2}+\dfrac{3}{2} \Rightarrow \dfrac{1}{2}x=4 \Rightarrow 2 \cdot \dfrac{1}{2}x=4 \cdot 2 \Rightarrow x=8$

55. $-\dfrac{3}{8}x+\dfrac{1}{4}=\dfrac{1}{8} \Rightarrow -\dfrac{3}{8}x=-\dfrac{1}{4}+\dfrac{1}{8} \Rightarrow -\dfrac{3}{8}x=-\dfrac{1}{8} \Rightarrow \left(-\dfrac{8}{3}\right)\left(-\dfrac{3}{8}\right)x=\left(-\dfrac{1}{8}\right)\left(-\dfrac{8}{3}\right) \Rightarrow x=\dfrac{1}{3}$

57. $4y-2(y+1)=0 \Rightarrow 4y-2y-2=0 \Rightarrow 2y-2=0 \Rightarrow 2y-2+2=0+2 \Rightarrow 2y=2 \Rightarrow \dfrac{2y}{2}=\dfrac{2}{2} \Rightarrow y=1$

59. $ax+b=0 \Rightarrow ax+b-b=0-b \Rightarrow ax=-b \Rightarrow \dfrac{ax}{a}=\dfrac{-b}{a} \Rightarrow x=-\dfrac{b}{a}$

61. $5x=5x+1 \Rightarrow 5x-5x=5x-5x+1 \Rightarrow 0=1$

Because the equation $0=1$ is always false, there are no solutions.

63. $8x=0 \Rightarrow \dfrac{8x}{8}=\dfrac{0}{8} \Rightarrow x=0$  Thus, there is one solution.

65. $4x=5(x+3)-x \Rightarrow 4x=5x+15-x \Rightarrow 4x=4x+15 \Rightarrow 4x-4x=4x-4x+15 \Rightarrow 0=15$

Because the equation $0=15$ is always false, there are no solutions.

67. $5(2x+7)-(10x+5)=30 \Rightarrow 10x+35-10x-5=30 \Rightarrow 30=30$

Since the equation $30=30$ is always true, there are infinitely many solutions.

69. $x-(3x+2)=15-2x \Rightarrow x-3x-2=15-2x \Rightarrow -2x-2=15-2x \Rightarrow$

$-2x+2x-2=15-2x+2x \Rightarrow -2=15$

Because the equation $-2=15$ is always false, there are no solutions.

71. (a) See Figure 71.

(b) Let $D$ represent the distance from home and $x$ represent the number of hours. Then $D=4+8x$.

(c) Substitute 3 for $x$. Then, $D=4+8(3)=28$ miles. This agrees with the value found in the table.

(d) Using the formula $D=4+8x$, substitute 22 for $D$. Then, $22=4+8x$. Then, solving for $x$:

$22-4=4-4+8x \Rightarrow 18=8x \Rightarrow \dfrac{18}{8}=\dfrac{8x}{8} \Rightarrow \dfrac{9}{4}=x \Rightarrow x=2.25$ hours. Thus, the bicyclist is 22

miles from home after 2 hours and 15 minutes.

| Hours ($x$) | 0 | 1 | 2 | 3 | 4 |
|---|---|---|---|---|---|
| Distance ($D$) | 4 | 12 | 20 | 28 | 36 |

Figure 71

73. Using the formula, substitute 1730 for $I$ and solve for $x$. $1730 = 241x - 482,440 \Rightarrow$

$$1730 + 482,440 = 241x - 482,440 + 482,440 \Rightarrow 484,170 = 241x \Rightarrow \frac{484,170}{241} = \frac{241x}{241} \Rightarrow x \approx 2009$$

75. Using the formula, substitute 1.5 for $N$ and solve for $x$.

$$1.5 = 0.03x - 58.62 \Rightarrow 1.5 + 58.62 = 0.03x - 58.62 + 58.62$$

$$\Rightarrow 60.12 = 0.03x \Rightarrow \frac{60.12}{0.03} = \frac{0.03x}{0.03} \Rightarrow x = 2004$$

77. Using the formula, substitute 2841 for $H$ and solve for $x$. $2841 = -33x + 69,105 \Rightarrow$

$$2841 - 69,105 = -33x - 69,105 + 69,105 \Rightarrow -66264 = -33x \Rightarrow \frac{-66264}{-33} = \frac{-33x}{-33} \Rightarrow 2008 = x$$

Thus, the number of hospitals reached 2841 in 2008.

## Checking Basic Concepts  Sections 2.1 and 2.2

1. (a)  $4x^3 - 2 = 0$ is not linear because it cannot be written in the form $ax + b = 0$. It has a non-zero

    term containing $x^3$.

   (b)  $2(x+1) = 4$ is a linear equation. Use the distributive property to obtain $2x + 2 = 4$, then

    $2x + 2 - 4 = 4 - 4 \Rightarrow 2x - 2 = 0$.  $a = 2$ and $b = -2$.

2. For $x = 3$, substitute 3 for $x$ and solve:  $4(3) - 3 = 12 - 3 = 9$

   For $x = 3.5$, substitute 3.5 for $x$ and solve:  $4(3.5) - 3 = 14 - 3 = 11$

   For $x = 4$, substitute 4 for $x$ and solve:  $4(4) - 3 = 16 - 3 = 13$

   For $x = 4.5$, substitute 4.5 for $x$ and solve:  $4(4.5) - 3 = 18 - 3 = 15$

   For $x = 5$, substitute 5 for $x$ and solve:  $4(5) - 3 = 20 - 3 = 17$

   See Figure 2. To solve $4x - 3 = 13$, the table tells us that when $x = 4$, $4x - 3 = 13$.

| $x$ | 3 | 3.5 | 4 | 4.5 | 5 |
|---|---|---|---|---|---|
| $4x - 3$ | 9 | 11 | 13 | 15 | 17 |

Figure 2

3. (a)  $x - 12 = 6 \Rightarrow x - 12 + 12 = 6 + 12 \Rightarrow x = 18$  To check the answer, substitute 18 for $x$ in the

    original equation $x - 12 = 6$.  $18 - 12 = 6 \Rightarrow 6 = 6$.  Since this is true $x = 18$ is correct.

   (b)  $\frac{3}{4}z = \frac{1}{8} \Rightarrow \frac{4}{3} \cdot \frac{3}{4}z = \frac{1}{8} \cdot \frac{4}{3} \Rightarrow z = \frac{4}{24} \Rightarrow z = \frac{1}{6}$

(c)    $0.6t + 0.4 = 2 \Rightarrow 0.6t + 0.4 - 0.4 = 2 - 0.4 \Rightarrow 0.6t = 1.6 \Rightarrow \dfrac{0.6t}{0.6} = \dfrac{1.6}{0.6} \Rightarrow t = 2.\overline{6}$

(d)    $5 - 2(x - 2) = 3(4 - x) \Rightarrow 5 - 2x + 4 = 12 - 3x \Rightarrow 9 - 2x = 12 - 3x \Rightarrow$

$9 - 2x + 3x = 12 - 3x + 3x \Rightarrow 9 + x = 12 \Rightarrow 9 - 9 + x = 12 - 9 \Rightarrow x = 3$

4. (a)    $x - 5 = 6x \Rightarrow x - x - 5 = 6x - x \Rightarrow -5 = 5x \Rightarrow \dfrac{-5}{5} = \dfrac{5x}{5} \Rightarrow -1 = x$.  Thus, the equation has

one solution.

(b)    $-2(x - 5) = 10 - 2x \Rightarrow -2x + 10 = 10 - 2x \Rightarrow -2x + 2x + 10 = 10 - 2x + 2x \Rightarrow 10 = 10$

Since $10 = 10$ is always true, the equation has infinitely many solutions.

(c)    $-(x - 1) = -x - 1 \Rightarrow -x + 1 = -x - 1 \Rightarrow -x + x + 1 = -x + x - 1 \Rightarrow 1 = -1$

Since this is never true, the equation has no solutions.

5. (a)    Let $D$ represent distance from home and $x$ represent hours driven.  Note that the driver is
initially 300 miles from home and that each hour driven the driver gets closer to home by 75
miles.  Thus, the formula is $D = 300 - 75x$.

(b)    Since the distance from home, when the driver is home, is 0, use the formula and set $D$ equal to
0.  Thus, $0 = 300 - 75x$.

(c)    $0 = 300 - 75x \Rightarrow 0 + 75x = 300 - 75x + 75x \Rightarrow 75x = 300 \Rightarrow \dfrac{75x}{75} = \dfrac{300}{75} \Rightarrow x = 4$ hours.

## Section 2.3  Introduction to Problem Solving

1. Check your solution.

3. $=$

5. $\dfrac{x}{100}$

7. left

9. $\dfrac{B - A}{A} \cdot 100$

11. $rt$

13. Let $t$ represent the number.  $2 + t = 12 \Rightarrow 2 - 2 + t = 12 - 2 \Rightarrow t = 10$

15. $\dfrac{x}{5} = x - 24 \Rightarrow \dfrac{x}{5} \cdot 5 = 5(x - 24) \Rightarrow x = 5x - 120 \Rightarrow x - 5x = 5x - 5x - 120 \Rightarrow -4x = -120 \Rightarrow$

$\dfrac{-4x}{-4} = \dfrac{-120}{-4} \Rightarrow x = 30$

17. $\dfrac{x + 5}{2} = 7 \Rightarrow \dfrac{x + 5}{2} \cdot 2 = 7 \cdot 2 \Rightarrow x + 5 = 14 \Rightarrow x + 5 - 5 = 14 - 5 \Rightarrow x = 9$

19. $\dfrac{x}{2} = 17 \Rightarrow \dfrac{x}{2} \cdot 2 = 17 \cdot 2 \Rightarrow x = 34$

21. Let the smallest natural number be represented by $x$. $x + (x+1) + (x+2) = 96 \Rightarrow 3x + 3 = 96 \Rightarrow$

    $3x + 3 - 3 = 96 - 3 \Rightarrow 3x = 93 \Rightarrow \dfrac{3x}{3} = \dfrac{93}{3} \Rightarrow x = 31$  Thus, the numbers are 31, 32 and 33.

23. $3x = 102 \Rightarrow \dfrac{3x}{3} = \dfrac{102}{3} \Rightarrow x = 34$

25. $5x = 2x + 24 \Rightarrow 5x - 2x = 2x - 2x + 24 \Rightarrow 3x - 24 \Rightarrow \dfrac{3x}{3} = \dfrac{24}{3} \Rightarrow x = 8$

27. $\dfrac{6x}{7} = 18 \Rightarrow \dfrac{6x}{7} \cdot 7 = 18 \cdot 7 \Rightarrow 6x = 126 \Rightarrow \dfrac{6x}{6} = \dfrac{126}{6} \Rightarrow x = 21$

29. Let $x$ represent the child's current age.

    $x + 10 = 2x + 3 \Rightarrow x - x + 10 = 2x - x + 3 \Rightarrow 10 = x + 3 \Rightarrow 10 - 3 = x + 3 - 3 \Rightarrow 7 = x \Rightarrow x = 7$

31. Let $x$ be the previous weight of the individual.

    $x - 30 = \dfrac{1}{3}x + 110 \Rightarrow 3(x - 30) = 3\left(\dfrac{1}{3}x + 110\right) \Rightarrow 3x - 90 = x + 330 \Rightarrow 3x - x - 90 = x - x + 330 \Rightarrow$

    $2x - 90 = 330 \Rightarrow 2x - 90 + 90 = 330 + 90 \Rightarrow 2x = 420 \Rightarrow \dfrac{2x}{2} = \dfrac{420}{2} \Rightarrow x = 210$ lb

33. Let $x$ be the number of waste sites in Washington.

    $24 = 2x - 2 \Rightarrow 24 + 2 = 2x - 2 + 2 \Rightarrow 26 = 2x \Rightarrow \dfrac{26}{2} = \dfrac{2x}{2} \Rightarrow 13 = x \Rightarrow x = 13$

35. Let $x$ be the number of reptiles on the endangered species list.

    $92 = 2x + 12 \Rightarrow 92 - 12 = 2x - 12 + 12 \Rightarrow 80 = 2x \Rightarrow \dfrac{80}{2} = \dfrac{2x}{2} \Rightarrow 40 = x \Rightarrow x = 40$

37. $x - 1562 = 250 \Rightarrow x - 1562 + 1562 = 250 + 1562 \Rightarrow x = 1812$ million

39. $2x - 4 = 70 \Rightarrow 2x - 4 + 4 = 70 + 4 \Rightarrow 2x = 74 \Rightarrow \dfrac{2x}{2} = \dfrac{74}{2} \Rightarrow x = 37$

    There were 37,000 cosmetic surgeries performed on persons under age 18.

41. $106 = 2(x + 5) + 2x \Rightarrow 106 = 2x + 10 + 2x \Rightarrow 106 = 4x + 10 \Rightarrow 106 - 10 = 4x + 10 - 10 \Rightarrow$

    $96 = 4x \Rightarrow \dfrac{96}{4} = \dfrac{4x}{4} \Rightarrow 24 = x \Rightarrow x = 24$

43. $62 = 2x + 2(x + 7) \Rightarrow 62 = 2x + 2x + 14 \Rightarrow 62 = 4x + 14 \Rightarrow 62 - 14 = 4x + 14 - 14 \Rightarrow$

    $48 = 4x \Rightarrow \dfrac{48}{4} = \dfrac{4x}{4} \Rightarrow 12 = x \Rightarrow x = 12$ and $x + 7 = 19$

    The length is 19 inches and the width is 12 inches.

45. $170 = x + (2x - 10) \Rightarrow 170 = 3x - 10 \Rightarrow 170 + 10 = 3x - 10 + 10 \Rightarrow$

    $180 = 3x \Rightarrow \dfrac{180}{3} = \dfrac{3x}{3} \Rightarrow 60 = x \Rightarrow x = 60$ and $2x - 10 = 110$

    Therefore, Facebook had 60 million users and MySpace had 110 million users.

47. $248,000 = x + (x - 52,000) \Rightarrow 248,000 = 2x - 52,000 \Rightarrow 248,000 + 52,000 = 2x \Rightarrow$

$300,000 = 2x \Rightarrow \dfrac{300,000}{2} = \dfrac{2x}{2} \Rightarrow 150,000 = x \Rightarrow x = 150,000$ and $x - 52,000 = 98,000$

In 2003 there were 150,000 troops and in 2010 there were 98,000 troops.

49. $37\% = \dfrac{37}{100}$          $37\% = 37 \times 0.01 = 0.37$

51. $148\% = \dfrac{148}{100} = \dfrac{37 \cdot 4}{25 \cdot 4} = \dfrac{37}{25}$          $148\% = 148 \times 0.01 = 1.48$

53. $6.9\% = \dfrac{6.9}{100} = \dfrac{6.9}{100} \cdot \dfrac{10}{10} = \dfrac{69}{1000}$          $6.9\% = 6.9 \times 0.01 = 0.069$

55. $0.05\% = \dfrac{0.05}{100} = \dfrac{0.05}{100} \cdot \dfrac{100}{100} = \dfrac{5}{10,000} = \dfrac{1 \cdot 5}{2000 \cdot 5} = \dfrac{1}{2000}$

$0.05\% = 0.05 \times 0.01 = 0.0005$

57. $0.45 = 0.45 \times 100\% = 45\%$

59. $1.8 = 1.8 \times 100\% = 180\%$

61. $0.006 = 0.006 \times 100\% = 0.6\%$

63. $\dfrac{2}{5} = 0.4 = 0.4 \times 100\% = 40\%$

65. $\dfrac{3}{4} = 0.75 = 0.75 \times 100\% = 75\%$

67. $\dfrac{5}{6} = 0.83\overline{3} = 0.83\overline{3} \times 100\% = 83.\overline{3}\%$

69. Let $B$ represent voters in 2008 and let $A$ represent voters in 1980. Then, the percent change in the

number of voters is $\dfrac{B - A}{A} = \dfrac{132.6 - 86.5}{86.5} = \dfrac{46.1}{86.5} \approx 0.533$ or about 53.3%.

71. Calculate the value of 4% of 950, then add the value to 950. Thus, $4\%$ of $950 = .04(950) = 38$.

Then, $950 + 38 = \$988$ per credit.

73. To calculate the total area of Wisconsin, let $x$ represent the unknown number and note that 13.6 is

38% of the unknown number. Then, $0.38x = 13.6 \Rightarrow \dfrac{0.38x}{0.38} = \dfrac{13.6}{0.38} \Rightarrow x \approx 35.8$. Thus, the total area

of Wisconsin is about 35.8 million acres.

75. The percent change from \$1.20 to \$1.50 is, $\dfrac{1.50 - 1.20}{1.20} \cdot 100 = 25$  Thus, the percent change

from \$1.20 to \$1.50 is 25%. The percent change from \$1.50 to \$1.20 is $\dfrac{1.20 - 1.50}{1.50} \cdot 100 = -20$

Thus, the percent change from \$1.50 to \$1.20 is −20%.

77. $d = rt \Rightarrow d = 4 \cdot 2 \Rightarrow d = 8$ miles

79. $d = rt \Rightarrow 1000 = r \cdot 50 \Rightarrow \dfrac{1000}{50} = \dfrac{r \cdot 50}{50} \Rightarrow 20 = r \Rightarrow r = 20$ feet/second

81. $d = rt \Rightarrow 200 = 40t \Rightarrow \dfrac{200}{40} = \dfrac{40t}{40} \Rightarrow 5 = t \Rightarrow t = 5$ hours

83. Given that the distance traveled ($d$) is 255 and that the time spent traveling ($t$) is 4.25 hours, calculate the speed of the car ($r$).

   Then, $d = rt \Rightarrow 255 = r \cdot 4.25 \Rightarrow \dfrac{255}{4.25} = \dfrac{r \cdot 4.25}{4.25} \Rightarrow 60 = r \Rightarrow r = 60$ miles/hour.

85. Let the slower runner be standing still. Then, the faster runner will be traveling at $0 + 2$ mph. Then, this problem is equivalent to solving how long it takes the faster runner to travel $\dfrac{3}{4}$ of a mile. Using the $d = rt$ formula: $\dfrac{3}{4} = 2t \Rightarrow \dfrac{3}{4} \cdot \dfrac{1}{2} = \dfrac{2}{1} \cdot \dfrac{1}{2} t \Rightarrow \dfrac{3}{8} = t \Rightarrow t = \dfrac{3}{8}$. So, in $\dfrac{3}{8}$ hour the faster runner will be $\dfrac{3}{4}$ mile ahead of the slower runner.

87. Let $t$ represent the amount of time spent running 5 mph. Since the total time spent running was 1.1 hours, then $1.1 - t$ represents the amount of time running at 8 mph. Using the $d = rt$ formula, the distance run will equal the sum of $5t$ and $8(1.1 - t)$.

   Thus, $7 = 5t + 8(1.1 - t) \Rightarrow 7 = 5t + 8.8 - 8t \Rightarrow 7 = 8.8 - 3t \Rightarrow$

   $3t + 7 = 8.8 - 3t + 3t \Rightarrow 3t + 7 = 8.8 \Rightarrow 3t + 7 - 7 = 8.8 - 7 \Rightarrow 3t = 1.8 \Rightarrow t = \dfrac{1.8}{3} = 0.6.$

   Therefore, the athlete ran at 5 mph for 0.6 hour and ran at 8 mph for $(1.1 - 0.6) = 0.5$ hour.

89. Since the plane is already 300 miles west of Chicago, it will have to fly $2175 - 300 = 1875$ miles to be 2175 miles west of Chicago. The plane is traveling at 500 mph. Then,

   $1875 = 500t \Rightarrow \dfrac{1875}{500} = \dfrac{500t}{500} \Rightarrow 3.75 = t \Rightarrow t = 3.75.$ Therefore, it will take the plane 3.75 hours to be 2175 miles west of Chicago.

91. Let $x$ represent the amount of water that should be added. Note that there is no salt in pure water and that we will add the amount of pure water to the 3% salt solution to obtain a 1.2% solution. Therefore, set up the equation so that the amount of salt on both sides of the equation is equal. Thus, $x(0.00) + 20(0.03) = (x + 20)(0.012) \Rightarrow 0.00x + 0.6 = 0.012x + 0.24 \Rightarrow$

   $0.6 - 0.24 = 0.012x + 0.24 - 0.24 \Rightarrow 0.36 = 0.012x \Rightarrow \dfrac{0.36}{0.012} = \dfrac{0.012x}{0.012} \Rightarrow 30 = x \Rightarrow x = 30.$

Therefore, 30 ounces of water should be added.

93. Let $x$ represent the amount of the loan at 6% and let $x+1000$ represent the amount of the loan at 5%. The total interest for one year is $215 and this is the sum of the interest paid on the two loans. Therefore: $0.06x+0.05(x+1000) = 215 \Rightarrow 0.06x+0.05x+50 = 215 \Rightarrow$

$0.11x+50-50 = 215-50 \Rightarrow 0.11x = 165 \Rightarrow \dfrac{0.11x}{0.11} = \dfrac{165}{0.11} \Rightarrow x = 1500$. Therefore, the amount of the

loan at 6% interest is $1500 and the amount of the loan at 5% interest is $1500+1000 = \$2500$.

95. Let $x$ represent the amount of 70% antifreeze. Then, the 45% antifreeze mixture is the sum of the 70% mixture and the 30% mixture. Therefore, $0.7x+10(0.3) = (x+10)(0.45) \Rightarrow$

$0.7x+3 = 0.45x+4.5 \Rightarrow 0.7x+3-3 = 0.45x+4.5-3 \Rightarrow$

$0.7x = 0.45x+1.5 \Rightarrow 0.7x-0.45x = 0.45x-0.45x+1.5 \Rightarrow 0.25x = 1.5 \Rightarrow \dfrac{0.25x}{0.25} = \dfrac{1.5}{0.25} \Rightarrow x = 6$.

Therefore, 6 gallons of 70% antifreeze should be mixed with 10 gallons of 30% antifreeze to obtain the 45% mixture.

97. Let $x$ represent the amount of 2.5% cream. Then, the 1% cream is the sum of the 2.5% cream and the base cream $(0\%)$. Therefore $0.025x+15(0) = (x+15)(0.01) \Rightarrow$

$0.025x = 0.01x+0.15 \Rightarrow 0.025x-0.01x = 0.01x-0.01x+0.15 \Rightarrow$

$0.015x = 0.15 \Rightarrow \dfrac{0.015x}{0.015} = \dfrac{0.15}{0.015} \Rightarrow x = 10$. Therefore, 10 grams of 2.5% hydrocortisone cream

should be mixed with the base to obtain the 1% cream.


## Section 2.4  Formulas

1. formula

3. $\dfrac{1}{2}bh$

5. 360

7. $lwh$

9. $2\pi r$

11. $\pi r^2 h$

13. $A = lw$. Thus, $A = 6 \cdot 3 = 18$ ft$^2$.

15. $A = \dfrac{1}{2}bh$. Thus, $A = \dfrac{1}{2} \cdot 6 \cdot 3 = 9$ in$^2$.

17. $A = \pi r^2$. Thus, $A = \pi 4^2 = 16\pi \approx 50.3$ cm$^2$.

19. $A = \dfrac{1}{2}(a+b)h$. Thus, $A = \dfrac{1}{2}(5+6)2 = \dfrac{1}{2}(11)2 = 11$ mm$^2$.

21. $A = lw$. Thus, $A = 13 \cdot 7 = 91$ in$^2$.

23. Convert yards to feet in order to keep the units consistent. Therefore, $7$ yards $= 7 \cdot 3 = 21$ feet.

$A = lw$. Thus, $A = 21 \cdot 5 = 105$ ft$^2$. Or, convert feet to yards to obtain $\dfrac{5}{3}$ yards. Then

$A = lw = 7 \cdot \dfrac{5}{3} = \dfrac{35}{3}$ yd$^2$.

25. $C = 2\pi r$. Because the circle has a diameter of 8 inches, the radius is $= \dfrac{8}{2} = 4$ inches. Thus,

$C = 2\pi r = 2\pi 4 = 8\pi \approx 25.1$ inches.

27. The total area of the lot is the sum of the area of the square and the area of the triangle. The area of

the square is $lw = 52 \cdot 52 = 2704$ ft$^2$. The area of the triangle is $\dfrac{1}{2}bh = \dfrac{1}{2} \cdot 73 \cdot 52 = 1898$ ft$^2$. Thus,

the area of the lot is $1898 + 2704 = 4602$ ft$^2$.

29. The sum of the angles of a triangle is 180°. Let the unknown angle be represented by $x$. Then,
$x + 75 + 40 = 180 \Rightarrow x + 75 + 40 - 75 - 40 = 180 - 75 - 40 \Rightarrow x = 65$. Thus, the third angle is 65°.

31. The sum of the angles of a triangle is 180°. Let the unknown angle be represented by $x$. Then,
$x + 23 + 76 = 180 \Rightarrow x + 23 + 76 - 23 - 76 = 180 - 23 - 76 \Rightarrow x = 81$. Thus, the third angle is 81°.

33. Because the sum of the angles of a triangle is 180, $x + 2x + 2x = 180 \Rightarrow 5x = 180 \Rightarrow \dfrac{5x}{5} = \dfrac{180}{5} \Rightarrow$

$x = 36$. Thus, the value of $x$ is 36°.

35. Let $x$ represent the largest angle. Then, $x + \dfrac{1}{3}x + \dfrac{1}{3}x = 180 \Rightarrow \dfrac{5}{3}x = 180 \Rightarrow \dfrac{3}{5} \cdot \dfrac{5}{3}x = \dfrac{180}{1} \cdot \dfrac{3}{5} \Rightarrow$

$x = \dfrac{540}{5} \Rightarrow x = 108$. Thus, the largest angle has a measure of 108° and the two small angles each

have measure $\dfrac{1}{3} \cdot \dfrac{108}{1} = \dfrac{108}{3} = 36°$.

37. $C = 2\pi r$. Since the diameter of the circle is 12 inches, the radius is $\dfrac{12}{2} = 6$ inches. Then,

$C = 2\pi 6 = 12\pi \approx 37.7$ inches. Then, $A = \pi r^2 = \pi 6^2 = 36\pi \approx 113.1$ in$^2$.

39. $C = 2\pi r$. Then, set $C$ equal to $2\pi$ and solve for $r$.

$2\pi = 2\pi r \Rightarrow \dfrac{2\pi}{2\pi} = \dfrac{2\pi r}{2\pi} \Rightarrow 1 = r \Rightarrow r = 1$ inch. Then, $A = \pi r^2$.

Because $r = 1$, substitute 1 for $r$ and solve for $A$.

$A = \pi(1)^2 = \pi$. Thus, the area is equal to $\pi$, which is approximately equal to $3.14$ in$^2$.

41. $V = lwh$. Thus, $V = 22 \cdot 12 \cdot 10 = 2640$ in$^3$. Surface area equals $2lw + 2lh + 2wh$.

Thus, $S = 2 \cdot 22 \cdot 12 + 2 \cdot 22 \cdot 10 + 2 \cdot 12 \cdot 10 = 528 + 440 + 240 = 1208$ in$^2$.

43. Convert yards to feet. Then, $\frac{2}{3}$ yard $= \frac{2}{3} \cdot 3 = 2$ feet. Then, $V = lwh = 2 \cdot \frac{2}{3} \cdot \frac{3}{2} = 2$ ft$^3$.  Surface area

   equals $2lw + 2lh + 2wh$.  Thus, $S = 2 \cdot 2 \cdot \frac{2}{3} + 2 \cdot 2 \cdot \frac{3}{2} + 2 \cdot \frac{2}{3} \cdot \frac{3}{2} = \frac{8}{3} + 6 + 2 = 8\frac{8}{3} = 10\frac{2}{3}$ ft$^2$.

45. $V = \pi r^2 h$. Thus, $V = \pi 2^2 \cdot 5 = \pi \cdot 4 \cdot 5 = 20\pi$ in$^3$.

47. Convert feet to inches to obtain $h = 2$ feet $= 2$ feet $\cdot \dfrac{12 \text{ inches}}{1 \text{ foot}} = 24$ inches.

   Then, $V = \pi r^2 h$. Thus, $V = \pi 5^2 \cdot 24 = \pi \cdot 25 \cdot 24 = 600\pi$ in$^3$.

49. (a)  $V = \pi r^2 h$. Thus, $V = \pi \left(\dfrac{3}{4}\right)^2 \left(\dfrac{5}{2}\right) = \pi \left(\dfrac{9}{16}\right)\left(\dfrac{5}{2}\right) = \pi \left(\dfrac{45}{32}\right) = \dfrac{45}{32}\pi \approx 4.4$ in$^3$.

   (b)  Because there is about 4.4 cubic inches of volume in the can, the number of fluid ounces is

   $(4.4)(0.554) \approx 2.4$ fl oz.

51. $9x + 3y = 6 \Rightarrow 9x - 9x + 3y = 6 - 9x \Rightarrow 3y = 6 - 9x \Rightarrow \dfrac{3y}{3} = \dfrac{6}{3} - \dfrac{9x}{3} \Rightarrow y = 2 - 3x \Rightarrow y = -3x + 2$

53. $4x + 3y = 12 \Rightarrow 4x - 4x + 3y = 12 - 4x \Rightarrow 3y = 12 - 4x \Rightarrow \dfrac{3y}{3} = \dfrac{12}{3} - \dfrac{4x}{3} \Rightarrow$

   $y = 4 - \dfrac{4}{3}x \Rightarrow y = -\dfrac{4}{3}x + 4$

55. The formula is given as $A = lw$.  To solve for $w$, proceed as follows: $A = lw \Rightarrow \dfrac{A}{l} = \dfrac{lw}{l} \Rightarrow w = \dfrac{A}{l}$.

57. The formula is given as $V = \pi r^2 h$.  To solve for $h$, proceed as follows: $V = \pi r^2 h \Rightarrow \dfrac{V}{\pi r^2} = \dfrac{\pi r^2 h}{\pi r^2} \Rightarrow$

   $h = \dfrac{V}{\pi r^2}$.

59. The formula is given as $A = \dfrac{1}{2}(a+b)h$  To solve for $a$, proceed as follows: $A = \dfrac{1}{2}(a+b)h \Rightarrow$

   $2A = \dfrac{1}{2}(2)(a+b)h \Rightarrow 2A = (a+b)h \Rightarrow \dfrac{2A}{h} = \dfrac{(a+b)h}{h} \Rightarrow \dfrac{2A}{h} = a+b \Rightarrow \dfrac{2A}{h} - b = a+b-b \Rightarrow$

   $a = \dfrac{2A}{h} - b$.

61. The formula is given as $V = lwh$.  To solve for $w$, proceed as follows: $V = lwh \Rightarrow \dfrac{V}{lh} = \dfrac{lwh}{lh} \Rightarrow$

   $w = \dfrac{V}{lh}$.

63. $s = \dfrac{a+b+c}{2} \Rightarrow 2s = \dfrac{a+b+c}{2} \cdot 2 \Rightarrow 2s = a+b+c \Rightarrow 2s - a - c = a+b+c-a-c \Rightarrow$

   $2s - a - c = b \Rightarrow b = 2s - a - c$

65. $\dfrac{a}{b}-\dfrac{c}{b}=1 \Rightarrow b\left(\dfrac{a}{b}-\dfrac{c}{b}\right)=1(b) \Rightarrow a-c=b \Rightarrow b=a-c$

67. $ab=cd+ad \Rightarrow ab-ad=cd+ad-ad \Rightarrow ab-ad=cd \Rightarrow a(b-d)=cd \Rightarrow$

$\dfrac{a(b-d)}{(b-d)}=\dfrac{cd}{(b-d)} \Rightarrow a=\dfrac{cd}{b-d}$

69. Because the perimeter equals the lengths of the four sides, $P=2W+2L.$ Thus, $P=2W+2L \Rightarrow$

$40=2(5)+2L \Rightarrow 40=10+2L \Rightarrow 40-10=10+2L-10 \Rightarrow 2L=30 \Rightarrow \dfrac{2L}{2}=\dfrac{30}{2} \Rightarrow L=15$ Thus, the

length of the rectangle is 15 inches.

71. The formula for GPA is given by $\dfrac{4a+3b+2c+d}{a+b+c+d+f}.$

$\dfrac{4(30)+3(45)+2(12)+1(4)}{30+45+12+4+4}=\dfrac{120+135+24+4}{95}=\dfrac{283}{95}\approx 2.98.$  Thus, the GPA is 2.98.

73. The formula for GPA is given by $\dfrac{4a+3b+2c+d}{a+b+c+d+f}.$

$\dfrac{4(0)+3(60)+2(80)+1(10)}{0+60+80+10+6}=\dfrac{0+180+160+10}{156}=\dfrac{350}{156}\approx 2.24.$  Thus, the GPA is 2.24.

75. To convert Celsius to Fahrenheit temperature, the formula given is $\dfrac{9}{5}C+32=F.$

$\dfrac{9}{5}(25)+32=F \Rightarrow \dfrac{225}{5}+\dfrac{160}{5}=F \Rightarrow \dfrac{385}{5}=F \Rightarrow F=77^\circ\text{F}.$

77. To convert Celsius to Fahrenheit temperature, the formula given is $\dfrac{9}{5}C+32=F.$

$\dfrac{9}{5}(-40)+32=F \Rightarrow \dfrac{-360}{5}+\dfrac{160}{5}=F \Rightarrow \dfrac{-200}{5}=F \Rightarrow F=-40^\circ\text{F}.$

79. To convert Fahrenheit to Celsius temperature, the formula given is $C=\dfrac{5}{9}(F-32).$

$C=\dfrac{5}{9}(23-32) \Rightarrow C=\dfrac{5}{9}(-9) \Rightarrow C=\dfrac{-45}{9} \Rightarrow C=-5^\circ\text{C}.$

81. To convert Fahrenheit to Celsius temperature, the formula given is $C=\dfrac{5}{9}(F-32).$

$C=\dfrac{5}{9}(-4-32) \Rightarrow C=\dfrac{5}{9}(-36) \Rightarrow C=\dfrac{-180}{9} \Rightarrow C=-20^\circ\text{C}.$

83. The formula for calculating gas mileage is $M = \dfrac{D}{G}$, where $D$ is the distance and $G$ is the gasoline

used. Therefore, $M = \dfrac{88,043 - 87,625}{38} \Rightarrow M = \dfrac{418}{38} \Rightarrow M = 11$ mpg.

85. $D = \dfrac{x}{5} \Rightarrow D = \dfrac{12}{5} \Rightarrow D = 2\dfrac{2}{5} = 2.4$ miles.

## Checking Basic Concepts  Sections 2.3 and 2.4

1. (a)  $3x = 36 \Rightarrow \dfrac{3x}{3} = \dfrac{36}{3} \Rightarrow x = 12$

   (b)  $35 - x = 43 \Rightarrow 35 - 35 - x = 43 - 35 \Rightarrow -x = 8 \Rightarrow -1(x) = 8(-1) \Rightarrow x = -8$

2. $x + (x+1) + (x+2) = -93 \Rightarrow 3x + 3 = -93 \Rightarrow 3x + 3 - 3 = -93 - 3 \Rightarrow 3x = -96 \Rightarrow$

   $\dfrac{3x}{3} = \dfrac{-96}{3} \Rightarrow x = -32$.  The three consecutive integers are $-32, -31, -30$.

3. $9.5\% = 0.095$

4. $\dfrac{5}{4} = 1\dfrac{1}{4} = 1.25 = 125\%$

5. Convert 38.8% to the decimal 0.388 and let $x$ represent the unknown rate. Therefore,

   $x - 0.388x = 46,357$.  Thus, $x - 0.388x = 46,357 \Rightarrow 0.612x = 46,357 \Rightarrow$

   $\dfrac{0.612x}{0.612} = \dfrac{46,357}{0.612} \Rightarrow x \approx 75,747$.  Thus, the rate in 2009 was about 75,747.

6. Use the formula $D = rt$, where $D$ is distance, $r$ is the speed and $t$ is the time. Thus, $390 = 60t \Rightarrow$

   $\dfrac{390}{60} = \dfrac{60t}{60} \Rightarrow 6.5 = t$.  Thus, the travel time is 6.5 hours.

7. Let $x$ represent the amount of the loan at 7% and $x + 2000$ represent the amount of the loan at

   6%. Thus, $0.07x + 0.06(x + 2000) = 510 \Rightarrow 0.07x + 0.06x + 120 = 510 \Rightarrow$

   $0.13x + 120 - 120 = 510 - 120 \Rightarrow 0.13x = 390 \Rightarrow \dfrac{0.13x}{0.13} = \dfrac{390}{0.13} \Rightarrow x = 3000$.  Thus, the loan at 7%

   was $3000 and the loan at 6% was $5000.

8. The formula for gas mileage is $M = \dfrac{D}{G}$, where $D$ is the distance and $G$ is the gasoline used.

   Therefore, $28 = \dfrac{504}{G} \Rightarrow 28G = \left(\dfrac{504}{G}\right)G \Rightarrow 28G = 504 \Rightarrow \dfrac{28G}{28} = \dfrac{504}{28} \Rightarrow G = 18$ gal.

9. The area of a triangle is given by $A = \frac{1}{2}bh.$  Thus, $A = \frac{1}{2}bh \Rightarrow 36 = \frac{1}{2}(6)h \Rightarrow 36 = 3h \Rightarrow \frac{36}{3} = \frac{3h}{3} \Rightarrow$

   $12 = h.$  Thus, the height of the triangle is 12 inches.

10. The area of a circle is given by $A = \pi r^2.$  Thus, $A = \pi r^2 \Rightarrow A = \pi(3)^2 \Rightarrow A = 9\pi \approx 28.3 \text{ ft}^2.$

    The circumference of a circle is given by $C = 2\pi r.$  Thus, $C = 2\pi r \Rightarrow C = 2\pi 3 \Rightarrow C = 6\pi \approx 18.8$ feet.

11. $x + 2x + 3x = 180 \Rightarrow 6x = 180 \Rightarrow \frac{6x}{6} = \frac{180}{6} \Rightarrow x = 30.$  Thus, the value of $x$ is $30°$.

12. $A = \pi r^2 + \pi r l \Rightarrow A - \pi r^2 = \pi r^2 - \pi r^2 + \pi r l \Rightarrow A - \pi r^2 = \pi r l \Rightarrow \frac{A - \pi r^2}{\pi r} = \frac{\pi r l}{\pi r} \Rightarrow l = \frac{A - \pi r^2}{\pi r}$

## Section 2.5  Linear Inequalities

1. $<\,;\,\le\,;\,>\,;\,\ge$

3. solution

5. True

7. number line

9. $<$

11.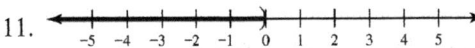

13.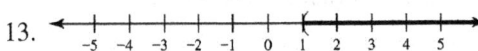

15.

17.

19. $x < 0$

21. $x \le 3$

23. $x \ge 10$

25. $[6, \infty)$

27. $(-2, \infty)$

29. $(-\infty, 7]$

31. Substitute $x$ with 4 in order to test the inequality. Thus, $x + 5 > 5 \Rightarrow 4 + 5 > 5 \Rightarrow 9 > 5.$  Because this inequality statement is true, $x = 4$ is a solution.

33. Substitute $x$ with 5 in order to test the inequality. Thus, $5(5) \ge 25 \Rightarrow 25 \ge 25.$  Because this inequality statement is true, $x = 5$ is a solution.

35. Substitute $y$ with $-3$ in order to test the inequality. Thus, $3(-3) + 5 \ge -8 \Rightarrow -9 + 5 \ge -8 \Rightarrow -4 \ge -8$ Because this inequality statement is true, $y = -3$ is a solution.

37. Substitute $z$ with $-4$ in order to test the inequality. Thus,

$5(-4+1) < 3(-4)-7 \Rightarrow 5(-3) < -12-7 \Rightarrow -15 < -19$ Because this inequality statement is not true,

$z = -4$ is not a solution.

39. $x > -2$

41. $x < 1$

43. To complete the table, insert the $x$ value into $-2x+6$ whenever there is a missing value in the

table. Thus, $-2x+6 \Rightarrow -2(2)+6 \Rightarrow -4+6 = 2; \ -2x+6 \Rightarrow -2(3)+6 \Rightarrow -6+6 = 0;$

$-2x+6 \Rightarrow -2(4)+6 \Rightarrow -8+6 = -2.$ Thus, the missing values in the table are 2, 0 and $-2$.

See Figure 43.

From the table, we see that $-2x+6 \le 0$ whenever $x \ge 3$. Thus, the solution to the inequality is $x \ge 3$.

| $x$ | 1 | 2 | 3 | 4 | 5 |
|---|---|---|---|---|---|
| $-2x+6$ | 4 | 2 | 0 | $-2$ | $-4$ |

Figure 43

45. To complete the table, insert the $x$ value into $5-x$ and $x+7$ whenever there is a missing value in the

table. $5-x \Rightarrow 5-(-2) = 7; \ 5-x \Rightarrow 5-(-1) = 6; \ 5-x \Rightarrow 5-(0) = 5.$

Thus, the missing values in the table that correspond to $5-x$ are 7, 6 and 5.

$x+7 \Rightarrow (-2)+7 = 5; \ x+7 \Rightarrow (-1)+7 = 6; \ x+7 \Rightarrow (0)+7 = 7.$

Thus, the missing values in the table that correspond to $x+7$ are 5, 6 and 7. See Figure 45.

From the table, we see that $5-x > x+7$ whenever $x < -1$. The solution to the inequality is $x < -1$.

| $x$ | $-3$ | $-2$ | $-1$ | 0 | 1 |
|---|---|---|---|---|---|
| $5-x$ | 8 | 7 | 6 | 5 | 4 |
| $x+7$ | 4 | 5 | 6 | 7 | 8 |

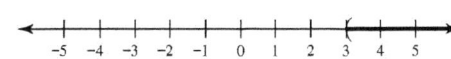

Figure 45          Figure 47

47. $x-3 > 0 \Rightarrow x-3+3 > 0+3 \Rightarrow x > 3.$ See Figure 47.

49. $3-y \le 5 \Rightarrow 3-5-y+y \le 5-5+y \Rightarrow -2 \le y \Rightarrow y \ge -2.$ See Figure 49.

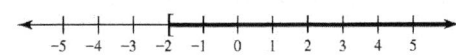

Figure 49

51. $12 < 4+z \Rightarrow 12-4 < 4-4+z \Rightarrow 8 < z \Rightarrow z > 8.$ See Figure 51.

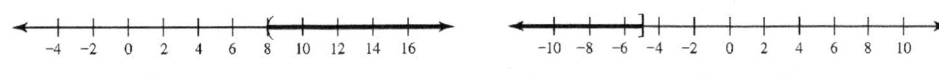

Figure 51          Figure 53

53. $5-2t \ge 10-t \Rightarrow 5-10-2t+2t \ge 10-10-t+2t \Rightarrow -5 \ge t \Rightarrow t \le -5.$ See Figure 53.

55. $2x < 10 \Rightarrow \dfrac{2x}{2} < \dfrac{10}{2} \Rightarrow x < 5.$ See Figure 55.

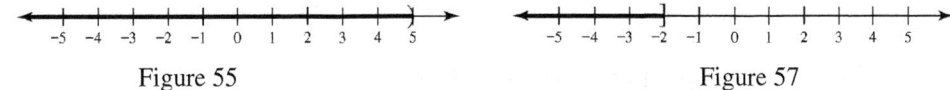

Figure 55          Figure 57

57. $-\dfrac{1}{2}t \geq 1 \Rightarrow -2 \cdot \dfrac{1}{-2}t \leq -2 \cdot 1 \Rightarrow t \leq -2$.  See Figure 57.

59. $\dfrac{3}{4} > -5y \Rightarrow -5y < \dfrac{3}{4} \Rightarrow \dfrac{-5y}{-5} > \dfrac{\frac{3}{4}}{-5} \Rightarrow y > \dfrac{3}{4} \cdot \left(-\dfrac{1}{5}\right) \Rightarrow y > -\dfrac{3}{20}$.  See Figure 59.

**Figure 59**                          **Figure 61**

61. $-\dfrac{2}{3} \leq \dfrac{1}{7}z \Rightarrow \dfrac{1}{7}z \geq -\dfrac{2}{3} \Rightarrow \dfrac{7}{1}\left(\dfrac{1}{7}z\right) \geq -\dfrac{2}{3}\left(\dfrac{7}{1}\right) \Rightarrow z \geq -\dfrac{14}{3}$.  See Figure 61.

63. $x+6 > 7 \Rightarrow x+6-6 > 7-6 \Rightarrow x > 1 \Rightarrow \{x \mid x > 1\}$

65. $-3x \leq 21 \Rightarrow \dfrac{-3x}{-3} \geq \dfrac{21}{-3} \Rightarrow x \geq -7 \Rightarrow \{x \mid x \geq -7\}$

67. $2x-3 < 9 \Rightarrow 2x-3+3 < 9+3 \Rightarrow 2x < 12 \Rightarrow \dfrac{2x}{2} < \dfrac{12}{2} \Rightarrow x < 6 \Rightarrow \{x \mid x < 6\}$

69. $3x+1 < 22 \Rightarrow 3x+1-1 < 22-1 \Rightarrow 3x < 21 \Rightarrow \dfrac{3x}{3} < \dfrac{21}{3} \Rightarrow x < 7$

71. $5-\dfrac{3}{4}x \geq 6 \Rightarrow 5-5-\dfrac{3}{4}x \geq 6-5 \Rightarrow -\dfrac{3}{4}x \geq 1 \Rightarrow -\dfrac{4}{3}\left(-\dfrac{3}{4}x\right) \leq 1\left(-\dfrac{4}{3}\right) \Rightarrow x \leq -\dfrac{4}{3}$

73. $45 > 6-2x \Rightarrow 6-2x < 45 \Rightarrow 6-6-2x < 45-6 \Rightarrow -2x < 39 \Rightarrow \dfrac{-2x}{-2} > \dfrac{39}{-2} \Rightarrow x > -\dfrac{39}{2}$

75. $5x-2 \leq 3x+1 \Rightarrow 5x-3x-2 \leq 3x-3x+1 \Rightarrow 2x-2 \leq 1 \Rightarrow 2x-2+2 \leq 1+2 \Rightarrow 2x \leq 3 \Rightarrow$

$\dfrac{2x}{2} \leq \dfrac{3}{2} \Rightarrow x \leq \dfrac{3}{2}$

77. $-x+24 < x+23 \Rightarrow -x-x+24 < x-x+23 \Rightarrow -2x+24 < 23 \Rightarrow -2x+24-24 < 23-24 \Rightarrow$

$-2x < -1 \Rightarrow \dfrac{-2x}{-2} > \dfrac{-1}{-2} \Rightarrow x > \dfrac{1}{2}$

79. $-(x+1) \geq 3(x-2) \Rightarrow -x-1 \geq 3x-6 \Rightarrow -x-3x-1 \geq 3x-3x-6 \Rightarrow -4x-1 \geq -6 \Rightarrow$

$-4x-1+1 \geq -6+1 \Rightarrow -4x \geq -5 \Rightarrow \dfrac{-4x}{-4} \leq \dfrac{-5}{-4} \Rightarrow x \leq \dfrac{5}{4}$

81. $3(2x+1) > -(5-3x) \Rightarrow 6x+3 > -5+3x \Rightarrow 6x-3x+3 > -5+3x-3x \Rightarrow 3x+3 > -5 \Rightarrow$

$3x+3-3 > -5-3 \Rightarrow 3x > -8 \Rightarrow \dfrac{3x}{3} > \dfrac{-8}{3} \Rightarrow x > -\dfrac{8}{3}$

83. $1.6x+0.4 \leq 0.4x \Rightarrow 1.6x-0.4x+0.4 \leq 0.4x-0.4x \Rightarrow 1.2x+0.4 \leq 0 \Rightarrow$

$1.2x+0.4-0.4 \leq 0-0.4 \Rightarrow 1.2x \leq -0.4 \Rightarrow \dfrac{1.2x}{1.2} \leq \dfrac{-0.4}{1.2} \Rightarrow x \leq -\dfrac{1}{3}$

85. $0.8x-0.5 < x+1-0.5x \Rightarrow 0.8x-0.5 < 0.5x+1 \Rightarrow 0.8x-0.5x-0.5 < 0.5x-0.5x+1 \Rightarrow$

$0.3x-0.5 < 1 \Rightarrow 0.3x-0.5+0.5 < 1+0.5 \Rightarrow 0.3x < 1.5 \Rightarrow \dfrac{0.3x}{0.3} < \dfrac{1.5}{0.3} \Rightarrow x < 5$

87. $-\dfrac{1}{2}\left(\dfrac{2}{3}x+4\right) \ge x \Rightarrow -\dfrac{1}{3}x-2 \ge x \Rightarrow -\dfrac{1}{3}x-x-2 \ge x-x \Rightarrow -\dfrac{4}{3}x-2 \ge 0 \Rightarrow$

$-\dfrac{4}{3}x-2+2 \ge 0+2 \Rightarrow -\dfrac{4}{3}x \ge 2 \Rightarrow -\dfrac{3}{4}\left(-\dfrac{4}{3}x\right) \le 2\left(-\dfrac{3}{4}\right) \Rightarrow x \le -\dfrac{3}{2}$

89. $\dfrac{3}{7}x+\dfrac{2}{7} > -\dfrac{1}{7}x-\dfrac{5}{14} \Rightarrow \dfrac{3}{7}x+\dfrac{1}{7}x+\dfrac{2}{7} > -\dfrac{1}{7}x+\dfrac{1}{7}x-\dfrac{5}{14} \Rightarrow \dfrac{4}{7}x+\dfrac{2}{7} > -\dfrac{5}{14} \Rightarrow$

$\dfrac{4}{7}x+\dfrac{2}{7}-\dfrac{2}{7} > -\dfrac{5}{14}-\dfrac{2}{7} \Rightarrow \dfrac{4}{7}x > -\dfrac{5}{14}-\dfrac{4}{14} \Rightarrow \dfrac{4}{7}x > -\dfrac{9}{14} \Rightarrow \dfrac{7}{4}\left(\dfrac{4}{7}x\right) > -\dfrac{9}{14}\left(\dfrac{7}{4}\right) \Rightarrow x > -\dfrac{63}{56} \Rightarrow x > -\dfrac{9}{8}$

91. $\dfrac{x}{3}+\dfrac{5x}{6} \le \dfrac{2}{3} \Rightarrow \dfrac{2x}{6}+\dfrac{5x}{6} \le \dfrac{2}{3} \Rightarrow \dfrac{7x}{6} \le \dfrac{2}{3} \Rightarrow 6\left(\dfrac{7x}{6}\right) \le \dfrac{2}{3}(6) \Rightarrow 7x \le 4 \Rightarrow \dfrac{7x}{7} \le \dfrac{4}{7} \Rightarrow x \le \dfrac{4}{7}$

93. $\dfrac{6x}{7} < \dfrac{1}{3}x+1 \Rightarrow \dfrac{6x}{7}-\dfrac{1}{3}x < \dfrac{1}{3}x-\dfrac{1}{3}x+1 \Rightarrow \dfrac{6x}{7}-\dfrac{x}{3} < 1 \Rightarrow \dfrac{18x}{21}-\dfrac{7x}{21} < 1 \Rightarrow \dfrac{11x}{21} < 1 \Rightarrow$

$21\left(\dfrac{11x}{21}\right) < 1(21) \Rightarrow 11x < 21 \Rightarrow \dfrac{11x}{11} < \dfrac{21}{11} \Rightarrow x < \dfrac{21}{11}$

95. $x > 60$

97. $x \ge 21$

99. $x > 40,000$

101. $x \le 70$

103. $2(x+5)+2x < 50 \Rightarrow 2x+10+2x < 50 \Rightarrow 4x+10 < 50 \Rightarrow 4x+10-10 < 50-10 \Rightarrow$

$4x < 40 \Rightarrow \dfrac{4x}{4} < \dfrac{40}{4} \Rightarrow x < 10$ feet.

105. Let $x$ represent the unknown test score. Then, $\dfrac{74+x}{2} \ge 80 \Rightarrow 2\left(\dfrac{74+x}{2}\right) \ge 80(2) \Rightarrow 74+x \ge 160 \Rightarrow$

$74-74+x \ge 160-74 \Rightarrow x \ge 86$. Thus, the student needs a score of 86 or more to maintain an average of at least 80.

107. Let $x$ represent the number of hours parked after the first half hour. We see that there is a $2.00 cost for the first half hour and $1.25 cost for each hour after that. Therefore,

$2+1.25x \le 8 \Rightarrow 2-2+1.25x \le 8-2 \Rightarrow 1.25x \le 6 \Rightarrow \dfrac{1.25x}{1.25} \le \dfrac{6}{1.25} \Rightarrow x \le 4.8$. This result would

indicate that the student can park for as long as 4.8 hours beyond the first half hour for $8.00. However, because a partial hour of parking is charged as a full hour, the longest amount of time that the student could park beyond the first hour is 4 hours for a total of 4.5 hours.

109. Let $x$ represent the number of days. Then, $25x+0.20(90)x \le 200 \Rightarrow 25x+18x \le 200 \Rightarrow$

$43x \leq 200 \Rightarrow \dfrac{43x}{43} \leq \dfrac{200}{43} \Rightarrow x \leq 4.65$. Because the car can not be rented for a partial day, the person

can rent the car for 4 days.

111. (a)    $C = 1.5x + 2000$

(b)    $R = 12x$

(c)    $P = 12x - (1.5x + 2000) \Rightarrow P = 10.5x - 2000$

(d)    To yield a positive profit, revenue must be greater than cost. Then, $12x > 1.5x + 2000 \Rightarrow$

$12x - 1.5x > 1.5x - 1.5x + 2000 \Rightarrow 10.5x > 2000 \Rightarrow \dfrac{10.5x}{10.5} > \dfrac{2000}{10.5} \Rightarrow x > 190.476$. Thus, 191 or

more compact discs must be sold to yield a profit.

113. (a)    Set the distances equal and then solve for $x$. Then, $\dfrac{1}{6}x = \dfrac{1}{8}x + 2 \Rightarrow 4x = 3x + 48 \Rightarrow x = 48$.

Thus, at 48 minutes the athletes are the same distance from the parking lot.

(b)    $\dfrac{1}{6}x > \dfrac{1}{8}x + 2 \Rightarrow 4x > 3x + 48 \Rightarrow x > 48$. Thus, after more than 48 minutes, the first athlete is

farther from the parking lot than the second athlete.

115. Because $T = 90 - 19x$, set an inequality statement with $T$ equal to 4.5 and solve. Then,

$90 - 19x < 4.5 \Rightarrow 90 - 90 - 19x < 4.5 - 90 \Rightarrow -19x < -85.5 \Rightarrow \dfrac{-19x}{-19} > \dfrac{-85.5}{-19} \Rightarrow x > 4.5$.

Thus, at altitudes more than 4.5 miles, the air temperature is less than 4.5°F.


## Checking Basic Concepts  Section 2.5

1.

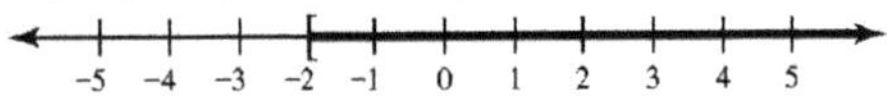

2.    $x < 1$

3.    Substitute $-3$ into the inequality. Thus $4(-3) - 5 \leq -15 \Rightarrow -12 - 5 + 5 \leq -15 \Rightarrow -17 \leq -15$.

Thus, $-3$ is a solution to the inequality.

4.    When $x = -2$, then $5 - 2(-2) = 5 + 4 = 9$;  When $x = -1$, then $5 - 2(-1) = 5 + 2 = 7$;

When $x = 0$, then $5 - 2(0) = 5 - 0 = 5$;  When $x = 1$, then $5 - 2(1) = 5 - 2 = 3$. Therefore, the

numbers that complete the table are 9, 7, 5 and 3. See Figure 4.

| $x$ | $-2$ | $-1$ | 0 | 1 | 2 |
|---|---|---|---|---|---|
| $5 - 2x$ | 9 | 7 | 5 | 3 | 1 |

Figure 4

From the table, we see that $5 - 2x \leq 7$ whenever $x \geq -1$.

Thus, the solution to the inequality is $x \geq -1$.

5. (a) $x+5>8 \Rightarrow x+5-5>8-5 \Rightarrow x>3$

(b) $-\dfrac{5}{7}x \le 25 \Rightarrow -\dfrac{7}{5}\left(-\dfrac{5}{7}x\right) \ge 25\left(-\dfrac{7}{5}\right) \Rightarrow x \ge -\dfrac{175}{5} \Rightarrow x \ge -35$

(c) $3x \ge -2(1-2x)+3 \Rightarrow 3x \ge -2+4x+3 \Rightarrow 3x \ge 1+4x \Rightarrow 3x-4x \ge 1+4x-4x \Rightarrow$

   $-x \ge 1 \Rightarrow -1(-x) \le -1(1) \Rightarrow x \le -1$

6. $x \le 12$

7. Let $l$ represent length and $w$ represent width. Then, $l=2w+5$. Therefore, $2(2w+5)+2w>88 \Rightarrow$

   $4w+10+2w>88 \Rightarrow 6w+10>88 \Rightarrow 6w+10-10>88-10 \Rightarrow 6w>78 \Rightarrow \dfrac{6w}{6}>\dfrac{78}{6} \Rightarrow w>13.$

   Thus, the possible widths must be more than 13 inches.

## Chapter 2  Review

1. $x+9=3 \Rightarrow x+9-9=3-9 \Rightarrow x=-6$  Check:  $-6+9=3 \Rightarrow 3=3$

2. $x-4=-2 \Rightarrow x-4+4=-2+4 \Rightarrow x=2$  Check:  $2-4=-2 \Rightarrow -2=-2$

3. $x-\dfrac{3}{4}=\dfrac{3}{2} \Rightarrow x-\dfrac{3}{4}+\dfrac{3}{4}=\dfrac{3}{2}+\dfrac{3}{4} \Rightarrow x=\dfrac{6}{4}+\dfrac{3}{4} \Rightarrow x=\dfrac{9}{4}$  Check:  $\dfrac{9}{4}-\dfrac{3}{4}=\dfrac{3}{2} \Rightarrow \dfrac{6}{4}=\dfrac{3}{2} \Rightarrow \dfrac{3}{2}=\dfrac{3}{2}$

4. $x+0.5=0 \Rightarrow x+0.5-0.5=0-0.5 \Rightarrow x=-0.5 \Rightarrow x=-\dfrac{1}{2}$

   Check:  $-\dfrac{1}{2}+0.5=0 \Rightarrow -0.5+0.5=0 \Rightarrow 0=0$

5. $4x=12 \Rightarrow \dfrac{4x}{4}=\dfrac{12}{4} \Rightarrow x=3$  Check:  $4(3)=12 \Rightarrow 12=12$

6. $3x=-7 \Rightarrow \dfrac{3x}{3}=\dfrac{-7}{3} \Rightarrow x=-\dfrac{7}{3}$  Check:  $3\left(-\dfrac{7}{3}\right)=-7 \Rightarrow -7=-7$

7. $-0.5x=1.25 \Rightarrow \dfrac{-0.5x}{-0.5}=\dfrac{1.25}{-0.5} \Rightarrow x=-2.5$  Check:  $-0.5(-2.5)=1.25 \Rightarrow 1.25=1.25$

8. $-\dfrac{1}{3}x=\dfrac{7}{6} \Rightarrow -\dfrac{3}{1}\left(-\dfrac{1}{3}x\right)=\dfrac{7}{6}\left(-\dfrac{3}{1}\right) \Rightarrow x=-\dfrac{21}{6} \Rightarrow x=-\dfrac{7}{2}$  Check:  $-\dfrac{1}{3}\left(-\dfrac{7}{2}\right)=\dfrac{7}{6} \Rightarrow \dfrac{7}{6}=\dfrac{7}{6}$

9. The equation $-4x+3=2$ is linear.  $-4x+3=2 \Rightarrow -4x+3-2=2-2 \Rightarrow -4x+1=0;$  $a=-4,\ b=1$

10. The equation $\dfrac{3}{8}x^2-x=\dfrac{1}{4}$ is not a linear equation because it cannot be written in the form

    $ax+b=0.$

11. $4x-5=3 \Rightarrow 4x-5+5=3+5 \Rightarrow 4x=8 \Rightarrow \dfrac{4x}{4}=\dfrac{8}{4} \Rightarrow x=2.$  To check the solution, substitute 2 for $x$

    in the original equation:  $4x-5=3 \Rightarrow 4(2)-5=3 \Rightarrow 8-5=3.$  Because this statement is true, the solution checks.

12. $7 - \frac{1}{2}x = -4 \Rightarrow 7 - 7 - \frac{1}{2}x = -4 - 7 \Rightarrow -\frac{1}{2}x = -11 \Rightarrow -\frac{2}{1}\left(-\frac{1}{2}x\right) = -\frac{11}{1}\left(-\frac{2}{1}\right) \Rightarrow x = 22.$ To check

the solution, substitute 22 for $x$ in the original equation:

$7 - \frac{1}{2}x = -4 \Rightarrow 7 - \frac{1}{2}(22) = -4 \Rightarrow 7 - 11 = -4.$ Because this statement is true, the solution checks.

13. $5(x - 3) = 12 \Rightarrow 5x - 15 = 12 \Rightarrow 5x - 15 + 15 = 12 + 15 \Rightarrow 5x = 27 \Rightarrow \frac{5x}{5} = \frac{27}{5} \Rightarrow x = \frac{27}{5}.$ To check the

solution, substitute $\frac{27}{5}$ for $x$ in the original equation:  $5\left(\frac{27}{5} - 3\right) = 12 \Rightarrow 5\left(\frac{27}{5} - \frac{15}{5}\right) = 12 \Rightarrow$

$5\left(\frac{12}{5}\right) = 12 \Rightarrow \frac{60}{5} = 12.$  Because this statement is true, the solution checks.

14. $3 + x = 2x - 4 \Rightarrow 3 + x - x = 2x - x - 4 \Rightarrow 3 = x - 4 \Rightarrow 3 + 4 = x - 4 + 4 \Rightarrow 7 = x \Rightarrow x = 7.$

To check the solution, substitute 7 for $x$ in the original equation:

$3 + x = 2x - 4 \Rightarrow 3 + 7 = 2(7) - 4 \Rightarrow 10 = 14 - 4 \Rightarrow 10 = 10.$  Because this statement is true, the

solution checks.

15. $2(x - 1) = 4(x + 3) \Rightarrow 2x - 2 = 4x + 12 \Rightarrow 2x - 2x - 2 = 4x - 2x + 12 \Rightarrow -2 = 2x + 12$

$\Rightarrow -2 - 12 = 2x + 12 - 12 \Rightarrow -14 = 2x \Rightarrow \frac{-14}{2} = \frac{2x}{2} \Rightarrow -7 = x \Rightarrow x = -7.$

To check the solution, substitute $-7$ for $x$ in the original equation:

$2(x - 1) = 4(x + 3) \Rightarrow 2(-7 - 1) = 4(-7 + 3) \Rightarrow 2(-8) = 4(-4) \Rightarrow -16 = -16.$  Because this statement

is true, the solution checks.

16. $1 - (x - 3) = 6 + 2x \Rightarrow 1 - x + 3 = 6 + 2x \Rightarrow 4 - x = 6 + 2x \Rightarrow 4 - x + x = 6 + 2x + x \Rightarrow$

$4 = 6 + 3x \Rightarrow 4 - 6 = 6 - 6 + 3x \Rightarrow -2 = 3x \Rightarrow \frac{-2}{3} = \frac{3x}{3} \Rightarrow -\frac{2}{3} = x \Rightarrow x = -\frac{2}{3}.$  To check the solution,

substitute $-\frac{2}{3}$ for $x$ in the original equation:  $1 - \left(-\frac{2}{3} - 3\right) = 6 + 2\left(-\frac{2}{3}\right) \Rightarrow$

$1 + \frac{2}{3} + 3 = 6 - \frac{4}{3} \Rightarrow 4\frac{2}{3} = 4\frac{2}{3}.$  Because this statement is true, the solution checks.

17. $3.4x - 4 = 5 - 0.6x \Rightarrow 3.4x - 4 + 4 = 5 + 4 - 0.6x \Rightarrow 3.4x = 9 - 0.6x \Rightarrow$

$3.4x + 0.6x = 9 - 0.6x + 0.6x \Rightarrow 4x = 9 \Rightarrow \frac{4x}{4} = \frac{9}{4} \Rightarrow x = \frac{9}{4}.$  To check the solution, substitute $\frac{9}{4}$ for

$x$ in the original equation:  $3.4\left(\frac{9}{4}\right) - 4 = 5 - 0.6\left(\frac{9}{4}\right) \Rightarrow 3.4(2.25) - 4 = 5 - 0.6(2.25) \Rightarrow$

$7.65 - 4 = 5 - 1.35 \Rightarrow 3.65 = 3.65.$  Because this statement is true, the solution checks.

18. $-\frac{1}{3}(3-6x)=-(x+2)+1 \Rightarrow -1+2x=-x-2+1 \Rightarrow 2x-1=-x-1 \Rightarrow$

$2x-1+1=-x-1+1 \Rightarrow 2x=-x \Rightarrow 2x+x=-x+x \Rightarrow 3x=0 \Rightarrow \frac{3x}{3}=\frac{0}{3} \Rightarrow x=0.$

To check the solution, substitute 0 for $x$ in the original equation: $-\frac{1}{3}(3-6(0))=-(0+2)+1 \Rightarrow$

$-\frac{1}{3}(3)=-2+1 \Rightarrow -1=-1.$ Because this statement is true, the solution checks.

19. $\frac{2}{3}x-\frac{1}{6}=\frac{5}{12} \Rightarrow \frac{2}{3}x-\frac{1}{6}+\frac{1}{6}=\frac{5}{12}+\frac{1}{6} \Rightarrow \frac{2}{3}x=\frac{5}{12}+\frac{2}{12} \Rightarrow \frac{2}{3}x=\frac{7}{12} \Rightarrow \frac{3}{2}\left(\frac{2}{3}x\right)=\frac{7}{12}\left(\frac{3}{2}\right) \Rightarrow$

$x=\frac{21}{24} \Rightarrow x=\frac{7}{8}.$ To check the solution, substitute $\frac{7}{8}$ for $x$ in the original equation:

$\frac{2}{3}\left(\frac{7}{8}\right)-\frac{1}{6}=\frac{5}{12} \Rightarrow \frac{14}{24}-\frac{4}{24}=\frac{10}{24} \Rightarrow \frac{10}{24}=\frac{10}{24}.$ Because this statement is true, the solution checks.

20. $2y-3(2-y)=5+y \Rightarrow 2y-6+3y=5+y \Rightarrow 5y-6=5+y \Rightarrow 5y-y-6=5+y-y \Rightarrow$

$4y-6=5 \Rightarrow 4y-6+6=5+6 \Rightarrow 4y=11 \Rightarrow \frac{4y}{4}=\frac{11}{4} \Rightarrow y=\frac{11}{4}.$ To check the solution, substitute

$\frac{11}{4}$ for $y$ in the original equation: $2\left(\frac{11}{4}\right)-3\left(2-\frac{11}{4}\right)=5+\frac{11}{4} \Rightarrow \frac{22}{4}-6+\frac{33}{4}=\frac{20}{4}+\frac{11}{4} \Rightarrow$

$\frac{22}{4}-\frac{24}{4}+\frac{33}{4}=\frac{20}{4}+\frac{11}{4} \Rightarrow \frac{31}{4}=\frac{31}{4}.$ Because this statement is true, the solution checks.

21. First, solve for $x$: $4(3x-2)=2(6x+5) \Rightarrow 12x-8=12x+10 \Rightarrow$

$12x-12x-8=12x-12x+10 \Rightarrow -8=10.$ Because this statement is not true, the equation has no

solutions.

22. First, solve for $x$: $5(3x-1)=15x-5 \Rightarrow 15x-5=15x-5.$ Because this statement is true for any

value of $x$, the equation has infinitely many solutions.

23. First, solve for $x$: $8x=5x+3x \Rightarrow 8x=8x.$ Because this statement is true for any value of $x$, the

equation has infinitely many solutions.

24. First solve for $x$: $9x-2=8x-2 \Rightarrow 9x-8x-2=8x-8x-2 \Rightarrow x-2=-2 \Rightarrow$

$x-2+2=-2+2 \Rightarrow x=0.$ Thus, there is one solution to the equation.

25. When $x=1.0,$ then $-2(1.0)+3=-2+3=1;$ When $x=1.5,$ then $-2(1.5)+3=-3+3=0;$

When $x=2.0,$ then $-2(2.0)+3=-4+3=-1;$ When $x=2.5,$ then $-2(2.5)+3=-5+3=-2;$

Thus, the missing values in the table are 1, 0, $-1$ and $-2.$ See Figure 25. From the table we see

that when $x=1.5,$ the value of $-2x+3$ is 0.

| $x$ | 0.5 | 1.0 | 1.5 | 2.0 | 2.5 |
|---|---|---|---|---|---|
| $-2x+3$ | 2 | 1 | 0 | $-1$ | $-2$ |

Figure 25

| $x$ | $-2$ | $-1$ | 0 | 1 | 2 |
|---|---|---|---|---|---|
| $-(x+1)+3$ | 4 | 3 | 2 | 1 | 0 |

Figure 26

26. When $x=-2$, then $-(-2+1)+3=2-1+3=4$;   When $x=-1$, then $-(-1+1)+3=1-1+3=3$;

When $x=0$, then $-(0+1)+3=0-1+3=2$;   When $x=1$, then $-(1+1)+3=-2+3=1$;

When $x=2$, then $-(2+1)+3=-3+3=0$;   Thus, the missing values in the table are

4, 3, 2, 1 and 0   See Figure 26.  From the table we see that when $x=0$, the value

of $-(x+1)+3$ is 2.

27. $6x=72 \Rightarrow \dfrac{6x}{6}=\dfrac{72}{6} \Rightarrow x=12$

28. $x+18=-23 \Rightarrow x+18-18=-23-18 \Rightarrow x=-41$

29. $2x-5=x+4 \Rightarrow 2x-5+5=x+4+5 \Rightarrow 2x=x+9 \Rightarrow 2x-x=x-x+9 \Rightarrow x=9$

30. $x+4=3x \Rightarrow x-3x+4=3x-3x \Rightarrow -2x+4=0 \Rightarrow -2x+4-4=0-4 \Rightarrow -2x=-4 \Rightarrow$

$\dfrac{-2x}{-2}=\dfrac{-4}{-2} \Rightarrow x=2$

31. $\dfrac{5x}{3}=15 \Rightarrow \dfrac{3}{5}\cdot\dfrac{5x}{3}=\dfrac{3}{5}\cdot\dfrac{15}{1} \Rightarrow x=9$

32. $x+(x+1)+(x+2)=-153 \Rightarrow 3x+3=-153 \Rightarrow 3x+3-3=-153-3 \Rightarrow 3x=-156 \Rightarrow$

$\dfrac{3x}{3}=\dfrac{-156}{3} \Rightarrow x=-52$.  The numbers are $-52$, $-51$ and $-50$.

33. $85\%=\dfrac{85}{100}=\dfrac{17}{20}$;  $85\%=0.85$

34. $5.6\%=\dfrac{5.6}{100}=\dfrac{56}{1000}=\dfrac{7}{125}$;  $5.6\%=0.056$

35. $0.03\%=\dfrac{0.03}{100}=\dfrac{3}{10,000}$;  $0.03\%=0.0003$

36. $342\%=\dfrac{342}{100}=\dfrac{171}{50}$;  $342\%=3.42$

37. $0.89=89\%$

38. $0.005=0.5\%$

39. $2.3=230\%$

40. $1=100\%$

41. $d=rt \Rightarrow d=8(3) \Rightarrow d=24$ miles.

42. $d=rt \Rightarrow d=70(55) \Rightarrow d=3850$ feet.

43. $d = rt \Rightarrow 500 = r(20) \Rightarrow \dfrac{500}{20} = r\dfrac{20}{20} \Rightarrow \dfrac{500}{20} = r \Rightarrow r = 25$ yd/sec.

44. $d = rt \Rightarrow 125 = 15t \Rightarrow \dfrac{125}{15} = \dfrac{15t}{15} \Rightarrow \dfrac{125}{15} = t \Rightarrow t = \dfrac{25}{3}$ hours.

45. The area of a triangle is $\dfrac{1}{2}(\text{base})(\text{height})$.  Thus, $A = \dfrac{1}{2}(b)(h) = \dfrac{1}{2}(5)(3) = 7.5$ m$^2$.

46. The area of a circle is $\pi r^2$ where $r$ represents radius.  Thus, $A = \pi r^2 = \pi\left(6^2\right) = 36\pi \approx 113.1$ ft$^2$.

47. The area of a rectangle is given as length ($l$) times width ($w$).  Thus,

    $A = lw = (36)(24) = 864$ in$^2$ or 6 ft$^2$.

48. $P = 2l + 2w$ where $P$ represents perimeter, $l$ represents the length of the base and $w$ represents the

    width.  Thus, $P = 2l + 2w = 2(13) + 2(7) = 26 + 14 = 40$ in.

49. The circumference of a circle is $2\pi r$, where $r$ represents radius.  Thus,

    $r = \dfrac{1}{2}(\text{diameter}) = \dfrac{1}{2}(18) = 9.$   $C = 2\pi r = 2\pi(9) = 18\pi \approx 56.5$ feet.

50. $A = \pi r^2$, where $A$ represents area and $r$ represents radius.  Thus, $A = \pi r^2 = \pi\left(5^2\right) = 25\pi \approx 78.5$ in$^2$.

51. The angles in a triangle must add up to $180°$.  Let $x$ represent the unknown angle.  Thus,

    $90 + 40 + x = 180 \Rightarrow 130 + x = 180 \Rightarrow 130 - 130 + x = 180 - 130 \Rightarrow x = 50°.$

52. The angles in a triangle must add up to $180°$.  Thus, $x + 3x + 4x = 180 \Rightarrow 8x = 180 \Rightarrow \dfrac{8x}{8} = \dfrac{180}{8} \Rightarrow$

    $x = 22.5°.$

53. $V = \pi r^2 h = \pi\left(5^2\right)(25) = \pi(25)(25) = 625\pi \approx 1963.5$ in$^3$.

54. First, convert height ($h$) and base ($b$) to inches.  $h = 5$ feet $\cdot \dfrac{12 \text{ inches}}{1 \text{ foot}} = 60$ inches and

    $b = 3$ feet $\cdot \dfrac{12 \text{ inches}}{1 \text{ foot}} = 36$ inches.

    Then, $A = \dfrac{1}{2}(a + b)h = \dfrac{1}{2}(36 + 18)60 = \dfrac{1}{2}(54)60 = (27)60 = 1620$ in$^2$.

    Or, convert the base in inches to feet.  $b = 18$ inches $\cdot \dfrac{1 \text{ foot}}{12 \text{ inches}} = 1.5$ feet.  Then,

    $A = \dfrac{1}{2}(a + b)h = \dfrac{1}{2}(3 + 1.5)5 = \dfrac{1}{2}(4.5)5 = (2.25)5 = 11.25$ ft$^2$.

55. The total area of the figure is the sum of the area of the triangle and the area of the rectangle.  The

    area of the triangle is $\dfrac{1}{2}bh = \dfrac{1}{2} \cdot 8 \cdot 6 = 24$ in$^2$.  The area of the rectangle is $lw = 25 \cdot 6 = 150$ in$^2$.

    Thus, the total area is $24 + 150 = 174$ in$^2$.

56. The total area of the figure is the sum of the area of the rectangle and the area of a circle. The area of

the rectangle is $lw = 12 \cdot 4 = 48 \text{ ft}^2.$ The area of the circle is $\pi r^2 = \pi\left(\dfrac{4}{2}\right)^2 = \pi(2)^2 = 4\pi \approx 12.6 \text{ ft}^2.$

Thus, the total area is about $48 + 12.6 = 60.6 \text{ ft}^2.$

57. $3x = 5 + y \Rightarrow 3x - 5 = 5 - 5 + y \Rightarrow 3x - 5 = y$

58. $16 = 2x + 2y \Rightarrow 16 - 2x = 2x - 2x + 2y \Rightarrow 16 - 2x = 2y \Rightarrow \dfrac{16 - 2x}{2} = \dfrac{2y}{2} \Rightarrow 8 - x = y \Rightarrow y = -x + 8$

59. $z = 2xy \Rightarrow \dfrac{z}{2x} = \dfrac{2xy}{2x} \Rightarrow \dfrac{z}{2x} = y \Rightarrow y = \dfrac{z}{2x}$

60. $S = \dfrac{a+b+c}{3} \Rightarrow 3S = \dfrac{a+b+c}{3} \cdot 3 \Rightarrow 3S = a+b+c \Rightarrow 3S - a - c = a + b + c - a - c \Rightarrow$

$3S - a - c = b \Rightarrow b = 3S - a - c$

61. $T = \dfrac{a}{3} + \dfrac{b}{4} \Rightarrow \dfrac{12T}{12} = \dfrac{4a}{12} + \dfrac{3b}{12} \Rightarrow 12\left(\dfrac{12T}{12}\right) = 12\left(\dfrac{4a}{12}\right) + 12\left(\dfrac{3b}{12}\right) \Rightarrow 12T = 4a + 3b \Rightarrow$

$12T - 4a = 4a - 4a + 3b \Rightarrow 12T - 4a = 3b \Rightarrow \dfrac{12T - 4a}{3} = \dfrac{3b}{3} \Rightarrow \dfrac{12T - 4a}{3} = b \Rightarrow b = \dfrac{12T - 4a}{3}$

62. $cd = ab + bc \Rightarrow cd - bc = ab + bc - bc \Rightarrow cd - bc = ab \Rightarrow c(d - b) = ab \Rightarrow$

$\dfrac{c(d-b)}{(d-b)} = \dfrac{ab}{(d-b)} \Rightarrow c = \dfrac{ab}{d-b}$

63. The formula for GPA is given by $\dfrac{4a + 3b + 2c + d}{a + b + c + d + f}.$

$\dfrac{4(20) + 3(25) + 2(12) + 4}{20 + 25 + 12 + 4 + 4} = \dfrac{80 + 75 + 24 + 4}{65} = \dfrac{183}{65} \approx 2.82.$ Thus, the GPA is 2.82.

64. The formula for GPA is given by $\dfrac{4a + 3b + 2c + d}{a + b + c + d + f}.$

$\dfrac{4(64) + 3(32) + 2(20) + 10}{64 + 32 + 20 + 10 + 3} = \dfrac{256 + 96 + 40 + 10}{129} = \dfrac{402}{129} \approx 3.12.$ Thus, the GPA is 3.12.

65. To convert Celsius to Fahrenheit temperature, the formula given is $\dfrac{9}{5}C + 32 = F.$

$\dfrac{9}{5}(15) + 32 = F \Rightarrow \dfrac{135}{5} + 32 = F \Rightarrow 27 + 32 = F \Rightarrow F = 59°\text{F}.$

66. To convert Fahrenheit to Celsius temperature, the formula given is $C = \dfrac{5}{9}(F - 32).$

$C = \dfrac{5}{9}(113 - 32) \Rightarrow C = \dfrac{5}{9}(81) \Rightarrow C = \dfrac{405}{9} \Rightarrow C = 45°\text{C}.$

67.

68.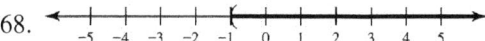

69. (number line from -5 to 5)

70. (number line from -5 to 5)

71. $x < 3$

72. $x \geq -1$

73. Substitute $-2$ for $x$ and check for accuracy: $1-(x+3) \geq x \Rightarrow 1-(-2+3) \geq -2 \Rightarrow 1-1 \geq -2 \Rightarrow$

$0 \geq -2$. Because this statement is true, $x = -2$ is a solution to the inequality.

74. Substitute $-1$ for $x$ and check for accuracy: $4(x+1) < -(5-x) \Rightarrow 4(-1+1) < -(5-(-1)) \Rightarrow$

$4(0) < -(6) \Rightarrow 0 < -6$. Because this statement is not true, $x = -1$ is not a solution to the inequality.

75. When $x = 1$, then $5 - x = 5 - 1 = 4$; When $x = 2$, then $5 - x = 5 - 2 = 3$;

When $x = 3$, then $5 - x = 5 - 3 = 2$; When $x = 4$, then $5 - x = 5 - 4 = 1$;

Thus, the missing values in the table are 4, 3, 2 and 1. See Figure 75. From the table we see that

$5 - x > 3$, when $x < 2$.

| $x$ | 0 | 1 | 2 | 3 | 4 |
|-----|---|---|---|---|---|
| $5 - x$ | 5 | 4 | 3 | 2 | 1 |

Figure 75

| $x$ | 1 | 1.5 | 2 | 2.5 | 3 |
|-----|---|-----|---|-----|---|
| $2x - 5$ | -3 | -2 | -1 | 0 | 1 |

Figure 76

76. When $x = 1.5$, then $2x - 5 = 2(1.5) - 5 = 3 - 5 = -2$; When $x = 2$, then $2x - 5 = 2(2) - 5 = 4 - 5 = -1$;

When $x = 2.5$, then $2x - 5 = 2(2.5) - 5 = 5 - 5 = 0$; When $x = 3$, then $2x - 5 = 2(3) - 5 = 6 - 5 = 1$;

Thus, the missing values in the table are $-2$, $-1$, 0 and 1. See Figure 76. From the table we see

that $2x - 5 \leq 0$ when $x \leq 2.5$.

77. $x - 3 > 0 \Rightarrow x - 3 + 3 > 0 + 3 \Rightarrow x > 3$

78. $-2x \leq 10 \Rightarrow \dfrac{-2x}{-2} \geq \dfrac{10}{-2} \Rightarrow x \geq -5$

79. $5 - 2x \geq 7 \Rightarrow 5 - 5 - 2x \geq 7 - 5 \Rightarrow -2x \geq 2 \Rightarrow \dfrac{-2x}{-2} \leq \dfrac{2}{-2} \Rightarrow x \leq -1$

80. $3(x-1) < 20 \Rightarrow 3x - 3 < 20 \Rightarrow 3x - 3 + 3 < 20 + 3 \Rightarrow 3x < 23 \Rightarrow \dfrac{3x}{3} < \dfrac{23}{3} \Rightarrow x < \dfrac{23}{3}$

81. $5x \leq 3 - (4x+2) \Rightarrow 5x \leq 3 - 4x - 2 \Rightarrow 5x + 4x \leq 3 - 4x + 4x - 2 \Rightarrow 9x \leq 1 \Rightarrow \dfrac{9x}{9} \leq \dfrac{1}{9} \Rightarrow x \leq \dfrac{1}{9}$

82. $3x - 2(4-x) \geq x + 1 \Rightarrow 3x - 8 + 2x \geq x + 1 \Rightarrow 5x - 8 + 8 \geq x + 1 + 8 \Rightarrow 5x \geq x + 9 \Rightarrow$

$5x - x \geq x - x + 9 \Rightarrow 4x \geq 9 \Rightarrow \dfrac{4x}{4} \geq \dfrac{9}{4} \Rightarrow x \geq \dfrac{9}{4}$

83. $x < 50$

84. $x \leq 45,000$

85. $x \geq 16$

86. $x < 1995$ or $x \leq 1994$

87. (a)  See Figure 87.

(b)  $R = 2 + \dfrac{3}{4}x$

(c)  At 5 PM, $x = 5$;  $R = 2 + \dfrac{3}{4}(5) = 2 + \dfrac{15}{4} = \dfrac{23}{4} = 5\dfrac{3}{4}$ inches.  This value does agree with the

table.

(d)  At 3: 45 PM, $x = 3.75$;  $R = 2 + \dfrac{3}{4}\left(3\dfrac{3}{4}\right) = 2 + \dfrac{3}{4}\left(\dfrac{15}{4}\right) = 2 + \dfrac{45}{16} = \dfrac{32}{16} + \dfrac{45}{16} = \dfrac{77}{16} = 4\dfrac{13}{16}$ inches.

| Time | 12:00 | 1:00 | 2:00 | 3:00 | 4:00 | 5:00 |
|------|-------|------|------|------|------|------|
| Rainfall (R) | 2 | 2.75 | 3.5 | 4.25 | 5 | 5.75 |

| Hours (x) | 1 | 2 | 3 | 4 | 5 |
|-----------|---|---|---|---|---|
| Distance (D) | 40 | 30 | 20 | 10 | 0 |

Figure 87                                   Figure 89

88. Let $x$ represent the cost of the laptop.  $0.05x = 106.25 \Rightarrow \dfrac{0.05x}{0.05} = \dfrac{106.25}{0.05} \Rightarrow x = 2125$.  Thus, the cost

of the laptop is \$2125.

89. (a)  See Figure 89.

(b)  $D = 50 - 10x$

(c)  $D = 50 - 10x = 50 - 10(3) = 50 - 30 = 20$ miles.  This value does agree with the table.

(d)  $D \geq 20$.  Thus, $50 - 10x \geq 20 \Rightarrow 50 - 50 - 10x \geq 20 - 50 \Rightarrow -10x \geq -30 \Rightarrow \dfrac{-10x}{-10} \leq \dfrac{-30}{-10} \Rightarrow$

$x \leq 3$.  Thus, the bicyclist was at least 20 miles from home when he had traveled for 3 or fewer

hours, or from noon to 3 PM.

90. First, subtract the smaller number from the larger to obtain the difference between them:

$625,000 - 468,500 = 156,500$.  Then, determine what percentage 156,500 is of 468,500  Do this by

dividing the smaller number by the larger:  $\dfrac{156,500}{468,500} \approx 0.334$.  Thus, there was about an 33.4%

change in master's degrees received between 2001 and 2008.

91. Use the distance $(d) =$ rate $(r) \times$ time $(t)$ formula.  Determine how long it takes the faster car to be

2 miles ahead of the slower car, let $(r + 12)$ be the rate of the faster car and $r$ be the rate of the

slower car.  $d = rt \Rightarrow 2 = (r + 12 - r)t \Rightarrow 2 = 12t \Rightarrow \dfrac{2}{12} = \dfrac{12t}{12} \Rightarrow \dfrac{1}{6} = t \Rightarrow t = \dfrac{1}{6}$ hour, or 10 minutes.

92. Perimeter $(P) = 2 \times$ width $(W) + 2 \times$ length $(L)$.  Then, $W = L - 10 \Rightarrow 2(L - 10) + 2L = 112 \Rightarrow$

$2L - 20 + 2L = 112 \Rightarrow 2L - 20 + 20 + 2L = 112 + 20 \Rightarrow 4L = 132 \Rightarrow \dfrac{4L}{4} = \dfrac{132}{4} \Rightarrow L = 33$.  Because the

length is 33, the width is $(L - 10) = 23$.  Thus, the dimensions are 33 by 23 inches.

93. Let $x$ represent the amount of water.  The amount of salt on one side of the equation must equal the

amount of salt on the other side. Thus, $100(0.03) + x(0.00) = (100 + x)(0.02) \Rightarrow 3 + 0 = 2 + 0.02x \Rightarrow$

$3 - 2 = 2 - 2 + 0.02x \Rightarrow 1 = 0.02x \Rightarrow \dfrac{1}{0.02} = \dfrac{0.02x}{0.02} \Rightarrow 50 = x \Rightarrow x = 50.$

Thus, 50 ml of water must be added.

94. Let $x$ represent the higher interest rate. Then, $800(x) + 500(x - 0.02) = 55 \Rightarrow$

$800x + 500x - 10 = 55 \Rightarrow 1300x - 10 + 10 = 55 + 10 \Rightarrow 1300x = 65 \Rightarrow \dfrac{1300x}{1300} = \dfrac{65}{1300} \Rightarrow x = 0.05.$

Thus, the interest rate on the $800 loan is 5% and the interest rate on the $500 loan is 3%.

95. Area $(A)$ of a triangle is $\dfrac{1}{2} \times$ base $(b) \times$ height $(h)$. Thus, $A = \dfrac{1}{2}bh \Rightarrow \dfrac{1}{2}bh \leq 100 \Rightarrow \dfrac{1}{2}b(8) \leq 100 \Rightarrow$

$4b \leq 100 \Rightarrow \dfrac{4b}{4} \leq \dfrac{100}{4} \Rightarrow b \leq 25.$ Therefore, the base must be 25 inches or less.

96. Let $x$ represent the unknown test score. Then, $\dfrac{75 + 91 + x}{3} \geq 80 \Rightarrow \dfrac{166 + x}{3} \geq 80 \Rightarrow$

$3\left(\dfrac{166 + x}{3}\right) \geq 3(80) \Rightarrow 166 + x \geq 240 \Rightarrow 166 - 166 + x \geq 240 - 166 \Rightarrow x \geq 74.$ Thus, the student must

score 74 or more.

97. Let $x$ represent the unknown number of hours after the first hour. Then, $2.25 + 1.25x = 9 \Rightarrow$

$2.25 - 2.25 + 1.25x = 9 - 2.25 \Rightarrow 1.25x = 6.75 \Rightarrow \dfrac{1.25x}{1.25} = \dfrac{6.75}{1.25} \Rightarrow x = 5.4.$ Because each partial hour

is charged as a full hour, the person can park for $1 + 5 = 6$ hours.

98. (a) $C = 150,000 + 85x$

(b) $R = 225x$

(c) $P = 225x - (150,000 + 85x) \Rightarrow P = 140x - 150,000$

(d) $140x - 150,000 < 0 \Rightarrow 140x - 150,000 + 150,000 < 0 + 150,000 \Rightarrow 140x < 150,000 \Rightarrow$

$\dfrac{140x}{140} < \dfrac{150,000}{140} \Rightarrow x < 1071.43.$ Therefore, if 1071 or fewer DVD players are sold, there will

be a loss.

## Chapter 2 Test

1. $9 = 3 - x \Rightarrow 9 - 3 = 3 - 3 - x \Rightarrow 6 = -x \Rightarrow 6(-1) = (-x)(-1) \Rightarrow -6 = x \Rightarrow x = -6$

To check the solution: $9 = 3 - (-6) \Rightarrow 9 = 9.$ The solution checks.

2. $4x - 3 = 7 \Rightarrow 4x - 3 + 3 = 7 + 3 \Rightarrow 4x = 10 \Rightarrow \dfrac{4x}{4} = \dfrac{10}{4} \Rightarrow x = \dfrac{5}{2}$

To check the solution: $4\left(\dfrac{5}{2}\right)-3=7 \Rightarrow \dfrac{20}{2}-\dfrac{6}{2}=7 \Rightarrow \dfrac{14}{2}=7 \Rightarrow 7=7$.  The solution checks.

3. $4x-(2-x)=-3(2x+6) \Rightarrow 4x-2+x=-6x-18 \Rightarrow 5x-2=-6x-18 \Rightarrow$

   $5x-2+2=-6x-18+2 \Rightarrow 5x=-6x-16 \Rightarrow 5x+6x=-6x+6x-16 \Rightarrow 11x=-16 \Rightarrow$

   $\dfrac{11x}{11}=\dfrac{-16}{11} \Rightarrow x=-\dfrac{16}{11}$.  To check the solution: $4\left(-\dfrac{16}{11}\right)-\left(2-\left(-\dfrac{16}{11}\right)\right)=-3\left(2\left(-\dfrac{16}{11}\right)+6\right) \Rightarrow$

   $-\dfrac{64}{11}-2-\dfrac{16}{11}=-6\left(-\dfrac{16}{11}\right)-18 \Rightarrow -\dfrac{64}{11}-\dfrac{22}{11}-\dfrac{16}{11}=\dfrac{96}{11}-\dfrac{198}{11} \Rightarrow -\dfrac{102}{11}=-\dfrac{102}{11}$.  The solution checks.

4. $\dfrac{1}{12}x-\dfrac{2}{3}=\dfrac{1}{2}\left(\dfrac{3}{4}-\dfrac{1}{3}x\right) \Rightarrow \dfrac{1}{12}x-\dfrac{2}{3}=\dfrac{3}{8}-\dfrac{1}{6}x \Rightarrow \dfrac{1}{12}x+\dfrac{1}{6}x-\dfrac{2}{3}=\dfrac{3}{8}-\dfrac{1}{6}x+\dfrac{1}{6}x \Rightarrow$

   $\dfrac{3}{12}x-\dfrac{2}{3}+\dfrac{2}{3}=\dfrac{3}{8}+\dfrac{2}{3} \Rightarrow \dfrac{3}{12}x=\dfrac{9}{24}+\dfrac{16}{24} \Rightarrow \dfrac{3}{12}x=\dfrac{25}{24} \Rightarrow \dfrac{12}{3}\left(\dfrac{3}{12}x\right)=\dfrac{12}{3}\left(\dfrac{25}{24}\right) \Rightarrow x=\dfrac{300}{72}=\dfrac{25}{6}$.

   To check the solution: $\dfrac{1}{12}\left(\dfrac{25}{6}\right)-\dfrac{2}{3}=\dfrac{1}{2}\left(\dfrac{3}{4}-\dfrac{1}{3}\left(\dfrac{25}{6}\right)\right) \Rightarrow \dfrac{25}{72}-\dfrac{48}{72}=\dfrac{1}{2}\left(\dfrac{27}{36}-\dfrac{50}{36}\right) \Rightarrow$

   $-\dfrac{23}{72}=\dfrac{1}{2}\left(-\dfrac{23}{36}\right) \Rightarrow -\dfrac{23}{72}=-\dfrac{23}{72}$.  The solution checks.

5. First, solve for $x$: $6(2x-1)=-4(3-3x) \Rightarrow 12x-6=-12+12x \Rightarrow 12x-12x-6=-12+12x-12x \Rightarrow$

   $-6=-12$.  Because this statement is not true, there are no solutions.

6. First, solve for $x$: $4(2x-1)=8x-4 \Rightarrow 8x-4=8x-4 \Rightarrow 8x-8x-4+4=8x-8x-4+4 \Rightarrow 0=0$

   Because this statement is always true, there are infinitely many solutions.

7. When $x=1$, then $6-2x=6-2(1)=4$;   When $x=2$, then $6-2x=6-2(2)=2$;

   When $x=3$, then $6-2x=6-2(3)=0$;   When $x=4$, then $6-2x=6-2(4)=-2$;

   Thus, the missing values in the table are 4, 2, 0 and $-2$.  See Figure 7. From the table we see that

   $6-2x=0$, when $x=3$.

| $x$ | 0 | 1 | 2 | 3 | 4 |
|-----|---|---|---|---|----|
| $6-2x$ | 6 | 4 | 2 | 0 | -2 |

   Figure 7

8. $x+(-7)=6 \Rightarrow x-7=6 \Rightarrow x-7+7=6+7 \Rightarrow x=13$

9. $2x+6=x-7 \Rightarrow 2x-x+6=x-x-7 \Rightarrow x+6=-7 \Rightarrow x+6-6=-7-6 \Rightarrow x=-13$

10. $x+(x+1)+(x+2)=336 \Rightarrow 3x+3=336 \Rightarrow 3x+3-3=336-3 \Rightarrow 3x=333 \Rightarrow$

    $\dfrac{3x}{3}=\dfrac{333}{3} \Rightarrow x=111$.  Thus, the three numbers are 111, 112 and 113.

11. $3.2\%=0.032; 3.2\%=\dfrac{3.2}{100}=\dfrac{32}{1000}=\dfrac{16}{500}=\dfrac{8}{250}=\dfrac{4}{125}$

12. $0.345=34.5\%$

13. $d = rt \Rightarrow 200 = r \cdot 4 \Rightarrow r = \dfrac{200}{4} = 50$   Thus, 50 ft/sec

14. Area $(A) = \dfrac{1}{2} \times$ base $(b) \times$ height $(h)$. Thus, $A = \dfrac{1}{2}bh = \dfrac{1}{2}(5)(3) = 7.5$ in$^2$.

15. Circumference of a circle is given as

$C = 2\pi r.$ Then, $C = 2\pi r = 2\pi \left(\dfrac{30}{2}\right) = 2\pi(15) = 30\pi \approx 94.2$ inches.

Area of a circle is given as $A = \pi r^2$. Then, $A = \pi r^2 = \pi(15)^2 = 225\pi \approx 706.9$ in$^2$.

16. The angles in a triangle must add up to 180°.

Then, $x + 2x + 3x = 180 \Rightarrow 6x = 180 \Rightarrow \dfrac{6x}{6} = \dfrac{180}{6} \Rightarrow x = 30.$  Thus, the angles are 30°, 60° and 90°.

17. $z = y - 3xy \Rightarrow z - y = y - y - 3xy \Rightarrow z - y = -3xy \Rightarrow \dfrac{z-y}{-3y} = \dfrac{-3xy}{-3y} \Rightarrow \dfrac{z-y}{-3y} = x \Rightarrow x = \dfrac{y-z}{3y}$

18. $R = \dfrac{x}{4} + \dfrac{y}{5} \Rightarrow 20R = 5x + 4y \Rightarrow 20R - 4y = 5x \Rightarrow x = \dfrac{20R - 4y}{5}$

19.

20. $x > 0$

21. $-3x + 9 \geq x - 15 \Rightarrow -3x + 9 - 9 \geq x - 15 - 9 \Rightarrow -3x \geq x - 24 \Rightarrow -3x - x \geq x - x - 24$

$\Rightarrow -4x \geq -24 \Rightarrow \dfrac{-4x}{-4} \leq \dfrac{-24}{-4} \Rightarrow x \leq 6$

22. $3(6 - 5x) < 20 - x \Rightarrow 18 - 15x < 20 - x \Rightarrow 18 - 18 - 15x < 20 - 18 - x \Rightarrow -15x < 2 - x \Rightarrow$

$-15x + x < 2 - x + x \Rightarrow -14x < 2 \Rightarrow \dfrac{-14x}{-14} > \dfrac{2}{-14} \Rightarrow x > -\dfrac{1}{7}$

23. (a)   $S = 5 + 2x,$ where $x$ represents hours past noon.

(b)   $x = 8.$ Thus, $S = 5 + 2(8) = 21$ inches.

(c)   $x = 6.25.$ Thus, $S = 5 + 2(6.25) = 17.5$ inches.

24. The amount of acid on the left side of the equation must equal the amount of acid on the right side of the equation. Let $x$ represent the unknown amount of water. Then,

$1000(0.45) + x(0) = (1000 + x)(0.15) \Rightarrow 450 = 150 + 0.15x \Rightarrow 450 - 150 = 150 - 150 + 0.15x \Rightarrow$

$300 = 0.15x \Rightarrow \dfrac{300}{0.15} = \dfrac{0.15x}{0.15} \Rightarrow 2000 = x \Rightarrow x = 2000.$ Thus, 2000 ml of water must be added.

25. Subtract the lesser amount from the larger amount and then calculate the percentage difference as

compared to the smaller amount. Then, $95 - 75 = 20; \dfrac{20}{75} \approx 0.267 \Rightarrow 26.7\%.$ Therefore, there will

be a 26.7% increase in cost from 2010 to 2015.

### Chapters 2 Extended Discovery Exercises

1. For the first hour, the distance traveled was $d = rt$ such that $d = (50)(1) = 50$ miles. For the second

   hour, the distance traveled was $d = rt$ such that $d = (70)(1) = 70$ miles.

   Thus, for the two hours $r = \dfrac{d}{t}$ such that $r = \dfrac{70+50}{1+1} = \dfrac{120}{2} = 60$. Thus, the average speed of the car

   was 60 mph.

2. Uphill, $t = \dfrac{d}{r}$ such that $t = \dfrac{1}{5} = \dfrac{1}{5}$ of an hour.

   Downhill, $t = \dfrac{d}{r}$ such that $t = \dfrac{1}{10} = \dfrac{1}{10}$ of an hour. Thus, the average speed

   $r = \dfrac{d}{t}$ is $r = \dfrac{1+1}{\frac{1}{5}+\frac{1}{10}} = \dfrac{2}{\frac{3}{10}} = \dfrac{20}{3} = 6.\overline{6}$ mph. *Answers may vary.*

3. For the first two miles, $t = \dfrac{d}{r}$ such that $t = \dfrac{2}{8} = \dfrac{1}{4}$ of an hour. For the third mile, $t = \dfrac{d}{t}$ such that

   $t = \dfrac{1}{10} = \dfrac{1}{10}$ of an hour. Thus, the average speed of the athlete is

   $r = \dfrac{d}{t} = \dfrac{3}{\frac{1}{4}+\frac{1}{10}} = \dfrac{3}{\frac{5}{20}+\frac{2}{20}} = \dfrac{3}{\frac{7}{20}} = \dfrac{60}{7} \approx 8.6$ mph.

4. Choose a distance of 400 miles as the distance between the two cities (the distance is arbitrary

   because any distance gives the same average speed). Then, the pilot flew at 200 mph for 1 hour and

   at 100 mph for 2 hours. Then, $r = \dfrac{d}{t} = \dfrac{400}{1+2} = \dfrac{400}{3} = 133.\overline{3}$. Thus, the average speed is $133.\overline{3}$ mph.

5. The lighter coin can be found in two weighings as follows: Place two coins on each pan of the

   balance and set three coins off to the side. Case 1: The pans balance and the lighter coin is one of the

   three coins that were set off to the side. Case 2: The pans do not balance and the lighter coin is one

   of the two coins on the higher pan. To find the lighter coin in Case 1, work only with the three

   remaining coins. Place one coin on each side of the balance and set one coin off to the side. If the

   pans do not balance, the lighter coin is the one on the higher pan. To find the lighter coin in Case 2,

   work with only the two coins from the higher pan. Place one coin on each side of the balance. The

   lighter coin is on the higher pan.

6. (a)   Surface area $(A) = 4\pi r^2 = 4\pi(3960)^2 \approx 197{,}060{,}797$ mi$^2$.

   (b)   $0.71(197{,}060{,}797) \approx 139{,}913{,}166$ mi$^2$.

(c) $\dfrac{680,000}{139,913,166} \approx 0.00486$ miles. To convert 0.00486 miles to feet:

$0.00486 \text{ mile} \cdot \dfrac{5280 \text{ feet}}{1 \text{ mile}} \approx 25.7 \text{ feet.}$

(d) They would be flooded.

(e) Divide the volume of the Antarctic ice cap by the surface area of the oceans:

$\dfrac{6,300,000}{139,913,166} \approx 0.045$ miles. To convert 0.045 miles to feet: $0.045 \text{ mile} \cdot \dfrac{5280 \text{ feet}}{1 \text{ mile}} \approx 237.7 \text{ feet.}$

## Chapters 1 and 2 Cumulative Review Exercises

1. The number 45 is a composite number because it has factors other than itself and 1; $45 = 3 \times 3 \times 5$

2. The number 37 is a prime number because its only factors are itself and 1.

3. $\dfrac{4}{3} \cdot \dfrac{3}{8} = \dfrac{4 \cdot 3}{3 \cdot 8} = \dfrac{12}{24} = \dfrac{1 \cdot 12}{2 \cdot 12} = \dfrac{1}{2} \cdot \dfrac{12}{12} = \dfrac{1}{2} \cdot 1 = \dfrac{1}{2}$

4. $\dfrac{2}{3} \div 6 = \dfrac{2}{3} \div \dfrac{6}{1} = \dfrac{2}{3} \cdot \dfrac{1}{6} = \dfrac{2 \cdot 1}{3 \cdot 6} = \dfrac{2}{18} = \dfrac{1 \cdot 2}{9 \cdot 2} = \dfrac{1}{9} \cdot \dfrac{2}{9} = \dfrac{1}{9} \cdot 1 = \dfrac{1}{9}$

5. $\dfrac{11}{12} - \dfrac{3}{8} = \dfrac{11}{12} \cdot \dfrac{2}{2} - \dfrac{3}{8} \cdot \dfrac{3}{3} = \dfrac{11 \cdot 2}{12 \cdot 2} - \dfrac{3 \cdot 3}{8 \cdot 3} = \dfrac{22}{24} - \dfrac{9}{24} = \dfrac{22 - 9}{24} = \dfrac{13}{24}$

6. $\dfrac{2}{3} + \dfrac{1}{5} = \dfrac{2}{3} \cdot \dfrac{5}{5} + \dfrac{1}{5} \cdot \dfrac{3}{3} = \dfrac{2 \cdot 5}{3 \cdot 5} + \dfrac{1 \cdot 3}{5 \cdot 3} = \dfrac{10}{15} + \dfrac{3}{15} = \dfrac{10 + 3}{15} = \dfrac{13}{15}$

7. $-1$ is a rational number, and is an integer.

8. $\sqrt{3}$ is an irrational number.

9. $15 - 4 \cdot 3 = 15 - 12 = 3$

10. $30 \div 6 \cdot 2 = 5 \cdot 2 = 10$

11. $23 - 4^2 \div 2 = 23 - 16 \div 2 = 23 - 8 = 15$

12. $11 - \dfrac{3+1}{6-4} = 11 - \dfrac{4}{2} = 11 - 2 = 9$

13. $5x^3 - x^3 = (5-1)x^3 = 4x^3$

14. $4 + 2x - 1 + 3x = 4 - 1 + 2x + 3x = 4 - 1 + (2+3)x = 3 + 5x$

15. $x - 3 = 11 \Rightarrow x - 3 + 3 = 11 + 3 \Rightarrow x = 14$

16. $4x - 6 = -22 \Rightarrow 4x - 6 + 6 = -22 + 6 \Rightarrow 4x = -16 \Rightarrow \dfrac{4x}{4} = \dfrac{-16}{4} \Rightarrow x = -4$

17. $5(6y + 2) = 25 \Rightarrow 5(6y) + 5(2) = 25 \Rightarrow 30y + 10 = 25 \Rightarrow 30y + 10 - 10 = 25 - 10 \Rightarrow 30y = 15 \Rightarrow$

$$\frac{30y}{30} = \frac{15}{30} \Rightarrow y = \frac{1}{2}$$

18. $11 - (y+2) = 3y+5 \Rightarrow 11 - y - 2 = 3y+5 \Rightarrow 11 - 2 - y = 3y+5$

    $\Rightarrow 9 - y = 3y+5 \Rightarrow 9 - y + y = 3y + y + 5 \Rightarrow 9 = 4y+5$

    $\Rightarrow 9 - 5 = 4y+5-5 \Rightarrow 4 = 4y \Rightarrow \frac{4}{4} = \frac{4y}{4} \Rightarrow 1 = y \Rightarrow y = 1$

19. First, solve for $x$: $6x+2 = 2(3x+1) \Rightarrow 6x+2 = 6x+2 \Rightarrow 6x - 6x + 2 = 6x - 6x + 2 \Rightarrow 2 = 2$. Because this statement is true for any value of $x$, the equation has infinitely many solutions.

20. First, solve for $x$: $2(3x-4) = 6(x-1) \Rightarrow 6x-8 = 6x-6 \Rightarrow 6x-6x-8 = 6x-6x-6 \Rightarrow -8 = -6$. Because this statement is not true, the equation has no solutions.

21. Let $x$ represent the first integer. Then $x+1$ and $x+2$ represent the other two integers. The sum of the integers is $x+x+1+x+2$, $x+x+1+x+2 = 90 \Rightarrow 3x+3 = 90 \Rightarrow 3x+3-3 = 90-3 \Rightarrow 3x = 87$

    $\Rightarrow \frac{3x}{3} = \frac{87}{3} \Rightarrow x = 29$, $x+1 = 30$, $x+2 = 31$. The integers are 29, 30, and 31.

22. $4.7\% = \frac{4.7}{100} = \frac{47}{1000} = 0.047$

23. $0.17 = 0.17 \times 100\% = 17\%$

24. Given that the distance traveled ($d$) is 325 miles and that the time spent traveling ($t$) is 5 hours, calculate the speed of the car ($r$).

    Then, $d = rt \Rightarrow 325 = r \cdot 5 \Rightarrow \frac{325}{5} = \frac{r \cdot 5}{5} \Rightarrow 65 = r \Rightarrow r = 65$ miles/hour.

25. The sum of the measures of the angles of a triangle is $180°$. Then

    $$2x+3x+4x = 180 \Rightarrow (2+3+4)x = 180 \Rightarrow 9x = 180 \Rightarrow \frac{9x}{9} = \frac{180}{9} \Rightarrow x = 20$$

26. Since the diameter of the circle is 10 inches, the radius is $\frac{10}{2} = 5$ inches. Then,

    $$A = \pi r^2 = \pi 5^2 = 25\pi \approx 78.5 \text{ in}^2.$$

27. $a = 3xy-4 \Rightarrow a+4 = 3xy-4+4 \Rightarrow a+4 = 3xy \Rightarrow \frac{a+4}{3y} = \frac{3xy}{3y} \Rightarrow \frac{a+4}{3y} = x$

28. $A = \frac{x+y+z}{3} \Rightarrow 3A = 3\left(\frac{x+y+z}{3}\right) \Rightarrow 3A = x+y+z \Rightarrow 3A-y-z = x+y-y+z-z \Rightarrow$

    $3A-y-z = x$

29. $7-3x > 4 \Rightarrow 7-7-3x > 4-7 \Rightarrow -3x > 4+(-7) \Rightarrow -3x > -3 \Rightarrow \frac{-3x}{-3} < \frac{-3}{-3} \Rightarrow x < 1$

30. $6x \le 5 - (x - 9) \Rightarrow 6x \le 5 - x + 9 \Rightarrow 6x \le 14 - x \Rightarrow 6x + x \le 14 - x + x \Rightarrow 7x \le 14 \Rightarrow \dfrac{7x}{7} \le \dfrac{14}{7} \Rightarrow x \le 2$

31. $I = 36Y$

32. Start with the initial balance, subtract the amounts withdrawn and add the amounts deposited:

$468 - 14 + 200 - 73 - 21 + 58 = 454 + 200 - 73 - 21 + 58 =$

$654 - 73 - 21 + 58 = 581 - 21 + 58 = 560 + 58 = \$618$

33. Let $x$ represent the amount of 4% acid. Then, the 6% acid solution is the sum of the 4% acid and the

10% acid. Therefore $0.04x + 0.10(150) = 0.06\,(x + 150) \Rightarrow 0.04x + 15 = 0.06x + 9$

$\Rightarrow 0.04x - 0.04x + 15 = 0.06x - 0.04x + 9 \Rightarrow 15 = 0.02x + 9 \Rightarrow 15 - 9 = 0.02x + 9 - 9 \Rightarrow 6 = 0.02x$

$\Rightarrow \dfrac{6}{0.02} = \dfrac{0.02x}{0.02} \Rightarrow 300 = x$

300 mL of 4% acid solution should be mixed with the 10% acid solution to dilute it to a 6% acid

solution.

34. Let $x$ represent the 2009 production level and the equation is $2x - 188 = 356$.

$2x - 188 = 356 \Rightarrow 2x - 188 + 188 = 356 + 188 \Rightarrow 2x = 544 \Rightarrow \dfrac{2x}{2} = \dfrac{544}{2} \Rightarrow x = 272$

The 2009 production level was 272 billion kilowatt-hours.

# Chapter 3: Graphing Equations

## Section 3.1: Introduction to Graphing

1. *xy*-plane

3. $(0, 0)$

5. False

7. scatterplot

9. $(-2, -2)$, $(-2, 2)$, $(0, 0)$, $(2, 2)$

11. $(-1, 0)$, $(0, -3)$, $(0, 2)$, $(2, 0)$

13. See Figure 13.  $(1, 3)$ QI,  $(0, -3)$ None,  $(-2, 2)$ QII

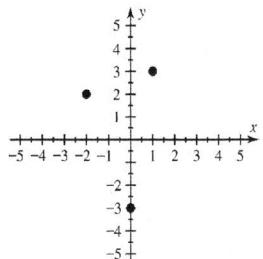

Figure 13                         Figure 15

15. See Figure 15.  $(0, 6)$ None,  $(8, -4)$ QIV,  $(-6, -6)$ QIII

17. (a)    Quadrant I

    (b)    Quadrant III

19. (a)    None, because the point is on the *x*-axis.

    (b)    Quadrant I

21. (a)    Quadrant II

    (b)    Quadrant IV

23. I and III

25. See Figure 25.

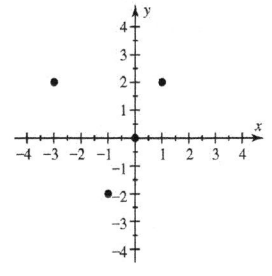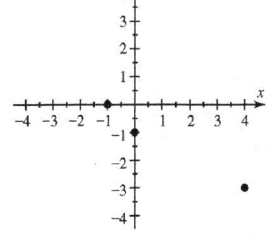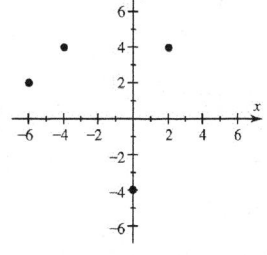

Figure 25                    Figure 27                    Figure 29

27. See Figure 27.

29. See Figure 29.

31. See Figure 31.

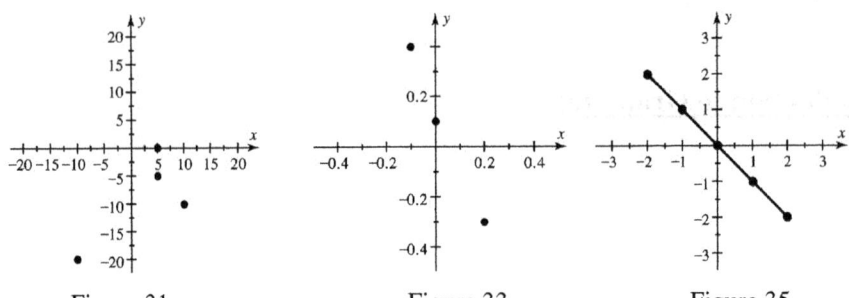

Figure 31          Figure 33          Figure 35

33. See Figure 33.

35. See Figure 35.

37. See Figure 37.

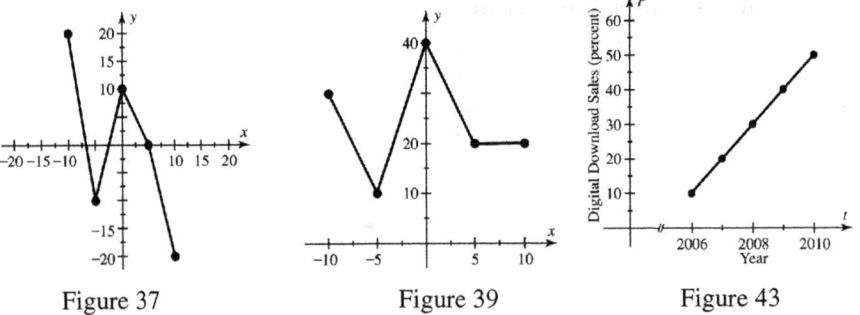

Figure 37          Figure 39          Figure 43

39. See Figure 39.

41. $(1960, 484)$, $(1980, 632)$, $(2000, 430)$, $(2010, 325)$ In 1960, there were 484 billion cigarettes

consumed in the U.S. *Answers may vary slightly.*

43. (a)   See Figure 43.

    (b)   The percentage of digital download sales increased.

45. (a)   See Figure 45.

    (b)   The number of welfare beneficiaries increased and then decreased.

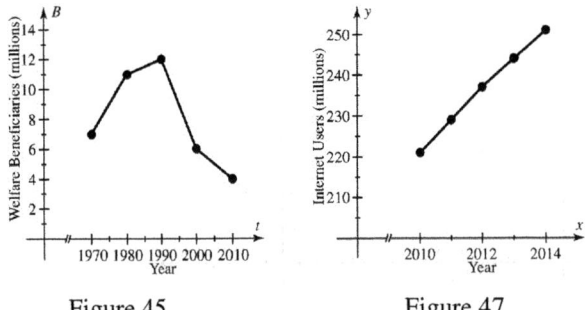

Figure 45          Figure 47

47. (a)   See Figure 47.

    (b)   The number of Internet users is increasing.

49. (a)    The rate decreased.

    (b)    From the graph, the infant mortality rate in 1990 was 9 per 1000 births.

    (c)    From the graph, the rate in 1970 was about 20 and the rate in 2010 was about 6. The percent

           change was $\dfrac{6-20}{20} \times 100 = -70\%$.

## Section 3.2  Linear Equations in Two Variables

1. True

3. ordered pair

5. False

7. graph

9. False

11. Substitute 5 for $x$ and 6 for $y$:  $y = x+1 \Rightarrow 6 = 5+1 \Rightarrow 6 = 6$.  This is a true statement, so the ordered

    pair (5, 6) is a solution.

13. Substitute 2 for $x$ and 13 for $y$:  $y = 4x+7 \Rightarrow 13 = 4(2)+7 \Rightarrow 13 = 8+7 \Rightarrow 13 = 15$.  This is not a true

    statement, so the ordered pair (2, 13) is not a solution.

15. Substitute $-2$ for $x$ and 3 for $y$:  $4x - y = -13 \Rightarrow 4(-2)-3 = -13 \Rightarrow -8-3 = -13 \Rightarrow -11 = -13$.

    This is not a true statement, so the ordered pair $(-2,\ 3)$ is not a solution.

17. Substitute $\dfrac{1}{2}$ for $x$ and 2 for $y$:  $y - 6x = -1 \Rightarrow 2 - 6\left(\dfrac{1}{2}\right) = -1 \Rightarrow 2 - 3 = -1 \Rightarrow -1 = -1$.  This is a

    true statement, so the ordered pair $\left(\dfrac{1}{2},\ 2\right)$ is a solution.

19. Substitute 100 for $x$ and 100 for $y$:  $0.31x - 0.42y = -9 \Rightarrow 0.31(100) - 0.42(100) = -9 \Rightarrow$

    $31 - 42 = -9 \Rightarrow -11 = -9$.  This is not a true statement, so the ordered pair (100, 100) is not a

    solution.

21. When $x = -1$: $y = 4x \Rightarrow y = 4(-1) \Rightarrow y = -4$;   when $x = 0$: $y = 4x \Rightarrow y = 4(0) \Rightarrow y = 0$;

    when $x = 1$: $y = 4x \Rightarrow y = 4(1) \Rightarrow y = 4$;   when $x = 2$: $y = 4x \Rightarrow y = 4(2) \Rightarrow y = 8$.

    Thus, the missing values in the table are $-4$, 0, 4 and 8.  See Figure 21.

    | $x$ | $-2$ | $-1$ | 0 | 1 | 2 |
    |-----|------|------|---|---|---|
    | $y$ | $-8$ | $-4$ | 0 | 4 | 8 |

    Figure 21

23. When $y = -2$: $3y + 2x = 6 \Rightarrow 3(-2) + 2x = 6 \Rightarrow -6 + 2x = 6 \Rightarrow -6 + 6 + 2x = 6 + 6 \Rightarrow$

    $2x = 12 \Rightarrow \dfrac{2x}{2} = \dfrac{12}{2} \Rightarrow x = 6$; when $y = 0$: $3y + 2x = 6 \Rightarrow 3(0) + 2x = 6 \Rightarrow 2x = 6 \Rightarrow \dfrac{2x}{2} = \dfrac{6}{2} \Rightarrow$

    $x = 3$; when $y = 2$: $3y + 2x = 6 \Rightarrow 3(2) + 2x = 6 \Rightarrow 6 + 2x = 6 \Rightarrow 6 - 6 + 2x = 6 - 6 \Rightarrow$

    $2x = 0 \Rightarrow \dfrac{2x}{2} = \dfrac{0}{2} \Rightarrow x = 0$; when $y = 4$: $3y + 2x = 6 \Rightarrow 3(4) + 2x = 6 \Rightarrow 12 + 2x = 6 \Rightarrow$

    $12 - 12 + 2x = 6 - 12 \Rightarrow 2x = -6 \Rightarrow \dfrac{2x}{2} = \dfrac{-6}{2} \Rightarrow x = -3$; when $y = 8$: $3y + 2x = 6 \Rightarrow$

    $3(8) + 2x = 6 \Rightarrow 24 + 2x = 6 \Rightarrow 24 - 24 + 2x = 6 - 24 \Rightarrow 2x = -18 \Rightarrow \dfrac{2x}{2} = \dfrac{-18}{2} \Rightarrow x = -9.$

    Thus, the missing values in the table are 6, 3, 0, $-3$ and $-9$. See Figure 23.

25. When $y = 0$: $y = x + 4 \Rightarrow 0 = x + 4 \Rightarrow 0 - 4 = x + 4 - 4 \Rightarrow -4 = x \Rightarrow x = -4$;

    when $y = 4$: $y = x + 4 \Rightarrow 4 = x + 4 \Rightarrow 4 - 4 = x + 4 - 4 \Rightarrow 0 = x \Rightarrow x = 0$;

    when $y = 8$: $y = x + 4 \Rightarrow 8 = x + 4 \Rightarrow 8 - 4 = x + 4 - 4 \Rightarrow 4 = x \Rightarrow x = 4$;

    when $y = 12$: $y = x + 4 \Rightarrow 12 = x + 4 \Rightarrow 12 - 4 = x + 4 - 4 \Rightarrow 8 = x \Rightarrow x = 8.$

    Thus, the missing values in the table are $-4$, 0, 4 and 8. See Figure 25.

| $x$ | 6 | 3 | 0 | $-3$ | $-9$ |
|---|---|---|---|---|---|
| $y$ | $-2$ | 0 | 2 | 4 | 8 |

| $x$ | $-8$ | $-4$ | 0 | 4 | 8 |
|---|---|---|---|---|---|
| $y$ | $-4$ | 0 | 4 | 8 | 12 |

Figure 23                                    Figure 25

27. When $x = -6$: $2x - 6y = 12 \Rightarrow 2(-6) - 6y = 12 \Rightarrow -12 - 6y = 12 \Rightarrow -12 + 12 - 6y = 12 + 12 \Rightarrow$

    $-6y = 24 \Rightarrow \dfrac{-6y}{-6} = \dfrac{24}{-6} \Rightarrow y = -4$;

    when $x = 0$: $2x - 6y = 12 \Rightarrow 2(0) - 6y = 12 \Rightarrow -6y = 12 \Rightarrow \dfrac{-6x}{-6} = \dfrac{12}{-6} \Rightarrow y = -2$;

    when $x = 6$: $2x - 6y = 12 \Rightarrow 2(6) - 6y = 12 \Rightarrow 12 - 12 - 6y + 1 = 12 - 12 \Rightarrow -6y = 0 \Rightarrow$

    $\dfrac{-6y}{-6} = \dfrac{0}{-6} \Rightarrow y = 0$;

    when $x = 12$: $2x - 6y = 12 \Rightarrow 2(12) - 6y = 12 \Rightarrow 24 - 24 - 6y = 12 - 24 \Rightarrow -6y = -12 \Rightarrow$

    $\dfrac{-6y}{-6} = \dfrac{-12}{-6} \Rightarrow y = 2.$

    when $x = 18$: $2x - 6y = 12 \Rightarrow 2(18) - 6y = 12 \Rightarrow 36 - 36 - 6y = 12 - 36 \Rightarrow -6y = -24 \Rightarrow$

    $\dfrac{-6y}{-6} = \dfrac{-24}{-6} \Rightarrow y = 4.$

    Thus, the missing values in the table are $-4, -2, 0, 2, 4$  See Figure 27.

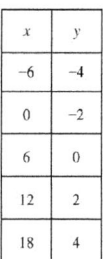

| $x$ | $y$ |
|-----|-----|
| -6  | -4  |
| 0   | -2  |
| 6   | 0   |
| 12  | 2   |
| 18  | 4   |

Figure 27

| $x$ | -3 | 0 | 3 | 6  |
|-----|----|---|---|----|
| $y$ | -9 | 0 | 9 | 18 |

Figure 29

29. See Figure 29.

31. See Figure 31.

| $x$ | 8  | 6 | 4 | 2 |
|-----|----|---|---|---|
| $y$ | -2 | 0 | 2 | 4 |

Figure 31

| $x$ | -8 | -4 | 0 | 4 |
|-----|----|----|---|---|
| $y$ | -2 | 0  | 2 | 4 |

Figure 33

| $x$ | $-\frac{1}{2}$ | $-\frac{1}{4}$ | 0 | $\frac{1}{4}$ |
|-----|----------------|----------------|---|---------------|
| $y$ | -2             | -1             | 0 | 1             |

Figure 35

33. See Figure 33. Tables provided are horizontal due to space restrictions.

35. See Figure 35. Tables provided are horizontal due to space restrictions.

37. They must be multiples of 5.

39.

| $x$ | -1 | 0 | 1  |
|-----|----|---|----|
| $y$ | 2  | 0 | -2 |

Table values may vary.

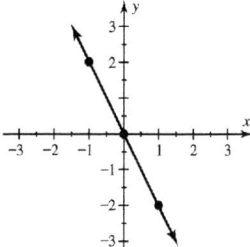

41.

| $x$ | 0 | 1 | 2 |
|-----|---|---|---|
| $y$ | 3 | 2 | 1 |

Table values may vary.

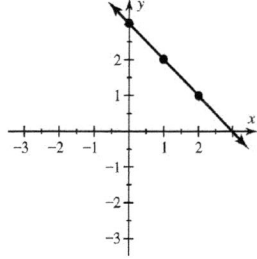

43.

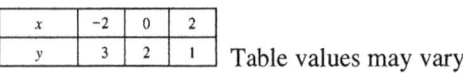 Table values may vary.

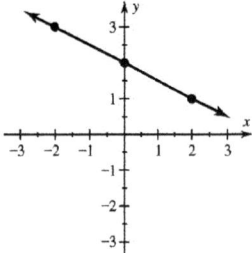

45. See Figure 45.

47. See Figure 47.

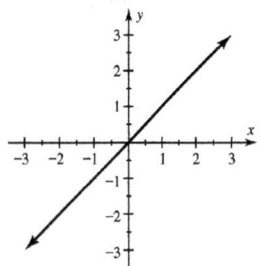

Figure 45

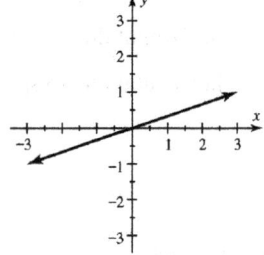

Figure 47

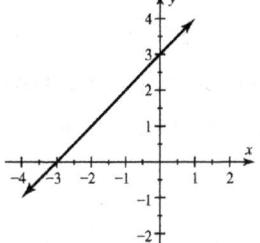

Figure 49

49. See Figure 49.

51. See Figure 51.

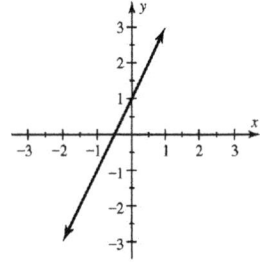

Figure 51

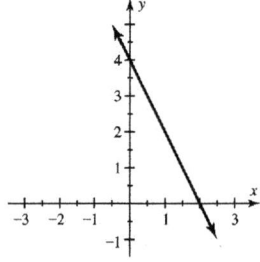

Figure 53

53. See Figure 53.

55. See Figure 55.

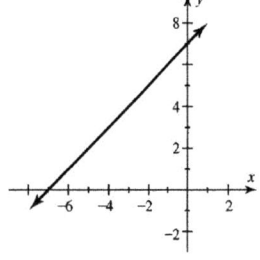

Figure 55

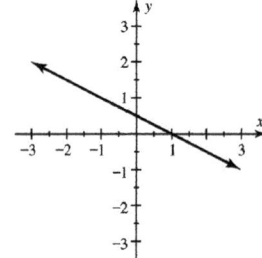

Figure 57

57. See Figure 57.

59. $2x + 3y = 6 \Rightarrow 3y = -2x + 6 \Rightarrow y = -\dfrac{2}{3}x + 2$  See Figure 59.

61. $x + 4y = 4 \Rightarrow 4y = -x + 4 \Rightarrow y = -\dfrac{1}{4}x + 1$  See Figure 61.

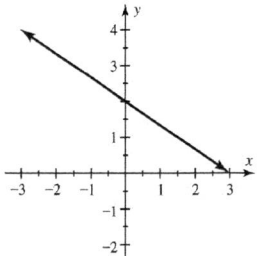
Figure 59

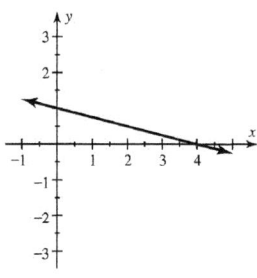
Figure 61

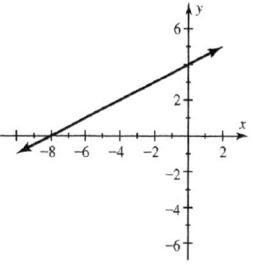
Figure 63

63. $-x + 2y = 8 \Rightarrow 2y = x + 8 \Rightarrow y = \dfrac{1}{2}x + 4$  See Figure 63.

65. $y - 2x = 7 \Rightarrow y = 2x + 7$  See Figure 65.

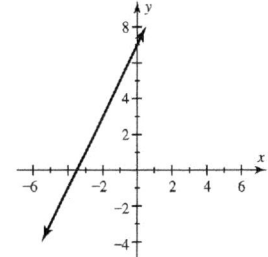
Figure 65

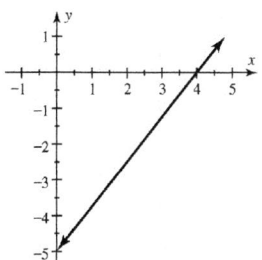
Figure 67

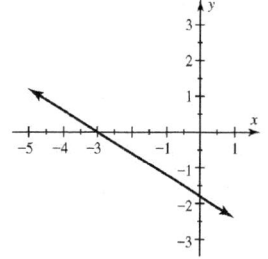
Figure 69

67. $5x - 4y = 20 \Rightarrow -4y = -5x + 20 \Rightarrow y = \dfrac{5}{4}x - 5$  See Figure 67.

69. $3x + 5y = -9 \Rightarrow 5y = -3x - 9 \Rightarrow y = -\dfrac{3}{5}x - \dfrac{9}{5}$  See Figure 69.

71. (a)

| $t$ | 2010 | 2020 | 2030 | 2040 | 2050 |
|-----|------|------|------|------|------|
| $P$ | 13.2 | 15.0 | 16.7 | 18.5 | 20.3 |

   (b)    From the table, the percentage is 16.7% in 2030.

73. (a)    See Figure 73.

   (b)    Set $A = 100$ and solve for $t$:  $A = 2.7t \Rightarrow 100 = 2.7t \Rightarrow \dfrac{100}{2.7} = \dfrac{2.7t}{2.7} \Rightarrow t \approx 37$.  Therefore, about

   37 days.

75. (a)    For year 2006, set $t = 2006$ and solve for $P$:  $P = 10t - 20{,}050 = 10(2006) - 20{,}050 =$

   $20{,}060 - 20{,}050 = 10$.  Therefore, in year 2006 it was 10%.

   For year 2010, set $t = 2010$ and solve for $P$:  $P = 10t - 20{,}050 = 10(2010) - 20{,}050 =$

   $20{,}100 - 20{,}050 = 50$.  Therefore, in year 2010 it was 50%.

(b)    See Figure 75.

(c)    *P* was 30% in year 2008.

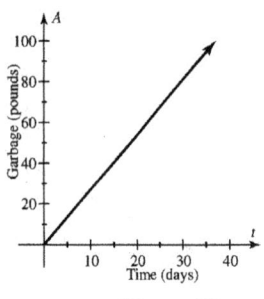

Figure 73

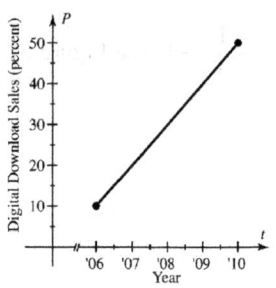

Figure 75

## Checking Basic Concepts  Sections 3.1 and 3.2

1. $(-2, 2)$, II; $(-1, -2)$, III; $(1, 3)$, I; $(3, 0)$, none.

2. See Figure 2.

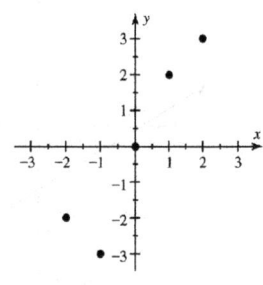

Figure 2

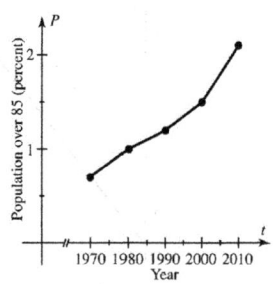

Figure 3

3. See Figure 3. The percentage increased.

4. Substitute $-2$ for $x$ and $-3$ for $y$: $-2x - y = 7 \Rightarrow -2(-2) - (-3) = 7 \Rightarrow 4 + 3 = 7 \Rightarrow 7 = 7$. This is a true statement, so the ordered pair $(-2, -3)$ is a solution.

5. When $x = -1$: $y = -2x + 1 = -2(-1) + 1 = 2 + 1 = 3$;

   when $x = 0$: $y = -2x + 1 = -2(0) + 1 = 1$;

   when $x = 1$: $y = -2x + 1 = -2(1) + 1 = -2 + 1 = -1$;

   when $x = 2$: $y = -2x + 1 = -2(2) + 1 = -4 + 1 = -3$.

   Thus, the missing values in the table are 3, 1, $-1$, $-3$. See Figure 5.

   | $x$ | $-2$ | $-1$ | 0 | 1 | 2 |
   |---|---|---|---|---|---|
   | $y$ | 5 | 3 | 1 | $-1$ | $-3$ |

   Figure 5

6. (a)    See Figure 6a.

   (b)    See Figure 6b.

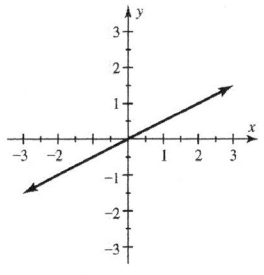

Figure 6a

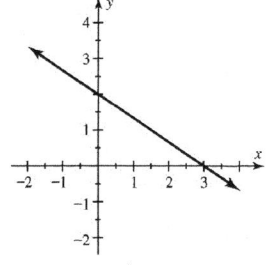

Figure 6b

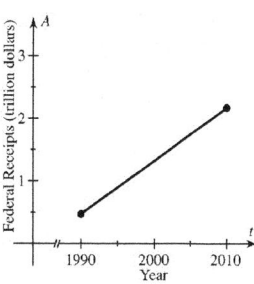

Figure 7

7. (a)   Set $t = 1990$ and solve for $A$:  $A = 0.085t - 168.68 = 0.085(1990) - 168.68 = 0.47$

Set $t = 2010$ and solve for $A$:  $A = 0.085t - 168.68 = 0.085(2010) - 168.68 = 2.17$

In 1990, receipts were \$0.47 trillion; in 2010, receipts were \$2.17 trillion.

(b)   See Figure 7.

(c)   $1.83 = 0.085t - 168.68 \Rightarrow 1.83 + 168.68 = 0.085t - 168.68 + 168.68 \Rightarrow 170.51 = 0.085t \Rightarrow$

$\dfrac{170.51}{0.085} = \dfrac{0.085t}{0.085} \Rightarrow 2006 = t$   The total receipts were \$1.83 trillion in 2006.

## Section 3.3  More Graphing of Lines

1. $x$-intercept

3. $y$-intercept

5. horizontal

7. vertical

9. Because the graph crosses the $x$-axis at $x = 3$, $x = 3$ is the $x$-intercept.

Because the graph crosses the $y$-axis at $y = -2$, $y = -2$ is the $y$-intercept.

11. Because the graph crosses the $x$-axis at $x = 0$, $x = 0$ is the $x$-intercept.

Because the graph crosses the $y$-axis at $y = 0$, $y = 0$ is the $y$-intercept.

13. Because the graph crosses the $x$-axis at $x = -2$ and at $x = 2$, $x = -2$ and $x = 2$ are the $x$-intercepts.

Because the graph crosses the $y$-axis at $y = 4$, $y = 4$ is the $y$-intercept.

15. Because the graph touches the $x$-axis at $x = 1$, $x = 1$ is the $x$-intercept.

Because the graph crosses the $y$-axis at $y = 1$, $y = 1$ is the $y$-intercept.

17. When $x = -2$:  $y = x + 2 = -2 + 2 = 0$;   when $x = -1$:  $y = x + 2 = -1 + 2 = 1$;

when $x = 0$:  $y = x + 2 = 0 + 2 = 2$;   when $x = 1$:  $y = x + 2 = 1 + 2 = 3$;

when $x = 2$:  $y = x + 2 = 2 + 2 = 4$.   Therefore, the missing values in the table are 0, 1, 2, 3 and 4.

See Figure 17. The $x$-intercept is at $x = -2$, because at $x = -2$ the value of $y$ *is* 0.

The $y$-intercept is at $y = 2$, because at $y = 2$ the value of $x$ is 0.

| $x$ | $-2$ | $-1$ | $0$ | $1$ | $2$ |
|---|---|---|---|---|---|
| $y$ | $0$ | $1$ | $2$ | $3$ | $4$ |

Figure 17

| $x$ | $-4$ | $-2$ | $0$ | $2$ | $4$ |
|---|---|---|---|---|---|
| $y$ | $-6$ | $-4$ | $-2$ | $0$ | $2$ |

Figure 19

19. When $x = -4$: $-x + y = -2 \Rightarrow -(-4) + y = -2 \Rightarrow 4 + y = -2 \Rightarrow 4 - 4 + y = -2 - 4 \Rightarrow y = -6$;

when $x = -2$: $-x + y = -2 \Rightarrow -(-2) + y = -2 \Rightarrow 2 + y = -2 \Rightarrow 2 - 2 + y = -2 - 2 \Rightarrow y = -4$;

when $x = 0$: $-x + y = -2 \Rightarrow -(0) + y = -2 \Rightarrow y = -2$;

when $x = 2$: $-x + y = -2 \Rightarrow -(2) + y = -2 \Rightarrow -2 + y = -2 \Rightarrow -2 + 2 + y = -2 + 2 \Rightarrow y = 0$;

when $x = 4$: $-x + y = -2 \Rightarrow -(4) + y = -2 \Rightarrow -4 + y = -2 \Rightarrow -4 + 4 + y = -2 + 4 \Rightarrow y = 2$.

Therefore, the missing values in the table are $-6$, $-4$, $-2$, 0 and 2. See Figure 19.

The $x$-intercept is at $x = 2$, because at $x = 2$ the value of $y$ is 0.

The $y$-intercept is at $y = -2$, because at $y = -2$ the value of $x$ is 0.

21. To find the $x$-intercept, set $y = 0$ and solve for $x$: $-2x + 3y = -6 \Rightarrow -2x + 3(0) = -6 \Rightarrow$

$-2x = -6 \Rightarrow \dfrac{-2x}{-2} = \dfrac{-6}{-2} \Rightarrow x = 3$. Therefore, the $x$-intercept is at $x = 3$.

To find the $y$-intercept, set $x = 0$ and solve for $y$:

$-2x + 3y = -6 \Rightarrow -2(0) + 3y = -6 \Rightarrow 3y = -6 \Rightarrow \dfrac{3y}{3} = \dfrac{-6}{3} \Rightarrow y = -2$. Therefore, the $y$-intercept is at

$y = -2$. See Figure 21.

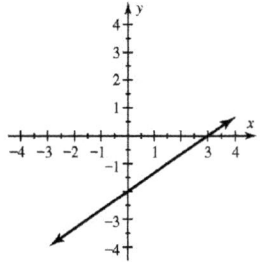

Figure 21

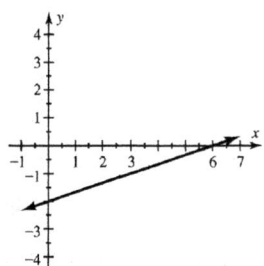

Figure 23

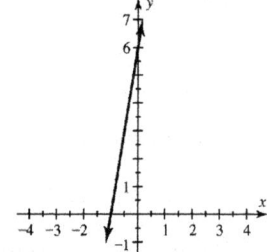

Figure 25

23. To find the $x$-intercept, set $y = 0$ and solve for $x$: $x - 3y = 6 \Rightarrow x - 3(0) = 6 \Rightarrow x = 6$.

Therefore, the $x$-intercept is at $x = 6$.

To find the $y$-intercept, set $x = 0$ and solve for $y$: $x - 3y = 6 \Rightarrow 0 - 3y = 6 \Rightarrow$

$-3y = 6 \Rightarrow \dfrac{-3y}{-3} = \dfrac{6}{-3} \Rightarrow y = -2$. Therefore, the $y$-intercept is at $y = -2$. See Figure 23.

25. To find the $x$-intercept, set $y = 0$ and solve for $x$: $6x - y = -6 \Rightarrow 6x - 0 = -6 \Rightarrow$

$6x = -6 \Rightarrow \dfrac{6x}{6} = \dfrac{-6}{6} \Rightarrow x = -1$. Therefore, the $x$-intercept is at $x = -1$.

To find the $y$-intercept, set $x = 0$ and solve for $y$: $6x - y = -6 \Rightarrow 6(0) - y = -6 \Rightarrow -y = -6 \Rightarrow$

$-1(-y) = -1(-6) \Rightarrow y = 6$. Therefore, the $y$-intercept is at $y = 6$. See Figure 25.

27. To find the $x$-intercept, set $y = 0$ and solve for $x$:  $3x + 7y = 21 \Rightarrow 3x + 7(0) = 21 \Rightarrow$

$3x = 21 \Rightarrow \dfrac{3x}{3} = \dfrac{21}{3} \Rightarrow x = 7$.  Therefore, the $x$-intercept is at $x = 7$.

To find the $y$-intercept, set $x = 0$ and solve for $y$:  $3x + 7y = 21 \Rightarrow 3(0) + 7y = 21 \Rightarrow$

$7y = 21 \Rightarrow \dfrac{7y}{7} = \dfrac{21}{7} \Rightarrow y = 3$.  Therefore, the $y$-intercept is at $y = 3$.  See Figure 27.

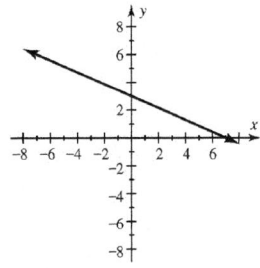
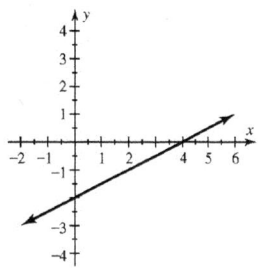

Figure 27                                    Figure 29                                    Figure 31

29. To find the $x$-intercept, set $y = 0$ and solve for $x$:  $40y - 30x = -120 \Rightarrow 40(0) - 30x = -120 \Rightarrow$

$-30x = -120 \Rightarrow \dfrac{-30x}{-30} = \dfrac{-120}{-30} \Rightarrow x = 4$.  Therefore, the $x$-intercept is at $x = 4$.

To find the $y$-intercept, set $x = 0$ and solve for $y$:  $40y - 30x = -120 \Rightarrow 40y - 30(0) = -120 \Rightarrow$

$40y = -120 \Rightarrow \dfrac{40y}{40} = \dfrac{-120}{40} \Rightarrow y = -3$.  Therefore, the $y$-intercept is at $y = -3$.  See Figure 29.

31. To find the $x$-intercept, set $y = 0$ and solve for $x$:  $\dfrac{1}{2}x - y = 2 \Rightarrow \dfrac{1}{2}x - 0 = 2 \Rightarrow$

$\dfrac{1}{2}x = 2 \Rightarrow 2\left(\dfrac{1}{2}x\right) = 2(2) \Rightarrow x = 4$.  Therefore, the $x$-intercept is at $x = 4$.

To find the $y$-intercept, set $x = 0$ and solve for $y$:  $\dfrac{1}{2}x - y = 2 \Rightarrow \dfrac{1}{2}(0) - y = 2 \Rightarrow$

$-y = 2 \Rightarrow -1(-y) = 2(-1) \Rightarrow y = -2$.  Therefore, the $y$-intercept is at $y = -2$.  See Figure 31.

33. To find the $x$-intercept, set $y = 0$ and solve for $x$:  $-\dfrac{x}{4} + \dfrac{y}{3} = 1 \Rightarrow -\dfrac{x}{4} + \dfrac{0}{3} = 1 \Rightarrow$

$-\dfrac{x}{4} = 1 \Rightarrow -4\left(-\dfrac{x}{4}\right) = 1(-4) \Rightarrow x = -4$.  Therefore, the $x$-intercept is at $x = -4$.

To find the $y$-intercept, set $x = 0$ and solve for $y$:  $-\dfrac{x}{4} + \dfrac{y}{3} = 1 \Rightarrow -\dfrac{0}{4} + \dfrac{y}{3} = 1 \Rightarrow$

$\dfrac{y}{3} = 1 \Rightarrow 3\left(\dfrac{y}{3}\right) = 3(1) \Rightarrow y = 3$.  Therefore, the $y$-intercept is at $y = 3$.  See Figure 33.

35. To find the $x$-intercept, set $y = 0$ and solve for $x$:  $\dfrac{x}{3} + \dfrac{y}{2} = 1 \Rightarrow \dfrac{x}{3} + \dfrac{0}{2} = 1 \Rightarrow$

$\dfrac{x}{3} = 1 \Rightarrow 3\left(\dfrac{x}{3}\right) = 1(3) \Rightarrow x = 3$.  Therefore, the $x$-intercept is at $x = 3$.

To find the $y$-intercept, set $x = 0$ and solve for $y$:  $\dfrac{x}{3} + \dfrac{y}{2} = 1 \Rightarrow \dfrac{0}{3} + \dfrac{y}{2} = 1 \Rightarrow$

$\dfrac{y}{2} = 1 \Rightarrow 2\left(\dfrac{y}{2}\right) = 2(1) \Rightarrow y = 2$.  Therefore, the $y$-intercept is at $y = 2$.  See Figure 35.

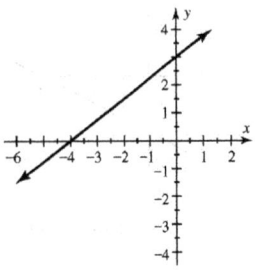

Figure 33                             Figure 35                             Figure 37

37. To find the $x$-intercept, set $y = 0$ and solve for $x$:

$$0.6y - 1.5x = 3 \Rightarrow 0.6(0) - 1.5x = 3 \Rightarrow -1.5x = 3 \Rightarrow \dfrac{-1.5x}{-1.5} = \dfrac{3}{-1.5} \Rightarrow x = -2.$$

Therefore, the $x$-intercept is at $x = -2$.  To find the $y$-intercept, set

$x = 0$ and solve for $y$:  $0.6y - 1.5x = 3 \Rightarrow 0.6y - 1.5(0) = 3 \Rightarrow 0.6y = 3 \Rightarrow$

$\dfrac{0.6y}{0.6} = \dfrac{3}{0.6} \Rightarrow y = 5$.  Therefore, the $y$-intercept is at $y = 5$.  See Figure 37.

39. To find the $x$-intercept, set $y = 0$ and solve for $x$:

$$Ax + By = C \Rightarrow Ax + B(0) = C \Rightarrow Ax = C \Rightarrow \dfrac{Ax}{A} = \dfrac{C}{A} \Rightarrow x = \dfrac{C}{A}.$$

Therefore, the $x$-intercept is at $x = \dfrac{C}{A}$.

To find the $y$-intercept, set $x = 0$ and solve for $y$:

$$Ax + By = C \Rightarrow A(0) + By = C \Rightarrow By = C \Rightarrow \dfrac{By}{B} = \dfrac{C}{B} \Rightarrow y = \dfrac{C}{B}.$$

Therefore, the $y$-intercept is at $y = \dfrac{C}{B}$.

41. Because $y = 1$ for every $x$-value, the graph is a horizontal line.  The equation for the line is $y = 1$.

43. Because $x = -6$ for every $y$-value, the graph is a vertical line.  The equation for the line is $x = -6$.

45. (a) See Figure 45a.

    (b) See Figure 45b.

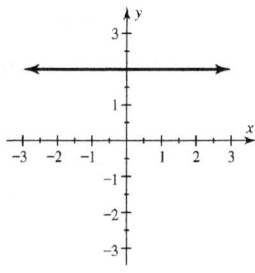

Figure 45a

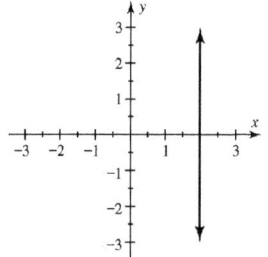

Figure 45b

47. (a) See Figure 47a.

    (b) See Figure 47b.

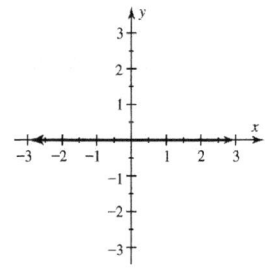

Figure 47a

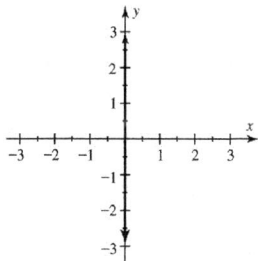

Figure 47b

49. (a) See Figure 49a.

    (b) See Figure 49b.

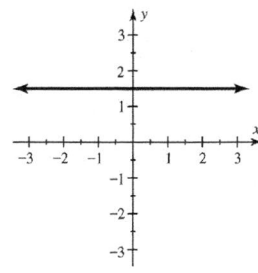

Figure 49a

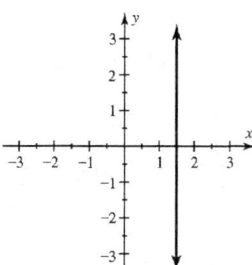

Figure 49b

51.  $y = 4$

53.  $x = -1$

55.  $y = -6$

57.  $x = 5$

59. A horizontal line that passes through (1, 2) has a $y$-value of 2 for every $x$-value. Therefore, the equation for the horizontal line is $y = 2$. A vertical line that passes through (1, 2) has an $x$-value of 1 for every $y$-value. Therefore, the equation for the vertical line is $x = 1$.

61. A horizontal line that passes through $\left(20, -45\right)$ has a $y$-value of $-45$ for every $x$-value. Therefore, the equation for the horizontal line is $y = -45$. A vertical line that passes through $\left(20, -45\right)$ has an $x$-value of $20$ for every $y$-value. Therefore, the equation for the vertical line is $x = 20$.

63. A horizontal line that passes through (0, 5) has a $y$-value of 5 for every $x$-value. Therefore, the equation for the horizontal line is $y = 5$. A vertical line that passes through (0, 5) has an $x$-value of 0 for every $y$-value. Therefore, the equation for the vertical line is $x = 0$.

65. A vertical line that passes through $(-1, 6)$ has an $x$-value of $-1$ for every $y$-value. Therefore, the equation for the vertical line is $x = -1$.

67. A horizontal line that passes through $\left( \dfrac{3}{4}, -\dfrac{5}{6} \right)$ has a $y$-value of $-\dfrac{5}{6}$ for every $x$-value. Therefore, the equation for the horizontal line is $y = -\dfrac{5}{6}$.

69. Because $y = \dfrac{1}{2}$ is the equation of a horizontal line, a line perpendicular to $y = \dfrac{1}{2}$ is a vertical line. Therefore, the equation of the vertical line passing through $(4, -9)$ is $x = 4$.

71. Because $x = 4$ is the equation of a vertical line, a line parallel to $x = 4$ is also a vertical line. Therefore, the equation of the vertical line passing through $\left( -\dfrac{2}{3}, \dfrac{1}{2} \right)$ is $x = -\dfrac{2}{3}$.

73. $y = 0$

75. (a) $y$-intercept, 200; $x$-intercept, 4.

    (b) The driver was initially 200 mi from home; the driver arrived home after 4 hr.

77. (a) To find the $v$-intercept, set $t = 0$ and solve for $v$: $v = 128 - 32t = 128 - 32(0) = 128$.

    To find the $t$-intercept, set $v = 0$ and solve for $t$: $v = 128 - 32t \Rightarrow 0 = 128 - 32t \Rightarrow$

$$0 + 32t = 128 - 32t + 32t \Rightarrow 32t = 128 \Rightarrow \frac{32t}{32} = \frac{128}{32} \Rightarrow t = 4. \text{ See Figure 77.}$$

    (b) The initial velocity was 128 ft/sec; the velocity after 4 seconds was 0.

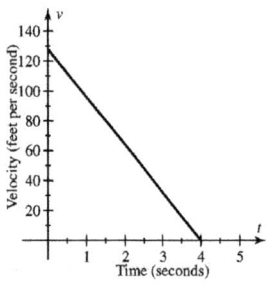

Figure 77

79. (a) $y$-intercept, 2000; $x$-intercept, 4.

    (b) The pool initially contained 2000 gal; the pool was empty after 4 hr.

## **Section 3.4  Slope and Rates of Change**

1. True

3. rise; run

5. vertical

7. positive

9. rate

11. Because the graph of the line rises as it goes from left to right, the slope is positive.

13. Because the graph of the line neither falls nor rises as it goes from left to right, the slope is zero.

15. Because the graph of the line falls as it goes from left to right, the slope is negative.

17. Because the graph is a vertical line, the run is 0.  Thus, the slope is undefined.

19. Because the graph is a horizontal line, the slope is 0.  Thus, the rise always equals 0.

21. A point on the graph is $(0, \ -2)$ and another point is $(2, \ 0)$.

    Let $x_1 = 0$, $x_2 = 2$, $y_1 = -2$ and $y_2 = 0$.  Then: $m \ (\text{slope}) = \dfrac{y_2 - y_1}{x_2 - x_1} = \dfrac{0 - (-2)}{2 - 0} = \dfrac{2}{2} = 1$.

    Thus, the slope is 1; the graph rises 1 unit for each unit of run.

23. A point on the graph is $(-1, -1)$ and another point is $(0, 1)$

    Let $x_1 = -1$, $x_2 = 0$, $y_1 = -1$ and $y_2 = 1$.   Then: $m \ (\text{slope}) = \dfrac{y_2 - y_1}{x_2 - x_1} = \dfrac{1 - (-1)}{0 - (-1)} = \dfrac{2}{1} = 2$ .

    Thus, the slope is 2; the graph rises 2 units for each unit of run.

25. Because the run always equals 0, the slope is undefined.

27. A point on the graph is $(0, \ 1)$ and another point is $(2, \ -2)$.

    Let $x_1 = 0$, $x_2 = 2$, $y_1 = 1$ and $y_2 = -2$.  Then: $m \ (\text{slope}) = \dfrac{y_2 - y_1}{x_2 - x_1} = \dfrac{-2 - 1}{2 - 0} = \dfrac{-3}{2} = -\dfrac{3}{2}$.

    Thus, the slope is $-\dfrac{3}{2}$; the graph falls 3 units for each 2 units of run.

29. A point on the graph is $(0, \ 4)$ and another point is $(4, \ 0)$.

    Let $x_1 = 0$, $x_2 = 4$, $y_1 = 4$ and $y_2 = 0$.  Then: $m \ (\text{slope}) = \dfrac{y_2 - y_1}{x_2 - x_1} = \dfrac{0 - 4}{4 - 0} = \dfrac{-4}{4} = -1$.

    Thus, the slope is $-1$; the graph falls 1 unit for each unit of run.

31. $m = \dfrac{y_2 - y_1}{x_2 - x_1} = \dfrac{4 - 2}{2 - 1} = \dfrac{2}{1} = 2$.  Thus, the slope of the line is 2.  See Figure 31.

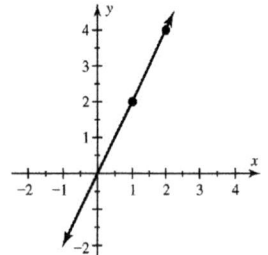

Figure 31

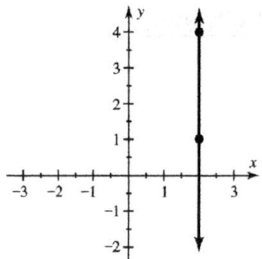

Figure 33

33. $m = \dfrac{y_2 - y_1}{x_2 - x_1} = \dfrac{4-1}{2-2} = \dfrac{3}{0}$. Because the denominator is 0, the slope is undefined. See Figure 33.

35. $m = \dfrac{y_2 - y_1}{x_2 - x_1} = \dfrac{5-3}{-2-1} = \dfrac{2}{-3} = -\dfrac{2}{3}$. Thus, the slope of the line is $-\dfrac{2}{3}$. See Figure 35.

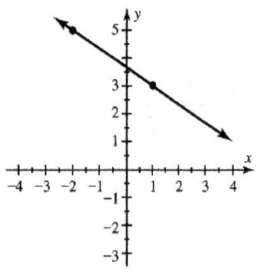

Figure 35

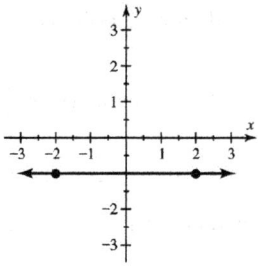

Figure 37

37. $m = \dfrac{y_2 - y_1}{x_2 - x_1} = \dfrac{-1-(-1)}{-2-2} = \dfrac{-1+1}{-2+(-2)} = \dfrac{0}{-4} = 0$. Thus, the slope of the line is 0. See Figure 37.

39. $m = \dfrac{y_2 - y_1}{x_2 - x_1} = \dfrac{-9-(-2)}{-3-4} = \dfrac{-7}{-7} = 1$. Thus, the slope of the line is 1

41. $m = \dfrac{y_2 - y_1}{x_2 - x_1} = \dfrac{-2-4}{4-(-3)} = \dfrac{-6}{7} = -\dfrac{6}{7}$. Thus, the slope of the line is $-\dfrac{6}{7}$.

43. $m = \dfrac{y_2 - y_1}{x_2 - x_1} = \dfrac{5-5}{2-(-3)} = \dfrac{0}{5} = 0$. Thus, the slope of the line is 0.

45. $m = \dfrac{y_2 - y_1}{x_2 - x_1} = \dfrac{-4-6}{-1-(-1)} = \dfrac{-10}{0}$. Because the denominator is 0, the slope is undefined.

47. $m = \dfrac{y_2 - y_1}{x_2 - x_1} = \dfrac{18-5}{2000-1980} = \dfrac{13}{20}$. Thus, the slope of the line is $\dfrac{13}{20}$

49. $m = \dfrac{y_2 - y_1}{x_2 - x_1} = \dfrac{10.6-6.1}{2000-1950} = \dfrac{4.5}{50} = \dfrac{9}{100}$. Thus, the slope of the line is $\dfrac{9}{100}$.

51. $m = \dfrac{y_2 - y_1}{x_2 - x_1} = \dfrac{\frac{3}{7}-\left(-\frac{2}{7}\right)}{-\frac{2}{3}-\frac{1}{3}} = \dfrac{\frac{5}{7}}{-\frac{3}{3}} = -\dfrac{5}{7}$. Thus, the slope of the line is $-\dfrac{5}{7}$.

53. $m = \dfrac{y_2 - y_1}{x_2 - x_1} = \dfrac{64-(-34)}{14-12} = \dfrac{98}{2} = 49$. Thus, the slope of the line is 49.

55. See Figure 55.

57. See Figure 57.

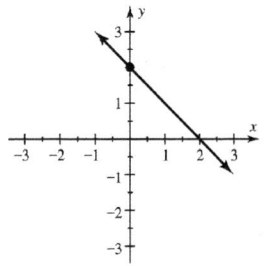

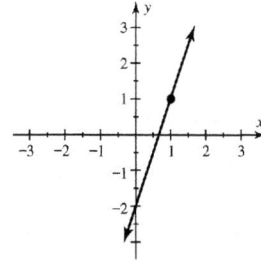

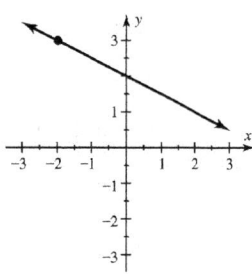

Figure 55                Figure 57                Figure 59

59. See Figure 59.

61. See Figure 61.

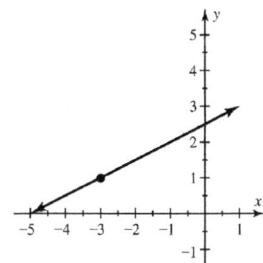

Figure 61

63. Because the slope is 2, the $y$-value increases by 2 when the $x$-value increases by 1. Thus,

when $x = 1$, $y = -4 + 2 = -2$; when $x = 2$, $y = -2 + 2 = 0$; when $x = 3$, $y = 0 + 2 = 2$. See Figure 63.

| $x$ | 0 | 1 | 2 | 3 |
|---|---|---|---|---|
| $y$ | -4 | -2 | 0 | 2 |

| $x$ | 1 | 2 | 3 | 4 |
|---|---|---|---|---|
| $y$ | 4 | 1 | -2 | -5 |

| $x$ | -4 | -2 | 0 | 2 |
|---|---|---|---|---|
| $y$ | 0 | 3 | 6 | 9 |

Figure 63                Figure 65                Figure 67

65. Because the slope is $-3$, the $y$-value decreases by 3 when the $x$-value increases by 1. Thus,

when $x = 2$, $y = 4 - 3 = 1$; when $x = 3$, $y = 1 - 3 = -2$; when $x = 4$, $y = -2 - 3 = -5$; See Figure 65.

67. Because the slope is $\dfrac{3}{2}$, the $y$-value increases by $\dfrac{3}{2}$ when the $x$-value increases by 1 and the $y$-value

increases by $2\left(\dfrac{3}{2}\right) = 3$ when the $x$-value increases by 2. Thus, when $x = -2$, $y = 0 + 3 = 3$;

when $x = 0$, $y = 3 + 3 = 6$; when $x = 2$, $y = 6 + 3 = 9$. See Figure 67.

69. Because zero gumballs cost zero cents, the graph must pass through (0, 0). c

71. The average cost of a new car has increased over the past 30 years. b

73. (a)    When $0 \le x \le 3$, the $y$-value increases by 1000 when $x$ increases by 1. $m_1 = 1000$

When $3 \le x \le 5$, the $y$-value decreases by 1000 when $x$ increases by 1. $m_2 = -1000$

(b)    $m_1 = 1000$: Water is being added to the pool at a rate of 1000 gallons per hour.

$m_2 = -1000$: Water is being removed from the pool at a rate of 1000 gallons per hour.

(c)    Initially the pool contained 2000 gallons of water. Over the first 3 hours, water was pumped into the pool at a rate of 1000 gallons per hour. For the next 2 hours, water was pumped out of the pool at a rate of 1000 gallons per hour.

75. (a)    When $0 \le x \le 2$, the $y$-value increases by 50 when $x$ increases by 1.  $m_1 = 50$

When $2 \le x \le 3$, the $y$-value remains constant when $x$ increases.  $m_2 = 0$

When $3 \le x \le 4$, the $y$-value decreases by 50 when $x$ increases by 1.  $m_3 = -50$

(b)    $m_1 = 50$:  The car is moving away from home at a rate of 50 mph.

$m_2 = 0$:  The car is not moving.

$m_3 = -50$:  The car is moving toward home at a rate of 50 mph.

(c)    Initially the car is at home. Over the first 2 hours, the car travels away from home at a rate of 50 mph. Then the car is parked for 1 hour.

Finally, the car travels toward home at a rate of 50 mph.

77.  $m^3/min$

79.  See Figure 79.

81.  See Figure 81.

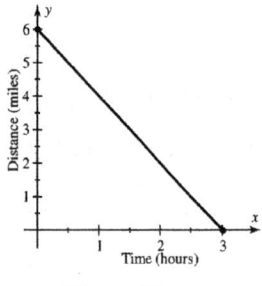                    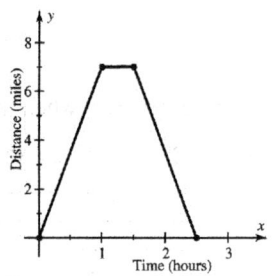

Figure 79                                       Figure 81

83. (a)    76,000,000

(b)    281,000,000

(c)    $m = \dfrac{y_2 - y_1}{x_2 - x_1} = \dfrac{281,000,000 - 76,000,000}{2000 - 1900} = \dfrac{205,000,000}{100} = 2,050,000 \Rightarrow 2,050,000$ people/yr

85. (a)    $m = \dfrac{y_2 - y_1}{x_2 - x_1} = \dfrac{148,000 - 25,600}{4 - 1} = \dfrac{122,400}{3} = 40,800$

(b)    The celebrity gained 40,800 followers each year, on average.

87. (a)    A point on the line is $(40, 1000)$ and another point is $(0, 0)$.

$m = \dfrac{y_2 - y_1}{x_2 - x_1} = \dfrac{1000 - 0}{40 - 0} = \dfrac{1000}{40} = 25$.  The slope of the line is 25.

(b)    The revenue is \$25 per flashdrive.

89. (a)    See Figure 89.

    (b)    $m_1 = \dfrac{y_2 - y_1}{x_2 - x_1} = \dfrac{100 - 0}{5 - 0} = \dfrac{100}{5} = 20$;   $m_2 = \dfrac{y_2 - y_1}{x_2 - x_1} = \dfrac{250 - 100}{10 - 5} = \dfrac{150}{5} = 30$;

    $m_3 = \dfrac{y_2 - y_1}{x_2 - x_1} = \dfrac{450 - 250}{15 - 10} = \dfrac{200}{5} = 40$.

    (c)    $m_1 = 20$:  Each mile between 0 and 5 miles is worth \$20/mile.

    $m_2 = 30$:  Each mile between 5 and 10 miles is worth \$30/mile.

    $m_3 = 40$:  Each mile between 10 and 15 miles is worth \$40/mile.

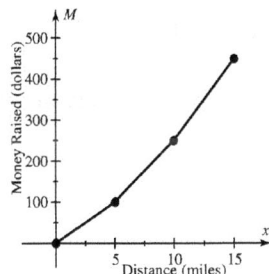

Figure 89

91. (a)    $m = \dfrac{y_2 - y_1}{x_2 - x_1} = \dfrac{50,000 - 42,000}{2008 - 2000} = \dfrac{8000}{8} = 1000$.   The slope of the line is 1000.

    (b)    Median family income increased on average by \$1000/year over this period of time.

    (c)    $42,000 + 14(1000) = \$56,000$

93. (a)    Because the slope of the line is negative, the ant is moving toward the stone.

    (b)    When the ant starts moving, $x = 0$, so $-2(0) + 10 = 0 + 10 = 10$.

    The ant is 10 feet from the stone.

    (c)    The slope is $-2$, so the ant is moving 2 ft/min.

    (d)    When the ant reaches the stone the distance from the stone is 0. So, solve the equation

    $0 = -2x + 10$ for $x$:  $0 - 10 = -2x + 10 - 10 \Rightarrow -10 = -2x \Rightarrow \dfrac{-10}{-2} = \dfrac{-2x}{-2} \Rightarrow 5 = x$

## Checking Basic Concepts  Sections 3.3 and 3.4

1. The line crosses the $x$-axis at $x = -2$;  therefore, the $x$-intercept is $-2$.

   The line crosses the $y$-axis at $y = 3$;  therefore, the $y$-intercept is 3.

2. $2x - y = 2 \Rightarrow 2x - y + y = 2 + y \Rightarrow 2x - 2 = 2 - 2 + y \Rightarrow y = 2x - 2$.

   When $x = -1$:  $y = 2(-1) - 2 \Rightarrow y = -2 - 2 \Rightarrow y = -4$;  When $x = 0$:  $y = 2x - 2 = 2(0) - 2 = -2$;

   When $x = 1$:  $y = 2x - 2 = 2(1) - 2 = 0$;  When $x = 2$:  $y = 2x - 2 = 2(2) - 2 = 2$.

   Therefore, the missing values in the table are $-4$, $-2$, 0, and 2.  See Figure 2.

When $y = 0$, $x = 1$. Therefore the $x$-intercept is $x = 1$.

When $x = 0$, $y = -2$. Therefore, the $y$-intercept is $y = -2$.

| $x$ | -2 | -1 | 0 | 1 | 2 |
|---|---|---|---|---|---|
| $y$ | -6 | -4 | -2 | 0 | 2 |

Figure 2

3.  (a)    To find the $x$-intercept, set $y = 0$ and solve for $x$:

$x - 2y = 6 \Rightarrow x - 2(0) = 6 \Rightarrow x = 6$. Therefore, the $x$-intercept is $x = 6$.

To find the $y$-intercept, set $x = 0$ and solve for $y$: $x - 2y = 6 \Rightarrow 0 - 2(y) = 6 \Rightarrow -2y = 6 \Rightarrow$

$\dfrac{-2y}{-2} = \dfrac{6}{-2} \Rightarrow y = -3$. Therefore, the $y$-intercept is $y = -3$. See Figure 3a.

(b)    $y = 2$ is the graph of a horizontal line. Therefore, the line does not cross the $x$-axis and does

not have an $x$-intercept. The line crosses the $y$-axis at $y = 2$, and therefore its $y$-intercept is at

$y = 2$. See Figure 3b.

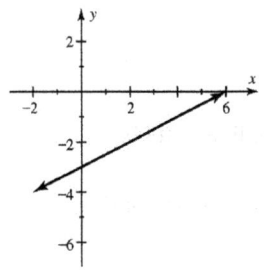      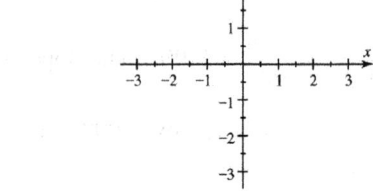      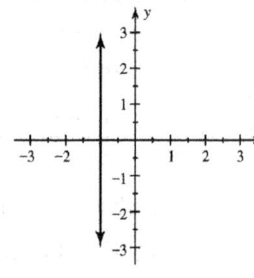

Figure 3a                        Figure 3b                        Figure 3c

(c)    $x = -1$ is the graph of a vertical line. Therefore, the line crosses the $x$-axis at $x = -1$ and

therefore its $x$-intercept is at $x = -1$. The line does not cross the $y$-axis, and therefore does not

have a $y$-intercept. See Figure 3c.

4.  A horizontal line passing through $(-2, 4)$ has the equation $y = 4$.

A vertical line passing through $(-2, 4)$ has the equation $x = -2$.

5.  (a)    $\dfrac{6 - 3}{2 - (-2)} = \dfrac{3}{4}$   Therefore, the slope of the line is $\dfrac{3}{4}$.

(b)    $\dfrac{3 - 3}{0 - (-5)} = \dfrac{0}{5} = 0$   Therefore, the slope of the line is 0.

(c)    $\dfrac{8 - 5}{1 - 1} = \dfrac{3}{0}$   Therefore, the slope of the line is undefined.

6.  A point on the line is $(2, 1)$ and another point is $(0, 2)$. Thus, $\dfrac{2 - 1}{0 - 2} = \dfrac{1}{-2} = -\dfrac{1}{2}$.

Therefore, the slope of the line is $-\dfrac{1}{2}$.

7.  See Figure 7.

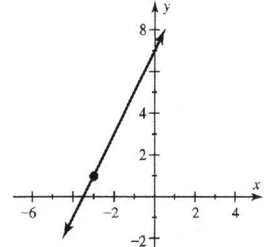

Figure 7

8.  (a)   $\dfrac{5-5}{1-0} = \dfrac{0}{1} = 0.$  Thus $m_1 = 0.$   $\dfrac{7-5}{2-1} = \dfrac{2}{1} = 2.$  Thus $m_2 = 2.$   $\dfrac{5-7}{5-2} = \dfrac{-2}{3} = -\dfrac{2}{3}.$  Thus $m_3 = -\dfrac{2}{3}.$

(b)   $m_1 = 0$:  before the rain, the depth of water in the pond does not change.

$m_2 = 2$:  while it rains, the depth of water in the pond increases by 2 feet for every hour of rain.

$m_3 = -\dfrac{2}{3}$:  after it stops raining, the depth of water in the pond decreases by $\dfrac{2}{3}$ feet per hour

until it reaches its original depth.

(c)   Initially the pond had a depth of 5 feet.  For the first hour, there was no change in the depth of

the pond.  For the next hour, the depth of the pond increased at a rate of 2 feet per hour to a

depth of 7 feet.  Finally, the depth of the pond decreased for 3 hours at a rate of $\dfrac{2}{3}$ foot per

hour until it was 5 feet deep.

## Section 3.5  Slope-Intercept Form

1.  $y = mx + b$

3.  $y$-intercept

5.  parallel

7.  perpendicular

9.  f

11.  a

13.  e

15.  The graph of the line shows the $y$-intercept at $-1$.  For every $x$-increase of 1, the $y$-value increases by

1, so the slope is 1.  Therefore:  $y = mx + b \Rightarrow y = x - 1.$

17.  The graph of the line shows the $y$-intercept at 1.  For every $x$-increase of 1, the $y$-value decreases by

2, so the slope is $-2$.  Therefore:  $y = mx + b \Rightarrow y = -2x + 1.$

19.  The graph of the line shows the $y$-intercept at 0.  For every $x$-increase of 1, the $y$-value decreases by

2, so the slope is $-2$.  Therefore:  $y = mx + b = -2x + 0 \Rightarrow y = -2x.$

21. The graph of the line shows the $y$-intercept at 2. For every $x$-increase of 4, the $y$-value increases by 3, so the slope is $\dfrac{3}{4}$. Therefore: $y = mx + b \Rightarrow y = \dfrac{3}{4}x + 2$.

23. See Figure 23. The slope-intercept form is $y = mx + b$, so with $m = 1$ and $b = 2$, $y = x + 2$.

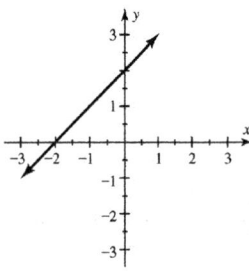

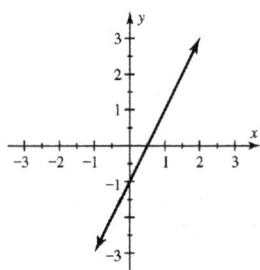

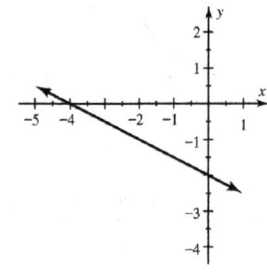

Figure 23                          Figure 25                          Figure 27

25. See Figure 25. The slope-intercept form is $y = mx + b$, so with $m = 2$ and $b = -1$, $y = 2x - 1$.

27. See Figure 27. The slope-intercept form is $y = mx + b$, so with $m = -\dfrac{1}{2}$ and $b = -2$, $y = -\dfrac{1}{2}x - 2$.

29. See Figure 29. $y = mx + b = \dfrac{1}{3}x + 0 \Rightarrow y = \dfrac{1}{3}x$.

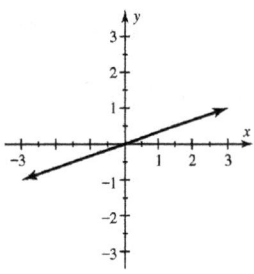

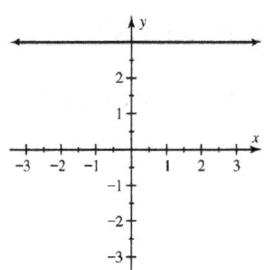

Figure 29                          Figure 31

31. See Figure 31. $y = mx + b \Rightarrow y = 0x + 3 \Rightarrow y = 3$

33. (a)   To write the equation in slope-intercept form, solve for $y$:

$x + y = 4 \Rightarrow x - x + y = 4 - x \Rightarrow y = 4 - x \Rightarrow y = -x + 4$

   (b)   Because the slope-intercept form is $y = mx + b$, the slope is $-1$ and the $y$-intercept is 4.

35. (a)   To write the equation in slope-intercept form, solve for $y$:

$2x + y = 4 \Rightarrow 2x - 2x + y = 4 - 2x \Rightarrow y = 4 - 2x \Rightarrow y = -2x + 4$

   (b)   Because the slope-intercept form is $y = mx + b$, the slope is $-2$ and the $y$-intercept is 4.

37. (a)   To write the equation in slope-intercept form, solve for $y$: $x - 2y = -4 \Rightarrow x - x - 2y = -4 - x \Rightarrow$

$-2y = -4 - x \Rightarrow -2y = -x - 4 \Rightarrow \dfrac{-2y}{-2} = \dfrac{-x - 4}{-2} \Rightarrow y = \dfrac{-x}{-2} - \dfrac{4}{-2} \Rightarrow y = \dfrac{1}{2}x + 2$

   (b)   Because the slope-intercept form is $y = mx + b$, the slope is $\dfrac{1}{2}$ and the $y$-intercept is 2.

39. (a)   To write the equation in slope-intercept form, solve for $y$:

$$2x - 3y = 6 \Rightarrow 2x - 2x - 3y = 6 - 2x \Rightarrow$$

$$-3y = 6 - 2x \Rightarrow -3y = -2x + 6 \Rightarrow \frac{-3y}{-3} = \frac{-2x + 6}{-3} \Rightarrow y = \frac{-2x}{-3} + \frac{6}{-3} \Rightarrow y = \frac{2}{3}x - 2$$

   (b)   Because the slope-intercept form is $y = mx + b$, the slope is $\frac{2}{3}$ and the $y$-intercept is $-2$.

41. (a)   To write the equation in slope-intercept form, solve for $y$:   $x = 4y - 6 \Rightarrow x - 4y = 4y - 4y - 6 \Rightarrow$

$$x - 4y = -6 \Rightarrow x - x - 4y = -6 - x \Rightarrow -4y = -x - 6 \Rightarrow \frac{-4y}{-4} = \frac{-x}{-4} - \frac{6}{-4} \Rightarrow y = \frac{1}{4}x + \frac{3}{2}$$

   (b)   Because the slope-intercept form is $y = mx + b$, the slope is $\frac{1}{4}$ and the $y$-intercept is $\frac{3}{2}$.

43. (a)   To write the equation in slope-intercept form, solve for $y$:

$$\frac{1}{2}x + \frac{3}{2}y = 1 \Rightarrow \frac{1}{2}x - \frac{1}{2}x + \frac{3}{2}y = 1 - \frac{1}{2}x \Rightarrow$$

$$\frac{3}{2}y = -\frac{1}{2}x + 1 \Rightarrow \frac{2}{3}\left(\frac{3}{2}y\right) = \frac{2}{3}\left(-\frac{1}{2}x + 1\right) \Rightarrow y = -\frac{1}{3}x + \frac{2}{3}$$

   (b)   Because the slope-intercept form is $y = mx + b$, the slope is $-\frac{1}{3}$ and the $y$-intercept is $\frac{2}{3}$.

45.  See Figure 45.

47.  See Figure 47.

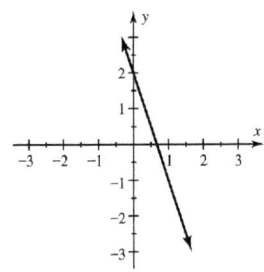

Figure 45

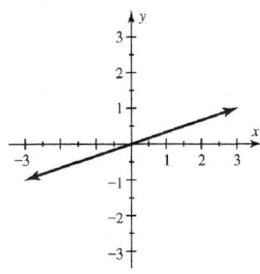

Figure 47

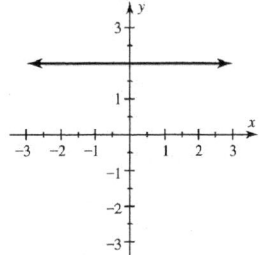

Figure 49

49.  See Figure 49.

51.  Solve for $y$:   $x + y = 3 \Rightarrow x - x + y = 3 - x \Rightarrow y = -x + 3$   See Figure 51.

53.  Solve for $y$:   $x + 2y = 2 \Rightarrow x - x + 2y = 2 - x \Rightarrow 2y = 2 - x \Rightarrow \frac{2y}{2} = \frac{2}{2} - \frac{x}{2} \Rightarrow y = -\frac{1}{2}x + 1$

See Figure 53.

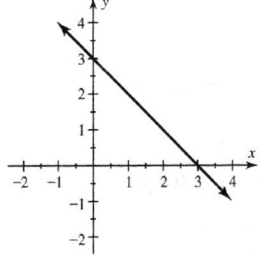

Figure 51

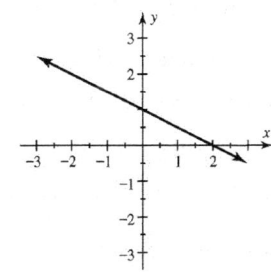

Figure 53

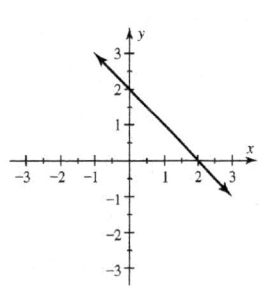

Figure 55

55. Solve for $y$: $x = 2 - y \Rightarrow x - x + y = 2 - x - y + y \Rightarrow y = 2 - x$   See Figure 55

57. For every increase in $x$ of 1, the $y$-value increases by 2. Therefore, the slope $m = 2$. When $x = 0$, $y = 2$. Therefore the $y$-intercept $b = 2$. Therefore: $y = mx + b \Rightarrow y = 2x + 2$.

59. For every increase in $x$ of 2, the $y$-value increases by 2. Therefore, the slope $m = 1$. When $x = 0$, $y = -2$. Therefore the $y$-intercept $b = -2$. Therefore: $y = mx + b \Rightarrow y = x - 2$.

61. $y = \dfrac{4}{7}x + 3$

63. Because the line parallel to $y = 3x + 1$ has slope 3, $m = 3$.

Because when $x = 0$, $y = 0$, the $y$-intercept $b = 0$. Therefore: $y = 3x$.

65. Solve $2x + 4y = 5$ for $y$: $2x + 4y = 5 \Rightarrow 2x - 2x + 4y = 5 - 2x \Rightarrow 4y = -2x + 5 \Rightarrow$

$\dfrac{4y}{4} = \dfrac{-2x}{4} + \dfrac{5}{4} \Rightarrow y = -\dfrac{1}{2}x + \dfrac{5}{4}$. Because the line parallel to $2x + 4y = 5$ has slope

$-\dfrac{1}{2}$, $m = -\dfrac{1}{2}$. Then, to find the $y$-intercept $b$, replace $x$ with 1 and $y$ with 2:

$y = -\dfrac{1}{2}x + b \Rightarrow 2 = -\dfrac{1}{2}(1) + b \Rightarrow 2 = -\dfrac{1}{2} + b \Rightarrow 2 + \dfrac{1}{2} = -\dfrac{1}{2} + \dfrac{1}{2} + b \Rightarrow \dfrac{5}{2} = b$.

Therefore, $y = -\dfrac{1}{2}x + \dfrac{5}{2}$.

67. Because the line $y = -\dfrac{1}{2}x - 3$` has slope $-\dfrac{1}{2}$, the line perpendicular to it has slope 2, the negative

reciprocal of $-\dfrac{1}{2}$. Because when $x = 0$, $y = 0$, the $y$-intercept is $b = 0$.

Therefore, $y = mx + b \Rightarrow y = 2x$.

69. Solve $x = -\dfrac{1}{3}y$ for $y$: $x = -\dfrac{1}{3}y \Rightarrow -3(x) = -3\left(-\dfrac{1}{3}y\right) \Rightarrow -3x = y \Rightarrow y = -3x$.

Because the line $y = -3x$ has slope $-3$, the line perpendicular to it has slope $\dfrac{1}{3}$, the negative

reciprocal of $-3$. Then, to find the $y$-intercept $b$, replace $y$ with 0 and $x$ with $-1$:

$0 = \dfrac{1}{3}(-1) + b \Rightarrow 0 = -\dfrac{1}{3} + b \Rightarrow 0 + \dfrac{1}{3} = -\dfrac{1}{3} + \dfrac{1}{3} + b \Rightarrow \dfrac{1}{3} = b$. Therefore, $y = \dfrac{1}{3}x + \dfrac{1}{3}$.

71. (a)    When $x = 0$, $y = 25$. Therefore, \$25.

    (b)    It costs $0.25x$ for each additional mile $x$. Therefore, 25 cents.

    (c)    When $x = 0$, $y = 25$ and the $y$-intercept is 25. This represents the fixed cost of renting a car.

    (d)    The slope is 0.25. This represents the cost per mile of renting the car.

73. (a)    Because $b = 3.95$ and $m = 0.07$, $y = 0.07x + 3.95$. Set $x = 50$ and solve for $y$:

        $y = 0.07(50) + 3.95 = 3.5 + 3.95 = 7.45$. Therefore, the charge is \$7.45.

    (b)    $C = 0.07x + 3.95$

(c)   Set $C = 8.64$ and solve for $x$: $8.64 = 0.07x + 3.95 \Rightarrow 8.64 - 3.95 = 0.07x \Rightarrow 4.69 = 0.07x \Rightarrow$

$$\frac{4.69}{0.07} = \frac{0.07x}{0.07} \Rightarrow x = 67 \text{ min.}$$

75. (a)   Because the $y$-value increases by $0.35 for every increase of 1 in the $x$-value, the slope

$m = 0.35$.   Because the fixed cost of \$164.30 represents the $y$-value when $x = 0$, $b = 164.3$.

(b)   The fixed cost of owning the car for one month.

## Section 3.6  Point-Slope Form

1.  True

3.  $y = mx + b$

5.  1; 3

7.  Yes; every nonvertical line has one slope and one $y$-intercept.

9.  Substitute 3 for $x$ and $-11$ for $y$:   $y + 1 = -2(x + 3) \Rightarrow -11 + 1 = -2(3 + 3) \Rightarrow -10 = 2(6) \Rightarrow -10 = -12$

Because this is not a true statement, the point $(3, -11)$ does not lie on the line.

11.  Substitute 0 for $x$ and 4 for $y$:   $y = \frac{1}{2}(x + 4) + 2 \Rightarrow 4 = \frac{1}{2}(0 + 4) + 2 \Rightarrow 4 = \frac{1}{2}(4) + 2 \Rightarrow$

$4 = 2 + 2 \Rightarrow 4 = 4$.   Because this is a true statement, the point $(0, 4)$ lies on the line.

13.  The points $(-2, 3)$ and $(0, -1)$ are on the line.  First, determine the slope:

$m = \frac{y_2 - y_1}{x_2 - x_1} = \frac{-1 - 3}{0 - (-2)} = \frac{-4}{2} = -2$.   Then substitute the $x$ and $y$ values of the point $(-2, 3)$ into the

point-slope form:   $y - y_1 = m(x - x_1) \Rightarrow y - 3 = m(x - (-2))$. Then, substitute the value of the slope

for $m$:   $y - 3 = -2(x + 2)$.

15.  The points $(1, 2)$ and $(-3, -1)$ are on the line.  First, determine the slope:

$m = \frac{y_2 - y_1}{x_2 - x_1} = \frac{-1 - 2}{-3 - 1} = \frac{-3}{-4} = \frac{3}{4}$.   Then substitute the $x$ and $y$ values of the point $(1, 2)$ into the point

slope form:   $y - y_1 = m(x - x_1) \Rightarrow y - 2 = m(x - 1)$. Then, substitute the value of the slope for

$m$:   $y - 2 = \frac{3}{4}(x - 1)$.

17.  The points $(3, -1)$ and $(-1, 1)$ are on the line.  First, determine the slope:

$m = \frac{y_2 - y_1}{x_2 - x_1} = \frac{1 - (-1)}{-1 - 3} = \frac{1 + 1}{-4} = \frac{2}{-4} = -\frac{1}{2}$.   Then substitute the $x$ and $y$ values of the point $(3, -1)$

and the value of the slope into the point-slope form:

$y - y_1 = m(x - x_1) \Rightarrow y - (-1) = -\frac{1}{2}(x - 3) \Rightarrow y + 1 = -\frac{1}{2}(x - 3)$.

19.  $y - y_1 = m(x - x_1) \Rightarrow y - 1 = 4(x - (-3)) \Rightarrow y - 1 = 4(x + 3)$

21. $y - y_1 = m(x - x_1) \Rightarrow y - (-3) = \dfrac{1}{2}(x - (-5)) \Rightarrow y + 3 = \dfrac{1}{2}(x + 5)$

23. $y - y_1 = m(x - x_1) \Rightarrow y - 30 = 1.5(x - 2010)$

25. First, determine the slope: $m = \dfrac{y_2 - y_1}{x_2 - x_1} = \dfrac{-3 - 4}{-1 - 2} = \dfrac{-7}{-3} = \dfrac{7}{3}$. Then, insert the values of the first point

    and the value of the slope into the point-slope form: $y - y_1 = m(x - x_1) \Rightarrow y - 4 = \dfrac{7}{3}(x - 2)$.

27. First determine the slope: $m = \dfrac{y_2 - y_1}{x_2 - x_1} = \dfrac{-3 - 0}{0 - 5} = \dfrac{-3}{-5} = \dfrac{3}{5}$. Then, insert the values of the first point

    and the value of the slope into the point-slope form:

    $y - y_1 = m(x - x_1) \Rightarrow y - 0 = \dfrac{3}{5}(x - 5) \Rightarrow y = \dfrac{3}{5}(x - 5)$.

29. First determine the slope: $m = \dfrac{y_2 - y_1}{x_2 - x_1} = \dfrac{65 - 15}{2013 - 2003} = \dfrac{50}{10} = 5$. Then, insert the values of the first

    point and the value of the slope into the point-slope form:

    $y - y_1 = m(x - x_1) \Rightarrow y - 15 = 5(x - 2003)$.

31. To convert to slope-intercept form, use the distributive property of multiplication and solve for $y$:

    $y - 4 = 3(x - 2) \Rightarrow y - 4 = 3x - 6 \Rightarrow y - 4 + 4 = 3x - 6 + 4 \Rightarrow y = 3x - 2$.

33. To convert to slope-intercept form, use the distributive property of multiplication and solve for $y$:

    $y + 2 = \dfrac{1}{3}(x + 6) \Rightarrow y + 2 = \dfrac{1}{3}x + 2 \Rightarrow y + 2 - 2 = \dfrac{1}{3}x + 2 - 2 \Rightarrow y = \dfrac{1}{3}x$.

35. To convert to slope-intercept form, use the distributive property of multiplication and solve for $y$:

    $y - \dfrac{3}{4} = \dfrac{2}{3}(x - 1) \Rightarrow y - \dfrac{3}{4} = \dfrac{2}{3}x - \dfrac{2}{3} \Rightarrow y - \dfrac{3}{4} + \dfrac{3}{4} = \dfrac{2}{3}x - \dfrac{2}{3} + \dfrac{3}{4} \Rightarrow y = \dfrac{2}{3}x - \dfrac{8}{12} + \dfrac{9}{12} \Rightarrow y = \dfrac{2}{3}x + \dfrac{1}{12}$.

37. Use the distributive property of multiplication and solve for $y$:

    $y = -2(x - 2) + 5 \Rightarrow y = -2x + 4 + 5 \Rightarrow y = -2x + 9$.

39. Use the distributive property of multiplication and solve for $y$:

    $y = \dfrac{3}{5}(x - 5) + 1 \Rightarrow y = \dfrac{3}{5}x - 3 + 1 \Rightarrow y = \dfrac{3}{5}x - 2$.

41. $y = -16(x + 1.5) + 5 \Rightarrow y = -16x - 24 + 5 \Rightarrow y = -16x - 19$

43. First, find the point-slope form for the line: $y - y_1 = m(x - x_1) \Rightarrow y - (-3) = -2(x - 4)$.

    To find the slope-intercept form, use the distributive property of multiplication, then solve for $y$:

    $y - (-3) = -2(x - 4) \Rightarrow y + 3 = -2x + 8 \Rightarrow y + 3 - 3 = -2x + 8 - 3 \Rightarrow y = -2x + 5$.

45. First, find the slope of the line: $m = \dfrac{y_2 - y_1}{x_2 - x_1} = \dfrac{-1 - (-2)}{2 - 3} = \dfrac{-1 + 2}{-1} = \dfrac{1}{-1} = -1.$ Then, find the point

slope form for the line by inserting the values of the point $(3, -2)$ and the value of the slope into

the point-slope form: $y - y_1 = m(x - x_1) \Rightarrow y - (-2) = -1(x - 3) \Rightarrow y + 2 = -1(x - 3).$ Then, use the

distributive property of multiplication and solve for $y$:

$y + 2 = -x + 3 \Rightarrow y + 2 - 2 = -x + 3 - 2 \Rightarrow y = -x + 1.$

47. The points $(3, 0)$ and $\left(0, \dfrac{1}{3}\right)$ are on the line. First find the slope of the line:

$m = \dfrac{y_2 - y_1}{x_2 - x_1} = \dfrac{\frac{1}{3} - 0}{0 - 3} = \dfrac{\frac{1}{3}}{-3} = \dfrac{1}{3}\left(-\dfrac{1}{3}\right) = -\dfrac{1}{9}.$ Then, insert the values of the point $(3, 0)$ and the value

of the slope into the point-slope form: $y - y_1 = m(x - x_1) \Rightarrow y - 0 = -\dfrac{1}{9}(x - 3).$ Then, use the

distributive property of multiplication and solve for $y$: $y = -\dfrac{1}{9}x + \dfrac{1}{3}.$

49. The line parallel to $y = 2x - 1$ has slope 2. Insert the values of the point $(2, -3)$ and the value of

the slope into the point-slope form:

$y - y_1 = m(x - x_1) \Rightarrow y - (-3) = 2(x - 2) \Rightarrow y + 3 = 2(x - 2).$ Then, convert to slope-intercept

form: $y + 3 = 2x - 4 \Rightarrow y + 3 - 3 = 2x - 4 - 3 \Rightarrow y = 2x - 7.$

51. The line perpendicular to $y = -\dfrac{1}{2}x + 3$ has slope 2. Insert the values of the point $(6, -3)$ and the

value of the slope into the point-slope form:

$y - y_1 = m(x - x_1) \Rightarrow y - (-3) = 2(x - 6) \Rightarrow y + 3 = 2(x - 6).$ Then, convert to slope-intercept

form: $y + 3 = 2x - 12 \Rightarrow y + 3 - 3 = 2x - 12 - 3 \Rightarrow y = 2x - 15.$

53. As $x$ increases by 1, $y$ decreases by 2. Thus, the slope is $-2$. Insert the values of the point $(1, -3)$

and the value of the slope into the point slope form:

$y - y_1 = m(x - x_1) \Rightarrow y - (-3) = -2(x - 1) \Rightarrow y + 3 = -2(x - 1).$ Then, convert to slope-intercept

form: $y + 3 = -2x + 2 \Rightarrow y + 3 - 3 = -2x + 2 - 3 \Rightarrow y = -2x - 1.$

55. As $x$ increases by 2, $y$ increases by 1. Thus, the slope is $\dfrac{1}{2}$. Insert the values of the point $(-1, -3)$

and the value of the slope into the point slope form:

$y - y_1 = m(x - x_1) \Rightarrow y - (-3) = \dfrac{1}{2}(x - (-1)) \Rightarrow y + 3 = \dfrac{1}{2}(x + 1).$ Then, convert to slope-intercept

form: $y + 3 = \dfrac{1}{2}x + \dfrac{1}{2} \Rightarrow y + 3 - 3 = \dfrac{1}{2}x + \dfrac{1}{2} - 3 \Rightarrow y = \dfrac{1}{2}x - \dfrac{5}{2}.$

57. Write the point-slope form in slope-intercept form:

$$y - y_1 = m(x - x_1) \Rightarrow y - y_1 = mx - mx_1 \Rightarrow y - y_1 + y_1 = mx - mx_1 + y_1 \Rightarrow y = mx - mx_1 + y_1.$$

so the $y$-intercept is $-mx_1 + y_1$.

59. (a)  Find the slope:  $m = \dfrac{y_2 - y_1}{x_2 - x_1} = \dfrac{15 - 40}{6 - 1} = \dfrac{-25}{5} = -5°$ F/hr.

   (b)  One point of the graph of the line is $(1,\ 40)$. First, find the point-slope form of the line:

   $$y - y_1 = m(x - x_1) \Rightarrow y - 40 = -5(x - 1).$$ Then convert to slope-intercept form:

   $$T = -5x + 5 + 40 \Rightarrow T = -5x + 45.$$ The temperature is decreasing at a rate of $5°$F per hour.

   (c)  To find the $x$-intercept, use the slope-intercept form of the line and set $T = 0$:  $T = -5x + 45 \Rightarrow$

   $$0 = -5x + 45 \Rightarrow 5x = 45 \Rightarrow x = 9.$$ Therefore, at 9 A.M. the temperature was $0°$F.

   (d)  See Figure 59.

   (e)  At 4 A.M., the temperature was $25°$F.

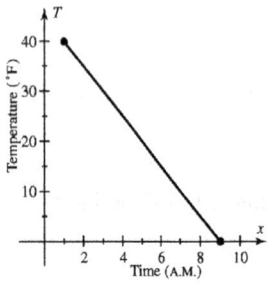

Figure 59

61. (a)  Because $y$ increases as $x$ increases, water is entering the tank. After 4 minutes, 300 gallons of water are in the tank.

   (b)  When $x = 0$, $y = 100$. therefore, the $y$-intercept is 100. This means that the tank held 100 gallons of water initially.

   (c)  First, find the slope:  $m = \dfrac{y_2 - y_1}{x_2 - x_1} = \dfrac{200 - 100}{2 - 0} = \dfrac{100}{2} = 50.$

   Then:  $y = mx + b \Rightarrow y = 50x + 100$;  the amount of water is increasing at a rate of 50 gal/min.

   (d)  $y = 50x + 100 \Rightarrow 500 = 50x + 100 \Rightarrow 400 = 50x \Rightarrow 8 = x$; The tank will be full after 8 minutes.

63. (a)  As $x$ increases, the $y$-value decreases. Because $y$ represents the distance from home, the person is traveling toward home.

   (b)  After 1 hour, the person is 250 miles from home; after 4 hours, the person is 100 miles from home.

   (c)  Because in 3 hours the person travels 150 miles, the driver is traveling at 50 mph.

   (d)  First, find the slope:  $m = \dfrac{y_2 - y_1}{x_2 - x_1} = \dfrac{100 - 250}{4 - 1} = \dfrac{-150}{3} = -50.$  Then, $y = -50x + b$. Because

   when $x = 0$, $y = 300$, the $y$-intercept $b$ is 300. Then, $y = -50x + 300$. The car is traveling toward home at 50 mph.

65. (a) In 2002, average home size was 2334 square feet.

    (b) Using the points (2002, 2334) and (2008, 2538) we have

$$m = \frac{y_2 - y_1}{x_2 - x_1} = \frac{2538 - 2334}{2008 - 2002} = \frac{204}{6} = 34.$$ Substituting the point (2002, 2334) and the slope of

34 into the point-slope equation we have $y - y_1 = m(x - x_1) \Rightarrow y - 2334 = 34(x - 2002)$.

    (c) According to the graph we have about 2500 square feet.

Using the equation: $y - 2334 = 34(2007 - 2002) \Rightarrow y - 2334 = 34(5) \Rightarrow y - 2334 = 170 \Rightarrow$

$y = 170 + 2334 \Rightarrow y = 2504$ square feet

    (d) Home size increased, on average, by 34 square feet per year.

67. (a) The rate of change is given as 515 children per year. Therefore, the slope of the line is 515.
    In 2006 there were 731 children adopted into the U.S. Therefore, one point on the line is
    (2006, 731). The point-slope equation is $y - y_1 = m(x - x_1) \Rightarrow y - 731 = 515(x - 2006)$.

    (b) $y - 731 = 515(2008 - 2006) \Rightarrow y - 731 = 515(2) \Rightarrow y - 731 = 1030 \Rightarrow$

$y = 731 + 1030 \Rightarrow y = 1761$   The result is 1761 children.

69. Cigarette consumption decreased, on average, by 10.33 billion cigarettes per year.

71. (a) The slope-intercept form is $y = mx + b$. Solve for $b$ by setting $y$ equal to 33, $m$ equal to 2.7,
    and $x$ equal to 2009:   $33 = 2.7(2000) + b \Rightarrow 33 = 5424.3 + b \Rightarrow -5391.3 = b$

Therefore,   $y = 2.7x - 5391.3$.

    (b) Substitute 2012 for $x$:   $y = 2.7(2012) - 5391.3 = 5432.4 - 5391.3 \Rightarrow y = 41.1$ million.

## Checking Basic Concepts  Sections 3.5 and 3.6

1. For every $x$-value increase of 1, the $y$-value decreases by 3.  Therefore, the slope is $-3$.  When
   $x = 0, y = 1$.  Thus, the $y$-intercept is 1.  Therefore:   $y = mx + b \Rightarrow y = -3x + 1$.

2. To convert to slope-intercept form, solve for $y$:   $4x - 5y = 20 \Rightarrow 4x - 4x - 5y = 20 - 4x \Rightarrow$

$-5y = -4x + 20 \Rightarrow \frac{-5y}{-5} = \frac{-4x}{-5} + \frac{20}{-5} \Rightarrow y = \frac{4}{5}x - 4.$  The slope is $\frac{4}{5}$ and the $y$-intercept is $-4$.

3. See Figure 3.

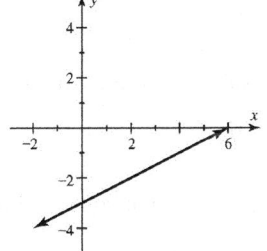

Figure 3

4. (a) The slope $m$ is 3 and the $y$-intercept $b$ is $-2$. Therefore: $y = mx + b \Rightarrow y = 3x - 2$.

(b) The slope of the line $y = \frac{2}{3}x$ is $\frac{2}{3}$. Therefore, the slope of the line perpendicular to $y = \frac{2}{3}x$

has slope $-\frac{3}{2}$. Because for every 2 units of increase in run the rise decreases by 3, the

$y$-intercept $b$ exists when $x = 0$. Thus, $b = 0$. Then: $y = mx + b \Rightarrow y = -\frac{3}{2}x + 0 \Rightarrow y = -\frac{3}{2}x$.

(c) First, find the slope: $m = \frac{y_2 - y_1}{x_2 - x_1} = \frac{3 - (-4)}{-2 - 1} = \frac{3 + 4}{-2 - 1} = \frac{7}{-3} = -\frac{7}{3}$. Then insert the values of the

point $(1, -4)$ and the value of the slope into the point-slope form: $y - y_1 = m(x - x_1) \Rightarrow$

$y - (-4) = -\frac{7}{3}(x - 1) \Rightarrow y + 4 = -\frac{7}{3}(x - 1)$. Then, convert to slope-intercept form:

$y + 4 = -\frac{7}{3}x + \frac{7}{3} \Rightarrow y + 4 - 4 = -\frac{7}{3}x + \frac{7}{3} - 4 \Rightarrow y = -\frac{7}{3}x + \frac{7}{3} - \frac{12}{3} \Rightarrow y = -\frac{7}{3}x - \frac{5}{3}$.

5. $y - y_1 = m(x - x_1) \Rightarrow y - 3 = -2(x - (-1)) \Rightarrow y - 3 = -2(x + 1)$

6. Use the distributive property of multiplication and solve for $y$:

$y + 3 = -2(x - 2) \Rightarrow y + 3 = -2x + 4 \Rightarrow y + 3 - 3 = -2x + 4 - 3 \Rightarrow y = -2x + 1$.

7. First, use the points $(-3, -3)$ and $(-1, 1)$ to find the slope: $m = \frac{y_2 - y_1}{x_2 - x_1} = \frac{1 - (-3)}{-1 - (-3)} = \frac{1 + 3}{-1 + 3} = \frac{4}{2} = 2$.

Use the point-slope form to write an equation for the line then convert to slope-intercept form:

$y - y_1 = m(x - x_1) \Rightarrow y - (-3) = 2(x - (-3)) \Rightarrow$

$y + 3 = 2(x + 3) \Rightarrow y + 3 = 2x + 6 \Rightarrow y + 3 - 3 = 2x + 6 - 3 \Rightarrow y = 2x + 3$

8. (a) First, find the slope: $m = \frac{y_2 - y_1}{x_2 - x_1} = \frac{12 - 36}{3 - 1} = \frac{-24}{2} = -12$. Then, insert the values for the point

$(1, 36)$ and the value of the slope into the point-slope form:

$y - y_1 = m(x - x_1) \Rightarrow y - 36 = -12(x - 1)$. Then convert to slope-intercept form:

$y - 36 = -12x + 12 \Rightarrow y - 36 + 36 = -12x + 12 + 36 \Rightarrow y = -12x + 48$.

(b) Because the slope is $-12$, the bicyclist gets 12 miles closer to home every hour ridden.

Therefore the bicyclist is traveling at 12 mph.

(c) The $y$-value will equal 0 when the rider reaches home. Therefore, substitute 0 for $y$ in the slope-intercept form and solve for $x$: $y = -12x + 48 \Rightarrow 0 = -12x + 48 \Rightarrow 0 - 48 = -12x + 48 - 48 \Rightarrow$

$-48 = -12x \Rightarrow \frac{-48}{-12} = \frac{-12x}{-12} \Rightarrow 4 = x \Rightarrow x = 4$. Therefore, the rider will arrive home at 4 pm.

(d) At noon, $x = 0$. Substitute 0 for $x$ in the slope-intercept form and solve for $y$:

$y = -12x + 48 \Rightarrow y = -12(0) + 48 \Rightarrow y = 48$. At noon the rider was 48 miles from home.

9. (a) At noon, $t = 0$. Substitute 0 for $t$ in the equation and solve for $S$:

   $S = 2t + 5 = 2(0) + 5 \Rightarrow S = 5$. Therefore, 5 inches of snow fell by noon.

   (b) Because the slope equals 2, the snow fell at a rate of 2 inches per hour in the afternoon.

   (c) The $S$-intercept is 5. This represents the total inches of snow that fell before noon.

   (d) The slope is 2. This represents that the rate of snowfall was 2 inches per hour.

## Section 3.7 Introduction to Modeling

1. linear

3. approximate

5. constant rate of change

7. f

9. a

11. e

13. For the linear equation $y = 2x + 2$, the $y$-intercept is 2 and the slope is 2. The ordered pairs in the table are modeled exactly by the linear equation.

15. For the linear equation $y = -4x$, the $y$-intercept is 0 and the slope is $-4$. The ordered pairs in the table are not modeled exactly by the linear equation.

17. For the linear equation $y = 1.4x - 4$, the $y$-intercept is $-4$ and the slope is 1.4. The ordered pairs in the table are not modeled exactly by the linear equation.

19. Because the line goes through the points on the graph, the linear model is exact. To find the equation

   of the line, first find the slope: $m = \dfrac{y_2 - y_1}{x_2 - x_1} = \dfrac{6-4}{4-3} = \dfrac{2}{1} = 2$. Then to find the $y$-intercept $b$, insert the

   values of the point (4, 6) and the value of the slope into the slope-intercept form and solve for $b$:

   $y = mx + b \Rightarrow 6 = 2(4) + b \Rightarrow 6 = 8 + b \Rightarrow 6 - 8 = 8 - 8 + b \Rightarrow -2 = b \Rightarrow b = -2$. Thus, the equation of

   the line is $y = 2x - 2$.

21. Because the line does not go through all the points on the graph, the linear model is approximate. To

   find the equation of the line, first find the slope: $m = \dfrac{y_2 - y_1}{x_2 - x_1} = \dfrac{6-2}{2-0} = \dfrac{4}{2} = 2$. Then to find the $y$

   intercept $b$, insert the values of the point (2, 6) and the value of the slope into the slope-intercept

   form and solve for $b$: $y = mx + b \Rightarrow 6 = 2(2) + b \Rightarrow$

   $6 = 4 + b \Rightarrow 6 - 4 = 4 - 4 + b \Rightarrow 2 = b \Rightarrow b = 2$. The equation of the line is $y = 2x + 2$.

23. Because the line does not go through all the points on the graph, the linear model is

   approximate. The equation of the line is $y = 2$ because the slope is 0 and because the $y$-intercept is

   at $y = 2$.

25. The initial value is 40, so the $y$-intercept $b$ is 40. The rate of increase is 5 per minute, so the slope is 5. Thus: $y = mx + b \Rightarrow y = 5x + 40$.

27. The initial value is $-5$, so the $y$-intercept $b$ is $-5$. The rate of decrease is 20 per day, so the slope is $-20$. Thus: $y = mx + b \Rightarrow y = -20x - 5$.

29. The initial value is 8, so the $y$-intercept $b$ is 8. Because $y$ remains constant, the slope is 0. Thus: $y = mx + b \Rightarrow y = 0(x) + 8 \Rightarrow y = 8$.

31. (a)    See Figure 31a. A line could pass through all five points.

    (b)    See Figure 31b.

    (c)    Because $y$ decreases by 2 for every increase in $x$, the slope $m$ is $-2$. Because when $x = 0$, $y = 4$, the $y$-intercept $b$ is 4. Thus: $y = mx + b \Rightarrow y = -2x + 4$.

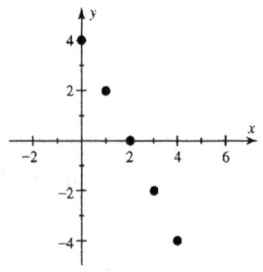    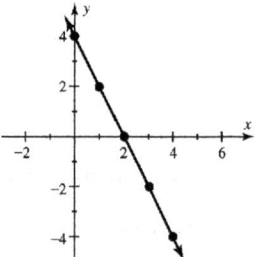

Figure 31a                           Figure 31b

33. (a)    See Figure 33a. A line could not pass through all five points.

    (b)    See Figure 33b.

    (c)    To find the equation of the line, first find the slope: $m = \dfrac{y_2 - y_1}{x_2 - x_1} = \dfrac{0 - 4}{0 - (-2)} = \dfrac{-4}{2} = -2$.

    Then, the line passes through the origin, so the $y$-intercept $b$ is 0.

    Thus: $y = mx + b \Rightarrow y = -2x + 0 \Rightarrow y = -2x$.

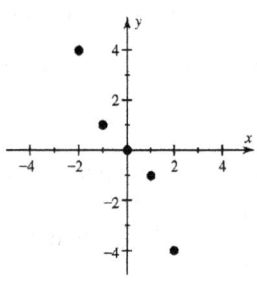    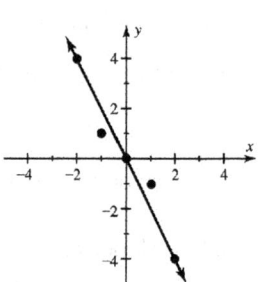

Figure 33a                           Figure 33b

35. (a)    See Figure 35a. A line could pass through all five points.

    (b)    See Figure 35b.

    (c)    To find the equation of the line, first find the slope: $m = \dfrac{y_2 - y_1}{x_2 - x_1} = \dfrac{-2 - 0}{0 - (-4)} = \dfrac{-2}{4} = -\dfrac{1}{2}$.

    Then, when $x = 0$, $y = -2$, the $y$-intercept $b$ is $-2$. Thus: $y = mx + b \Rightarrow y = -\dfrac{1}{2}x - 2$.

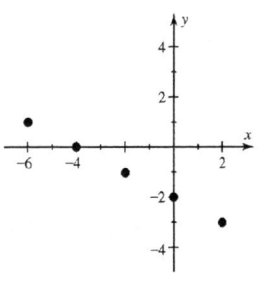

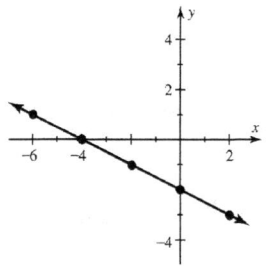

Figure 35a                                      Figure 35b

37. (a)   See Figure 37a. A line could not pass through all five points.

   (b)   See Figure 37b.

   (c)   To find the equation of the line, first find the slope: $m = \dfrac{y_2 - y_1}{x_2 - x_1} = \dfrac{0 - (-2)}{3 - (-3)} = \dfrac{2}{6} = \dfrac{1}{3}$.

        Then, when $x = 0$, $y = -1$, the $y$-intercept $b$ is $-1$. Thus: $y = mx + b \Rightarrow y = \dfrac{1}{3}x - 1$.

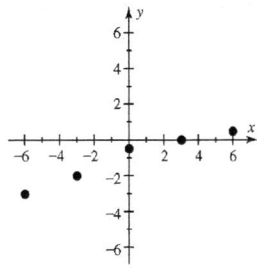

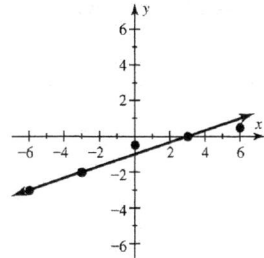

Figure 37a                                      Figure 37b

39. The initial value is 200, so the $g$-intercept $b$ is 200. The barrel is being filled at a rate of 5 gallons per minute, so the slope $m$ is 5. Thus: $g = mt + b \Rightarrow g = 5t + 200$, where $g$ represents gallons of water and $t$ represents time in minutes.

41. The initial value is 5, so the $d$-intercept $b$ is 5. The athlete is jogging at 6 miles per hour, so the slope $m$ is 6. Thus: $d = mt + b \Rightarrow d = 6t + 5$, where $d$ represents distance in miles and $t$ represents time in hours.

43. The initial value is 200, so the $p$-intercept $b$ is 200. The worker is being paid \$8 per hour, so the slope $m$ is 8. Thus: $p = mt + b \Rightarrow p = 8t + 200$, where $p$ represents total pay in dollars and $t$ represents time in hours.

45. The initial value is 5, so the $r$-intercept $b$ is 5. The carpenter is shingling roofs at a rate of 1 per day, so the slope $m$ is 1. Thus: $r = mt + b \Rightarrow r = t + 5$, where $r$ represents total number of roofs shingled and $t$ represents time in days.

47. (a)   A point on the graph of the volume of the glacier is (1912, 5) and another point is (2002, 1). Because there is assumed to be a constant melt rate, to find the yearly rate find the slope of the line connecting the two points:

$$m = \frac{y_2 - y_1}{x_2 - x_1} = \frac{1 - 5}{2002 - 1912} = \frac{-4}{90} = -\frac{2}{45} \text{ acres per year.}$$

(b)   The slope $m$ is $-\dfrac{2}{45}$. The initial value $b$ is 5. Thus: $A = mt + b \Rightarrow A = -\dfrac{2}{45}t + 5$, where $A$

represents acres of glacier and $t$ represents time in years.

49. (a)   First, use the first and last data points to determine slope:

$m = \dfrac{y_2 - y_1}{x_2 - x_1} = \dfrac{208 - 172}{2009 - 2003} = \dfrac{36}{6} = 6$. Then, insert the values of the first point into the point

slope form: $y - y_1 = m(x - x_1) \Rightarrow y - 172 = 6(x - 2003)$. Then, convert to slope-intercept

form: $y - 172 = 6x - 12,018 \Rightarrow y - 172 + 172 = 6x - 12,018 + 172 \Rightarrow y = 6x - 11,846$

(b)   Set $x = 2013$ and solve for $y$: $y = 6(2013) - 11,846 = 12,078 - 11,846 = 232$. Thus, the prison

population in 2013 will be about 232,000.

51. (a)   See Figure 51a.

(b)   See Figure 51b.

(c)   $m = \dfrac{y_2 - y_1}{x_2 - x_1} = \dfrac{120 - 60}{6 - 3} = \dfrac{60}{3} = 20$. For every gallon of gasoline, the car travels 20 miles.

(d)   $y - 60 = 20(x - 3) \Rightarrow y - 60 = 20x - 60 \Rightarrow y = 20x$.

(e)   $y = 20x = 20(7) = 140$ miles.

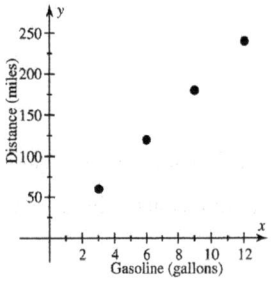

Figure 51a

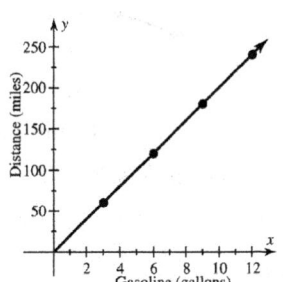

Figure 51b

## Checking Basic Concepts  Section 3.7

1. The ordered pairs in the table indicate that the $y$-intercept is 10 and the slope is $-5$. Thus, the line

$y = -5x + 10$ models exactly the ordered pairs.

2. Because the line does not pass through each point on the graph, the linear model is approximate. For

each increase in run of 1, the value of the rise increases by 1, so the slope $m$ is 1. When

$x = 0$, $y = -1$, so the $y$-intercept $b$ is $-1$. Thus: $y = mx + b \Rightarrow y = x - 1$.

3. (a)   The initial value $b$ is 50 and the slope $m$ is 10. Thus: $y = 10x + 50$.

(b)   The initial value $b$ is 200 and the slope $m$ is $-2$. Thus: $y = -2x + 200$.

4. (a)   See Figure 4a. A line could pass through all four points.

(b)   See Figure 4b.

(c)   The y-intercept $b$ is 1 and the slope $m$ is $\dfrac{1-2}{0-(-2)} = \dfrac{-1}{2} = -\dfrac{1}{2}$.   Thus: $y = -\dfrac{1}{2}x + 1$.

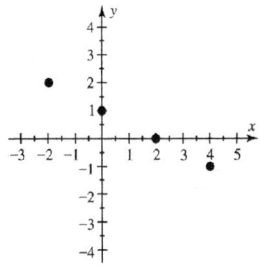    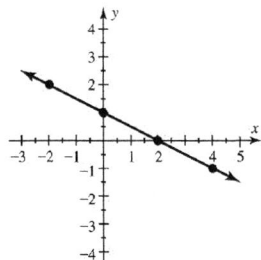

Figure 4a                          Figure 4b

5. (a)   The slope $m$ is 0.075 and the y-intercept is 0.  Thus, $T = 0.075x$.

(b)   2013 is 68 years after 1945, so substitute 68 for $x$ and solve for $T$. $T = 0.075\,(68) = 5.1$. The

temperature increase is 5.1° F.

## Chapter3  Review

1. $(-2, 0)$: none;  $(-1, 2)$: Quadrant II;  $(0, 0)$: none;  $(1, -2)$: Quadrant IV; $(1, 3)$: Quadrant I

2. See Figure 2.

3. (a)   $(-4, 3)$ QII

   (b)   $\left(\dfrac{1}{3}, -\dfrac{1}{2}\right)$ QIV

4. (a)   $(0, 3.2)$ None

   (b)   $(-5, -1.7)$ QIII

5. See Figure 5.

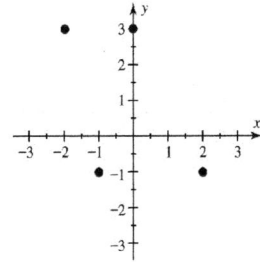         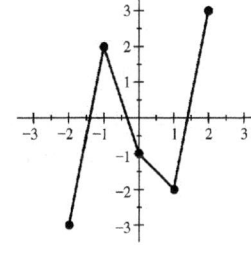         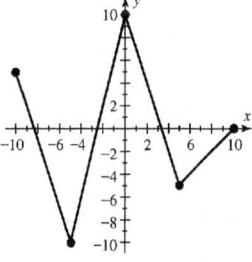

Figure 2                          Figure 5                          Figure 6

6. See Figure 6.

7. Substitute 6 for $x$ and 3 for $y$:  $y = x - 3 \Rightarrow 3 = 6 - 3 \Rightarrow 3 = 3$.  Because this is a true statement, the

ordered pair (6, 3) is a solution.

8. Substitute $-2$ for $x$ and 1 for $y$:  $y = 5 - 2x \Rightarrow 1 = 5 - 2(-2) \Rightarrow 1 = 5 + 4 \Rightarrow 1 = 9$.  Because this is not

a true statement, the ordered pair $(-2, 1)$ is not a solution.

9. Substitute $-1$ for $x$ and 6 for $y$: $3x - y = 3 \Rightarrow 3(-1) - 6 = 3 \Rightarrow -3 - 6 = 3 \Rightarrow -9 = 3$. Because this is

   not a true statement, the ordered pair $(-1, 6)$ is not a solution.

10. Substitute $-4$ for $x$ and $-3$ for $y$: $\frac{1}{2}x + 2y = -8 \Rightarrow \frac{1}{2}(-4) + 2(-3) = -8 \Rightarrow -2 - 6 = -8 \Rightarrow -8 = -8$.

    Because this is a true statement, the ordered pair $(-4, -3)$ is a solution.

11. For $x = -2$: $y = -3x = -3(-2) \Rightarrow y = 6$; for $x = -1$: $y = -3x = -3(-1) \Rightarrow y = 3$;

    for $x = 0$: $y = -3x = -3(0) \Rightarrow y = 0$; for $x = 1$: $y = -3x = -3(1) \Rightarrow y = -3$;

    for $x = 2$: $y = -3x = -3(2) \Rightarrow y = -6$. Therefore, the missing values in the table are

    6, 3, 0, $-3$, and $-6$. See Figure 11.

12. For $y = -3$: $2x + y = 5 \Rightarrow 2x + (-3) = 5 \Rightarrow 2x - 3 = 5 \Rightarrow 2x + 3 - 3 = 5 + 3 \Rightarrow 2x = 8 \Rightarrow$

    $\frac{2x}{2} = \frac{8}{2} \Rightarrow x = 4$; for $y = -1$: $2x + y = 5 \Rightarrow 2x + (-1) = 5 \Rightarrow 2x - 1 = 5 \Rightarrow 2x - 1 + 1 = 5 + 1 \Rightarrow$

    $2x = 6 \Rightarrow \frac{2x}{2} = \frac{6}{2} \Rightarrow x = 3$; for $y = 0$: $2x + y = 5 \Rightarrow 2x + 0 = 5 \Rightarrow 2x = 5 \Rightarrow \frac{2x}{2} = \frac{5}{2} \Rightarrow x = 2.5$;

    for $y = 1$: $2x + y = 5 \Rightarrow 2x + 1 = 5 \Rightarrow 2x + 1 - 1 = 5 - 1 \Rightarrow 2x = 4 \Rightarrow \frac{2x}{2} = \frac{4}{2} \Rightarrow x = 2$;

    for $y = 3$: $2x + y = 5 \Rightarrow 2x + 3 = 5 \Rightarrow 2x + 3 - 3 = 5 - 3 \Rightarrow 2x = 2 \Rightarrow \frac{2x}{2} = \frac{2}{2} \Rightarrow x = 1$.

    Therefore, the missing values in the tables are 4, 3, 2.5, 2, and 1. See Figure 12.

| $x$ | $-2$ | $-1$ | 0 | 1 | 2 |
|---|---|---|---|---|---|
| $y$ | 6 | 3 | 0 | $-3$ | $-6$ |

Figure 11

| $x$ | 4 | 3 | 2.5 | 2 | 1 |
|---|---|---|---|---|---|
| $y$ | $-3$ | $-1$ | 0 | 1 | 3 |

Figure 12

| $x$ | $-2$ | 0 | 2 | 4 |
|---|---|---|---|---|
| $y$ | $-4$ | 2 | 8 | 14 |

Figure 13

13. See Figure 13.

14. See Figure 14.

15. See Figure 15.

| $x$ | 1 | 2 | 3 | 4 |
|---|---|---|---|---|
| $y$ | 6 | 5 | 4 | 3 |

Figure 14

| $x$ | $-0.5$ | 0 | 0.5 | 1 |
|---|---|---|---|---|
| $y$ | $-1$ | 0 | 1 | 2 |

Figure 15

| $x$ | $-1$ | $-3$ | $-5$ | $-7$ |
|---|---|---|---|---|
| $y$ | 1 | 2 | 3 | 4 |

Figure 16

16. See Figure 16.

17. See Figure 17.

18. See Figure 18.

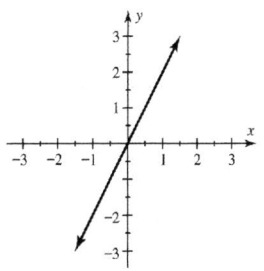

Figure 17

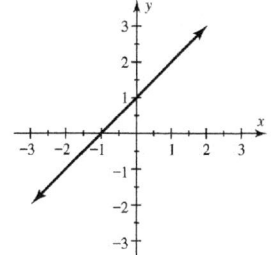

Figure 18

19. See Figure 19.

20. See Figure 20.

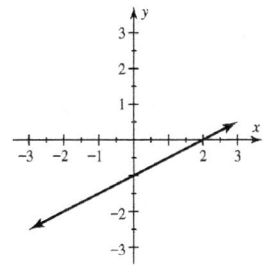

Figure 19

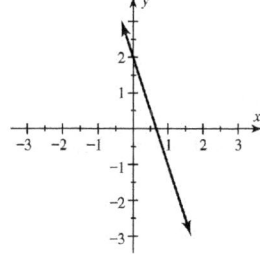

Figure 20

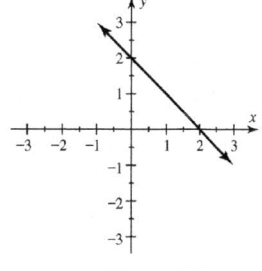

Figure 21

21. See Figure 21.

22. See Figure 22.

23. See Figure 23.

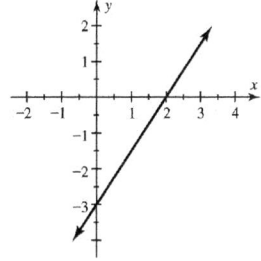

Figure 22

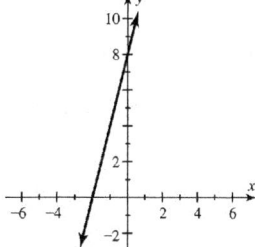

Figure 23

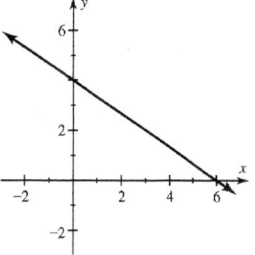

Figure 24

24. See Figure 24.

25. The line crosses the $x$-axis at $x = 3$, therefore the $x$-intercept is 3. The line crosses the $y$-axis at $y = -2$, therefore the $y$-intercept is $-2$.

26. The graph crosses the $x$-axis at $x = -2$ and $x = 2$. Therefore, the $x$-intercepts are 2, $-2$. The graph crosses the $y$-axis at $y = -4$, therefore, the $y$-intercept is $-4$.

27. For $x = -2$: $y = 2 - x = 2 - (-2) \Rightarrow y = 4$; for $x = -1$: $y = 2 - x = 2 - (-1) = 2 + 1 \Rightarrow y = 3$;

for $x = 0$: $y = 2 - x = 2 - 0 \Rightarrow y = 2$; for $x = 1$: $y = 2 - x = 2 - 1 \Rightarrow y = 1$;

for $x = 2$: $y = 2 - x = 2 - 2 \Rightarrow y = 0$. Therefore, the missing values in the table are 4, 3, 2, 1 ,0.

See Figure 27. When $y = 0$, $x = 2$. Therefore the $x$-intercept is 2. When $x = 0$, $y = 2$. Therefore, the $y$-intercept is 2.

| $x$ | $-2$ | $-1$ | 0 | 1 | 2 |
|---|---|---|---|---|---|
| $y$ | 4 | 3 | 2 | 1 | 0 |

Figure 27

| $x$ | $-4$ | $-2$ | 0 | 2 | 4 |
|---|---|---|---|---|---|
| $y$ | $-4$ | $-3$ | $-2$ | $-1$ | 0 |

Figure 28

28. For $x = -4$: $x - 2y = 4 \Rightarrow -4 - 2y = 4 \Rightarrow -4 + 4 - 2y = 4 + 4 \Rightarrow -2y = 8 \Rightarrow \dfrac{-2y}{-2} = \dfrac{8}{-2} \Rightarrow$

    $y = -4$; for $x = -2$: $x - 2y = 4 \Rightarrow -2 - 2y = 4 \Rightarrow -2 + 2 - 2y = 4 + 2 \Rightarrow -2y = 6 \Rightarrow$

    $\dfrac{-2y}{-2} = \dfrac{6}{-2} \Rightarrow y = -3$; for $x = 0$: $x - 2y = 4 \Rightarrow 0 - 2y = 4 \Rightarrow -2y = 4 \Rightarrow \dfrac{-2y}{-2} = \dfrac{4}{-2} \Rightarrow y = -2$;

    for $x = 2$: $x - 2y = 4 \Rightarrow 2 - 2y = 4 \Rightarrow 2 - 2 - 2y = 4 - 2 \Rightarrow -2y = 2 \Rightarrow \dfrac{-2y}{-2} = \dfrac{2}{-2} \Rightarrow y = -1$;

    for $x = 4$: $x - 2y = 4 \Rightarrow 4 - 2y = 4 \Rightarrow 4 - 4 - 2y = 4 - 4 \Rightarrow -2y = 0 \Rightarrow \dfrac{-2y}{-2} = \dfrac{0}{-2} \Rightarrow y = 0$.

    Therefore, the missing values in the table are $-4$, $-3$, $-2$, $-1$, 0. See Figure 28. When
    $y = 0$, $x = 4$. Therefore, the $x$-intercept is 4. When $x = 0$, $y = -2$. Therefore, the $y$-intercept is $-2$.

29. To find the $x$-intercept, set $y = 0$: $2x - 3y = 6 \Rightarrow 2x - 3(0) = 6 \Rightarrow 2x = 6 \Rightarrow \dfrac{2x}{2} = \dfrac{6}{2} \Rightarrow x = 3$.

    Therefore, the $x$-intercept is 3. To find the $y$-intercept, set $x = 0$: $2x - 3y = 6 \Rightarrow$

    $2(0) - 3y = 6 \Rightarrow -3y = 6 \Rightarrow \dfrac{-3y}{-3} = \dfrac{6}{-3} \Rightarrow y = -2$. Therefore, the $y$-intercept is $-2$. See Figure 29.

30. Set $y = 0$ and solve for $x$: $5x - y = 5 \Rightarrow 5x - 0 = 5 \Rightarrow 5x = 5 \Rightarrow \dfrac{5x}{5} = \dfrac{5}{5} \Rightarrow x = 1$. Therefore, the

    $x$-intercept is 1. Then set $x = 0$ and solve for $y$: $5x - y = 5 \Rightarrow 5(0) - y = 5 \Rightarrow -y = 5 \Rightarrow$

    $-1(-y) = -1(5) \Rightarrow y = -5$. Therefore, the $y$-intercept is $-5$. See Figure 30.

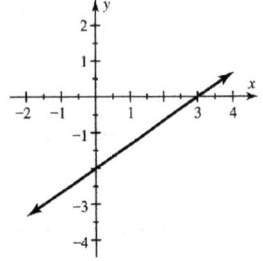

Figure 29

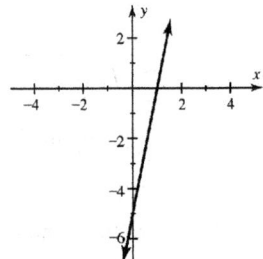

Figure 30

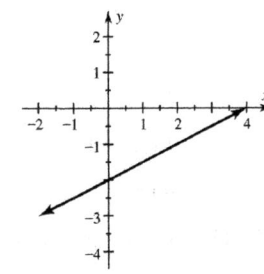

Figure 31

31. Set $y = 0$ and solve for $x$: $0.1x - 0.2y = 0.4 \Rightarrow 0.1x - 0.2(0) = 0.4 \Rightarrow 0.1x = 0.4 \Rightarrow \dfrac{0.1x}{0.1} = \dfrac{0.4}{0.1} \Rightarrow$

    $x = 4$. Therefore the $x$-intercept is 4. Then set $x = 0$ and solve for $y$: $0.1x - 0.2y = 0.4 \Rightarrow$

    $0.1(0) - 0.2y = 0.4 \Rightarrow -0.2y = 0.4 \Rightarrow \dfrac{-0.2y}{-0.2} = \dfrac{0.4}{-0.2} \Rightarrow y = -2$. Therefore, the $y$-intercept is $-2$.

    See Figure 31.

32. Set $y = 0$ and solve for $x$: $\dfrac{x}{2} + \dfrac{y}{3} = 1 \Rightarrow \dfrac{x}{2} + \dfrac{0}{3} = 1 \Rightarrow \dfrac{x}{2} = 1 \Rightarrow 2\left(\dfrac{x}{2}\right) = 2(1) \Rightarrow x = 2.$

Therefore the $x$-intercept is 2. Then set $x = 0$ and solve for $y$: $\dfrac{x}{2} + \dfrac{y}{3} = 1 \Rightarrow \dfrac{0}{2} + \dfrac{y}{3} = 1 \Rightarrow \dfrac{y}{3} = 1 \Rightarrow$

$3\left(\dfrac{y}{3}\right) = 3(1) \Rightarrow y = 3.$ Therefore, the $y$-intercept is 3. See Figure 32.

33. Because the $y$-value is constant, the slope $m = 0$. Because when $x = 0$, $y = 1$, the $y$-intercept $b = 1$.

Thus: $y = mx + b \Rightarrow y = 0x + 1 \Rightarrow y = 1.$

34. Because the $x$-value is a constant 3, the graph is a vertical line with slope $m$ undefined. Thus, $x = 3$.

35. (a) See Figure 35a.

    (b) See Figure 35b.

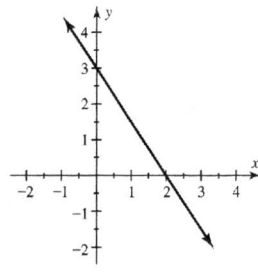    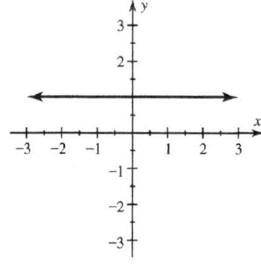    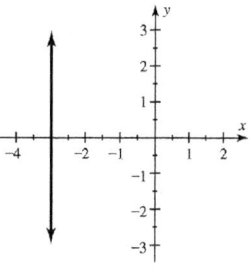

      Figure 32             Figure 35a           Figure 35b

36. $x = -1$; $y = 1$

37. horizontal line: $y = 3$; vertical line: $x = -2$

38. The line that is perpendicular to the horizontal line $y = -\dfrac{1}{2}$ is a vertical line in the form $x = k$, where

$k$ is the $x$-intercept. Since we know the graph passes through the point $(4, 1)$, then the equation of the

line is $x = 4$.

39. The line that is parallel to the horizontal line $y = 3$ is a horizontal line in the form $y = b$, where

$b$ is the $y$-intercept. Since we know the graph passes through the point $(-6, -5)$, then the equation of

the line is $y = -5$.

40. (a) When $x = 0$, $y = 90$. Therefore, the $y$-intercept is 90. When $y = 0$, $x = 3$. Therefore, the $x$-

    intercept is 3.

    (b) The driver is initially 90 miles from home; the driver arrives home after 3 hours.

41. $m = \dfrac{y_2 - y_1}{x_2 - x_1} = \dfrac{7 - 3}{4 - 2} = \dfrac{4}{2} = 2.$

42. $m = \dfrac{y_2 - y_1}{x_2 - x_1} = \dfrac{-1 - 1}{2 - (-3)} = \dfrac{-2}{2 + 3} = \dfrac{-2}{5} = -\dfrac{2}{5}.$

43. $m = \dfrac{y_2 - y_1}{x_2 - x_1} = \dfrac{1 - 1}{5 - 2} = \dfrac{0}{3} = 0.$

44.  $m = \dfrac{y_2 - y_1}{x_2 - x_1} = \dfrac{10 - 6}{-5 - (-5)} = \dfrac{4}{-5 + 5} = \dfrac{4}{0}$.  Thus, undefined.

45.  The points $(1, 3)$ and $(0, 0)$ are on the graph.  Insert these two points into the slope formula:

$$m = \dfrac{y_2 - y_1}{x_2 - x_1} = \dfrac{0 - 3}{0 - 1} = \dfrac{-3}{-1} = 3.$$

46.  The points $(-2, 3)$ and $(0, 2)$ are on the graph.  Insert these two points into the slope formula:

$$m = \dfrac{y_2 - y_1}{x_2 - x_1} = \dfrac{2 - 3}{0 - (-2)} = \dfrac{-1}{0 + 2} = \dfrac{-1}{2} = -\dfrac{1}{2}.$$

47.  (a)    See Figure 47.

     (b)    The slope is $-2$ and the $y$-intercept is 0.

48.  (a)    See Figure 48.

     (b)    The slope is 1 and the $y$-intercept is $-1$.

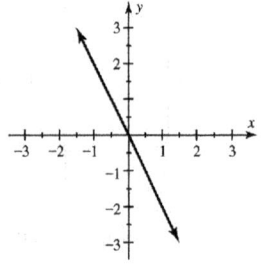

Figure 47

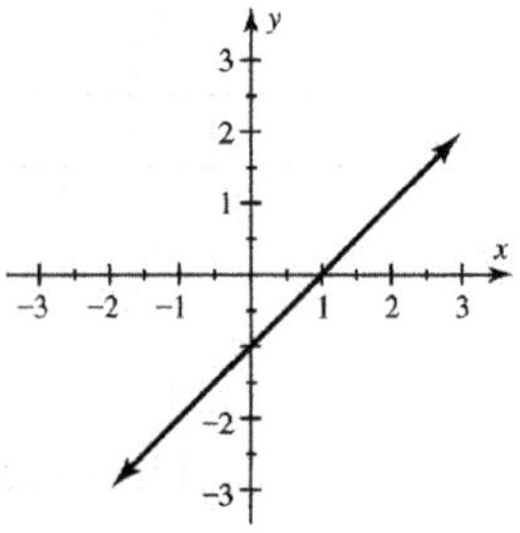

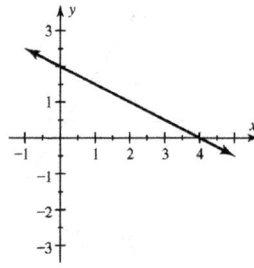

Figure 48

Figure 49

49.  (a)    See Figure 49.

     (b)    Solve for $y$:

$$x + 2y = 4 \Rightarrow x - x + 2y = 4 - x \Rightarrow 2y = -x + 4 \Rightarrow \dfrac{1}{2}(2y) = \dfrac{1}{2}(-x + 4) \Rightarrow y = -\dfrac{1}{2}x + 2.$$

Therefore, the slope is $-\dfrac{1}{2}$ and the $y$-intercept is 2.

50.  (a)    See Figure 50.

     (b)    Solve for $y$:

$$2x - 3y = -6 \Rightarrow 2x - 2x - 3y = -6 - 2x \Rightarrow -3y = -2x - 6 \Rightarrow -\dfrac{1}{3}(-3y) = -\dfrac{1}{3}(-2x - 6) \Rightarrow$$

$y = \dfrac{2}{3}x + 2$.  Therefore, the slope is $\dfrac{2}{3}$ and the $y$-intercept is 2.

51.  See Figure 51.

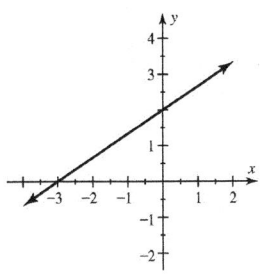

Figure 50

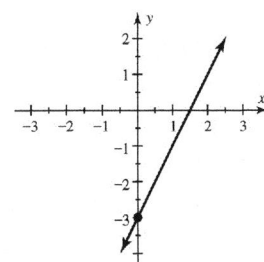

Figure 51

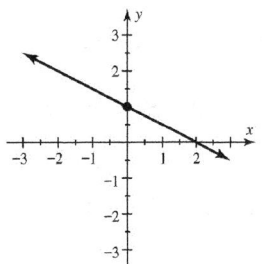

Figure 52

52. See Figure 52.

53. See Figure 53.

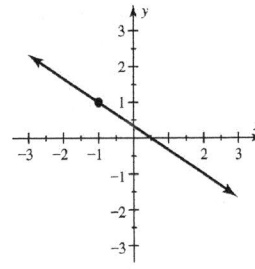

Figure 53

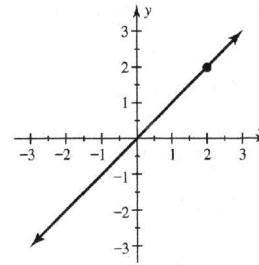

Figure 54

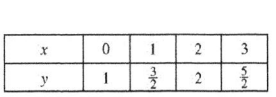

Figure 55

54. See Figure 54.

55. Because the slope is $\dfrac{1}{2}$, for every increase of 1 in the run, the rise increases by $\dfrac{1}{2}$. Thus, when

$x = 1$, $y = 1 + \dfrac{1}{2} = \dfrac{3}{2}$. When $x = 2$, $y = \dfrac{3}{2} + \dfrac{1}{2} = \dfrac{4}{2} = 2$. When $x = 3$, $y = 2 + \dfrac{1}{2} = \dfrac{4}{2} + \dfrac{1}{2} = \dfrac{5}{2}$.

See Figure 55.

56. Since the cost of buying coffee drinks increases as the number of drinks purchased increases the slope will be positive.

57. For every increase of 1 in the run, the rise increases by 1. Therefore, the slope $m$ is 1. When $x = 0$, $y = 1$. Therefore, the $y$-intercept is 1. Thus, $y = mx + b \Rightarrow y = 1x + 1 \Rightarrow y = x + 1$.

58. For every increase of 1 in the run, the rise decreases by 2. Therefore, the slope $m$ is $-2$. When $x = 0$, $y = 2$. Therefore, the $y$-intercept is 2. Thus, $y = mx + b \Rightarrow y = -2x + 2$.

59. See Figure 59. $y = mx + b \Rightarrow y = 2x - 2$

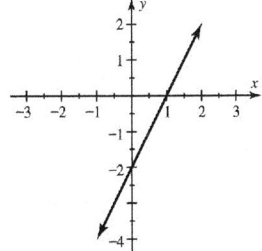

Figure 59

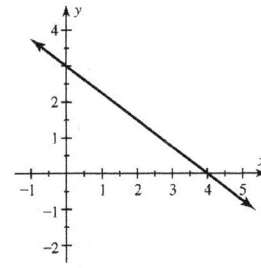

Figure 60

60. See Figure 60. $y = mx + b \Rightarrow y = -\dfrac{3}{4}x + 3$

61. (a)    Solve for $y$: $x + y = 3 \Rightarrow x - x + y = 3 - x \Rightarrow y = -x + 3$.

   (b)    The slope is $-1$ and the $y$-intercept is 3.

62. (a)    Solve for $y$: $-3x + 2y = -6 \Rightarrow -3x + 3x + 2y = -6 + 3x \Rightarrow 2y = 3x - 6 \Rightarrow$

        $\dfrac{1}{2}(2y) = \dfrac{1}{2}(3x - 6) \Rightarrow y = \dfrac{3}{2}x - 3$.

   (b)    The slope is $\dfrac{3}{2}$ and the $y$-intercept is $-3$.

63. (a)    Solve for $y$: $20x - 10y = 200 \Rightarrow 20x - 20x - 10y = 200 - 20x \Rightarrow -10y = -20x + 200 \Rightarrow$

        $-\dfrac{1}{10}(-10y) = -\dfrac{1}{10}(-20x + 200) \Rightarrow y = 2x - 20$.

   (b)    The slope is 2 and the $y$-intercept is $-20$.

64. (a)    Solve for $y$: $5x - 6y = 30 \Rightarrow 5x - 5x - 6y = 30 - 5x \Rightarrow -6y = -5x + 30 \Rightarrow$

        $-\dfrac{1}{6}(-6y) = -\dfrac{1}{6}(-5x + 30) \Rightarrow y = \dfrac{5}{6}x - 5$.

   (b)    The slope is $\dfrac{5}{6}$ and the $y$-intercept is $-5$.

65. See Figure 65.

66. See Figure 66.

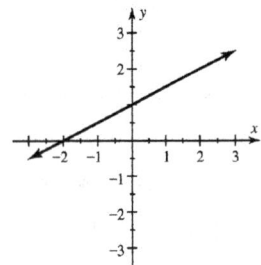 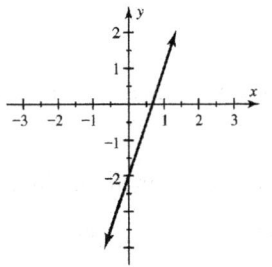 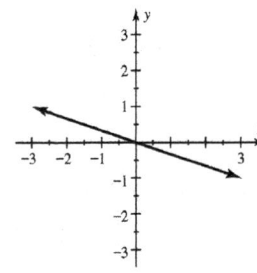

    Figure 65                      Figure 66                      Figure 67

67. See Figure 67.

68. See Figure 68.

69. See Figure 69.

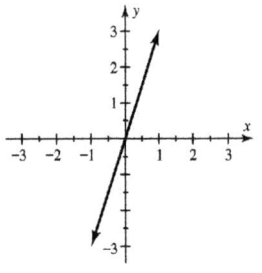 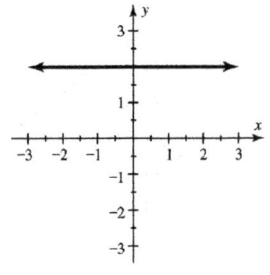 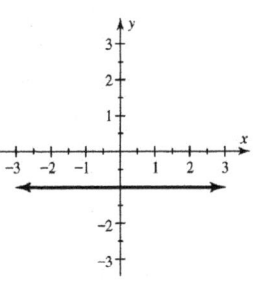

    Figure 68                      Figure 69                      Figure 70

70. See Figure 70.

71. See Figure 71.

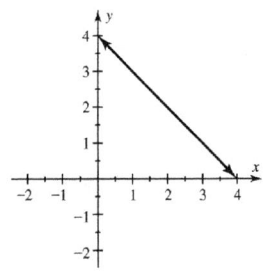

Figure 71                                     Figure 72

72. See Figure 72.

73. When the run increases by 1, the rise increases by 5. Thus, the slope $m$ is 5. When

$x = 0$, $y = -5$. Therefore, the $y$-intercept $b$ is $-5$. Thus, $y = mx + b \Rightarrow y = 5x - 5$.

74. When the run increases by 1, the rise decreases by 2. Thus, the slope $m$ is $-2$. When

$x = 0$, $y = 0$. Therefore, the $y$-intercept $b$ is 0. Thus, $y = mx + b \Rightarrow y = -2x + 0 \Rightarrow y = -2x$.

75. $y = mx + b \Rightarrow y = -\dfrac{5}{6}x + 2$.

76. The line parallel to $y = -2x + 1$ has slope $-2$. Insert the values of the point $(1, \ -5)$ and the value

of the slope into the slope-intercept form and solve for $b$:

$y = mx + b \Rightarrow -5 = -2(1) + b \Rightarrow -5 = -2 + b \Rightarrow -5 + 2 = -2 + 2 + b \Rightarrow -3 = b \Rightarrow b = -3$. Thus, the

equation of the line is $y = -2x - 3$.

77. The line perpendicular to $y = -\dfrac{3}{2}x$ has slope $\dfrac{2}{3}$, the negative reciprocal of $-\dfrac{3}{2}$. Insert the values of

the point $(3, \ 0)$ and the value of the slope into the slope-intercept form and solve for $b$:

$y = mx + b \Rightarrow 0 = \dfrac{2}{3}(3) + b \Rightarrow 0 = 2 + b \Rightarrow 0 - 2 = 2 - 2 + b \Rightarrow -2 = b \Rightarrow b = -2$. Thus, $y = \dfrac{2}{3}x - 2$.

78. The line perpendicular to $y = 5x - 3$ has slope $-\dfrac{1}{5}$, the negative reciprocal of 5. Insert the values of

the point $(0, \ -2)$ and the value of the slope into the slope-intercept form and solve for $b$:

$y = mx + b \Rightarrow -2 = -\dfrac{1}{5}(0) + b \Rightarrow -2 = 0 + b \Rightarrow -2 = b \Rightarrow b = -2$. Thus, $y = -\dfrac{1}{5}x - 2$.

79. Substitute $-3$ for $x$ and 1 for $y$: $y - 1 = 2(x + 3) \Rightarrow 1 - 1 = 2(-3 + 3) \Rightarrow 0 = 2(0) \Rightarrow 0 = 0$.

Because this is a true statement, the point $(-3, 1)$ lies on the line.

80. Substitute 3 for $x$ and $-8$ for $y$: $y = -3(x - 1) + 2 \Rightarrow -8 = -3(3 - 1) + 2 \Rightarrow -8 = -3(2) + 2 \Rightarrow$

$-8 = -6 + 2 \Rightarrow -8 = -4$. Because this is not a true statement, the point $(3, \ -8)$ is not on the line.

81. Substitute the values of the point $(1, 2)$ and the value of the slope into the point-slope form

$$y - y_1 = m(x - x_1) \Rightarrow y - 2 = 5(x - 1)$$

82. Substitute the values of the point $(3, -5)$ and the value of the slope into the point-slope form

$$y - y_1 = m(x - x_1) \Rightarrow y - (-5) = 20(x - 3) \Rightarrow y + 5 = 20(x - 3)$$

83. Find the slope: $m = \dfrac{y_2 - y_1}{x_2 - x_1} = \dfrac{-1-1}{1-(-2)} = \dfrac{-2}{1+2} = \dfrac{-2}{3} = -\dfrac{2}{3}.$  Then, insert the values of the point

$(-2, 1)$ and the value of the slope into the point-slope form:

$$y - y_1 = m(x - x_1) \Rightarrow \quad y - 1 = -\frac{2}{3}\big(x - (-2)\big) \Rightarrow \quad y - 1 = -\frac{2}{3}(x + 2).$$

84. Find the slope: $m = \dfrac{y_2 - y_1}{x_2 - x_1} = \dfrac{30-(-30)}{40-20} = \dfrac{60}{20} = 3.$   Then, insert the values of the point $(20, -30)$

and the value of the slope into the point-slope form: $y - y_1 = m(x - x_1) \Rightarrow y - (-30) = 3(x - 20) \Rightarrow$

$y + 30 = 3(x - 20).$

85. One point on the line is $(3, 0)$ and another point is $(0, -4)$.  Find the slope:

$$m = \frac{y_2 - y_1}{x_2 - x_1} = \frac{-4-0}{0-3} = \frac{-4}{-3} = \frac{4}{3}.$$  Then, insert the values of the point $(3, 0)$ and the value of the

slope into the point-slope form: $y - y_1 = m(x - x_1) \Rightarrow \quad y - 0 = \dfrac{4}{3}(x - 3) \Rightarrow y = \dfrac{4}{3}(x - 3)$

86. One point on the line is $\left(\dfrac{1}{2}, 0\right)$ and another point is $(0, -1)$. Find the slope:

$$m = \frac{-1-0}{0-\frac{1}{2}} = \frac{-1}{-\frac{1}{2}} = -2(-1) = 2.$$  Then, insert the values of the point $\left(\dfrac{1}{2}, 0\right)$ and the value of the

slope into the point-slope form: $y - y_1 = m(x - x_1) \Rightarrow \quad y - 0 = 2\left(x - \dfrac{1}{2}\right) \Rightarrow y = 2\left(x - \dfrac{1}{2}\right)$

87. The line parallel to $y = 2x$ has slope 2.  Insert the values of the point $(5, 7)$ and the value of the

slope into the point-slope form: $y - y_1 = m(x - x_1) \Rightarrow \quad y - 7 = 2(x - 5).$

88. First, find the slope by converting to slope-intercept form:  $y - 4 = \dfrac{3}{2}(x + 1) \Rightarrow y - 4 = \dfrac{3}{2}x + \dfrac{3}{2} \Rightarrow$

$y - 4 + 4 = \dfrac{3}{2}x + \dfrac{3}{2} + 4 \Rightarrow y = \dfrac{3}{2}x + \dfrac{3}{2} + \dfrac{8}{2} \Rightarrow y = \dfrac{3}{2}x + \dfrac{11}{2}.$  The line perpendicular to $y = \dfrac{3}{2}x + \dfrac{11}{2}$ has

slope $-\dfrac{2}{3},$ the negative reciprocal of $\dfrac{3}{2}.$  Insert the values of the point $(-1, 0)$ and the value of the

slope into the point-slope form:

$$y - y_1 = m(x - x_1) \Rightarrow y - 0 = -\frac{2}{3}\big(x - (-1)\big) \Rightarrow y - 0 = -\frac{2}{3}(x + 1) \Rightarrow y = -\frac{2}{3}(x + 1)$$

89. Use the distributive property of multiplication and solve for $y$:

$$y - 2 = 3(x+1) \Rightarrow y - 2 = 3x + 3 \Rightarrow y - 2 + 2 = 3x + 3 + 2 \Rightarrow y = 3x + 5.$$

90. Use the distributive property of multiplication and solve for $y$:

$$y - 9 = \frac{1}{3}(x - 6) \Rightarrow y - 9 = \frac{1}{3}x - 2 \Rightarrow y - 9 + 9 = \frac{1}{3}x - 2 + 9 \Rightarrow y = \frac{1}{3}x + 7.$$

91. Use the distributive property of multiplication:

$$y = 2(x+3) + 5 \Rightarrow y = 2x + 6 + 5 \Rightarrow y = 2x + 11.$$

92. Use the distributive property of multiplication:

$$y = -\frac{1}{4}(x-8) + 1 \Rightarrow y = -\frac{1}{4}x + 2 + 1 \Rightarrow y = -\frac{1}{4}x + 3.$$

93. Because a straight line would not pass through all the points, no.

94. Because the line does not pass through all the points, the linear model is approximate. When the run increases by 1, the rise decreases by 1. Therefore, the slope is $-1$. When $x = 0$, $y = 5$. Thus the $y$-intercept is 5. Therefore: $y = mx + b \Rightarrow y = -1x + 5 \Rightarrow y = -x + 5$.

95. Because the initial value is 40, $b = 40$. Because $y$ decreases at a rate of 2 pounds per minute, the slope is $-2$. Therefore: $y = mx + b \Rightarrow y = -2x + 40$.

96. Because the initial value is 200, $b = 200$. Because the rate of increase is 20 gallons per hour, the slope is 20. Therefore: $y = mx + b \Rightarrow y = 20x + 200$.

97. Because the initial value is 50, $b = 50$. Because $y$ remains constant, the slope is 0. Therefore: $y = 50$.

98. Because the initial value is $-20$, $b = -20$. Because the rise is 5 feet per second, the slope is 5. Therefore: $y = 5x - 20$.

99. (a) See Figure 99a. The line could pass through all five points.

(b) See Figure 99b.

(c) The initial value is 10, and the slope is $-4$. Thus, $y = -4x + 10$.

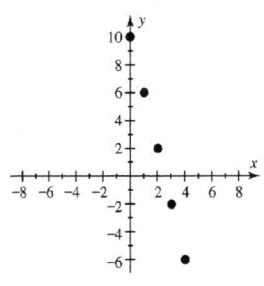

Figure 99a

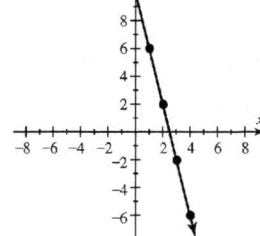

Figure 99b

100. (a) See Figure 100a. The line could not pass through all five points.

(b) See Figure 100b.

(c) $\dfrac{3-1}{0-(-4)} = \dfrac{2}{4} = \dfrac{1}{2}$, so the slope is $\dfrac{1}{2}$. The $y$-intercept is 3, so $y = \dfrac{1}{2}x + 3$.

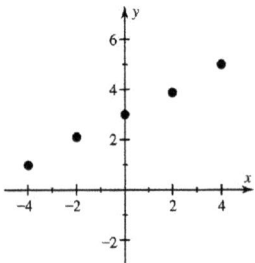

Figure 100a

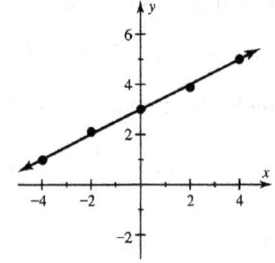

Figure 100b

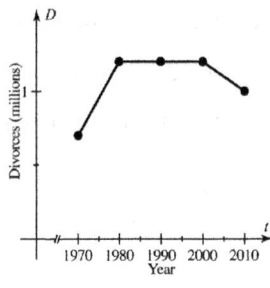

Figure 101

101. (a)   See Figure 101.

 (b)   The number of divorces increased between 1970 and 1980, then remained unchanged from
        1980 to 2000, then decreased between 2000 and 2010.

102. (a)   $G = 100t$

 (b)   See Figure 102.

 (c)   When $G = 5000$, $t = 50$.  Therefore, 50 days.

103. (a)   See Figure 103.

 (b)   $v$-intercept, 160; $t$-intercept, 5;  the initial velocity was 160 ft/sec. and the velocity after 5
        seconds was 0.

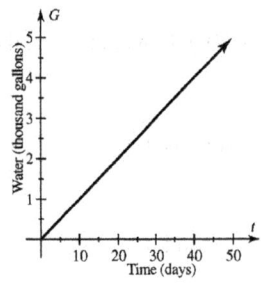

Figure 102

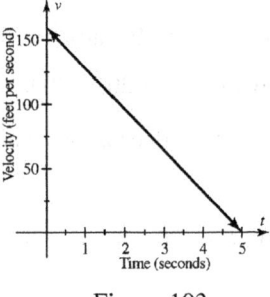

Figure 103

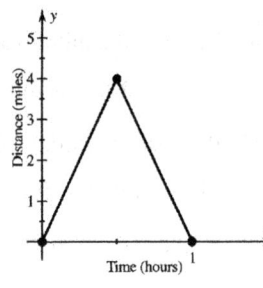

Figure 105

104. (a)   $m_1 = 0$;  $m_2 = -1500$;  $m_3 = 500$

 (b)   $m_1 = 0$: The population remained unchanged.  $m_2 = -1500$: The population decreased at a rate
        of 1500 insects per week.  $m_3 = 500$: The population increased at a rate of 500 insects per
        week.

 (c)   For the first week the population did not change from the initial value of 4000.  Over the next
        two weeks the population decreased at a rate of 1500 insects per week until it reached 1000.
        Finally, the population increased at a rate of 500 per week for two weeks, reaching 2000.

105. See Figure 105.

106. (a)   $\dfrac{16,100-19,100}{2010-1985} = \dfrac{-3000}{25} = -120$

 (b)   The number of nursing homes decreased on average at a rate of 120/yr.

107. (a)    Because the initial value is 35, \$35.

(b)    Because the slope is 0.2, each additional mile costs 20¢.

(c)    35; the fixed cost of renting the car.

(d)    The slope is 0.2; the cost of each mile driven.

108. (a)    The person is driving toward home; the slope is negative.

(b)    After 1 hour the car is 200 miles from home, after 3 hours the car is 100 miles from home.

(c)    The initial value is 250, so the $y$-intercept $b$ is 250. For each increase of 1 in run, the rise decreases by 50, so the slope is $-50$. Thus, $y = -50x + 250$. The car is moving toward home at 50 mph.

(d)    When time = 2, the distance from home is 150 miles.

$$y = -50x + 250 = -50(2) + 250 = -100 + 250 = 150 \text{ miles}$$

109. (a)    See Figure 109.

(b)    Find the slope: $\dfrac{600 - 400}{1980 - 1970} = \dfrac{200}{10} = 20$. Then, insert the values of the point (1970, 400) and

the value of the slope into the point-slope form: $y - y_1 = m(x - x_1) \Rightarrow y - 400 = 20(x - 1970)$.

Then, convert to slope-intercept form:

$$y - 400 = 20x - 39,400 \Rightarrow y - 400 + 400 = 20x - 39,400 + 400 \Rightarrow y = 20x - 39,000; \text{ the}$$

number of icebergs increased at an average rate of 20 per year.

(c)    Because you could draw a straight line through all the points on the graph, yes.

(d)    $I = 20(2005) - 39,000 = 1100$ icebergs

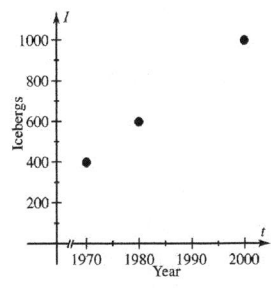

Figure 109

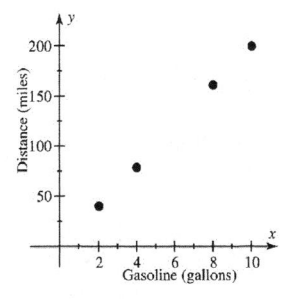

Figure 110a

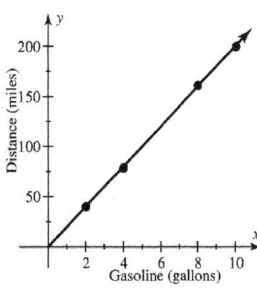

Figure 110b

110. (a)    See Figure 110a.

(b)    See Figure 110b.

(c)    $\dfrac{200 - 40}{10 - 2} = \dfrac{160}{8} = 20$; The mileage is 20 miles per gallon.

(d)    $y = 20x$; it is not an exact model because when $x = 4$, $y \neq 20(4)$, $y = 79$

(e)    $y = 20(9) = 180$; about 180 miles.

## Chapter 3 Test

1. $(-2,-2)$: Quadrant III; $(-2, 1)$: Quadrant II; $(0, -2)$: none; $(1, 0)$: none; $(2, 3)$: Quadrant I;

   $(3, -1)$:  Quadrant IV

2. See Figure 2.

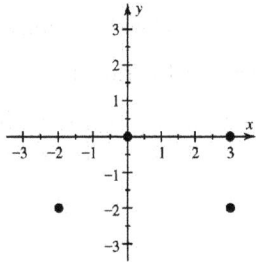    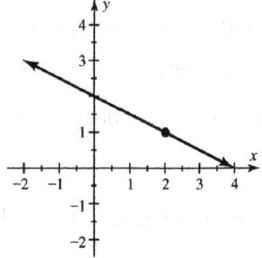    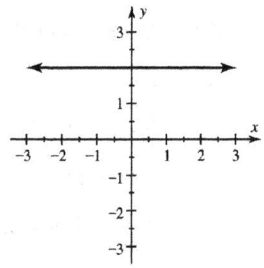

Figure 2                     Figure 5                     Figure 7

3. For $x = -2$: $y = 2x - 4 = 2(-2) - 4 = -4 - 4 \Rightarrow y = -8$.

   For $x = -1$: $y = 2x - 4 = 2(-1) - 4 = -2 - 4 \Rightarrow y = -6$.

   For $x = 0$: $y = 2x - 4 = 2(0) - 4 \Rightarrow y = -4$.

   For $x = 1$: $y = 2x - 4 = 2(1) - 4 = 2 - 4 \Rightarrow y = -2$.

   For $x = 2$: $y = 2x - 4 = 2(2) - 4 = 4 - 4 \Rightarrow y = 0$.  See Figure 3.

   The $x$-intercept is 2 and the $y$-intercept is $-4$.

   | $x$ | $-2$ | $-1$ | 0 | 1 | 2 |
   |---|---|---|---|---|---|
   | $y$ | $-8$ | $-6$ | $-4$ | $-2$ | 0 |

   Figure 3

4. Let $x = 1$ and $y = -3$ in the equation $2x - y = 5$.   $2(1) - (-3) \overset{?}{=} 5 \Rightarrow 2 + 3 \overset{?}{=} 5 \Rightarrow 5 = 5$

   Yes, the ordered pair $(1, -3)$ is a solution.

5. See Figure 5.

6. To find the $x$-intercept, set $y = 0$ and solve for $x$:

   $5x - 3y = 15 \Rightarrow 5x - 3(0) = 15 \Rightarrow 5x = 15 \Rightarrow \dfrac{5x}{5} = \dfrac{15}{5} \Rightarrow x = 3$. Therefore, the $x$-intercept is at $x = 3$.

   To find the $y$-intercept, set $x = 0$ and solve for $y$:   $5x - 3y = 15 \Rightarrow 5(0) - 3y = 15 \Rightarrow$

   $-3y = 15 \Rightarrow \dfrac{-3y}{-3} = \dfrac{15}{-3} \Rightarrow y = -5$. Therefore, the $y$-intercept is at $y = -5$.

7. See Figure 7.

8. See Figure 8.

9. See Figure 9.

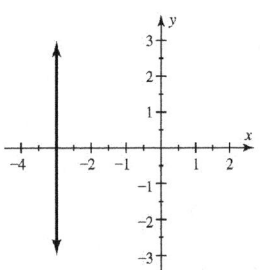

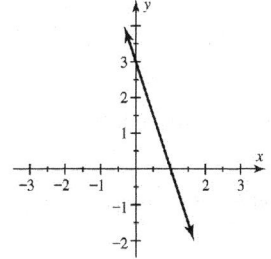

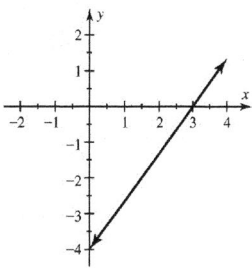

Figure 8                              Figure 9                              Figure 10

10. See Figure 10.

11. For every increase of 1 in the run, the rise decreases by 1, so the slope is $-1$. When $x = 0$, $y = 2$, so the $y$-intercept is 2. Therefore, the equation of the line is $y = -x + 2$.

12. Solve for $y$: $-4x + 2y = 1 \Rightarrow -4x + 4x + 2y = 1 + 4x \Rightarrow 2y = 4x + 1 \Rightarrow \dfrac{2y}{2} = \dfrac{4x}{2} + \dfrac{1}{2} \Rightarrow$

    $y = 2x + \dfrac{1}{2}$. The slope is 2 and the $y$-intercept is $\dfrac{1}{2}$.

13. The equation of a horizontal line through $(1, -5)$ is $y = -5$. The equation of a vertical line through

    $(1, -5)$ is $x = 1$.

14. $m = \dfrac{y_2 - y_1}{x_2 - x_1} = \dfrac{1 - 3}{5 - (-4)} = \dfrac{-2}{9} = -\dfrac{2}{9}$

15. $y = -\dfrac{4}{3}x - 5$

16. The line parallel to $y = 3x - 1$ has slope 3. Insert the values of the point $(2, -5)$ and the value of

    the slope into the point-slope form:

    $y - y_1 = m(x - x_1) \Rightarrow y - (-5) = 3(x - 2) \Rightarrow y + 5 = 3(x - 2)$. Then, convert to slope-intercept

    form:   $y + 5 = 3x - 6 \Rightarrow y + 5 - 5 = 3x - 6 - 5 \Rightarrow y = 3x - 11$.

17. The line perpendicular to $y = \dfrac{1}{3}x$ has slope $-3$, the negative reciprocal to $\dfrac{1}{3}$. Insert the values of

    the point $(1, 2)$ and the value of the slope into the point-slope form:

    $y - y_1 = m(x - x_1) \Rightarrow y - 2 = -3(x - 1)$. Then, convert to slope-intercept

    form:   $y - 2 = -3x + 3 \Rightarrow y - 2 + 2 = -3x + 3 + 2 \Rightarrow y = -3x + 5$.

18. Find the slope:   $\dfrac{-1 - 2}{2 - (-4)} = \dfrac{-3}{6} = -\dfrac{1}{2}$. Then, insert the values of the point $(-4, 2)$ and the value of

    the slope into the point-slope form:   $y - 2 = -\dfrac{1}{2}(x - (-4)) \Rightarrow y - 2 = -\dfrac{1}{2}(x + 4)$. Then, convert to

    slope-intercept form:   $y - 2 = -\dfrac{1}{2}x - 2 \Rightarrow y - 2 + 2 = -\dfrac{1}{2}x - 2 + 2 \Rightarrow y = -\dfrac{1}{2}x$.

19. $y-3=\dfrac{1}{2}(x+4)\Rightarrow y-3=\dfrac{1}{2}x+2\Rightarrow y-3+3=\dfrac{1}{2}x+2+3\Rightarrow y=\dfrac{1}{2}x+5$

20. Use the first and last points in the table to find the slope. $m=\dfrac{4-(-8)}{2-(-2)}=\dfrac{12}{4}=3$

    Substitute the last point and the slope into the point-slope equation. $y-4=3(x-2)$

    Solve for $y$: $y-4=3(x-2)\Rightarrow y-4=3x-6\Rightarrow y=3x-6+4\Rightarrow y=3x-2$

21. Since parallel lines have equal slopes, the slope of the line that is parallel to the line $y=-3x$ is $-3$.

    Substitute the given point and the slope into the point-slope equation.

    $y-7=-3(x-(-2))\Rightarrow y-7=3(x+2)$

22. The linear model is approximate, because it does not go through all the points exactly. For every
    increase of 1 in the run, the increase in rise is 1, so the slope is 1. Insert the values of the point (3, 2)
    and the value of the slope into the point-slope form: $y-2=1(x-3)$. Then, convert to slope
    intercept form: $y-2=x-3\Rightarrow y-2+2=x-3+2\Rightarrow y=x-1$.

23. See Figure 23.

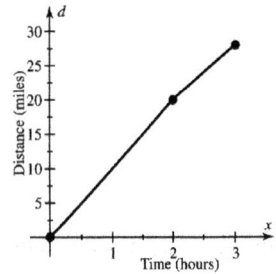

    Figure 23

24. (a)    $m_1=2;\ m_2=-9;\ m_3=2;\ m_4=5$

    (b)    $m_1=2$: The population increased at a rate of 2000 fish per year. $m_2=-9$: The population
           decreased at a rate of 9000 fish per year. $m_3=2$: The population increased at a rate of 2000
           fish per year. $m_4=5$: The population increased at a rate of 5000 fish per year.

    (c)    For the first year the population increased from an initial value of 8000 to 10,000 at a rate of
           2000 fish per year. During the second year the population dropped dramatically to 1000 at a
           rate of 9000 fish per year. Over the third year the population grew to 3000 at a rate of 2000 fish
           per year. Finally, over the fourth year, the population grew at a rate of 5000 fish per year to
           reach 8000.

25. $N=100x+2000$

## Chapter 3 Extended and Discovery Exercises

1. (a)  See Figure 1.

   (b)  Insert the values of the points (2007, 23.5) and (1993, 20.5) into the slope formula:

   $\dfrac{23.5 - 20.5}{2007 - 1993} = \dfrac{3}{14} \approx 0.21.$ Then insert the values of the point (1993, 20.5) and the value of the

   slope into the point-slope form:

   $P - 20.5 = 0.21(x - 1993) \Rightarrow P - 20.5 + 20.5 = 0.21(x - 1993) + 20.5$

   $\Rightarrow P = 0.21(x - 1993) + 20.5.$  *Answers may vary.*

   (c)  Substitute 2010 for $x$ and solve for $P$:   $P = 0.21(2010 - 1993) + 20.5 \Rightarrow P = 0.21(17) + 20.5$

   $\Rightarrow P = 3.57 + 20.5 \Rightarrow P = 24.07 \Rightarrow P \approx 24.1$ percent.

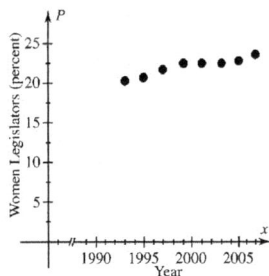

Figure 1

2. (a)  See Figure 2.

   (b)  Find the slope:  $\dfrac{309 - 203}{2010 - 1970} = \dfrac{106}{40} = 2.65.$  Then, insert the values of the point

   (1970, 203) and the value of the slope into the point-slope form:  $P - 203 = 2.65(x - 1970) \Rightarrow$

   $P - 203 + 203 = 2.65(x - 1970) + 203 \Rightarrow P = 2.65(x - 1970) + 203.$  *Answers may vary.*

   (c)  Substitute 2005 for $x$ in the equation and solve for $P$:   $P = 2.65(2005 - 1970) + 203 \Rightarrow$

   $P = 2.65(35) + 203 \Rightarrow P \approx 296$ million.

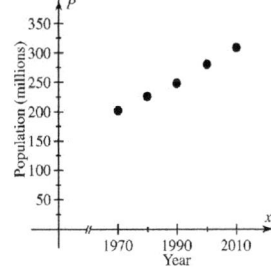

Figure 2

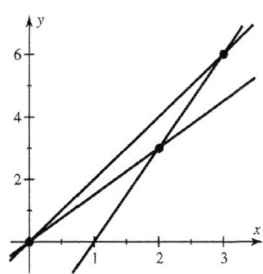

Figure 3

3. (a) The slope of the line connecting the points (0, 0) and (2, 3) is $\dfrac{3}{2}$, and the $y$-intercept is 0.

Therefore the equation of the line is $y = \dfrac{3}{2}x$. The slope of the line connection the points (2, 3) and (3, 6) is 3. Therefore the point-slope form is: $y - 3 = 3(x - 2)$. The slope-intercept form is thus: $y - 3 + 3 = 3x - 6 + 3 \Rightarrow y = 3x - 3$. The slope of the line connecting the points (3, 6) and (0, 0) is 2. The $y$-intercept is 0. Therefore the equation of the line is: $y = 2x$.

   (b) See Figure 3. A triangle is formed by the line segments connecting the three points.

4. (a) $y = -x + 1$

   (b) See Figure 4. A parallelogram is formed.

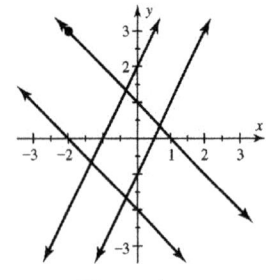

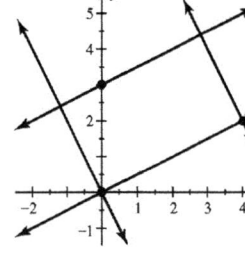

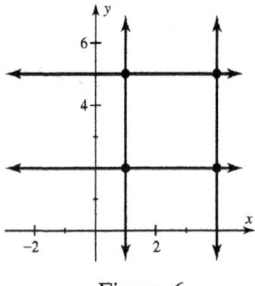

Figure 4                          Figure 5                          Figure 6

5. (a) $y = \dfrac{1}{2}x$; $y = \dfrac{1}{2}x + 3$; $y = -2x$; $y = -2x + 10$

   (b) See Figure 5. A rectangle is formed.

6. (a) (1, 5)

   (b) $x = 1$; $x = 4$; $y = 2$; $y = 5$

   (c) See Figure 6. A square is formed.

## Chapters 1-3  Cumulative Review Exercises

1. Composite; $40 = 2 \cdot 2 \cdot 2 \cdot 5$

2. Prime

3. $n + 10$

4. $n^2 - 2$

5. $\dfrac{3}{4} \div \dfrac{9}{8} = \dfrac{3}{4} \cdot \dfrac{8}{9} = \dfrac{24}{36} = \dfrac{12 \cdot 2}{12 \cdot 3} = \dfrac{2}{3}$

6. $\dfrac{7}{10} - \dfrac{2}{15} = \dfrac{7 \cdot 3}{10 \cdot 3} - \dfrac{2 \cdot 2}{15 \cdot 2} = \dfrac{21}{30} - \dfrac{4}{30} = \dfrac{21 - 4}{30} = \dfrac{17}{30}$

7. $20 - 2 \cdot 3 = 20 - (2 \cdot 3) = 20 - 6 = 14$

8. $-3^2 = -\left(3^2\right) = -(9) = -9$

9. Rational

10. Irrational, because $\sqrt{3}$ cannot be expressed as a fraction.

11. $3 + 4x - 2 + 3x = 3 - 2 + 4x + 3x = 1 + 7x = 7x + 1$

12. $2(x-1) - (x+2) = 2x - 2 - x - 2 = 2x - x - 2 - 2 = x - 4$

13. $3t - 5 = 1 \Rightarrow 3t - 5 + 5 = 1 + 5 \Rightarrow 3t = 6 \Rightarrow \dfrac{3t}{3} = \dfrac{6}{3} \Rightarrow t = 2$. Replace $t$ with 2 in the original

   equation: $3t - 5 = 1 \Rightarrow 3(2) - 5 = 1 \Rightarrow 6 - 5 = 1 \Rightarrow 1 = 1$. The solution checks.

14. $2(x-3) = -6 - x \Rightarrow 2x - 6 = -6 - x \Rightarrow 2x + x - 6 = -6 - x + x \Rightarrow 3x - 6 = -6$

   $\Rightarrow 3x - 6 + 6 = -6 + 6 \Rightarrow 3x = 0 \Rightarrow \dfrac{3x}{3} = \dfrac{0}{3} \Rightarrow x = 0$. Replace $x$ with 0 in the original equation:

   $2(x-3) = -6 - x \Rightarrow 2(0-3) = -6 - 0 \Rightarrow 2(-3) = -6 \Rightarrow -6 = -6$. The solution checks.

15. Area of a rectangle is width multiplied by length, $A = LW$. Replace $W$ with 18 and $L$ with 40 and

   solve for $A$: $A = LW \Rightarrow 40 \cdot 18 = 720 \text{ ft}^2$.

16. Area of a triangle is $\dfrac{1}{2}$ times base times height, $A = \dfrac{1}{2}bh$. Replace $b$ with 8 and $h$ with 5 and solve

   for $A$: $A = \dfrac{1}{2}bh \Rightarrow A = \dfrac{1}{2}(8)(5) = \dfrac{1}{2}(40) = 20 \text{ ft}^2$.

17. When $x = -2$: $6 - 2x = 6 - 2(-2) = 6 - (-4) = 6 + 4 = 10$;

   when $x = -1$: $6 - 2x = 6 - 2(-1) = 6 - (-2) = 6 + 2 = 8$;

   when $x = 0$: $6 - 2x = 6 - 2(0) = 6 - 0 = 6$; when $x = 1$: $6 - 2x = 6 - 2(1) = 6 - 2 = 4$;

   when $x = 2$: $6 - 2x = 6 - 2(2) = 6 - 4 = 2$; thus, the missing values in the table are 10, 8, 6, 4, and 2.

   See Figure 17. From the table, $6 - 2x = 4$, when $x = 1$.

   | $x$ | -2 | -1 | 0 | 1 | 2 |
   |---|---|---|---|---|---|
   | $y$ | 10 | 8 | 6 | 4 | 2 |

   Figure 17

18. $2n + 2 = n - 5 \Rightarrow 2n - n + 2 = n - n - 5 \Rightarrow n + 2 = -5 \Rightarrow n + 2 - 2 = -5 - 2 \Rightarrow n = -7$

19. $d = rt \Rightarrow 80 = 10t \Rightarrow \dfrac{80}{10} = \dfrac{10t}{10} \Rightarrow 8 = t \Rightarrow t = 8$ hours.

20. $x < 2$

21. $3 - 6x < 3 \Rightarrow 3 - 3 - 6x < 3 - 3 \Rightarrow -6x < 0 \Rightarrow \dfrac{-6x}{-6} > \dfrac{0}{-6} \Rightarrow x > 0; \{x | x > 0\}$

22. $2x \le 1 - (2x - 1) \Rightarrow 2x \le 1 - 2x + 1 \Rightarrow 2x + 2x \le 1 - 2x + 2x + 1 \Rightarrow 4x \le 2 \Rightarrow \dfrac{4x}{4} \le \dfrac{2}{4} \Rightarrow$

$x \le \dfrac{1}{2}; \quad \left\{ x \,\middle|\, x \le \dfrac{1}{2} \right\}$

23. $(-2, -3)$: Quadrant III; $(0, 3)$: none; $(2, -2)$: Quadrant IV; $(2, 1)$: Quadrant I

24. Solve the equation for $y$: $x + 2y = 4 \Rightarrow x - x + 2y = 4 - x \Rightarrow 2y = 4 - x \Rightarrow \dfrac{2y}{2} = \dfrac{4}{2} - \dfrac{x}{2} \Rightarrow$

$y = 2 - \dfrac{x}{2}$. Then, when $x = -2$: $y = 2 - \dfrac{x}{2} \Rightarrow y = 2 - \dfrac{-2}{2} = 2 - (-1) = 2 + 1 = 3$;

when $x = -1$: $y = 2 - \dfrac{x}{2} \Rightarrow y = 2 - \left( \dfrac{-1}{2} \right) = 2 + \dfrac{1}{2} = \dfrac{5}{2} = 2.5$;

when $x = 0$: $y = 2 - \dfrac{x}{2} \Rightarrow y = 2 - \dfrac{0}{2} = 2 - 0 = 2$; when $x = 1$: $y = 2 - \dfrac{x}{2} \Rightarrow y = 2 - \dfrac{1}{2} = \dfrac{3}{2} = 1.5$;

when $x = 2$: $y = 2 - \dfrac{x}{2} \Rightarrow y = 2 - \dfrac{2}{2} = 2 - 1 = 1$. Thus, the missing values in the table are 3, 2.5, 2,

1.5, and 1. See Figure 24.

| $x$ | $-2$ | $-1$ | 0 | 1 | 2 |
|---|---|---|---|---|---|
| $y$ | 3 | 2.5 | 2 | 1.5 | 1 |

Figure 24

25. See Figure 25.

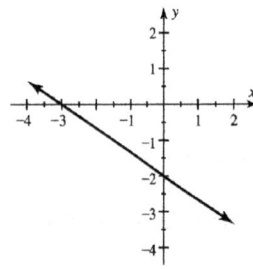

Figure 25

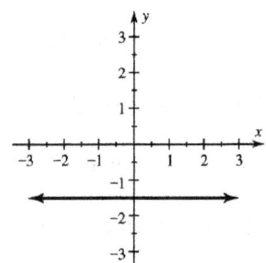

Figure 26

26. See Figure 26.

27. To find the $x$-intercept, set $y = 0$ and solve for $x$:

$-4x + 5y = 40 \Rightarrow -4x + 5(0) = 40 \Rightarrow -4x + 0 = 40 \Rightarrow -4x = 40 \Rightarrow \dfrac{-4x}{-4} = \dfrac{40}{-4} \Rightarrow x = -10.$

Thus, the $x$-intercept is $-10$. To find the $y$-intercept, set $x = 0$ and solve for $y$:

$-4x + 5y = 40 \Rightarrow -4(0) + 5y = 40 \Rightarrow 0 + 5y = 40 \Rightarrow 5y = 40 \Rightarrow \dfrac{5y}{5} = \dfrac{40}{5} \Rightarrow y = 8.$

Thus, the $y$-intercept is 8.

28. $x = \dfrac{3}{2}; \quad y = -2$

29. When $x$ increases by 1, $y$ increases by 3; therefore the slope is 3. When $x = 0$, $y = -3$; therefore the $y$-intercept is $-3$. Thus, the slope-intercept form of the line is $y = 3x - 3$.

30. Solve for $y$: $3x - 5y = 15 \Rightarrow 3x - 3x - 5y = 15 - 3x \Rightarrow -5y = -3x + 15 \Rightarrow$

$\dfrac{-5y}{-5} = \dfrac{-3x}{-5} + \dfrac{15}{-5} \Rightarrow y = \dfrac{3}{5}x - 3$. See Figure 30.

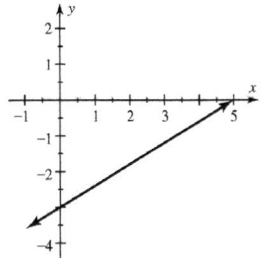

      Figure 30

31. First, solve for $y$: $3x - 2y = 6 \Rightarrow 3x - 3x - 2y = 6 - 3x \Rightarrow -2y = -3x + 6 \Rightarrow \dfrac{-2y}{-2} = \dfrac{-3x}{-2} + \dfrac{6}{-2} \Rightarrow$

$y = \dfrac{3}{2}x - 3$. The line perpendicular to $y = \dfrac{3}{2}x - 3$ has slope $-\dfrac{2}{3}$, the negative reciprocal of

$\dfrac{3}{2}$. Insert the values of the point $(0, -3)$ and the value of the slope into the point-slope

form: $y - (-3) = -\dfrac{2}{3}(x - 0) \Rightarrow y + 3 + (-3) = -\dfrac{2}{3}x - 3 \Rightarrow y = -\dfrac{2}{3}x - 3$.

32. First, find the slope: $\dfrac{-3 - 3}{2 - (-1)} = \dfrac{-6}{2 + 1} = \dfrac{-6}{3} = -2$. Then, insert the values of the point $(-1, 3)$ and the

value of the slope into the point-slope form: $y - 3 = -2(x - (-1)) \Rightarrow y - 3 = -2(x + 1) \Rightarrow$

$y - 3 = -2x - 2 \Rightarrow y - 3 + 3 = -2x - 2 + 3 \Rightarrow y = -2x + 1$.

33. $y = 5x + 100$

34. $I = 5000x + 20{,}000$

35. $C = 8x$

36. 9 out of 10, or $\dfrac{9}{10}$, of the portion of mail consisting of first-class mail and periodicals were first class

mail. Thus, $\dfrac{9}{10} \cdot \dfrac{11}{20} = \dfrac{99}{200}$ represents the fraction of all mail that was first-class mail.

37. Let $x$ represent the amount of money invested at 3%. Then, $2x$ represents the amount of money invested at 4%.

$0.03x + 0.04(2x) = 110 \Rightarrow 0.03x + 0.08x = 110 \Rightarrow 0.11x = 110 \Rightarrow \dfrac{0.11x}{0.11} = \dfrac{110}{0.11} \Rightarrow x = 1000$.

Thus, $1000 is invested at 3% and $2000 is invested at 4%.

38. (a)    Replace $x$ with 200:   $C = 0.3(200) + 25 = 60 + 25 = \$85$.

    (b)    Because the initial value is 25, it costs \$25 to rent the car but not drive it.

    (c)    Because the slope is 0.3, it costs 30¢ to drive the car 1 additional mile.

# Chapter 4: Systems of Linear Equations in Two Variables

### Section 4.1 Solving Systems of Linear Equations Graphically and Numerically

1. ordered

3. no solutions; one solution; infinitely many

5. inconsistent

7. dependent

9. the same

11. Graph the system $y = 2$, $y = 2x$. See Figure 11.  The intersection $x = 1$ is the solution.

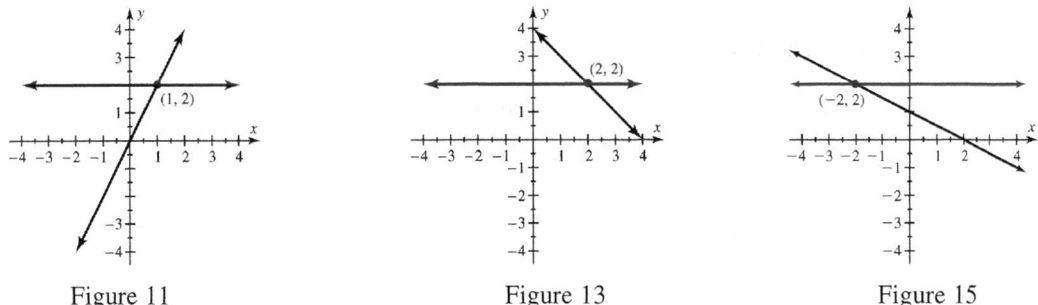

| Figure 11 | Figure 13 | Figure 15 |

13. Graph the system $y = 2$, $y = 4 - x \Rightarrow y = -x + 4$. See Figure 13.  The intersection $x = 2$ is the

   solution.

15. Graph the system $y = 2$, $y = -\dfrac{1}{2}x + 1$. See Figure 15.  The intersection $x = -2$ is the solution.

17. Graph the system $y = 2$, $y = 2x + y = 6 \Rightarrow y = -2x + 6$. See Figure 17. The intersection $x = 2$ is

   the solution.

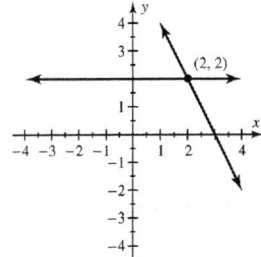

Figure 17

19. Since the lines intersect at one point there is one solution, the system is consistent, and the equations

   are independent.

21. Since the lines intersect at an infinite number of points there are infinitely many solutions, the system

   is consistent, and the equations are dependent.

23. Since the lines are parallel there is no solution and the system is inconsistent.

25. For the system $x + y = 2$ and $x - y = 0$, $(1, 1)$ is the solution because $(1) + (1) = 2$ and $(1) - (1) = 0$.

27. For the system $2x + 3y = -5$ and $4x - 5y = 23$, $(2, -3)$ is the solution because

$2(2) + 3(-3) = 4 + (-9) = -5$ and $4(2) - 5(-3) = 8 - (-15) = 8 + 15 = 23$.

29. For the system $-5x + 5y = -10$ and $4x + 9y = 8$, $(2, 0)$ is the solution because

$-5(2) + 5(0) = -10 + 0 = -10$ and $4(2) + 9(0) = 8 + 0 = 8$.

31. The intersection and solution is $(2, 1)$, $y = 1$ and $2 + 1 = 3$.

33. The intersection and solution is $(3, 2)$, $3 + 2 = 5$ and $3 - 2 = 1$.

35. The intersection and solution is $(-1, 1)$, $-(-1) + 2(1) = 1 + 2 = 3$ and $2(-1) + 3(1) = -2 + 3 = 1$.

37. Since $y = 4$ for both equations when $x = 2$, the solution is $(2, 4)$.

39. Since $y = 1$ for both equations when $x = 3$, the solution is $(3, 1)$.

41. See Figure 41. $Y_1 = Y_2 = 3$ when $x = 1$, the solution is $(1, 3)$.

| $x$ | 0 | 1 | 2 | 3 |
|---|---|---|---|---|
| $y = x + 2$ | 2 | 3 | 4 | 5 |
| $y = 4 - x$ | 4 | 3 | 2 | 1 |

Figure 41

43. (a)   Graph the system. See Figure 43a. The intersection and solution is $(-1, 1)$.

(b)   See Figure 43b. Since $y = 1$ for both equations, when $x = -1$ the solution is $(-1, 1)$.

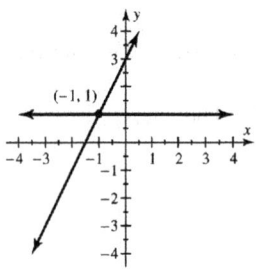

Figure 43a

| $x$ | $-2$ | $-1$ | 0 | 1 |
|---|---|---|---|---|
| $y = 2x + 3$ | $-1$ | 1 | 3 | 5 |
| $y = 1$ | 1 | 1 | 1 | 1 |

Figure 43b

45. (a)   Graph the system. See Figure 45a. The intersection and solution is $(3, 1)$.

(b)   See Figure 45b. Since $y = 1$ for both equations, when $x = 3$ the solution is $(3, 1)$.

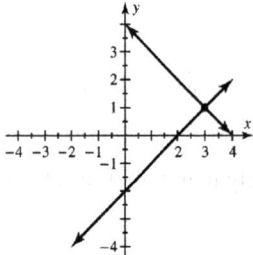

Figure 45a

| $x$ | 2 | 3 | 4 | 5 |
|---|---|---|---|---|
| $y = 4 - x$ | 2 | 1 | 0 | $-1$ |
| $y = x - 2$ | 0 | 1 | 2 | 3 |

Figure 45b

47. (a)   Graph the system. See Figure 47a. The intersection and solution is $(1, 3)$.

    (b)   See Figure 47b. Since $y = 3$ for both equations, when $x = 1$ the solution is $(1, 3)$.

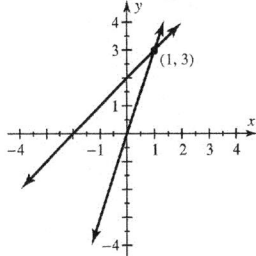

| $x$ | $-1$ | 0 | 1 | 2 |
|---|---|---|---|---|
| $y = 3x$ | $-3$ | 0 | 3 | 6 |
| $y = x + 2$ | 1 | 2 | 3 | 4 |

Figure 47a                    Figure 47b

49. Graph the system. See Figure 49. The intersection and solution is $(-1, -2)$.

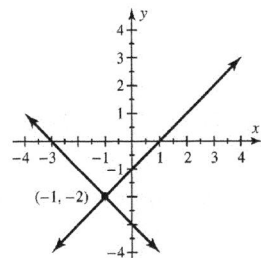

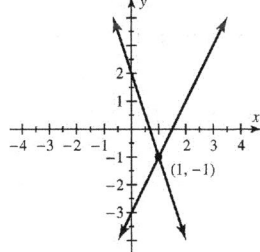

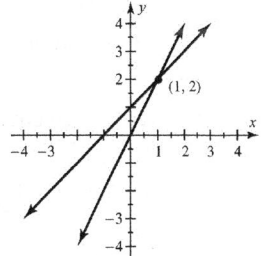

Figure 49                          Figure 51                          Figure 53

51. Graph the system. See Figure 51. The intersection and solution is $(1, -1)$.

53. Graph the system. See Figure 53. The intersection and solution is $(1, 2)$.

55. Graph the system. See Figure 55. The intersection and solution is $(3, 2)$.

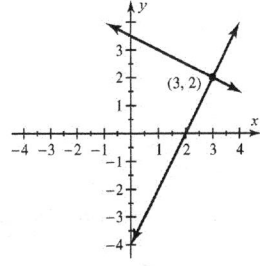

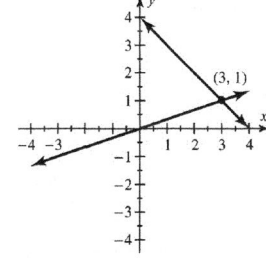

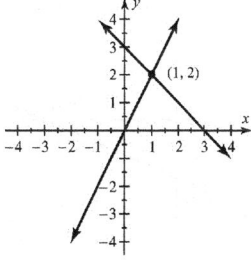

Figure 55                          Figure 57                          Figure 59

57. Graph the system. See Figure 57. The intersection and solution is $(3, 1)$.

59. Graph the system. See Figure 59. The intersection and solution is $(1, 2)$.

61. (a)   Let $x$ and $y$ represent the numbers. Then $x + y = 4$ and $x - y = 0$.

    (b)   Write each equation in slope-intercept form: $y = -x + 4$ and $y = x$. The graphs of these lines

          are shown in Figure 61, and they intersect at the point $(2, 2)$. The unknown numbers are 2

          and 2.

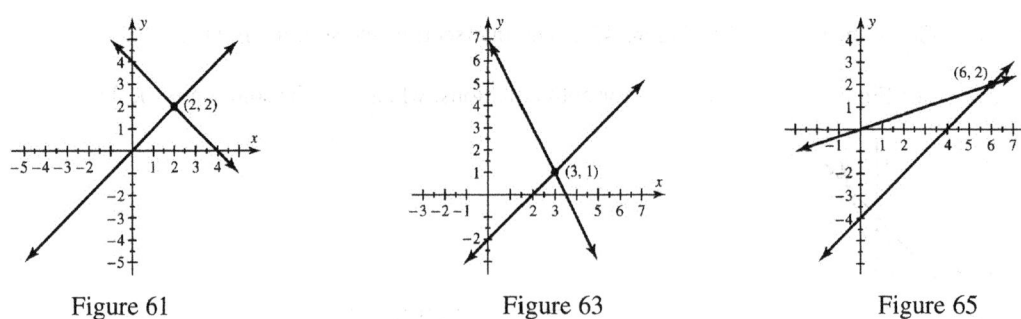

Figure 61                          Figure 63                          Figure 65

63. (a)    Let $x$ and $y$ represent the numbers. Then $2x + y = 7$ and $x - y = 2$.

    (b)    Write each equation in slope-intercept form: $y = -2x + 7$ and $y = x - 2$. The graphs of these

           lines are shown in Figure 63, and they intersect at the point $(3, 1)$. The unknown numbers are

           3 and 1.

65. (a)    Let $x$ and $y$ represent the numbers. Then $x = 3y$ and $x - y = 4$.

    (b)    Write each equation in slope-intercept form: $y = \dfrac{1}{3}x$ and $y = x - 4$. The graphs of these lines

           are shown in Figure 65, and they intersect at the point $(6, 2)$. The unknown numbers are 6

           and 2.

67. (a)    $x = $ miles, $C = $ cost, then $C = 0.5x + 50$.

    (b)    Graph $C = 0.5x + 50$ and $C = 80$. See Figure 67b.

           The intersection and solution $(60, 80) \Rightarrow 60$ miles.

    (c)    See Figure 67c. Since $C = 80$ for both equations when $x = 60$,

           the solution is $(60, 80) \Rightarrow 60$ miles.

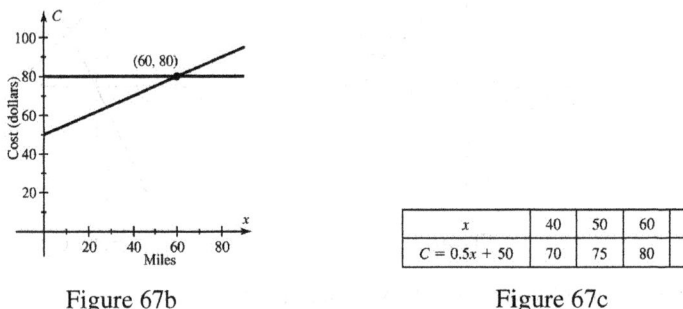

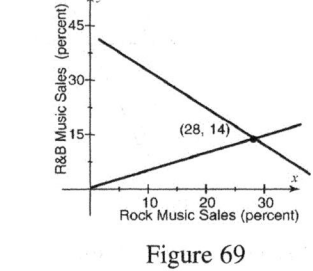

| $x$ | 40 | 50 | 60 | 70 |
|---|---|---|---|---|
| $C = 0.5x + 50$ | 70 | 75 | 80 | 85 |

Figure 67b                          Figure 67c                          Figure 69

69. (a)    $x = \%$ rock music, $y = \%$ R&B music, then $x + y = 42$ and $x = 2y$.

    (b)    Graph $x + y = 42$ and $x = 2y$. See Figure 69. The intersection and solution is $(28, 14)$.

           Rock music accounted for 28% of all music sales and R&B music accounted for 14%.

71. (a)    Let $x = $ length, $y = $ width, then $2x + 2y = 28$ and $x = y + 4$ or $x - y = 4$.

    (b)    Graph $2x + 2y = 28$ and $x - y = 4$. See Figure 71.

           The intersection and solution is $(9, 5) \Rightarrow 9$ in.$\times 5$ in.

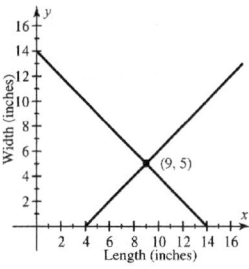

Figure 71

## Section 4.2 Solving Systems of Linear Equations by Substitution

1. exact

3. Because $1 = 1$ is true, it has infinitely many solutions.

5. Substituting $y = 2x$ into the first equation yields the following: $x + (2x) = 9 \Rightarrow 3x = 9 \Rightarrow x = 3$ and

   so $y = 2(3) \Rightarrow y = 6$. The solution is $(3, 6)$.

7. Substituting $x = 2y$ into the first equation yields the following: $(2y) + 2y = 4 \Rightarrow 4y = 4 \Rightarrow y = 1$

   and so $x = 2(1) \Rightarrow x = 2$. The solution is $(2, 1)$.

9. Substituting $y = x + 1$ into the first equation yields the following: $2x + (x + 1) = -2 \Rightarrow 3x + 1 = -2 \Rightarrow$

   $3x = -3 \Rightarrow x = -1$ and so $y = -1 + 1 \Rightarrow y = 0$. The solution is $(-1, 0)$.

11. Substituting $x = y + 3$ into the first equation yields the following: $(y + 3) + 3y = 3 \Rightarrow 4y + 3 = 3 \Rightarrow$

    $4y = 0 \Rightarrow y = 0$ and so $x = 0 + 3 \Rightarrow x = 3$. The solution is $(3, 0)$.

13. Substituting $y = 2x - 1$ into the first equation yields the following: $3x + 2(2x - 1) = \dfrac{3}{2} \Rightarrow$

    $3x + 4x - 2 = \dfrac{3}{2} \Rightarrow 7x - 2 = \dfrac{3}{2} \Rightarrow 7x = \dfrac{7}{2} \Rightarrow x = \dfrac{1}{2}$ and so $y = 2\left(\dfrac{1}{2}\right) - 1 \Rightarrow y = 0$.

    The solution is $\left(\dfrac{1}{2},\ 0\right)$.

15. Substituting $x = 2 - \dfrac{1}{2}y$ into the first equation yields the following: $2\left(2 - \dfrac{1}{2}y\right) - 3y = -12 \Rightarrow$

    $4 - y - 3y = -12 \Rightarrow 4 - 4y = -12 \Rightarrow -4y = -16 \Rightarrow y = 4$ and so $x = 2 - \dfrac{1}{2}(4) \Rightarrow x = 2 - 2 \Rightarrow x = 0$.

    The solution is $(0, 4)$.

17. Note that $3x - y = 1 \Rightarrow -y = -3x + 1 \Rightarrow y = 3x - 1$, substituting $y = 3x - 1$ into the first equation

    yields the following: $2x - 3(3x - 1) = -4 \Rightarrow 2x - 9x + 3 = -4 \Rightarrow$

    $-7x + 3 = -4 \Rightarrow -7x = -7 \Rightarrow x = 1$ and so $y = 3(1) - 1 \Rightarrow y = 2$. The solution is $(1, 2)$.

19. Note that $x - 5y = 26 \Rightarrow x = 5y + 26$, substituting $x = 5y + 26$ into the second equation yields the

following: $2(5y + 26) + 6y = -12 \Rightarrow 10y + 52 + 6y = -12 \Rightarrow 16y + 52 = -12 \Rightarrow 16y = -64 \Rightarrow$

$y = -4$ and so $x = 5(-4) + 26 \Rightarrow x = 6$. The solution is $(6, -4)$.

21. Note that $y - 3z = 13 \Rightarrow y = 3z + 13$, substituting $y = 3z + 13$ into the first equation yields the

following:

$\dfrac{1}{2}(3z + 13) - z = 5 \Rightarrow \dfrac{3}{2}z + \dfrac{13}{2} - z = 5 \Rightarrow \dfrac{1}{2}z + \dfrac{13}{2} = 5 \Rightarrow \dfrac{1}{2}z = -\dfrac{3}{2} \Rightarrow z = -3$ and so

$y = 3(-3) + 13 \Rightarrow y = -9 + 13 \Rightarrow y = 4$. The solution is $(4, -3)$.

23. Note that $r + 60t = -29 \Rightarrow r = -60t - 29$, substituting $r = -60t - 29$ into the first equation yields the

following: $10(-60t - 29) - 20t = 20 \Rightarrow -600t - 290 - 20t = 20 \Rightarrow -620t - 290 = 20 \Rightarrow$

$-620t = 310 \Rightarrow t = -\dfrac{1}{2}$ and so $r = -60\left(-\dfrac{1}{2}\right) - 29 \Rightarrow r = 30 - 29 \Rightarrow r = 1$. The solution is $\left(1, -\dfrac{1}{2}\right)$.

25. Note that $3x + 2y = 9 \Rightarrow 3x = -2y + 9 \Rightarrow x = -\dfrac{2}{3}y + 3$, substituting $x = -\dfrac{2}{3}y + 3$ into the second

equation yields the following: $2\left(-\dfrac{2}{3}y + 3\right) - 3y = -7 \Rightarrow -\dfrac{4}{3}y + 6 - 3y = -7 \Rightarrow -\dfrac{13}{3}y + 6 = -7 \Rightarrow$

$-\dfrac{13}{3}y = -13 \Rightarrow y = 3$ and so $x = -\dfrac{2}{3}(3) + 3 \Rightarrow x = -2 + 3 \Rightarrow x = 1$. The solution is $(1, 3)$.

27. Note that $2a - 3b = 6 \Rightarrow 2a = 3b + 6 \Rightarrow a = \dfrac{3}{2}b + 3$, substituting $a = \dfrac{3}{2}b + 3$ into the second equation

yields the following: $-5\left(\dfrac{3}{2}b + 3\right) + 4b = -8 \Rightarrow -\dfrac{15}{2}b - 15 + 4b = -8 \Rightarrow -\dfrac{7}{2}b - 15 = -8 \Rightarrow$

$-\dfrac{7}{2}b = 7 \Rightarrow b = -2$ and so $a = \dfrac{3}{2}(-2) + 3 \Rightarrow a = -3 + 3 \Rightarrow a = 0$. The solution is $(0, -2)$.

29. Note that $2x - \dfrac{1}{2}y = 3 \Rightarrow -\dfrac{1}{2}y = -2x + 3 \Rightarrow y = 4x - 6$, substituting $y = 4x - 6$ into the first equation

yields the following:

$-\dfrac{1}{2}x + 3(4x - 6) = 5 \Rightarrow -\dfrac{1}{2}x + 12x - 18 = 5 \Rightarrow \dfrac{23}{2}x - 18 = 5 \Rightarrow \dfrac{23}{2}x = 23 \Rightarrow x = 2$ and so

$y = 4(2) - 6 \Rightarrow y = 8 - 6 \Rightarrow y = 2$. The solution is $(2, 2)$.

31. Note that $-8a + 2b = 34 \Rightarrow 2b = 8a + 34 \Rightarrow b = 4a + 17$, substituting

$b = 4a + 17$ into the first equation yields the following:

$3a + 5(4a + 17) = 16 \Rightarrow 3a + 20a + 85 = 16 \Rightarrow 23a = -69 \Rightarrow a = -3$ and so

$b = 4(-3) + 17 \Rightarrow b = -12 + 17 \Rightarrow b = 5$. The solution is $(-3, 5)$.

33. Note that $x + y = 7 \Rightarrow y = -x + 7$, substituting $y = -x + 7$ into the first equation yields the

following: $x + (-x + 7) = 9 \Rightarrow 7 = 9$, which is false $\Rightarrow$ no solutions.

35. Note that $x - y = 4 \Rightarrow x = y + 4$, substituting $x = y + 4$ into the second equation yields the

following: $2(y + 4) - 2y = 8 \Rightarrow 2y + 8 - 2y = 8 \Rightarrow 8 = 8$, which is true $\Rightarrow$ infinitely many solutions.

37. Note that $x + y = 4 \Rightarrow y = -x + 4$, substituting $y = -x + 4$ into the second equation yields the

following: $x - (-x + 4) = 2 \Rightarrow 2x = 6 \Rightarrow x = 3$ and so $y = -(3) + 4 \Rightarrow y = 1$. The solution is (3, 1).

39. Note that $-x + y = -7 \Rightarrow y = x - 7$, substituting $y = x - 7$ into the first equation yields the

following: $x - (x - 7) = 7 \Rightarrow 7 = 7$, which is true $\Rightarrow$ infinitely many solutions.

41. Note that $u - 2v = 5 \Rightarrow u = 2v + 5$, substituting $u = 2v + 5$ into the second equation yields the

following: $2(2v + 5) - 4v = -2 \Rightarrow 4v + 10 - 4v = -2 \Rightarrow 10 = -2$, which is false $\Rightarrow$ no solutions.

43. Note that $r - 3t = -5 \Rightarrow r = 3t - 5$, substituting $r = 3t - 5$ into the first equation yields the

following: $2(3t - 5) + 3t = 1 \Rightarrow 6t - 10 + 3t = 1 \Rightarrow 9t - 10 = 1 \Rightarrow 9t = 11 \Rightarrow t = \dfrac{11}{9}$ and so

$r = 3\left(\dfrac{11}{9}\right) - 5 \Rightarrow r = \dfrac{11}{3} - 5 \Rightarrow r = -\dfrac{4}{3}$. The solution is $\left(-\dfrac{4}{3}, \dfrac{11}{9}\right)$.

45. Substituting $y = 5x$ into the second equation yields the following: $5x = -3x \Rightarrow 8x = 0 \Rightarrow x = 0$, and

so $y = 5(0) \Rightarrow y = 0$. The solution is $(0, 0)$.

47. Substituting $5a = 4 - b$ into the second equation yields the following: $4 - b = 3 - b \Rightarrow 4 = 3$, which is

false $\Rightarrow$ no solutions.

49. Note that $2x + 4y = 0 \Rightarrow 2x = -4y \Rightarrow x = -2y$, substituting $x = -2y$ into the second equation

yields the following: $3(-2y) + 6y = 5 \Rightarrow -6y + 6y = 5 \Rightarrow 0 = 5$, which is false $\Rightarrow$ no solutions.

51. They are a single line.

53. (a)    $2L + 2W = 72$ and $L = W + 10$

    (b)    Substituting $L = W + 10$ into the first equation yields the following: $2(W + 10) + 2W = 72 \Rightarrow$

           $2W + 20 + 2W = 72 \Rightarrow 4W = 52 \Rightarrow W = 13$ and so $L = 13 + 10 \Rightarrow L = 23$. The solution is

           (23, 13). Checking: $2(23) + 2(13) = 46 + 26 = 72$, yes. $23 = 13 + 10$, yes.

55. (a)    $x + y = 90$ and $x = \dfrac{1}{2}y$

    (b)    Substituting $x = \dfrac{1}{2}y$ into the first equation yields the following:

           $\dfrac{1}{2}y + y = 90 \Rightarrow \dfrac{3}{2}y = 90 \Rightarrow y = 60$ and so $x = \dfrac{1}{2}(60) \Rightarrow x = 30$. The solution is (30, 60).

    (c)    Graph the system. See Figure 55. The lines intersect at the solution (30, 60).

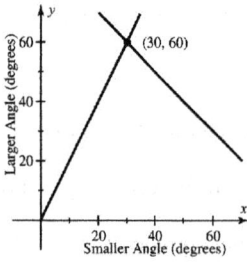

Figure 55

57. (a)    $x - y = 21$ and $y = 0.86x$

     (b)    Substituting $y = 0.86 x$ into the first equation yields the following: $x - 0.86x = 21 \Rightarrow$

            $0.14x = 21 \Rightarrow x = 150$ and so $150 - y = 21 \Rightarrow y = 129$. The solution is $(150, \ 129)$.

59. Let $L = $ length and $W = $ width, then let $L = W + 44$ and $2L + 2W = 288$, substituting $L = W + 44$ into

     the second equation yields the following: $2(W + 44) + 2W = 288 \Rightarrow 2W + 88 + 2W = 288 \Rightarrow$

     $4W = 200 \Rightarrow W = 50$ and so $L = 50 + 44 \Rightarrow L = 94$. The solution is $(94, 50)$ or $94 \ \text{ft} \times 50 \ \text{ft}$.

61. Let $x = $ larger number and $y = $ smaller number, then $x + y = 70$ and $x = 3y + 2$, substituting

     $x = 3y + 2$ into the first equation yields the following: $(3y + 2) + y = 70 \Rightarrow 4y + 2 = 70 \Rightarrow$

     $4y = 68 \Rightarrow y = 17$ and so $x = 3(17) + 2 \Rightarrow x = 51 + 2 \Rightarrow x = 53$. The numbers are 17 and 53.

63. Let $x = $ speed of tugboat and $y = $ speed of current, then $15x - 15y = 120$ and $10x + 10y = 120$. Note

     that $10x + 10y = 120 \Rightarrow 10y = -10x + 120 \Rightarrow y = -x + 12$, substituting $y = -x + 12$ into the first

     equation yields the following: $15x - 15(-x + 12) = 120 \Rightarrow 15x + 15x - 180 = 120 \Rightarrow$

     $30x = 300 \Rightarrow x = 10$ and so $y = -10 + 12 \Rightarrow y = 2$. The tugboat speed is 10 mph and the current's

     speed is 2 mph.

65. Let $x = $ liters of 20% solution and $y = $ liters of 50% solution, then $x + y = 10$ and

     $0.2x + 0.5y = 0.4(10) \Rightarrow 0.2x + 0.5y = 4$. Note that $x + y = 10 \Rightarrow y = -x + 10$, substituting

     $y = -x + 10$ into the second equation yields the following: $0.2x + 0.5(-x + 10) = 4 \Rightarrow$

     $0.2x - 0.5x + 5 = 4 \Rightarrow -0.3x = -1 \Rightarrow x = 3.\overline{3}$ and so $3.\overline{3} + y = 10 \Rightarrow y = 6.\overline{6}$. The amounts

     are: $3.\overline{3}$ liters of 20% solution and $6.\overline{6}$ liters of 50% solution.

67. Let $x = \text{mi}^2$ of Lake Superior and $y = \text{mi}^2$ of Lake Michigan, then $x + y = 54,000$ and $x = y + 10,000$,

     substituting $x = y + 10,000$ into the first equation yields the following:

     $(y + 10,000) + y = 54,000 \Rightarrow 2y + 10,000 = 54,000 \Rightarrow 2y = 44,000 \Rightarrow y = 22,000$ and so

     $x = 22,000 + 10,000 \Rightarrow x = 32,000$.

     Lake Superior has $32,000 \ \text{mi}^2$ and Lake Michigan has $22,000 \ \text{mi}^2$.

## Checking Basic Concepts  Sections 4.1 and 4.2

1. (a)   Graph the system $y = 2$ and $y = 1 - \frac{1}{2}x$.  See Figure 1a.

   The intersection and solution is $(-2,\ 2) \Rightarrow x = -2$.

   (b)   Graph the system $y = 2$ and $2x - 3y = 6$.  See Figure 1b.

   The intersection and solution is $(6,\ 2) \Rightarrow x = 6$.

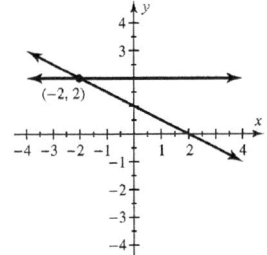          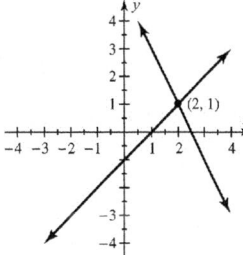

   Figure 1a                      Figure 1b                      Figure 3

2. $(4, 2)$ is the solution because $2(4) - 5(2) = 8 - 10 = -2$ and $3(4) + 2(2) = 12 + 4 = 16$.

3. Graph the system.  See Figure 3. The intersection and solution is $(2,\ 1)$.

   Checking: $2 - 1 = 1$ (yes) and $2(2) + 1 = 5$ (yes).

4. (a)   Substituting $y = 2 - x$ into the first equation yields the following:  $x + (2 - x) = -1 \Rightarrow 2 = -1$

   which is false $\Rightarrow$ no solution.

   (b)   Note that $-x + y = -2 \Rightarrow y = x - 2$, substituting $y = x - 2$ into the first equation yields the

   following: $4x - (x - 2) = 5 \Rightarrow 3x + 2 = 5 \Rightarrow 3x = 3 \Rightarrow x = 1$ and so $y = 1 - 2 \Rightarrow y = -1$.  The

   solution is $(1,\ -1)$, one solution.

   (c)   Note that $x + 2y = 3 \Rightarrow x = -2y + 3$, substituting $x = -2y + 3$ into the second equation yields

   the following: $-(-2y + 3) - 2y = -3 \Rightarrow 2y - 3 - 2y = -3 \Rightarrow -3 = -3$, which is true $\Rightarrow$

   infinitely many solutions.

5. (a)   $x + y = 300,\ 150x + 200y = 55,000$

   (b)   Note that $x + y = 300 \Rightarrow y = -x + 300$, substituting $y = -x + 300$ into the second equation

   yields the following:

   $150x + 200(-x + 300) = 55,000 \Rightarrow 150x - 200x + 60,000 = 55,000 \Rightarrow -50x = -5000 \Rightarrow$

   $x = 100$ and so $y = -100 + 300 \Rightarrow y = 200$.  The solution is $(100, 200)$.  Checking the answer

   $100 + 200 = 300$ (yes) and $150(100) + 200(200) = 15,000 + 40,000 = 55,000$ (true).

## Section 4.3 Solving Systems of Linear Equations by Elimination

1. Substitution; elimination

3. $=$

5. It has infinitely many solutions.

7. Adding the two equations will eliminate the variable $y$.

$$\begin{array}{r} x - y = 0 \\ x + y = 2 \\ \hline 2x = 2 \end{array}$$

Thus, $x = 1$. And so $1 + y = 2 \Rightarrow y = 1$. The solution is $(1, 1)$.

The result is supported by the graph's intersection point of $(1, 1)$.

9. Adding the two equations will eliminate the variable $y$.

$$\begin{array}{r} 2x + 3y = -1 \\ 2x - 3y = -7 \\ \hline 4x = -8 \end{array}$$

Thus, $x = -2$. And so $2(-2) + 3y = -1 \Rightarrow -4 + 3y = -1 \Rightarrow 3y = 3 \Rightarrow y = 1$.

The solution is $(-2, 1)$. The result is supported by the graph's intersection point

of $(-2, 1)$.

11. Multiplying the first equation by $-1$ and adding the two equations will eliminate both variables.

$$\begin{array}{r} -x - y = -3 \\ x + y = -1 \\ \hline 0 = -4 \end{array}$$

This is false $\Rightarrow$ no solution. This result is supported by the graph's parallel lines

which has no solution.

13. Multiplying the second equation by $-2$ and adding the two equations will eliminate both variables.

$$\begin{array}{r} 2x + 2y = 6 \\ -2x - 2y = -6 \\ \hline 0 = 0 \end{array}$$

Which is true $\Rightarrow$ there are infinitely many solutions. This result is supported by

the graph having two lines that coincide.

15. Adding the two equations will eliminate the variable $y$.

$$\begin{array}{r} x + y = 7 \\ x - y = 5 \\ \hline 2x = 12 \end{array}$$

Thus, $x = 6$. And so $6 + y = 7 \Rightarrow y = 1$. The solution is $(6, 1)$.

17. Adding the two equations will eliminate the variable $x$.

$$\begin{array}{r} -x + y = 5 \\ x + y = 3 \\ \hline 2y = 8 \end{array}$$

Thus, $y = 4$. And so $x + 4 = 3 \Rightarrow x = -1$. The solution is $(-1, 4)$.

19. Adding the two equations will eliminate the variable $y$.

$$2x + y = 8$$
$$\underline{3x - y = 2}$$
$$5x = 10$$     Thus, $x = 2$. And so $2(2) + y = 8 \Rightarrow 4 + y = 8 \Rightarrow y = 4$. The solution is $(2, 4)$.

21. Adding the two equations will eliminate the variable $x$.

$$-2x + y = -3$$
$$\underline{2x - 4y = 0}$$
$$-3y = -3$$     Thus, $y = 1$. And so $-2x + 1 = -3 \Rightarrow -2x = -4 \Rightarrow x = 2$. The solution is $(2, 1)$.

23. Adding the two equations will eliminate the variable $y$.

$$\frac{1}{2}x - y = 3$$
$$\underline{\frac{3}{2}x + y = 5}$$
$$2x = 8$$     Thus, $x = 4$. And so $\frac{1}{2}(4) - y = 3 \Rightarrow 2 - y = 3 \Rightarrow -y = 1 \Rightarrow y = -1$.

The solution is $(4, -1)$.

25. Multiplying the second equation by $-1$ and adding the two equations will eliminate the variable $a$.

$$a + 6b = 2$$
$$\underline{-a - 3b = 1}$$
$$3b = 3$$     Thus, $b = 1$. And so $a + 6(1) = 2 \Rightarrow a = -4$. The solution is $(-4, 1)$.

27. Multiplying the first equation by $-1$ and adding the two equations will eliminate the variable $a$.

$$-5a + 6b = 2$$
$$\underline{5a + 5b = 9}$$
$$11b = 11$$     Thus, $b = 1$. And so $5a + 5(1) = 9 \Rightarrow 5a = 4 \Rightarrow a = \frac{4}{5}$. The solution is $\left(\frac{4}{5}, 1\right)$.

29. Multiplying the second equation by $-2$ and adding the two equations will eliminate the variable $v$.

$$3u + 2v = -16$$
$$\underline{-4u - 2v = 18}$$
$$-u = 2$$     Thus $u = -2$. And so $3(-2) + 2v = -16 \Rightarrow -6 + 2v = -16 \Rightarrow 2v = -10 \Rightarrow v = -5$.

The solution is $(-2, -5)$.

31. Multiplying the first equation by 2, the second equation by 5 and adding the two equations will eliminate the variable $x$.

$$10x - 14y = 10$$
$$\underline{-10x + 10y = -10}$$
$$-4y = 0$$     Thus $y = 0$. And so $10x - 14(0) = 10 \Rightarrow 10x = 10 \Rightarrow x = 1$. The solution is $(1, 0)$.

33. Multiplying the second equation by $-5$, the first equation by 3 and adding the two equations will eliminate the variable $x$.

$$15x - 9y = 12$$
$$\underline{-15x - 10y = -50}$$
$$-19y = -38$$   Thus $y = 2$. And so $5x - 3(2) = 4 \Rightarrow 5x - 6 = 4 \Rightarrow 5x = 10 \Rightarrow x = 2$.

The solution is $(2, 2)$.

35. Multiplying the first equation by 2 and adding the two equations will eliminate the variable $x$.

$$-10x - 20y = -44$$
$$\underline{10x + 15y = 35}$$
$$-5y = -9$$   Thus $y = \dfrac{9}{5}$. And so $10x + 15\left(\dfrac{9}{5}\right) = 35 \Rightarrow 10x + 27 = 35 \Rightarrow 10x = 8 \Rightarrow x = \dfrac{4}{5}$.

The solution is $\left(\dfrac{4}{5}, \dfrac{9}{5}\right)$.

37. Since $y = 2$, when $x = 3$ for both equations $\Rightarrow$ the solution is $(3, 2)$.

39. Since $y = 1$, when $x = 0$ for both equations $\Rightarrow$ the solution is $(0, 1)$.

41. (a)   Adding the two equations will eliminate the variable $y$.

$$2x + y = 5$$
$$\underline{x - y = 1}$$
$$3x = 6$$   Thus, $x = 2$. And so $2 - y = 1 \Rightarrow y = 1$. The solution is $(2, 1)$.

(b)   Graph the system. See Figure 41b. The intersection and solution is $(2, 1)$.

(c)   See Figure 41c. Since $y = 1$ when $x = 2$ for both equations the solution is $(2, 1)$.

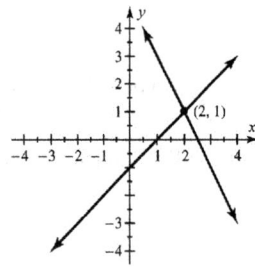

| $x$ | 0 | 1 | 2 | 3 |
|---|---|---|---|---|
| $y = 5 - 2x$ | 5 | 3 | 1 | $-1$ |
| $y = x - 1$ | $-1$ | 0 | 1 | 2 |

Figure 41b                    Figure 41c

43. (a)   Multiplying the second equation by $-1$ and adding the two equations will eliminate the variable $y$.

$$2x + y = 5$$
$$\underline{-x - y = -1}$$
$$x = 4$$   And so $-4 - y = -1 \Rightarrow -y = 3 \Rightarrow y = -3$. The solution is $(4, -3)$.

(b)   Graph the system. See Figure 43b. The intersection and solution is $(4, -3)$.

(c)   See Figure 43c. Since $y = -3$ when $x = 4$ for both equations the solution is $(4, -3)$.

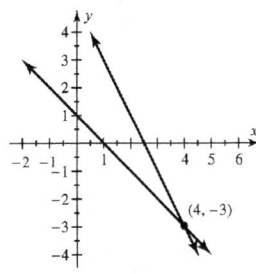

Figure 43b

| $x$ | 2 | 3 | 4 | 5 |
|---|---|---|---|---|
| $y = 5 - 2x$ | 1 | $-1$ | $-3$ | $-5$ |
| $y = 1 - x$ | $-1$ | $-2$ | $-3$ | $-4$ |

Figure 43c

45. (a) Multiplying the second equation by 3 and adding the two equations will eliminate the variable $x$.

$$6x + 3y = 6$$
$$\underline{-6x + 6y = -6}$$
$$9y = 0$$

Thus, $y = 0$. And so $6x + 3(0) = 6 \Rightarrow 6x = 6 \Rightarrow x = 1$. The solution is $(1, 0)$.

(b) Graph the system. See Figure 45b. The intersection and solution is $(1, 0)$.

(c) See Figure 45c. Since $y = 0$ when $x = 1$ for both equations the solution is $(1, 0)$.

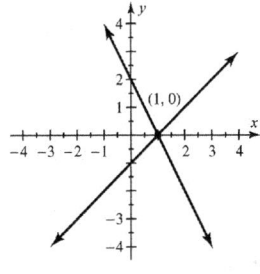

Figure 45b

| $x$ | $-1$ | 0 | 1 | 2 |
|---|---|---|---|---|
| $y = 2 - 2x$ | 4 | 2 | 0 | $-2$ |
| $y = x - 1$ | $-2$ | $-1$ | 0 | 1 |

Figure 45c

47. Multiplying the second equation by 2 and adding the two equations will eliminate both variables.

$$2x - 2y = 4$$
$$\underline{-2x + 2y = -4}$$
$$0 = 0$$

This is always true $\Rightarrow$ infinitely many solutions. See Figure 47.

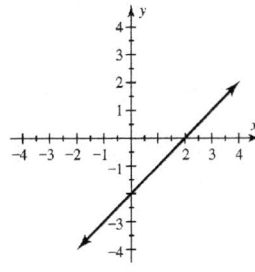

Figure 47

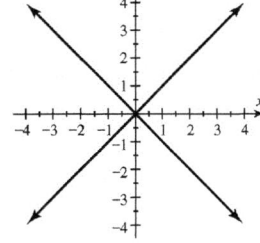

Figure 49

49. Adding the two equations will eliminate the variable $y$.

$$x - y = 0$$
$$\underline{x + y = 0}$$
$$2x = 0$$

Thus, $x = 0$. And so $0 - y = 0 \Rightarrow -y = 0 \Rightarrow y = 0$. The solution is $(0, 0) \Rightarrow$ one solution. See Figure 49

51. Multiplying the first equation by $-1$ and adding the two equations will eliminate both variables.

$$-x + y = -4$$
$$\underline{x - y = 1}$$
$$0 = -3$$

This is never true $\Rightarrow$ no solutions. See Figure 51.

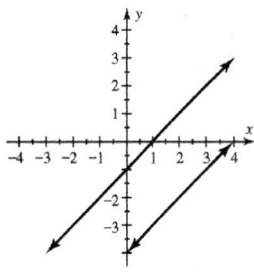
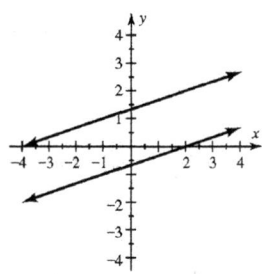
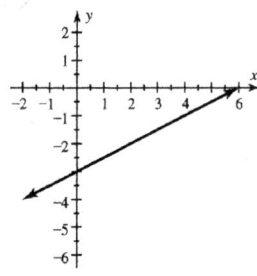

Figure 51          Figure 53          Figure 55

53. Adding the two equations will eliminate both variables.

$$x - 3y = 2$$
$$\underline{-x + 3y = 4}$$
$$0 = 6$$

This is never true $\Rightarrow$ no solutions. See Figure 53.

55. Multiplying the first equation by 3, the second by $-2$ and adding the two equations will eliminate both variables.

$$12x - 24y = 72$$
$$\underline{-12x + 24y = -72}$$
$$0 = 0$$

This is always true $\Rightarrow$ infinitely many solutions. See Figure 55.

57. Let $x =$ skin cancer in men and let $y =$ skin cancer in women. Then

$x + y = 68,000$ and $x = y + 10,000$. Note that $x = y + 10,000 \Rightarrow x - y = 10,000$. Adding the two equations will eliminate the variable $y$.

$$x + y = 68,000$$
$$\underline{x - y = 10,000}$$
$$2x = 78,000$$

Thus $x = 39,000$. And so $39,000 - y = 10,000 \Rightarrow -y = -29,000 \Rightarrow y = 29,000$.

Therefore men: 39,000; Women: 29,000.

59. Let $x =$ minutes on a stationary bike and let $y =$ minutes on a stair climber. Then $x + y = 30$ and $9x + 11.5y = 300$. Multiplying the first equation by $-9$ and adding the two equations will eliminate the variable $x$.

$$-9x - 9y = -270$$
$$\underline{9x + 11.5y = 300}$$
$$2.5y = 30$$

Thus $y = 12$. And so $x + 12 = 30 \Rightarrow x = 18$. Therefore bicycle: 18 minutes; stair climber: 12 minutes.

61. Let $x =$ speed of riverboat and let $y =$ speed of the current. Then $8x+8y = 64$ and $16x-16y = 64$.

    Multiplying the first equation by 2 and adding the two equations will eliminate the variable $y$.

    $16x+16y = 128$
    $16x-16y = 64$
    ───────────
    $32x = 192$     Thus $x = 6$. So $8(6)+8y = 64 \Rightarrow 48+8y = 64 \Rightarrow 8y = 16 \Rightarrow y = 2$.

                     Therefore current: 2 mph; boat: 6 mph.

63. Let $x =$ amount of money invested at 3% and let $y =$ amount of money invested at 5%. Then

    $x+y = 5000$ and $0.03x+0.05y = 210$. Multiplying the second equation by $-20$ and adding the

    two equations will eliminate the variable $y$.

    $x+y = \phantom{-}5000$
    $-0.6x-y = -4200$
    ────────────
    $0.4x = \phantom{-}800$     Thus $x = 2000$. So $2000+y = 5000 \Rightarrow y = 3000$.

                  Therefore \$2000 invested at 3%; \$3000 invested at 5%.

65. Let $x =$ one of two integers and let $y =$ the other of two integers. Then $x+y = -17$ and

    $x-y = -69$. Adding the two equations will eliminate the variable $y$.

    $x+y = -17$
    $x-y = -69$     Thus $x = -43$. So $-43+y = -17 \Rightarrow y = 26$. Therefore the numbers are
    ──────────
    $2x = -86$     $-43$ and 26.

67. (a) The graph's intersection is $20\,\text{in.}\times 40\,\text{in.}$

    (b) Note that $L = 2W \Rightarrow -2W+L = 0$, using this and $2W+2L = 120$ and adding the two

    equations will eliminate the variable $W$.

    $-2W+L = 0$
    $2W+2L = 120$     Thus
    ───────────
    $3L = 120$     $L = 40$. So $2W+2(40) = 120 \Rightarrow 2W+80 = 120 \Rightarrow 2W = 40 \Rightarrow W = 20$.

                  Therefore $L = 40$ and $W = 20$ so $20\,\text{in.}\times 40\,\text{in.}$

## Section 4.4  Systems of Linear Inequalities

1. inequality

3. All points below and including the line $y = k$

5. All points above and including the line $y = x$

7. dashed

9. $Ax+By = C$

11. intersect

13. Yes, $3 > 2$ is true.

15. No, $0 \geq 2$ is false.

17. No, $4 \geq 5$ is false.

19. Yes, $0 < 3 - 1 \Rightarrow 0 < 2$ is true.

21. Yes, $-2 + 6 \leq 4 \Rightarrow 4 \leq 4$ is true.

23. No, $2(-1) + (-1) \geq -1 \Rightarrow -2 + (-1) \geq -1 \Rightarrow -3 \geq -1$ is false.

25. $x > 1$

27. $y \geq 2$

29. $y < x$

31. $-x + y \leq 1$

33. See Figure 33.

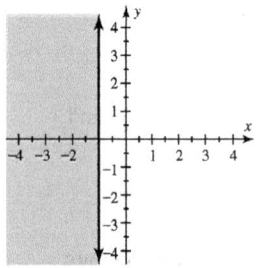

Figure 33

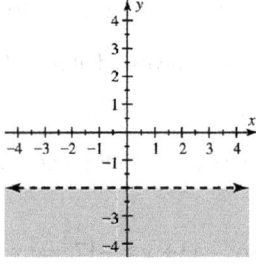

Figure 35

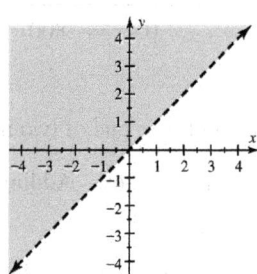

Figure 37

35. See Figure 35.

37. See Figure 37.

39. See Figure 39.

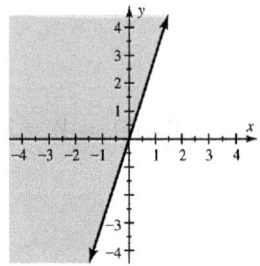

Figure 39

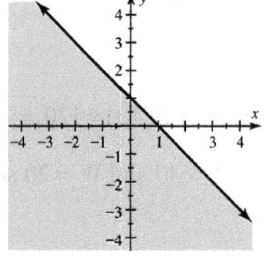

Figure 41

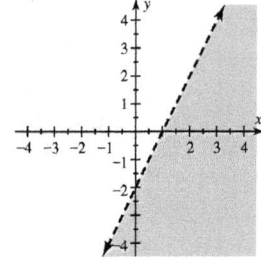

Figure 43

41. See Figure 41.

43. See Figure 43.

45. Yes, because $3 - 1 < 3 \Rightarrow 2 < 3$ is true and $3 + 1 > 3 \Rightarrow 4 > 3$ is true.

47. No, because $3(-2) - 2(3) \geq 1 \Rightarrow -6 - 6 \geq 1 \Rightarrow -12 \geq 1$ is false.

49. Yes, because $4 - 2(-2) \geq 8 \Rightarrow 4 + 4 \geq 8 \Rightarrow 8 \geq 8$ is true and $-2(4) - 5(-2) > 0 \Rightarrow -8 + 10 > 0 \Rightarrow$

$2 > 0$ is true.

51. The region containing $(1, 2)$, because $1 \leq 2$ is true and $1 + 2 \geq 2 \Rightarrow 3 \geq 2$ is true.

53. The region containing $(1, 0)$, because $1 + 0 \leq 3 \Rightarrow 1 \leq 3$ is true and $0 \leq 2(1) \Rightarrow 0 \leq 2$ is true.

55. See Figure 55.

57. See Figure 57.

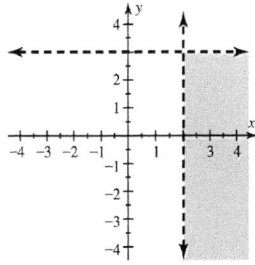

Figure 55

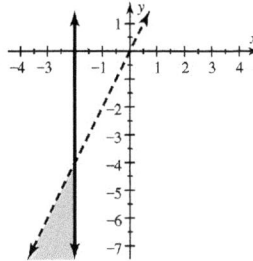

Figure 57

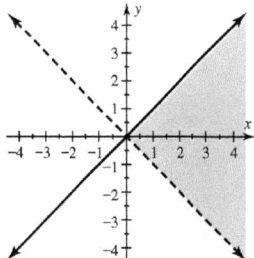

Figure 59

59. See Figure 59.

61. See Figure 61.

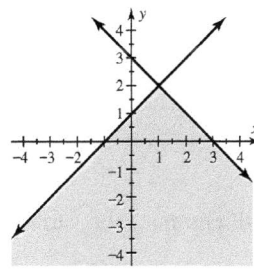

Figure 61

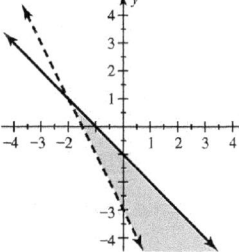

Figure 63

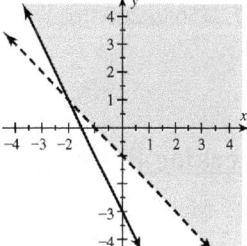

Figure 65

63. See Figure 63.

65. See Figure 65.

67. See Figure 67.

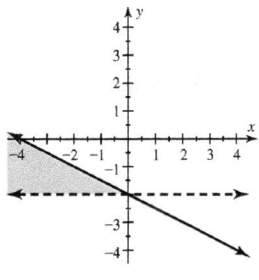

Figure 67

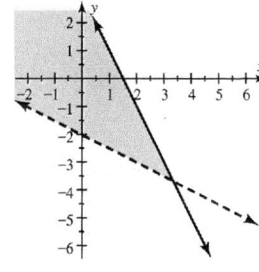

Figure 69

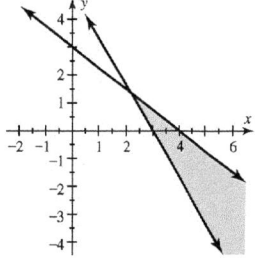

Figure 71

69. See Figure 69.

71. See Figure 71.

73. Graph the system, $M \geq 2V$ and $M + V < 90$. See Figure 73.

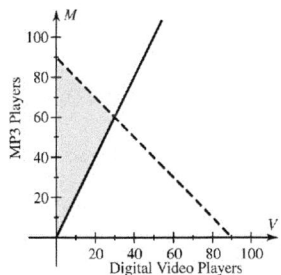

Figure 73

75. (a)    $R = 220 - 20 \Rightarrow R = 200$ bpm  for a 20 year old person and $R = 220 - 70 \Rightarrow R = 150$ bpm for

a 70 year old person.

(b)    See Figure 75.

(c)    The shaded region represents possible heart rates for ages 20 to 70.

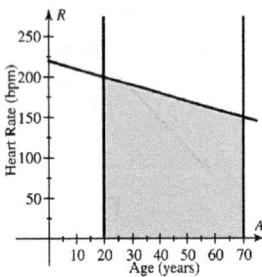

Figure 75

77. 150 to 200 lbs.

## Checking Basic Concepts  Sections 4.3 and 4.4

1. Multiplying the second equation by $-2$ and adding the two equations will eliminate the variable $x$.

$$2x + 3y = 5$$
$$\underline{-2x + 14y = 12}$$
$$17y = 17$$    Thus $y = 1$. So $x - 7(1) = -6 \Rightarrow x = 1$. The solution is $(1,\ 1)$.

2. (a)    Multiplying the first equation by $-1$ and adding the two equations will eliminate the variable $x$.

$$-x - y = 1$$
$$\underline{x - 2y = 2}$$
$$-3y = 3$$    Thus $y = -1$. So $x - 2(-1) = 2 \Rightarrow x + 2 = 2 \Rightarrow x = 0$. The solution is $(0,\ -1)$.

There is one solution.

(b)    Adding the two equations  will eliminate both variables.

$$5x - 6y = 4$$
$$\underline{-5x + 6y = 1}$$
$$0 = 5$$    This is false, therefore there are no solutions.

(c)    Multiplying the first equation by $-2$ and adding the two equations will eliminate both variables.

$$-2x + 6y = 0$$
$$\underline{2x - 6y = 0}$$
$$0 = 0$$    This is true, therefore there are infinitely many solutions.

3. Substituting $y = 2x$ into the first equation yields the following:  $-2x + (2x) = 0 \Rightarrow 0 = 0$. This is

true $\Rightarrow$ infinitely many solutions.  Graph the system. See Figure 3a. Lines coincide $\Rightarrow$ infinitely

many solutions. Numerically: See Figure 3b. Since for all values of $x$ both equations produce the

same solutions $\Rightarrow$ infinitely many solutions.

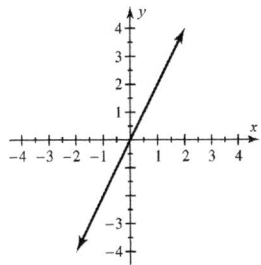

Figure 3a

| $x$ | 0 | 1 | 2 | 3 |
|---|---|---|---|---|
| $y = 2x$ | 0 | 2 | 4 | 6 |
| $-2x + y = 0$ | 0 | 2 | 4 | 6 |

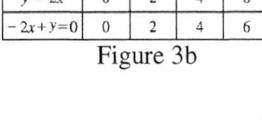

Figure 3b

4. (a) See Figure 4a.

   (b) See Figure 4b.

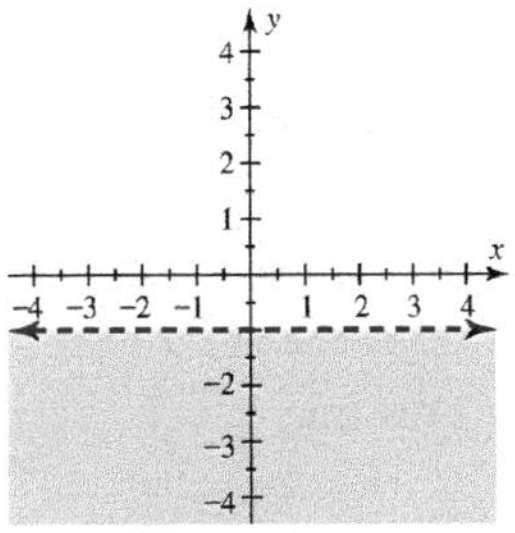

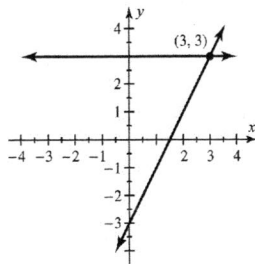

Figure 4a

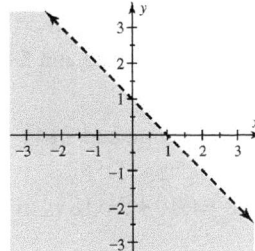

Figure 4b

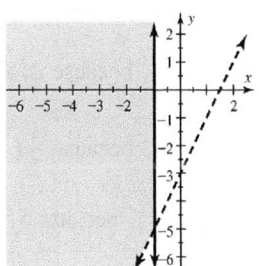

Figure 5

5. See Figure 5.
6. (a) $x + y = 11$

   $x = y + 5$ or $x - y = 5$.

   (b) Substituting $x = y + 5$ into the first equation yields the following: $(y+5) + y = 11 \Rightarrow 2y + 5 = 11 \Rightarrow 2y = 6 \Rightarrow y = 3$. And so $x + 3 = 11 \Rightarrow x = 8$. The solution is (8, 3) or New York had a population of 8 million people and Chicago had a population of 3 million people.

## Chapter 4 Review

1. Graph the system $y = 3$ and $y = 2x - 3$. See Figure 1. The intersection and solution is $(3, 3) \Rightarrow x = 3$.

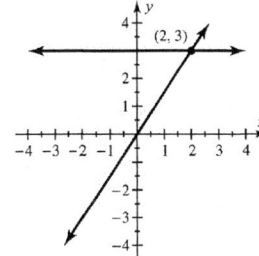

Figure 1

Figure 2

2. Graph the system $y = 3$ and $y = \frac{3}{2}x$. See Figure 2. The intersection and solution is $(2,\ 3) \Rightarrow x = 2$.

3. (a)   Parallel lines so no solutions.

    (b)   No solutions so inconsistent.

4. (a)   An intersection so one solution.

    (b)   One solution is consistent and independent.

5. (a)   Coinciding lines so infinitely many solutions.

    (b)   Infinitely many solutions is consistent and dependent.

6. (a)   Parallel lines so no solutions.

    (b)   No solutions is inconsistent.

7. $(1, 2)$, because $1 + 2(2) = 5 \Rightarrow 1 + 4 = 5$ is true and $1 - (2) = -1$ is true.

8. $(5, 2)$, because $2(5) - 2 = 8 \Rightarrow 10 - 2 = 8$ is true and $5 + 3(2) = 11 \Rightarrow 5 + 6 = 11$ is true.

9. $(4, 3)$, because $\dfrac{1}{2}(4) = 3 - 1 \Rightarrow 2 = 2$ is true and $2(4) = 3(3) - 1 \Rightarrow 8 = 9 - 1$ is true.

10. $(2, -4)$, because $5(2) - 2(-4) = 18 \Rightarrow 10 + 8 = 18$ is true and $-4 = -2(2)$ is true.

11. The intersection and solution is $(2, 2)$. Checking: $2 = 2$ is true and $-2(2) + 2 = -2 \Rightarrow -4 + 2 = -2$

    is true.

12. The intersection and solution is $(1, 2)$. Checking: $1 + 2 = 3$ is true and $2 = 2(1)$ is true.

13. Since when $x = 2$, $y = 6$ for both equations $\Rightarrow$ the solution is $(2, 6)$.

14. Since when $x = 1$, $y = 1$ for both equations $\Rightarrow$ the solution is $(1, 1)$.

15. Graph the system. See Figure 15. The intersection and solution is $(4, -3)$.

16. Graph the system. See Figure 16. The intersection and solution is $(1, 2)$.

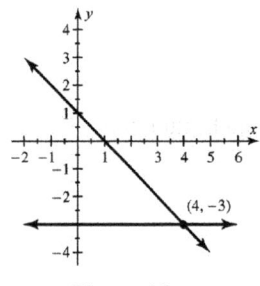   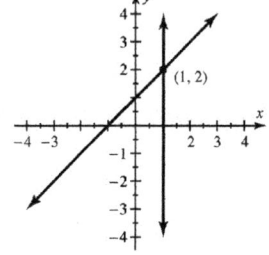   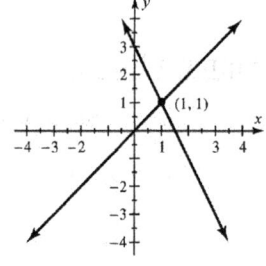

Figure 15                    Figure 16                    Figure 17

17. Graph the system. See Figure 17. The intersection and solution is $(1, 1)$.

18. Graph the system. See Figure 18. The intersection and solution is $(1, 2)$.

19. Graph the system. See Figure 19. The intersection and solution is $(1, 1)$.

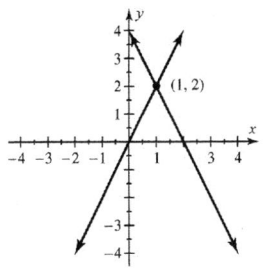

Figure 18

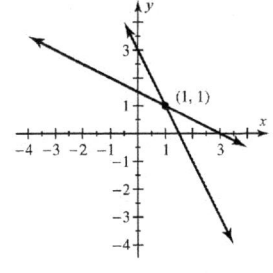

Figure 19

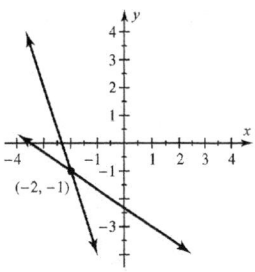

Figure 20

20. Graph the system. See Figure 20. The intersection and solution is $(-2, -1)$.

21. Substituting $y = 3x$ into the first equation yields the following: $x + (3x) = 8 \Rightarrow 4x = 8 \Rightarrow x = 2$,

    and so $y = 3(2) \Rightarrow y = 6$. The solution is $(2,\ 6)$.

22. Substituting $y = -5x$ into the first equation yields the following:

    $x - 2(-5x) = 22 \Rightarrow x + 10x = 22 \Rightarrow 11x = 22 \Rightarrow x = 2$, and so

    $y = -5(2) \Rightarrow y = -10$. The solution is $(2, -10)$.

23. Note that $x + 3y = 1 \Rightarrow x = -3y + 1$. Substituting $x = -3y + 1$ into the second equation yields the

    following: $-2(-3y + 1) + 2y = 6 \Rightarrow 6y - 2 + 2y = 6 \Rightarrow 8y - 2 = 6 \Rightarrow 8y = 8 \Rightarrow y = 1$, and so

    $x = -3(1) + 1 \Rightarrow x = -2$. The solution is $(-2, 1)$.

24. Note that $2x - y = -4 \Rightarrow -y = -2x - 4 \Rightarrow y = 2x + 4$. Substituting $y = 2x + 4$ into the first

    equation yields the following:

    $3x - 2(2x + 4) = -4 \Rightarrow 3x - 4x - 8 = -4 \Rightarrow -x - 8 = -4 \Rightarrow -x = 4 \Rightarrow x = -4$, and so

    $y = 2(-4) + 4 \Rightarrow y = -8 + 4 \Rightarrow y = -4$. The solution is $(-4, -4)$.

25. Substituting $y = -x$ into the first equation yields the following: $x + (-x) = 2 \Rightarrow 0 = 2$. This is not

    true $\Rightarrow$ no solutions.

26. Substituting $x + y = -2 \Rightarrow y = -x - 2$ into the second equation yields the following:

    $x + (-x - 2) = 3 \Rightarrow -2 = 3$. This is not true $\Rightarrow$ no solutions.

27. Note that $x - 2y = -2 \Rightarrow x = 2y - 2$. Substituting $x = 2y - 2$ into the first equation yields the

    following: $-(2y - 2) + 2y = 2 \Rightarrow -2y + 2 + 2y = 2 \Rightarrow 2 = 2$. This is true $\Rightarrow$ infinitely many

    solutions.

28. Note that $-x - y = -2 \Rightarrow -y = x - 2 \Rightarrow y = -x + 2$. Substituting $y = -x + 2$ into the second

    equation yields the following: $2x - (-x + 2) = 1 \Rightarrow 2x + x - 2 = 1 \Rightarrow 3x = 3 \Rightarrow x = 1$, and so

    $y = -1 + 2 \Rightarrow y = 1$. The solution is $(1,\ 1)$.

29. The lines intersect at the solution (2, 1). Adding the two equations will eliminate the variable $y$.

$$x + y = 3$$
$$\underline{x - y = 1}$$
$$2x = 4$$    Thus $x = 2$. So $2 + y = 3 \Rightarrow y = 1$. The solution is (2, 1).

30. The lines intersect at the solution $(-1,\ 2)$. Multiplying the second equation by $-2$ and adding the two equations will eliminate the variable $x$.

$$2x + 3y = 4$$
$$\underline{-2x + 4y = 10}$$
$$7y = 14$$

   Thus $y = 2$. So $2x + 3(2) = 4 \Rightarrow 2x + 6 = 4 \Rightarrow 2x = -2 \Rightarrow x = -1$. The solution is $(-1,\ 2)$.

31. Adding the two equations will eliminate the variable $y$.

$$x + y = 10$$
$$\underline{x - y = 12}$$
$$2x = 22$$    Thus $x = 11$. So $11 + y = 10 \Rightarrow y = -1$. The solution is $(11, -1)$.

32. Adding the two equations will eliminate the variable $y$.

$$2x - y = 2$$
$$\underline{3x + y = 3}$$
$$5x = 5$$    Thus $x = 1$. So $2(1) - y = 2 \Rightarrow 2 - y = 2 \Rightarrow -y = 0 \Rightarrow y = 0$. The solution is (1, 0).

33. Multiplying the second equation by 2 and adding the two equations will eliminate the variable $x$.

$$-2x + 2y = -1$$
$$\underline{2x - 6y = -6}$$
$$-4y = -7$$    Thus $y = \frac{7}{4}$. So $-2x + 2\left(\frac{7}{4}\right) = -1 \Rightarrow -2x + \frac{14}{4} = -1 \Rightarrow -2x = -\frac{18}{4} \Rightarrow x = \frac{9}{4}$.

   The solution is $\left(\frac{9}{4}, \frac{7}{4}\right)$.

34. Multiplying the first equation by $-1$ and adding the two equations will eliminate the variable $x$.

$$-2x + 5y = 0$$
$$\underline{2x + 4y = 9}$$
$$9y = 9$$    Thus $y = 1$. So $2x + 4(1) = 9 \Rightarrow 2x + 4 = 9 \Rightarrow 2x = 5 \Rightarrow x = \frac{5}{2}$.

   The solution is $\left(\frac{5}{2}, 1\right)$.

35. Multiplying the first equation by 2 and adding the two equations will eliminate the variable $b$.

$$4a + 2b = 6$$
$$\underline{-3a - 2b = -1}$$
$$a = 5$$    And so $4(5) + 2b = 6 \Rightarrow 20 + 2b = 6 \Rightarrow 2b = -14 \Rightarrow b = -7$. The solution is $(5, -7)$.

36. Multiplying the first equation by $-3$ and adding the two equations will eliminate the variable $a$.

$$-3a + 9b = -6$$
$$\underline{3a + b = 26}$$
$$10b = 20 \qquad \text{Thus } b = 2. \text{ And so } 3a + 2 = 26 \Rightarrow 3a = 24 \Rightarrow a = 8. \text{ The solution is } (8,\ 2).$$

37. Multiplying the first equation by 2, the second equation by 5 and adding the two equations will eliminate the variable $r$.

$$10r + 6t = -2$$
$$\underline{-10r - 25t = -55}$$
$$-19t = -57 \qquad \text{Thus } t = 3. \text{ So } 10r + 6(3) = -2 \Rightarrow 10r + 18 = -2 \Rightarrow 10r = -20 \Rightarrow r = -2.$$

The solution is $(-2,\ 3)$.

38. Multiplying the first equation by 7, the second equation by 2 and adding the two equations will eliminate the variable $t$.

$$35r + 14t = 35$$
$$\underline{6r - 14t = 6}$$
$$41r = 41 \qquad \text{Thus } r = 1. \text{ So } 6(1) - 14t = 6 \Rightarrow 6 - 14t = 6 \Rightarrow -14t = 0 \Rightarrow t = 0.$$

The solution is $(1,\ 0)$.

39. (a)   Adding the two equations will eliminate the variable $y$.

$$3x + y = 6$$
$$\underline{x - y = -2}$$
$$4x = 4 \qquad \text{Thus } x = 1. \text{ So } 1 - y = -2 \Rightarrow -y = -3 \Rightarrow y = 3. \text{ The solution is } (1, 3).$$

(b)   See Figure 39b. The intersection and solution is $(1, 3)$.

(c)   See Figure 39c. Since when $x = 1$, $y = 3$ for both equations the solution is $(1, 3)$.

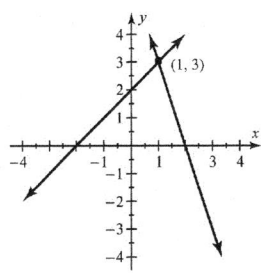

| $x$ | $-1$ | 0 | 1 | 2 |
|---|---|---|---|---|
| $y = 6 - 3x$ | 9 | 6 | 3 | 0 |
| $y = x + 2$ | 1 | 2 | 3 | 4 |

Figure 39b                                              Figure 39c

40. (a)   Multiplying the second equation by 2 and adding the two equations will eliminate the variable $x$.

$$2x + y = 3$$
$$\underline{-2x + 4y = -8}$$
$$5y = -5 \qquad \text{Thus } y = -1. \text{ So } 2x - 1 = 3 \Rightarrow 2x = 4 \Rightarrow x = 2. \text{ The solution is } (2, -1).$$

(b)   See Figure 40b. The intersection and solution is $(2, -1)$.

(c)   See Figure 40c. Since when $x = 2$, $y = -1$ for both equations the solution is $(2, -1)$.

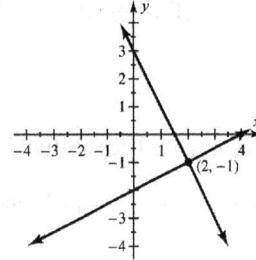

Figure 40b

| $x$ | $-2$ | $0$ | $2$ | $4$ |
|---|---|---|---|---|
| $y = 3 - 2x$ | 7 | 3 | $-1$ | $-5$ |
| $y = (x - 4)/2$ | $-3$ | $-2$ | $-1$ | 0 |

Figure 40

41. Adding the two equations will eliminate both variables.

$$x - y = 5$$
$$\underline{-x + y = -5}$$
$$0 = 0 \quad \text{Since this is true, there are infinitely many solutions.}$$

42. Multiplying the second equation by 3 and adding the two equations will eliminate both variables.

$$3x - 3y = 0$$
$$\underline{-3x + 3y = 0}$$
$$0 = 0 \quad \text{Since this is true, there are infinitely many solutions.}$$

43. Adding the two equations will eliminate both variables.

$$-2x + y = 3$$
$$\underline{2x - y = 3}$$
$$0 = 6 \quad \text{Since this is false, there are no solutions.}$$

44. Adding the two equations will eliminate the variable $y$.

$$-2x + y = 2$$
$$\underline{3x - y = 3}$$
$$x = 5$$

So $3(5) - y = 3 \Rightarrow 15 - y = 3 \Rightarrow -y = -12 \Rightarrow y = 12$. The solution is $(5, 12)$,

therefore there is one solution.

45. Yes, because $-3 \leq 2$ is true.

46. No, because $-1 > -1$ is false.

47. No, because $1 + 2 < -2$ is false.

48. Yes, because $2(1) - 3(-4) \geq 2 \Rightarrow 2 + 12 \geq 2$ is true.

49. $y > 1$

50. $y \leq 2x + 1$

51. See Figure 51.

52. See Figure 52.

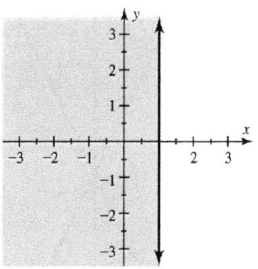

Figure 51

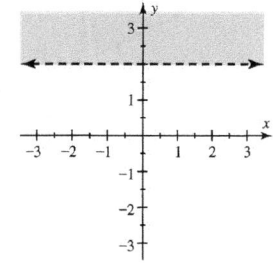

Figure 52

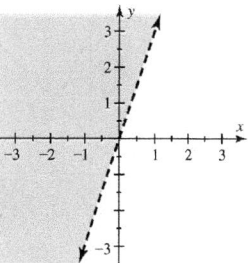

Figure 53

53. See Figure 53.

54. See Figure 54.

55. See Figure 55.

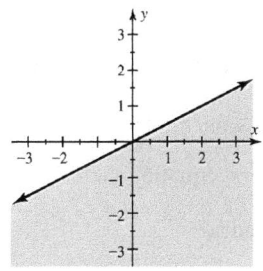

Figure 54

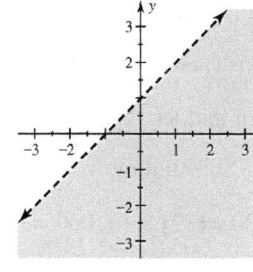

Figure 55

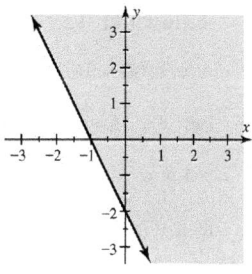

Figure 56

56. See Figure 56.

57. Yes, because $1-2(-2)>3 \Rightarrow 1+4>3$ is true and $2(1)+(-2)<3 \Rightarrow 0<3$ is true.

58. No, because $4-(-3)\geq 1 \Rightarrow 4+3\geq 1$ is true but $4(4)+3(-3)\leq 4 \Rightarrow 16-9\leq 4$ is false.

59. The region containing $(2,-2)$, because $-2\leq 1$ is true and $2(2)+(-2)\geq -1 \Rightarrow 4-2\geq -1$ is true.

60. The region containing $(1,\ 3)$, because $3\geq 1$ is true and $1+3\geq 2$ is true.

61. See Figure 61.

62. See Figure 62.

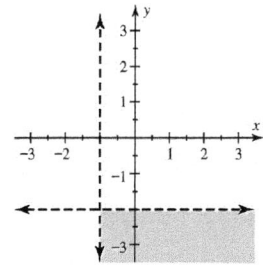

Figure 61

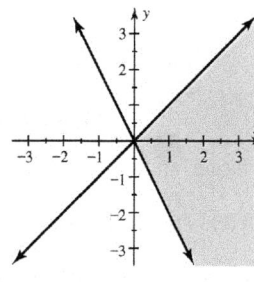

Figure 62

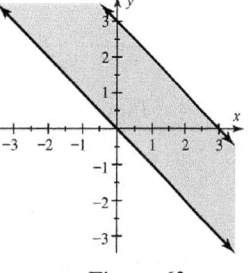

Figure 63

63. See Figure 63.

64. See Figure 64.

65. See Figure 65.

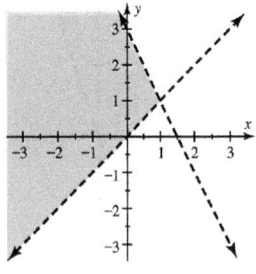

Figure 64

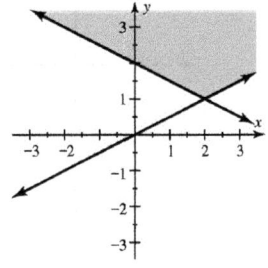

Figure 65

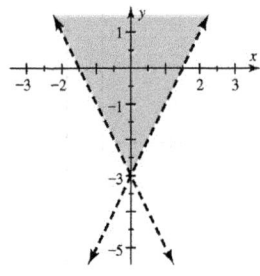

Figure 66

66. See Figure 66.

67. Let $x$ = number of traffic fatalities in 1912 and let $y$ = number of traffic fatalities in 2008. Then
$12x = y$ and $y = x + 34,100$. Substituting $12x = y$ into the second equation yields the
following: $12x = x + 34,100 \Rightarrow 11x = 34,100 \Rightarrow x = 3100$, and so
$y = 3100 + 34,100 \Rightarrow y = 37,200$. There were 3100 deaths in 1912 and 37,200 deaths in 2008.

68. Let $x$ = cases of lymphoma in men and let $y$ = cases of lymphoma in women. Then
$x + y = 74,000$ and $x = y + 6000$. Substituting $x = y + 6000$ into the first equation yields the
following: $(y + 6000) + y = 74,000 \Rightarrow 2y = 68,000 \Rightarrow y = 34,000$, and so
$x = 34,000 + 6000 \Rightarrow x = 40,000$. Therefore men will have 40,000 cases of lymphoma
reported and women will have 34,000 cases of lymphoma reported.

69. (a)  $C = 0.2x + 40$

    (b)  See Figure 69b. The intersection and solution is (250, 90) or 250 miles.

    (c)  See Figure 69c. Since when $x = 250$, $C = 90$ the solution is $(250,\ 90) \Rightarrow 250$ miles.

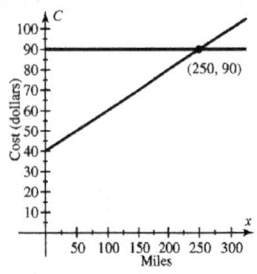

Figure 69b

| $x$ | 150 | 200 | 250 | 300 |
|---|---|---|---|---|
| $C = 0.2x + 40$ | 70 | 80 | 90 | 100 |

Figure 69c

70. Let $x$ = smaller angle and let $y$ = larger angle. Then $x + y = 180$ and $x = y - 30$. Substituting
$x = y - 30$ into the first equation yields the following: $(y - 30) + y = 180 \Rightarrow 2y = 210 \Rightarrow$
$y = 105$, and so $x = 105 - 30 \Rightarrow x = 75$. The angles are 75° and 105°.

71. (a)  $2x + y = 180$ and $2x = y + 40$

    (b)  Note that $2x + y = 180 \Rightarrow y = -2x + 180$. Substituting $y = -2x + 180$ into the second
        equation yields the following: $2x = (-2x + 180) + 40 \Rightarrow 2x = -2x + 220 \Rightarrow$
        $4x = 220 \Rightarrow x = 55$, and so $y = -2(55) + 180 \Rightarrow y = -110 + 180 \Rightarrow y = 70$.

        The solution is $(55,\ 70)$ or two angles at 55° and one at 70°.

(c)    Note that $2x = y + 40 \Rightarrow 2x - y = 40$. Adding the two equations will eliminate the variable $y$.

$2x + y = 180$

$\underline{2x - y = 40}$

$4x = 220$    Thus $x = 55$. So $2(55) = y + 40 \Rightarrow 110 = y + 40 \Rightarrow y = 70$. The solution is

(55, 70) or two angles at $55°$ and one angle at $70°$.

72. Let $x =$ garden's width and let $y =$ garden's length. Then $2x + 2y = 88$ and $y = x + 4$. Substituting

$y = x + 4$ into the first equation yields the following:

$2x + 2(x + 4) = 88 \Rightarrow 2x + 2x + 8 = 88 \Rightarrow 4x + 8 = 88 \Rightarrow 4x = 80 \Rightarrow x = 20$, and

so $y = 20 + 4 \Rightarrow y = 24$. The solution is $(20, 24)$.

The dimensions of the garden are $20$ feet $\times$ $24$ feet.

73. (a)    Let $x =$ number of $80 rooms and let $y =$ number of $120 rooms. Then $x + y = 10$ and

$80x + 120y = 920$.

(b)    Multiplying the first equation by $-80$ and adding the two equations will eliminate the

variable $x$.

$-80x - 80y = -800$

$\underline{80x + 120y = 920}$

$40y = 120$    Thus $y = 3$. So $x + 3 = 10 \Rightarrow x = 7$. The solution is $(7, 3)$ or 7 $80

rooms and 3 $120 rooms.

74. Let $x =$ pounds of $2 candy and let $y =$ pounds of $3 candy. Then $x + y = 18$

and $2x + 3y = 47$. Multiplying the first equation by $-2$ and adding the two equations will eliminate

the variable $x$.

$-2x - 2y = -36$

$\underline{2x + 3y = 47}$

$y = 11$    So $x + 11 = 18 \Rightarrow x = 7$. The solution is $(7, 11)$ or 7 pounds of $2 candy and

11 pounds of $3 candy.

75. Let $x =$ minutes on the stationary bike and let $y =$ minutes on the stair climber. Then $x + y = 60$

and $9x + 11y = 590$. Multiplying the first equation by $-9$ and adding the two equations will

eliminate the variable $x$.

$-9x - 9y = -540$

$\underline{9x + 11y = 590}$

$2y = 50$    Thus $y = 25$. So $x + 25 = 60 \Rightarrow x = 35$. The solution is $(35, 25)$ or 35

minutes on the bike and 25 minutes on the stair climber.

76. Let $x =$ speed of the boat and let $y =$ speed of the current. Then $10x + 10y = 140$ and

$14x - 14y = 140$. Multiplying the first equation by 7, the second by 5 and adding the two equations

will eliminate the variable $y$.

$$70x + 70y = 980$$
$$\underline{70x - 70y = 700}$$
$$140x = 1680$$

Thus

$x = 12$. So $10(12) + 10y = 140 \Rightarrow 120 + 10y = 140 \Rightarrow 10y = 20 \Rightarrow y = 2$.

The solution is $(12,\ 2)$. The current is 2 mph.

77. (a) The intersection and solution is approximately $16$ feet $\times 24$ feet.

(b) Using the system $2L + 2W = 80$ and $L = \dfrac{3}{2}W$ and substituting $L = \dfrac{3}{2}W$ into the first equation

yields the following: $2\left(\dfrac{3}{2}W\right) + 2W = 80 \Rightarrow 3W + 2W = 80 \Rightarrow 5W = 80 \Rightarrow W = 16$, and so

$L = \dfrac{3}{2}(16) \Rightarrow L = 24$. The dimensions are $16$ feet $\times 24$ feet.

78. Let $W =$ number of wheels made and let $T =$ number of trailers made. Then

$W + T \le 30$ and $W \ge 2T$. Graph this system and shade. See Figure 78.

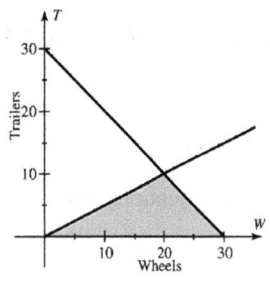

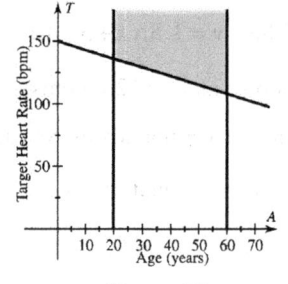

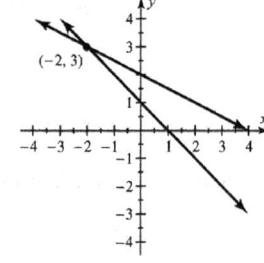

Figure 78                    Figure 79                    Figure 4

79. (a) $T = 150 - 0.7(20) \Rightarrow T = 150 - 14 \Rightarrow T = 136$ bpm for a 20 year old person and

$T = 150 - 0.7(60) \Rightarrow T = 150 - 42 \Rightarrow T = 108$ bpm for a 60 year old person.

(b) See Figure 79.

(c) The shaded region represents target heart rates above 70% of the maximum heart rate for ages 20 to 60.

## Chapter 4 Test

1. $(1,\ 2)$, because $3(1) + 2(2) = 7 \Rightarrow 3 + 4 = 7 \Rightarrow$ is true and $2(1) - 2 = 0$ is true.

2. The intersection and solution is $(-2, -1)$. Checking: $-2 + 4(-1) = -6 \Rightarrow -2 + (-4) = -6$ is true and

$2(-2) + (-1) = -5 \Rightarrow -4 + (-1) = -5$.

3. Since when $x = -1$, $y = -2$ for both equations, the solution is $(-1, -2)$.

4. See Figure 4. The intersection and solution is $(-2, 3)$.

5. Substituting $y = 3x$ into the first equation yields the following:

   $3x + 2(3x) = 9 \Rightarrow 3x + 6x = 9 \Rightarrow 9x = 9 \Rightarrow x = 1$, and so $y = 3(1) \Rightarrow y = 3$.

   The solution is $(1, 3)$.

6. (a) Note that $x + 3y = 5 \Rightarrow x = -3y + 5$. Substituting $x = -3y + 5$ into the second equation yields

   the following:

   $3(-3y + 5) - 2y = 4 \Rightarrow -9y + 15 - 2y = 4 \Rightarrow -11y + 15 = 4 \Rightarrow -11y = -11 \Rightarrow y = 1$, and so

   $x = -3(1) + 5 \Rightarrow x = -3 + 5 \Rightarrow x = 2$. The solution is $(2, 1) \Rightarrow$ one solution $\Rightarrow$ consistent.

   (b) Note that $2x - y = -4 \Rightarrow -y = -2x - 4 \Rightarrow y = 2x + 4$. Substituting $y = 2x + 4$ into the first

   equation yields the following:

   $-x + \frac{1}{2}(2x + 4) = 12 \Rightarrow -x + x + 2 = 12 \Rightarrow 2 = 12$. This is false $\Rightarrow$ no solutions $\Rightarrow$ inconsistent.

7. There is only one line, which indicates that the graphs are identical, or coincide.

   (a) There are infinitely many solutions.

   (b) The system is consistent and the equations are dependent.

8. The lines are parallel.

   (a) There are no solutions.

   (b) The system is inconsistent.

9. Adding the two equations will eliminate the variable $y$.

   $\begin{aligned} x + 2y &= 5 \\ 3x - 2y &= -17 \\ \hline 4x &= -12 \end{aligned}$  Thus $x = -3$. So $-3 + 2y = 5 \Rightarrow 2y = 8 \Rightarrow y = 4$. The solution is $(-3, 4)$.

10. Multiplying the second equation by 2 and adding the two equations will eliminate both variables.

    $\begin{aligned} 2x - 2y &= 3 \\ -2x + 2y &= 10 \\ \hline 0 &= 13 \end{aligned}$  This is false $\Rightarrow$ no solutions.

11. Multiply the first equation by 3 and then add.

    $\begin{aligned} 3x - 6y &= 9 \\ -3x + 6y &= -9 \\ \hline 0 &= 0 \end{aligned}$

    The equation $0 = 0$ is always true, which indicates that the system has infinitely many solutions.

12. Multiply the first equation by 2, the second equation by 3, and then add.

$$8x + 6y = 10$$
$$\underline{9x - 6y = -27}$$
$$17x \phantom{+6y} = -17, \quad \text{or} \quad x = -1$$

Find $y$ by substituting $-1$ for $x$ in the first given equation, $4x + 3y = 5$.

$$4(-1) + 3y = 5 \Rightarrow -4 + 3y = 5 \Rightarrow -4 + 4 + 3y = 5 + 4 \Rightarrow 3y = 9 \Rightarrow \frac{3y}{3} = \frac{9}{3} \Rightarrow y = 3$$

The solution is $(-1, 3)$.

13. $y \le -\dfrac{1}{2}x$

14. Point $(4, -3)$ is a solution because $2(4) + (-3) = 8 + (-3) = 5 > 3$ and $4 - (-3) = 7 \ge 7$.

15. See Figure 15.

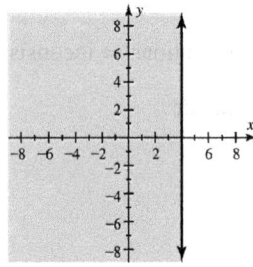

Figure 15               Figure 16

16. See Figure 16.

17. Substitute the test point $(0, 0)$ into both inequalities as follows:

$(0) + (0) \le 5 \Rightarrow 0 \le 5$. This is a true statement and we will shade below the blue line to include the

point $(0, 0)$.

$(0) - (0) \ge 1 \Rightarrow 0 \ge 1$. This is a false statement and we will shade below the red line to exclude the

point $(0, 0)$. Therefore, the region that will include both shaded areas will be area 4.

18. See Figure 18.

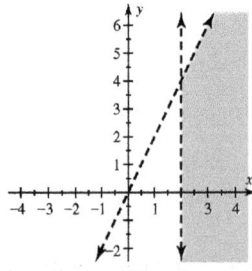

Figure 18               Figure 19

19. See Figure 19.

20. Let $x$ = taxes collected in 2009 and let $y$ = taxes collected in 2010. Then

$x + y = 8$ and $y = x + 0.6$. Substituting $y = x + 0.6$ into the first equation yields the following:

$x + (x + 0.6) = 8 \Rightarrow 2x + 0.6 = 8 \Rightarrow 2x = 7.4 \Rightarrow x = 3.7$, and $y = 3.7 + 0.6 \Rightarrow y = 4.3$. The

solution is (3.7, 4.3) or \$3.7 trillion collected in 2009 and \$4.3 trillion collected in 2010.

21. Let $x$ = liters of 20% solution and let $y$ = liters of 60% solution. Then $x + y = 10 \Rightarrow y = 10 - x$

and $0.2x + 0.6y = 0.3(10)$. Substituting $y = 10 - x$ into the second equation yields the following:

$0.2x + 0.6(10 - x) = 3 \Rightarrow 0.2x + 6 - 0.6x = 3 \Rightarrow -0.4x = -3 \Rightarrow x = 7.5$, and $y = 10 - 7.5 \Rightarrow$

$y = 2.5$  The solution is (7.5, 2.5) or 7.5 L of 20% solution and 2.5 L of 60% solution.

22. Let $x$ = hrs jogged at 6 mph and let $y$ = hrs jogged at 9 mph. Then $x + y = 1$ and

$6x + 9y = 7$. Multiplying the first equation by $-6$ and adding the two equations will eliminate the

variable $x$.

$$\begin{array}{r} -6x - 6y = -6 \\ 6x + 9y = 7 \\ \hline 3y = 1 \end{array}$$

Thus $y = \dfrac{1}{3}$. So $x + \dfrac{1}{3} = 1 \Rightarrow x = \dfrac{2}{3}$. The solution is

$\left( \dfrac{2}{3}, \dfrac{1}{3} \right)$ or $\dfrac{2}{3}$ hour at 6 miles per hour and $\dfrac{1}{3}$ hour at 9 miles per hour.

## Chapters 4 Extended and Discovery Exercises

1  The forest has higher precipitation for higher temperatures than grasslands $\Rightarrow 7P - 5T \geq -70$.

2. The deserts have less precipitation for high temperatures than grasslands $\Rightarrow 35P - 3T \leq 140$.

3. The grassland is lower precipitation for higher temperatures than forests and higher than desert

$\Rightarrow 7P - 5T \leq -70$ and $35P - 3T \geq 140$.

4. From the graph, grasslands would be the type of plant growth, and

$7(14) - 5(50) \leq -70 \Rightarrow 98 - 250 \leq -70$  is true and

$35(14) - 3(50) \geq 140 \Rightarrow 490 - 150 \geq 140$ is true $\Rightarrow$ yes.

5. $-0.5$; average the $x$-values $-1$ and $0$ because the solution $y$-value $0$ is half way between their

corresponding $y$-values $-1$ and $1$.

6. $0.5$; average the $x$-values $0$ and $1$ because the solution $y$-value $5$ is half way between their

corresponding $y$-values $3$ and $7$.

7. $1.5$; average the $x$-values $1$ and $2$ because the solution $y$-value $3.75$ is half way between their

corresponding $y$-values $3.5$ and $4$.

8. $\frac{1}{3}$; because 0 is one-third of the way between the $y$-values $-1$ and $2$, you must choose a value one third of the way between the corresponding $x$-values 0 and 1.

## Chapters 1-4  Cumulative Review Exercises

1. $120 = 2 \cdot 2 \cdot 2 \cdot 3 \cdot 5$

2. (a)  $2^3 \div \frac{5+7}{9-3} = 8 \div \frac{12}{6} = 8 \div 2 = 4$

   (b)  $-\frac{2}{5} \cdot (5-25) = -\frac{2}{5} \cdot (-20) = \frac{-2}{5} \cdot \frac{-20}{1} = \frac{(-2)(-20)}{(5)(1)} = \frac{40}{5} = 8$

3. (a)  $-6.9$ is a rational number.

   (b)  $\sqrt{14}$ is an irrational number.

4. (a)  $|-5| = 5$ and $-5 < 5$ so $-5 < |-5|$.

   (b)  $|7| = 7$ and $|-1| = 1$ and $7 > 1$ so $|7| > |-1|$.

5. (a)  $5x^2 - x^2 = (5-1)x^2 = 4x^2$

   (b)  $3 - 2x + 7x - 5 = -2x + 7x + 3 - 5 = (-2 + 7)x + (3 + -5) = 5x - 2$

6. $5(2x+1) = 7+x \Rightarrow 10x+5 = 7+x \Rightarrow 10x - x + 5 = 7 + x - x$

   $\Rightarrow 9x + 5 = 7 \Rightarrow 9x + 5 - 5 = 7 - 5 \Rightarrow 9x = 2 \Rightarrow \frac{9x}{9} = \frac{2}{9} \Rightarrow x = \frac{2}{9}$

7. $1 - (x+1) = x - 1 \Rightarrow 1 - x - 1 = x - 1 \Rightarrow -x = x - 1 \Rightarrow -x - x = x - x - 1 \Rightarrow -2x = -1$

   $\Rightarrow \frac{-2x}{-2} = \frac{-1}{-2} \Rightarrow x = \frac{1}{2}$

8. First, solve for $x$: $2(5x+1) = 10x - 3 \Rightarrow 10x + 2 = 10x - 3 \Rightarrow 10x - 10x + 2 = 10x - 10x - 3 \Rightarrow 2 = -3$.
   Because this statement is not true, the equation has no solutions

9. Let $x$ represent the smallest integer.

   $x + (x+1) + (x+2) + (x+3) = 50 \Rightarrow 4x + 6 = 50 \Rightarrow 4x + 6 - 6 = 50 - 6 \Rightarrow 4x = 44 \Rightarrow$

   $\frac{4x}{4} = \frac{44}{4} \Rightarrow x = 11$  Thus, the numbers are 11, 12, 13, and 14.

10. Convert feet to inches to keep the units consistent. Thus, 1 foot = 1 foot $\cdot \dfrac{12 \text{ inches}}{1 \text{ foot}} = 12$ inches.

   Then, $A = LW = 36 \cdot 12 = 432 \text{ in}^2$. Or convert inches to feet to obtain 36 inches $\cdot \dfrac{1 \text{ foot}}{12 \text{ inches}} = 3$ feet.

   Then, $A = LW = 3 \cdot 1 = 3 \text{ ft}^2$.

11. The formula is given as $W = 3x - 7y$. To solve for $x$, proceed as follows:

$$W = 3x - 7y \Rightarrow W + 7y = 3x \Rightarrow \frac{W+7y}{3} = \frac{3x}{3} \Rightarrow x = \frac{W+7y}{3}$$

12. $3 - (2x - 7) \le 8x \Rightarrow 3 - 2x + 7 \le 8x \Rightarrow 10 - 2x \le 8x \Rightarrow 10 - 2x + 2x \le 8x + 2x$

$$\Rightarrow 10 \le 10x \Rightarrow \frac{10}{10} \le \frac{10x}{10} \Rightarrow 1 \le x \Rightarrow x \ge 1$$

13. See Figure 13.

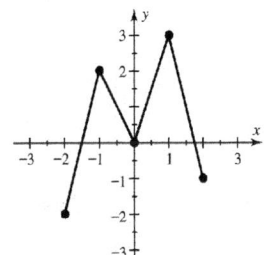

Figure 13

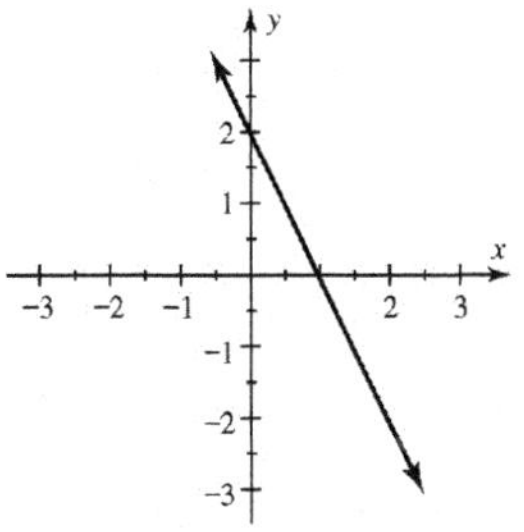

Figure 14a

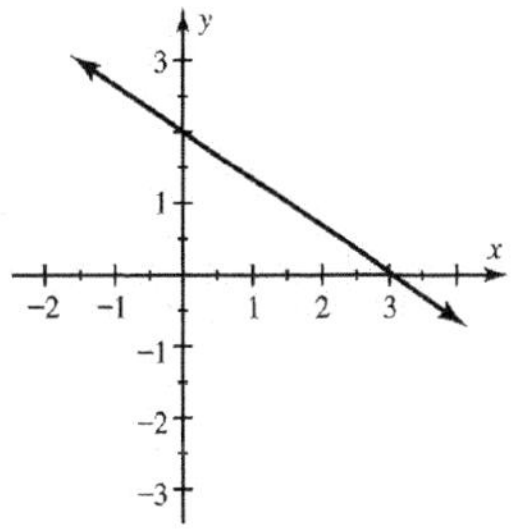

Figure 14b

14. (a)    See Figure 14a.

    (b)    See Figure 14b.

15. To find the $x$-intercept, set $y = 0$ and solve for $x$:

$$4x - y = 8 \Rightarrow 4x - 0 = 8 \Rightarrow 4x = 8 \Rightarrow \frac{4x}{4} = \frac{8}{4} \Rightarrow x = 2. \text{ Therefore, the } x\text{-intercept is at } x = 2.$$

To find the $y$-intercept, set $x = 0$ and solve for $y$:

$$4x - y = 8 \Rightarrow 4(0) - y = 8 \Rightarrow 0 - y = 8 \Rightarrow -y = 8 \Rightarrow y = -8. \text{ Therefore, the } y\text{-intercept is at } y = -8.$$

16. (a)    See Figure 16a.

    (b)    See Figure 16b.

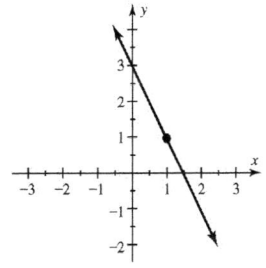

Figure 16a

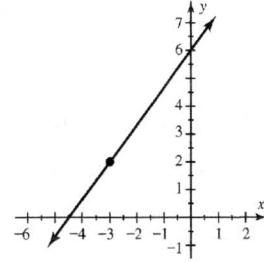

Figure 16b

17. (a)    For every increase of 1 in the run, the rise decreases by $\frac{1}{2}$. Therefore, the slope $m$ is $-\frac{1}{2}$.

When $x = 0$, $y = -1$. Therefore, the $y$-intercept is $-1$. Thus, $y = mx + b \Rightarrow y = -\frac{1}{2}x - 1$.

    (b)    For every increase of 1 in the run, the rise increases by 4. Therefore, the slope $m$ is 4. When

$x = 0$, $y = -3$. Therefore, the $y$-intercept is $-3$. Thus, $y = mx + b \Rightarrow y = 4x - 3$.

18. Find the slope: $m = \dfrac{y_2 - y_1}{x_2 - x_1} = \dfrac{4-2}{-4-4} = \dfrac{2}{-8} = -\dfrac{1}{4}$. Then, insert the values of the point (4, 2) and the

    value of the slope into the point-slope form:

$$y - y_1 = m(x - x_1) \Rightarrow y - 2 = -\frac{1}{4}(x-4) \Rightarrow y - 2 = -\frac{1}{4}x + 1 \Rightarrow y - 2 + 2 = -\frac{1}{4}x + 1 + 2 \Rightarrow y = -\frac{1}{4}x + 3$$

19. Solve the given equation for $y$:

$$2x - 6y = 7 \Rightarrow -2x + 2x - 6y = -2x + 7 \Rightarrow -6y = -2x + 7 \Rightarrow \frac{-6y}{-6} = \frac{-2x}{-6} + \frac{7}{-6} \Rightarrow y = \frac{1}{3}x - \frac{7}{6}.$$ The

    slope of this line is $\dfrac{1}{3}$, so the slope of a line perpendicular to it is $-3$. Insert the values of the point

    (1, 1) and the value of the slope into the point-slope form:

$$y - y_1 = m(x - x_1) \Rightarrow y - 1 = -3(x - 1) \Rightarrow y - 1 = -3x + 3 \Rightarrow y - 1 + 1 = -3x + 3 + 1 \Rightarrow y = -3x + 4.$$

20. (4, 4), because $3(4) - 4 = 8 \Rightarrow 12 - 4 = 8$ is true and $2(4) + 4 = 12 \Rightarrow 8 + 4 = 12$ is true.

21. Multiplying the second equation by $-1$ and adding the two equations will eliminate the variable $y$.

$$\begin{array}{r} x - y = 4 \\ 2x + y = -1 \\ \hline 3x = 3 \end{array}$$     Thus $x = 1$. So $1 - y = 4 \Rightarrow y = -3$.

         The solution is $(1, -3)$.

22. Multiplying the first equation by 2 and adding the two equations will eliminate both variables.

$$\begin{array}{r} 4x + 6y = 8 \\ -4x - 6y = 7 \\ \hline 0 = 15 \end{array}$$     Since this is false, there are no solutions.

23. Multiplying the first equation by 5 and adding the two equations will eliminate both variables.

$$\begin{array}{r} 15x - 20y = 40 \\ -15x + 20y = -40 \\ \hline 0 = 0 \end{array}$$     Since this is true, there are infinitely many solutions.

24. Multiplying the first equation by 3, the second equation by 2, and adding the two equations will

    eliminate the variable $y$.

$$\begin{array}{r} 21x + 6y = -9 \\ -10x - 6y = -2 \\ \hline 11x \qquad\;\; = -11 \end{array}$$

         Thus $x = -1$. So $7(-1) + 2y = -3 \Rightarrow -7 + 2y = -3 \Rightarrow 2y = 4 \Rightarrow y = 2$. The

         solution is $(-1, 2)$.

25. See Figure 25.

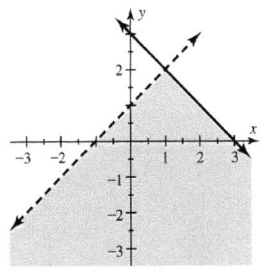

Figure 25

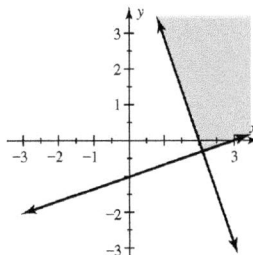

Figure 26

26. See Figure 26.

27. $F = 16.5R$

28. The temperature change is the difference between the two temperatures: $83 - (-11) = 83 + 11 = 94°C$.

29. The sales tax \$17.15 is 7% of the cost of the camera, $c$. $17.15 = 0.07c \Rightarrow \dfrac{17.15}{0.07} = \dfrac{0.07c}{0.07} \Rightarrow 245 = c$

   The camera cost \$245.

30. The increase will be 9% of \$145 per credit, and $0.09(145) = 13.05$. Thus, tuition is increasing by

   \$13.05 per credit so the new tuition will be $\$145 + \$13.05 = \$158.05$ per credit.

31. The truck uses 3 gallons per hour, so the slope $m$ is –3. The initial amount is 30 gallons, so the

   $y$-intercept is $b = 30$. Thus $y = mx + b \Rightarrow G = -3x + 30$.

32. Let $x$ represent the amount borrowed at 5%. Then the amount borrowed at 6% is $2400 - x$. The total

   interest for one year will equal the sum of the interest for each loan. Therefore:

   $0.05x + 0.06(2400 - x) = 132 \Rightarrow 0.05x + 144 - 0.06x = 132 \Rightarrow -0.01x + 144 = 132$

   $\Rightarrow -0.01x + 144 - 144 = 132 - 144 \Rightarrow -0.01x = -12 \dfrac{-0.01x}{-0.01} = \dfrac{-12}{-0.01} \Rightarrow x = 1200,\ 2400 - x = 1200.$

   Therefore, the amount borrowed at 5% interest is \$1200 and the amount borrowed at 6% interest is

   \$1200.

# Chapter 5  Polynomials and Exponents

## Section 5.1 Rules for Exponents

1.  base; exponent

3.  $a^{m+n}$

5.  $a^n b^n$

7.  $8^2 = 8 \cdot 8 = 64$

9.  $(-2)^3 = (-2) \cdot (-2) \cdot (-2) = -8$

11.  $-2^3 = -(2 \cdot 2 \cdot 2) = -8$

13.  $6^0 = 1$

15.  $3 + \dfrac{4^2}{2} = 3 + \dfrac{(4 \cdot 4)}{2} = 3 + \dfrac{16}{2} = 3 + 8 = 11$

17.  $4 \cdot \left(\dfrac{1}{2}\right)^3 = 4 \cdot \dfrac{1 \cdot 1 \cdot 1}{(2 \cdot 2 \cdot 2)} = 4 \cdot \dfrac{1}{8} = \dfrac{4}{8} = \dfrac{1}{2}$

19.  $3 \cdot 3^2 = 3^1 \cdot 3^2 = 3^{1+2} = 3^3 = 27$

21.  $4^2 \cdot 4^6 = 4^{2+6} = 4^8 = 65,536$

23.  $x^3 \cdot x^6 = x^{3+6} = x^9$

25.  $x^2 x^2 x^2 = x^{2+2+2} = x^6$

27.  $4x^2 \cdot 5x^5 = 20x^{2+5} = 20x^7$

29.  $3\left(-xy^3\right)\left(x^2 y\right) = 3\left(-x^{1+2} y^{3+1}\right) = 3\left(-x^3 y^4\right) = -3x^3 y^4$

31.  $\left(2^3\right)^2 = 2^{2 \cdot 3} = 2^6 = 64$

33.  $\left(n^3\right)^4 = n^{3 \cdot 4} = n^{12}$

35.  $x\left(x^3\right)^2 = x\left(x^{3 \cdot 2}\right) = x\left(x^6\right) = x^{1+6} = x^7$

37.  $(-7b)^2 = (-7) \cdot (-7) \cdot (b \cdot b) = 49b^2$

39.  $(ab)^3 = (a \cdot a \cdot a)(b \cdot b \cdot b) = a^3 b^3$

41.  $\left(2x^2\right)^0 = (2)^0 \left(x^2\right)^0 = 1x^0 = 1 \cdot 1 = 1$

43.  $\left(-4b^2\right)^3 = (-4) \cdot (-4) \cdot (-4)\left(b^{2 \cdot 3}\right) = -64b^6$

45. $\left(x^2 y^3\right)^7 = \left(x^{2 \cdot 7} y^{3 \cdot 7}\right) = x^{14} y^{21}$

47. $\left(y^3\right)^2 \left(x^4 y\right)^3 = \left(y^{3 \cdot 2}\right)\left(x^{4 \cdot 3} y^{1 \cdot 3}\right) = \left(y^6\right)\left(x^{12} y^3\right) = x^{12} y^{6+3} = x^{12} y^9$

49. $\left(a^2 b\right)^2 \left(a^2 b^2\right)^3 = \left(a^{2 \cdot 2} b^{1 \cdot 2}\right)\left(a^{2 \cdot 3} b^{2 \cdot 3}\right) = \left(a^4 b^2\right)\left(a^6 b^6\right) = a^{4+6} b^{2+6} = a^{10} b^8$

51. $\left(\dfrac{1}{3}\right)^3 = \dfrac{(1 \cdot 1 \cdot 1)}{(3 \cdot 3 \cdot 3)} = \dfrac{1}{27}$

53. $\left(\dfrac{a}{b}\right)^5 = \dfrac{(a \cdot a \cdot a \cdot a \cdot a)}{(b \cdot b \cdot b \cdot b \cdot b)} = \dfrac{a^5}{b^5}$

55. $\left(\dfrac{x-y}{3}\right)^3 = \dfrac{(x-y)^3}{3^3} = \dfrac{(x-y)^3}{3 \cdot 3 \cdot 3} = \dfrac{(x-y)^3}{27}$

57. $\left(\dfrac{5}{a+b}\right)^2 = \dfrac{5^2}{(a+b)^2} = \dfrac{5 \cdot 5}{(a+b)^2} = \dfrac{25}{(a+b)^2}$

59. $\left(\dfrac{2x}{5}\right)^3 = \dfrac{(2 \cdot 2 \cdot 2)\left(x^{1 \cdot 3}\right)}{(5 \cdot 5 \cdot 5)} = \dfrac{8x^3}{125}$

61. $\left(\dfrac{3x^2}{5y^4}\right)^3 = \dfrac{(3 \cdot 3 \cdot 3)\left(x^{2 \cdot 3}\right)}{(5 \cdot 5 \cdot 5)\left(y^{4 \cdot 3}\right)} = \dfrac{27x^6}{125y^{12}}$

63. $(x+y)(x+y)^3 = (x+y)^{1+3} = (x+y)^4$

65. $(a+b)^2 (a+b)^3 = (a+b)^{2+3} = (a+b)^5$

67. $6\left(x^4 y^6\right)^0 = 6\left(x^{4 \cdot 0} y^{6 \cdot 0}\right) = 6\left(x^0 y^0\right) = 6 \cdot 1 = 6$

69. $a\left(a^2 + 2b^2\right) = a^{1+2} + 2ab^2 = a^3 + 2ab^2$

71. $3a^3 \left(4a^2 + 2b\right) = 3 \cdot 4a^{3+2} + 3 \cdot 2a^3 b = 12a^5 + 6a^3 b$

73. $(r+t)(rt) = \left(r^{1+1} t\right) + \left(r\, t^{1+1}\right) = r^2 t + rt^2$

75. For example, $a = 3$, $b = 1$, $m = 1$, and $n = 0$. $a^m \cdot b^n = 3^1 \cdot 1^0 = 3 \cdot 1 = 3$ and

   $(ab)^{m+n} = (3 \cdot 1)^{1+0} = 3^1 = 3$. *Answers may vary.*

77. $2x^2 \cdot 5x^2 = 10x^{2+2} = 10x^4$

79. $\pi\left(3x^2\right)^2 = \pi(3 \cdot 3)\left(x^{2 \cdot 2}\right) = \pi\left(9x^4\right) = 9\pi x^4$

81. $4x \cdot 2x \cdot x = 8x^{1+1+1} = 8x^3$

83. $1000(1+0.05)^3 = 1000(1.05)^3 = 1000(1.05 \cdot 1.05 \cdot 1.05) = 1000(1.157625) \approx \$1157.63$

85. (a)  The tripling will be represented by the 3 and the two times will be represented by the 2 in the expression $3^2$.

    (b)  $3^2 \cdot 8000 = 9 \cdot 8000 = 72,000$  The result is $72,000.

## Section 5.2 Addition and Subtraction of Polynomials

1. monomial

3. degree

5. binomial

7. like

9. opposite

11. Because $x^2 \Rightarrow$ the degree is 2; the coefficient is 3.

13. Because $ab = a^1b^1$ and $1+1=2 \Rightarrow$ the degree is 2; $-ab = -1ab \Rightarrow$ the coefficient is $-1$.

15. Because $rt = r^1t^1$ and $1+1=2 \Rightarrow$ the degree is 2; the coefficient is $-5$.

17. Because there are no variables the degree is 0; the coefficient is 6.

19. Yes it is a polynomial; 1 term $-x$; one variable $x$; $x^1 \Rightarrow$ degree is 1.

21. Yes it is a polynomial; 3 terms $4x^2, -5x,$ and 9; one variable $x$; $x^2 \Rightarrow$ degree is 2.

23. Not a polynomial because it has a variable in the denominator.

25. Not a polynomial because it has negative exponents on the variables.

27. Yes it is a polynomial; 1 term $-2^3a^4bc$; three variables $a, b,$ and $c$; $-2^3a^4bc$ and $4+1+1=6 \Rightarrow$ degree is 6.

29. Yes, $5x + (-4x) = (5-4)x = 1x = x$.

31. Yes, $x^3 + (-6x^3) = [1+(-6)]x^3 = -5x^3$.

33. No, $x$ and $xy$ are not like terms.

35. Yes $\Rightarrow ab + ba = (1+1)ab = 2ab$.

37. $(3x+5) + (-4x+4) = 3x + (-4x) + 5 + 4 = (3-4)x + (5+4) = -x+9$

39. $(3x^2 + 4x + 1) + (x^2 + 4x) = 3x^2 + x^2 + 4x + 4x + 1 = (3+1)x^2 + (4+4)x + 1 = 4x^2 + 8x + 1$

41. $(y^3 + 3y^2 - 5) + (3y^3 + 4y - 4) = 3y^3 + y^3 + 3y^2 + 4y + (-5) + (-4) =$

    $(3+1)y^3 + 3y^2 + 4y + (-5-4) = 4y^3 + 3y^2 + 4y - 9$

43. $(-xy+5) + (5xy-4) = 5xy + (-xy) + 5 + (-4) = (5-1)xy + (5-4) = 4xy + 1$

45. $\left(a^3b^2 + a^2b^3\right) + \left(a^2b^3 - a^3b^2\right) = a^3b^2 + \left(-a^3b^2\right) + a^2b^3 + a^2b^3 = (1-1)a^3b^2 + (1+1)a^2b^3 =$

$0a^3b^2 + 2a^2b^3 = 2a^2b^3$

47. $\begin{array}{r} 4x^2 - 2x + 1 \\ +5x^2 + 3x - 7 \\ \hline 9x^2 + \ x - 6 \end{array}$

49. $\begin{array}{r} -x^2 + \ x + 0 \\ + 2x^2 - 8x - 1 \\ \hline x^2 - 7x - 1 \end{array}$

51. $-5x^2$

53. $-3a^2 + a - 4$

55. $2t^2 + 3t - 4$

57. $(3x+1) - (-x+3) = (3x+1) + (x-3) = (3+1)x + (1-3) = 4x - 2$

59. $\left(-x^2 + 6x\right) - \left(2x^2 + x - 2\right) = \left(-x^2 + 6x\right) + \left(-2x^2 - x + 2\right) = (-1-2)x^2 + (6-1)x + (2) = -3x^2 + 5x + 2$

61. $\left(z^3 - 2z^2 - z\right) - \left(4z^2 + 5z + 1\right) = \left(z^3 - 2z^2 - z\right) + \left(-4z^2 - 5z - 1\right) =$

$z^3 + (-2-4)z^2 + (-1-5)z - 1 = z^3 - 6z^2 - 6z - 1$

63. $\left(4xy + x^2y^2\right) - \left(xy - x^2y^2\right) = \left(4xy + x^2y^2\right) + \left(-xy + x^2y^2\right) = (4-1)xy + (1+1)x^2y^2 = 3xy + 2x^2y^2$

65. $\left(ab^2\right) - \left(ab^2 + a^3b\right) = \left(ab^2\right) + \left(-ab^2 - a^3b\right) = (1-1)ab^2 - a^3b = 0ab^2 - a^3b = -a^3b$

67. $\left(x^2 + 2x - 3\right) - \left(2x^2 + 7x + 1\right) = \quad \begin{array}{r} x^2 + 2x - 3 \\ +\left(-2x^2\right) - 7x - 1 \\ \hline -x^2 - 5x - 4 \end{array}$

69. $\left(3x^3 - 2x\right) - \left(5x^3 + 4x + 2\right) = \quad \begin{array}{r} 3x^3 \ - 2x + 0 \\ +\left(-5x^3\right) - 4x - 2 \\ \hline -2x^3 - 6x - 2 \end{array}$

71. (a)   Let $t = 0$, then $1.6t^2 - 28t + 200 = 1.6\left(0^2\right) - 28(0) + 200 = 0 - 0 + 200 = 200$ bpm.

(b)   Let $t = 5$, then $1.6t^2 - 28t + 200 = 1.6\left(5^2\right) - 28(5) + 200 = 1.6(25) - 28(5) + 200 =$

$40 - 140 + 200 = 100$ bpm.

(c)   It decreases quickly at first and then more slowly.

73. $z^2 + z^2 = (1+1)z^2 = 2z^2$;   Let $z = 10$ in., then $2\left(10^2\right) = 2(100) = 200$ in$^2$

75. $2x \cdot x + x \cdot x = 2x^2 + x^2$ or $3x^2$; Let $x = 6$ feet, then $3x^2 = 3\left(6^2\right) = 3(36) = 108$ ft$^2$

77. $\pi x^2 + \pi y^2$; Let $x = 2$ feet and $y = 3$ feet, then $\pi x^2 + \pi y^2 = \pi\left(2^2\right) + \pi\left(3^2\right) = \pi(4) + \pi(9) = 13\pi$ ft$^2$

79. (a)   Slope $m = \dfrac{P_2 - P_1}{t_2 - t_1}$, so $m_1 = \dfrac{5-4}{1987-1974} = \dfrac{1}{13} \Rightarrow m_1 = 0.077$; $m_2 = \dfrac{6-5}{1999-1987} = \dfrac{1}{12} \Rightarrow$

   $m_2 = 0.083$; $m_3 = \dfrac{7-6}{2012-1999} = \dfrac{1}{13} \Rightarrow m_3 = 0.077$. A line is a reasonable estimate, but it is

   not exact.

   (b)   For the given years, the polynomial gives a reasonable estimate.

## Checking Basic Concepts  Sections 5.1 and 5.2

1. (a)   $-5^2 = -(5 \cdot 5) = -25$

   (b)   $3^2 - 2^3 = (3 \cdot 3) - (2 \cdot 2 \cdot 2) = 9 - 8 = 1$

2. (a)   $10^3 \cdot 10^5 = 10^{3+5} = 10^8$

   (b)   $\left(3x^2\right)\left(-4x^5\right) = -12x^{2+5} = -12x^7$

   (c)   $\left(a^3 b\right)^2 = a^{3 \cdot 2} b^{1 \cdot 2} = a^6 b^2$

   (d)   $\left(\dfrac{x}{z^3}\right)^4 = \dfrac{x^{1 \cdot 4}}{z^{3 \cdot 4}} = \dfrac{x^4}{z^{12}}$

3. (a)   $\left(4y^3\right)^0 = 1$

   (b)   $\left(x^3\right)^2 \left(3x^4\right)^2 = x^{3 \cdot 2} \cdot 3^2 \cdot x^{4 \cdot 2} = 9x^6 \cdot x^8 = 9x^{6+8} = 9x^{14}$

   (c)   $2a^2\left(5a^3 - 7\right) = 2 \cdot 5a^{2+3} - 2 \cdot 7a^2 = 10a^5 - 14a^2$

4.  3 terms $5x^3 y$, $-2x^2 y$, and 5;  2 variables $x$ and $y$; $x^3 y = x^3 y^1$, $3 + 1 = 4 \Rightarrow$ the degree is 4.

5. (a)   Let length $= 2w$, width $= w$, and height $= h$, then $2w \cdot w \cdot h = 2w^2 h$.

   (b)   Let $w = 12$ and $h = 10$, then $2w^2 h = 2\left(12^2\right)(10) = 2(144)(10) = 2880$ in$^3$.

6. (a)   $\left(2a^2 + 3a - 1\right) + \left(a^2 - 3a + 7\right) = 2a^2 + a^2 + 3a + (-3a) + (-1) + 7 =$

   $(2+1)a^2 + (3-3)a + (-1+7) = 3a^2 + 0a + 6 = 3a^2 + 6$

   (b)   $\left(4z^3 + 5z\right) - \left(2z^3 - 2z + 8\right) = \left(4z^3 + 5z\right) + \left(-2z^3 + 2z - 8\right) = (4-2)z^3 + (5+2)z - 8 =$

   $2z^3 + 7z - 8$

(c)    $\left(x^2 + 2xy + y^2\right) - \left(x^2 - 2xy + y^2\right) = \left(x^2 + 2xy + y^2\right) + \left(-x^2 + 2xy - y^2\right) =$

$(1-1)x^2 + (2+2)xy + (1-1)y^2 = 0x^2 + 4xy + 0y^2 = 4xy$

## Section 5.3 Multiplication of Polynomials

1. The product rule

3. term; term

5. $x^2 \cdot x^5 = x^{2+5} = x^7$

7. $-3a \cdot 4a = (-3)(4)aa = -12a^{1+1} = -12a^2$

9. $4x^3 \cdot 5x^2 = (4)(5)x^3 x^2 = 20x^{3+2} = 20x^5$

11. $xy^2 \cdot 4xy = (1)(4)xxy^2 y = 4x^{1+1}y^{2+1} = 4x^2 y^3$

13. $\left(-3xy^2\right)\left(4x^2 y\right) = (-3)(4)xx^2 y^2 y = -12x^{1+2}y^{2+1} = -12x^3 y^3$

15. $3(x+4) = 3 \cdot x + 3 \cdot 4 = 3x + 12$

17. $-5(9x+1) = -5 \cdot 9x + (-5 \cdot 1) = -45x - 5$

19. $(4-z)z = 4 \cdot z + (-z \cdot z) = 4z - z^2$

21. $-y(5+3y) = -y \cdot 5 + (-y \cdot 3y) = -5y - 3y^2$

23. $3x\left(5x^2 - 4\right) = 3x \cdot 5x^2 + 3x \cdot (-4) = (3)(5)xx^2 - 12x = 15x^{1+2} - 12x = 15x^3 - 12x$

25. $(6x-6)x^2 = 6x \cdot x^2 + (-6) \cdot x^2 = 6x^{1+2} - 6x^2 = 6x^3 - 6x^2$

27. $-8\left(4t^2 + t + 1\right) = -8 \cdot 4t^2 + (-8)(t) + (-8)(1) = -32t^2 - 8t - 8$

29. $n^2\left(-5n^2 + n - 2\right) = \left(n^2\right)\left(-5n^2\right) + \left(n^2\right)(n) + \left(n^2\right)(-2) = -5n^{2+2} + n^{2+1} - 2n^2 = -5n^4 + n^3 - 2n^2$

31. $xy(x+y) = xy \cdot x + xy \cdot y = x^{1+1}y + xy^{1+1} = x^2 y + xy^2$

33. $x^2\left(x^2 y - xy^2\right) = x^2 \cdot x^2 y + \left(x^2\right)\left(-xy^2\right) = x^{2+2}y - x^{2+1}y^2 = x^4 y - x^3 y^2$

35. $-ab\left(a^3 - 2b^3\right) = -ab \cdot a^3 + (-ab)\left(-2b^3\right) = -a^{1+3}b + 2ab^{3+1} = -a^4 b + 2ab^4$

37. (a)    Area $= x^2 + 3x$. See Figure 37.

     (b)    $x(x+3) = x \cdot x + x \cdot 3 = x^{1+1} + 3x = x^2 + 3x$

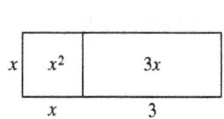

Figure 37

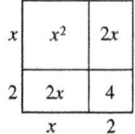

Figure 39

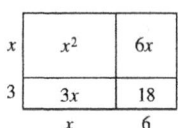

Figure 41

39. (a)   Area $= x^2 + 2x + 2x + 4 = x^2 + 4x + 4$. See Figure 39.

    (b)   $(x+2)(x+2) = x \cdot x + x \cdot 2 + 2 \cdot x + 2 \cdot 2 = x^{1+1} + 2x + 2x + 4 = x^2 + 4x + 4$

41. (a)   Area $= x^2 + 3x + 6x + 18 = x^2 + 9x + 18$.  See Figure 41.

    (b)   $(x+3)(x+6) = x \cdot x + x \cdot 6 + 3 \cdot x + 3 \cdot 6 = x^{1+1} + 6x + 3x + 18 = x^2 + 9x + 18$

43. $(x+3)(x+5) = x \cdot x + x \cdot 5 + 3 \cdot x + 3 \cdot 5 = x^{1+1} + 5x + 3x + 15 = x^2 + 8x + 15$

45. $(x-8)(x-9) = x \cdot x + (x)(-9) + (-8)(x) + (-8)(-9) = x^{1+1} + (-9x) + (-8x) + 72 = x^2 - 17x + 72$

47. $(3z-2)(2z-5) = 3z \cdot 2z + (3z)(-5) + (-2)(2z) + (-2)(-5) = 6z^{1+1} + (-15z) + (-4z) + 10 =$

    $6z^2 - 19z + 10$

49. $(8b-1)(8b+1) = 8b \cdot 8b + 8b \cdot 1 + (-1)(8b) + (-1)(1) = 64b^{1+1} + 8b - 8b - 1 = 64b^2 - 1$

51. $(10y+7)(y-1) = 10y \cdot y + (10y)(-1) + 7 \cdot y + (7)(-1) = 10y^{1+1} - 10y + 7y - 7 = 10y^2 - 3y - 7$

53. $(5-3a)(1-2a) = 5 \cdot 1 + (5)(-2a) + (-3a)(1) + (-3a)(-2a) = 5 + (-10a) + (-3a) + 6a^{1+1} =$

    $5 - 13a + 6a^2$

55. $(1-3x)(1+3x) = 1 \cdot 1 + 1 \cdot 3x + (-3x)(1) + (-3x)(3x) = 1 + 3x - 3x - 9x^{1+1} = 1 - 9x^2$

57. $(x-1)(x^2+1) = x \cdot x^2 + x \cdot 1 + (-1)(x^2) + (-1)(1) = x^{1+2} + x - x^2 - 1 = x^3 - x^2 + x - 1$

59. $(x^2+4)(4x-3) = x^2 \cdot 4x + (x^2)(-3) + 4 \cdot 4x + (4)(-3) = 4x^{2+1} - 3x^2 + 16x - 12 =$

    $4x^3 - 3x^2 + 16x - 12$

61. $(2n+1)(n^2+3) = 2n \cdot n^2 + 2n \cdot 3 + 1 \cdot n^2 + 1 \cdot 3 = 2n^{1+2} + 6n + n^2 + 3 = 2n^3 + n^2 + 6n + 3$

63. $(m+1)(m^2+3m+1) = m \cdot m^2 + m \cdot 3m + m \cdot 1 + 1 \cdot m^2 + 1 \cdot 3m + 1 \cdot 1 =$

    $m^{1+2} + 3m^{1+1} + m + m^2 + 3m + 1 = m^3 + 3m^2 + m + m^2 + 3m + 1 = m^3 + 4m^2 + 4m + 1$

65. $(3x-2)(2x^2-x+4) = 3x \cdot 2x^2 + (3x)(-x) + 3x \cdot 4 + (-2)(2x^2) + (-2)(-x) + (-2)(4) =$

    $6x^{1+2} - 3x^{1+1} + 12x - 4x^2 + 2x - 8 = 6x^3 - 3x^2 + 12x - 4x^2 + 2x - 8 = 6x^3 - 7x^2 + 14x - 8$

67. $(x+1)(x^2-x+1) = x \cdot x^2 + (x)(-x) + x \cdot 1 + 1 \cdot x^2 + (1)(-x) + 1 \cdot 1 =$

    $x^{1+2} - x^{1+1} + x + x^2 - x + 1 = x^3 - x^2 + x + x^2 - x + 1 = x^3 + 1$

69. $(4b^2+3b+7)(b^2+3) = 4b^2 \cdot b^2 + 4b^2 \cdot 3 + 3b \cdot b^2 + 3b \cdot 3 + 7 \cdot b^2 + 7 \cdot 3 =$

    $4b^{2+2} + 12b^2 + 3b^{1+2} + 9b + 7b^2 + 21 = 4b^4 + 12b^2 + 3b^3 + 9b + 7b^2 + 21 =$

    $4b^4 + 3b^3 + 19b^2 + 9b + 21$

71. 
$$
\begin{array}{r}
x^2 - 3x + 1 \\
x + 2 \\
\hline
2x^2 - 6x + 2 \\
x^3 - 3x^2 + x \\
\hline
x^3 - x^2 - 5x + 2
\end{array}
$$

73. 
$$
\begin{array}{r}
a^2 + 2a + 4 \\
a - 2 \\
\hline
-2a^2 - 4a - 8 \\
a^3 + 2a^2 + 4a \\
\hline
a^3 \qquad\quad - 8
\end{array}
$$

75. 
$$
\begin{array}{r}
3x^2 - x + 1 \\
2x^2 + 1 \\
\hline
3x^2 - x + 1 \\
6x^4 - 2x^3 + 2x^2 \\
\hline
6x^4 - 2x^3 + 5x^2 - x + 1
\end{array}
$$

77. Every term in the first polynomial must be multiplied by every term in the second polynomial, so there are $m \cdot n$ products, hence $m \cdot n$ terms.

79. (a)　Let $h = $ height, $h + 2 = $ length, and $h - 4 = $ width, then $(h - 4)(h + 2) =$

$h \cdot h + h \cdot 2 + (-4)(h) + (-4)(2) = h^{1+1} + 2h - 4h - 8 = h^2 - 2h - 8 \Rightarrow (h)\left(h^2 - 2h - 8\right) =$

$h \cdot h^2 + (h)(-2h) + (h)(-8) = h^{1+2} - 2h^{1+1} - 8h = h^3 - 2h^2 - 8h.$

(b)　$h^3 - 2h^2 - 8h = 25^3 - 2(25)^2 - 8(25) = 15{,}625 - 2(625) - 200 = 15{,}625 - 1250 - 200 =$

$14{,}175 \text{ in}^3$

81. (a)　$x(50 - x) = x \cdot 50 + (x)(-x) = 50x - x^{1+1} = 50x - x^2$

(b)　$50x - x^2 = 50(25) - \left(25^2\right) = 1250 - 625 = 625$

83. $(x + 1)(x + 1) = x \cdot x + x \cdot 1 + 1 \cdot x + 1 \cdot 1 = x^{1+1} + x + x + 1 = x^2 + 2x + 1 \Rightarrow 1 \text{ side} = x^2 + 2x + 1.$

A cube has all sides the same $\Rightarrow 6 \text{ sides} = (6)\left(x^2 + 2x + 1\right) = 6 \cdot x^2 + 6 \cdot 2x + 6 \cdot 1 = 6x^2 + 12x + 6.$

85. (a)　$t(64 - 16t) = t \cdot 64 + (t)(-16t) = 64t - 16t^{1+1} = 64t - 16t^2$

(b)　$64t - 16t^2 = 64(2) - 16(2)^2 = 128 - 16(4) = 128 - 64 = 64;$

$t(64 - 16t) = 2(64 - 16(2)) = 2(64 - 32) = 2(32) = 64$

(c)　Yes; yes

## **Section 5.4 Special Products**

1. $(a+b)(a-b) = (a)^2 - (b)^2 = a^2 - b^2$

3. $(a-b)^2 = (a)^2 - 2(a)(b) + (b)^2 = a^2 - 2ab + b^2$

5. False

7. $(x-3)(x+3) = (x)^2 - (3)^2 = x^2 - 9$

9. $(4x-1)(4x+1) = (4x)^2 - (1)^2 = 16x^2 - 1$

11. $(1+2a)(1-2a) = (1)^2 - (2a)^2 = 1 - 4a^2$

13. $(2x+3y)(2x-3y) = (2x)^2 - (3y)^2 = 4x^2 - 9y^2$

15. $(ab-5)(ab+5) = (ab)^2 - (5)^2 = a^2b^2 - 25$

17. $(a^2-b^2)(a^2+b^2) = (a^2)^2 - (b^2)^2 = a^4 - b^4$

19. $(x^3-y^3)(x^3+y^3) = (x^3)^2 - (y^3)^2 = x^6 - y^6$

21. $101 \cdot 99 = (100+1)(100-1) = (100)^2 - (1)^2 = 10,000 - 1 = 9999$

23. $23 \cdot 17 = (20+3)(20-3) = (20)^2 - (3)^2 = 400 - 9 = 391$

25. $90 \cdot 110 = (100-10)(100+10) = (100)^2 - (10)^2 = 10,000 - 100 = 9900$

27. $(a-2)^2 = (a)^2 - 2(a)(2) + (2)^2 = a^2 - 4a + 4$

29. $(2x+3)^2 = (2x)^2 + 2(2x)(3) + (3)^2 = 4x^2 + 12x + 9$

31. $(3b+5)^2 = (3b)^2 + 2(3b)(5) + (5)^2 = 9b^2 + 30b + 25$

33. $\left(\dfrac{3}{4}a - 4\right)^2 = \left(\dfrac{3}{4}a\right)^2 - 2\left(\dfrac{3}{4}a\right)(4) + (4)^2 = \dfrac{9}{16}a^2 - 6a + 16$

35. $(1-b)^2 = (1)^2 - 2(1)(b) + (b)^2 = 1 - 2b + b^2$

37. $(5+y^3)^2 = (5)^2 + 2(5)(y^3) + (y^3)^2 = 25 + 10y^3 + y^6$

39. $(a^2+b)^2 = (a^2)^2 + 2(a^2)(b) + (b)^2 = a^4 + 2a^2b + b^2$

41. $(a+1)^3 = (a+1)(a+1)^2 = (a+1)(a^2 + 2a + 1) =$

    $a \cdot a^2 + a \cdot 2a + a \cdot 1 + 1 \cdot a^2 + 1 \cdot 2a + 1 \cdot 1 = a^3 + 2a^2 + a + a^2 + 2a + 1 = a^3 + 3a^2 + 3a + 1$

43. $(x-2)^3 = (x-2)(x-2)^2 = (x-2)(x^2 - 4x + 4) =$

    $x \cdot x^2 + x \cdot (-4x) + x \cdot 4 + (-2) \cdot x^2 + (-2)(-4x) + (-2)4 =$

    $x^3 + (-4x^2) + 4x + (-2x^2) + 8x + (-8) = x^3 - 6x^2 + 12x - 8$

45. $(2x+1)^3 = (2x+1)(2x+1)^2 = (2x+1)(4x^2+4x+1) =$

     $2x\cdot 4x^2 + 2x\cdot 4x + 2x\cdot 1 + 1\cdot 4x^2 + 1\cdot 4x + 1\cdot 1 = 8x^3 + 8x^2 + 2x + 4x^2 + 4x + 1 = 8x^3 + 12x^2 + 6x + 1$

47. $(6u-1)^3 = (6u-1)(6u-1)^2 = (6u-1)(36u^2-12u+1) =$

     $6u\cdot 36u^2 + 6u\cdot(-12u) + 6u\cdot 1 + (-1)\cdot 36u^2 + (-1)\cdot(-12u) + (-1)\cdot 1 =$

     $216u^3 + (-72u^2) + 6u + (-36u^2) + 12u + (-1) = 216u^3 - 108u^2 + 18u - 1$

49. $4(5x+9) = 4\cdot 5x + 4\cdot 9 = 20x + 36$

51. $(x-5)(x+7) = x\cdot x + x\cdot 7 + (-5)(x) + (-5)(7) = x^2 + 2x - 35$

53. $(3x-5)^2 = (3x)^2 - 2(3x)(5) + (-5)^2 = 9x^2 - 30x + 25$

55. $(5x+3)(5x+4) = 5x\cdot 5x + 5x\cdot 4 + 3\cdot 5x + 3\cdot 4 = 25x^2 + 35x + 12$

57. $(4b-5)(4b+5) = (4b)^2 - (5)^2 = 16b^2 - 25$

59. $-5x(4x^2-7x+2) = (-5x)(4x^2) + (-5x)(-7x) + (-5x)(2) = -20x^3 + 35x^2 - 10x$

61. $(4-a)^3 = (4-a)(4-a)^2 = (4-a)(16-8a+a^2) =$

     $4\cdot 16 + (4)(-8a) + 4\cdot a^2 + (-a)(16) + (-a)(-8a) + (-a)(a^2) = 64 - 48a + 12a^2 - a^3$

63. $x(x+3)^2 = x(x^2+6x+9) = x^3 + 6x^2 + 9x$

65. $(x+2)(x-2)(x+1)(x-1) = (x^2-4)(x^2-1) = x^4 - x^2 - 4x^2 + 4 = x^4 - 5x^2 + 4$

67. $(a^n+b^n)(a^n-b^n) = (a^n)^2 - (b^n)^2 = a^{2n} - b^{2n} = a^{2n} - b^{2n}$

69. (a)   $(x+2)(x+2) = (x)^2 + 2(x)(2) + (2)^2 = x^2 + 4x + 4$

    (b)   $x\cdot x + 2\cdot x + x\cdot 2 + 2\cdot 2 = x^2 + 2x + 2x + 4 = x^2 + 4x + 4$

71. (a)   $(2x+3)(2x+3) = (2x)^2 + 2(2x)(3) + (3)^2 = 4x^2 + 12x + 9$

    (b)   $2x\cdot 2x + 3\cdot 2x + 2x\cdot 3 + 3\cdot 3 = 4x^2 + 6x + 6x + 9 = 4x^2 + 12x + 9$

73. (a)   $6(x+5)^2 = 6(x^2+10x+25) = 6x^2 + 60x + 150$

    (b)   $(x+5)^3 = (x+5)(x+5)^2 = (x+5)(x^2+10x+25) =$

        $x\cdot x^2 + x\cdot 10x + x\cdot 25 + 5\cdot x^2 + 5\cdot 10x + 5\cdot 25 = x^3 + 15x^2 + 75x + 125$

75. (a)   $(1+x)^2 = (1)^2 + 2(1)(x) + (x)^2 = 1 + 2x + x^2$

    (b)   $1 + 2x + x^2 = 1 + 2(0.10) + (0.10)^2 = 1 + 0.20 + 0.01 = 1.21;$ the money increases by 1.21 times in

       2 years if the interest rate is 10%.

77. (a)  $(1-x)^2 = (1)^2 - 2(1)(x) + (x)^2 = 1 - 2x + x^2$

   (b)  $1 - 2x + x^2 = 1 - 2(0.50) + (0.50)^2 = 1 - 1 + 0.25 = 0.25$;  if the chance of rain on each day is

   50%, then there is a 25% chance of no rain on either day.

79. (a)  $(z+16)^2 - (z)^2 = (z^2 + 32z + 256) - (z^2) = 32z + 256$

   (b)  $32z + 256 = 32(60) + 256 = 1920 + 256 = 2176$;  the area of an 8-foot-wide sidewalk around a

   $60\,\text{ft} \times 60\,\text{ft}$  pool is  $2176\ \text{ft}^2$.


## Checking Basic Concepts  Sections 5.3 and 5.4

1. (a)  $\left(-3xy^4\right)\left(5x^2 y\right) = (-3)(5)xx^2 y^4 y = -15x^3 y^5$

   (b)  $-x(6-4x) = (-x)(6) + (-x)(-4x) = -6x + 4x^2$

   (c)  $3ab\left(a^2 - 2ab + b^2\right) = 3ab \cdot a^2 + (3ab)(-2ab) + 3ab \cdot b^2 = 3a^3 b - 6a^2 b^2 + 3ab^3$

2. (a)  $(x+3)(4x-3) = x \cdot 4x + (x)(-3) + 3 \cdot 4x + (3)(-3) = 4x^2 + 9x - 9$

   (b)  $\left(x^2 - 1\right)\left(2x^2 + 2\right) = x^2 \cdot 2x^2 + x^2 \cdot 2 + (-1)\left(2x^2\right) + (-1)(2) = 2x^4 - 2$

   (c)  $(x+y)\left(x^2 - xy + y^2\right) = x \cdot x^2 + (x)(-xy) + x \cdot y^2 + y \cdot x^2 + (y)(-xy) + y \cdot y^2 = x^3 + y^3$

3. (a)  $(5x+2)(5x-2) = (5x)^2 - (2)^2 = 25x^2 - 4$

   (b)  $(x+3)^2 = (x)^2 + 2(x)(3) + (3)^2 = x^2 + 6x + 9$

   (c)  $(2-7x)^2 = (2)^2 - 2(2)(7x) + (7x)^2 = 4 - 28x + 49x^2$

   (d)  $(t+2)^3 = (t+2)(t+2)^2 = (t+2)\left(t^2 + 4t + 4\right) = t \cdot t^2 + t \cdot 4t + t \cdot 4 + 2 \cdot t^2 + 2 \cdot 4t + 2 \cdot 4 =$

   $t^3 + 6t^2 + 12t + 8$

4. (a)  $(m+5)^2 = (m)^2 + 2(5)(m) + (5)^2 = m^2 + 10m + 25$

   (b)  $m \cdot m + m \cdot 5 + 5 \cdot m + 5 \cdot 5 = m^2 + 5m + 5m + 25 = m^2 + 10m + 25$


## Section 5.5 Integer Exponents and the Quotient Rule

1.  $a^{-n} = \dfrac{1}{a^n}$

3.  $\dfrac{a^m}{a^n} = a^{m-n}$

5. $\left(\dfrac{a}{b}\right)^{-n} = \left(\dfrac{b}{a}\right)^{n}$

7. (a) $\quad 4^{-1} = \dfrac{1}{4^{1}} = \dfrac{1}{4}$

(b) $\quad \left(\dfrac{1}{3}\right)^{-2} = \left(\dfrac{3}{1}\right)^{2} = \dfrac{3^{2}}{1^{2}} = \dfrac{3 \cdot 3}{1 \cdot 1} = \dfrac{9}{1} = 9$

9. (a) $\quad 2^{3} \cdot 2^{-2} = 2^{3+(-2)} = 2^{1} = 2$

(b) $\quad 10^{-1} \cdot 10^{-2} = 10^{-1+(-2)} = 10^{-3} = \dfrac{1}{10^{3}} = \dfrac{1}{10 \cdot 10 \cdot 10} = \dfrac{1}{1000}$

11. (a) $\quad 3^{-2} \cdot 3^{-1} \cdot 3^{-1} = 3^{-2+(-1)+(-1)} = 3^{-4} = \dfrac{1}{3^{4}} = \dfrac{1}{3 \cdot 3 \cdot 3 \cdot 3} = \dfrac{1}{81}$

(b) $\quad \left(2^{3}\right)^{-1} = 2^{(3)(-1)} = 2^{-3} = \dfrac{1}{2^{3}} = \dfrac{1}{2 \cdot 2 \cdot 2} = \dfrac{1}{8}$

13. (a) $\quad \left(3^{2}4^{3}\right)^{-1} = 3^{(2)(-1)}4^{(3)(-1)} = 3^{-2}4^{-3} = \dfrac{1}{3^{2}4^{3}} = \dfrac{1}{(3 \cdot 3)(4 \cdot 4 \cdot 4)} = \dfrac{1}{9 \cdot 64} = \dfrac{1}{576}$

(b) $\quad \dfrac{4^{5}}{4^{2}} = 4^{5-2} = 4^{3} = 4 \cdot 4 \cdot 4 = 64$

15. (a) $\quad \dfrac{1^{9}}{1^{7}} = 1^{9-7} = 1^{2} = 1 \cdot 1 = 1$

(b) $\quad \dfrac{1}{4^{-3}} = 4^{3} = 4 \cdot 4 \cdot 4 = 64$

17. (a) $\quad \dfrac{5^{-2}}{5^{-4}} = 5^{-2-(-4)} = 5^{2} = 5 \cdot 5 = 25$

(b) $\quad \left(\dfrac{2}{7}\right)^{-2} = \left(\dfrac{7}{2}\right)^{2} = \dfrac{7^{2}}{2^{2}} = \dfrac{7 \cdot 7}{2 \cdot 2} = \dfrac{49}{4}$

19. (a) $\quad x^{-1} = \dfrac{1}{x}$

(b) $\quad a^{-4} = \dfrac{1}{a^{4}}$

21. (a) $\quad x^{-2} \cdot x^{-1} \cdot x = x^{-2+(-1)+1} = x^{-2} = \dfrac{1}{x^{2}}$

(b) $\quad a^{-5} \cdot a^{-2} \cdot a^{-1} = a^{-5+(-2)+(-1)} = a^{-8} = \dfrac{1}{a^{8}}$

23. (a) $\quad x^{2}y^{-3}x^{-5}y^{6} = x^{2+(-5)}y^{-3+6} = x^{-3}y^{3} = \dfrac{y^{3}}{x^{3}}$

(b) $\quad (xy)^{-3} = x^{-3}y^{-3} = \dfrac{1}{x^{3}y^{3}}$

25. (a)  $(2t)^{-4} = 2^{-4}t^{-4} = \dfrac{1}{2^4 t^4} = \dfrac{1}{16t^4}$

    (b)  $(x+1)^{-7} = \dfrac{1}{(x+1)^7}$

27. (a)  $\left(a^{-2}\right)^{-4} = a^{(-2)(-4)} = a^8$

    (b)  $\left(rt^3\right)^{-2} = r^{-2}t^{(3)(-2)} = r^{-2}t^{-6} = \dfrac{1}{r^2 t^6}$

29. (a)  $(ab)^2\left(a^2\right)^{-3} = \left(a^2 b^2\right)\left(a^{(2)(-3)}\right) = \left(a^2 b^2\right)\left(a^{-6}\right) = a^{2+(-6)}b^2 = a^{-4}b^2 = \dfrac{b^2}{a^4}$

    (b)  $\dfrac{x^4}{x^2} = x^{4-2} = x^2$

31. (a)  $\dfrac{a^{10}}{a^{-3}} = a^{10-(-3)} = a^{13}$

    (b)  $\dfrac{4z}{2z^4} = \dfrac{4}{2}\cdot\dfrac{z}{z^4} = 2z^{1-4} = 2z^{-3} = \dfrac{2}{z^3}$

33. (a)  $\dfrac{-4xy^5}{6x^3 y^2} = \dfrac{-4}{6}\cdot\dfrac{xy^5}{x^3 y^2} = \dfrac{-2}{3}x^{1-3}y^{5-2} = \dfrac{-2}{3}x^{-2}y^3 = -\dfrac{2y^3}{3x^2}$

    (b)  $\dfrac{x^{-4}}{x^{-1}} = x^{-4-(-1)} = x^{-3} = \dfrac{1}{x^3}$

35. (a)  $\dfrac{10b^{-4}}{5b^{-5}} = \dfrac{10}{5}\cdot\dfrac{b^{-4}}{b^{-5}} = 2b^{-4-(-5)} = 2b$

    (b)  $\left(\dfrac{a}{b}\right)^3 = \dfrac{a^3}{b^3}$

37. (a)  $\dfrac{6x^2 y^{-4}}{18x^{-5}y^4} = \dfrac{6}{18}\cdot\dfrac{x^2 y^{-4}}{x^{-5}y^4} = \dfrac{1}{3}x^{2-(-5)}y^{-4-4} = \dfrac{x^7 y^{-8}}{3} = \dfrac{x^7}{3y^8}$

    (b)  $\dfrac{16a^{-3}b^{-5}}{4a^{-8}b} = \dfrac{16}{4}\cdot\dfrac{a^{-3}b^{-5}}{a^{-8}b} = 4a^{-3-(-8)}b^{-5-1} = \dfrac{4a^5 b^{-6}}{1} = \dfrac{4a^5}{b^6}$

39. (a)  $\dfrac{1}{y^{-5}} = y^5$

    (b)  $\dfrac{4}{2t^{-3}} = \dfrac{4}{2}\cdot\dfrac{1}{t^{-3}} = 2\cdot t^3 = 2t^3$

41. (a)  $\dfrac{3a^4}{\left(2a^{-2}\right)^3} = \dfrac{3a^4}{8a^{-6}} = \dfrac{3}{8}a^{4-(-6)} = \dfrac{3}{8}a^{10}$

    (b)  $\dfrac{\left(2b^5\right)^{-3}}{4b^{-6}} = \dfrac{1}{4b^{-6}\cdot\left(2b^5\right)^3} = \dfrac{1}{4\cdot 8\cdot b^{-6}\cdot b^{15}} = \dfrac{1}{32b^{-6+15}} = \dfrac{1}{32b^9}$

43. (a) $\dfrac{1}{(xy)^{-2}} = \dfrac{1}{x^{-2}y^{-2}} = x^2 y^2$

(b) $\dfrac{1}{\left(a^2 b\right)^{-3}} = \dfrac{1}{a^{(2)(-3)}b^{-3}} = \dfrac{1}{a^{-6}b^{-3}} = a^6 b^3$

45. (a) $\dfrac{\left(3m^4 n\right)^{-2}}{\left(2mn^{-2}\right)^3} = \dfrac{1}{\left(2mn^{-2}\right)^3 \cdot \left(3m^4 n\right)^2} = \dfrac{1}{8 \cdot 9 \cdot m^3 \cdot m^8 \cdot n^{-6} \cdot n^2} = \dfrac{1}{72 m^{3+8} n^{-6+2}} = \dfrac{1}{72 m^{11} n^{-4}} = \dfrac{n^4}{72 m^{11}}$

(b) $\dfrac{\left(-4x^4 y\right)^2}{\left(xy^{-5}\right)^{-3}} = \dfrac{\left(-4x^4 y\right)^2 \cdot \left(xy^{-5}\right)^3}{1} = \dfrac{16 \cdot x^8 \cdot x^3 \cdot y^2 \cdot y^{-15}}{1} = \dfrac{16}{1} \cdot x^{8+3} \cdot y^{2+(-15)} = \dfrac{16 x^{11} y^{-13}}{1} = \dfrac{16 x^{11}}{y^{13}}$

47. (a) $\left(\dfrac{a}{b}\right)^{-2} = \left(\dfrac{b}{a}\right)^2 = \dfrac{b^2}{a^2}$

(b) $\left(\dfrac{u}{4v}\right)^{-1} = \left(\dfrac{4v}{u}\right)^1 = \dfrac{4v}{u}$

49. (a) $\left(\dfrac{3a^4 b}{2ab^{-2}}\right)^{-2} = \left(\dfrac{2ab^{-2}}{3a^4 b}\right)^2 = \dfrac{4a^2 b^{-4}}{9a^8 b^2} = \dfrac{4}{9} a^{2-8} b^{(-4)-2} = \dfrac{4a^{-6} b^{-6}}{9} = \dfrac{4}{9a^6 b^6}$

(b) $\left(\dfrac{4m^4 n}{5m^{-3} n^2}\right)^2 = \dfrac{16 m^8 n^2}{25 m^{-6} n^4} = \dfrac{16}{25} m^{8-(-6)} \cdot n^{2-4} = \dfrac{16 m^{14} n^{-2}}{25} = \dfrac{16 m^{14}}{25 n^2}$

51. $\dfrac{a^n}{a^m} = a^{n-m} = a^{-1(m-n)} = (a^{m-n})^{-1} = \dfrac{1}{a^{m-n}}$

53. Thousand

55. Billion

57. Hundredth

59. Move the decimal point three places to the right, $2 \times 10^3 = 2000$.

61. Move the decimal point four places to the right, $4.5 \times 10^4 = 45,000$.

63. Move the decimal point three places to the left, $8 \times 10^{-3} = 0.008$.

65. Move the decimal point four places to the left, $4.56 \times 10^{-4} = 0.000456$.

67. Move the decimal point seven places to the right, $3.9 \times 10^7 = 39,000,000$.

69. Move the decimal point five places to the right, $-5 \times 10^5 = -500,000$.

71. Move the decimal point three places to the left, $2000 = 2 \times 10^3$.

73. Move the decimal point two places to the left, $567 = 5.67 \times 10^2$.

75. Move the decimal point seven places to the left, $12,000,000 = 1.2 \times 10^7$.

77. Move the decimal point three places to the right, $0.004 = 4 \times 10^{-3}$.

79. Move the decimal point four places to the right, $0.000895 = 8.95 \times 10^{-4}$.

81. Move the decimal point two places to the right, $-0.05 = -5 \times 10^{-2}$.

83. $\left(5 \times 10^3\right)\left(3 \times 10^2\right) = (5 \cdot 3) \times \left(10^3 \cdot 10^2\right) = 15 \times 10^{3+2} = 15 \times 10^5 = 1.5 \times 10^6$;

    Move the decimal point six places to the right, $1.5 \times 10^6 = 1,500,000$.

85. $\left(-3 \times 10^{-3}\right)\left(5 \times 10^2\right) = (-3 \cdot 5) \times \left(10^{-3} \cdot 10^2\right) = -15 \times 10^{-3+2} = -15 \times 10^{-1} = -1.5 \times 10^0 = -1.5$

87. $\dfrac{4 \times 10^5}{2 \times 10^2} = \dfrac{4}{2} \cdot \dfrac{10^5}{10^2} = 2 \times 10^{5-2} = 2 \times 10^3$;

    Move the decimal point three places to the right, $2 \times 10^3 = 2000$.

89. $\dfrac{8 \times 10^{-6}}{4 \times 10^{-3}} = \dfrac{8}{4} \cdot \dfrac{10^{-6}}{10^{-3}} = 2 \times 10^{-6-(-3)} = 2 \times 10^{-3}$;

    Move the decimal point three places to the left, $2 \times 10^{-3} = 0.002$.

91. (a)   $\left(1.86 \times 10^5\right)\left(3.15 \times 10^7\right) = (1.86 \cdot 3.15) \times \left(10^5 \cdot 10^7\right) \approx 5.859 \times 10^{5+7} \approx 5.859 \times 10^{12}$ miles.

    (b)   $\left(5.859 \times 10^{12}\right)(4.27) = (5.859 \cdot 4.27) \times \left(10^{12}\right) \approx 25 \times 10^{12} \approx 2.5 \times 10^{13}$ miles.

93. $\dfrac{10^5 \cdot \pi \, (\text{light years})}{1} \div \dfrac{2 \times 10^8 \, (\text{years})}{1} \cdot \dfrac{5.859 \times 10^{12} \text{ miles}}{1} \approx 9.2 \times 10^9$ miles.

95. $1 \times 10^{100}$

97. (a)   Move the decimal point 13 places to the left, $12,460,000,000,000 = 1.246 \times 10^{13}$.

    (b)   $\dfrac{1.246 \times 10^{13}}{2.98 \times 10^8} = \dfrac{1.246}{2.98} \cdot \dfrac{10^{13}}{10^8} \approx 0.41812 \times 10^{13-8} = 0.41812 \times 10^5 = \$41,812$


## Section 5.6 Division of Polynomials

1. $\dfrac{a+b}{d} = \dfrac{a}{d} + \dfrac{b}{d}$

3. term

5. False

7. $(x+1) \cdot \left(2x^2 - 2x + 1\right) + 4 = 2x^3 - x + 5$

9. $\dfrac{6x^2}{3x} = \dfrac{6}{3} \cdot \dfrac{x^2}{x} = 2x^{2-1} = 2x$;   Checking: $3x \cdot 2x = (3 \cdot 2)(x \cdot x) = 6x^2$

11. $\dfrac{z^4 + z^3}{z} = \dfrac{z^4}{z} + \dfrac{z^3}{z} = z^{4-1} + z^{3-1} = z^3 + z^2$;   Checking: $(z)\left(z^3 + z^2\right) = \left(z^3\right)(z) + \left(z^2\right)(z) = z^4 + z^3$

13. $\dfrac{a^5 - 6a^3}{2a^3} = \dfrac{a^5}{2a^3} - \dfrac{6a^3}{2a^3} = \dfrac{1}{2}a^{5-3} - 3a^{3-3} = \dfrac{a^2}{2} - 3a^0 = \dfrac{a^2}{2} - 3$

   Checking: $(2a^3)\left(\dfrac{a^2}{2} - 3\right) = (2a^3)\left(\dfrac{a^2}{2}\right) - (2a^3)(3) = \dfrac{2a^{3+2}}{2} - 6a^3 = a^5 - 6a^3$

15. $\dfrac{y + 6y^2}{3y^3} = \dfrac{y}{3y^3} + \dfrac{6y^2}{3y^3} = \dfrac{1}{3}y^{1-3} + \dfrac{6}{3}y^{2-3} = \dfrac{y^{-2}}{3} + 2y^{-1} = \dfrac{1}{3y^2} + \dfrac{2}{y}$

   Checking: $3y^3\left(\dfrac{1}{3y^2} + \dfrac{2}{y}\right) = \dfrac{3y^3}{3y^2} + \dfrac{6y^3}{y} = \dfrac{3}{3}y^{3-2} + 6y^{3-1} = y + 6y^2$

17. $\dfrac{4x - 7x^4}{x^2} = \dfrac{4x}{x^2} - \dfrac{7x^4}{x^2} = 4x^{1-2} - 7x^{4-2} = 4x^{-1} - 7x^2 = \dfrac{4}{x} - 7x^2$

19. $\dfrac{6y^2 + 3y}{3y^3} = \dfrac{6y^2}{3y^3} + \dfrac{3y}{3y^3} = \dfrac{6}{3}y^{2-3} + \dfrac{3}{3}y^{1-3} = 2y^{-1} + y^{-2} = \dfrac{2}{y} + \dfrac{1}{y^2}$

21. $\dfrac{9x^4 - 3x + 6}{3x} = \dfrac{9x^4}{3x} - \dfrac{3x}{3x} + \dfrac{6}{3x} = 3x^{4-1} - 1 + \dfrac{2}{x} = 3x^3 - 1 + \dfrac{2}{x}$

23. $\dfrac{12y^4 - 3y^2 + 6y}{3y^2} = \dfrac{12y^4}{3y^2} - \dfrac{3y^2}{3y^2} + \dfrac{6y}{3y^2} = 4y^{4-2} - 1 + 2y^{1-2} = 4y^2 - 1 + 2y^{-1} = 4y^2 - 1 + \dfrac{2}{y}$

25. $\dfrac{15m^4 - 10m^3 + 20m^2}{5m^2} = \dfrac{15m^4}{5m^2} - \dfrac{10m^3}{5m^2} + \dfrac{20m^2}{5m^2} = 3m^{4-2} - 2m^{3-2} + 4m^{2-2} = 3m^2 - 2m + 4$

27.
$$
\begin{array}{r}
2x+1 \\
x-2{\overline{\smash{\big)}\,2x^2-3x+1}} \\
\underline{2x^2-4x}\phantom{xxxx} \\
x+1 \\
\underline{x-2} \\
3
\end{array}
$$
The solution is: $2x + 1 + \dfrac{3}{x - 2}$

   Check: $(x - 2)(2x + 1) + 3 = x \cdot 2x + x \cdot 1 + (-2) \cdot 2x + (-2) \cdot 1 + 3 = 2x^2 + x - 4x - 2 + 3 = 2x^2 - 3x + 1$

29.
$$
\begin{array}{r}
x+1 \\
x+1{\overline{\smash{\big)}\,x^2+2x+1}} \\
\underline{x^2+\phantom{2}x}\phantom{xxx} \\
x+1 \\
\underline{x+1} \\
0
\end{array}
$$
The solution is: $x + 1$

   Check: $(x + 1)(x + 1) = (x)^2 + (2)(x)(1) + (1)^2 = x^2 + 2x + 1$

31. $x-1\overline{)x^3-x^2+x-2}$      The solution is: $x^2+1+\dfrac{-1}{x-1}$

$\phantom{x-1)}\dfrac{x^2+1}{\phantom{x^3-x^2+x-2}}$

$\phantom{x-1)}\underline{x^3-x^2}$

$\phantom{x-1)x^3-x^2}x-2$

$\phantom{x-1)x^3-x^2}\underline{x-1}$

$\phantom{x-1)x^3-x^2x}-1$

Check: $(x-1)(x^2+1)+(-1)=x\cdot x^2+x\cdot 1+(-1)(x^2)+(-1)(1)+(-1)=$

$x^3+x-x^2-1-1=x^3-x^2+x-2$

33. $x-2\overline{)x^3+\phantom{ }x^2-7x+2}$      The solution is: $x^2+3x-1$

$\phantom{x-2)}\dfrac{x^2+3x-1}{\phantom{x^3+x^2-7x+2}}$

$\phantom{x-2)}\underline{x^3-2x^2}$

$\phantom{x-2)x}3x^2-7x$

$\phantom{x-2)x}\underline{3x^2-6x}$

$\phantom{x-2)x3x^2}-x+2$

$\phantom{x-2)x3x^2}\underline{-x+2}$

$\phantom{x-2)x3x^2-x+}0$

Check: $(x-2)(x^2+3x-1)=x\cdot x^2+x\cdot 3x-x\cdot 1-2\cdot x^2-2\cdot 3x-2(-1)=$

$x^3+3x^2-x-2x^2-6x+2=x^3+x^2-7x+2$

35. $4x+1\overline{)4x^3-3x^2+7x+3}$      The solution is: $x^2-x+2+\dfrac{1}{4x+1}$

$\phantom{4x+1)}\dfrac{x^2-x+2}{\phantom{4x^3-3x^2+7x+3}}$

$\phantom{4x+1)}\underline{4x^3+\phantom{ }x^2}$

$\phantom{4x+1)4}-4x^2+7x$

$\phantom{4x+1)4}\underline{-4x^2-\phantom{ }x}$

$\phantom{4x+1)4-4x^2}8x+3$

$\phantom{4x+1)4-4x^2}\underline{8x+2}$

$\phantom{4x+1)4-4x^2+8x}1$

37. $x-2\overline{)x^3-0x^2-\phantom{ }x+2}$      The solution is: $x^2+2x+3+\dfrac{8}{x-2}$

$\phantom{x-2)}\dfrac{x^2+2x+3}{\phantom{x^3-0x^2-x+2}}$

$\phantom{x-2)}\underline{x^3-2x^2}$

$\phantom{x-2)x}2x^2-\phantom{ }x$

$\phantom{x-2)x}\underline{2x^2-4x}$

$\phantom{x-2)x2x^2}3x+2$

$\phantom{x-2)x2x^2}\underline{3x-6}$

$\phantom{x-2)x2x^2+3x}8$

39.   $x-1\overline{\smash{\big)}\,3x^3+0x^2+0x+2}$    overset $3x^2+3x+3$      The solution is: $3x^2+3x+3+\dfrac{5}{x-1}$

$\underline{3x^3-3x^2}$

$3x^2+0x$

$\underline{3x^2-3x}$

$3x+2$

$\underline{3x-3}$

$5$

41.  $x^2+1\overline{\smash{\big)}\,x^3+3x^2+0x+1}$    overset $x+3$      The solution is: $x+3+\dfrac{-x-2}{x^2+1}$

$\underline{x^3+0x^2+\ x}$

$3x^2-\ x+1$

$\underline{3x^2+0x+3}$

$-x-2$

43.  $x^2-x+1\overline{\smash{\big)}\,x^3+0x^2+0x+1}$   overset $x+1$   The solution is: $x+1$

$\underline{x^3-\ x^2+\ x}$

$x^2-\ x+1$

$\underline{x^2-\ x+1}$

$0$

45.  $x+2\overline{\smash{\big)}\,x^3+0x^2+0x+8}$    overset $x^2-2x+4$      The solution is: $x^2-2x+4$

$\underline{x^3+2x^2}$

$-2x^2+0x$

$\underline{-2x^2-4x}$

$4x+8$

$\underline{4x+8}$

$0$

47.  They are the same.

49.  $2x\cdot l=8x^2 \Rightarrow l=\dfrac{8x^2}{2x} \Rightarrow l=4x^{2-1} \Rightarrow l=4x$

51.  $2x^2\cdot H=2x^3+4x^2 \Rightarrow H=\dfrac{2x^3+4x^2}{2x^2} \Rightarrow H=\dfrac{2x^3}{2x^2}+\dfrac{4x^2}{2x^2}=x^{3-2}+2x^{2-2}=x+2.$ See Figure 51.

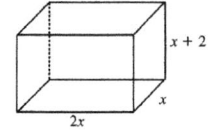

Figure 51

## Checking Basic Concepts  Sections 5.5 and 5.6

1. (a) $9^{-2} = \dfrac{1}{9^2} = \dfrac{1}{9 \cdot 9} = \dfrac{1}{81}$

   (b) $\dfrac{3x^{-3}}{6x^4} = \dfrac{1}{2}x^{-3-4} = \dfrac{1}{2}x^{-7} = \dfrac{1}{2x^7}$

   (c) $\left(4ab^{-4}\right)^{-2} = 4^{-2}a^{-2}b^8 = \dfrac{b^8}{16a^2}$

2. (a) $\dfrac{1}{z^{-5}} = z^5$

   (b) $\dfrac{x^{-3}}{y^{-6}} = \dfrac{y^6}{x^3}$

   (c) $\left(\dfrac{3}{x^2}\right)^{-3} = \left(\dfrac{x^2}{3}\right)^3 = \dfrac{x^{2 \cdot 3}}{3^3} = \dfrac{x^6}{27}$

3. (a) Move the decimal point four places to the left: $45,000 = 4.5 \times 10^4$.

   (b) Move the decimal point four places to the right: $0.000234 = 2.34 \times 10^{-4}$.

   (c) Move the decimal point two places to the right: $0.01 = 1 \times 10^{-2}$.

4. (a) Move the decimal point four places to the right: $4.71 \times 10^4 = 47,100$.

   (b) Move the decimal point three places to the left: $6 \times 10^{-3} = 0.006$.

5. $\dfrac{25a^4 - 15a^3}{5a^3} = \dfrac{25a^4}{5a^3} - \dfrac{15a^3}{5a^3} = 5a^{4-3} - 3a^{3-3} = 5a - 3$

6.
$$
\begin{array}{r}
3x + 2 \\
x - 1 \overline{\smash{\big)}\, 3x^2 -\ x - 4} \\
\underline{3x^2 - 3x} \\
2x - 4 \\
\underline{2x - 2} \\
-2
\end{array}
$$

The quotient is: $3x + 2$. The remainder is: $-2$

7.
$$
\begin{array}{r}
x^2\ + 2x + 1 \\
x^2 + 0x - 3 \overline{\smash{\big)}\, x^4 + 2x^3 - 2x^2 - 5x - 2} \\
\underline{x^4 + 0x^3 - 3x^2} \\
2x^3 +\ x^2 - 5x \\
\underline{2x^3 + 0x^2 - 6x} \\
x^2 + x - 2 \\
\underline{x^2 + 0x - 3} \\
x + 1
\end{array}
$$

The quotient is: $x^2\ + 2x + 1$. The remainder is: $x+1$

8. (a)   Move the decimal point seven places to the left: $93,000,000 = 9.3 \times 10^7$.

   (b)   $\dfrac{9.3 \times 10^7}{1.86 \times 10^5} = \dfrac{9.3}{1.86} \cdot \dfrac{10^7}{10^5} \approx 5 \times 10^{7-5} = 5 \times 10^2 = 500$ seconds. (8 minutes 20 seconds)

## Chapter 5 Review

1. $5^3 = 5 \cdot 5 \cdot 5 = 125$

2. $-3^4 = -(3 \cdot 3 \cdot 3 \cdot 3) = -81$

3. $4(-2)^0 = 4(1) = 4$

4. $3 + 3^2 - 3^0 = 3 + (3 \cdot 3) - 1 = 3 + 9 - 1 = 11$

5. $\dfrac{-5^2}{5} = \dfrac{-(5 \cdot 5)}{5} = \dfrac{-25}{5} = -5$

6. $\left(\dfrac{-5}{5}\right)^2 = \dfrac{(-5)(-5)}{(5)(5)} = \dfrac{25}{25} = 1$

7. $6^2 \cdot 6^3 = 6^{2+3} = 6^5$

8. $10^5 \cdot 10^7 = 10^{5+7} = 10^{12}$

9. $z^4 \cdot z^5 = z^{4+5} = z^9$

10. $y^2 \cdot y \cdot y^3 = y^{2+1+3} = y^6$

11. $5x^2 \cdot 6x^7 = 5 \cdot 6 \cdot x^{2+7} = 30x^9$

12. $\left(ab^3\right)\left(a^3b\right) = a^{1+3}b^{3+1} = a^4b^4$

13. $\left(2^5\right)^2 = 2^{5 \cdot 2} = 2^{10}$

14. $\left(m^4\right)^5 = m^{4 \cdot 5} = m^{20}$

15. $(ab)^3 = a^3b^3$

16. $\left(x^2y^3\right)^4 = x^{2 \cdot 4}y^{3 \cdot 4} = x^8y^{12}$

17. $(xy)^3(x^2y^4)^2 = (x^3y^3)(x^{2 \cdot 2}y^{4 \cdot 2}) = (x^3y^3)(x^4y^8) = x^{3+4}y^{3+8} = x^7y^{11}$

18. $\left(a^2b^9\right)^0 = 1$

19. $(r-t)^4(r-t)^5 = (r-t)^{4+5} = (r-t)^9$

20. $(a+b)^2(a+b)^4 = (a+b)^{2+4} = (a+b)^6$

21. $\left(\dfrac{3}{x-y}\right)^2 = \dfrac{3^2}{(x-y)^2} = \dfrac{3\cdot 3}{(x-y)^2} = \dfrac{9}{(x-y)^2}$

22. $\left(\dfrac{x+y}{2}\right)^3 = \dfrac{(x+y)^3}{2^3} = \dfrac{(x+y)^3}{2\cdot 2\cdot 2} = \dfrac{(x+y)^3}{8}$

23. $2x^2(3x-5) = 2\cdot 3x^2\cdot x - 2\cdot 5\cdot x^2 = 6x^{2+1} - 10x^2 = 6x^3 - 10x^2$

24. $3x\left(4x+x^3\right) = 3\cdot 4\cdot x\cdot x + 3\cdot x\cdot x^3 = 12x^{1+1} + 3x^{1+3} = 12x^2 + 3x^4$

25. degree is 7; coefficient is 6

26. degree is 5; coefficient is $-1$

27. Yes, it is a polynomial; 1 term: $8y$; 1 variable: $y$; $y^1$ so it is a 1st degree polynomial.

28. Yes, it is a polynomial;

    4 terms: $8x^3, -3x^2, x$ and $-5$; 1 variables: $x$; $x^3$ so it is a 3rd degree polynomial.

29. Yes, it is a polynomial;

    3 terms: $a^2, 2ab$ and $b^2$; 2 variables: $a$ and $b$; $a^2$ so it is a 2nd degree polynomial.

30. No, it is not a polynomial, because there are variables in the denominator.

31. $\begin{array}{r} 3x^2 + 4x + 8 \\ + 2x^2 - 5x - 5 \\ \hline 5x^2 - x + 3 \end{array}$

32. $-6x^2 + 3x + 7$

33. $(4x-3)+(-x+7) = 4x+(-x)+(-3)+7 = 3x+4$

34. $\left(3x^2-1\right)-\left(5x^2+12\right) = \left(3x^2-1\right)+\left(-5x^2-12\right) = 3x^2+\left(-5x^2\right)+(-1)+(-12) = -2x^2 - 13$

35. $\left(x^2+5x+6\right)-\left(3x^2-4x+1\right) = \left(x^2+5x+6\right)+\left(-3x^2+4x-1\right) =$

    $x^2+\left(-3x^2\right)+5x+4x+6+(-1) = -2x^2 + 9x + 5$

36. $\left(x^2+3x-5\right)+\left(2x^2-5x-1\right) = x^2+2x^2+3x-5x-5+(-1) = 3x^2-2x-6$

37. $\left(a^3+4a^2\right)+\left(a^3-5a^2+7a\right) = a^3+a^3+4a^2+\left(-5a^2\right)+7a = 2a^3-a^2+7a$

38. $\left(4x^3-2x+6\right)-\left(4x^3-6\right) = \left(4x^3-2x+6\right)+\left(-4x^3+6\right) = 4x^3+\left(-4x^3\right)-2x+6+6 = -2x+12$

39. $\left(xy+y^2\right)+\left(4y^2-4xy\right) = y^2+4y^2+xy+(-4xy) = 5y^2-3xy$

40. $\left(7x^2+2xy+y^2\right)-\left(7x^2-2xy+y^2\right) = \left(7x^2+2xy+y^2\right)+\left(-7x^2+2xy-y^2\right) =$

    $7x^2+\left(-7x^2\right)+2xy+2xy+y^2+\left(-y^2\right) = 4xy$

41. $-x^2\cdot x^3 = -x^{2+3} = -x^5$

42. $-\left(r^2t^3\right)(rt) = -\left(r^{2+1}t^{3+1}\right) = -r^3t^4$

43. $-3(2t-5) = (-3)(2t)+(-3)(-5) = -6t+15$

44. $2y(1-6y) = 2y\cdot 1 + 2y(-6y) = 2y - 12y^2$

45. $6x^3\left(3x^2+5x\right) = 6x^3\cdot 3x^2 + 6x^3\cdot 5x = 18x^{3+2} + 30x^{3+1} = 18x^5 + 30x^4$

46. $-x\left(x^2-2x+9\right) = (-x)\left(x^2\right) + (-x)(-2x) + (-x)(9) = -x^{1+2} + 2x^{1+1} - 9x = -x^3 + 2x^2 - 9x$

47. $-ab\left(a^2-2ab+b^2\right) = -ab\cdot a^2 + (-ab)(-2ab) + (-ab)\left(b^2\right) = -a^3b + 2a^2b^2 - ab^3$

48. $(a-2)(a+5) = a\cdot a + a\cdot 5 + (-2)(a) + (-2)(5) = a^2 + 5a - 2a - 10 = a^2 + 3a - 10$

49. $(8x-3)(x+2) = 8x\cdot x + 8x\cdot 2 + (-3)(x) + (-3)(2) = 8x^2 + 16x - 3x - 6 = 8x^2 + 13x - 6$

50. $(2x-1)(1-x) = 2x\cdot 1 + (2x)(-x) + (-1)(1) + (-1)(-x) = 2x - 2x^2 - 1 + x = -2x^2 + 3x - 1$

51. $\left(y^2+1\right)(2y+1) = y^2\cdot 2y + y^2\cdot 1 + 1\cdot 2y + 1\cdot 1 = 2y^3 + y^2 + 2y + 1$

52. $\left(y^2-1\right)\left(2y^2+1\right) = y^2\cdot 2y^2 + y^2\cdot 1 + (-1)\left(2y^2\right) + (-1)(1) = 2y^4 + y^2 - 2y^2 - 1 = 2y^4 - y^2 - 1$

53. $(z+1)(z^2-z+1) = z\cdot z^2 + (z)(-z) + z\cdot 1 + 1\cdot z^2 + (1)(-z) + 1\cdot 1 = z^3 - z^2 + z + z^2 - z + 1 = z^3 + 1$

54. $(4z-3)(z^2-3z+1) = 4z\cdot z^2 + (4z)(-3z) + 4z\cdot 1 + (-3)\left(z^2\right) + (-3)(-3z) + (-3)(1) =$

    $4z^3 - 12z^2 + 4z - 3z^2 + 9z - 3 = 4z^3 - 15z^2 + 13z - 3$

55. (a)　　$z^2 + z$; See Figure 55.

　　(b)　　$z(z+1) = z\cdot z + z\cdot 1 = z^2 + z$

56. (a)　　$2x^2 + 4x$; See Figure 56.

　　(b)　　$2x(x+2) = 2x\cdot x + 2x\cdot 2 = 2x^2 + 4x$

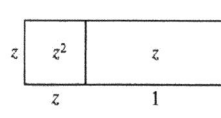
Figure 55

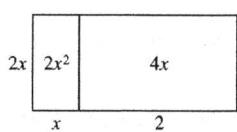
Figure 56

57. $(z+2)(z-2) = (z)^2 - (2)^2 = z^2 - 4$

58. $(5z-9)(5z+9) = (5z)^2 - (9)^2 = 25z^2 - 81$

59. $(1-3y)(1+3y) = (1)^2 - (3y)^2 = 1 - 9y^2$

60. $(5x+4y)(5x-4y) = (5x)^2 - (4y)^2 = 25x^2 - 16y^2$

61. $(rt+1)(rt-1) = (rt)^2 - (1)^2 = r^2t^2 - 1$

62. $\left(2m^2 - n^2\right)\left(2m^2 + n^2\right) = \left(2m^2\right)^2 - \left(n^2\right)^2 = 4m^4 - n^4$

63. $(x+1)^2 = (x)^2 + (2)(x)(1) + (1)^2 = x^2 + 2x + 1$

64. $(4x+3)^2 = (4x)^2 + (2)(4x)(3) + (3)^2 = 16x^2 + 24x + 9$

65. $(y-3)^2 = (y)^2 - (2)(y)(3) + (3)^2 = y^2 - 6y + 9$

66. $(2y-5)^2 = (2y)^2 - (2)(2y)(5) + (5)^2 = 4y^2 - 20y + 25$

67. $(4+a)^2 = (4)^2 + (2)(4)(a) + (a)^2 = 16 + 8a + a^2$

68. $(4-a)^2 = (4)^2 - (2)(4)(a) + (a)^2 = 16 - 8a + a^2$

69. $\left(x^2 + y^2\right)^2 = \left(x^2\right)^2 + (2)\left(x^2\right)\left(y^2\right) + \left(y^2\right)^2 = x^4 + 2x^2y^2 + y^4$

70. $(xy-2)^2 = (xy)^2 - (2)(xy)(2) + (2)^2 = x^2y^2 - 4xy + 4$

71. $(z+5)^3 = (z+5)(z+5)^2 = (z+5)\left(z^2 + 10z + 25\right) = z \cdot z^2 + z \cdot 10z + z \cdot 25 + 5 \cdot z^2 + 5 \cdot 10z + 5 \cdot 25 =$

    $z^3 + 10z^2 + 25z + 5z^2 + 50z + 125 = z^3 + 15z^2 + 75z + 125$

72. $(2z-1)^3 = (2z-1)(2z-1)^2 = (2z-1)\left(4z^2 - 4z + 1\right) =$

    $2z \cdot 4z^2 + (2z)(-4z) + 2z \cdot 1 + (-1)\left(4z^2\right) + (-1)(-4z) + (-1)(1) = 8z^3 - 8z^2 + 2z - 4z^2 + 4z - 1 =$

    $8z^3 - 12z^2 + 6z - 1$

73. $59 \cdot 61 = (60-1)(60+1) = 60^2 - 1^2 = 3600 - 1 = 3599$

74. $22 \cdot 18 = (20+2)(20-2) = 20^2 - 2^2 = 400 - 4 = 396$

75. $9^{-1} = \dfrac{1}{9}$

76. $3^{-2} = \dfrac{1}{3^2} = \dfrac{1}{3 \cdot 3} = \dfrac{1}{9}$

77. $4^3 \cdot 4^{-2} = 4^{3+(-2)} = 4^1 = 4$

78. $10^{-6} \cdot 10^3 = 10^{-6+3} = 10^{-3} = \dfrac{1}{10^3} = \dfrac{1}{10 \cdot 10 \cdot 10} = \dfrac{1}{1000}$

79. $\dfrac{1}{6^{-2}} = 6^2 = 36$

80. $\dfrac{5^7}{5^9} = 5^{7-9} = 5^{-2} = \dfrac{1}{5^2} = \dfrac{1}{5 \cdot 5} = \dfrac{1}{25}$

81. $\left(3^{-1} \cdot 2^2\right)^{-2} = 3^2 \cdot 2^{-4} = 9 \cdot \dfrac{1}{2^4} = 9 \cdot \dfrac{1}{16} = \dfrac{9}{16}$

82. $\left(2^{-4} \cdot 5^3\right)^0 = 2^0 \cdot 5^0 = 1 \cdot 1 = 1$

83. $z^{-2} = \dfrac{1}{z^2}$

84. $y^{-4} = \dfrac{1}{y^4}$

85. $a^{-4} \cdot a^2 = a^{-4+2} = a^{-2} = \dfrac{1}{a^2}$

86. $x^2 \cdot x^{-5} \cdot x = x^{2+(-5)+1} = x^{-2} = \dfrac{1}{x^2}$

87. $(2t)^{-2} = 2^{-2}t^{-2} = \dfrac{1}{2^2 t^2} = \dfrac{1}{4t^2}$

88. $\left(ab^2\right)^{-3} = a^{-3}b^{(2)(-3)} = a^{-3}b^{-6} = \dfrac{1}{a^3 b^6}$

89. $(xy)^{-2}(x^{-2}y)^{-1} = (x^{-2}y^{-2})(x^{(-2)(-1)}y^{-1}) = x^{-2+2}y^{-2+(-1)} = y^{-3} = \dfrac{1}{y^3}$

90. $\dfrac{x^6}{x^2} = x^{6-2} = x^4$

91. $\dfrac{4x}{2x^4} = 2x^{1-4} = 2x^{-3} = \dfrac{2}{x^3}$

92. $\dfrac{20x^5 y^3}{30xy^6} = \dfrac{2}{3}x^{5-1}y^{3-6} = \dfrac{2}{3}x^4 y^{-3} = \dfrac{2x^4}{3y^3}$

93. $\left(\dfrac{a}{b}\right)^5 = \dfrac{a^5}{b^5}$

94. $\dfrac{4}{t^{-4}} = 4t^4$

95. $\dfrac{\left(3m^3 n\right)^{-2}}{\left(2m^2 n^{-3}\right)^3} = \dfrac{3^{-2}m^{-6}n^{-2}}{2^3 m^6 n^{-9}} = \dfrac{1}{9}\dfrac{1}{8} \cdot m^{-6-6} \cdot n^{-2-(-9)} = \dfrac{1}{72}m^{-12}n^7 = \dfrac{n^7}{72m^{12}}$

96. $\left(\dfrac{x^{-4}y^2}{3xy^{-3}}\right)^{-2} = \left(\dfrac{3xy^{-3}}{x^{-4}y^2}\right)^2 = \dfrac{3^2 x^2 y^{-6}}{x^{-8}y^4} = 9x^{2-(-8)}y^{-6-4} = 9x^{10}y^{-10} = \dfrac{9x^{10}}{y^{10}}$

97. $\left(\dfrac{x}{y}\right)^{-2} = \left(\dfrac{y}{x}\right)^2 = \dfrac{y^2}{x^2}$

98. $\left(\dfrac{3u}{2v}\right)^{-1} = \left(\dfrac{2v}{3u}\right)^1 = \dfrac{2v}{3u}$

99. Move the decimal point two places to the right, $6 \times 10^2 = 600$.

100. Move the decimal point four places to the right, $5.24 \times 10^4 = 52,400$.

101. Move the decimal point three places to the left, $3.7 \times 10^{-3} = 0.0037$.

102. Move the decimal point two places to the left, $6.234 \times 10^{-2} = 0.06234$.

103. Move the decimal point four places to the left, $10,000 = 1 \times 10^4$.

104. Move the decimal point seven places to the left, $56,100,000 = 5.61 \times 10^7$.

105. Move the decimal point five places to the right, $0.000054 = 5.4 \times 10^{-5}$.

106. Move the decimal point three places to the right, $0.001 = 1 \times 10^{-3}$.

107. $\left(4 \times 10^2\right)\left(6 \times 10^4\right) = (4 \cdot 6) \times \left(10^2 \cdot 10^4\right) = 24 \times 10^6 = 2.4 \times 10^7$;

Move the decimal point seven places to the right, $2.4 \times 10^7 = 24,000,000$.

108. $\left(\dfrac{8 \times 10^3}{4 \times 10^4}\right) = \dfrac{8}{4} \times \dfrac{10^3}{10^4} = 2 \times 10^{3-4} = 2 \times 10^{-1}$;

Move the decimal point one place to the left, $2 \times 10^{-1} = 0.2$.

109. $\dfrac{5x^2 + 3x}{3x} = \dfrac{5x^2}{3x} + \dfrac{3x}{3x} = \dfrac{5}{3}x^{2-1} + 1 = \dfrac{5}{3}x + 1$; Checking: $3x\left(\dfrac{5}{3}x + 1\right) = 3x \cdot \dfrac{5}{3}x + 3x \cdot 1 = 5x^2 + 3x$

110. $\dfrac{6b^4 - 4b^2 + 2}{2b^2} = \dfrac{6b^4}{2b^2} - \dfrac{4b^2}{2b^2} + \dfrac{2}{2b^2} = 3b^{4-2} - 2b^{2-2} + \dfrac{1}{b^2} = 3b^2 - 2 + \dfrac{1}{b^2}$;

Checking: $\left(2b^2\right)\left(3b^2 - 2 + \dfrac{1}{b^2}\right) = 2b^2 \cdot 3b^2 + \left(2b^2\right)(-2) + 2b^2 \cdot \dfrac{1}{b^2} = 6b^4 - 4b^2 + 2$

111.
$$
\begin{array}{r}
3x + 2 \\
x-1{\overline{\smash{\big)}\,3x^2 -\ \ x + 2}} \\
\underline{3x^2 -\ 3x\phantom{ + 2}} \\
2x + 2 \\
\underline{2x - 2} \\
4
\end{array}
$$
The solution is: $3x + 2 + \dfrac{4}{x-1}$

Check: $(x-1)(3x+2) + 4 = x \cdot 3x + x \cdot 2 + (-1)(3x) + (-1)(2) + 4 = 3x^2 + 2x - 3x - 2 + 4 = 3x^2 - x + 2$

112.
$$
\begin{array}{r}
3x - 4 \\
3x+2{\overline{\smash{\big)}\,9x^2 - 6x - 2}} \\
\underline{9x^2 + 6x\phantom{ - 2}} \\
-12x - 2 \\
\underline{-12x - 8} \\
6
\end{array}
$$
The solution is: $3x - 4 + \dfrac{6}{3x+2}$

Check: $(3x+2)(3x-4) + 6 = 3x \cdot 3x + (3x)(-4) + 2 \cdot 3x + (2)(-4) + 6 =$

$9x^2 - 12x + 6x - 8 + 6 = 9x^2 - 6x - 2$

113.
$$\begin{array}{r} x^2-3x-1 \\ 4x+1{\overline{\smash{\big)}\,4x^3-11x^2-7x-1}} \end{array}$$
The solution is: $x^2-3x-1$

$$\underline{4x^3+\ \ x^2}$$
$$-12x^2-7x$$
$$\underline{-12x^2-3x}$$
$$-4x-1$$
$$\underline{-4x-1}$$
$$0$$

Check: $(4x+1)(x^2-3x-1)=4x\cdot x^2+(4x)(-3x)+(4x)(-1)+1\cdot x^2+(1)(-3x)+(1)(-1)=$

$4x^3-12x^2-4x+x^2-3x-1=4x^3-11x^2-7x-1$

114.
$$\begin{array}{r} x^2 \\ 2x-1{\overline{\smash{\big)}\,2x^3-x^2-1}} \end{array}$$
The solution is: $x^2+\dfrac{-1}{2x-1}$

$$\underline{2x^3-x^2}$$
$$0-1$$
$$-1$$

Check: $(x^2)(2x-1)+(-1)=x^2\cdot 2x+(x^2)(-1)+(-1)=2x^3-x^2-1$

115.
$$\begin{array}{r} x-1 \\ x^2+1{\overline{\smash{\big)}\,x^3-x^2-\ \ x+1}} \end{array}$$
The solution is: $x-1+\dfrac{-2x+2}{x^2+1}$

$$\underline{x^3+0x^2+\ \ x}$$
$$-x^2-2x+1$$
$$\underline{-x^2+0x-1}$$
$$-2x+2$$

Check: $(x-1)(x^2+1)+(-2x+2)=x\cdot x^2+x\cdot1+(-1)(x^2)+(-1)(1)+(-2x+2)=$

$x^3+x-x^2-1-2x+2=x^3-x^2-x+1$

116.
$$\begin{array}{r} x^2+2x+5 \\ x^2+x+1{\overline{\smash{\big)}\,x^4+3x^3+8x^2+7x+5}} \end{array}$$
The solution is: $x^2+2x+5$

$$\underline{x^4+\ \ x^3+\ \ x^2}$$
$$2x^3+7x^2+7x$$
$$\underline{2x^3+2x^2+2x}$$
$$5x^2+5x+5$$
$$\underline{5x^2+5x+5}$$
$$0$$

Check: $(x^2+x+1)(x^2+2x+5)=x^2\cdot x^2+x^2\cdot2x+x^2\cdot5+x\cdot x^2+x\cdot2x+x\cdot5+1\cdot x^2+1\cdot2x+1\cdot5=$

$x^4+2x^3+5x^2+x^3+2x^2+5x+x^2+2x+5=x^4+3x^3+8x^2+7x+5$

117. (a) $\dfrac{1}{2}t^2 + 60 \Rightarrow \dfrac{1}{2}(0^2) + 60 = 60$ bpm

(b) $\dfrac{1}{2}t^2 + 60 \Rightarrow \dfrac{1}{2}(10^2) + 60 = 50 + 60 = 110$ bpm

(c) It increases.

118. $L \times W = 2xy = 1$ rectangle $\Rightarrow$ 3 rectangles $= 3 \cdot 2xy = 6xy$;

for $x = 3$ ft, $y = 4$ ft, $6xy = 6(3)(4) = 72$ ft$^2$

119. $5 \cdot 3z = 15z$, $5 \cdot 2z = 10z$, $2z \cdot (3z + 2z) = 6z^2 + 4z^2 \Rightarrow 15z + 10z + 6z^2 + 4z^2 = 10z^2 + 25z$; for

$z = 6$ in. $\Rightarrow 10z^2 + 25z = 10(6)^2 + 25(6) = 10(36) + 150 = 360 + 150 = 510$ in$^2$

120. $\left(x^2 y\right)\left(x^2 y\right) = x^{2+2} y^{1+1} = x^4 y^2$

121. $P(1 + 0.06)^3 = P(1.06)^3 = P(1.191016)$; let $P = \$700 \Rightarrow \$700(1.191016) = \$833.71$.

122. $\dfrac{4}{3}\pi(x+2)^3 = \dfrac{4}{3}\pi(x+2)(x+2)^2 = \dfrac{4}{3}\pi(x+2)\left(x^2 + 4x + 4\right) =$

$\dfrac{4}{3}\pi\left(x \cdot x^2 + x \cdot 4x + x \cdot 4 + 2 \cdot x^2 + 2 \cdot 4x + 2 \cdot 4\right) = \dfrac{4}{3}\pi\left(x^3 + 4x^2 + 4x + 2x^2 + 8x + 8\right) =$

$\dfrac{4}{3}\pi\left(x^3 + 6x^2 + 12x + 8\right) = \dfrac{4}{3}\pi \cdot x^3 + \dfrac{4}{3}\pi \cdot 6x^2 + \dfrac{4}{3}\pi \cdot 12x + \dfrac{4}{3}\pi \cdot 8 = \dfrac{4}{3}\pi x^3 + 8\pi x^2 + 16\pi x + \dfrac{32}{3}\pi$

123. (a) $t(96 - 16t) = 96t - 16t^2$

(b) $t = 2 \Rightarrow 96t - 16t^2 = 96(2) - 16(2)^2 = 192 - 16(4) = 192 - 64 = 128$, after 2 seconds the ball is

128 feet high.

124. (a) If $P = 2L + 2W$ then $1200 = 2L + 2W \Rightarrow 1200 - 2L = 2W \Rightarrow 600 - L = W$. Now $A = L \cdot W$ so

$A = L \cdot (600 - L) \Rightarrow A = 600L - L^2$.

(b) $L = 50 \Rightarrow 600L - L^2 = 600(50) - (50)^2 = 30{,}000 - 2500 = 27{,}500$. A rectangular building

with a perimeter of 1200 ft and a side of length 50 ft has an area of $27{,}500$ ft$^2$.

125. (a) $(x+5)(x+5) = (x)^2 + 2(x)(5) + (5)^2 = x^2 + 10x + 25$

(b) $x \cdot x + 5 \cdot x + x \cdot 5 + 5 \cdot 5 = x^2 + 5x + 5x + 25 = x^2 + 10x + 25$

126. (a) $(x+4)(x+4) - (x-4)(x-4) = x^2 + 8x + 16 - \left(x^2 - 8x + 16\right) = 16x$

(b) $x = 100 \Rightarrow 16x = 16(100) = 1600$

127. 2.19 trillion $= 2.19 \times 10^{12}$, 249 million $= 2.49 \times 10^8$

$\dfrac{2.19 \times 10^{12}}{2.49 \times 10^8} = \dfrac{2.19}{2.49} \times \dfrac{10^{12}}{10^8} = 0.8795 \times 10^{12-8} = 0.8795 \times 10^4 = 8.795 \times 10^3 \approx \$8{,}795$/person.

128.  $239 \text{ million} = 2.39 \times 10^8$; $(2.31)(2.39 \times 10^8) = (2.31)(2.39) \times 10^8 = 5.5209 \times 10^8 \text{ gal} \approx$

$552,090,000 \text{ gal}$.

## Chapter 5 Test

1. (a)  $-5^0 = -1 \cdot 5^0 = -1 \cdot 1 = -1$

   (b)  $-9^2 = -1 \cdot 9^2 = -1 \cdot 81 = -81$

2. (a)  $-4^2 + 10 = -(4 \cdot 4) + 10 = -16 + 10 = -6$

   (b)  $8^{-2} = \dfrac{1}{8^2} = \dfrac{1}{8 \cdot 8} = \dfrac{1}{64}$

   (c)  $\dfrac{1}{2^{-3}} = 2^3 = 2 \cdot 2 \cdot 2 = 8$

   (d)  $-3x^0 = -3(1) = -3$

3. The polynomial has 3 terms: $5x^2, -3xy, -7y^3$; 2 variables $x$ and $y$; $y^3 \Rightarrow$ 3rd degree.

4.  $x^3 - 4x + 8$

5.  $(-3x + 4) + (7x + 2) = (-3x) + (7x) + 4 + 2 = 4x + 6$

6.  $(y^2 - 2y + 6) - (4y^3 + 5) = (y^3 - 2y + 6) + (-4y^3 - 5) = y^3 + (-4y^3) + (-2y) + 6 + (-5) = -3y^3 - 2y + 1$

7.  $\left(5x^2 - x + 3\right) - \left(4x^2 - 2x + 10\right) = \left(5x^2 - x + 3\right) + \left(-4x^2 + 2x - 10\right) =$

$5x^2 + \left(-4x^2\right) + (-x) + 2x + 3 + (-10) = x^2 + x - 7$

8.  $\left(a^3 + 5ab\right) + \left(3a^3 - 3ab\right) = a^3 + 3a^3 + 5ab + (-3ab) = 4a^3 + 2ab$

9.  $6y^4 \cdot 4y^7 = 6 \cdot 4 y^{4+7} = 24y^{11}$

10.  $(a^2 b^3)^2 (ab^2) = a^{2 \cdot 2} b^{3 \cdot 2} ab^2 = a^{4+1} b^{6+2} = a^5 b^8$

11.  $x^7 \cdot x^{-3} = x^{7+(-3)} = x^4$

12.  $(a^{-1} b^2)^{-3} = a^{(-1)(-3)} b^{(2)(-3)} = a^3 b^{-6} = \dfrac{a^3}{b^6}$

13.  $ab\left(a^2 - b^2\right) = ab \cdot a^2 + ab\left(-b^2\right) = a^3 b - ab^3$

14.  $\left(\dfrac{3a^2}{2b^{-3}}\right)^{-2} = \dfrac{3^{-2} a^{(2)(-2)}}{2^{-2} b^{(-3)(-2)}} = \dfrac{2^2}{3^2 a^4 b^6} = \dfrac{4}{9a^4 b^6}$

15.  $\dfrac{12xy^4}{6x^2 y} = 2x^{1-2} y^{4-1} = 2x^{-1} y^3 = \dfrac{2y^3}{x}$

16. $\left(\dfrac{2}{a+b}\right)^4 = \dfrac{2^4}{(a+b)^4} = \dfrac{16}{(a+b)^4}$

17. $3x^2\left(4x^3-6x+1\right) = 3x^2 \cdot 4x^3 + \left(3x^2\right)(-6x) + 3x^2 \cdot 1 = 12x^5 - 18x^3 + 3x^2$

18. $(z-3)(2z+4) = z \cdot 2z + z \cdot 4 + (-3)(2z) + (-3)(4) = 2z^2 + 4z - 6z - 12 = 2z^2 - 2z - 12$

19. $\left(7y^2-3\right)\left(7y^2+3\right) = \left(7y^2\right)^2 - (3)^2 = 49y^4 - 9$

20. $(3x-2)^2 = (3x)^2 - (2)(3x)(2) + (2)^2 = 9x^2 - 12x + 4$

21. $(m+3)^3 = (m+3)(m+3)^2 = (m+3)(m^2+6m+9) = m \cdot m^2 + m \cdot 6m + m \cdot 9 + 3 \cdot m^2 + 3 \cdot 6m + 3 \cdot 9 =$

  $m^3 + 6m^2 + 9m + 3m^2 + 18m + 27 = m^3 + 9m^2 + 27m + 27$

22. $(y+2)(y^2-2y+3) = (y+2)(y^2) + (y+2)(-2y) + (y+2)(3)$

  $= y \cdot y^2 + 2 \cdot y^2 + y(-2y) + 2(-2y) + y \cdot 3 + 2 \cdot 3$

  $= y^3 + 2y^2 + (-2y^2) + (-4y) + 3y + 6 = y^3 - y + 6$

23. $78 \cdot 82 = (80-2)(80+2) = (80)^2 - (2)^2 = 6400 - 4 = 6396$

24. Move the decimal point three places to the left, $6.1 \times 10^{-3} = 0.0061$.

25. Move the decimal point three places to the left, $5410 = 5.41 \times 10^3$.

26. $\dfrac{9x^3 - 6x^2 + 3x}{3x^2} = \dfrac{9x^3}{3x^2} - \dfrac{6x^2}{3x^2} + \dfrac{3x}{3x^2} = 3x^{3-2} - 2x^{2-2} + \dfrac{1}{x} = 3x - 2 + \dfrac{1}{x}$

27.
$$\begin{array}{r} x^2 - x + 1 \\ x+2\overline{\smash{)}\,x^3 + x^2 - x + 1} \\ \underline{x^3 + 2x^2} \\ -x^2 - x \\ \underline{-x^2 - 2x} \\ x + 1 \\ \underline{x + 2} \\ -1 \end{array}$$

  The solution is: $x^2 - x + 1 + \dfrac{-1}{x+2}$

28. (a) $20t$

  (b) $2t + 2000$

  (c) $(20t) - (2t + 2000) = 20t - 2t - 2000 = 18t - 2000$; profit from selling $t$ tickets.

29. $(3x)(2x) + (3x)(2x) = 6x^2 + 6x^2 = 12x^2$; For $x = 10$ feet $\Rightarrow 12x^2 = 12(10)^2 = 12(100) = 1200 \text{ ft}^2$

30. $(3x)(x+3)(x+6) = (3x)(x \cdot x + x \cdot 6 + 3 \cdot x + 3 \cdot 6) = (3x)\left(x^2 + 9x + 18\right) =$

  $3x \cdot x^2 + 3x \cdot 9x + 3x \cdot 18 = 3x^3 + 27x^2 + 54x$

31. (a)  $t(88-16t)=t\cdot 88-t\cdot 16t=88t-16t^2$

    (b)  $t=3 \Rightarrow 88t-16t^2 = 88(3)-16(3)^2 = 264-16(9)=264-144=120$

    After 3 seconds the ball is 120 feet high.

## Chapter 5  Extended and Discovery Exercises

1. Conjecture: Make the power of the 10's the same and you can add or subtract the lead numbers and multiply by that power of 10. $(4\times10^3)+(3\times10^3)=(4+3)\times10^3 = 7\times10^3$; Answer checks.

2. Conjecture: Make the power of the 10's the same and you can add or subtract the lead numbers and multiply by that power of 10. $(5\times10^{-2})-(2\times10^{-2})=(5-2)\times10^{-2} = 3\times10^{-2}$; Answer checks.

3. Conjecture: Make the power of the 10's the same and you can add or subtract the lead numbers and multiply by that power of 10. $(1.2\times10^4)-(3\times10^3)=(12\times10^3)-(3\times10^3)=(12-3)\times10^3 =9\times10^3$; Answer checks.

4. Conjecture: Make the power of the 10's the same and you can add or subtract the lead numbers and multiply by that power of 10.

    $(2\times10^2)+(6\times10^1)=(2\times10^2)+(0.6\times10^2)=(2+0.6)\times10^2 = 2.6\times10^2$; Answer checks.

5. Conjecture: Make the power of the 10's the same and you can add or subtract the lead numbers and multiply by that power of 10.

    $(2\times10^{-1})+(4\times10^{-2})=(2\times10^{-1})+(0.4\times10^{-1})=(2+0.4)\times10^{-1} = 2.4\times10^{-1}$; Answer checks.

6. Conjecture: Make the power of the 10's the same and you can add or subtract the lead numbers and multiply by that power of 10.

    $(2\times10^{-3})-(5\times10^{-2})=(0.2\times10^{-2})-(5\times10^{-2})=(0.2-5)\times10^{-2} = -4.8\times10^{-2}$; Answer checks.

7. (a)  The length of the box is $30-2x$, the width is $20-2x$, and the height is $x$.

    If $V = L\times W\times H$ then, $V = (30-2x)(20-2x)(x)=$

    $(600-60x-40x+4x^2)(x)=(4x^2-100x+600)(x)=\ 4x^3-100x^2+600x$

    (b)  Let $x=4$, then $4(4)^3-100(4)^2+600(4)=4(64)-100(16)+600(4)=$

    $256-1600+2400 = 1056\ \text{in}^3$

8. (a)  The length and width would be $25-2x$ and the height would be $x$. Therefore the base is equal

    to $(25-2x)^2 = 625-100x+4x^2$ and the four sides are each

    $(25-2x)(x)=25x-2x^2$. 4 sides $= 100x-8x^2$.

    Adding, $625-100x+4x^2+100x-8x^2 = 625-4x^2$.

(b) Let $x = 3$, then $625 - 4(3)^2 = 625 - 4(9) = 625 - 36 = 589$ in$^2$.

9. See Figure 9a & 9b. No, not equal for all values of $x$; $3x(4 - 5x) = 12x - 15x^2$

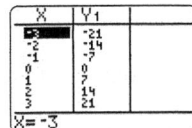

      Figure 9a          Figure 9b

10. See Figure 10a & 10b. No, not equal for all values of $x$; $(x - 1)^2 = x^2 - 2x + 1$

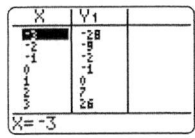

      Figure 10a         Figure 10b

11. See Figure 11a & 11b. Yes, equal for all values of $x$.

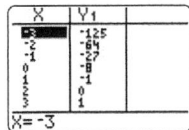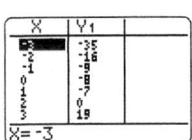

      Figure 11a         Figure 11b

12. See Figure 12a & 12b. No, not equal for all values of $x$; $(x - 2)^3 = (x - 2)(x - 2)^2 =$

$$(x - 2)(x^2 - 4x + 4) = x^3 - 4x^2 + 4x - 2x^2 + 8x - 8 = x^3 - 6x^2 + 12x - 8$$

      Figure 12a         Figure 12b

## Chapters 1-5  Cumulative Review Exercises

1. (a)    $18 - 2 \cdot 5 = 18 - 10 = 8$

   (b)    $42 \div 7 + 2 = 6 + 2 = 8$

2. (a)    $21 - (-8) = 21 + 8 = 29$

   (b)    $-\dfrac{7}{3} \div \left(-\dfrac{14}{9}\right) = -\dfrac{7}{3} \cdot -\dfrac{9}{14} = \dfrac{7 \cdot 9}{3 \cdot 14} = \dfrac{7 \cdot 3 \cdot 3}{3 \cdot 2 \cdot 7} = \dfrac{3}{2}$

3. (a)    $(x - 3) + x = 4 + x \Rightarrow x + x - 3 = 4 + x \Rightarrow 2x - 3 = 4 + x$

       $\Rightarrow 2x - x - 3 = 4 + x - x \Rightarrow x - 3 = 4 \Rightarrow x - 3 + 3 = 4 + 3 \Rightarrow x = 7$

   (b)    $2(5x - 4) = 1 + 10x \Rightarrow 10x - 8 = 1 + 10x \Rightarrow 10x - 10x - 8 = 1 + 10x - 10x \Rightarrow -8 = 1$. Because this

       statement is not true, the equation has no solutions.

4. (a)  $2 + 6x = 2(3x + 1) \Rightarrow 2 + 6x = 6x + 2 \Rightarrow 2 + 6x = 2 + 6x$. Because this statement is true for any

value for $x$ the equation has infinitely many solutions.

   (b)  $11x - 9 = -31 \Rightarrow 11x - 9 + 9 = -31 + 9 \Rightarrow 11x = -22 \Rightarrow \dfrac{11x}{11} = \dfrac{-22}{11} \Rightarrow x = -2$

5. Use the distance $(d) = $ rate $(r) \times $ time $(t)$ formula. 4 hours 30 minutes equals 4.5 hours. Thus,

$$d = rt \Rightarrow 306 = r(4.5) \Rightarrow \frac{306}{4.5} = \frac{4.5r}{4.5} \Rightarrow 68 = r, \text{ or } 68 \text{ mph.}$$

6. (a)  $42\% = \dfrac{42}{100} = \dfrac{21 \cdot 2}{50 \cdot 2} = \dfrac{21}{50}$

   (b)  $0.076 = \dfrac{76}{1000} = \dfrac{19 \cdot 4}{250 \cdot 4} = \dfrac{19}{250}$

7.

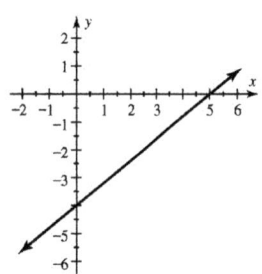

8.

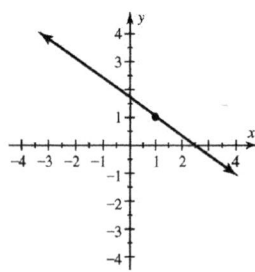

9. For every increase of 1 in the run, the rise decreases by 2. Therefore, the slope $m$ is $-2$. When

$x = 0$, $y = 1$. Therefore, the $y$-intercept is 1. Thus $y = mx + b \Rightarrow y = -2x + 1$.

10. To find the $x$-intercept, set $y = 0$: $2y = 3x - 6 \Rightarrow 2(0) = 3x - 6 \Rightarrow$

$0 = 3x - 6 \Rightarrow 3x = 6 \Rightarrow \dfrac{3x}{3} = \dfrac{6}{3} \Rightarrow x = 2$. Therefore, the $x$-intercept is 2.

To find the $y$-intercept, set $x = 0$: $2y = 3x - 6 \Rightarrow 2y = 3(0) - 6 \Rightarrow 2y = -6 \Rightarrow \dfrac{2y}{2} = \dfrac{-6}{2} \Rightarrow y = -3$.

Therefore, the $y$-intercept is $-3$.

11. $3x - 6y = 7 \Rightarrow -6y = -3x + 7 \Rightarrow y = \dfrac{1}{2}x - \dfrac{7}{6}$. The line parallel to $y = \dfrac{1}{2}x - \dfrac{7}{6}$ has slope $\dfrac{1}{2}$. Insert the

values of the point $(2, -3)$ and the value of the slope into the slope-intercept form and solve for $b$:

$$y = mx + b \Rightarrow -3 = \frac{1}{2}(2) + b \Rightarrow -3 = 1 + b \Rightarrow -3 - 1 = 1 - 1 + b \Rightarrow -4 = b. \text{ Thus, } y = \frac{1}{2}x - 4.$$

12. Find the slope: $m = \dfrac{y_2 - y_1}{x_2 - x_1} = \dfrac{4 - (-5)}{1 - (-2)} = \dfrac{4 + 5}{1 + 2} = \dfrac{9}{3} = 3$. Then, insert the values of the point $(1, 4)$ and

the value of the slope into the point-slope form: $y - y_1 = m(x - x_1) \Rightarrow y - 4 = 3(x - 1)$. Then, convert

to slope-intercept form: $y - 4 = 3(x - 1) \Rightarrow y - 4 = 3x - 3 \Rightarrow y - 4 + 4 = 3x - 3 + 4 \Rightarrow y = 3x + 1$.

13. Multiplying the first equation by $-2$ and adding the two equations will eliminate both variables.

$$-8x - 6y = 12$$
$$\underline{8x + 6y = 12}$$
$$0 = 24 \qquad \text{Since this is false, there are no solutions.}$$

14. Note that $x - 3y = 5 \Rightarrow x = 3y + 5$. Substituting $x = 3y + 5$ into the second equation yields the

following: $3(3y + 5) + y = 5 \Rightarrow 9y + 15 + y = 5 \Rightarrow 10y + 15 = 5 \Rightarrow 10y = -10 \Rightarrow \dfrac{10y}{10} = \dfrac{-10}{10} \Rightarrow y = -1,$

and so $x = 3(-1) + 5 \Rightarrow x = 2$. The solution is $(2, -1)$.

15. Multiplying the first equation by 3 and adding the two equations will eliminate both variables.

$$3x + 12y = -24$$
$$\underline{-3x - 12y = \phantom{-}24}$$
$$0 = 0 \qquad \text{Since this is true, there are infinitely many solutions.}$$

16. Note that $x - 5y = 30 \Rightarrow x = 5y + 30$. Substituting $x = 5y + 30$ into the second equation yields the

following: $2(5y + 30) + y = -6 \Rightarrow 10y + 60 + y = -6 \Rightarrow 11y + 60 = -6$

$\Rightarrow 11y = -66 \Rightarrow \dfrac{11y}{11} = \dfrac{-66}{11} \Rightarrow y = -6,$ and so $x = 5(-6) + 30 = 0$. The solution is $(0, -6)$.

17. See Figure 17.

18. See Figure 18.

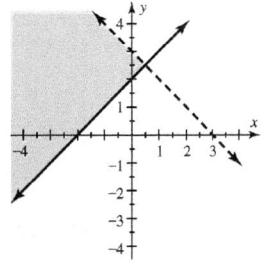

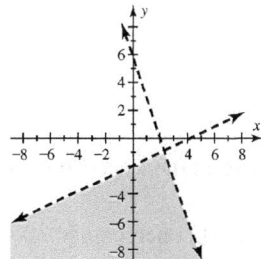

Figure 17                                Figure 18

19. (a) $\quad 3x^2 \cdot 5x^3 = (3 \cdot 5)x^{2+3} = 15x^5$

(b) $\quad (x^3 y)^2 (x^4 y^5) = (x^{3 \cdot 2} y^2)(x^4 y^5) = x^{6+4} y^{2+5} = x^{10} y^7$

20. (a) $\quad (5x^2 - 3x + 4) - (3x^2 - 2x + 1) = 5x^2 + (-3x^2) + (-3x) + 2x + 4 + (-1) = 2x^2 - x + 3$

(b) $\quad (7a^3 - 4a^2 - 5) + (5a^3 + 4a^2 + a) = 7a^3 + 5a^3 + (-4a^2) + 4a^2 + a + (-5) = 12a^3 + a - 5$

21. (a)   $(2x+3)(x-7) = 2x \cdot x + 2x(-7) + 3 \cdot x + 3(-7) = 2x^2 - 14x + 3x - 21 = 2x^2 - 11x - 21$

    (b)   $(y+3)(y^2 - 3y - 1) = y \cdot y^2 + y(-3y) + y(-1) + 3 \cdot y^2 + 3(-3y) + 3(-1)$

          $= y^3 - 3y^2 - y + 3y^2 - 9y - 3 = y^3 - 10y - 3$

    (c)   $(4x+7)(4x-7) = (4x)^2 - (7)^2 = 16x^2 - 49$

    (d)   $(5a+3)^2 = (5a)^2 + 2(5a)(3) + (3)^2 = 25a^2 + 30a + 9$

22. (a)   $x^{-5} \cdot x^3 \cdot x = x^{-5+3+1} = x^{-1} = \dfrac{1}{x}$

    (b)   $\left(\dfrac{2}{x^3}\right)^{-3} = \left(\dfrac{x^3}{2}\right)^3 = \dfrac{x^{3 \cdot 3}}{2^3} = \dfrac{x^9}{2 \cdot 2 \cdot 2} = \dfrac{x^9}{8}$

    (c)   $\dfrac{3x^2 y^{-1}}{6x^{-2} y} = \dfrac{3}{6} x^{2-(-2)} y^{-1-1} = \dfrac{1}{2} x^4 y^{-2} = \dfrac{x^4}{2y^2}$

    (d)   $(xy^{-2})^3 (x^{-2}y)^{-2} = (x^3 y^{-2 \cdot 3})(x^{(-2)(-2)} y^{-2}) = (x^3 y^{-6})(x^4 y^{-2}) = x^{3+4} y^{-6+(-2)} = x^7 y^{-8} = \dfrac{x^7}{y^8}$

23. Move the decimal point 10 places to the left, $24{,}000{,}000{,}000 = 2.4 \times 10^{10}$.

24. Move the decimal point 7 places to the left, $4.71 \times 10^{-7} = 0.000000471$.

25. (a)   $\dfrac{8x^3 - 2x}{2x} = \dfrac{8x^3}{2x} - \dfrac{2x}{2x} = 4x^{3-1} - 1 = 4x^2 - 1$

    (b)   
    $$\begin{array}{r}
    2x - 5 \\
    x+3 \overline{)\, 2x^2 + x - 14} \\
    \underline{2x^2 + 6x} \\
    -5x - 14 \\
    \underline{-5x - 15} \\
    1
    \end{array}$$
    The solution is $2x - 5 + \dfrac{1}{x+3}$.

26. First, subtract the smaller number from the larger to obtain the difference between them:

    $1200 - 900 = 300$. Then, determine what percentage 300 is of 1200: $\dfrac{300}{1200} = 0.25$. Thus the percent

    change from \$1200 to \$900 is $-25\%$.

27. Let $x$ represent the amount of 3% acid. Then, the 5% acid solution is the sum of the 3% acid and 6%

    acid solutions. Therefore $0.03x + 0.06(400) = 0.05(x+400) \Rightarrow 0.03x + 24 = 0.05x + 20$

    $\Rightarrow 0.03x - 0.03x + 24 = 0.05x - 0.03x + 20 \Rightarrow 24 = 0.02x + 20 \Rightarrow 24 - 20 = 0.02x + 20 - 20$

    $\Rightarrow 4 = 0.02x \Rightarrow \dfrac{4}{0.02} = \dfrac{0.02x}{0.02} \Rightarrow 200 = x$

    200 ml of 3% acid should be added to the 6% acid solution to dilute it to a 5% acid solution.

28. (a)   $(x+5)(2x) = x \cdot 2x + 5 \cdot 2x = 2x^{1+1} + 10x = 2x^2 + 10x$

   (b)   $(x+2)(x+5) = x \cdot x + x \cdot 5 + 2 \cdot x + 2 \cdot 5 = x^{1+1} + 5x + 2x + 10 = x^2 + 7x + 10$

   (c)   $(x+2)(2x) = x \cdot 2x + 2 \cdot 2x = 2x^{1+1} + 4x = 2x^2 + 4x$

   (d)   $3 \text{ sides} = 2x^2 + 10x + x^2 + 7x + 10 + 2x^2 + 4x$

   $= 2x^2 + x^2 + 2x^2 + 10x + 7x + 4x + 10 = 5x^2 + 21x + 10;$

   $6 \text{ sides} = 2(5x^2 + 21x + 10) = 2 \cdot 5x^2 + 2 \cdot 21x + 2 \cdot 10 = 10x^2 + 42x + 20$

# Chapter 6: Factoring Polynomials and Solving Equations

## Section 6.1 Introduction to Factoring

1. factor

3. multiplying

5. greatest common factor (GCF)

7. Factors of $2x^2$ are 1, 2, $x$, $2x$, $x^2$ and $2x^2$; factors of $4x$ are 1, 2, 4, $x$, $2x$ and $4x$. Therefore, common factors are 1, 2, $x$ and $2x$.

9. $2x = 2 \cdot x$ and $4 = 2 \cdot 2 \Rightarrow 2(x+2)$. See Figure 9.

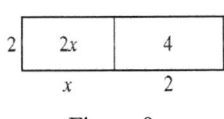

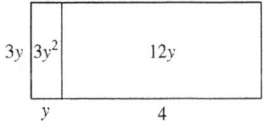

Figure 9                              Figure 11                              Figure 13

11. $z^2 = z \cdot z$ and $4z = 4 \cdot z \Rightarrow z(z+4)$. See Figure 11.

13. $3y^2 = 3y \cdot y$ and $12y = 3y \cdot 4 \Rightarrow 3y(y+4)$. See Figure 13.

15. In the expression $3x^2 + 9x$, the terms $3x^2$ and $9x$ both contain a common factor of $3x$ because $3x^2 = 3x \cdot x$ and $9x = 3x \cdot 3$. Therefore $3x^2 + 9x = 3x(x+3)$.

17. In the expression $4y^3 - 2y^2$, the terms $4y^3$ and $-2y^2$ both contain a common factor of $2y^2$ because $4y^3 = 2y^2 \cdot 2y$ and $-2y^2 = 2y^2(-1)$. Therefore $4y^3 - 2y^2 = 2y^2(2y-1)$.

19. In the expression $2z^3 + 8z^2 - 4z$, the terms $2z^3, 8z^2$, and $-4z$ all contain a common factor of $2z$ because $2z^3 = 2z \cdot z^2$, $8z^2 = 2z \cdot 4z$, and $-4z = 2z(-2)$. Therefore $2z^3 + 8z^2 - 4z = 2z(z^2 + 4z - 2)$.

21. In the expression $6x^2 y - 3xy^2$, the terms $6x^2 y$ and $-3xy^2$ both contain a common factor of $3xy$ because $6x^2 y = 3xy \cdot 2x$ and $-3xy^2 = 3xy(-y)$. Therefore $6x^2 y - 3xy^2 = 3xy(2x - y)$.

23. $6x - 18x^2$; because $6x = 2 \cdot 3 \cdot x$ and $18x^2 = 3 \cdot 3 \cdot 2 \cdot x \cdot x$, common factors are $2 \cdot 3 \cdot x \Rightarrow$ GCF $= 6x$ and $\Rightarrow 6x(1 - 3x)$.

25. $8y^3 - 12y^2$; because $8y^3 = 2 \cdot 2 \cdot 2 \cdot y \cdot y \cdot y$ and $12y^2 = 2 \cdot 2 \cdot 3 \cdot y \cdot y$, common factors are $2 \cdot 2 \cdot y \cdot y \Rightarrow$ GCF $= 4y^2$ and $\Rightarrow 4y^2(2y - 3)$.

27. $6z^3 + 3z^2 + 9z$; because $6z^3 = 2 \cdot 3 \cdot z \cdot z \cdot z$ and $3z^2 = 3 \cdot z \cdot z$ and $9z = 3 \cdot 3 \cdot z$, common factors are $3 \cdot z \Rightarrow$ GCF $= 3z$ and $\Rightarrow 3z(2z^2 + z + 3)$.

29. $x^4 - 5x^3 - 4x^2$; because $x^4 = x \cdot x \cdot x \cdot x$ and $5x^3 = 5 \cdot x \cdot x \cdot x$ and $4x^2 = 2 \cdot 2 \cdot x \cdot x$, common factors

   are $x \cdot x \Rightarrow$ GCF $= x^2$ and $\Rightarrow x^2(x^2 - 5x - 4)$.

31. $5y^5 + 10y^4 - 15y^3 + 10y^2$; because $5y^5 = 5 \cdot y \cdot y \cdot y \cdot y \cdot y$ and $10y^4 = 2 \cdot 5 \cdot y \cdot y \cdot y \cdot y$ and

   $15y^3 = 3 \cdot 5 \cdot y \cdot y \cdot y$ and $10y^2 = 2 \cdot 5 \cdot y \cdot y$, common factors are $5 \cdot y \cdot y \Rightarrow$

   GCF $= 5y^2$ and $\Rightarrow 5y^2(y^3 + 2y^2 - 3y + 2)$.

33. $xy + xz$; because $xy = x \cdot y$ and $xz = x \cdot z$, common factors are $x \Rightarrow$ GCF $= x$ and $\Rightarrow x(y + z)$.

35. $ab^2 - a^2b$; because $ab^2 = a \cdot b \cdot b$ and $a^2b = a \cdot a \cdot b$, common factors are $a \cdot b \Rightarrow$

   GCF $= ab$ and $\Rightarrow ab(b - a)$.

37. $5x^2y^4 + 10x^3y^3$; because $5x^2y^4 = 5 \cdot x \cdot x \cdot y \cdot y \cdot y \cdot y$ and $10x^3y^3 = 2 \cdot 5 \cdot x \cdot x \cdot x \cdot y \cdot y \cdot y$, common

   factors are $5 \cdot x \cdot x \cdot y \cdot y \cdot y \Rightarrow$ GCF $= 5x^2y^3$ and $\Rightarrow 5x^2y^3(y + 2x)$.

39. $a^2b + ab^2 + ab$; because $a^2b = a \cdot a \cdot b$ and $ab^2 = a \cdot b \cdot b$ and $ab = a \cdot b$, common factors $= a \cdot b \Rightarrow$

   GCF $= ab$ and $\Rightarrow ab(a + b + 1)$.

41. $x(x+1) - 2(x+1)$ has common binomial $(x+1) \Rightarrow (x-2)(x+1)$

43. $(z+5)z + (z+5)4$ has common binomial $(z+5) \Rightarrow (z+4)(z+5)$

45. $4x^3(x-5) - 2x(x-5)$ has common binomial $(x-5) \Rightarrow (4x^3 - 2x)(x-5) \Rightarrow 2x(2x^2 - 1)(x-5)$

47. $x^3 + 2x^2 + 3x + 6$ by associative property $= (x^3 + 2x^2) + (3x + 6) \Rightarrow$

   common factors $= x^2(x+2) + 3(x+2) \Rightarrow (x^2 + 3)(x+2)$

49. $2y^3 + y^2 + 2y + 1$ by associative property $= (2y^3 + y^2) + (2y + 1) \Rightarrow$

   common factors $= y^2(2y+1) + (2y+1) \Rightarrow (y^2 + 1)(2y+1)$

51. $2z^3 - 6z^2 + 5z - 15$ by associative property $= (2z^3 - 6z^2) + (5z - 15) \Rightarrow$

   common factors $= 2z^2(z-3) + 5(z-3) \Rightarrow (2z^2 + 5)(z-3)$

53. $4t^3 - 20t^2 + 3t - 15$ by associative property $= (4t^3 - 20t^2) + (3t - 15) \Rightarrow$

   common factors $= 4t^2(t-5) + 3(t-5) \Rightarrow (4t^2 + 3)(t-5)$

55. $9r^3 + 6r^2 - 6r - 4$ by associative property $= (9r^3 + 6r^2) - (6r + 4) \Rightarrow$

   common factors $= 3r^2(3r+2) - 2(3r+2) \Rightarrow (3r^2 - 2)(3r+2)$

57. $7x^3 + 21x^2 - 2x - 6$ by associative property $= \left(7x^3 + 21x^2\right) - \left(2x + 6\right) \Rightarrow$

common factors $= 7x^2\left(x + 3\right) - 2\left(x + 3\right) \Rightarrow \left(7x^2 - 2\right)\left(x + 3\right)$

59. $2y^3 - 7y^2 - 4y + 14$ by associative property $= \left(2y^3 - 7y^2\right) - \left(4y - 14\right) \Rightarrow$

common factors $= y^2\left(2y - 7\right) - 2\left(2y - 7\right) \Rightarrow \left(y^2 - 2\right)\left(2y - 7\right)$

61. $z^3 - 4z^2 - 7z + 28$ by associative property $= \left(z^3 - 4z^2\right) - \left(7z - 28\right) \Rightarrow$

common factors $= z^2\left(z - 4\right) - 7\left(z - 4\right) \Rightarrow \left(z^2 - 7\right)\left(z - 4\right)$

63. $2x^4 - 3x^3 + 4x - 6$ by associative property $= \left(2x^4 - 3x^3\right) + \left(4x - 6\right) \Rightarrow$

common factors $= x^3\left(2x - 3\right) + 2\left(2x - 3\right) \Rightarrow \left(x^3 + 2\right)\left(2x - 3\right)$

65. $ax + bx + ay + by$ by associative property $= \left(ax + bx\right) + \left(ay + by\right) \Rightarrow$

common factors $= x\left(a + b\right) + y\left(a + b\right) \Rightarrow \left(x + y\right)\left(a + b\right)$

67. $3x^3 + 6x^2 + 3x + 6$ has GCF $3 \Rightarrow 3(x^3 + 2x^2 + x + 2) \Rightarrow$ by associative

property $= 3[(x^3 + 2x^2) + (x + 2)] \Rightarrow$ common factors $= 3[x^2(x + 2) + 1(x + 2)] \Rightarrow 3(x^2 + 1)(x + 2)$

69. $6y^4 - 24y^3 - 2y^2 + 8y$ has GCF $2y \Rightarrow 2y(3y^3 - 12y^2 - y + 4) \Rightarrow$ by associative

property $= 2y[(3y^3 - 12y^2) - (y - 4)] \Rightarrow$ common factors $=$

$2y[3y^2(y - 4) - 1(y - 4)] \Rightarrow 2y(3y^2 - 1)(y - 4)$

71. $x^5 + 2x^4 - 3x^3 - 6x^2$ has GCF $x^2 \Rightarrow x^2(x^3 + 2x^2 - 3x - 6) \Rightarrow$ by associative

property $= x^2[(x^3 + 2x^2) - (3x + 6)] \Rightarrow$ common factors $=$

$x^2[x^2(x + 2) - 3(x + 2)] \Rightarrow x^2(x^2 - 3)(x + 2)$

73. $4x^5 + 2x^4 - 12x^3 - 6x^2$ has GCF $2x^2 \Rightarrow 2x^2(2x^3 + x^2 - 6x - 3) \Rightarrow$ by associative

property $= 2x^2[(2x^3 + x^2) - (6x + 3)] \Rightarrow$ common factors $=$

$2x^2[x^2(2x + 1) - 3(2x + 1)] \Rightarrow 2x^2(x^2 - 3)(2x + 1)$

75. $x^3y + x^2y^2 - 2x^2y - 2xy^2$ has GCF $xy \Rightarrow xy(x^2 + xy - 2x - 2y) \Rightarrow$ by associative

property $= xy[(x^2 + xy) - (2x + 2y)] \Rightarrow$ common factors $= xy[x(x + y) - 2(x + y)] \Rightarrow xy(x - 2)(x + y)$

77. $2x^3y^3 - 2x^4y^2 + 4x^2y^3 - 4x^3y^2$ has GCF $2x^2y^2 \Rightarrow 2x^2y^2(xy - x^2 + 2y - 2x) \Rightarrow$ by associative

property $= 2x^2y^2[(xy - x^2) + (2y - 2x)] \Rightarrow$ common factors $=$

$2x^2y^2[x(y - x) + 2(y - x)] \Rightarrow 2x^2y^2(y - x)(x + 2)$

79. $ax^2 + bx + c = a\left(x^2 + \dfrac{b}{a}x + \dfrac{c}{a}\right)$ because $a \cdot x^2 = ax^2$, $a \cdot \dfrac{b}{a}x = bx$, and $a \cdot \dfrac{c}{a} = c$.

81. (a)    $80t = 2 \cdot 2 \cdot 2 \cdot 2 \cdot 5 \cdot t$ and $16t^2 = 2 \cdot 2 \cdot 2 \cdot 2 \cdot t \cdot t$, common factors are

    $2 \cdot 2 \cdot 2 \cdot 2 \cdot t \Rightarrow \text{GCF} = 2 \cdot 2 \cdot 2 \cdot 2 \cdot t = 16t$.

    (b)    $80t - 16t^2 = 16t(5 - t)$

83. The area of the large rectangle is computed as $2x \cdot x = 2x^2$ and the area of the small rectangle is

    computed as $4 \cdot x = 4x$. The area of the shaded region will be represented as $2x^2 - 4x$. The GCF of

    the expression $2x^2 - 4x$ is $2x$ and the resulting factored polynomial is $2x(x - 2)$.

85. The area of the large rectangle is computed as $4y \cdot 2y = 8y^2$ and the area of the small triangles is

    computed as $2 \cdot \dfrac{1}{2}x \cdot y = xy$. The area of the shaded region will be represented as $8y^2 - xy$. The

    GCF of the expression $8y^2 - xy$ is $y$ and the resulting factored polynomial is $y(8y - x)$.

87. (a)    $x = 3$, $4x^3 - 60x^2 + 200x \Rightarrow 4(3)^3 - 60(3)^2 + 200(3) = 108 - 540 + 600 = 168 \Rightarrow V = 168 \text{ in}^3$.

    (b)    $4x^3 = 2 \cdot 2 \cdot x \cdot x \cdot x$ and $60x^2 = 2 \cdot 2 \cdot 3 \cdot 5 \cdot x \cdot x$ and $200x = 2 \cdot 2 \cdot 2 \cdot 5 \cdot 5 \cdot x$, common factors

    are $2 \cdot 2 \cdot x \Rightarrow \text{GCF} = 4x \Rightarrow 4x\left(x^2 - 15x + 50\right)$.

## Section 6.2  Factoring Trinomials I  *(x²+bx+c)*

1.  1

3.  $x^2 + bx + c$, so $c$ is the third term $\Rightarrow mn = c$ and $b$ is the second term $\Rightarrow m + n = b$.

5.  Factors of 28 are 1, 28; 2, 14; 4, 7 and only $4 + 7 = 11 \Rightarrow 4$, 7.

7.  Factors of $-30$ are $-1, 30; -2, 15; -3, 10; -5, 6; 1, -30; 2, -15; 3, -10; 5, -6$ and only

    $3 + (-10) = -7 \Rightarrow 3, -10$.

9.  Factors of $-50$ are $-1, 50; 1, -50; -2, 25; 2, -25; -5, 10; 5, -10$ and only

    $-5 + 10 = 5 \Rightarrow -5, 10$.

11. Factors of 28 are 1, 28; $-1, -28$; 2, 14; $-2, -14$; 4, 7; $-4, -7$ and only

    $(-4) + (-7) = -11 \Rightarrow -4, -7$.

13. Factors of 2 with sum of 3 are 1 and $2 \Rightarrow x^2 + 3x + 2 = (x + 1)(x + 2)$.

15. Factors of 4 with sum of 4 are 2 and $2 \Rightarrow y^2 + 4y + 4 = (y + 2)(y + 2)$.

17. The only factors of 7 are 1 and 7, whose sum is 8, not 3. The polynomial $z^2 + 3z + 7$ is prime.

19. Factors of 15 with sum of 8 are 3 and $5 \Rightarrow x^2 + 8x + 15 = (x + 3)(x + 5)$.

21. Factors of 36 with sum of 13 are 4 and 9 $\Rightarrow m^2+13m+36=(m+4)(m+9)$.

23. Factors of 100 with sum of 20 are 10 and 10 $\Rightarrow n^2+20n+100=(n+10)(n+10)$.

25. Factors of 5 with sum of −6 are −1 and $-5 \Rightarrow x^2-6x+5=(x-1)(x-5)$.

27. Factors of 12 with sum of −7 are −3 and $-4 \Rightarrow y^2-7y+12=(y-3)(y-4)$.

29. Factors of 40 with sum of −13 are −5 and $-8 \Rightarrow z^2-13z+40=(z-5)(z-8)$.

31. Factors of 63 with sum of −16 are −7 and $-9 \Rightarrow a^2-16a+63=(a-7)(a-9)$.

33. The factors of 10 are $\pm1, \pm10, \pm2, \pm5$, but none of these has a pair has sum −6. The polynomial

    $y^2-6y+10$ is prime.

35. Factors of 125 with sum of −30 are −5 and $-25 \Rightarrow b^2-30b+125=(b-5)(b-25)$.

37. Factors of −90 with sum of 13 are −5 and $18 \Rightarrow x^2+13x-90=(x-5)(x+18)$.

39. Factors of −45 with sum of 4 are −5 and $9 \Rightarrow m^2+4m-45=(m-5)(m+9)$.

41. The factors of −63 are −1, 63; 1, −63; 3, −21; −3, 21; 7, −9; and −7, 9 but none of these pairs of

    factors has sum 16. The polynomial $a^2+16a-63$ is prime.

43. Factors of −200 with sum of 10 are −10 and $20 \Rightarrow n^2+10n-200=(n-10)(n+20)$.

45. Factors of −23 with sum of 22 are −1 and $23 \Rightarrow x^2+22x-23=(x-1)(x+23)$.

47. Factors of −32 with sum of 4 are −4 and $8 \Rightarrow a^2+4a-32=(a-4)(a+8)$.

49. Factors of −20 with sum of −1 are 4 and $-5 \Rightarrow b^2-b-20 \Rightarrow (b+4)(b-5)$.

51. The factors of −22 are 1, −22; −1, 22; −2, 11; and 2, −11 but none of these pairs of factors has sum

    −14. The polynomial $m^2-14m-22$ is prime.

53. Factors of −72 with sum of −1 are 8 and $-9 \Rightarrow x^2-x-72 \Rightarrow (x+8)(x-9)$.

55. Factors of −34 with sum of −15 are 2 and $-17 \Rightarrow y^2-15y-34 \Rightarrow (y+2)(y-17)$.

57. Factors of −66 with sum of −5 are 6 and $-11 \Rightarrow z^2-5z-66 \Rightarrow (z+6)(z-11)$.

59. The GCF of $5x^2-10x-40$ is $5 \Rightarrow 5(x^2-2x-8)$. Factors of −8 with sum of −2 are −4 and

    $2 \Rightarrow 5(x^2-2x-8) \Rightarrow 5(x-4)(x+2)$.

61. The GCF of $-3m^2-9m+12$ is $-3 \Rightarrow -3(m^2+3m-4)$. Factors of −4 with sum of 3 are 4 and

    $-1 \Rightarrow -3(m^2+3m-4) \Rightarrow -3(m+4)(m-1)$.

63. The GCF of $y^3-7y^2+10y$ is $y \Rightarrow y(y^2-7y+10)$. Factors of 10 with sum of −7 are −5 and

    $-2 \Rightarrow y(y^2-7y+10) \Rightarrow y(y-5)(y-2)$.

65. The GCF of $-x^3 - 2x^2 + 15x$ is $-x \Rightarrow -x(x^2 + 2x - 15)$. Factors of $-15$ with sum of 2 are 5 and

$-3 \Rightarrow -x(x^2 + 2x - 15) \Rightarrow -x(x+5)(x-3)$.

67. The GCF of $3a^3 + 21a^2 + 18a$ is $3a \Rightarrow 3a(a^2 + 7a + 6)$. Factors of 6 with sum of 7 are 6 and

$1 \Rightarrow 3a(a^2 + 7a + 6) \Rightarrow 3a(a+6)(a+1)$.

69. The GCF of $-2x^3 + 6x^2 - 8x$ is $-2x \Rightarrow -2x(x^2 - 3x + 4)$. Factors of 4 are $\pm 1, \pm 4, \pm 2$, but neither of

these pairs has sum $-3$. The polynomial $x^2 - 3x + 4$ is prime, so factored form of the original

polynomial is $-2x(x^2 - 3x + 4)$.

71. The GCF of $2m^4 - 10m^3 - 28m^2$ is $2m^2 \Rightarrow 2m^2(m^2 - 5m - 14)$. Factors of $-14$ with sum of $-5$ are $-7$

and $2 \Rightarrow 2m^2(m^2 - 5m - 14) \Rightarrow 2m^2(m-7)(m+2)$.

73. The GCF of $-3x^4 + 3x^3 + 6x^2$ is $-3x^2 \Rightarrow -3x^2(x^2 - x - 2)$. Factors of $-2$ with sum of $-1$ are $-2$ and

$1 \Rightarrow -3x^2(x^2 - x - 2) \Rightarrow -3x^2(x-2)(x+1)$.

75. $5 + 6x + x^2$ in standard form is $x^2 + 6x + 5$ and factors of 5 with a sum of $6 \Rightarrow 1, 5 \Rightarrow (x+5)(x+1)$.

77. $3 - 4x + x^2$ in standard form is $x^2 - 4x + 3$ and factors of 3 with a sum of $-4 \Rightarrow -1, -3 \Rightarrow$

$(x-1)(x-3)$.

79. Using the hint, the answer will be in the form $(m-x)(n+x)$. We need to find $m$ and $n$. Factors of

12 with a difference of 4 are 6 and $2 \Rightarrow 12 + 4x - x^2 = (6-x)(2+x)$.

81. Factors of 32 with a difference of $-4$ are 8 and $4 \Rightarrow 32 - 4x - x^2 = (8+x)(4-x)$.

83. Factors of $k$ with a sum of $k+1$ are $k$ and $1 \Rightarrow x^2 + (k+1)x + k \Rightarrow (x+1)(x+k)$.

85. $x^2 + 2x + 1 \Rightarrow (x+1)(x+1) \Rightarrow L \cdot W = x^2 + 2x + 1 \Rightarrow L = x+1$. See Figure 85.

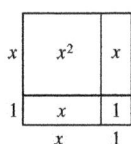
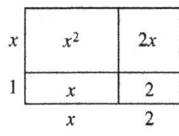

Figure 85                    Figure 87

87. $x^2 + 3x + 2 = (x+2)(x+1) = L \cdot W \Rightarrow L = x+2$ or $x+1$. See Figure 87.

89. $6x^2 + 12x + 6$ divided by 6 $\left(\text{for 6 surfaces}\right) = x^2 + 2x + 1 = (x+1)(x+1) \Rightarrow$ each side is $x+1$.

91. Add the four regions: $x^2 + 2x + 6x + 12 = x^2 + 8x + 12 = (x+2)(x+6)$.

## Checking Basic Concepts  Sections 6.1 and 6.2

1. $8x^3 = 2 \cdot 2 \cdot 2 \cdot x \cdot x \cdot x$ and $12x^2 = 2 \cdot 2 \cdot 3 \cdot x \cdot x$ and $24x = 2 \cdot 2 \cdot 2 \cdot 3 \cdot x \Rightarrow GCF = 2 \cdot 2 \cdot x = 4x.$

2. GCF of $12z^3$ and $18z^2$ is $6z^2 \Rightarrow 6z^2(2z-3).$

3. (a)   $6y(y-2)+5(y-2)=(6y+5)(y-2)$

   (b)   $\left(2x^3+x^2\right)+(10x+5)=x^2(2x+1)+5(2x+1)=\left(x^2+5\right)(2x+1)$

   (c)   GCF is $4 \Rightarrow 4z^3-12z^2+4z-12 \Rightarrow 4(z^3-3z^2+z-3) \Rightarrow 4[(z^3-3z^2)+(z-3)] \Rightarrow$

   $4[z^2(z-3)+1(z-3)] \Rightarrow 4(z^2+1)(z-3)$

4. (a)   Factors of 8 with a sum of 6 are 2 and $4 \Rightarrow (x+2)(x+4).$

   (b)   Factors of $-42$ with a sum of $-1$ are $-7$ and $6 \Rightarrow (x-7)(x+6).$

   (c)   The factors of $-5$ are $1, -5$ and $-1, 5$ but neither of these pairs has sum 3. The polynomial

   $a^2+3a-5$ is prime.

   (d)   The GCF of $4a^3+20a^2+24a$ is $4a \Rightarrow 4a(a^2+5a+6) \Rightarrow 4a(a+2)(a+3)$

5. $x^2+5x+5x+25 \Rightarrow x^2+10x+25 \Rightarrow (x+5)(x+5)$

## Section 6.3  Factoring Trinomials II $(ax^2+bx+c)$

1. $ac$; $b$

3. By $ax^2+bx+c$, if $a>0$, $b>0$ and $c>0$ then: $+$ ; $+$

5. By $ax^2+bx+c$, if $a>0$, $b<0$ and $c>0$ then: $-$ ; $-$

7. If $(4x+a)(b+2)=4x^2+11x+6$, then by FOIL: $4x \cdot b = 4x^2 \Rightarrow b=x$ and $a \cdot 2 = 6 \Rightarrow a=3.$

9. If $(2x-a)(2x+b)=4x^2+4x-3$, then by FOIL: -$a \cdot b = -3 \Rightarrow ab=3$ and $-2ax+2bx=4x.$

   If $a=1$, then $b=3$ and $-2x+6x=4x$ .

11. If $(a+7)(2x-b)=6x^2+11x-7$, then by FOIL: $2x \cdot a = 6x^2 \Rightarrow a=3x$ and $-b \cdot (7)=-7 \Rightarrow b=1.$

13. If $(3x-a)(2x+b)=6x^2-x-15$, then by FOIL: $-a \cdot b = -15 \Rightarrow a \cdot b=15$ and $-2xa+3xb = -x$

    If $a=5$ then $b=3$ and we have $-2x(5)+3x(3) = -10x+9x=-x$. Therefore, $a=5$ and $b=3$

15. Using factoring by grouping: For $2x^2+7x+3$, $m \cdot n = a \cdot c = 2 \cdot 3 = 6$ and $m+n=b=7 \Rightarrow$

    $m=6$, $n=1 \Rightarrow 2x^2+6x+x+3 \Rightarrow 2x(x+3)+(x+3) \Rightarrow (2x+1)(x+3).$

17. To factor $3y^2+2y+4$, find numbers $m$ and $n$ such that $mn=3 \cdot 4 = 12$ and $m+n=2$. Because no

    such numbers exist the polynomial is prime.

19. Using factoring by grouping: For $3x^2 + 4x + 1$, $m \cdot n = a \cdot c = 3 \cdot 1 = 3$ and $m + n = b = 4 \Rightarrow$ $m = 3$, $n = 1 \Rightarrow 3x^2 + 3x + x + 1 \Rightarrow 3x(x+1) + (x+1) \Rightarrow (3x+1)(x+1)$.

21. Using factoring by grouping: For $6x^2 + 11x + 3$, $m \cdot n = a \cdot c = 6 \cdot 3 = 18$ and $m + n = b = 11 \Rightarrow$ $m = 9$, $n = 2 \Rightarrow 6x^2 + 9x + 2x + 3 \Rightarrow 3x(2x+3) + (2x+3) \Rightarrow (3x+1)(2x+3)$.

23. Using factoring by grouping: For $5x^2 - 11x + 2$, $m \cdot n = a \cdot c = 5 \cdot 2 = 10$ and $m + n = b = -11 \Rightarrow$ $m = -10$, $n = -1 \Rightarrow 5x^2 - 10x - x + 2 \Rightarrow 5x(x-2) - (x-2) \Rightarrow (5x-1)(x-2)$.

25. Using factoring by grouping: For $2y^2 - 7y + 5$, $m \cdot n = a \cdot c = 2 \cdot 5 = 10$ and $m + n = b = -7 \Rightarrow$ $m = -5$, $n = -2 \Rightarrow 2y^2 - 5y - 2y + 5 \Rightarrow y(2y-5) - (2y-5) \Rightarrow (y-1)(2y-5)$.

27. To factor $3m^2 - 11m - 6$, find numbers $m$ and $n$ such that $mn = 3(-6) = -18$ and $m + n = -11$. Because no such numbers exist the polynomial is prime.

29. Using factoring by grouping: For $7z^2 - 37z + 10$, $m \cdot n = a \cdot c = 7 \cdot 10 = 70$ and $m + n = b = -37 \Rightarrow$ $m = -35$, $n = -2 \Rightarrow 7z^2 - 35z - 2z + 10 \Rightarrow 7z(z-5) - 2(z-5) \Rightarrow (7z-2)(z-5)$.

31. Using factoring by grouping: For $3t^2 - 7t - 6$, $m \cdot n = a \cdot c = 3 \cdot (-6) = -18$ and $m + n = b = -7 \Rightarrow$ $m = -9$, $n = 2 \Rightarrow 3t^2 - 9t + 2t - 6 \Rightarrow 3t(t-3) + 2(t-3) \Rightarrow (3t+2)(t-3)$.

33. Using factoring by grouping: For $15r^2 + r - 6$, $m \cdot n = a \cdot c = 15 \cdot (-6) = -90$ and $m + n = b = 1 \Rightarrow$ $m = 10$, $n = -9 \Rightarrow 15r^2 + 10r - 9r - 6 \Rightarrow 5r(3r+2) - 3(3r+2) \Rightarrow (5r-3)(3r+2)$.

35. Using factoring by grouping: For $24m^2 - 23m - 12$, $m \cdot n = a \cdot c = 24 \cdot (-12) = -288$ and $m + n = b = -23 \Rightarrow m = -32$, $n = 9 \Rightarrow 24m^2 - 32m + 9m - 12 \Rightarrow$ $8m(3m-4) + 3(3m-4) \Rightarrow (8m+3)(3m-4)$.

37. Using factoring by grouping: For $25x^2 + 5x - 2$, $m \cdot n = a \cdot c = 25 \cdot (-2) = -50$ and $m + n = b = 5 \Rightarrow$ $m = 10$, $n = -5 \Rightarrow 25x^2 + 10x - 5x - 2 \Rightarrow 5x(5x+2) - (5x+2) \Rightarrow (5x-1)(5x+2)$.

39. Using factoring by grouping: For $6x^2 + 11x - 2$, $m \cdot n = a \cdot c = 6 \cdot (-2) = -12$ and $m + n = b = 11 \Rightarrow$ $m = 12$, $n = -1 \Rightarrow 6x^2 + 12x - x - 2 \Rightarrow 6x(x+2) - (x+2) \Rightarrow (6x-1)(x+2)$.

41. To factor $15y^2 - 7y + 2$, find numbers $m$ and $n$ such that $mn = 15 \cdot 2 = 30$ and $m + n = -7$. Because no such numbers exist the polynomial is prime.

43. Using factoring by grouping: For $21n^2 + 4n - 1$, $m \cdot n = a \cdot c = 21 \cdot (-1) = -21$ and $m + n = b = 4 \Rightarrow$ $m = 7$, $n = -3 \Rightarrow 21n^2 + 7n - 3n - 1 \Rightarrow 7n(3n+1) - (3n+1) \Rightarrow (7n-1)(3n+1)$.

45. Using factoring by grouping: For $14y^2+23y+3$, $m \cdot n = a \cdot c = 14 \cdot 3 = 42$ and $m+n=b=23 \Rightarrow$

    $m=21$, $n=2 \Rightarrow 14y^2+21y+2y+3 \Rightarrow 7y(2y+3)+(2y+3) \Rightarrow (7y+1)(2y+3)$.

47. Using factoring by grouping: For $28z^2-25z+3$, $m \cdot n = a \cdot c = 28 \cdot 3 = 84$ and $m+n=b=-25 \Rightarrow$

    $m=-21$, $n=-4 \Rightarrow 28z^2-21z-4z+3 \Rightarrow 7z(4z-3)-(4z-3) \Rightarrow (7z-1)(4z-3)$.

49. Using factoring by grouping: For $30x^2-29x+6$, $m \cdot n = a \cdot c = 30 \cdot 6 = 180$ and $m+n=b=-29 \Rightarrow$

    $m=-20$, $n=-9 \Rightarrow 30x^2-20x-9x+6 \Rightarrow 10x(3x-2)-3(3x-2) \Rightarrow (10x-3)(3x-2)$.

51. To factor $20a^2+18a-5$, find numbers $m$ and $n$ such that $mn=20(-5)=-100$ and $m+n=18$.

    Because no such numbers exist the polynomial is prime.

53. Using factoring by grouping: For $18t^2+23t-6$, $m \cdot n = a \cdot c = 18 \cdot (-6) = -108$ and $m+n=b=23 \Rightarrow$

    $m=27$, $n=-4 \Rightarrow 18t^2+27t-4t-6 \Rightarrow 9t(2t+3)-2(2t+3) \Rightarrow (9t-2)(2t+3)$.

55. The GCF of $12a^2+12a-9$ is $3 \Rightarrow 3(4a^2+4a-3)$. Now use factoring by grouping: For

    $4a^2+4a-3$, $m \cdot n = a \cdot c = 4(-3) = -12$ and $m+n=b=4 \Rightarrow m=6$, $n=-2 \Rightarrow 3(4a^2+4a-3)$

    $\Rightarrow 3[(4a^2+6a)+(-2a-3)] \Rightarrow 3[2a(2a+3)-1(2a+3)] \Rightarrow 3(2a-1)(2a+3)$.

57. The GCF of $12y^3-11y^2+2y$ is $y \Rightarrow y(12y^2-11y+2)$. Now use factoring by grouping: For

    $12y^2-11y+2$, $m \cdot n = a \cdot c = 12 \cdot 2 = 24$ and $m+n=b=-11 \Rightarrow m=-3$, $n=-8 \Rightarrow$

    $y(12y^2-11y+2) \Rightarrow y[(12y^2-3y)+(-8y+2)] \Rightarrow y[3y(4y-1)-2(4y-1)] \Rightarrow y(3y-2)(4y-1)$.

59. The GCF of $24x^3-30x^2+9x$ is $3x \Rightarrow 3x(8x^2-10x+3)$. Now use factoring by grouping: For

    $8x^2-10x+3$, $m \cdot n = a \cdot c = 8 \cdot 3 = 24$ and $m+n=b=-10 \Rightarrow m=-6$, $n=-4 \Rightarrow$

    $3x(8x^2-10x+3) \Rightarrow 3x[(8x^2-6x)+(-4x+3)] \Rightarrow 3x[2x(4x-3)-1(4x-3)] \Rightarrow 3x(2x-1)(4x-3)$.

61. The GCF of $8x^4-6x^3+2x^2$ is $2x^2 \Rightarrow 2x^2(4x^2-3x+1)$. Now use factoring by grouping: For

    $4x^2-3x+1$, $m \cdot n = a \cdot c = 4 \cdot 1 = 4$ and $m+n=b=-3$, but no such $m$ and $n$ exist. Thus $4x^2-3x+1$

    is prime, and the factored form of the original polynomial is $2x^2(4x^2-3x+1)$.

63. The GCF of $28x^4+56x^3+21x^2$ is $7x^2 \Rightarrow 7x^2(4x^2+8x+3)$. Now use factoring by grouping: For

    $4x^2+8x+3$, $m \cdot n = a \cdot c = 4 \cdot 3 = 12$ and $m+n=b=8 \Rightarrow m=2$, $n=6 \Rightarrow 7x^2(4x^2+8x+3) \Rightarrow$

    $7x^2[(4x^2+2x)+(6x+3)] \Rightarrow 7x^2[2x(2x+1)+3(2x+1)] \Rightarrow 7x^2(2x+3)(2x+1)$.

65. Find numbers $m$ and $n$ such that $m \cdot n = 3k$ and $m+n=3k+1 \Rightarrow m=3k$, $n=1 \Rightarrow 3x^2+(3k+1)x+k$

    $\Rightarrow (3x^2+3kx)+(x+k) \Rightarrow 3x(x+k)+1(x+k) \Rightarrow (3x+1)(x+k)$.

67. Put $2+15x+7x^2$ in standard form $7x^2+15x+2$, $m \cdot n = a \cdot c = 7 \cdot 2 = 14$ and $m+n=b=15 \Rightarrow$

    $m=14$, $n=1 \Rightarrow 7x^2+14x+x+2 \Rightarrow 7x(x+2)+(x+2) \Rightarrow (7x+1)(x+2)$.

69. Put $2-5x+2x^2$ in standard form $2x^2-5x+2$, $m \cdot n = a \cdot c = 2 \cdot 2 = 4$ and $m+n = b = -5 \Rightarrow$

     $m = -4$, $n = -1 \Rightarrow 2x^2-4x-x+2 \Rightarrow 2x(x-2)-(x-2) \Rightarrow (2x-1)(x-2)$.

71. Put $3-2x-8x^2$ in standard form $-8x^2-2x+3 = -(8x^2+2x-3)$, $m \cdot n = a \cdot c = 8 \cdot (-3) = -24$ and

     $m+n = b = 2 \Rightarrow m = 6$, $n = -4 \Rightarrow -(8x^2+6x-4x-3) \Rightarrow -[2x(4x+3)-(4x+3)] \Rightarrow$

     $-(2x-1)(4x+3)$

73. $-2x^2-7x+15 = -\left(2x^2+7x-15\right)$, $m \cdot n = a \cdot c = 2 \cdot (-15) = -30$ and $m+n = b = 7 \Rightarrow$

     $m = 10$, $n = -3 \Rightarrow -\left(2x^2+10x-3x-15\right) \Rightarrow -\left[2x(x+5)-3(x+5)\right] \Rightarrow -(2x-3)(x+5)$.

75. $-5x^2+14x+3 = -\left(5x^2-14x-3\right)$, $m \cdot n = a \cdot c = 5 \cdot (-3) = -15$ and $m+n = b = -14 \Rightarrow$

     $m = -15$, $n = 1 \Rightarrow -\left(5x^2-15x+x-3\right) \Rightarrow -\left[5x(x-3)+(x-3)\right] \Rightarrow -(5x+1)(x-3)$.

77. $6x^2+7x+2$, $m \cdot n = a \cdot c = 6 \cdot 2 = 12$ and $m+n = b = 7 \Rightarrow m = 4$, $n = 3 \Rightarrow$

     $\left(6x^2+4x+3x+2\right) \Rightarrow 2x(3x+2)+(3x+2) \Rightarrow (2x+1)(3x+2)$. See Figure 77.

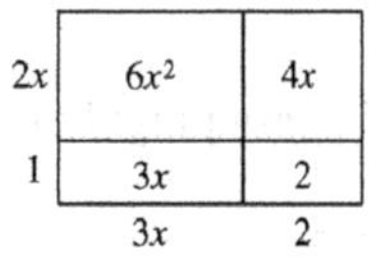

Figure 77

79. $\left(2x^2+6x+x+3\right) \Rightarrow 2x(x+3)+(x+3) \Rightarrow (2x+1)(x+3)$.


## Section 6.4  Special Types of Factoring

1. $(a-b)(a+b)$

3. False

5. $a^2+2ab+b^2 = (a+b)^2$

7. $(x+3)^2 = x^2+6x+9 \Rightarrow 6x$

9. $a^3+b^3 = (a+b)\left(a^2-ab+b^2\right)$

11. $8x^3+27y^3 = (2x)^3+(3y)^3 \Rightarrow a = 2x$ and $b = 3y$.

13. $y^3-8 = (y-2)(y+2y+4) \Rightarrow -;+$

15. $x^2-1 = (x)^2-(1)^2 \Rightarrow (x-1)(x+1)$

17. $z^2-100 = (z)^2-(10)^2 \Rightarrow (z-10)(z+10)$

19. $4y^2 - 1 = (2y)^2 - (1)^2 \Rightarrow (2y-1)(2y+1)$

21. $36z^2 - 25 = (6z)^2 - (5)^2 \Rightarrow (6z-5)(6z+5)$

23. $9 - x^2 = (3)^2 - (x)^2 \Rightarrow (3-x)(3+x)$

25. $1 - 9y^2 = (1)^2 - (3y)^2 \Rightarrow (1-3y)(1+3y)$

27. $4a^2 - 9b^2 = (2a)^2 - (3b)^2 \Rightarrow (2a-3b)(2a+3b)$

29. $36m^2 - 25n^2 = (6m)^2 - (5n)^2 \Rightarrow (6m-5n)(6m+5n)$

31. $81r^2 - 49t^2 = (9r)^2 - (7t)^2 \Rightarrow (9r-7t)(9r+7t)$

33. $x^2 + 8x + 16$ is a perfect square trinomial,

$a^2 = x^2$ so $a = x$, $b^2 = 4^2$ so $b = 4$, $2ab = 8x$, the middle term $\Rightarrow (a+b)^2 \Rightarrow (x+4)^2$.

35. $z^2 + 12z + 25$ is not a perfect square trinomial because

$a^2 = z^2$ so $a = z$, $b^2 = 5^2$ so $b = 5$, $2ab = 10z \neq 12z$, the middle term and also, FOIL does not work,

therefore: Not possible.

37. $x^2 - 6x + 9$ is a perfect square trinomial, $a^2 = x^2$ so $a = x$, $b^2 = (\pm 3)^2$ so $b = \pm 3$, using $b = -3$,

$2ab = -6x$, the middle term $\Rightarrow (a+b)^2 \Rightarrow (x-3)^2$.

39. $9y^2 + 6y + 1$ is a perfect square trinomial, $a^2 = (3y)^2$ so $a = 3y$, $b^2 = 1^2$ so $b = 1$, $2ab = 6y$,

the middle term $\Rightarrow (a+b)^2 \Rightarrow (3y+1)^2$.

41. $4z^2 - 4z + 1$ is a perfect square trinomial, $a^2 = (2z)^2$ so $a = 2z$, $b^2 = (\pm 1)^2$ so $b = \pm 1$,

using $b = -1$, $2ab = -4z$, the middle term $\Rightarrow (a+b)^2 \Rightarrow (2z-1)^2$.

43. $9t^2 + 16t + 4$ is not a perfect square trinomial because $a^2 = (3t)^2$ so $a = 3t$, $b^2 = (\pm 2)^2$ so $b = \pm 2$.

$2ab = \pm 12t \neq 16t$, the middle term and also, FOIL does not work, therefore: Not possible.

45. $9x^2 + 30x + 25$ is a perfect square trinomial, $a^2 = (3x)^2$ so $a = 3x$, $b^2 = (5)^2$ so $b = 5$,

$2ab = 30x$, the middle term $\Rightarrow (a+b)^2 \Rightarrow (3x+5)^2$.

47. $4a^2 - 36a + 81$ is a perfect square trinomial,

$a^2 = (2a)^2$ so $a = 2a$, $b^2 = (\pm 9)^2$ so $b = \pm 9$, using $b = -9$, $2ab = -36a$, the middle term $\Rightarrow$

$(a+b)^2 \Rightarrow (2a-9)^2$.

49. $x^2 + 2xy + y^2$ is a perfect square trinomial, $a^2 = x^2$ so $a = x$, $b^2 = y^2$ so $b = y$,

$2ab = 2xy$, the middle term $\Rightarrow (a+b)^2 \Rightarrow (x+y)^2$.

51. $r^2 - 10rt + 25t^2$ is a perfect square trinomial,

$a^2 = r^2$ so $a = r$, $b^2 = (\pm 5t)^2$ so $b = \pm 5t$, using $b = -5t$, $2ab = -10rt$, the middle term $\Rightarrow$

$(a+b)^2 \Rightarrow (r-5t)^2$.

53. $4y^2 - 10yz + 9z^2$ is not a perfect square trinomial because

$a^2 = (2y)^2$ so $a = 2y$, $b^2 = (\pm 3z)^2$ so $b = \pm 3z$, $2ab = \pm 12yz \neq -10yz$, the middle term and also,

FOIL does not work, therefore: Not possible.

55. Using the sum of cubes for $z^3 + 1$, $a^3 = z^3$ so $a = z$, $b^3 = 1^3$ so $b = 1 \Rightarrow$

$(a+b)(a^2 - ab + b^2) \Rightarrow (z+1)(z^2 - z + 1)$.

57. Using the sum of cubes for $x^3 + 64$, $a^3 = x^3$ so $a = x$, $b^3 = 4^3$ so $b = 4 \Rightarrow$

$(a+b)(a^2 - ab + b^2) \Rightarrow (x+4)(x^2 - 4x + 16)$.

59. Using the difference of cubes for $y^3 - 8$, $a^3 = y^3$ so $a = y$, $b^3 = 2^3$ so $b = 2 \Rightarrow$

$(a-b)(a^2 + ab + b^2) \Rightarrow (y-2)(y^2 + 2y + 4)$.

61. Using the difference of cubes for $n^3 - 1$, $a^3 = n^3$ so $a = n$, $b^3 = 1^3$ so $b = 1 \Rightarrow$

$(a-b)(a^2 + ab + b^2) \Rightarrow (n-1)(n^2 + n + 1)$.

63. Using the sum of cubes for $8x^3 + 1$, $a^3 = (2x)^3$ so $a = 2x$, $b^3 = 1^3$ so $b = 1 \Rightarrow$

$(a+b)(a^2 - ab + b^2) \Rightarrow (2x+1)(4x^2 - 2x + 1)$.

65. Using the difference of cubes for $m^3 - 64n^3$, $a^3 = m^3$ so $a = m$, $b^3 = (4n)^3$ so $b = 4n \Rightarrow$

$(a-b)(a^2 + ab + b^2) \Rightarrow (m - 4n)(m^2 + 4mn + 16n^2)$.

67. Using the sum of cubes for $8x^3 + 125y^3$, $a^3 = (2x)^3$ so $a = 2x$, $b^3 = (5y)^3$ so $b = 5y \Rightarrow$

$(a+b)(a^2 - ab + b^2) \Rightarrow (2x+5y)(4x^2 - 10xy + 25y^2)$.

69. $4x^2 - 16 = 4(x^2 - 4) = 4((x)^2 - (2)^2) \Rightarrow 4(x-2)(x+2)$

71. $2y^2 - 28y + 98 = 2(y^2 - 14y + 49)$; $y^2 - 14y + 49$ is a perfect square trinomial, $a^2 = y^2$ so

$a = y$, $b^2 = 49$ so $b = \pm 7$, using $b = -7$, $2ab = -14y$, the middle term $\Rightarrow (a+b)^2 \Rightarrow (y-7)^2$.

Therefore $2(y^2 - 14y + 49) = 2(y-7)^2$.

73. $5z^3 + 40 = 5(z^3 + 8)$; $z^3 + 8$ is a sum of cubes with $a = z$ and

$b = 2 \Rightarrow (a+b)(a^2 - ab + b^2) \Rightarrow (z+2)(z^2 - 2z + 4)$. Therefore $5z^3 + 40 = 5(z+2)(z^2 - 2z + 4)$.

75. $x^3 y - xy^3 = xy(x^2 - y^2) = xy(x-y)(x+y)$

77. $2m^3 - 10m^2 + 18m = 2m(m^2 - 5m + 9)$; $m^2 - 5m + 9$ is not a perfect square trinomial because

$a^2 = m^2$ so $a = m$, $b^2 = 9$ so $b = \pm 3$, $2ab = \pm 6m \neq -5m$, the middle term, and FOIL does not work,

therefore $2m^3 - 10m^2 + 18m = 2m\,(m^2 - 5m + 9)$.

79. $700x^4 - 63x^2 y^2 = 7x^2(100x^2 - 9y^2) = 7x^2((10x)^2 - (3y)^2) = 7x^2\,(10x - 3y)(10x + 3y)$

81. Using the sum of cubes for $16a^3 + 2b^3 = 2\,(8a^3 + b^3)$, $a^3 = (2a)^3$ so $a = 2a$, $b^3 = b^3$ so

$b = b \Rightarrow (a + b)(a^2 - ab + b^2) \Rightarrow 2\,(2a + b)(4a^2 - 2ab + b^2)$.

83. $4b^4 + 24b^3 + 36b^2 = 4b^2(b^2 + 6b + 9)$; $b^2 + 6b + 9$ is a perfect square trinomial, $a^2 = b^2$ so

$a = b$, $b^2 = 9$ so $b = 3$, $2ab = 6b$, the middle term $\Rightarrow (a + b)^2 \Rightarrow (b + 3)^2$. Therefore

$4b^4 + 24b^3 + 36b^2 = 4b^2(b + 3)^2$.

85. Using the difference of cubes for $500r^3 - 32t^3 = 4(125r^3 - 8t^3)$, $a^3 = (5r)^3$ so $a = 5r$, $b^3 = (2t)^3$ so

$b = 2t \Rightarrow (a - b)(a^2 + ab + b^2) \Rightarrow 4(5r - 2t)(25r^2 + 10rt + 4t^2)$

87. Sides must be the same $\Rightarrow$ perfect square trinomial,

$4x^2 + 12x + 9$, $a^2 = (2x)^2$ so $a = 2x$, $b^2 = 3^2$ so $b = 3$,

$2ab = 12x$ the middle term $\Rightarrow (2x + 3)^2$.  See Figure 87. Length $= 2x + 3$.

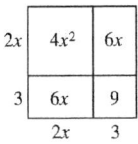

Figure 87

## Checking Basic Concepts  Sections 6.3 and 6.4

1. (a)  $2x^2 - 5x - 12$ (factor by grouping), $m \cdot n = -24$, $m + n = -5 \Rightarrow m = -8$, $n = 3 \Rightarrow$

$2x^2 - 8x + 3x - 12 \Rightarrow 2x(x - 4) + 3(x - 4) \Rightarrow (2x + 3)(x - 4)$.

(b)  $6x^2 + 17x - 14$ (factor by grouping), $m \cdot n = -84$, $m + n = 17 \Rightarrow m = 21$, $n = -4 \Rightarrow$

$6x^2 + 21x - 4x - 14 \Rightarrow 3x(2x + 7) - 2(2x + 7) \Rightarrow (3x - 2)(2x + 7)$.

2. (a)  To factor $3y^2 + 4y - 2$, find numbers $m$ and $n$ such that $mn = 3(-2) = -6$ and $m + n = 4$.

Because no such numbers exist the polynomial is prime.

(b)  $6y^3 - 10y^2 - 4y = 2y\,(3y^2 - 5y - 2)$; $m \cdot n = a \cdot c = 3(-2) = -6$ and

$m + n = b = -5 \Rightarrow m = -6$, $n = 1 \Rightarrow 2y\,(3y^2 - 5y - 2) \Rightarrow 2y\,[(3y^2 - 6y) + (y - 2)]$

$\Rightarrow 2y\,[3y\,(y - 2) + 1(y - 2)] \Rightarrow 2y\,(3y + 1)(y - 2)$

3. $3x^2 + 2x + 9x + 6 \Rightarrow x(3x + 2) + 3(3x + 2) \Rightarrow (x + 3)(3x + 2)$

4. (a)  $z^2 - 64$ (difference of squares), $a^2 = z^2$ so $a = z, b^2 = 8^2$ so $b = 8 \Rightarrow (z-8)(z+8)$.

   (b)  $9r^2 - 4t^2$ (difference of squares), $a^2 = (3r)^2$ so $a = 3r, b^2 = (2t)^2$ so
        $b = 2t \Rightarrow (3r + 2t)(3r - 2t)$.

5. (a)  $x^2 + 12x + 36$ (FOIL in reverse) $\Rightarrow (x+6)(x+6) \Rightarrow (x+6)^2$.

   (b)  $9a^2 - 12ab + 4b^2$ (perfect square trinomial), $a^2 = (3a)^2$ so $a = 3a, b^2 = (\pm 2b)^2$ so $b = \pm 2b$,
        using $b = -2b, 2ab = -12ab$ the second term $\Rightarrow (3a - 2b)^2$.

6. (a)  $m^3 - 27$ (difference of cubes), $a^3 = m^3$ so $a = m, b^3 = 3^3$ so $b = 3 \Rightarrow (m-3)(m^2 + 3m + 9)$.

   (b)  $125n^3 + 27$ (sum of cubes), $a^3 = (5n)^3$ so $a = 5n, b^3 = 3^3$ so
        $b = 3 \Rightarrow (5n + 3)(25n^2 - 15n + 9)$.

7. (a)  $16x^2 - 4 = 4(4x^2 - 1) = 4\left((2x)^2 - (1)^2\right) = 4(2x-1)(2x+1)$

   (b)  $3y^4 + 24y = 3y(y^3 + 8)$ (sum of cubes), $a^3 = y^3$ so $a = y, b^3 = 2^3$ so
        $b = 2 \Rightarrow 3y(y+2)(y^2 - 2y + 4)$.

## Section 6.5  Summary of Factoring

1. greatest common factor

3. grouping

5. No; a sum of squares cannot be factored.

7. square, grouping, FOIL

9. $4x - 2 = 2(2x - 1)$

11. $2y^2 - 4y + 4 = 2(y^2 - 2y + 2)$

13. $z^2 - 4 = (z-2)(z+2)$

15. $a^3 + 8 = (a+2)(a^2 - 2a + 4)$

17. $4b^2 - 12b + 9 = (2b - 3)^2$

19. $m^2 + 9$ is a sum of squares and cannot be factored.

21. $x^3 - x^2 + 5x - 5 = x^2(x-1) + 5(x-1) = (x^2 + 5)(x-1)$

23. $y^2 - 5y + 4 = (y-4)(y-1)$

25. $x^3 + 4x^2 - 9x - 36 = x^2(x+4) - 9(x+4) = (x^2 - 9)(x+4) = (x-3)(x+3)(x+4)$

27. $8a^3 - 64 = 8(a^3 - 8) = 8(a-2)(a^2 + 2a + 4)$

29. $12x^4 - 18x^3 + 4x^2 - 6x = 2x(6x^3 - 9x^2 + 2x - 3) = 2x[3x^2(2x-3) + 1(2x-3)] = 2x(3x^2 + 1)(2x-3)$

31. $54t^4 + 16t = 2t(27t^3 + 8) = 2t(3t + 2)(9t^2 - 6t + 4)$

33. $2r^3 + 6r^2 - 2r - 6 = 2r^2(r + 3) - 2(r + 3) = (2r^2 - 2)(r + 3) = 2(r^2 - 1)(r + 3) = 2(r - 1)(r + 1)(r + 3)$

35. $6z^4 - 21z^3 - 45z^2 = 3z^2(2z^2 - 7z - 15) = 3z^2(2z + 3)(z - 5)$

37. $12b^4 - 10b^3 + 2b^2 = 2b^2(6b^2 - 5b + 1) = 2b^2(3b - 1)(2b - 1)$

39. $6y^2z - 24z^3 = 6z(y^2 - 4z^2) = 6z(y - 2z)(y + 2z)$

41. $3x^2y - 30xy + 75y = 3y(x^2 - 10x + 25) = 3y(x - 5)^2$

43. $27m^3 - 8n^3 = (3m - 2n)(9m^2 + 6mn + 4n^2)$

45. $3x^5 - 12x^3 - 3x^2 + 12 = 3(x^5 - 4x^3 - x^2 + 4) = 3[x^3(x^2 - 4) - 1(x^2 - 4)] = 3(x^3 - 1)(x^2 - 4)$

$= 3(x - 1)(x^2 + x + 1)(x - 2)(x + 2)$

47. $5a^2 - 27a - 18 = 5a^2 - 30a + 3a - 18 = 5a(a - 6) + 3(a - 6) = (5a + 3)(a - 6)$

49. $3rt^2 + 33rt + 90r = 3r(t^2 + 11t + 30) = 3r(t + 5)(t + 6)$

51. $9b^3 + 6b^2 + 12b + 8 = 3b^2(3b + 2) + 4(3b + 2) = (3b^2 + 4)(3b + 2)$

53. $6n^3 + 2n^2 - 10n = 2n(3n^2 + n - 5)$

55. $4x^2 - 36y^2 = 4(x^2 - 9y^2) = 4(x - 3y)(x + 3y)$

57. $2a^3 - 16a^2 + 32a = 2a(a^2 - 8a + 16) = 2a(a - 4)^2$

59. $32xy^3 + 4x = 4x(8y^3 + 1) = 4x(2y + 1)(4y^2 - 2y + 1)$

61. $8b^4 + 24b^3 - 2b^2 - 6b = 2b(4b^3 + 12b^2 - b - 3)$

$= 2b[4b^2(b + 3) - 1(b + 3)] = 2b(4b^2 - 1)(b + 3) = 2b(2b - 1)(2b + 1)(b + 3)$

63. Area of one square $= \dfrac{1}{3}(27x^2 + 18x + 3) = 9x^2 + 6x + 1$; sides must be the same $\Rightarrow$ perfect square

trinomial, $9x^2 + 6x + 1 = (3x + 1)^2 \Rightarrow$ one side length is $3x + 1$.


## Section 6.6  Solving Equations by Factoring I (Quadratics)

1. $0, 0$

3. $2x = 0, \; x + 6 = 0$

5. Apply the zero-product property by setting $x + 5 = 0$ and $x - 4 = 0$.

7. zero

9. $ax^2 + bx + c = 0$ with $a \neq 0$.

11. descending

13. $x = 0$

15. $2x = 0$ or $x + 8 = 0$, so $x = -8,\ 0$.

17. $y - 1 = 0$ or $y - 2 = 0$, so $y = 1,\ 2$.

19. $2z - 1 = 0$ or $4z - 3 = 0$, then $2z - 1 = 0 \Rightarrow 2z = 1 \Rightarrow z = \dfrac{1}{2}$ or $4z - 3 = 0 \Rightarrow 4z = 3 \Rightarrow$

     $z = \dfrac{3}{4}$, so $z = \dfrac{1}{2}, \dfrac{3}{4}$.

21. $1 - 3n = 0$ or $3 - 7n = 0$, then $1 - 3n = 0 \Rightarrow -3n = -1 \Rightarrow n = \dfrac{-1}{-3} = \dfrac{1}{3}$ or $3 - 7n = 0 \Rightarrow$

     $-7n = -3 \Rightarrow n = \dfrac{-3}{-7} = \dfrac{3}{7}$, so $n = \dfrac{1}{3}, \dfrac{3}{7}$.

23. $x = 0$ or $x - 5 = 0$ or $x - 8 = 0$, then $x - 5 = 0 \Rightarrow x = 5$ or $x - 8 = 0 \Rightarrow x = 8$, so $x = 0,\ 5,\ 8$.

25. $x^2 - x = 0 \Rightarrow x(x - 1) = 0$, then $x = 0$ or $x - 1 = 0 \Rightarrow x = 1$, so $x = 0, 1$.

27. $z^2 - 5z = 0 \Rightarrow z(z - 5) = 0$, then $z = 0$ or $z - 5 = 0 \Rightarrow z = 5$, so $z = 0,\ 5$.

29. $10y^2 + 15y = 0 \Rightarrow 5y(2y + 3) = 0$, then $5y = 0 \Rightarrow y = 0$ or $2y + 3 = 0 \Rightarrow 2y = -3 \Rightarrow$

     $y = \dfrac{-3}{2}$, so $y = 0, \dfrac{-3}{2}$.

31. $x^2 - 1 = 0 \Rightarrow (x + 1)(x - 1)$ then $x + 1 = 0 \Rightarrow x = -1$ or $x - 1 = 0 \Rightarrow x = 1$ so $x = -1, 1$.

33. $4n^2 - 1 = 0 \Rightarrow (2n + 1)(2n - 1) = 0$, then $2n + 1 = 0 \Rightarrow 2n = -1 \Rightarrow n = \dfrac{-1}{2}$ or $2n - 1 = 0 \Rightarrow$

     $2n = 1 \Rightarrow n = \dfrac{1}{2}$, so $n = -\dfrac{1}{2}, \dfrac{1}{2}$.

35. $z^2 + 3z + 2 = 0 \Rightarrow (z + 2)(z + 1) = 0$, then $z + 2 = 0 \Rightarrow z = -2$ or $z + 1 = 0 \Rightarrow z = -1$, so $z = -2, -1$.

37. $x^2 - 12x + 35 = 0 \Rightarrow (x - 7)(x - 5) = 0$, then $x - 7 = 0 \Rightarrow x = 7$ or $x - 5 = 0 \Rightarrow x = 5$, so $x = 5,\ 7$.

39. $2b^2 + 3b - 2 = 0 \Rightarrow (2b - 1)(b + 2) = 0$, then $2b - 1 = 0 \Rightarrow 2b = 1 \Rightarrow$

     $b = \dfrac{1}{2}$ or $b + 2 = 0 \Rightarrow b = -2$, so $b = -2, \dfrac{1}{2}$.

41. $6y^2 + 19y + 10 = 0$, $\left(\text{factor by grouping}\right)$, $m \cdot n = 60$, $m + n = 19 \Rightarrow m = 15,\ n = 4 \Rightarrow$

     $6y^2 + 15y + 4y + 10 = 0 \Rightarrow 3y(2y + 5) + 2(2y + 5) = 0 \Rightarrow (3y + 2)(2y + 5) = 0$, then $3y + 2 = 0 \Rightarrow$

     $3y = -2 \Rightarrow y = \dfrac{-2}{3}$ or $2y + 5 = 0 \Rightarrow 2y = -5 \Rightarrow y = \dfrac{-5}{2}$, so $y = -\dfrac{5}{2}, -\dfrac{2}{3}$.

43. $x^2 = 25 \Rightarrow x^2 - 25 = 0 \Rightarrow (x + 5)(x - 5) = 0$, then $x + 5 = 0 \Rightarrow x = -5$ or $x - 5 = 0 \Rightarrow$

     $x = 5$, so $x = -5,\ 5$.

45. $t^2 = 5t \Rightarrow t^2 - 5t = 0 \Rightarrow t(t - 5) = 0$, then $t = 0$ or $t - 5 = 0 \Rightarrow t = 5$, so $t = 0,\ 5$.

47. $3m^2 = -9m \Rightarrow 3m^2 + 9m = 0 \Rightarrow 3m(m+3) = 0$, then $3m = 0 \Rightarrow m = 0$ or $m+3 = 0 \Rightarrow$

    $m = -3$, so $m = -3, \ 0$.

49. $x^2 = 5x + 6 \Rightarrow x^2 - 5x - 6 = 0 \Rightarrow (x+1)(x-6) = 0$, then $x+1 = 0 \Rightarrow x = -1$ or $x-6 = 0 \Rightarrow$

    $x = 6$, so $x = -1, \ 6$.

51. $12z^2 = 5 - 4z \Rightarrow 12z^2 + 4z - 5 = 0$, (factor by grouping), $m \cdot n = -60$, $m+n = 4 \Rightarrow$

    $m = 10, \ n = -6 \Rightarrow 12z^2 + 10z - 6z - 5 = 0 \Rightarrow 2z(6z+5) - (6z+5) = 0 \Rightarrow (2z-1)(6z+5) = 0$,

    then $2z - 1 = 0 \Rightarrow 2z = 1 \Rightarrow z = \dfrac{1}{2}$ or $6z + 5 = 0 \Rightarrow 6z = -5 \Rightarrow z = \dfrac{-5}{6}$, so $z = -\dfrac{5}{6}, \dfrac{1}{2}$.

53. $t(t+1) = 2 \Rightarrow t^2 + t = 2 \Rightarrow t^2 + t - 2 = 0 \Rightarrow (t+2)(t-1) = 0$, then $t+2 = 0 \Rightarrow$

    $t = -2$ or $t - 1 = 0 \Rightarrow t = 1$, so $t = -2, \ 1$.

55. $x(2x+5) = 3 \Rightarrow 2x^2 + 5x = 3 \Rightarrow 2x^2 + 5x - 3 = 0 \Rightarrow (2x-1)(x+3) = 0$, then $2x-1 = 0 \Rightarrow$

    $2x = 1 \Rightarrow x = \dfrac{1}{2}$ or $x+3 = 0 \Rightarrow x = -3$, so $x = -3, \dfrac{1}{2}$.

57. $12x^2 + 12x = -3 \Rightarrow 12x^2 + 12x + 3 = 0 \Rightarrow 3(4x^2 + 4x + 1) = 0 \Rightarrow 3(2x+1)^2 = 0$, then

    $2x + 1 = 0 \Rightarrow 2x = -1 \Rightarrow x = -\dfrac{1}{2}$.

59. $30y^2 + 50y + 20 = 0 \Rightarrow 10(3y^2 + 5y + 2) = 0 \Rightarrow 10(3y+2)(y+1) = 0$, then

    $3y + 2 = 0 \Rightarrow 3y = -2 \Rightarrow y = -\dfrac{2}{3}$ or $y + 1 = 0 \Rightarrow y = -1$, so $y = -1, -\dfrac{2}{3}$.

61. Both sides are the same so $x^2 = 144 \Rightarrow x^2 - 144 = 0 \Rightarrow (x+12)(x-12) = 0$, then $x+12 = 0 \Rightarrow$

    $x = -12$ or $x - 12 = 0 \Rightarrow x = 12$, so $x = -12, \ 12$, but length is not negative so $x = 12$ feet.

63. Area of circle $-$ circumference of circle $= 8\pi \Rightarrow \pi r^2 - 2\pi r = 8\pi \Rightarrow r^2 - 2r = 8 \Rightarrow$

    $r^2 - 2r - 8 = 0 \Rightarrow (r-4)(r+2) = 0$, then $r - 4 = 0 \Rightarrow r = 4$ or $r + 2 = 0 \Rightarrow r = -2$, so $r = -2, \ 4$,

    but radius cannot be $-2$ so $r = 4$.

65. Using Pythagorean Theorem

    $a^2 + b^2 = c^2, \ (x-1)^2 + x^2 = (x+1)^2 \Rightarrow x^2 - 2x + 1 + x^2 = x^2 + 2x + 1 \Rightarrow$

    $2x^2 - 2x + 1 = x^2 + 2x + 1 \Rightarrow x^2 - 4x = 0 \Rightarrow x(x-4) = 0$, then $x = 0$ or $x - 4 = 0 \Rightarrow$

    $x = 4$, so $x = 0, \ 4$, but length cannot be 0 so $x = 4$.

67. (a)   Height is zero at ground,

    $0 = 96t - 16t^2 \Rightarrow 0 = 16t(6-t)$, then $16t = 0 \Rightarrow t = 0$ or $6 - t = 0 \Rightarrow$

    $-t = -6 \Rightarrow t = 6$, so $t = 0, \ 6$, but $t = 0$ is when it was first hit, so after it was hit, it took 6

    seconds to hit the ground.

(b)    3 seconds.  See Figure 67.

| Time ($t$) | 0 | 1 | 2 | 3 | 4 | 5 | 6 |
|---|---|---|---|---|---|---|---|
| Height ($h$) | 0 | 80 | 128 | 144 | 128 | 80 | 0 |

Figure 67

69. (a)    For $x = 30$, $D = \dfrac{1}{11}(30)^2 \Rightarrow D = \dfrac{1}{11}(900) \Rightarrow D \approx 81.8$ feet.

For $x = 60$, $D = \dfrac{1}{11}(60)^2 \Rightarrow D = \dfrac{1}{11}(3600) \Rightarrow D = 327.3$ feet.

When the speed doubles, the braking distance quadruples.

(b)    $33 = \dfrac{1}{11}x^2 \Rightarrow 363 = x^2 \Rightarrow x \approx 19$ mph.

(c)    See Figure 69.  About 19 miles per hour: yes.

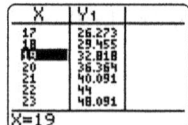          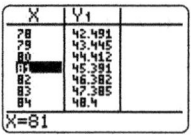

Figure 69                      Figure 71

71. (a)    In 1930, $x = 30 \Rightarrow W = \dfrac{19}{3125}(30)^2 + \dfrac{11}{2} \Rightarrow W = \dfrac{19}{3125}(900) + \dfrac{11}{2} \Rightarrow$

$W = 10.972$ million $\approx 11$ million.

In 2000, $x = 100 \Rightarrow W = \dfrac{19}{3125}(100)^2 + \dfrac{11}{2} \Rightarrow W = \dfrac{19}{3125}(10,000) + \dfrac{11}{2} \Rightarrow$

$W = 66.3$ million $\approx 66$ million.

(b)    See Figure 71.  About 1981.

73.  Width $= x$, Length $= x + 10$, $x(x + 10) = 2000 \Rightarrow x^2 + 10x = 2000 \Rightarrow x^2 + 10x - 2000 = 0 \Rightarrow$

$(x + 50)(x - 40) = 0$, then $x + 50 = 0 \Rightarrow x = -50$ or $x - 40 = 0 \Rightarrow x = 40$, so $x = -50$, 40, but we

can not have $-50$ pixels, so $x = 40 \Rightarrow$ width 40 and length $40 + 10 = 50 \Rightarrow 40 \times 50$ pixels.

## Checking Basic Concepts  Sections 6.5 and 6.6

1. (a)    $9a^2 - 18a + 27 = 9(a^2 - 2a + 3)$

(b)    $7xy^2 + 28x = 7x\,(y^2 + 4)$

2. (a)    $6z^4 - 28z^3 + 16z^2 = 2z^2(3z^2 - 14z + 8)\; = 2z^2(3z - 2)(z - 4)$

(b)    $2r^2t^2 - 18r^2 = 2r^2(t^2 - 9) = 2r^2(t - 3)(t + 3)$

3. (a)    $36x^3 - 48x^2 + 16x = 4x\,(9x^2 - 12x + 4)\; = 4x\,(3x - 2)^2$

(b)    $24b^3 - 81 = 3(8b^3 - 27) = 3(2b - 3)(4b^2 + 6b + 9)$

4. (a)    $4y^2 - 6y = 0 \Rightarrow 2y(2y - 3) = 0$, then $2y = 0$ or $2y - 3 = 0 \Rightarrow y = 0, \dfrac{3}{2}$

   (b)    $5z^2 + 2z = 3 \Rightarrow 5z^2 + 2z - 3 = 0 \Rightarrow (5z - 3)(z + 1) = 0$, then $5z - 3 = 0 \Rightarrow 5z = 3 \Rightarrow z = \dfrac{3}{5}$ or

          $z + 1 = 0$, so $z = -1, \dfrac{3}{5}$

5. Symbolic: $x^2 + 2x - 3 = 0 \Rightarrow (x + 3)(x - 1) = 0$, then $x + 3 = 0$ or $x - 1 = 0$, so $x = -3, 1$

   Numerical: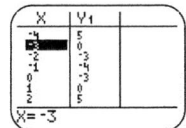

6. $88t - 16t^2 = 0 \Rightarrow -8t(2t - 11) = 0$, then $-8t = 0 \Rightarrow t = 0$ or $2t - 11 = 0 \Rightarrow 2t = 11 \Rightarrow t = \dfrac{11}{2}$,

   so $t = 0, \dfrac{11}{2}$; $t = 0$ is when the ball was hit so the golf ball will hit the ground 5.5 seconds after it

   was hit.

## Section 6.7  Solving Equations by Factoring II (Higher Degree)

1. GCF

3. factors

5. $(z^2 + 1)(z^2 + 2)$

7. Yes, $x^2 + 1$ cannot be factored further.

9. Factor out $x^2$.

11. $5x^2 - 5x - 30 \Rightarrow 5(x^2 - x - 6) \Rightarrow 5(x + 2)(x - 3)$

13. $-4y^2 - 32y - 48 \Rightarrow -4(y^2 + 8y + 12) \Rightarrow -4(y + 6)(y + 2)$

15. $-20z^2 - 110z - 50 \Rightarrow -10(2z^2 + 11z + 5) \Rightarrow -10(2z + 1)(z + 5)$

17. $60 - 64t - 28t^2 \Rightarrow -4(7t^2 + 16t - 15) \Rightarrow -4(7t - 5)(t + 3)$

19. $r^3 - r \Rightarrow r(r^2 - 1) \Rightarrow r(r + 1)(r - 1)$

21. $3x^3 + 3x^2 - 18x \Rightarrow 3x(x^2 + x - 6) \Rightarrow 3x(x + 3)(x - 2)$

23. $72z^3 + 12z^2 - 24z \Rightarrow 12z(6z^2 + z - 2) \Rightarrow 12z(3z + 2)(2z - 1)$

25. $x^4 - 4x^2 \Rightarrow x^2(x^2 - 4) \Rightarrow x^2(x + 2)(x - 2)$

27. $t^4 + t^3 - 2t^2 \Rightarrow t^2\left(t^2 + t - 2\right) \Rightarrow t^2\left(t+2\right)(t-1)$

29. $x^4 - 5x^2 + 6 \Rightarrow \left(x^2 - 2\right)\left(x^2 - 3\right)$

31. $2x^4 + 7x^2 + 3 \Rightarrow \left(2x^2 + 1\right)\left(x^2 + 3\right)$

33. $y^4 + 6y^2 + 9 \Rightarrow \left(y^2 + 3\right)\left(y^2 + 3\right) \Rightarrow \left(y^2 + 3\right)^2$

35. $x^4 - 9 \Rightarrow \left(x^2 + 3\right)\left(x^2 - 3\right)$

37. $x^4 - 81 \Rightarrow \left(x^2 + 9\right)\left(x^2 - 9\right) \Rightarrow \left(x^2 + 9\right)(x+3)(x-3)$

39. $z^5 + 2z^4 + z^3 \Rightarrow z^3(z^2 + 2z + 1) \Rightarrow z^3(z+1)(z+1) \Rightarrow z^3(z+1)^2$

41. $2x^2 + xy - y^2 \Rightarrow (2x - y)(x + y)$

43. $a^4 - 2a^2b^2 + b^4 \Rightarrow \left(a^2 - b^2\right)\left(a^2 - b^2\right) \Rightarrow (a+b)(a-b)(a+b)(a-b) \Rightarrow (a+b)^2(a-b)^2$

45. $x^3 - xy^2 \Rightarrow x\left(x^2 - y^2\right) \Rightarrow x(x+y)(x-y)$

47. $4x^3 + 4x^2y + xy^2 \Rightarrow x\left(4x^2 + 4xy + y^2\right) \Rightarrow x(2x+y)(2x+y) \Rightarrow x(2x+y)^2$

49. (a)   $x^3 - 4x \Rightarrow x\left(x^2 - 4\right) \Rightarrow x(x+2)(x-2)$

    (b)   $x(x+2)(x-2) = 0$, then $x = 0$ or $x+2 = 0$ or $x - 2 = 0 \Rightarrow x = -2,\ 0,\ 2.$

51. (a)   $2y^3 - 6y^2 - 36y \Rightarrow 2y\left(y^2 - 3y - 18\right) \Rightarrow 2y(y-6)(y+3)$

    (b)   $2y(y-6)(y+3) = 0$, then $2y = 0$ or $y - 6 = 0$ or $y + 3 = 0 \Rightarrow y = -3,\ 0,\ 6.$

53. (a)   $x^3 - x^2 + 4x - 4 \Rightarrow x^2(x-1) + 4(x-1) \Rightarrow \left(x^2 + 4\right)(x-1)$

    (b)   $\left(x^2 + 4\right)(x-1) = 0$, then $\left(x^2 + 4\right) = 0$ or $(x-1) = 0 \Rightarrow x = 1$ and there is no real number that

    makes $\left(x^2 + 4\right) = 0.$

55. $3x^2 + 33x + 72 = 0 \Rightarrow 3\left(x^2 + 11x + 24\right) = 0 \Rightarrow 3(x+8)(x+3) = 0,$ then $x + 8 = 0$ or $x + 3 = 0 \Rightarrow$

    $x = -8,\ -3.$

57. $25x^2 = 50x + 75 \Rightarrow 25x^2 - 50x - 75 = 0 \Rightarrow 25\left(x^2 - 2x - 3\right) = 0 \Rightarrow 25(x-3)(x+1) = 0,$

    then $x - 3 = 0$ or $x + 1 = 0 \Rightarrow x = -1,\ 3.$

59. $y^3 - 3y^2 - 4y = 0 \Rightarrow y\left(y^2 - 3y - 4\right) = 0 \Rightarrow y(y-4)(y+1) = 0,$ then $y = 0$ or $y - 4 = 0$ or

    $y + 1 = 0 \Rightarrow y = -1,\ 0,\ 4.$

61. $3z^3 + 6z^2 = 72z \Rightarrow 3z^3 + 6z^2 - 72z = 0 \Rightarrow 3z\left(z^2 + 2z - 24\right) = 0 \Rightarrow 3z(z+6)(z-4) = 0,$ then

    $3z = 0$ or $z + 6 = 0$ or $z - 4 = 0 \Rightarrow z = -6,\ 0,\ 4.$

63. $x^4 - 36x^2 = 0 \Rightarrow x^2(x^2 - 36) = 0 \Rightarrow x^2(x+6)(x-6) = 0$, then $x^2 = 0$ or $x + 6 = 0$ or $x - 6 = 0 \Rightarrow$

   $x = -6, 0, 6$.

65. $r^4 + 6r^3 = 7r^2 \Rightarrow r^4 + 6r^3 - 7r^2 = 0 \Rightarrow r^2(r^2 + 6r - 7) = 0 \Rightarrow r^2(r+7)(r-1) = 0$, then $r^2 = 0$ or

   $r + 7 = 0$ or $r - 1 = 0 \Rightarrow r = -7, 0, 1$.

67. $x^4 - 13x^2 = -36 \Rightarrow x^4 - 13x^2 + 36 = 0 \Rightarrow (x^2 - 9)(x^2 - 4) = 0 \Rightarrow (x+3)(x-3)(x+2)(x-2) = 0$,

   then $x + 3 = 0$ or $x - 3 = 0$ or $x + 2 = 0$ or $x - 2 = 0 \Rightarrow x = -3, -2, 2, 3$.

69. $x^4 + 1 = 2x^2 \Rightarrow x^4 - 2x^2 + 1 = 0 \Rightarrow (x^2 - 1)(x^2 - 1) = 0 \Rightarrow (x+1)(x-1)(x+1)(x-1) = 0$, then

   $x + 1 = 0$ or $x - 1 = 0 \Rightarrow x = -1, 1$.

71. $a^4 = 81 \Rightarrow a^4 - 81 = 0 \Rightarrow (a^2 + 9)(a^2 - 9) = 0 \Rightarrow (a^2 + 9)(a+3)(a-3) = 0$, then $a^2 + 9 = 0$ or

   $a + 3 = 0$ or $a - 3 = 0$, but $a^2 + 9 = 0$ does not produce real solutions so $a = -3, 3$.

73. $x^3 - 2x^2 - x + 2 = 0 \Rightarrow x^2(x-2) - (x-2) = 0 \Rightarrow (x^2 - 1)(x-2) = 0 \Rightarrow$

   $(x+1)(x-1)(x-2) = 0$, then $x + 1 = 0$ or $x - 1 = 0$ or $x - 2 = 0 \Rightarrow x = -1, 1, 2$.

75. $x^3 - 5x^2 + x - 5 = 0 \Rightarrow x^2(x-5) + (x-5) = 0 \Rightarrow (x^2 + 1)(x-5) = 0 \Rightarrow$ then $x^2 + 1 = 0$ or

   $x - 5 = 0$ but $x^2 + 1 = 0$ does not produce real solutions so $x = 5$.

77. (a)   $x < 7.5$ *in.* because the width is 15 inches.

   (b)   Surface area of rectangle with 4 cut-out pieces is $15 \times 20 = 300$.  Area of cut-out pieces =

   $x^2 \cdot 4$ pieces $= 4x^2 \Rightarrow$ the surface area is $300 - 4x^2$.

   (c)   $300 - 4x^2 = 275 \Rightarrow 0 = 4x^2 - 25 \Rightarrow (2x+5)(2x-5) = 0$, then $2x + 5 = 0$ or $2x - 5 = 0$, but

   $2x + 5 = 0$ produces a negative length which we can not have so $2x - 5 = 0 \Rightarrow 2x = 5 \Rightarrow$

   $x = \dfrac{5}{2} \Rightarrow x = 2.5$ inches.

79. In 1990, $x = 30 \Rightarrow 0.0013(30)^3 - 0.085(30)^2 + 1.6(30) + 12 = 35.1 - 76.5 + 48 + 12 = 18.6$ trillion ft$^3$

81. Factoring is very difficult.


## Checking Basic Concepts  Section 6.7

1. (a)   $3x^2 - 6x - 24 = 3(x^2 - 2x - 8) = 3(x-4)(x+2)$

   (b)   $-10y^2 + 5y + 5 = -5(2y^2 - y - 1) = -5(2y+1)(y-1)$

2. (a)   $z^4 - 25 = (z^2 - 5)(z^2 + 5)$

(b)  $7t^4 - 7 = 7\left(t^4 - 1\right) = 7\left(t^2 - 1\right)\left(t^2 + 1\right) = 7(t-1)(t+1)\left(t^2 + 1\right)$

3. (a)  $x^4 - 8x^2 + 16 \Rightarrow \left(x^2 - 4\right)\left(x^2 - 4\right) \Rightarrow (x+2)(x-2)(x+2)(x-2) \Rightarrow (x+2)^2 (x-2)^2$

(b)  $2y^3 + 17y^2 - 30y \Rightarrow y\left(2y^2 + 17y - 30\right) \Rightarrow y(2y-3)(y+10)$

4. $t^4 + t^3 = 12t^2 \Rightarrow t^4 + t^3 - 12t^2 = 0 \Rightarrow t^2\left(t^2 + t - 12\right) = 0 \Rightarrow t^2(t+4)(t-3) = 0$, then $t^2 = 0$ or

$t + 4 = 0$ or $t - 3 = 0 \Rightarrow t = -4, 0, 3.$

5. $x^3 - 3x^2 + 2x - 6 = 0 \Rightarrow x^2(x-3) + 2(x-3) = 0 \Rightarrow \left(x^2 + 2\right)(x-3) = 0$, then $x^2 + 2 = 0 \Rightarrow x^2 = -2$

(not possible for real numbers) or $x - 3 = 0 \Rightarrow x = 3.$

## Chapter 6 Review

1. Factors of $5x = 5 \cdot x \Rightarrow$ factors of $15 = 3 \cdot 5 \Rightarrow$ GCF=5; $5(x-3)$

2. Factors of $y^2 = y \cdot y \Rightarrow$ factors of $2y = 2 \cdot y \Rightarrow$ GCF=y; $y(y+2)$

3. Factors of $8z^3 = 2 \cdot 2 \cdot 2 \cdot z \cdot z \cdot z \Rightarrow$ factors of $4z^2 = 2 \cdot 2 \cdot z \cdot z \Rightarrow$ GCF $= 2 \cdot 2 \cdot z \cdot z = 4z^2$;

   $4z^2 (2z-1)$

4. Factors of $6x^4 = 2 \cdot 3 \cdot x \cdot x \cdot x \cdot x \Rightarrow$ factors of $3x^3 = 3 \cdot x \cdot x \cdot x \Rightarrow$

   factors of $12x^2 = 2 \cdot 2 \cdot 3 \cdot x \cdot x \Rightarrow$ GCF $= 3 \cdot x \cdot x = 3x^2$; $3x^2\left(2x^2 + x - 4\right).$

5. Factors of $9xy = 3 \cdot 3 \cdot x \cdot y \Rightarrow$ factors of $15yz^2 = 3 \cdot 5 \cdot y \cdot z \cdot z \Rightarrow$ GCF=3$\cdot y = 3y$; $3y(3x + 5z^2).$

6. Factors of $a^2b^3 = a \cdot a \cdot b \cdot b \cdot b \Rightarrow$ factors of $a^3b^2 = a \cdot a \cdot a \cdot b \cdot b \Rightarrow$ GCF $= a \cdot a \cdot b \cdot b = a^2b^2$;

   $a^2b^2 (b+a).$

7. $x(x+2) - 3(x+2) = (x-3)(x+2)$

8. $y^2(x-5) + 3y(x-5) = \left(y^2 + 3y\right)(x-5) = y(y+3)(x-5)$

9. $z^3 - 2z^2 + 5z - 10 = z^2(z-2) + 5(z-2) = \left(z^2 + 5\right)(z-2)$

10. $t^3 + t^2 + 8t + 8 = t^2(t+1) + 8(t+1) = \left(t^2 + 8\right)(t+1)$

11. $x^3 - 3x^2 + 6x - 18 = x^2(x-3) + 6(x-3) = \left(x^2 + 6\right)(x-3)$

12. $ax + bx - ay - by = x(a+b) - y(a+b) = (x-y)(a+b)$

13. $x^5 + 3x^4 - 2x^3 - 6x^2 = x^2\left(x^3 + 3x^2 - 2x - 6\right) = x^2\left[x^2(x+3) - 2(x+3)\right] = x^2\left(x^2 - 2\right)(x+3)$

14. $2y^4 + 6y^3 + 2y^2 + 6y = 2y\left(y^3 + 3y^2 + y + 3\right) = 2y\left[y^2(y+3) + 1(y+3)\right] = 2y\left(y^2 + 1\right)(y+3)$

15. $4, 5$

16. $-3, 7$

17. $-9, -4$

18. $-25, 4$

19. Product $-12$, sum $-1$ is $-4, 3 \Rightarrow (x-4)(x+3)$.

20. Product $24$, sum $10$ is $4, 6 \Rightarrow (x+4)(x+6)$.

21. Product $-16$, sum $6$ is $-2, 8 \Rightarrow (x-2)(x+8)$.

22. Product $-42$, sum $-1$ is $-7, 6 \Rightarrow (x-7)(x+6)$.

23. Product $-6$, sum $4 \Rightarrow$ prime

24. Product $8$, sum $-5 \Rightarrow$ prime

25. Product $-3$, sum $2$ is $-1, 3 \Rightarrow (x-1)(x+3)$.

26. Product $120$, sum $22$ is $10, 12 \Rightarrow (x+10)(x+12)$.

27. $2x^3 + 6x^2 - 20x = 2x(x^2 + 3x - 10)$; product $-10$, sum $3$ is $5, -2 \Rightarrow 2x(x+5)(x-2)$.

28. $x^4 - 3x^3 - 28x^2 = x^2(x^2 - 3x - 28)$; product $-28$, sum $-3$ is $-7, 4 \Rightarrow x^2(x-7)(x+4)$.

29. Product $10$, sum $-7$ is $-5, -2 \Rightarrow (2-x)(5-x)$.

30. Product $24$, sum $2$ is $6, -4 \Rightarrow (6-x)(4+x)$.

31. First factor out the GCF of $-2$.  $-2(x^2 + 2x - 15)$  Product $-15$, sum $2$ is $5, -3 \Rightarrow -2(x+5)(x-3)$

32. First factor out the GCF of $-x$.  $-x(x^2 + 9x - 10)$  Product $-10$, sum $9$ is $10, -1 \Rightarrow -x(x+10)(x-1)$

33. Using FOIL:  $9x^2 + 3x - 2 = (3x+2)(3x-1)$.

34. Using FOIL:  $2x^2 + 3x - 5 = (2x+5)(x-1)$.

35. Using FOIL:  $3x^2 + 14x + 15 = (3x+5)(x+3)$.

36. Using FOIL:  $35x^2 - 2x - 1 = (5x-1)(7x+1)$.

37. Using FOIL:  $3x^2 + 4x - 5 \Rightarrow$ prime

38. Using FOIL:  $4x^2 - 12x - 5 \Rightarrow$ prime

39. Using FOIL:  $24x^2 - 7x - 5 = (8x-5)(3x+1)$.

40. Using FOIL:  $4x^2 + 33x - 27 = (x+9)(4x-3)$.

41. $12x^3 + 48x^2 + 21x = 3x(4x^2 + 16x + 7)$; using FOIL: $3x(2x+7)(2x+1)$.

42. $8x^4 + 14x^3 - 30x^2 = 2x^2(4x^2 + 7x - 15)$; using FOIL: $2x^2(x+3)(4x-5)$.

43. Using FOIL: $12 - 5x - 2x^2 = (3 - 2x)(4 + x)$.

44. Using FOIL: $1 + 3x - 10x^2 = (1 + 5x)(1 - 2x)$.

45. Difference of squares, $z^2 - 4 = (z + 2)(z - 2)$.

46. Difference of squares, $9z^2 - 64 = (3z + 8)(3z - 8)$.

47. Difference of squares, $36 - y^2 = (6 + y)(6 - y)$.

48. Difference of squares, $100a^2 - 81b^2 = (10a + 9b)(10a - 9b)$.

49. Perfect square trinomial,

    $a^2 = x^2$ so $a = x$, $b^2 = 7^2$ so $b = 7$, then $2ab = 2 \cdot 7 \cdot x = 14x$ the middle term $\Rightarrow (x + 7)^2$.

50. Perfect square trinomial,

    $a^2 = x^2$ so $a = x$, $b^2 = (\pm 5)^2$ so $b = \pm 5$, using $b = -5$, then $2ab = 2 \cdot (-5)x =$

    $-10x$ the middle term $\Rightarrow (x - 5)^2$.

51. Perfect square trinomial, $a^2 = (2x)^2$ so $a = 2x$, $b^2 = (\pm 3)^2$ so $b = \pm 3$, using $b = -3$, then

    $2ab = 2 \cdot (-3)2x = -12x$ the middle term $\Rightarrow (2x - 3)^2$.

52. Perfect square trinomial, $a^2 = (3x)^2$ so $a = 3x$, $b^2 = 8^2$ so $b = 8$,

    then $2ab = 2 \cdot 8 \cdot 3x = 48x$ the middle term $\Rightarrow (3x + 8)^2$.

53. Difference of cubes, $8t^3 - 1$, $a^3 = (2t)^3$ so $a = 2t$, $b^3 = 1^3$ so $b = 1 \Rightarrow (2t - 1)(4t^2 + 2t + 1)$.

54. Sum of cubes, $27r^3 + 8t^3$, $a^3 = (3r)^3$ so $a = 3r$, $b^3 = (2t)^3$ so $b = 2t \Rightarrow (3r + 2t)(9r^2 - 6rt + 4t^2)$.

55. $2x^3 - 50x = 2x(x^2 - 25) = 2x(x - 5)(x + 5)$ (difference of squares).

56. $24x^3 + 81 = 3(8x^3 + 27) = 3(2x + 3)(4x^2 - 6x + 9)$ (sum of cubes).

57. $2x^3 + 28x^2 + 98x = 2x(x^2 + 14x + 49) = 2x(x + 7)^2$ (perfect square trinomial).

58. $2x^4 - 128x = 2x(x^3 - 64) = 2x(x - 4)(x^2 + 4x + 16)$ (difference of cubes).

59. $12x - 8 = 4(3x - 2)$

60. $6x^3 + 9x^2 = 3x^2(2x + 3)$

61. $9y^2 - 6y + 6 = 3(3y^2 - 2y + 2)$

62. $yz^2 - 9y = y(z^2 - 9) = y(z - 3)(z + 3)$

63. $x^4 + 7x^3 - 4x^2 - 28x = x(x^3 + 7x^2 - 4x - 28) = x[x^2(x + 7) - 4(x + 7)] = x(x^2 - 4)(x + 7)$

    $= x(x - 2)(x + 2)(x + 7)$

64. $12x^3 + 36x^2 + 27x = 3x(4x^2 + 12x + 9) = 3x(2x+3)^2$

65. $3ab^3 - 24a = 3a(b^3 - 8) = 3a(b-2)(b^2 + 2b + 4)$

66. $5x^3 + 20x = 5x(x^2 + 4)$

67. $24x^3 - 6xy^2 = 6x(4x^2 - y^2) = 6x(2x - y)(2x + y)$

68. $x^3 y + 27y = y(x^3 + 27) = y(x+3)(x^2 - 3x + 9)$

69. $m = 0$ or $n = 0$

70. $y = 0$

71. $(4x-3)(x+9) = 0$, then $4x - 3 = 0 \Rightarrow 4x = 3 \Rightarrow x = \dfrac{3}{4}$ or $x + 9 = 0 \Rightarrow x = -9$, so $x = -9, \dfrac{3}{4}$.

72. $(1-4x)(6+5x) = 0$, then $1 - 4x = 0 \Rightarrow -4x = -1 \Rightarrow x = \dfrac{-1}{-4} \Rightarrow x = \dfrac{1}{4}$ or $6 + 5x = 0 \Rightarrow 5x = -6 \Rightarrow$

$x = \dfrac{-6}{5}$, so $x = -\dfrac{6}{5}, \dfrac{1}{4}$.

73. $z(z-1)(z-2) = 0$, then $z = 0$ or $z - 1 = 0 \Rightarrow z = 1$ or $z - 2 = 0 \Rightarrow z = 2$, so $z = 0, 1, 2$.

74. $z^2 - 7z = 0 \Rightarrow z(z-7) = 0$, then $z = 0$ or $z - 7 = 0 \Rightarrow z = 7$, so $z = 0, 7$.

75. $y^2 - 64 = 0$, difference of squares $(y+8)(y-8) = 0$, then $y + 8 = 0 \Rightarrow y = -8$ or $y - 8 = 0 \Rightarrow$

$y = 8$, so $y = -8, 8$.

76. $y^2 + 9y + 14 = 0 \Rightarrow (y+2)(y+7) = 0$, then $y + 2 = 0 \Rightarrow y = -2$ or $y + 7 = 0 \Rightarrow$

$y = -7$, so $y = -2, -7$.

77. $x^2 = x + 6 \Rightarrow x^2 - x - 6 = 0 \Rightarrow (x-3)(x+2) = 0$, then $x - 3 = 0 \Rightarrow x = 3$ or $x + 2 = 0 \Rightarrow$

$x = -2$, so $x = -2, 3$.

78. $10x^2 + 11x = 6 \Rightarrow 10x^2 + 11x - 6 = 0 \Rightarrow (2x+3)(5x-2) = 0$, then $2x + 3 = 0 \Rightarrow 2x = -3 \Rightarrow$

$x = \dfrac{-3}{2}$ or $5x - 2 = 0 \Rightarrow 5x = 2 \Rightarrow x = \dfrac{2}{5}$, so $x = -\dfrac{3}{2}, \dfrac{2}{5}$.

79. $t(t-14) = 72 \Rightarrow t^2 - 14t = 72 \Rightarrow t^2 - 14t - 72 = 0 \Rightarrow (t-18)(t+4) = 0$, then $t - 18 = 0 \Rightarrow$

$t = 18$ or $t + 4 = 0 \Rightarrow t = -4$, so $t = -4, 18$.

80. $t(2t-1) = 10 \Rightarrow 2t^2 - t = 10 \Rightarrow 2t^2 - t - 10 = 0 \Rightarrow (2t-5)(t+2) = 0$, then $2t - 5 = 0 \Rightarrow$

$2t = 5 \Rightarrow t = \dfrac{5}{2}$ or $t + 2 = 0 \Rightarrow t = -2$, so $t = -2, \dfrac{5}{2}$.

81. $5x^2 - 15x - 50 \Rightarrow 5(x^2 - 3x - 10) \Rightarrow 5(x-5)(x+2)$

82. $-3x^2 - 6x + 45 \Rightarrow -3(x^2 + 2x - 15) \Rightarrow -3(x+5)(x-3)$

83. $y^3 - 4y \Rightarrow y(y^2 - 4) \Rightarrow y(y+2)(y-2)$

84. $3y^3 + 6y^2 - 9y \Rightarrow 3y(y^2 + 2y - 3) \Rightarrow 3y(y+3)(y-1)$

85. $2z^4 + 14z^3 + 20z^2 \Rightarrow 2z^2(z^2 + 7z + 10) \Rightarrow 2z^2(z+2)(z+5)$

86. $8z^4 - 32z^2 \Rightarrow 8z^2(z^2 - 4) \Rightarrow 8z^2(z+2)(z-2)$

87. $x^4 - 6x^2 + 9 \Rightarrow (x^2 - 3)(x^2 - 3) \Rightarrow (x^2 - 3)^2$

88. $2x^4 - 15x^2 - 27 \Rightarrow (2x^2 + 3)(x^2 - 9) \Rightarrow (2x^2 + 3)(x+3)(x-3)$

89. $a^2 + 10ab + 25b^2 \Rightarrow (a + 5b)(a + 5b) \Rightarrow (a + 5b)^2$

90. $x^3 - xy^2 \Rightarrow x(x^2 - y^2) \Rightarrow x(x+y)(x-y)$

91. $16x^2 - 72x - 40 = 0 \Rightarrow 8(2x^2 - 9x - 5) = 0 \Rightarrow 8(2x+1)(x-5) = 0$, then $2x+1 = 0 \Rightarrow 2x = -1 \Rightarrow$

   $x = -\dfrac{1}{2}$ or $x - 5 = 0 \Rightarrow x = 5$, so $x = -\dfrac{1}{2}, 5$.

92. $2x^3 - 11x^2 + 15x = 0 \Rightarrow x(2x^2 - 11x + 15) = 0 \Rightarrow x(2x - 5)(x - 3) = 0$, then $x = 0$ or $2x - 5 = 0 \Rightarrow$

   $2x = 5 \Rightarrow x = \dfrac{5}{2}$ or $x - 3 = 0 \Rightarrow x = 3$, so $x = 0, \dfrac{5}{2}, 3$.

93. $t^3 - 25t = 0 \Rightarrow t(t^2 - 25) = 0 \Rightarrow t(t+5)(t-5) = 0$, then $t = 0$ or $t + 5 = 0 \Rightarrow$

   $t = -5$ or $t - 5 = 0 \Rightarrow t = 5$, so $t = -5, 0, 5$.

94. $t^4 - 7t^3 + 12t^2 = 0 \Rightarrow t^2(t^2 - 7t + 12) = 0 \Rightarrow t^2(t-3)(t-4) = 0$, then $t^2 = 0 \Rightarrow t = 0$ or $t - 3 = 0 \Rightarrow$

   $t = 3$ or $t - 4 = 0 \Rightarrow t = 4$, so $t = 0, 3, 4$.

95. $z^4 + 16 = 8z^2 \Rightarrow z^4 - 8z^2 + 16 = 0 \Rightarrow (z^2 - 4)(z^2 - 4) = 0 \Rightarrow (z+2)(z-2)(z+2)(z-2) = 0$, then

   $z + 2 = 0 \Rightarrow z = -2$ or $z - 2 = 0 \Rightarrow z = 2$, so $z = -2, 2$.

96. $z^4 - 256 = 0 \Rightarrow (z^2 + 16)(z^2 - 16) = 0 \Rightarrow (z^2 + 16)(z+4)(z-4) = 0$, then $z^2 + 16 = 0$ which

   produces no real solutions, or $z + 4 = 0 \Rightarrow z = -4$ or $z - 4 = 0 \Rightarrow z = 4$, so $z = -4, 4$.

97. $y^3 = -64 \Rightarrow y^3 + 64 = 0 \Rightarrow (y+4)(y^2 - 4y + 16) = 0$, then $y + 4 = 0 \Rightarrow$

   $y = -4$ or $y^2 - 4y + 16 = 0$ which produces no real solutions, so $y = -4$.

98. $y^3 - y^2 - y + 1 = 0 \Rightarrow y^2(y-1) - 1(y-1) = 0 \Rightarrow (y^2 - 1)(y-1) = 0 \Rightarrow$

   $(y+1)(y-1)(y-1) = 0$, then $y + 1 = 0 \Rightarrow y = -1$ or $y - 1 = 0 \Rightarrow y = 1$, so $y = -1, 1$.

99. The sides of a square are equal so the trinomial must be a perfect square trinomial,

$a^2 = (3x)^2$ so $a = 3x$, $b^2 = 7^2$ so $b = 7$, then $2ab = 2 \cdot 3x \cdot 7 = 42x$ the middle term,

so $(3x+7)^2$, so each side is $3x+7$. See Figure 99.

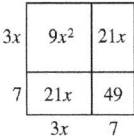

Figure 99    Figure 100

100. $x^2 + 6x + 5 = 0 \Rightarrow (x+5)(x+1) = 0$ so the sides are $(x+5)$ by $(x+1)$. See Figure 100.

101. A cube has six sides, so $(6x^2 + 12x + 6) \div 6 = x^2 + 2x + 1$, the area of each side $\Rightarrow (x+1)(x+1) \Rightarrow$

each side is $(x+1)$.

102. $x^2 + 3x + x + 3 = x^2 + 4x + 3 \Rightarrow (x+3)(x+1)$

103. $2x^2 + 3x + 12x + 18 = 2x^2 + 15x + 18 \Rightarrow (2x+3)(x+6)$

104. $(\text{area})\pi r^2 = 2r\pi(\text{circumference}) \Rightarrow r^2 = 2r \Rightarrow r = 2$

105. $x(x+7) = 120 \Rightarrow x^2 + 7x - 120 = 0 \Rightarrow (x+15)(x-8) = 0$, $x = -15$ (not possible), 8, so rectangle

$8 \times 15$ ft.

106. $100 = -16t^2 + 80t + 4 \Rightarrow 16t^2 - 80t + 96 = 0 \Rightarrow 16(t^2 - 5t + 6) = 0 \Rightarrow 16(t-2)(t-3) = 0$, so

$t = 2, 3$ so at 2 seconds and 3 seconds.

107. (a)   $D = \frac{1}{9}(45)^2 + \frac{11}{3}(45) \Rightarrow D = \frac{1}{9}(2025) + \frac{11}{3}(45) \Rightarrow D = 225 + 165 \Rightarrow D = 390$ feet.

(b)   $80 = \frac{1}{9}x^2 + \frac{11}{3}x \Rightarrow \frac{1}{9}x^2 + \frac{11}{3}x - 80 = 0 \Rightarrow$ multiply by 9 $\Rightarrow x^2 + 33x - 720 = 0 \Rightarrow$

$(x+48)(x-15) = 0$, so $D = -48$ (not possible) or 15 mph.

(c)   See Figure 107. 15 mph, yes.

108. (a)   $R = 100(200 - 100) \Rightarrow R = 100(100) \Rightarrow R = \$10,000$

(b)   $7500 = p(200 - p) \Rightarrow 7500 = 200p - p^2 \Rightarrow p^2 - 200p + 7500 = 0 \Rightarrow$

$(p-50)(p-150) = 0 \Rightarrow p = 50, 150 \Rightarrow \$50$ or $\$150$.

(c)   See Figure 108. $\$50$ or $\$150$, yes.

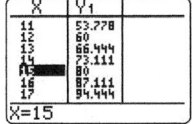

Figure 107

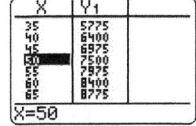

Figure 108

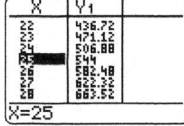

Figure 109

109. (a)    $Y = 20$ for 1970, $N = 0.68(20)^2 + 3.8(20) + 24 \Rightarrow N = 272 + 76 + 24 \Rightarrow N = 372$ million.

     (b)    See Figure 109. 1975.

110. $x(x+30) = 4000 \Rightarrow x^2 + 30x - 4000 = 0 \Rightarrow (x+80)(x-50) = 0$, so $x = -80$ (not possible)

     or 50, so 50×80 pixels.

111. (a)    Bottom $(50-2x)(40-2x) = 2000 - 80x - 100x + 4x^2 = 2000 - 180x + 4x^2$;

          2 sides $x(50-2x) \Rightarrow (50x - 2x^2)(2) = 100x - 4x^2$;

          2 sides $x(40-2x) \Rightarrow (40x - 2x^2)(2) = 80x - 4x^2$.

          Add all sides $2000 - 180x + 4x^2 + 100x - 4x^2 + 80x - 4x^2 = 2000 - 4x^2$.

     (b)    $1900 = 2000 - 4x^2 \Rightarrow -100 = -4x^2 \Rightarrow 25 = x^2 \Rightarrow x = 5$ inches.

112. Since we have two cubes first factor out the common factor of 2

     $12x^2 + 48x + 48 = 2(6x^2 + 24x + 24)$. The surface area of one of the cubes will be given as

     $6x^2 + 24x + 24$. The formula for the surface area of a cube is $S = 6s^2$. Simplify as follows:

     $S = 6s^2 = 6x^2 + 24x + 24 = 6(x^2 + 4x + 4) = 6(x+2)(x+2) = 6(x+2)^2$. Therefore, the length of the

     side of the cube is $x + 2$.

## Chapter 6 Test

1. $4x^2 y = 2 \cdot 2 \cdot x \cdot x \cdot y; 20xy^2 = 2 \cdot 2 \cdot 5 \cdot x \cdot y \cdot y; 12xy = 2 \cdot 2 \cdot 3 \cdot x \cdot y \Rightarrow$

     GCF $= 2 \cdot 2 \cdot x \cdot y = 4xy \Rightarrow 4x^2 y - 20xy^2 + 12xy = 4xy(x - 5y + 3)$

2. $9a^3 b^2 = 3 \cdot 3 \cdot a \cdot a \cdot a \cdot b \cdot b; 3a^2 b^2 = 3 \cdot a \cdot a \cdot b \cdot b \Rightarrow$ GCF $= 3 \cdot a \cdot a \cdot b \cdot b = 3a^2 b^2 \Rightarrow$

     $9a^3 b^2 + 3a^2 b^2 = 3a^2 b^2 (3a + 1)$

3. $ay + by + az + bz \Rightarrow y(a+b) + z(a+b) \Rightarrow (y+z)(a+b)$

4. $3x^3 + x^2 - 15x - 5 \Rightarrow x^2(3x+1) - 5(3x+1) \Rightarrow (x^2 - 5)(3x+1)$

5. $y^2 + 4y - 12 = (y+6)(y-2)$

6. $4x^2 + 20x + 25 = (2x+5)(2x+5) = (2x+5)^2$

7. $4z^2 - 19z + 12 = (4z-3)(z-4)$

8. $21 - 17t + 2t^2 = 2t^2 - 17t + 21 = (2t-3)(t-7)$

9. $x^2 + 7x - 10 \Rightarrow$ prime

10. $3y^2 + 4y + 2 \Rightarrow$ prime

11. $6x^3 + 3x^2 - 3x \Rightarrow 3x\left(2x^2 + x - 1\right) \Rightarrow 3x(2x-1)(x+1)$

12. $2z^4 - 12z^2 - 54 \Rightarrow 2\left(z^4 - 6z^2 - 27\right) \Rightarrow 2\left(z^2 - 9\right)\left(z^2 + 3\right) \Rightarrow 2(z+3)(z-3)\left(z^2 + 3\right)$

13. $36y^3 - 100y = 4y\left(9y^2 - 25\right) = 4y(3y-5)(3y+5)$

14. $7x^4 + 56x = 7x\left(x^3 + 8\right) = 7x(x+2)\left(x^2 - 2x + 4\right)$

15. $16a^4 + 24a^3 + 9a^2 = a^2\left(16a^2 + 24a + 9\right) = a^2\left(4a+3\right)^2$

16. $2b^4 - 32 = 2\left(b^4 - 16\right) = 2\left(b^2 - 4\right)\left(b^2 + 4\right) = 2(b-2)(b+2)\left(b^2 + 4\right)$

17. $x^2 - 16 = 0 \Rightarrow (x+4)(x-4) = 0$, then $x + 4 = 0$ or $x - 4 = 0$, so $x = -4, 4$.

18. $y^2 = y + 20 \Rightarrow y^2 - y - 20 = 0 \Rightarrow (y-5)(y+4) = 0$, then $y - 5 = 0$ or $y + 4 = 0$, so $y = -4, 5$.

19. $9z^2 + 16 = 24z \Rightarrow 9z^2 - 24z + 16 = 0$ is a perfect square trinomial $a^2 = \left(3z\right)^2$ so $a = 3z$,

    $b^2 = \left(\pm 4\right)^2$ so $b = \pm 4$, using $b = -4$, $2ab = 2 \cdot 3z \cdot -4 = -24z$ the middle term $\Rightarrow \left(3z - 4\right)^2$,

    then $3z - 4 = 0 \Rightarrow 3z = 4 \Rightarrow z = \dfrac{4}{3}$.

20. $x(x-5) = 66 \Rightarrow x^2 - 5x - 66 = 0 \Rightarrow (x-11)(x+6) = 0$, then $x - 11 = 0$ or $x + 6 = 0$, so $x = -6, 11$

21. $y^3 = 9y \Rightarrow y^3 - 9y = 0 \Rightarrow y\left(y^2 - 9\right) = 0 \Rightarrow y(y+3)(y-3) = 0$, then

    $y = 0$ or $y + 3 = 0$ or $y - 3 = 0$, so $y = -3, 0, 3$.

22. $x^4 - 5x^2 + 4 = 0 \Rightarrow \left(x^2 - 4\right)\left(x^2 - 1\right) = 0 \Rightarrow (x+2)(x-2)(x+1)(x-1) = 0$, then $x + 2 = 0$ or

    $x - 2 = 0$ or $x + 1 = 0$ or $x - 1 = 0$, so $x = -2, -1, 1, 2$.

23. In a square the sides must be equal so $9x^2 + 30x + 25$ must be a perfect square trinomial

    $a^2 = \left(3x\right)^2$ so $a = 3x$, $b^2 = 5^2$ so $b = 5$, $2ab = 2\left(3x\right)5 = 30x$ the middle term $\Rightarrow$

    $\left(3x + 5\right)^2 \Rightarrow$ each side is $3x + 5$.

24. $x^2 + 3x + 2x + 6 = x^2 + 5x + 6 = (x+2)(x+3)$

25. (a)    $D = \dfrac{1}{11}\left(55\right)^2 = \dfrac{1}{11}(3025) = 275$ ft.

    (b)    $99 = \dfrac{1}{11}x^2 \Rightarrow \dfrac{1}{11}x^2 - 99 = 0 \Rightarrow x^2 - 1089 = 0 \Rightarrow (x+33)(x-33) = 0$, then $x + 33 = 0$ or

    $x - 33 = 0$, so $x = -33$ (not possible) or 33, so 33 mph.

26. $36 = -16t^2 + 48t + 4 \Rightarrow 16t^2 - 48t + 32 = 0 \Rightarrow 16\left(t^2 - 3t + 2\right) = 0 \Rightarrow 16(t-2)(t-1) = 0$, then

$t - 2 = 0$ or $t - 1 = 0$, so $t = 1$, 2 so 1 sec. or 2 sec.

## Chapter 6  Extended and Discovery Exercises

1. $(x+2)(x+3)$

2. $(x+4)(x+5)$

3. $(x-5)(x-6)$

4. $(x+2)(x-5)$

5. $(x+6)(x-2)$

6. $(x-4)(x-4)$

7. $x^2 - 5 = x^2 - \left(\sqrt{5}\right)^2 = \left(x + \sqrt{5}\right)\left(x - \sqrt{5}\right)$

8. $y^2 - 7 = y^2 - \left(\sqrt{7}\right)^2 = \left(y + \sqrt{7}\right)\left(y - \sqrt{7}\right)$

9. $3z^2 - 25 = \left(\sqrt{3}z\right)^2 - (5)^2 = \left(\sqrt{3}z + 5\right)\left(\sqrt{3}z - 5\right)$

10. $7t^2 - 2 = \left(\sqrt{7}t\right)^2 - \left(\sqrt{2}\right)^2 = \left(\sqrt{7}t + \sqrt{2}\right)\left(\sqrt{7}t - \sqrt{2}\right)$

11. $x - 4 = \left(\sqrt{x}\right)^2 - (2)^2 = \left(\sqrt{x} + 2\right)\left(\sqrt{x} - 2\right)$

12. $x - 7 = \left(\sqrt{x}\right)^2 - \left(\sqrt{7}\right)^2 = \left(\sqrt{x} + \sqrt{7}\right)\left(\sqrt{x} - \sqrt{7}\right)$

13. $x^2 - 3 = 0 \Rightarrow x^2 - \left(\sqrt{3}\right)^2 = 0 \Rightarrow \left(x + \sqrt{3}\right)\left(x - \sqrt{3}\right) \Rightarrow x + \sqrt{3} = 0$ or $x - \sqrt{3} = 0 \Rightarrow x = -\sqrt{3}, \sqrt{3}$

14. $y^2 - 7 = 0 \Rightarrow y^2 - \left(\sqrt{7}\right)^2 = 0 \Rightarrow \left(y + \sqrt{7}\right)\left(y - \sqrt{7}\right) \Rightarrow y + \sqrt{7} = 0$ or $y - \sqrt{7} = 0 \Rightarrow y = -\sqrt{7}, \sqrt{7}$

15. $3x^2 - 25 = 0 \Rightarrow \left(\sqrt{3}x\right)^2 - (5)^2 = 0 \Rightarrow \left(\sqrt{3}x + 5\right)\left(\sqrt{3}x - 5\right) = 0 \Rightarrow \sqrt{3}x + 5 = 0$ or $\sqrt{3}x - 5 = 0 \Rightarrow$

$\sqrt{3}x = -5$ or $\sqrt{3}x = 5 \Rightarrow x = \dfrac{-5}{\sqrt{3}}$ or $x = \dfrac{5}{\sqrt{3}}$

16. $7x^2 - 11 = 0 \Rightarrow \left(\sqrt{7}x\right)^2 - \left(\sqrt{11}\right)^2 = 0 \Rightarrow \left(\sqrt{7}x + \sqrt{11}\right)\left(\sqrt{7}x - \sqrt{11}\right) = 0 \Rightarrow \sqrt{7}x + \sqrt{11} = 0$ or

$\sqrt{7}x - \sqrt{11} = 0 \Rightarrow \sqrt{7}x = -\sqrt{11}$ or $\sqrt{7}x = \sqrt{11} \Rightarrow x = \dfrac{-\sqrt{11}}{\sqrt{7}}$ or $x = \dfrac{\sqrt{11}}{\sqrt{7}}$

17. $x^4 - 9 = 0 \Rightarrow \left(x^2\right)^2 - (3)^2 = 0 \Rightarrow \left(x^2 + 3\right)\left(x^2 - 3\right) = 0 \Rightarrow x^2 + 3 = 0$ or $x^2 - 3 = 0 \Rightarrow$ only

$x^2 - \left(\sqrt{3}\right)^2 = 0 \Rightarrow \left(x + \sqrt{3}\right)\left(x - \sqrt{3}\right) = 0 \Rightarrow x + \sqrt{3} = 0$ or $x - \sqrt{3} = 0 \Rightarrow x = -\sqrt{3}$ or $x = \sqrt{3}$

18. $x^4 - 25 = 0 \Rightarrow \left(x^2\right)^2 - (5)^2 = 0 \Rightarrow \left(x^2 + 5\right)\left(x^2 - 5\right) = 0 \Rightarrow x^2 + 5 = 0$ or $x^2 - 5 = 0 \Rightarrow$ only

$x^2 - \left(\sqrt{5}\right)^2 = 0 \Rightarrow \left(x + \sqrt{5}\right)\left(x - \sqrt{5}\right) = 0 \Rightarrow x + \sqrt{5} = 0$ or $x - \sqrt{5} = 0 \Rightarrow x = -\sqrt{5}$ or $x = \sqrt{5}$

## Chapters 1–6 Cumulative Review Exercises

1. $\dfrac{3}{5} \cdot \dfrac{15}{21} = \dfrac{3}{5} \cdot \dfrac{5}{7} = \dfrac{3}{7}$

2. $\dfrac{4}{5} - \dfrac{1}{10} = \dfrac{8}{10} - \dfrac{1}{10} = \dfrac{7}{10}$

3. $26 - 3 \cdot 6 \div 2 = 26 - 18 \div 2 = 26 - 9 = 17$

4. $-2^2 + \dfrac{3+2}{8+2} = -\left(2^2\right) + \dfrac{5}{10} = -4 + \dfrac{1}{2} = -\dfrac{8}{2} + \dfrac{1}{2} = -\dfrac{7}{2}$

5. See Figure 5. $2x + 3 = 5$ when $x = 1$.

| $x$ | -2 | -1 | 0 | 1 | 2 |
|-----|----|----|---|---|---|
| $2x + 3$ | -1 | 1 | 3 | 5 | 7 |

   Figure 5

6. $3n - 5 = n - 7$; $3n - 5 = n - 7 \Rightarrow 3n - n = -7 + 5 \Rightarrow 2n = -2 \Rightarrow n = -1$

7. $\dfrac{5.7}{100} = \dfrac{57}{1000}$; $0.057$

8. $P = 2W + 2L \Rightarrow P - 2L = 2W \Rightarrow \dfrac{P - 2L}{2} = W \Rightarrow W = \dfrac{P - 2L}{2}$

9. $5 - 3z < -1 \Rightarrow -3z < -6 \Rightarrow z > 2$

10. See Figure 10.

11. See Figure 11.

   $x$-intercept: $0 = 3x - 2 \Rightarrow 2 = 3x \Rightarrow x = \dfrac{2}{3}$; $y$-intercept: $y = 3(0) - 2 \Rightarrow y = 0 - 2 \Rightarrow y = -2$

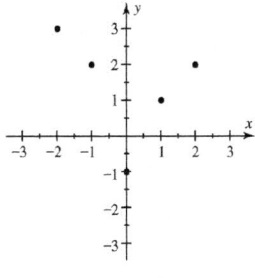
   Figure 10

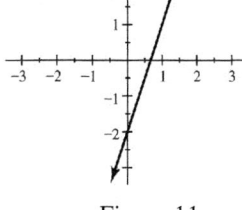
   Figure 11

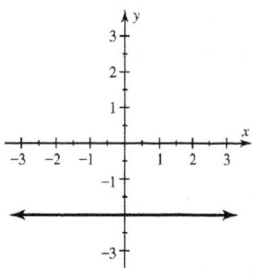
   Figure 12

12. See Figure 12. $x$-intercept: None; $y$-intercept: $y = -2$.

13. $2x - 3y = -6 \Rightarrow -3y = -2x - 6 \Rightarrow y = \dfrac{2}{3}x + 2,$ since perpendicular lines have slopes that are

negative reciprocals, $m = -\dfrac{3}{2},$ now using $(1, 2)$ and $y - y_1 = m(x - x_1) \Rightarrow y - 2 = -\dfrac{3}{2}(x - 1) \Rightarrow$

$y - 2 = -\dfrac{3}{2}x + \dfrac{3}{2} \Rightarrow y = -\dfrac{3}{2}x + \dfrac{7}{2}.$

14. $m = \dfrac{5 - 1}{1 - (-2)} = \dfrac{4}{3},$ using $(-2, 1)$ and $y - y_1 = m(x - x_1) \Rightarrow y - 1 = \dfrac{4}{3}\left[x - (-2)\right] \Rightarrow$

$y - 1 = \dfrac{4}{3}x + \dfrac{8}{3} \Rightarrow y = \dfrac{4}{3}x + \dfrac{11}{3}.$

15. $x$-intercept: $x = -1 \Rightarrow (-1, 0);$ $y$-intercept: $y = -2 \Rightarrow (0, -2);$ $m = \dfrac{-2 - 0}{0 - (-1)} = \dfrac{-2}{1} = -2;$

$y = mx + b \Rightarrow y = -2x - 2.$

16. $(1, 2); x + y = 3 \Rightarrow 1 + 2 = 3,$ Yes; $-2x + y = 0 \Rightarrow -2(1) + 2 = 0 \Rightarrow -2 + 2 = 0,$ Yes

17. Using substitution $y = -1$ into $2x + y = 1 \Rightarrow 2x + (-1) = 1 \Rightarrow 2x = 2 \Rightarrow x = 1 \Rightarrow (1, -1).$

18. Using substitution $5x + y = -5 \Rightarrow y = -5x - 5,$ substituting into $-x + 2y = 12 \Rightarrow$

$-x + 2(-5x - 5) = 12 \Rightarrow -x - 10x - 10 = 12 \Rightarrow -11x = 22 \Rightarrow x = -2$ and $y = -5(-2) - 5 \Rightarrow$

$y = 10 - 5 \Rightarrow y = 5 \Rightarrow (-2, 5).$

19. See Figure 19.

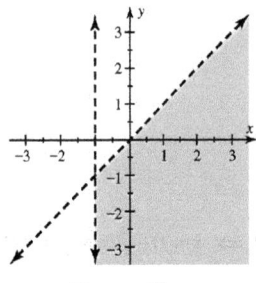    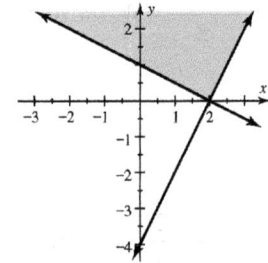

Figure 19                          Figure 20

20. See Figure 20. $2x - y \le 4 \Rightarrow -y \le -2x + 4 \Rightarrow y \ge 2x - 4$ and $x + 2y \ge 2 \Rightarrow 2y \ge -x + 2 \Rightarrow$

$y \ge -\dfrac{1}{2}x + 1.$

21. $-2^4 = -(2^4) = -16$

22. $(xy)^0 = 1$

23. $(xy)^4 \left(x^3 y^{-4}\right)^2 = x^4 \cdot y^4 \cdot x^6 \cdot y^{-8} = x^{10} y^{-4} = \dfrac{x^{10}}{y^4}$

24. $7x^3 \left(-2x^2 + 3x\right) = -14x^5 + 21x^4$

25. $a^{-4} \cdot a^2 = a^{-2} = \dfrac{1}{a^2}$

26. $\left(2t^3\right)^{-2} = 2^{-2} \cdot t^{-6} = \dfrac{1}{2^2} \cdot \dfrac{1}{t^6} = \dfrac{1}{4t^6}$

27. $(xy)^{-3}\left(x^{-1}y^2\right)^{-1} = x^{-3}y^{-3} \cdot x^1 \cdot y^{-2} = x^{-2} \cdot y^{-5} = \dfrac{1}{x^2 y^5}$

28. $\left(\dfrac{2x}{y^{-2}}\right)^5 = \dfrac{2^5 x^5}{y^{-10}} = \dfrac{32 x^5}{y^{-10}} = 32x^5 y^{10}$

29. $\dfrac{6x^3 + 12x^2}{3x} = \dfrac{6x^3}{3x} + \dfrac{12x^2}{3x} = 2x^2 + 4x$

30.
$$\begin{array}{r} 3x \phantom{+0x^2-x+1} \\ x^2+1\overline{)3x^3 + 0x^2 - x + 1} \\ \underline{3x^3 + 0x^2 + 3x} \phantom{+1} \\ -4x + 1 \end{array}$$

The solution is $3x + \dfrac{-4x+1}{x^2+1}$

31. FOIL, $x^2 + 3x - 28 = (x-4)(x+7)$

32. FOIL, $6y^2 + y - 12 = (2y+3)(3y-4)$

33. Difference of squares, $25x^2 - 4y^2 = (5x+2y)(5x-2y)$

34. FOIL, $64x^2 - 16x + 1 = (8x-1)(8x-1) = (8x-1)^2$

35. Difference of cubes, $27t^3 - 8 = (3t-2)\left(9t^2 + 6t + 4\right)$

36. $-4x^2 + 4x + 24 = -4\left(x^2 - x - 6\right) \Rightarrow$ FOIL, $-4(x-3)(x+2)$

37. $y^4 = 25y^2 \Rightarrow y^4 - 25y^2 = 0 \Rightarrow y^2\left(y^2 - 25\right) = 0 \Rightarrow y^2(y+5)(y-5) = 0$, then $y^2 = 0 \Rightarrow y = 0$ or

   $y + 5 = 0 \Rightarrow y = -5$ or $y - 5 = 0 \Rightarrow y = 5$, so $y = -5, 0, 5$.

38. $8z^2 + 8z - 16 = 0 \Rightarrow 8\left(z^2 + z - 2\right) = 0 \Rightarrow 8(z+2)(z-1) = 0$, then $z + 2 = 0 \Rightarrow z = -2$ or

   $z - 1 = 0 \Rightarrow z = 1$, so $z = -2, 1$.

39. $4z^3 = 49z \Rightarrow 4z^3 - 49z = 0 \Rightarrow z\left(4z^2 - 49\right) = 0 \Rightarrow z(2z+7)(2z-7) = 0$, then $z = 0$ or

   $2z + 7 = 0 \Rightarrow 2z = -7 \Rightarrow z = \dfrac{-7}{2}$ or $2z - 7 = 0 \Rightarrow 2z = 7 \Rightarrow z = \dfrac{7}{2}$, so $z = \dfrac{-7}{2}, 0, \dfrac{7}{2}$.

40. $x^4 - 18x^2 + 81 = 0 \Rightarrow \left(x^2 - 9\right)\left(x^2 - 9\right) = 0 \Rightarrow (x+3)(x-3)(x+3)(x-3) = 0$, then

   $x + 3 = 0 \Rightarrow x = -3$ or $x - 3 = 0 \Rightarrow x = 3$, so $x = -3, 3$.

41. (a)   $10x + 8x = 18x$

   (b)   $18x = 900 \Rightarrow x = 50$ min.

42. $1\dfrac{3}{4}+2\dfrac{1}{2}+2\dfrac{2}{3}=1\dfrac{9}{12}+2\dfrac{6}{12}+2\dfrac{8}{12}=5\dfrac{23}{12}=6\dfrac{11}{12}$ miles.

43. Let $x$ represent the minutes cross-country skiing and $60-x$ represent minutes running.

$12(x)+9(60-x)=615 \Rightarrow 12x+540-9x=615 \Rightarrow 3x=75 \Rightarrow x=25$, so 25 minutes skiing and

35 minutes running.

44. (a)   $C=0.25x+20$

   (b)   $100=0.25x+20 \Rightarrow 80=0.25x \Rightarrow x=320$ or 320 miles.

45. (a)   $x=$ one angle $\Rightarrow x+x+y=180 \Rightarrow 2x+y=180$ and $2x=y+20 \Rightarrow 2x-y=20$

   (b)   Using elimination,   $2x+y=180$
   $$\underline{\phantom{xxxx}2x-y=20\phantom{xxxx}}$$
   $$4x=200 \Rightarrow x=50 \Rightarrow 2(50)+y=180 \Rightarrow y=80.$$

   The result is 50, 50 and 80 degrees.

46. $3(3x)(2y)=18xy; 18(2)(3)=108$ yd$^2$.

47. $x^2+12x+36=(x+6)(x+6)=(x+6)^2 \Rightarrow$ each side is $x+6$

48. (a)   $0=64t-16t^2 \Rightarrow -16t(t-4)=0$, then $-16t=0 \Rightarrow t=0$ (when hit) or $t-4=0 \Rightarrow t=4$ sec

   (b)   $48=64t-16t^2 \Rightarrow 16t^2-64t+48=0 \Rightarrow 16(t^2-4t+3)=0 \Rightarrow 16(t-3)(t-1)=0$, then

   $t-3=0 \Rightarrow t=3$ or $t-1=0 \Rightarrow t=1$, so at 1 and 3 sec.

# Chapter 7: Rational Expressions

## Section 7.1: Introduction to Rational Expressions

1. $\dfrac{P}{Q}$; polynomials

3. denominator

5. $\dfrac{a}{b}$

7. $x = -7, \dfrac{3}{x} = \dfrac{3}{-7} = -\dfrac{3}{7}$

9. $x = -4, -\dfrac{x}{x-5} = -\dfrac{-4}{-4-5} = \dfrac{-(-4)}{-9} = \dfrac{4}{-9} = -\dfrac{4}{9}$

11. $y = -2, \dfrac{y+1}{y^2} = \dfrac{-2+1}{(-2)^2} = \dfrac{-1}{4} = -\dfrac{1}{4}$

13. $z = -2, \dfrac{7z}{z^2-4} = \dfrac{7(-2)}{(-2)^2-4} = \dfrac{-14}{4-4} = \dfrac{-14}{0} \Rightarrow$ undefined

15. $t = -2, \dfrac{5}{3t+6} = \dfrac{5}{3(-2)+6} = \dfrac{5}{-6+6} = \dfrac{5}{0} \Rightarrow$ undefined

17. $x = -2, \dfrac{4-x}{x-4} = \dfrac{4-(-2)}{-2-4} = \dfrac{6}{-6} = -1$

19. $x = 0, -\dfrac{6-x}{x-6} = -\dfrac{6-0}{0-6} = -\dfrac{6}{-6} = \dfrac{-(6)}{-6} = \dfrac{-6}{-6} = 1$

21. $x = -2, \dfrac{x}{x+1} = \dfrac{-2}{-2+1} = \dfrac{-2}{-1} = 2;$   $x = -1, \dfrac{x}{x+1} = \dfrac{-1}{-1+1} = \dfrac{-1}{0} \Rightarrow$ undefined;

    $x = 0, \dfrac{x}{x+1} = \dfrac{0}{0+1} = \dfrac{0}{1} = 0;$   $x = 1, \dfrac{x}{x+1} = \dfrac{1}{1+1} = \dfrac{1}{2};$   $x = 2, \dfrac{x}{x+1} = \dfrac{2}{2+1} = \dfrac{2}{3}.$   See Figure 21.

| $x$ | -2 | -1 | 0 | 1 | 2 |
|---|---|---|---|---|---|
| $\frac{x}{x+1}$ | 2 | — | 0 | $\frac{1}{2}$ | $\frac{2}{3}$ |

Figure 21

23. $x = -2$, $\dfrac{3x}{2x^2+1} = \dfrac{3(-2)}{2(-2)^2+1} = \dfrac{-6}{2(4)+1} = \dfrac{-6}{8+1} = \dfrac{-6}{9} = \dfrac{-2}{3} = -\dfrac{2}{3}$;

$x = -1$, $\dfrac{3x}{2x^2+1} = \dfrac{3(-1)}{2(-1)^2+1} = \dfrac{-3}{2(1)+1} = \dfrac{-3}{2+1} = \dfrac{-3}{3} = -1$;

$x = 0$, $\dfrac{3x}{2x^2+1} = \dfrac{3(0)}{2(0)^2+1} = \dfrac{0}{2(0)+1} = \dfrac{0}{0+1} = \dfrac{0}{1} = 0$;

$x = 1$, $\dfrac{3x}{2x^2+1} = \dfrac{3(1)}{2(1)^2+1} = \dfrac{3}{2(1)+1} = \dfrac{3}{2+1} = \dfrac{3}{3} = 1$;

$x = 2$, $\dfrac{3x}{2x^2+1} = \dfrac{3(2)}{2(2)^2+1} = \dfrac{6}{2(4)+1} = \dfrac{6}{8+1} = \dfrac{6}{9} = \dfrac{2}{3}$. See Figure 23.

| $x$ | $-2$ | $-1$ | $0$ | $1$ | $2$ |
|---|---|---|---|---|---|
| $\frac{3x}{2x^2+1}$ | $-\frac{2}{3}$ | $-1$ | $0$ | $1$ | $\frac{2}{3}$ |

Figure 23

25. A rational expression is undefined when the denominator $= 0$.   $x = 0$

27. A rational expression is undefined when the denominator $= 0$.   $z - 3 = 0 \Rightarrow z = 3$

29. A rational expression is undefined when the denominator $= 0$.   $5y + 4 = 0 \Rightarrow 5y = -4 \Rightarrow y = \dfrac{-4}{5}$

31. A rational expression is undefined when the denominator $= 0$.   $t^2 + 1 = 0 \Rightarrow t^2 = -1$  which is impossible, so none.

33. A rational expression is undefined when the denominator $= 0$.   $x^2 - 25 = 0 \Rightarrow x^2 = 25 \Rightarrow x = -5, 5$

35. A rational expression is undefined when the denominator $= 0$.

$x^2 + 5x + 6 = 0 \Rightarrow (x+3)(x+2) = 0 \Rightarrow x + 3 = 0 \Rightarrow x = -3$ or $x + 2 = 0 \Rightarrow x = -2$, so $x = -3, -2$

37. A rational expression is undefined when the denominator $= 0$.

$2z^2 - 7z + 5 = 0 \Rightarrow (2z-5)(z-1) = 0 \Rightarrow 2z - 5 = 0 \Rightarrow 2z = 5 \Rightarrow z = \dfrac{5}{2}$ or $z - 1 = 0 \Rightarrow z = 1$, so $z = 1, \dfrac{5}{2}$

39. $\dfrac{12}{18} = \dfrac{2}{3} \cdot \dfrac{6}{6} = \dfrac{2}{3} \cdot 1 = \dfrac{2}{3}$

41. $\dfrac{24}{48} = \dfrac{1}{2} \cdot \dfrac{24}{24} = \dfrac{1}{2} \cdot 1 = \dfrac{1}{2}$

43. $-\dfrac{6}{15} = -\dfrac{2}{5} \cdot \dfrac{3}{3} = -\dfrac{2}{5} \cdot 1 = -\dfrac{2}{5}$

45. $-\dfrac{25}{75} = -\dfrac{1}{3} \cdot \dfrac{25}{25} = -\dfrac{1}{3} \cdot 1 = -\dfrac{1}{3}$

47. (a) $\dfrac{8}{16} = \dfrac{1}{2} \cdot \dfrac{8}{8} = \dfrac{1}{2} \cdot 1 = \dfrac{1}{2}$

(b)  $\dfrac{x+2}{2x+4} = \dfrac{x+2}{2(x+2)} = \dfrac{1}{2} \cdot \dfrac{x+2}{x+2} = \dfrac{1}{2} \cdot 1 = \dfrac{1}{2}$

49. (a)  $\dfrac{7-3}{3-7} = \dfrac{4}{-4} = -1$

(b)  $\dfrac{7-x}{x-7} = \dfrac{7-x}{-1(7-x)} = -1 \cdot \dfrac{7-x}{7-x} = -1 \cdot 1 = -1$

51.  $\dfrac{5x^4}{10x^6} = \dfrac{1}{2} x^{4-6} = \dfrac{1}{2} x^{-2} = \dfrac{1}{2x^2}$

53.  $\dfrac{8xy^3}{6x^2 y^2} = \dfrac{4}{3} x^{1-2} y^{3-2} = \dfrac{4}{3} x^{-1} y = \dfrac{4y}{3x}$

55.  $\dfrac{x+4}{2x+8} = \dfrac{x+4}{2(x+4)} = \dfrac{1}{2} \cdot \dfrac{x+4}{x+4} = \dfrac{1}{2} \cdot 1 = \dfrac{1}{2}$

57.  $\dfrac{3z-9}{5z-15} = \dfrac{3(z-3)}{5(z-3)} = \dfrac{3}{5} \cdot \dfrac{z-3}{z-3} = \dfrac{3}{5} \cdot 1 = \dfrac{3}{5}$

59.  $\dfrac{(x+1)(x-1)}{(x+6)(x-1)} = \dfrac{x+1}{x+6} \cdot \dfrac{x-1}{x-1} = \dfrac{x+1}{x+6} \cdot 1 = \dfrac{x+1}{x+6}$

61.  $\dfrac{(5y+3)(2y-1)}{(2y-1)(y+2)} = \dfrac{(5y+3)(2y-1)}{(y+2)(2y-1)} = \dfrac{5y+3}{y+2} \cdot \dfrac{2y-1}{2y-1} = \dfrac{5y+3}{y+2} \cdot 1 = \dfrac{5y+3}{y+2}$

63.  $\dfrac{x-7}{7-x} = \dfrac{x-7}{-1(-7+x)} = \dfrac{x-7}{-1(x-7)} = -1$

65.  $\dfrac{a-b}{b-a} = \dfrac{a-b}{-1(-b+a)} = \dfrac{a-b}{-1(a-b)} = -1$

67.  $\dfrac{-6-x}{18+3x} = \dfrac{-1(6+x)}{3(6+x)} = -\dfrac{1}{3} \cdot \dfrac{6+x}{6+x} = -\dfrac{1}{3} \cdot 1 = -\dfrac{1}{3}$

69.  $\dfrac{x+1}{-2x-2} = \dfrac{1(x+1)}{-2(x+1)} = -\dfrac{1}{2} \cdot \dfrac{x+1}{x+1} = -\dfrac{1}{2} \cdot 1 = -\dfrac{1}{2}$

71.  $-\dfrac{9-x}{x-9} = \dfrac{x-9}{x-9} = 1$

73.  $\dfrac{(3x+5)(x-1)}{(3x-5)(1-x)} = \dfrac{3x+5}{3x-5} \cdot \dfrac{(x-1)}{-1(-1+x)} = \dfrac{3x+5}{3x-5} \cdot \dfrac{x-1}{-1(x-1)} = \dfrac{3x+5}{3x-5} \cdot -1 = -\dfrac{3x+5}{3x-5}$

75.  $\dfrac{n^2-n}{n^2-5n} = \dfrac{n(n-1)}{n(n-5)} = \dfrac{n}{n} \cdot \dfrac{n-1}{n-5} = 1 \cdot \dfrac{n-1}{n-5} = \dfrac{n-1}{n-5}$

77.  $\dfrac{x^2-3x}{6x-18} = \dfrac{x(x-3)}{6(x-3)} = \dfrac{x}{6} \cdot \dfrac{x-3}{x-3} = \dfrac{x}{6} \cdot 1 = \dfrac{x}{6}$

79. $\dfrac{z^2-3z+2}{z^2-4z+3} = \dfrac{(z-2)(z-1)}{(z-3)(z-1)} = \dfrac{z-2}{z-3} \cdot \dfrac{z-1}{z-1} = \dfrac{z-2}{z-3} \cdot 1 = \dfrac{z-2}{z-3}$

81. $\dfrac{2x^2+7x-4}{6x^2+x-2} = \dfrac{(x+4)(2x-1)}{(3x+2)(2x-1)} = \dfrac{x+4}{3x+2} \cdot \dfrac{2x-1}{2x-1} = \dfrac{x+4}{3x+2} \cdot 1 = \dfrac{x+4}{3x+2}$

83. $\dfrac{x-3}{3x^2-11x+6} = \dfrac{x-3}{(x-3)(3x-2)} = \dfrac{x-3}{x-3} \cdot \dfrac{1}{3x-2} = 1 \cdot \dfrac{1}{3x-2} = \dfrac{1}{3x-2}$

85. $-\dfrac{a-9}{9-a} = \dfrac{-(a-9)}{9-a} = \dfrac{-a+9}{9-a} = \dfrac{9-a}{9-a} = 1$

87. $\dfrac{-2x-1}{4x+2} = \dfrac{-2x-1}{2(2x+1)} = \dfrac{-1(2x+1)}{2(2x+1)} = -\dfrac{1}{2} \cdot \dfrac{2x+1}{2x+1} = -\dfrac{1}{2} \cdot 1 = -\dfrac{1}{2}$

89. $-1, \dfrac{x-a}{a-x} = \dfrac{1(x-a)}{a-x} = \dfrac{-1(-x+a)}{a-x} = \dfrac{-1(a-x)}{a-x} = -1$

91. (a)   Equation

   (b)   $x+1=7 \Rightarrow x+1-1=7-1 \Rightarrow x=6$

93. (a)   Expression

   (b)   $\dfrac{x}{x(x+1)} = \dfrac{x}{x} \cdot \dfrac{1}{x+1} = 1 \cdot \dfrac{1}{x+1} = \dfrac{1}{x+1}$

95. (a)   Expression

   (b)   $\dfrac{x^2-4}{x+2} = \dfrac{(x-2)(x+2)}{x+2} = \dfrac{x-2}{1} \cdot \dfrac{x+2}{x+2} = (x-2) \cdot 1 = x-2$

97. (a)   Equation

   (b)   $\dfrac{x}{2(1+3)} = 1 \Rightarrow \dfrac{x}{2(4)} = 1 \Rightarrow \dfrac{x}{8} = 1 \Rightarrow 8 \cdot \dfrac{x}{8} = 8 \cdot 1 \Rightarrow x=8$

99. (a)   For $x=3, T = \dfrac{1}{5-x} \Rightarrow T = \dfrac{1}{5-3} = \dfrac{1}{2}$;   When traffic arrives at an average rate of 3

   vehicles/min., the average wait is one-half minute.

   (b)   For $x=2, T = \dfrac{1}{5-x} \Rightarrow T = \dfrac{1}{5-2} = \dfrac{1}{3}$; For $x=4, T = \dfrac{1}{5-x} \Rightarrow T = \dfrac{1}{5-4} = \dfrac{1}{1} = 1$;

   For $x=4.5, T = \dfrac{1}{5-x} \Rightarrow T = \dfrac{1}{5-4.5} = \dfrac{1}{0.5} = 2$;

   For $x=4.9, T = \dfrac{1}{5-x} \Rightarrow T = \dfrac{1}{5-4.9} = \dfrac{1}{0.1} = 10$;

   For $x=4.99, T = \dfrac{1}{5-x} \Rightarrow T = \dfrac{1}{5-4.99} = \dfrac{1}{0.01} = 100$;   See Figure 99.

   As $x$ nears 5 vehicles/min., a small increase in $x$ increases the wait dramatically.

   | $x$ | 2 | 4 | 4.5 | 4.9 | 4.99 |
   |---|---|---|---|---|---|
   | $T$ | $\frac{1}{3}$ | 1 | 2 | 10 | 100 |

   Figure 99

101. (a)    For $x = 0$, $P = \dfrac{7x+3}{x+6} \Rightarrow P = \dfrac{7(0)+3}{0+6} = \dfrac{3}{6} = \dfrac{1}{2} = 0.5$

    For $x = 12$, $P = \dfrac{7x+3}{x+6} \Rightarrow P = \dfrac{7(12)+3}{12+6} = \dfrac{87}{18} = 4.8\overline{3}$

    For $x = 36$, $P = \dfrac{7x+3}{x+6} \Rightarrow P = \dfrac{7(36)+3}{36+6} = \dfrac{255}{42} \approx 6.07$

    For $x = 72$, $P = \dfrac{7x+3}{x+6} \Rightarrow P = \dfrac{7(72)+3}{72+6} = \dfrac{507}{78} = 6.5$

    (b)    Since the population is given in hundreds and $P = 0.5$ when $x = 0$, then the initial frog
    population is 50.

    (c)    The population increased quickly at first, but then leveled off.

103.  $\dfrac{1}{2}$

105. (a)    $\dfrac{3}{n}$

    (b)    $\dfrac{n-3}{n}$; $n = 100$, $\dfrac{n-3}{n} = \dfrac{100-3}{100} = \dfrac{97}{100}$;    There is a 97% chance that a winning ball will not be

    drawn.

107. (a)    $\dfrac{360}{60} = 6$ hours

    (b)    $\dfrac{M}{60}$

109. (a)    $T = \dfrac{1}{5-x}$, $5 - x = 0 \Rightarrow -x = -5 \Rightarrow x = 5$

    (b)    As the average arrival rate nears 5 cars/min., a small increase in $x$ increases the waiting time
    dramatically.

## Section 7.2 Multiplication and Division of Rational Expressions

1. numerators; denominators

3.  $\dfrac{AC}{BD}$

5.  $\dfrac{1}{2} \cdot \dfrac{4}{5} = \dfrac{4}{10} = \dfrac{2}{5}$

7.  $\dfrac{3}{7} \cdot 4 = \dfrac{3}{7} \cdot \dfrac{4}{1} = \dfrac{12}{7}$

9.  $\dfrac{5}{4} \cdot \dfrac{8}{15} = \dfrac{5 \cdot 2 \cdot 4}{4 \cdot 3 \cdot 5} = \dfrac{5}{5} \cdot \dfrac{4}{4} \cdot \dfrac{2}{3} = 1 \cdot 1 \cdot \dfrac{2}{3} = \dfrac{2}{3}$

11. $\dfrac{1}{3} \cdot \dfrac{2}{3} \cdot \dfrac{9}{11} = \dfrac{2}{9} \cdot \dfrac{9}{11} = \dfrac{18}{99} = \dfrac{2}{11}$

13. $\dfrac{2}{3} \div \dfrac{1}{6} = \dfrac{2}{3} \cdot \dfrac{6}{1} = \dfrac{12}{3} = 4$

15. $\dfrac{8}{9} \div \dfrac{5}{3} = \dfrac{8}{9} \cdot \dfrac{3}{5} = \dfrac{24}{45} = \dfrac{8}{15}$

17. $8 \div \dfrac{4}{5} = \dfrac{8}{1} \cdot \dfrac{5}{4} = \dfrac{40}{4} = 10$

19. $\dfrac{4}{5} \div \dfrac{2}{3} \div \dfrac{1}{2} = \dfrac{4}{5} \cdot \dfrac{3}{2} \div \dfrac{1}{2} = \dfrac{12}{10} \div \dfrac{1}{2} = \dfrac{6}{5} \cdot \dfrac{2}{1} = \dfrac{12}{5}$

21. $\dfrac{x+5}{x+5} = 1$

23. $\dfrac{(z+1)(z+2)}{(z+4)(z+2)} = \dfrac{z+1}{z+4} \cdot \dfrac{z+2}{z+2} = \dfrac{z+1}{z+4} \cdot 1 = \dfrac{z+1}{z+4}$

25. $\dfrac{8y(y+7)}{12y(y+7)} = \dfrac{8}{12} \cdot \dfrac{y}{y} \cdot \dfrac{y+7}{y+7} = \dfrac{8}{12} \cdot 1 \cdot 1 = \dfrac{8}{12} = \dfrac{2}{3}$

27. $\dfrac{x(x+2)(x+3)}{x(x-2)(x+3)} = \dfrac{x}{x} \cdot \dfrac{x+2}{x-2} \cdot \dfrac{x+3}{x+3} = 1 \cdot \dfrac{x+2}{x-2} \cdot 1 = \dfrac{x+2}{x-2}$

29. $\dfrac{8}{x} \cdot \dfrac{x+1}{x} = \dfrac{8(x+1)}{x^2}$

31. $\dfrac{8+x}{x} \cdot \dfrac{x-3}{x+8} = \dfrac{8+x}{8+x} \cdot \dfrac{x-3}{x} = 1 \cdot \dfrac{x-3}{x} = \dfrac{x-3}{x}$

33. $\dfrac{z+3}{z+4} \cdot \dfrac{z+4}{z-7} = \dfrac{z+4}{z+4} \cdot \dfrac{z+3}{z-7} = 1 \cdot \dfrac{z+3}{z-7} = \dfrac{z+3}{z-7}$

35. $\dfrac{5x+1}{3x+2} \cdot \dfrac{3x+2}{5x+1} = \dfrac{5x+1}{5x+1} \cdot \dfrac{3x+2}{3x+2} = 1 \cdot 1 = 1$

37. $\dfrac{(t+1)^2}{t+2} \cdot \dfrac{(t+2)^2}{t+1} = \dfrac{(t+1)(t+1)}{t+2} \cdot \dfrac{(t+2)(t+2)}{t+1} = \dfrac{t+1}{t+1} \cdot \dfrac{t+1}{1} \cdot \dfrac{t+2}{t+2} \cdot \dfrac{t+2}{1} =$

$1 \cdot (t+1) \cdot 1 \cdot (t+2) = (t+1)(t+2)$

39. $\dfrac{x^2}{x^2+4} \cdot \dfrac{x+4}{x} = \dfrac{x^2}{x} \cdot \dfrac{x+4}{x^2+4} = \dfrac{x}{x} \cdot \dfrac{x(x+4)}{x^2+4} = 1 \cdot \dfrac{x(x+4)}{x^2+4} = \dfrac{x(x+4)}{x^2+4}$

41. $\dfrac{(z^2-1)}{(z^2-4)} \cdot \dfrac{z-2}{z+1} = \dfrac{(z-1)(z+1)}{(z-2)(z+2)} \cdot \dfrac{z-2}{z+1} = \dfrac{z-1}{z+2} \cdot \dfrac{z+1}{z+1} \cdot \dfrac{z-2}{z-2} = \dfrac{z-1}{z+2} \cdot 1 \cdot 1 = \dfrac{z-1}{z+2}$

43. $\dfrac{y^2-2y}{y^2-1} \cdot \dfrac{y+1}{y-2} = \dfrac{y(y-2)}{(y-1)(y+1)} \cdot \dfrac{y+1}{y-2} = \dfrac{y}{y-1} \cdot \dfrac{y+1}{y+1} \cdot \dfrac{y-2}{y-2} = \dfrac{y}{y-1} \cdot 1 \cdot 1 = \dfrac{y}{y-1}$

45. $\dfrac{2x^2-x-3}{3x^2-8x-3}\cdot\dfrac{3x+1}{2x-3}=\dfrac{(2x-3)(x+1)}{(3x+1)(x-3)}\cdot\dfrac{3x+1}{2x-3}=\dfrac{2x-3}{2x-3}\cdot\dfrac{3x+1}{3x+1}\cdot\dfrac{x+1}{x-3}=1\cdot1\cdot\dfrac{x+1}{x-3}=\dfrac{x+1}{x-3}$

47. $\dfrac{(x-3)^3}{x^2-2x+1}\cdot\dfrac{x-1}{(x-3)^2}=\dfrac{(x-3)(x-3)^2}{(x-1)(x-1)}\cdot\dfrac{x-1}{(x-3)^2}=\dfrac{x-3}{x-1}\cdot\dfrac{(x-3)^2}{(x-3)^2}\cdot\dfrac{x-1}{x-1}=\dfrac{x-3}{x-1}\cdot1\cdot1=\dfrac{x-3}{x-1}$

49. $\dfrac{2}{x}\div\dfrac{2x+3}{x}=\dfrac{2}{x}\cdot\dfrac{x}{2x+3}=\dfrac{x}{x}\cdot\dfrac{2}{2x+3}=1\cdot\dfrac{2}{2x+3}=\dfrac{2}{2x+3}$

51. $\dfrac{x-2}{3x}\div\dfrac{2-x}{6x}=\dfrac{x-2}{3x}\cdot\dfrac{6x}{2-x}=\dfrac{x-2}{-1(-2+x)}\cdot\dfrac{6x}{3x}=\dfrac{x-2}{-1(-2+x)}\cdot2=-1\cdot2=-2$

53. $\dfrac{z+2}{z+1}\div\dfrac{z+2}{z-1}=\dfrac{z+2}{z+1}\cdot\dfrac{z-1}{z+2}=\dfrac{z+2}{z+2}\cdot\dfrac{z-1}{z+1}=1\cdot\dfrac{z-1}{z+1}=\dfrac{z-1}{z+1}$

55. $\dfrac{3y+4}{2y+1}\div\dfrac{3y+4}{y+2}=\dfrac{3y+4}{2y+1}\cdot\dfrac{y+2}{3y+4}=\dfrac{3y+4}{3y+4}\cdot\dfrac{y+2}{2y+1}=1\cdot\dfrac{y+2}{2y+1}=\dfrac{y+2}{2y+1}$

57. $\dfrac{t^2-1}{t^2+1}\div\dfrac{t+1}{4}=\dfrac{t^2-1}{t^2+1}\cdot\dfrac{4}{t+1}=\dfrac{(t+1)(t-1)}{t^2+1}\cdot\dfrac{4}{t+1}=\dfrac{t+1}{t+1}\cdot\dfrac{4(t-1)}{t^2+1}=1\cdot\dfrac{4(t-1)}{t^2+1}=\dfrac{4(t-1)}{t^2+1}$

59. $\dfrac{y^2-9}{y^2-25}\div\dfrac{y+3}{y+5}=\dfrac{y^2-9}{y^2-25}\cdot\dfrac{y+5}{y+3}=\dfrac{(y-3)(y+3)}{(y-5)(y+5)}\cdot\dfrac{y+5}{y+3}=\dfrac{y-3}{y-5}\cdot\dfrac{y+3}{y+3}\cdot\dfrac{y+5}{y+5}=\dfrac{y-3}{y-5}\cdot1\cdot1=\dfrac{y-3}{y-5}$

61. $\dfrac{2x^2-4x}{2x-1}\div\dfrac{x-2}{2x-1}=\dfrac{2x^2-4x}{2x-1}\cdot\dfrac{2x-1}{x-2}=\dfrac{2x(x-2)}{x-2}\cdot\dfrac{2x-1}{2x-1}=2x\cdot1\cdot1=2x$

63. $\dfrac{2z^2-5z-3}{z^2+z-20}\div\dfrac{z-3}{z-4}=\dfrac{(2z+1)(z-3)}{(z+5)(z-4)}\cdot\dfrac{z-4}{z-3}=\dfrac{2z+1}{z+5}\cdot\dfrac{z-3}{z-3}\cdot\dfrac{z-4}{z-4}=\dfrac{2z+1}{z+5}\cdot1\cdot1=\dfrac{2z+1}{z+5}$

65. $\dfrac{t^2-1}{t^2+5t-6}\div(t+1)=\dfrac{(t+1)(t-1)}{(t-1)(t+6)}\cdot\dfrac{1}{(t+1)}=\dfrac{t+1}{t+1}\cdot\dfrac{t-1}{t-1}\cdot\dfrac{1}{t+6}=1\cdot1\cdot\dfrac{1}{t+6}=\dfrac{1}{t+6}$

67. $\dfrac{a-b}{a+b}\div\dfrac{a-b}{2a+3b}=\dfrac{a-b}{a+b}\cdot\dfrac{2a+3b}{a-b}=\dfrac{a-b}{a-b}\cdot\dfrac{2a+3b}{a+b}=1\cdot\dfrac{2a+3b}{a+b}=\dfrac{2a+3b}{a+b}$

69. $\dfrac{x-y}{x^2+2xy+y^2}\div\dfrac{1}{(x+y)^2}=\dfrac{x-y}{(x+y)^2}\cdot\dfrac{(x+y)^2}{1}=\dfrac{x-y}{1}\cdot\dfrac{(x+y)^2}{(x+y)^2}=(x-y)\cdot1=x-y$

71. $\dfrac{a-b}{b-c}\cdot\dfrac{c-b}{b-a}=\dfrac{a-b}{b-c}\cdot\dfrac{-1(b-c)}{-1(a-b)}=\dfrac{-1}{-1}\cdot\dfrac{a-b}{a-b}\cdot\dfrac{b-c}{b-c}=1\cdot1\cdot1=1$

73. (a) $\quad D=\dfrac{900}{30}\cdot\dfrac{1}{x}=30\cdot\dfrac{1}{x}=\dfrac{30}{x}\Rightarrow D=\dfrac{30}{x}$

   (b) $\quad x=0.1,\ D=\dfrac{30}{0.1}=300$ ft; $x=0.4,\ D=\dfrac{30}{0.4}=75$ ft; stopping distance on dry pavement is 75 ft,

   one-fourth as far as the stopping distance on the icy road.

75. (a) $\quad\dfrac{1}{n}\cdot\dfrac{n}{n+1}=\dfrac{n}{n}\cdot\dfrac{1}{n+1}=1\cdot\dfrac{1}{n+1}=\dfrac{1}{n+1}$

(b) $\quad \dfrac{1}{n+1} \Rightarrow \dfrac{1}{99+1} = \dfrac{1}{100}$

## Checking Basic Concepts  Sections 7.1 and 7.2

1. For $x = -1,\ \dfrac{3}{x^2 - 1} = \dfrac{3}{(-1)^2 - 1} = \dfrac{3}{1-1} = \dfrac{3}{0} \Rightarrow$ undefined.  For $x = 3,\ \dfrac{3}{x^2 - 1} = \dfrac{3}{(3)^2 - 1} = \dfrac{3}{9-1} = \dfrac{3}{8}$

2. (a) $\quad \dfrac{6x^3 y^2}{15 x^2 y^3} = \dfrac{2}{5} x y^{-1} = \dfrac{2x}{5y}$

   (b) $\quad \dfrac{5x - 15}{x - 3} = \dfrac{5(x - 3)}{x - 3} = \dfrac{5}{1} \cdot \dfrac{x - 3}{x - 3} = 5 \cdot 1 = 5$

   (c) $\quad \dfrac{x^2 - x - 6}{x^2 + x - 12} = \dfrac{(x+2)(x-3)}{(x+4)(x-3)} = \dfrac{x+2}{x+4} \cdot \dfrac{x-3}{x-3} = \dfrac{x+2}{x+4} \cdot 1 = \dfrac{x+2}{x+4}$

3. (a) $\quad \dfrac{4}{3x} \cdot \dfrac{2x}{6} = \dfrac{8x}{18x} = \dfrac{8}{18} \cdot \dfrac{x}{x} = \dfrac{8}{18} \cdot 1 = \dfrac{8}{18} = \dfrac{4}{9}$

   (b) $\quad \dfrac{2x+4}{x^2 - 1} \cdot \dfrac{x+1}{x+2} = \dfrac{2(x+2)}{(x-1)(x+1)} \cdot \dfrac{x+1}{x+2} = \dfrac{2}{x-1} \cdot \dfrac{x+2}{x+2} \cdot \dfrac{x+1}{x+1} = \dfrac{2}{x-1} \cdot 1 \cdot 1 = \dfrac{2}{x-1}$

4. (a) $\quad \dfrac{7}{3z^2} \div \dfrac{14}{5z^3} = \dfrac{7}{3z^2} \cdot \dfrac{5z^3}{14} = \dfrac{35z^3}{42z^2} = \dfrac{35z}{42} \cdot \dfrac{z^2}{z^2} = \dfrac{35z}{42} \cdot 1 = \dfrac{35z}{42} = \dfrac{5z}{6}$

   (b) $\quad \dfrac{x^2 + x}{x - 3} \div \dfrac{x}{x - 3} = \dfrac{x(x+1)}{x-3} \cdot \dfrac{x-3}{x} = \dfrac{x}{x} \cdot \dfrac{x-3}{x-3} \cdot \dfrac{x+1}{1} = 1 \cdot 1 \cdot (x+1) = x+1$

5. (a) For $x = 0.5,\ T = \dfrac{1}{2-x} \Rightarrow T = \dfrac{1}{2-0.5} = \dfrac{1}{1.5} = \dfrac{2}{3}$;  For $x = 1,\ T = \dfrac{1}{2-x} \Rightarrow T = \dfrac{1}{2-1} = \dfrac{1}{1} = 1$;

   For $x = 1.5,\ T = \dfrac{1}{2-x} \Rightarrow T = \dfrac{1}{2-1.5} = \dfrac{1}{0.5} = 2$;

   For $x = 1.9,\ T = \dfrac{1}{2-x} \Rightarrow T = \dfrac{1}{2-1.9} = \dfrac{1}{0.1} = 10$;  See Figure 5.

   (b) As $x$ nears 2 customers/min., a small increase in $x$ increases the wait dramatically.

| $x$ | 0.5 | 1.0 | 1.5 | 1.9 |
|-----|-----|-----|-----|-----|
| $T$ | $\frac{2}{3}$ | 1 | 2 | 10 |

Figure 5

## Section 7.3  Addition and Subtraction with Like Denominators

1. numerators; denominators

3. $\dfrac{A+B}{C}$

5. $\dfrac{1}{2}+\dfrac{1}{2}=\dfrac{1+1}{2}=\dfrac{2}{2}=1$

7. $\dfrac{4}{5}+\dfrac{2}{5}=\dfrac{4+2}{5}=\dfrac{6}{5}$

9. $\dfrac{1}{6}+\dfrac{5}{6}=\dfrac{1+5}{6}=\dfrac{6}{6}=1$

11. $\dfrac{4}{7}-\dfrac{1}{7}=\dfrac{4-1}{7}=\dfrac{3}{7}$

13. $\dfrac{7}{8}-\dfrac{3}{8}=\dfrac{7-3}{8}=\dfrac{4}{8}=\dfrac{1}{2}$

15. $\dfrac{11}{12}-\dfrac{5}{12}=\dfrac{11-5}{12}=\dfrac{6}{12}=\dfrac{1}{2}$

17. $\dfrac{7}{15}+\dfrac{4}{15}-\dfrac{1}{15}=\dfrac{7+4-1}{15}=\dfrac{11-1}{15}=\dfrac{10}{15}=\dfrac{2}{3}$

19. $\dfrac{2}{x}+\dfrac{1}{x}=\dfrac{2+1}{x}=\dfrac{3}{x}$

21. $\dfrac{7+2x}{4x}-\dfrac{7}{4x}=\dfrac{(7+2x)-7}{4x}=\dfrac{2x}{4x}=\dfrac{1}{2}\cdot\dfrac{x}{x}=\dfrac{1}{2}\cdot 1=\dfrac{1}{2}$

23. $\dfrac{y+3}{y-3}+\dfrac{2y-12}{y-3}=\dfrac{(y+3)+(2y-12)}{y-3}=\dfrac{3y-9}{y-3}=\dfrac{3(y-3)}{y-3}=3\cdot\dfrac{y-3}{y-3}=3\cdot 1=3$

25. $\dfrac{x}{x-3}+\dfrac{-3}{x-3}=\dfrac{x-3}{x-3}=1$

27. $\dfrac{5z}{4z+3}-\dfrac{z}{4z+3}=\dfrac{5z-z}{4z+3}=\dfrac{4z}{4z+3}$

29. $\dfrac{t+5}{t+6}+\dfrac{t+7}{t+6}=\dfrac{(t+5)+(t+7)}{t+6}=\dfrac{2t+12}{t+6}=\dfrac{2(t+6)}{t+6}=2\cdot\dfrac{t+6}{t+6}=2\cdot 1=2$

31. $\dfrac{5x}{2x+3}-\dfrac{3x-3}{2x+3}=\dfrac{5x-(3x-3)}{2x+3}=\dfrac{2x+3}{2x+3}=1$

33. $\dfrac{x-4}{x^2-x}+\dfrac{4}{x^2-x}=\dfrac{x-4+4}{x^2-x}=\dfrac{x}{x(x-1)}=\dfrac{x}{x}\cdot\dfrac{1}{x-1}=1\cdot\dfrac{1}{x-1}=\dfrac{1}{x-1}$

35. $\dfrac{z^2-1}{z-2}+\dfrac{3-3z}{z-2}=\dfrac{z^2-1+3-3z}{z-2}=\dfrac{z^2-3z+2}{z-2}=\dfrac{(z-1)(z-2)}{z-2}=\dfrac{z-1}{1}\cdot\dfrac{z-2}{z-2}=\dfrac{z-1}{1}\cdot 1=z-1$

37. $\dfrac{x^2+4x-1}{4x+2}-\dfrac{x^2-4x-5}{4x+2}=\dfrac{(x^2+4x-1)-(x^2-4x-5)}{4x+2}=\dfrac{8x+4}{4x+2}=\dfrac{2(4x+2)}{4x+2}=2\cdot\dfrac{4x+2}{4x+2}=2\cdot 1=2$

39. $\dfrac{3y}{5}+\dfrac{2y-5}{5}=\dfrac{3y+(2y-5)}{5}=\dfrac{5y-5}{5}=\dfrac{5(y-1)}{5}=\dfrac{5}{5}\cdot(y-1)=1\cdot(y-1)=y-1$

41. $\dfrac{x+y}{4} + \dfrac{x-y}{4} = \dfrac{(x+y)+(x-y)}{4} = \dfrac{2x}{4} = \dfrac{1x}{2} = \dfrac{x}{2}$

43. $\dfrac{z^2+4}{z-2} - \dfrac{4z}{z-2} = \dfrac{(z^2+4)-4z}{z-2} = \dfrac{z^2-4z+4}{z-2} = \dfrac{(z-2)(z-2)}{z-2} = \dfrac{z-2}{z-2} \cdot \dfrac{z-2}{1} = 1 \cdot (z-2) = z-2$

45. $\dfrac{2x^2-5x}{2x+1} - \dfrac{3}{2x+1} = \dfrac{(2x^2-5x)-3}{2x+1} = \dfrac{2x^2-5x-3}{2x+1} = \dfrac{(2x+1)(x-3)}{2x+1} = \dfrac{2x+1}{2x+1} \cdot \dfrac{x-3}{1} = 1 \cdot (x-3) = x-3$

47. $\dfrac{3n}{2n^2-n+5} + \dfrac{4n}{2n^2-n+5} = \dfrac{3n+4n}{2n^2-n+5} = \dfrac{7n}{2n^2-n+5}$

49. $\dfrac{1}{x+3} + \dfrac{2}{x+3} + \dfrac{3}{x+3} = \dfrac{1+2+3}{x+3} = \dfrac{6}{x+3}$

51. $\dfrac{8}{ab} + \dfrac{1}{ab} = \dfrac{8+1}{ab} = \dfrac{9}{ab}$

53. $\dfrac{x}{(x+y)^2} + \dfrac{y}{(x+y)^2} = \dfrac{x+y}{(x+y)^2} = \dfrac{x+y}{x+y} \cdot \dfrac{1}{x+y} = 1 \cdot \dfrac{1}{x+y} = \dfrac{1}{x+y}$

55. $\dfrac{5}{x-y} + \dfrac{-5}{y-x} = \dfrac{5}{x-y} + \dfrac{-5}{-(-y+x)} = \dfrac{5}{x-y} + \dfrac{5}{x-y} = \dfrac{10}{x-y}$

57. $\dfrac{8}{a-b} + \dfrac{8}{b-a} = \dfrac{8}{a-b} + \dfrac{8}{-1(a-b)} = \dfrac{8}{a-b} + \dfrac{-8}{a-b} = \dfrac{8-8}{a-b} = \dfrac{0}{a-b} = 0$

59. $\dfrac{a+b}{4a} - \dfrac{a-b}{4a} = \dfrac{a+b-a+b}{4a} = \dfrac{2b}{4a} = \dfrac{2}{2} \cdot \dfrac{b}{2a} = 1 \cdot \dfrac{b}{2a} = \dfrac{b}{2a}$

61. $\dfrac{x}{x+y} + \dfrac{y}{x+y} = \dfrac{x+y}{x+y} = 1$

63. $\dfrac{a^2}{a+b} - \dfrac{b^2}{a+b} = \dfrac{a^2-b^2}{a+b} = \dfrac{(a+b)(a-b)}{a+b} = \dfrac{a+b}{a+b} \cdot \dfrac{a-b}{1} = 1 \cdot (a-b) = a-b$

65. $\dfrac{2x-5}{2x^2+5x+2} + \dfrac{6}{2x^2+5x+2} = \dfrac{2x-5+6}{2x^2+5x+2} = \dfrac{2x+1}{(2x+1)(x+2)} = \dfrac{2x+1}{2x+1} \cdot \dfrac{1}{x+2} = \dfrac{1}{x+2}$

67. $\dfrac{3x+7}{3x^2-2x-5} - \dfrac{2x+6}{3x^2-2x-5} = \dfrac{3x+7-2x-6}{3x^2-2x-5} = \dfrac{x+1}{(3x-5)(x+1)} = \dfrac{x+1}{x+1} \cdot \dfrac{1}{3x-5} = \dfrac{1}{3x-5}$

69. $\dfrac{4x^2}{2x+3y} - \dfrac{9y^2}{2x+3y} = \dfrac{4x^2-9y^2}{2x+3y} = \dfrac{(2x+3y)(2x-3y)}{2x+3y} = \dfrac{2x+3y}{2x+3y} \cdot \dfrac{2x-3y}{1} = 1 \cdot (2x-3y) = 2x-3y$

71. $\dfrac{2}{3+x} + \dfrac{3}{3+x} = \dfrac{5}{10} \Rightarrow \dfrac{2+3}{3+x} = \dfrac{5}{10} \Rightarrow \dfrac{5}{3+x} = \dfrac{5}{10} \Rightarrow 3+x = 10 \Rightarrow x=7$

73. (a) $\dfrac{6}{n+1} + \dfrac{5}{n+1} + \dfrac{3}{n+1} = \dfrac{6+5+3}{n+1} = \dfrac{14}{n+1}$

     (b) For $n = 99$, $\dfrac{14}{n+1} \Rightarrow \dfrac{14}{99+1} = \dfrac{14}{100} = \dfrac{7}{50}$; there are 7 chances in 50 that a defective battery is

     chosen.

## Section 7.4 Addition and Subtraction with Unlike Denominators

1. Examples include 36 and 54. *Answers may vary.*

3. $\dfrac{3}{3}$

5. Factoring 4 and 6 completely yields:  $2^2$ and $2 \cdot 3$, a list of factors is $2^2 \cdot 3 = 12$.

7. 2 and 3 are both prime $\Rightarrow$ a list of factors is $2 \cdot 3 = 6$.

9. Factoring 10 and 15 completely yields:  $2 \cdot 5$ and $3 \cdot 5$, a list of factors is $2 \cdot 3 \cdot 5 = 30$.

11. Factoring 24 and 36 completely yields:  $2^3 \cdot 3$ and $2^2 \cdot 3^2$, a list of factors is $2^3 \cdot 3^2 = 72$.

13. Factoring $4x$ and $6x$ completely yields:  $2^2 \cdot x$ and $2 \cdot 3 \cdot x$, a list of factors is $2^2 \cdot 3 \cdot x = 12x$.

15. Factoring $5x$ and $10x^2$ completely yields:  $5 \cdot x$ and $2 \cdot 5 \cdot x^2$, a list of factors is $2 \cdot 5 \cdot x^2 = 10x^2$.

17. Both $x$ and $x+1$ are prime $\Rightarrow$ a list of factors is $x(x+1)$.

19. Both $2x+1$ and $x+3$ are prime $\Rightarrow$ a list of factors is $(2x+1)(x+3)$.

21. Factoring $x^2 - x$ and $x^2 + x$ completely yields:

    $x(x-1)$ and $x(x+1)$, a list of factors is $x(x-1)(x+1)$.

23. Both $(x-8)^2$ and $(x-8)(x+1)$ are factored completely $\Rightarrow$ a list of factors is $(x-8)^2(x+1)$.

25. Factoring $4x^2 - 1$ and $2x+1$ completely yields:  $(2x+1)(2x-1)$ and $(2x+1)$, a list of factors is

    $(2x+1)(2x-1)$.

27. Factoring $x^2 - 1$ and $x+1$ completely yields:  $(x+1)(x-1)$ and $(x+1)$, a list of factors is

    $(x+1)(x-1)$.

29. Factoring $2x^2 + 7x + 6$ and $x^2 + 5x + 6$ completely yields:  $(2x+3)(x+2)$ and $(x+2)(x+3)$,

    a list of factors is $(2x+3)(x+2)(x+3)$.

31. Factoring $3y^2 + 6y$ and $3y^3 + 3y^2 - 6y$ completely yields:  $3y(y+2)$ and $3y(y+2)(y-1)$, a list of

    factors is $3y(y+2)(y-1)$.

33. $\dfrac{1}{3},\, D = 9 \Rightarrow \dfrac{1}{3} \cdot \dfrac{3}{3} = \dfrac{3}{9}$

35. $\dfrac{5}{7},\, D = 21 \Rightarrow \dfrac{5}{7} \cdot \dfrac{3}{3} = \dfrac{15}{21}$

37. $\dfrac{1}{4x},\, D = 8x^3 \Rightarrow \dfrac{1}{4x} \cdot \dfrac{2x^2}{2x^2} = \dfrac{2x^2}{8x^3}$

39. $\dfrac{1}{x+2},\, D = x^2 - 4 \Rightarrow \dfrac{1}{x+2} \cdot \dfrac{x-2}{x-2} = \dfrac{x-2}{x^2-4}$

41. $\dfrac{1}{x+1}$, $D = x^2 + x \Rightarrow \dfrac{1}{x+1} \cdot \dfrac{x}{x} = \dfrac{x}{x^2+x}$

43. $\dfrac{2x}{x+1}$, $D = x^2 + 2x + 1 \Rightarrow \dfrac{2x}{x+1} \cdot \dfrac{x+1}{x+1} = \dfrac{2x^2+2x}{x^2+2x+1}$

45. $\dfrac{4}{5} + \dfrac{1}{2} = \dfrac{8}{10} + \dfrac{5}{10} = \dfrac{13}{10}$

47. $\dfrac{5}{9} - \dfrac{1}{3} = \dfrac{5}{9} - \dfrac{3}{9} = \dfrac{2}{9}$

49. $\dfrac{4}{25} + \dfrac{2}{5} = \dfrac{4}{25} + \dfrac{10}{25} = \dfrac{14}{25}$

51. $\dfrac{1}{5} + \dfrac{3}{4} - \dfrac{1}{2} = \dfrac{4}{20} + \dfrac{15}{20} - \dfrac{10}{20} = \dfrac{9}{20}$

53. $\dfrac{1}{3x} + \dfrac{3}{4x} = \dfrac{4}{12x} + \dfrac{9}{12x} = \dfrac{13}{12x}$

55. $\dfrac{5}{z^2} - \dfrac{7}{z^3} = \dfrac{5z}{z^3} - \dfrac{7}{z^3} = \dfrac{5z-7}{z^3}$

57. $\dfrac{1}{x} - \dfrac{1}{y} = \dfrac{y}{xy} - \dfrac{x}{xy} = \dfrac{y-x}{xy}$

59. $\dfrac{a}{b} + \dfrac{b}{a} = \dfrac{a^2}{ab} + \dfrac{b^2}{ab} = \dfrac{a^2+b^2}{ab}$

61. $\dfrac{1}{2x+4} + \dfrac{3}{x+2} = \dfrac{1}{2(x+2)} + \dfrac{3}{x+2} \cdot \dfrac{2}{2} = \dfrac{1}{2(x+2)} + \dfrac{6}{2(x+2)} = \dfrac{7}{2(x+2)}$

63. $\dfrac{2}{t-2} - \dfrac{1}{t} = \dfrac{2}{t-2} \cdot \dfrac{t}{t} - \dfrac{1}{t} \cdot \dfrac{t-2}{t-2} = \dfrac{2t}{t(t-2)} - \dfrac{t-2}{t(t-2)} = \dfrac{t+2}{t(t-2)}$

65. $\dfrac{5}{n-1} + \dfrac{n}{n+1} = \dfrac{5}{n-1} \cdot \dfrac{n+1}{n+1} + \dfrac{n}{n+1} \cdot \dfrac{n-1}{n-1} = \dfrac{5n+5}{(n-1)(n+1)} + \dfrac{n^2-n}{(n-1)(n+1)} = \dfrac{n^2+4n+5}{(n-1)(n+1)}$

67. $\dfrac{3}{x-3} + \dfrac{6}{3-x} = \dfrac{3}{x-3} + \dfrac{6}{-(x-3)} = \dfrac{3}{x-3} - \dfrac{6}{x-3} = -\dfrac{3}{x-3}$

69. $\dfrac{1}{5k-1} + \dfrac{1}{1-5k} = \dfrac{1}{5k-1} + \dfrac{1}{-(5k-1)} = \dfrac{1}{5k-1} - \dfrac{1}{5k-1} = 0$

71. $\dfrac{2x}{(x-1)^2} + \dfrac{4}{x-1} = \dfrac{2x}{(x-1)^2} + \dfrac{4}{x-1} \cdot \dfrac{x-1}{x-1} = \dfrac{2x}{(x-1)^2} + \dfrac{4x-4}{(x-1)^2} = \dfrac{6x-4}{(x-1)^2}$

73. $\dfrac{2y}{y(2y-1)} + \dfrac{1}{2y-1} = \dfrac{2y}{y(2y-1)} + \dfrac{1}{2y-1} \cdot \dfrac{y}{y} = \dfrac{2y}{y(2y-1)} + \dfrac{y}{y(2y-1)} = \dfrac{3y}{y(2y-1)} = \dfrac{3}{2y-1} \cdot \dfrac{y}{y} =$

$\dfrac{3}{2y-1} \cdot 1 = \dfrac{3}{2y-1}$

75. $\dfrac{1}{x+2}-\dfrac{1}{x^2+2x}=\dfrac{1}{x+2}\cdot\dfrac{x}{x}-\dfrac{1}{x(x+2)}=\dfrac{x}{x(x+2)}-\dfrac{1}{x(x+2)}=\dfrac{x-1}{x(x+2)}$

77. $\dfrac{3}{x-2}-\dfrac{1}{x^2-4}=\dfrac{3}{x-2}\cdot\dfrac{x+2}{x+2}-\dfrac{1}{(x-2)(x+2)}=\dfrac{3(x+2)}{(x-2)(x+2)}-\dfrac{1}{(x-2)(x+2)}$

$=\dfrac{3x+6-1}{(x-2)(x+2)}=\dfrac{3x+5}{(x-2)(x+2)}$

79. $\dfrac{2}{x^2-3x}-\dfrac{1}{x^2+3x}=\dfrac{2}{x(x-3)}\cdot\dfrac{x+3}{x+3}-\dfrac{1}{x(x+3)}\cdot\dfrac{x-3}{x-3}$

$=\dfrac{2(x+3)}{x(x-3)(x+3)}-\dfrac{x-3}{x(x-3)(x+3)}=\dfrac{2x+6-x+3}{x(x-3)(x+3)}=\dfrac{x+9}{x(x-3)(x+3)}$

81. $\dfrac{1}{x-2}-\dfrac{1}{x+2}+\dfrac{1}{x}=\dfrac{1}{x-2}\cdot\dfrac{x(x+2)}{x(x+2)}-\dfrac{1}{x+2}\cdot\dfrac{x(x-2)}{x(x-2)}+\dfrac{1}{x}\cdot\dfrac{(x-2)(x+2)}{(x-2)(x+2)}$

$=\dfrac{x(x+2)-x(x-2)+(x-2)(x+2)}{x(x-2)(x+2)}=\dfrac{x^2+2x-x^2+2x+x^2-4}{x(x-2)(x+2)}=\dfrac{x^2+4x-4}{x(x-2)(x+2)}$

83. $\dfrac{x}{x^2+4x+4}+\dfrac{1}{x+2}=\dfrac{x}{(x+2)^2}+\dfrac{1}{x+2}\cdot\dfrac{x+2}{x+2}=\dfrac{x}{(x+2)^2}+\dfrac{x+2}{(x+2)^2}=\dfrac{2x+2}{(x+2)^2}$

85. $\dfrac{x}{(x+1)(x+2)}-\dfrac{1}{(x+2)(x+3)}=\dfrac{x}{(x+1)(x+2)}\cdot\dfrac{x+3}{x+3}-\dfrac{1}{(x+2)(x+3)}\cdot\dfrac{x+1}{x+1}=$

$\dfrac{x^2+3x}{(x+1)(x+2)(x+3)}-\dfrac{x+1}{(x+1)(x+2)(x+3)}=\dfrac{x^2+2x-1}{(x+1)(x+2)(x+3)}$

87. $\dfrac{1}{a+b}-\dfrac{1}{a-b}=\dfrac{1}{a+b}\cdot\dfrac{a-b}{a-b}-\dfrac{1}{a-b}\cdot\dfrac{a+b}{a+b}=\dfrac{a-b}{(a+b)(a-b)}-\dfrac{a+b}{(a+b)(a-b)}=-\dfrac{2b}{(a+b)(a-b)}$

89. $\dfrac{r}{r-t}+\dfrac{t}{t-r}-1=\dfrac{r}{r-t}+\dfrac{t}{-(r-t)}-1=\dfrac{r}{r-t}-\dfrac{t}{r-t}-1=\dfrac{r-t}{r-t}-1=1-1=0$

91. $\dfrac{1}{2a}+\dfrac{1}{3a}+\dfrac{1}{4a}=\dfrac{1}{2a}\cdot\dfrac{6}{6}+\dfrac{1}{3a}\cdot\dfrac{4}{4}+\dfrac{1}{4a}\cdot\dfrac{3}{3}=\dfrac{6}{12a}+\dfrac{4}{12a}+\dfrac{3}{12a}=\dfrac{13}{12a}$

93. $\dfrac{2}{x-y}+\dfrac{3}{y-x}+\dfrac{1}{x-y}=\dfrac{2}{x-y}+\dfrac{3}{-(x-y)}+\dfrac{1}{x-y}=\dfrac{2}{x-y}-\dfrac{3}{x-y}+\dfrac{1}{x-y}=\dfrac{0}{x-y}=0$

95. $\dfrac{3}{x-3}-\dfrac{3}{x^2-3x}-\dfrac{6}{x(x-3)}=\dfrac{3}{x-3}\cdot\dfrac{x}{x}-\dfrac{3}{x(x-3)}-\dfrac{6}{x(x-3)}=\dfrac{3x}{x(x-3)}-\dfrac{3}{x(x-3)}-\dfrac{6}{x(x-3)}=$

$\dfrac{3x-9}{x(x-3)}=\dfrac{3(x-3)}{x(x-3)}=\dfrac{3}{x}\cdot\dfrac{x-3}{x-3}=\dfrac{3}{x}\cdot 1=\dfrac{3}{x}$

97. $x+\dfrac{1}{x-1}-\dfrac{1}{x+1}=\dfrac{x(x-1)(x+1)}{(x-1)(x+1)}+\dfrac{1(x+1)}{(x-1)(x+1)}-\dfrac{1(x-1)}{(x-1)(x+1)}=\dfrac{x^3-x+x+1-x+1}{(x-1)(x+1)}=\dfrac{x^3-x+2}{(x-1)(x+1)}$

99. $\dfrac{2x+1}{x-1} - \dfrac{3}{x+1} + \dfrac{x}{x-1} = \dfrac{(2x+1)(x+1)}{(x-1)(x+1)} - \dfrac{3(x-1)}{(x-1)(x+1)} + \dfrac{x(x+1)}{(x-1)(x+1)}$

$= \dfrac{2x^2+3x+1-3x+3+x^2+x}{(x-1)(x+1)} = \dfrac{3x^2+x+4}{(x-1)(x+1)}$

101. $R=120, S=200;\ \dfrac{1}{R} + \dfrac{1}{S} \Rightarrow \dfrac{1}{120} + \dfrac{1}{200} = \dfrac{1}{120}\cdot\dfrac{5}{5} + \dfrac{1}{200}\cdot\dfrac{3}{3} = \dfrac{5}{600} + \dfrac{3}{600} = \dfrac{8}{600} = \dfrac{1}{75}$ and

$\dfrac{S+R}{RS} \Rightarrow \dfrac{120+200}{120(200)} = \dfrac{320}{24,000} = \dfrac{1}{75}$. Yes, they are the same.

103. $\dfrac{1}{F} - \dfrac{1}{D} = \dfrac{1}{F}\cdot\dfrac{D}{D} - \dfrac{1}{D}\cdot\dfrac{F}{F} = \dfrac{D}{FD} - \dfrac{F}{FD} = \dfrac{D-F}{FD}$

## Checking Basic Concepts  Sections 7.3 and 7.4

1. (a) $\dfrac{x}{x+2} + \dfrac{2}{x+2} = \dfrac{x+2}{x+2} = 1$

   (b) $\dfrac{2}{3x} - \dfrac{x}{3x} = \dfrac{2-x}{3x}$

   (c) $\dfrac{z^2+z}{z+2} + \dfrac{z}{z+2} = \dfrac{z^2+2z}{z+2} = \dfrac{z(z+2)}{z+2} = \dfrac{z}{1}\cdot\dfrac{z+2}{z+2} = \dfrac{z}{1}\cdot 1 = z$

2. (a) Factoring $3x$ and $5x$ completely yields: $3\cdot x$ and $5\cdot x$, a list of factors is $3\cdot 5\cdot x = 15x$.

   (b) Factoring $4x$ and $x^2+x$ completely yields:

   $2^2\cdot x$ and $x(x+1)$, a list of factors is $2^2\cdot x\cdot(x+1) = 4x(x+1)$.

   (c) Both $x+1$ and $x-1$ are factored completely, a list of factors is $(x+1)(x-1)$.

3. (a) $\dfrac{1}{x+1} + \dfrac{5}{x} = \dfrac{1}{x+1}\cdot\dfrac{x}{x} + \dfrac{5}{x}\cdot\dfrac{x+1}{x+1} = \dfrac{x}{x(x+1)} + \dfrac{5x+5}{x(x+1)} = \dfrac{6x+5}{x(x+1)}$

   (b) $\dfrac{5}{x-3} + \dfrac{1}{3-x} = \dfrac{5}{x-3} + \dfrac{1}{-(x-3)} = \dfrac{5}{x-3} - \dfrac{1}{x-3} = \dfrac{4}{x-3}$

   (c) $\dfrac{-4}{4x+2} - \dfrac{x+2}{2x+1} = \dfrac{-4}{2(2x+1)} - \dfrac{x+2}{2x+1} = \dfrac{-2}{(2x+1)} - \dfrac{x+2}{2x+1} = \dfrac{-2-x-2}{(2x+1)} = -\dfrac{x+4}{2x+1}$

4. $\dfrac{a}{a-b} - \dfrac{b}{a+b} = \dfrac{a}{a-b}\cdot\dfrac{a+b}{a+b} - \dfrac{b}{a+b}\cdot\dfrac{a-b}{a-b} = \dfrac{a^2+ab}{(a-b)(a+b)} - \dfrac{ab-b^2}{(a-b)(a+b)} = \dfrac{a^2+b^2}{(a-b)(a+b)}$

## Section 7.5 Complex Fractions

1. $\dfrac{\frac{1}{2}}{\frac{3}{4}} = \dfrac{1}{2} \cdot \dfrac{4}{3} = \dfrac{4}{6} = \dfrac{2}{3}$

3. fractions

5. $\dfrac{\frac{a}{b}}{\frac{c}{d}}$

7. For $\dfrac{\frac{x}{5} - \frac{1}{6}}{\frac{2}{15} - 3x}$, the denominators 5, 6 and 15 have prime factorization of 5, $2 \cdot 3$ and $3 \cdot 5$,

   therefore $2 \cdot 3 \cdot 5 = 30$.

9. For $\dfrac{\frac{2}{x+1} - x}{\frac{2}{x-1} + x}$, the denominators $x+1$ and $x-1$ are prime, therefore $(x+1)(x-1)$.

11. For $\dfrac{\frac{1}{2x-1} - \frac{1}{2x+1}}{\frac{x+1}{x}}$, the denominators $2x-1$, $2x+1$ and $x$ are prime, therefore $x(2x-1)(2x+1)$.

13. $\dfrac{\frac{2}{3}}{\frac{5}{6}} = \dfrac{2}{3} \div \dfrac{5}{6} = \dfrac{2}{3} \cdot \dfrac{6}{5} = \dfrac{12}{15} = \dfrac{4}{5}$

15. $\dfrac{2\frac{1}{2}}{1\frac{3}{4}} = \dfrac{\frac{5}{2}}{\frac{7}{4}} = \dfrac{5}{2} \div \dfrac{7}{4} = \dfrac{5}{2} \cdot \dfrac{4}{7} = \dfrac{20}{14} = \dfrac{10}{7}$

17. $\dfrac{1\frac{1}{2}}{2\frac{1}{3}} = \dfrac{\frac{3}{2}}{\frac{7}{3}} = \dfrac{3}{2} \div \dfrac{7}{3} = \dfrac{3}{2} \cdot \dfrac{3}{7} = \dfrac{9}{14}$

19. $\dfrac{\frac{r}{t}}{\frac{2r}{t}} = \dfrac{r}{t} \div \dfrac{2r}{t} = \dfrac{r}{t} \cdot \dfrac{t}{2r} = \dfrac{t}{t} \cdot \dfrac{1}{2} \cdot \dfrac{r}{r} = 1 \cdot \dfrac{1}{2} \cdot 1 = \dfrac{1}{2}$

21. $\dfrac{\frac{6}{x}}{\frac{2}{y}} = \dfrac{6}{x} \div \dfrac{2}{y} = \dfrac{6}{x} \cdot \dfrac{y}{2} = \dfrac{6y}{2x} = \dfrac{3y}{x}$

23. $\dfrac{\frac{6}{m-2}}{\frac{2}{m-2}} = \dfrac{6}{m-2} \div \dfrac{2}{m-2} = \dfrac{6}{m-2} \cdot \dfrac{m-2}{2} = \dfrac{m-2}{m-2} \cdot \dfrac{6}{2} = 1 \cdot 3 = 3$

25. $\dfrac{\frac{p+1}{p}}{\frac{p+2}{p}} = \dfrac{p+1}{p} \div \dfrac{p+2}{p} = \dfrac{p+1}{p} \cdot \dfrac{p}{p+2} = \dfrac{p}{p} \cdot \dfrac{p+1}{p+2} = 1 \cdot \dfrac{p+1}{p+2} = \dfrac{p+1}{p+2}$

27. $\dfrac{\frac{5}{z^2-1}}{\frac{z}{z^2-1}} = \dfrac{5}{z^2-1} \div \dfrac{z}{z^2-1} = \dfrac{5}{z^2-1} \cdot \dfrac{z^2-1}{z} = \dfrac{z^2-1}{z^2-1} \cdot \dfrac{5}{z} = 1 \cdot \dfrac{5}{z} = \dfrac{5}{z}$

29. $\dfrac{\frac{y}{y^2-9}}{\frac{1}{y+3}} = \dfrac{y}{y^2-9} \div \dfrac{1}{y+3} = \dfrac{y}{y^2-9} \cdot \dfrac{y+3}{1} = \dfrac{y}{(y+3)(y-3)} \cdot \dfrac{y+3}{1} = \dfrac{y}{y-3} \cdot \dfrac{y+3}{y+3} = \dfrac{y}{y-3} \cdot 1 = \dfrac{y}{y-3}$

31. $\dfrac{x-\dfrac{1}{x}}{x+\dfrac{1}{x}} = \dfrac{x-\dfrac{1}{x}}{x+\dfrac{1}{x}} \cdot \dfrac{x}{x} = \dfrac{x^2-1}{x^2+1}$

33. $\dfrac{x}{\dfrac{2}{x}+\dfrac{1}{x}} = \dfrac{x}{\dfrac{2}{x}+\dfrac{1}{x}} \cdot \dfrac{x}{x} = \dfrac{x^2}{2+1} = \dfrac{x^2}{3}$

35. $\dfrac{\dfrac{3}{x+1}}{\dfrac{4}{x+1}-\dfrac{1}{x+1}} = \dfrac{\dfrac{3}{x+1}}{\dfrac{3}{x+1}} = \dfrac{3}{x+1} \div \dfrac{3}{x+1} = \dfrac{3}{x+1} \cdot \dfrac{x+1}{3} = \dfrac{x+1}{x+1} \cdot \dfrac{3}{3} = 1 \cdot 1 = 1$

37. $\dfrac{\dfrac{1}{m^2n}+\dfrac{1}{mn^2}}{\dfrac{1}{m^2n}-\dfrac{1}{mn^2}} = \dfrac{\dfrac{1}{m^2n}+\dfrac{1}{mn^2}}{\dfrac{1}{m^2n}-\dfrac{1}{mn^2}} \cdot \dfrac{m^2n^2}{m^2n^2} = \dfrac{\dfrac{m^2n^2}{m^2n}+\dfrac{m^2n^2}{mn^2}}{\dfrac{m^2n^2}{m^2n}-\dfrac{m^2n^2}{mn^2}} = \dfrac{\left(\dfrac{m^2}{m^2}\cdot\dfrac{n}{n}\cdot n\right)+\left(m\cdot\dfrac{m}{m}\cdot\dfrac{n^2}{n^2}\right)}{\left(\dfrac{m^2}{m^2}\cdot\dfrac{n}{n}\cdot n\right)-\left(m\cdot\dfrac{m}{m}\cdot\dfrac{n^2}{n^2}\right)} = \dfrac{(1\cdot1\cdot n)+(m\cdot1\cdot1)}{(1\cdot1\cdot n)-(m\cdot1\cdot1)} = \dfrac{n+m}{n-m}$

39. $\dfrac{\dfrac{1}{2x}+\dfrac{1}{y}}{\dfrac{1}{y}-\dfrac{1}{2x}} = \dfrac{\dfrac{1}{2x}+\dfrac{1}{y}}{\dfrac{1}{y}-\dfrac{1}{2x}} \cdot \dfrac{2xy}{2xy} = \dfrac{\dfrac{2xy}{2x}+\dfrac{2xy}{y}}{\dfrac{2xy}{y}-\dfrac{2xy}{2x}} = \dfrac{\dfrac{2x}{2x}\cdot y+2x\cdot\dfrac{y}{y}}{2x\cdot\dfrac{y}{y}-\dfrac{2x}{2x}\cdot y} = \dfrac{1\cdot y+2x\cdot1}{2x\cdot1-1\cdot y} = \dfrac{2x+y}{2x-y}$

41. $\dfrac{\dfrac{1}{ab}+\dfrac{1}{a}}{\dfrac{1}{ab}-\dfrac{1}{b}} = \dfrac{\dfrac{1}{ab}+\dfrac{b}{ab}}{\dfrac{1}{ab}-\dfrac{a}{ab}} = \dfrac{\dfrac{1+b}{ab}}{\dfrac{1-a}{ab}} = \dfrac{1+b}{ab} \div \dfrac{1-a}{ab} = \dfrac{1+b}{ab} \cdot \dfrac{ab}{1-a} = \dfrac{ab}{ab} \cdot \dfrac{1+b}{1-a} = 1 \cdot \dfrac{1+b}{1-a} = \dfrac{1+b}{1-a}$

43. $\dfrac{\dfrac{2}{q}-\dfrac{1}{q+1}}{\dfrac{1}{q+1}} = \dfrac{\dfrac{2}{q}-\dfrac{1}{q+1}}{\dfrac{1}{q+1}} \cdot \dfrac{q(q+1)}{q(q+1)} = \dfrac{\dfrac{2q(q+1)}{q}-\dfrac{q(q+1)}{q+1}}{\dfrac{q(q+1)}{q+1}} = \dfrac{\dfrac{q}{q}\cdot2(q+1)-q\cdot\dfrac{q+1}{q+1}}{q\cdot\dfrac{q+1}{q+1}} = \dfrac{2q+2-q}{q} = \dfrac{q+2}{q}$

45. $\dfrac{\dfrac{1}{x+1}+\dfrac{1}{x+2}}{\dfrac{1}{x+1}-\dfrac{1}{x+2}} = \dfrac{\dfrac{1}{x+1}+\dfrac{1}{x+2}}{\dfrac{1}{x+1}-\dfrac{1}{x+2}} \cdot \dfrac{(x+1)(x+2)}{(x+1)(x+2)} = \dfrac{\dfrac{(x+1)(x+2)}{x+1}+\dfrac{(x+1)(x+2)}{x+2}}{\dfrac{(x+1)(x+2)}{x+1}-\dfrac{(x+1)(x+2)}{x+2}} =$

$\dfrac{\dfrac{x+1}{x+1}\cdot(x+2)+\dfrac{x+2}{x+2}\cdot(x+1)}{\dfrac{x+1}{x+1}\cdot(x+2)-\dfrac{x+2}{x+2}\cdot(x+1)} = \dfrac{1\cdot(x+2)+1\cdot(x+1)}{1\cdot(x+2)-1\cdot(x+1)} = \dfrac{x+2+x+1}{x+2-x-1} = \dfrac{2x+3}{1} = 2x+3$

47. $\dfrac{\dfrac{1}{2x-1}-\dfrac{1}{2x+1}}{\dfrac{x+1}{x}} = \dfrac{\dfrac{2x+1}{(2x-1)(2x+1)}-\dfrac{2x-1}{(2x-1)(2x+1)}}{\dfrac{x+1}{x}} = \dfrac{\dfrac{2}{(2x-1)(2x+1)}}{\dfrac{x+1}{x}} =$

$\dfrac{2}{(2x-1)(2x+1)} \div \dfrac{x+1}{x} = \dfrac{2}{(2x-1)(2x+1)} \cdot \dfrac{x}{x+1} = \dfrac{2x}{(x+1)(2x-1)(2x+1)}$

49. $\dfrac{\dfrac{1}{ab^2}-\dfrac{1}{a^2b}}{\dfrac{1}{b}-\dfrac{1}{a}} = \dfrac{\dfrac{a}{a^2b^2}-\dfrac{b}{a^2b^2}}{\dfrac{a}{ab}-\dfrac{b}{ab}} = \dfrac{\dfrac{a-b}{a^2b^2}}{\dfrac{a-b}{ab}} = \dfrac{a-b}{a^2b^2} \div \dfrac{a-b}{ab} = \dfrac{a-b}{a^2b^2} \cdot \dfrac{ab}{a-b} = \dfrac{a-b}{a-b} \cdot \dfrac{ab}{a^2b^2} = 1 \cdot \dfrac{1}{ab} = \dfrac{1}{ab}$

51. $\dfrac{1}{a^{-1}+b^{-1}} = \dfrac{1}{\dfrac{1}{a}+\dfrac{1}{b}} = \dfrac{1}{\dfrac{1}{a}\cdot\dfrac{b}{b}+\dfrac{1}{b}\cdot\dfrac{a}{a}} = \dfrac{1}{\dfrac{b}{ab}+\dfrac{a}{ab}} = \dfrac{1}{\dfrac{a+b}{ab}} = 1 \div \dfrac{a+b}{ab} = 1 \cdot \dfrac{ab}{a+b} = \dfrac{ab}{a+b}$

53. $\dfrac{P\left(1+\dfrac{r}{26}\right)^{52}-P}{\dfrac{r}{26}}$

55. $R = \dfrac{1}{\dfrac{1}{T}+\dfrac{1}{S}} = \dfrac{1}{\dfrac{1}{T}+\dfrac{1}{S}} \cdot \dfrac{ST}{ST} = \dfrac{ST}{\dfrac{ST}{T}+\dfrac{ST}{S}} = \dfrac{ST}{S\cdot\dfrac{T}{T}+\dfrac{S}{S}\cdot T} = \dfrac{ST}{S\cdot1+1\cdot T} = \dfrac{ST}{S+T}$

## Section 7.6 Rational Equations and Formulas

1. rational

3. $ad = bc$; $b$; $d$

5. $12x$

7. $\dfrac{x}{2} = \dfrac{3}{4} \Rightarrow x \cdot 4 = 2 \cdot 3 \Rightarrow 4x = 6 \Rightarrow \dfrac{4x}{4} = \dfrac{6}{4} \Rightarrow x = \dfrac{6}{4} \Rightarrow x = \dfrac{3}{2}$

   Check: $\dfrac{\frac{3}{2}}{2} = \dfrac{3}{4} \Rightarrow \dfrac{3}{2} \div \dfrac{2}{1} = \dfrac{3}{2} \cdot \dfrac{1}{2} = \dfrac{3}{4}$

9. $\dfrac{3}{z} = \dfrac{6}{5} \Rightarrow 3 \cdot 5 = z \cdot 6 \Rightarrow 15 = 6z \Rightarrow \dfrac{15}{6} = \dfrac{6z}{6} \Rightarrow z = \dfrac{15}{6} \Rightarrow z = \dfrac{5}{2}$   Check: $\dfrac{3}{\frac{5}{2}} = \dfrac{6}{5} \Rightarrow \dfrac{3}{1} \div \dfrac{5}{2} = \dfrac{3}{1} \cdot \dfrac{2}{5} = \dfrac{6}{5}$

11. $\dfrac{3y}{4} = \dfrac{7y}{2} \Rightarrow 3y \cdot 2 = 4 \cdot 7y \Rightarrow 6y = 28y \Rightarrow -22y = 0 \Rightarrow \dfrac{-22y}{-22} = \dfrac{0}{-22} \Rightarrow y = 0$

   Check: $\dfrac{3(0)}{4} = \dfrac{7(0)}{2} \Rightarrow \dfrac{0}{4} = \dfrac{0}{2} \Rightarrow 0 = 0$

13. $\dfrac{2}{3} = \dfrac{1}{2x+1} \Rightarrow 2(2x+1) = 3 \cdot 1 \Rightarrow 4x + 2 = 3 \Rightarrow 4x = 1 \Rightarrow \dfrac{4x}{4} = \dfrac{1}{4} \Rightarrow x = \dfrac{1}{4}$

   Check: $\dfrac{2}{3} = \dfrac{1}{2\left(\frac{1}{4}\right)+1} \Rightarrow \dfrac{1}{\frac{2}{4}+1} = \dfrac{1}{\frac{6}{4}} = \dfrac{1}{1} \div \dfrac{6}{4} = \dfrac{1}{1} \cdot \dfrac{4}{6} = \dfrac{4}{6} = \dfrac{2}{3}$

15. $\dfrac{5}{2x} = \dfrac{8}{x+2} \Rightarrow 5(x+2) = 2x \cdot 8 \Rightarrow 5x + 10 = 16x \Rightarrow -11x = -10 \Rightarrow \dfrac{-11x}{-11} = \dfrac{-10}{-11} \Rightarrow x = \dfrac{10}{11}$

   Check: $\dfrac{5}{2\left(\frac{10}{11}\right)} = \dfrac{8}{\frac{10}{11}+2} \Rightarrow \dfrac{5}{\frac{20}{11}} = \dfrac{8}{\frac{32}{11}} \Rightarrow \dfrac{5}{1} \div \dfrac{20}{11} = \dfrac{8}{1} \div \dfrac{32}{11} \Rightarrow \dfrac{5}{1} \cdot \dfrac{11}{20} = \dfrac{8}{1} \cdot \dfrac{11}{32} \Rightarrow \dfrac{55}{20} = \dfrac{88}{32} \Rightarrow \dfrac{11}{4} = \dfrac{11}{4}$

17. $\dfrac{1}{z-1} = \dfrac{2}{z+1} \Rightarrow 1 \cdot (z+1) = (z-1) \cdot 2 \Rightarrow z + 1 = 2z - 2 \Rightarrow 3 = z \Rightarrow z = 3$

   Check: $\dfrac{1}{3-1} = \dfrac{2}{3+1} \Rightarrow \dfrac{1}{2} = \dfrac{2}{4} \Rightarrow \dfrac{1}{2} = \dfrac{1}{2}$

19. $\dfrac{3}{n+5} = \dfrac{2}{n-5} \Rightarrow 3 \cdot (n-5) = (n+5) \cdot 2 \Rightarrow 3n - 15 = 2n + 10 \Rightarrow n = 25$

   Check: $\dfrac{3}{25+5} = \dfrac{2}{25-5} \Rightarrow \dfrac{3}{30} = \dfrac{2}{20} \Rightarrow \dfrac{1}{10} = \dfrac{1}{10}$

21. $\dfrac{m}{m-1} = \dfrac{5}{4} \Rightarrow m \cdot 4 = (m-1) \cdot 5 \Rightarrow 4m = 5m - 5 \Rightarrow -m = -5 \Rightarrow \dfrac{-m}{-1} = \dfrac{-5}{-1} \Rightarrow m = 5$

   Check: $\dfrac{5}{5-1} = \dfrac{5}{4} \Rightarrow \dfrac{5}{4} = \dfrac{5}{4}$

23. $\dfrac{5x}{5-x} = \dfrac{1}{3} \Rightarrow 5x \cdot 3 = (5-x) \cdot 1 \Rightarrow 15x = 5 - x \Rightarrow 16x = 5 \Rightarrow \dfrac{16x}{16} = \dfrac{5}{16} \Rightarrow x = \dfrac{5}{16}$

Check: $\dfrac{5\left(\frac{5}{16}\right)}{5-\frac{5}{16}} = \dfrac{1}{3} \Rightarrow \dfrac{\frac{25}{16}}{\frac{75}{16}} = \dfrac{25}{16} \div \dfrac{75}{16} = \dfrac{25}{16} \cdot \dfrac{16}{75} = \dfrac{16}{16} \cdot \dfrac{25}{75} = 1 \cdot \dfrac{1}{3} = \dfrac{1}{3}$

25. $\dfrac{6}{5-2x} = \dfrac{2}{1} \Rightarrow 6 \cdot 1 = (5-2x) \cdot 2 \Rightarrow 6 = 10 - 4x \Rightarrow -4 = -4x \Rightarrow \dfrac{-4}{-4} = \dfrac{-4x}{-4} \Rightarrow 1 = x \Rightarrow x = 1$

Check: $\dfrac{6}{5-2(1)} = 2 \Rightarrow \dfrac{6}{5-2} = \dfrac{6}{3} \Rightarrow 2 = 2$

27. $\dfrac{2x}{2x+1} = \dfrac{-1}{2x+1} \Rightarrow 2x(2x+1) = -1(2x+1) \Rightarrow 4x^2 + 2x = -2x - 1$

$\Rightarrow 4x^2 + 4x + 1 = 0 \Rightarrow (2x+1)^2 = 0 \Rightarrow 2x+1 = 0 \Rightarrow 2x = -1 \Rightarrow x = -\dfrac{1}{2}$

Check: $\dfrac{2\left(-\frac{1}{2}\right)}{2\left(-\frac{1}{2}\right)+1} \overset{2}{=} \dfrac{-1}{2\left(-\frac{1}{2}\right)+1} \Rightarrow \dfrac{-1}{-1+1} \overset{2}{=} \dfrac{-1}{-1+1}$, false because the denominators are 0 so both

expressions are undefined. There are no solutions because $-\dfrac{1}{2}$ is extraneous.

29. $\dfrac{1}{1-x} = \dfrac{3}{1+x} \Rightarrow 1 \cdot (1+x) = (1-x) \cdot 3 \Rightarrow 1 + x = 3 - 3x \Rightarrow 4x = 2 \Rightarrow \dfrac{4x}{4} = \dfrac{2}{4} \Rightarrow x = \dfrac{1}{2}$

Check: $\dfrac{1}{1-\frac{1}{2}} = \dfrac{3}{1+\frac{1}{2}} \Rightarrow \dfrac{1}{\frac{1}{2}} = \dfrac{3}{\frac{3}{2}} \Rightarrow \dfrac{1}{1} \div \dfrac{1}{2} = \dfrac{3}{1} \div \dfrac{3}{2} \Rightarrow \dfrac{1}{1} \cdot \dfrac{2}{1} = \dfrac{3}{1} \cdot \dfrac{2}{3} \Rightarrow 2 = \dfrac{6}{3} \Rightarrow 2 = 2$

31. $\dfrac{1}{z+2} = \dfrac{-z}{1} \Rightarrow 1 \cdot 1 = -z(z+2) \Rightarrow 1 = -z^2 - 2z \Rightarrow z^2 + 2z + 1 = 0 \Rightarrow (z+1)^2 = 0 \Rightarrow z = -1$

Check: $\dfrac{1}{-1+2} = \dfrac{-(-1)}{1} \Rightarrow \dfrac{1}{1} = \dfrac{1}{1} \Rightarrow 1 = 1$

33. $\dfrac{-1}{2x+5} = \dfrac{x}{3} \Rightarrow -1 \cdot 3 = (2x+5) \cdot x \Rightarrow -3 = 2x^2 + 5x \Rightarrow 0 = 2x^2 + 5x + 3 \Rightarrow$

$0 = (2x+3)(x+1) \Rightarrow 0 = 2x+3 \Rightarrow -\dfrac{3}{2} = \dfrac{2}{2}x \Rightarrow -\dfrac{3}{2} = x, \ 0 = x+1 \Rightarrow -1 = x \Rightarrow x = -\dfrac{3}{2}, \ -1$

Check: $\dfrac{-1}{2\left(\frac{-3}{2}\right)+5} = \dfrac{\frac{-3}{2}}{3} \Rightarrow \dfrac{-1}{\frac{4}{2}} = \dfrac{\frac{-3}{2}}{3} \Rightarrow \dfrac{-1}{1} \div \dfrac{4}{2} = \dfrac{-3}{2} \div \dfrac{3}{1} \Rightarrow \dfrac{-1}{1} \cdot \dfrac{2}{4} = \dfrac{-3}{2} \cdot \dfrac{1}{3} \Rightarrow \dfrac{-2}{4} = \dfrac{-3}{6} \Rightarrow$

$\dfrac{-1}{2} = \dfrac{-1}{2} ; \ \dfrac{-1}{2(-1)+5} = \dfrac{-1}{3} \Rightarrow \dfrac{-1}{-2+5} \Rightarrow \dfrac{-1}{3}$

35. $\dfrac{x}{2} + \dfrac{x}{4} = 3 \Rightarrow \dfrac{x(4)}{2} + \dfrac{x(4)}{4} = 3(4) \Rightarrow \dfrac{4x}{2} + \dfrac{4x}{4} = 12 \Rightarrow 2x + x = 12 \Rightarrow 3x = 12 \Rightarrow \dfrac{3x}{3} = \dfrac{12}{3} \Rightarrow x = 4$

Check: $\dfrac{4}{2} + \dfrac{4}{4} = 3 \Rightarrow 2 + 1 = 3 \Rightarrow 3 = 3$

37. $\dfrac{3x}{4} - \dfrac{x}{2} = 1 \Rightarrow \dfrac{3x(4)}{4} - \dfrac{x(4)}{2} = 1(4) \Rightarrow \dfrac{12x}{4} - \dfrac{4x}{2} = 4 \Rightarrow 3x - 2x = 4 \Rightarrow x = 4$

Check: $\dfrac{3(4)}{4} - \dfrac{4}{2} = 1 \Rightarrow \dfrac{12}{4} - \dfrac{4}{2} = 1 \Rightarrow 3 - 2 = 1 \Rightarrow 1 = 1$

39. $\dfrac{4}{t+1} + \dfrac{1}{t+1} = -1 \Rightarrow \dfrac{5}{t+1} = -1 \Rightarrow (t+1) \cdot \dfrac{5}{t+1} = -1 \cdot (t+1) \Rightarrow 5 = -t - 1 \Rightarrow 6 = -t \Rightarrow \dfrac{6}{-1} = \dfrac{-t}{-1} \Rightarrow t = -6$

Check: $\dfrac{4}{t+1} + \dfrac{1}{t+1} = -1 \Rightarrow \dfrac{5}{t+1} = -1 \Rightarrow \dfrac{5}{-6+1} = -1 \Rightarrow \dfrac{5}{-5} = -1 \Rightarrow -1 = -1$

41. $\dfrac{1}{x} + \dfrac{2}{x} = \dfrac{1}{2} \Rightarrow \dfrac{3}{x} = \dfrac{1}{2} \Rightarrow x \cdot \dfrac{3}{x} = \dfrac{1}{2} \cdot x \Rightarrow 3 = \dfrac{1}{2}x \Rightarrow 3 \cdot 2 = \dfrac{1}{2}x \cdot 2 \Rightarrow 6 = x \Rightarrow x = 6$

Check: $\dfrac{1}{6} + \dfrac{2}{6} = \dfrac{1}{2} \Rightarrow \dfrac{3}{6} = \dfrac{1}{2} \Rightarrow \dfrac{1}{2} = \dfrac{1}{2}$

43. $\dfrac{2}{x-1} + 1 = \dfrac{4}{x^2-1} \Rightarrow \dfrac{2+x-1}{x-1} = \dfrac{4}{x^2-1} \Rightarrow \dfrac{x+1}{x-1} = \dfrac{4}{(x-1)(x+1)}$

$\Rightarrow (x+1)(x-1)(x+1) = 4(x-1) \Rightarrow \dfrac{(x+1)(x-1)(x+1)}{(x-1)} = \dfrac{4(x-1)}{x-1}$

$\Rightarrow (x+1)(x+1) = 4 \Rightarrow x^2 + 2x + 1 = 4 \Rightarrow x^2 + 2x - 3 = 0$

$\Rightarrow (x+3)(x-1) = 0 \Rightarrow x+3 = 0, \ x = -3 \text{ and } x-1 = 0, \ x = 1$

Check: $\dfrac{2}{-3-1} + 1 = \dfrac{4}{(-3)^2 - 1} \Rightarrow \dfrac{2}{-4} + \dfrac{-4}{-4} = \dfrac{4}{9-1} \Rightarrow \dfrac{-2}{-4} = \dfrac{4}{8} \Rightarrow \dfrac{1}{2} = \dfrac{1}{2}$

$\dfrac{2}{1-1} + 1 \overset{?}{=} \dfrac{4}{1-1}$, false because the denominators are 0 so both expressions are undefined.

The only solution is $-3$ because 1 is extraneous.

45. $\dfrac{1}{x+2} = \dfrac{4}{4-x^2} - 1 \Rightarrow \dfrac{1}{x+2} = \dfrac{4-(4-x^2)}{4-x^2} \Rightarrow (4-x^2) \cdot 1 = (x+2)(4-4+x^2) \Rightarrow 4 - x^2 = (x+2)(x^2)$

$\Rightarrow (2-x)(2+x) = (x+2)(x^2) \Rightarrow \dfrac{(2-x)(2+x)}{x+2} = \dfrac{(x+2)(x^2)}{x+2}$

$\Rightarrow 2 - x = x^2 \Rightarrow x^2 + x - 2 = 0 \Rightarrow (x-1)(x+2) = 0 \Rightarrow x-1 = 0, \ x = 1 \text{ and } x+2 = 0, \ x = -2$

Check: $-2$ is extraneous because $\dfrac{1}{-2+2}$ is undefined. $\dfrac{1}{1+2} = \dfrac{4}{4-(1)^2} - 1 \Rightarrow \dfrac{1}{3} = \dfrac{4}{3} - \dfrac{3}{3} \Rightarrow \dfrac{1}{3} = \dfrac{1}{3}$

47. $\dfrac{5}{4z} - \dfrac{2}{3z} = 1 \Rightarrow \dfrac{5(12z)}{4z} - \dfrac{2(12z)}{3z} = 1(12z) \Rightarrow \dfrac{60z}{4z} - \dfrac{24z}{3z} = 12z \Rightarrow 15 - 8 = 12z \Rightarrow \dfrac{7}{12} = \dfrac{12z}{12} \Rightarrow$

$\dfrac{7}{12} = z \Rightarrow z = \dfrac{7}{12}$  Check: $\dfrac{5}{4\left(\frac{7}{12}\right)} - \dfrac{2}{3\left(\frac{7}{12}\right)} = 1 \Rightarrow \dfrac{5}{\frac{7}{3}} - \dfrac{2}{\frac{7}{4}} = 1 \Rightarrow \dfrac{5}{1} \div \dfrac{7}{3} - \dfrac{2}{1} \div \dfrac{7}{4} = 1 \Rightarrow$

$\dfrac{5}{1} \cdot \dfrac{3}{7} - \dfrac{2}{1} \cdot \dfrac{4}{7} = 1 \Rightarrow \dfrac{15}{7} - \dfrac{8}{7} = 1 \Rightarrow \dfrac{7}{7} = 1 \Rightarrow 1 = 1$

49. $\dfrac{4}{y-1}+\dfrac{1}{y}=\dfrac{6}{5} \Rightarrow \dfrac{4(5)(y)(y-1)}{y-1}+\dfrac{1(5)(y)(y-1)}{y}=\dfrac{6(5)(y)(y-1)}{5} \Rightarrow$

$\dfrac{20y(y-1)}{y-1}+\dfrac{5y(y-1)}{y}=\dfrac{\left(6y^2-6y\right)(5)}{5} \Rightarrow 20y+5y-5=6y^2-6y \Rightarrow 25y-5=6y^2-6y \Rightarrow$

$6y^2-31y+5=0 \Rightarrow (6y-1)(y-5)=0 \Rightarrow 6y-1=0 \Rightarrow 6y=1 \Rightarrow y=\dfrac{1}{6},\ y-5=0 \Rightarrow$

$y=5 \Rightarrow y=\dfrac{1}{6},\ 5$

Check: $\dfrac{4}{y-1}+\dfrac{1}{y}=\dfrac{6}{5} \Rightarrow \dfrac{4}{\frac{1}{6}-1}+\dfrac{1}{\frac{1}{6}}=\dfrac{6}{5} \Rightarrow \dfrac{4}{-\frac{5}{6}}+\dfrac{1}{\frac{1}{6}}=\dfrac{6}{5} \Rightarrow \dfrac{4}{1}\cdot-\dfrac{6}{5}+\dfrac{1}{1}\cdot\dfrac{6}{1}=\dfrac{6}{5} \Rightarrow -\dfrac{24}{5}+\dfrac{6}{1}=\dfrac{6}{5} \Rightarrow$

$-\dfrac{24}{5}+\dfrac{30}{5}=\dfrac{6}{5} \Rightarrow \dfrac{6}{5}=\dfrac{6}{5},\ \dfrac{4}{5-1}+\dfrac{1}{5}=\dfrac{6}{5} \Rightarrow \dfrac{4}{4}+\dfrac{1}{5}=\dfrac{6}{5} \Rightarrow \dfrac{20}{20}+\dfrac{4}{20}=\dfrac{24}{20} \Rightarrow \dfrac{24}{20}=\dfrac{24}{20} \Rightarrow \dfrac{6}{5}=\dfrac{6}{5}$

51. $\dfrac{1}{2x}-\dfrac{1}{x+3}=0 \Rightarrow \dfrac{1(2x)(x+3)}{2x}-\dfrac{1(2x)(x+3)}{x+3}=0(2x)(x+3) \Rightarrow \dfrac{(x+3)(2x)}{2x}-\dfrac{2x(x+3)}{x+3}=0 \Rightarrow$

$x+3-2x=0 \Rightarrow -x+3=0 \Rightarrow -x=-3 \Rightarrow x=3$

Check: $\dfrac{1}{2(3)}-\dfrac{1}{3+3}=0 \Rightarrow \dfrac{1}{6}-\dfrac{1}{6}=0 \Rightarrow 0=0$

53. $\dfrac{1}{x-1}+\dfrac{1}{x+1}=\dfrac{2}{x^2-1} \Rightarrow \dfrac{1}{x-1}\cdot\dfrac{x+1}{x+1}+\dfrac{1}{x+1}\cdot\dfrac{x-1}{x-1}=\dfrac{2}{x^2-1}$

$\Rightarrow \dfrac{x+1+x-1}{x^2-1}=\dfrac{2}{x^2-1} \Rightarrow \dfrac{2x}{x^2-1}=\dfrac{2}{x^2-1} \Rightarrow 2x(x^2-1)=2(x^2-1)$

$\Rightarrow \dfrac{2x(x^2-1)}{2(x^2-1)}=\dfrac{2(x^2-1)}{2(x^2-1)} \Rightarrow x=1$, extraneous because $\dfrac{1}{1-1}$ is undefined. There are no solutions.

55. $\dfrac{1}{x-2}+\dfrac{1}{x+2}=\dfrac{6}{x^2-4} \Rightarrow \dfrac{1(x-2)(x+2)}{x-2}+\dfrac{1(x-2)(x+2)}{x+2}=\dfrac{6(x-2)(x+2)}{(x-2)(x+2)} \Rightarrow$

$\dfrac{(x+2)(x-2)}{x-2}+\dfrac{(x-2)(x+2)}{x+2}=\dfrac{6(x-2)(x+2)}{(x-2)(x+2)} \Rightarrow x+2+x-2=6 \Rightarrow 2x=6 \Rightarrow x=\dfrac{6}{2} \Rightarrow x=3$

Check: $\dfrac{1}{3-2}+\dfrac{1}{3+2}=\dfrac{6}{(3)^2-4} \Rightarrow \dfrac{1}{1}+\dfrac{1}{5}=\dfrac{6}{5} \Rightarrow \dfrac{5}{5}+\dfrac{1}{5}=\dfrac{6}{5} \Rightarrow \dfrac{6}{5}=\dfrac{6}{5}$

57. $\dfrac{1}{p+1}+\dfrac{1}{p+2}=\dfrac{1}{p^2+3p+2} \Rightarrow \dfrac{1}{p+1}\cdot\dfrac{p+2}{p+2}+\dfrac{1}{p+2}\cdot\dfrac{p+1}{p+1}=\dfrac{1}{(p+1)(p+2)} \Rightarrow$

$\dfrac{p+2}{(p+1)(p+2)}+\dfrac{p+1}{(p+1)(p+2)}=\dfrac{1}{(p+1)(p+2)} \Rightarrow 2p+3=1 \Rightarrow 2p=-2 \Rightarrow$

$p=-1$, but $p=-1$ makes $\dfrac{1}{p+1}=\dfrac{1}{0}$ which is undefined $\Rightarrow$ no solution

59. $\dfrac{1}{x-2}+\dfrac{3}{2x-4}=\dfrac{6}{3x-6} \Rightarrow \dfrac{1}{x-2}+\dfrac{3}{2(x-2)}=\dfrac{6}{3(x-2)} \Rightarrow$

$\dfrac{1}{x-2}\cdot\dfrac{6}{6}+\dfrac{3}{2(x-2)}\cdot\dfrac{3}{3}=\dfrac{6}{3(x-2)}\cdot\dfrac{2}{2} \Rightarrow \dfrac{6}{6(x-2)}+\dfrac{9}{6(x-2)}=\dfrac{12}{6(x-2)} \Rightarrow 6+9=12 \Rightarrow$

$15=12$, which is false $\Rightarrow$ no solution

61. $\dfrac{1}{r^2-r-2}+\dfrac{2}{r^2-2r}=\dfrac{1}{r^2+r} \Rightarrow \dfrac{1(r)(r-2)(r+1)}{(r-2)(r+1)}+\dfrac{2(r)(r-2)(r+1)}{r(r-2)}=\dfrac{1(r)(r-2)(r+1)}{r(r+1)} \Rightarrow$

$\dfrac{r(r-2)(r+1)}{(r-2)(r+1)}+\dfrac{(2r+2)(r)(r-2)}{r(r-2)}=\dfrac{(r-2)(r)(r+1)}{r(r+1)} \Rightarrow r+2r+2=r-2 \Rightarrow 3r+2=r-2 \Rightarrow$

$3r=r-4 \Rightarrow 2r=-4 \Rightarrow r=\dfrac{-4}{2} \Rightarrow r=-2$  Check: $\dfrac{1}{(-2)^2-(-2)-2}+\dfrac{2}{(-2)^2-2(-2)}$

$=\dfrac{1}{(-2)^2+(-2)} \Rightarrow \dfrac{1}{4}+\dfrac{2}{8}=\dfrac{1}{2} \Rightarrow \dfrac{1}{4}+\dfrac{1}{4}=\dfrac{1}{2} \Rightarrow \dfrac{2}{4}=\dfrac{1}{2} \Rightarrow \dfrac{1}{2}=\dfrac{1}{2}$

63. Expression; $\dfrac{1}{x}-\dfrac{1-x}{x}=\dfrac{1-1+x}{x}=\dfrac{x}{x}=1$; when $x=2$, the expression has the value 1.

65. Equation; $\dfrac{1}{2x}-\dfrac{1}{4x}=\dfrac{1}{8} \Rightarrow \dfrac{2}{4x}-\dfrac{1}{4x}=\dfrac{1}{8} \Rightarrow \dfrac{1}{4x}=\dfrac{1}{8} \Rightarrow 4x=8 \Rightarrow x=2$

67. Equation; $\dfrac{x+1}{x-1}=\dfrac{2x-3}{2x-5} \Rightarrow (2x-5)(x+1)=(2x-3)(x-1)$

$\Rightarrow 2x^2-3x-5=2x^2-5x+3 \Rightarrow 2x-8=0 \Rightarrow 2x=8 \Rightarrow x=4$

69. Expression; $\dfrac{4x+4}{x+2}+\dfrac{x^2}{x+2}=\dfrac{x^2+4x+4}{x+2}=\dfrac{(x+2)^2}{x+2}=x+2$;  $x=2, 2+2=4$

71. The graphs intersect at $\left(-\frac{1}{2},-2\right)$ and $\left(\frac{1}{2},2\right)$. Check: $\dfrac{1}{-\frac{1}{2}}=4\left(-\dfrac{1}{2}\right) \Rightarrow -2=-2; \dfrac{1}{\frac{1}{2}}=4\left(\dfrac{1}{2}\right) \Rightarrow 2=2$

The solutions are $-\dfrac{1}{2},\dfrac{1}{2}$.

73. The graphs intersect at $(-1, 2)$ and $\left(\frac{1}{2},-1\right)$. Check: $\dfrac{-1-1}{-1}=-2(-1) \Rightarrow \dfrac{-2}{-1}=2 \Rightarrow 2=2$;

$\dfrac{\frac{1}{2}-1}{\frac{1}{2}}=-2\left(\frac{1}{2}\right) \Rightarrow \dfrac{-\frac{1}{2}}{\frac{1}{2}}=-1 \Rightarrow -1=-1$  The solutions are $-1,\dfrac{1}{2}$.

75. (a)    $-3, 1$; $y_1=\dfrac{3}{x}$, $y_2=x+2$.  See Figure 75a.

(b)    See Figure 75b.  $-3, 1$

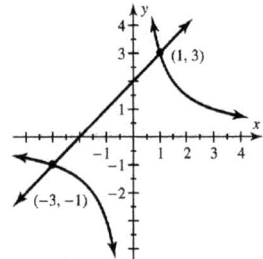

Figure 75a

| $x$ | $-3$ | $-1$ | $1$ | $3$ |
|---|---|---|---|---|
| $y = 3/x$ | $-1$ | $-3$ | $3$ | $1$ |
| $y = x + 2$ | $-1$ | $1$ | $3$ | $5$ |

Figure 75b

77. (a)  $-1;\ y_1 = \dfrac{3x}{2},\ y_2 = \dfrac{1}{2}x - 1.$  See Figure 77a.

(b)  See Figure 77b.  $-1$

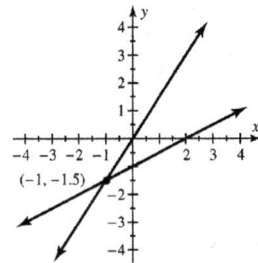

Figure 77a

| $x$ | $-2$ | $-1$ | $1$ | $2$ |
|---|---|---|---|---|
| $y = 3x/2$ | $-3$ | $-1.5$ | $1.5$ | $3$ |
| $y = x/2 - 1$ | $-2$ | $-1.5$ | $-0.5$ | $0$ |

Figure 77b

79. (a)  $2;\ y_1 = \dfrac{3}{x-1},\ y_2 = 3.$  See Figure 79a.

(b)  See Figure 79b.  $2$

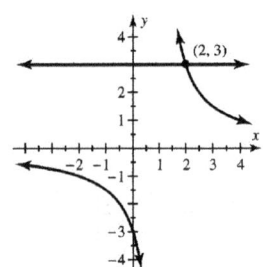

Figure 79a

| $x$ | $-2$ | $0$ | $2$ | $4$ |
|---|---|---|---|---|
| $y = 3/(x-1)$ | $-1$ | $-3$ | $3$ | $1$ |
| $y = 3$ | $3$ | $3$ | $3$ | $3$ |

Figure 79b

81. (a)  $-2,\ 2;\ y_1 = \dfrac{4}{x^2},\ y_2 = 1.$  See Figure 81a.

(b)  See Figure 81b.  $-2,\ 2$

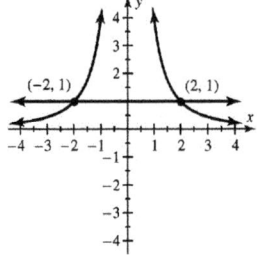

Figure 81a

| $x$ | $-2$ | $-1$ | $1$ | $2$ |
|---|---|---|---|---|
| $y = 4/x^2$ | $1$ | $4$ | $4$ | $1$ |
| $y = 1$ | $1$ | $1$ | $1$ | $1$ |

Figure 81b

83.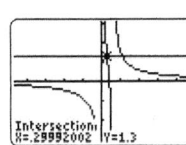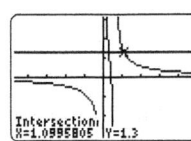

The solutions are 0.300 and 1.100.

85.

The solution is 1.084.

87. $m = \dfrac{F}{a}$ for $a \Rightarrow a \cdot m = \dfrac{F}{a} \cdot a \Rightarrow am = F \Rightarrow \dfrac{am}{m} = \dfrac{F}{m} \Rightarrow a = \dfrac{F}{m}$

89. $I = \dfrac{V}{R+r}$ for $r \Rightarrow I(R+r) = \dfrac{V}{R+r} \cdot (R+r) \Rightarrow I(R+r) = V \Rightarrow \dfrac{I(R+r)}{I} = \dfrac{V}{I} \Rightarrow$

$R + r = \dfrac{V}{I} \Rightarrow r = \dfrac{V}{I} - R$

91. $h = \dfrac{2A}{b}$ for $b \Rightarrow b \cdot h = \dfrac{2A}{b} \cdot b \Rightarrow bh = 2A \Rightarrow \dfrac{bh}{h} = \dfrac{2A}{h} \Rightarrow b = \dfrac{2A}{h}$

93. $\dfrac{3}{k} = \dfrac{z}{z+5}$ for $z \Rightarrow 3(z+5) = kz \Rightarrow 3z + 15 = kz \Rightarrow 15 = kz - 3z \Rightarrow 15 = z(k-3) \Rightarrow z = \dfrac{15}{k-3}$

95. $T = \dfrac{ab}{a+b}$ for $b \Rightarrow T(a+b) = ab \cdot 1 \Rightarrow aT + bT = ab \Rightarrow aT = ab - bT \Rightarrow aT = b(a-T) \Rightarrow b = \dfrac{aT}{a-T}$

97. $\dfrac{3}{k} = \dfrac{1}{x} - \dfrac{2}{y}$ for $x \Rightarrow \dfrac{3}{k} \cdot \dfrac{xy}{xy} = \dfrac{1}{x} \cdot \dfrac{ky}{ky} - \dfrac{2}{y} \cdot \dfrac{xk}{xk} \Rightarrow \dfrac{3xy}{kxy} = \dfrac{ky}{kxy} - \dfrac{2xk}{kxy} \Rightarrow 3xy = ky - 2xk \Rightarrow$

$3xy + 2xk = ky \Rightarrow x(3y+2k) = ky \Rightarrow x = \dfrac{ky}{3y+2k}$

99. $\dfrac{1}{10-x} = 1 \Rightarrow (10-x) \cdot \dfrac{1}{10-x} = 1(10-x) \Rightarrow 1 = 10 - x \Rightarrow x = 9$ cars/minute.

101. $\dfrac{t}{4} + \dfrac{t}{3} = 1 \Rightarrow \dfrac{t(12)}{4} + \dfrac{t(12)}{3} = 12 \Rightarrow \dfrac{12t}{4} + \dfrac{12t}{3} = 12 \Rightarrow 3t + 4t = 12 \Rightarrow 7t = 12 \Rightarrow t = \dfrac{12}{7} \Rightarrow t \approx 1.7$ hr

103. $\dfrac{d}{8} + \dfrac{d}{4} = 1 \Rightarrow \dfrac{(8)d}{8} + \dfrac{(8)d}{4} = 8 \Rightarrow \dfrac{8d}{8} + \dfrac{8d}{4} = 8 \Rightarrow d + 2d = 8 \Rightarrow 3d = 8 \Rightarrow d = \dfrac{8}{3} \Rightarrow d \approx 2.7$ days

105. Let $x$ represent the speed of the teammate. Then $x+2$ represents the speed of the winner. The

finishing time for the teammate is $\dfrac{6}{x}$ and the finishing time for the winner is $\dfrac{6}{x+2}$. Since the

winner's time is 2 minutes, or $\dfrac{1}{30}$ hour, faster than the teammate's time, the needed equation is

$\dfrac{6}{x} - \dfrac{6}{x+2} = \dfrac{1}{30}$. Multiply each term by the LCD, $30x(x+2)$. $\dfrac{6}{x} - \dfrac{6}{x+2} = \dfrac{1}{30} \Rightarrow$

$$30x(x+2)\cdot\frac{6}{x}-30x(x+2)\cdot\frac{6}{x+2}=30x(x+2)\cdot\frac{1}{30}\Rightarrow180x+360-180x=x^2+2x\Rightarrow$$

$$x^2+2x-360=0\Rightarrow(x+20)(x-18)=0\Rightarrow x=-20\text{ or }18$$  Since $x$ must be positive, the teammate's

speed is 18 mph and the winner's speed is $18+2=20$ mph.

107. (a)   $m=-0.05,\ B=\dfrac{30}{0.3+(-0.05)}\Rightarrow B=\dfrac{30}{0.25}\Rightarrow B=120$, the braking distance is 120 feet when

slope of the road is $-0.05$.

(b)   $B=150,\ 150=\dfrac{30}{0.3+m}\Rightarrow(0.3+m)(150)=30\Rightarrow0.3+m=\dfrac{30}{150}\Rightarrow0.3+m=.2\Rightarrow m=-0.1$

109.  $\dfrac{36}{x-3}=\dfrac{54}{x+3}\Rightarrow36x+108=54x-162\Rightarrow270=18x\Rightarrow15=x\Rightarrow x=15$ mph

111.  $\dfrac{450}{x-50}=\dfrac{750}{x+50}\Rightarrow450x+22,500=750x-37,500\Rightarrow60,000=300x\Rightarrow200=x\Rightarrow x=200$ mph

113.  Let $x=$ jogging in mph and $x+5=$ running in mph, then $x\cdot(t+1)=10$ and $(x+5)t=10$, so

$$t+1=\frac{10}{x}\Rightarrow t=\frac{10}{x}-1\text{ and }t=\frac{10}{x+5}.$$

Now,  $\dfrac{10}{x}-1=\dfrac{10}{x+5}\Rightarrow\dfrac{10}{x}\cdot\dfrac{x+5}{x+5}-\dfrac{1}{1}\cdot\dfrac{x(x+5)}{x(x+5)}=\dfrac{10}{x+5}\cdot\dfrac{x}{x}\Rightarrow10(x+5)-x(x+5)=10x\Rightarrow$

$$10x+50-x^2-5x=10x\Rightarrow x^2+5x-50=0\Rightarrow(x+10)(x-5)=0,\text{ so }x=-10,\ 5\text{ but }-10\text{ cannot be}$$

the rate so $x=5$.  So 10 mph running and 5 mph jogging.

## Checking Basic Concepts  Sections 7.5 and 7.6

1. (a)  $\dfrac{\frac{x}{3}}{\frac{2x}{5}}=\dfrac{x}{3}\div\dfrac{2x}{5}=\dfrac{x}{3}\cdot\dfrac{5}{2x}=\dfrac{5x}{6x}=\dfrac{5}{6}\cdot\dfrac{x}{x}=\dfrac{5}{6}\cdot1=\dfrac{5}{6}$

(b)  $\dfrac{\frac{2}{2x}-\frac{1}{3x}}{6x}=\dfrac{\frac{6}{6x}-\frac{2}{6x}}{6x}=\dfrac{\frac{4}{6x}}{6x}=\dfrac{4}{6x}\div\dfrac{6x}{1}=\dfrac{4}{6x}\cdot\dfrac{1}{6x}=\dfrac{4}{36x^2}=\dfrac{1}{9x^2}$

(c)  $\dfrac{\frac{1}{a}-\frac{1}{b}}{\frac{1}{a}+\frac{1}{b}}=\dfrac{\frac{b}{ab}-\frac{a}{ab}}{\frac{b}{ab}+\frac{a}{ab}}=\dfrac{\frac{b-a}{ab}}{\frac{b+a}{ab}}=\dfrac{b-a}{ab}\div\dfrac{b+a}{ab}=\dfrac{b-a}{ab}\cdot\dfrac{ab}{b+a}=\dfrac{ab}{ab}\cdot\dfrac{b-a}{b+a}=1\cdot\dfrac{b-a}{b+a}=\dfrac{b-a}{b+a}$

(d)  $\dfrac{\frac{1}{r^2}-\frac{1}{t^2}}{\frac{2}{r}-\frac{2}{t}}=\dfrac{\frac{t^2}{r^2t^2}-\frac{r^2}{r^2t^2}}{\frac{2t}{rt}-\frac{2r}{rt}}=\dfrac{\frac{t^2-r^2}{r^2t^2}}{\frac{2t-2r}{rt}}=\dfrac{t^2-r^2}{r^2t^2}\div\dfrac{2t-2r}{rt}=\dfrac{t^2-r^2}{r^2t^2}\cdot\dfrac{rt}{2t-2r}=\dfrac{rt}{r^2t^2}\cdot\dfrac{t^2-r^2}{2t-2r}=$

$$\dfrac{1}{2rt}\cdot\dfrac{(t+r)(t-r)}{t-r}=\dfrac{t+r}{2rt}=\dfrac{r+t}{2rt}$$

2. (a) $\dfrac{1}{2x} = \dfrac{3}{x+1} \Rightarrow 6x = x+1 \Rightarrow 5x = 1 \Rightarrow x = \dfrac{1}{5}$

Check: $\dfrac{1}{2\left(\frac{1}{5}\right)} = \dfrac{3}{\left(\frac{1}{5}\right)+1} \Rightarrow \dfrac{1}{\frac{2}{5}} = \dfrac{3}{\frac{6}{5}} \Rightarrow \dfrac{1}{5} \cdot \dfrac{5}{2} = \dfrac{3}{5} \cdot \dfrac{5}{6} \Rightarrow \dfrac{5}{2} = \dfrac{15}{6} \Rightarrow \dfrac{5}{2} = \dfrac{5}{2}$

(b) $\dfrac{x}{2x+3} = \dfrac{4}{5} \Rightarrow 5x = 8x+12 \Rightarrow -3x = 12 \Rightarrow x = \dfrac{12}{-3} \Rightarrow x = -4$

Check: $\dfrac{-4}{2(-4)+3} = \dfrac{4}{5} \Rightarrow \dfrac{-4}{-5} = \dfrac{4}{5} \Rightarrow \dfrac{4}{5} = \dfrac{4}{5}$

(c) $\dfrac{1}{2x} + \dfrac{3}{2x} = 1 \Rightarrow \dfrac{4}{2x} = 1 \Rightarrow 2x = 4 \Rightarrow x = \dfrac{4}{2} \Rightarrow x = 2$

Check: $\dfrac{1}{2(2)} + \dfrac{3}{2(2)} = 1 \Rightarrow \dfrac{1}{4} + \dfrac{3}{4} = 1 \Rightarrow \dfrac{4}{4} = 1 \Rightarrow 1 = 1$

(d) $\dfrac{3}{x+1} - \dfrac{2}{x} = -2 \Rightarrow \dfrac{3}{x+1} \cdot \dfrac{x}{x} - \dfrac{2}{x} \cdot \dfrac{x+1}{x+1} = \dfrac{-2}{1} \cdot \dfrac{x(x+1)}{x(x+1)} \Rightarrow 3x - 2x - 2 = -2x^2 - 2x \Rightarrow$

$x - 2 = -2x^2 - 2x \Rightarrow 2x^2 + 3x - 2 = 0 \Rightarrow (2x-1)(x+2) = 0$, so $x = \dfrac{1}{2},\ -2$

Check: $\dfrac{3}{\frac{1}{2}+1} - \dfrac{2}{\frac{1}{2}} = -2 \Rightarrow \dfrac{3}{\frac{3}{2}} - \dfrac{2}{\frac{1}{2}} = -2 \Rightarrow \dfrac{6}{3} - \dfrac{12}{3} = -2 \Rightarrow \dfrac{-6}{3} = -2 \Rightarrow -2 = -2$

$\dfrac{3}{-2+1} - \dfrac{2}{-2} = -2 \Rightarrow \dfrac{3}{-1} - \dfrac{2}{-2} = -2 \Rightarrow -3 - (-1) = -2 \Rightarrow -2 = -2$

(e) $\dfrac{1}{x-1} = \dfrac{2}{x^2-1} - \dfrac{1}{2} \Rightarrow \dfrac{1}{x-1} = \dfrac{2}{x^2-1} \cdot \dfrac{2}{2} - \dfrac{1}{2} \cdot \dfrac{x^2-1}{x^2-1} \Rightarrow \dfrac{1}{x-1} = \dfrac{4-x^2+1}{2\left(x^2-1\right)} \Rightarrow$

$2\left(x^2-1\right) = (x-1)\left(5-x^2\right) \Rightarrow \dfrac{2(x-1)(x+1)}{x-1} = \dfrac{(x-1)\left(5-x^2\right)}{x-1} \Rightarrow 2(x+1) = \left(5-x^2\right)$

$\Rightarrow 2x+2 = 5-x^2 \Rightarrow x^2+2x-3 = 0 \Rightarrow (x+3)(x-1) = 0$

$\Rightarrow x+3 = 0,\ x = -3$ and $x-1 = 0,\ x = 1$

Check: $\dfrac{1}{-3-1} = \dfrac{2}{(-3)^2-1} - \dfrac{1}{2} \Rightarrow \dfrac{1}{-4} = \dfrac{2}{8} - \dfrac{1}{2} \Rightarrow -\dfrac{1}{4} = \dfrac{1}{4} - \dfrac{2}{4} \Rightarrow -\dfrac{1}{4} = -\dfrac{1}{4}$; 1 is an extraneous

solution because $\dfrac{1}{1-1}$ is undefined. The solution is $-3$.

3 (a) $\dfrac{ax}{2} - 3y = b$ for $x \Rightarrow \dfrac{ax}{2} = b+3y \Rightarrow ax = 2(b+3y) \Rightarrow x = \dfrac{2(b+3y)}{a}$

(b) $\dfrac{1}{2m-1} = \dfrac{k}{m}$ for $m \Rightarrow m \cdot 1 = k(2m-1) \Rightarrow m = 2km - k \Rightarrow m - 2km = -k \Rightarrow$

$m(1-2k) = -k \Rightarrow m = \dfrac{-k}{1-2k} \Rightarrow m = \dfrac{-k}{-(2k-1)} \Rightarrow m = \dfrac{k}{2k-1}$

4.  (a)   $m = 0.1, \; D = \dfrac{120}{0.3 + 0.1} \Rightarrow D = \dfrac{120}{0.4} \Rightarrow D = 300;$  when the slope of the hill is 0.1, the braking

distance is 300 ft.

(b)   $D = 200, \; \; 200 = \dfrac{120}{0.3 + m} \Rightarrow 200(0.3 + m) = 120 \Rightarrow 0.3 + m = \dfrac{120}{200} \Rightarrow 0.3 + m = 0.6 \Rightarrow$

$m = 0.3;$  the braking distance is 200 ft when the slope of the road is 0.3.


## Section 7.7 Proportions and Variation

1.  A statement that two ratios are equal

3.  It doubles

5.  constant

7.  Directly; doubling the number being fed doubles the bill.

9.  inverse

11.  $\dfrac{x}{24} = \dfrac{5}{8} \Rightarrow 120 = 8x \Rightarrow \dfrac{120}{8} = x \Rightarrow x = 15$

13.  $\dfrac{14}{x} = \dfrac{2}{3} \Rightarrow 42 = 2x \Rightarrow \dfrac{42}{2} = x \Rightarrow x = 21$

15.  $\dfrac{3}{16} = \dfrac{h}{256} \Rightarrow 768 = 16h \Rightarrow \dfrac{768}{16} = h \Rightarrow h = 48$

17.  $\dfrac{3}{4} = \dfrac{2x}{7} \Rightarrow 21 = 8x \Rightarrow \dfrac{21}{8} = x \Rightarrow x = \dfrac{21}{8}$

19.  $\dfrac{x}{6} = \dfrac{8}{3x} \Rightarrow 3x^2 = 48 \Rightarrow x^2 = 16 \Rightarrow x^2 - 16 = 0 \Rightarrow (x - 4)(x + 4) = 0$

$\Rightarrow x - 4 = 0, x = 4 \text{ and } x + 4 = 0, \; x = -4$

21.  $\dfrac{x}{7} = \dfrac{7}{4x} \Rightarrow 4x^2 = 49 \Rightarrow x^2 = \dfrac{49}{4} \Rightarrow x^2 - \dfrac{49}{4} = 0$

$\Rightarrow \left(x - \dfrac{7}{2}\right)\left(x + \dfrac{7}{2}\right) = 0 \Rightarrow x - \dfrac{7}{2} = 0, \; x = \dfrac{7}{2} \text{ and } x + \dfrac{7}{2} = 0, \; x = -\dfrac{7}{2}$

23.  $\dfrac{a}{b} = \dfrac{c}{d} \Rightarrow ad = bc \Rightarrow \dfrac{ad}{c} = b \text{ or } b = \dfrac{ad}{c}$

25.  (a)   $\dfrac{5}{8} = \dfrac{9}{x}$

(b)   $\dfrac{5}{8} = \dfrac{9}{x} \Rightarrow 72 = 5x \Rightarrow \dfrac{72}{5} = x \Rightarrow x = \dfrac{72}{5}$

27.  (a)   $\dfrac{4}{8} = \dfrac{10}{x}$

(b)  $\dfrac{4}{8} = \dfrac{10}{x} \Rightarrow 4x = 80 \Rightarrow x = \dfrac{80}{4} \Rightarrow x = 20$

29. (a)  $\dfrac{98}{7} = \dfrac{x}{11}$

    (b)  $\dfrac{98}{7} = \dfrac{x}{11} \Rightarrow 1078 = 7x \Rightarrow \dfrac{1078}{7} = x \Rightarrow x = \$154$

31. (a)  $\dfrac{3}{750} = \dfrac{7}{x}$

    (b)  $\dfrac{3}{750} = \dfrac{7}{x} \Rightarrow 3x = 5250 \Rightarrow x = \dfrac{5250}{3} \Rightarrow x = 1750$ songs

33. (a)  $k = \dfrac{y}{x},\ y = 4,\ x = 2 \Rightarrow k = \dfrac{4}{2} = 2$

    (b)  $y = kx,\ k = 2,\ x = 6 \Rightarrow y = (2)(6) = 12$

35. (a)  $k = \dfrac{y}{x},\ y = 3,\ x = 2 \Rightarrow k = \dfrac{3}{2}$

    (b)  $y = kx,\ k = \dfrac{3}{2},\ x = 6 \Rightarrow y = \left(\dfrac{3}{2}\right)(6) = 9$

37. (a)  $k = \dfrac{y}{x},\ y = -60,\ x = 8 \Rightarrow k = \dfrac{-60}{8} = -\dfrac{15}{2}$

    (b)  $y = kx,\ k = -\dfrac{15}{2},\ x = 6 \Rightarrow y = \left(-\dfrac{15}{2}\right)(6) = -45$

39. (a)  $k = yx,\ y = 6,\ x = 4 \Rightarrow k = 6 \cdot 4 = 24$

    (b)  $y = \dfrac{k}{x},\ k = 24,\ x = 8 \Rightarrow y = \dfrac{24}{8} = 3$

41. (a)  $k = yx,\ y = 80\ x = \dfrac{1}{2} \Rightarrow k = 80 \cdot \dfrac{1}{2} = \dfrac{80}{2} = 40$

    (b)  $y = \dfrac{k}{x},\ k = 40,\ x = 8 \Rightarrow y = \dfrac{40}{8} = 5$

43. (a)  $k = yx,\ y = 20,\ x = 20 \Rightarrow k = 20 \cdot 20 = 400$

    (b)  $y = \dfrac{k}{x},\ k = 400,\ x = 8 \Rightarrow y = \dfrac{400}{8} = 50$

45. (a)  $z = kxy \Rightarrow 6 = 3(8)k \Rightarrow 6 = 24k \Rightarrow k = 0.25$

    (b)  $z = 0.25xy \Rightarrow z = 0.25(5)(7) = 8.75$

47. (a)  $z = kxy \Rightarrow 5775 = 25(21)k \Rightarrow 5775 = 525k \Rightarrow k = 11$

    (b)  $z = 11xy \Rightarrow z = 11(5)(7) = 385$

49. (a)   $z = kxy \Rightarrow 25 = \dfrac{1}{2}(5)k \Rightarrow 25 = \dfrac{5}{2}k \Rightarrow k = 10$

    (b)   $z = 10xy \Rightarrow z = 10(5)(7) = 350$

51. (a)   Direct, because as $x$ increases, $y$ increases and $k = \dfrac{y}{x} \Rightarrow k = \dfrac{3}{2} = \dfrac{4.5}{3} = \dfrac{6}{4}$, etc.

    (b)   $k = \dfrac{y}{x} \Rightarrow k = \dfrac{3}{2} \Rightarrow y = kx \Rightarrow y = \dfrac{3}{2}x$

    (c)   See Figure 51.

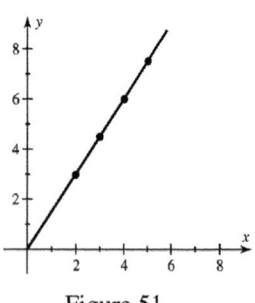
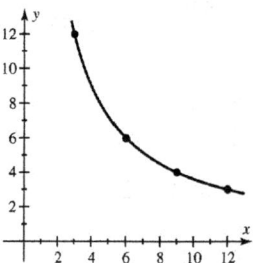

Figure 51                                    Figure 53

53. (a)   Inverse, because as $x$ increases, $y$ decreases and $k = xy \Rightarrow k = 3 \cdot 12 = 6 \cdot 6 = 9 \cdot 4$, etc.

    (b)   $k = xy \Rightarrow k = 3 \cdot 12 = 36 \Rightarrow y = \dfrac{k}{x} \Rightarrow y = \dfrac{36}{x}$

    (c)   See Figure 53.

55. (a)   Neither, because neither the products $k = xy \Rightarrow 4(10) \neq 6(20)$ nor the ratios , but

          $k = \dfrac{y}{x} \Rightarrow \dfrac{10}{4} \neq \dfrac{40}{20}$ are constant.

    (b)   NA

    (c)   NA

57. Direct, $k = \dfrac{y}{x} \Rightarrow k = \dfrac{6}{3} \Rightarrow k = 2$

59. Neither

61. Inverse, $k = xy \Rightarrow k = 4 \cdot 2 \Rightarrow k = 8$

63. $\dfrac{85}{750} = \dfrac{x}{420} \Rightarrow 750x = 35,700 \Rightarrow x = \dfrac{35,700}{750} \Rightarrow x = 47.6$ minutes

65. $\dfrac{8}{1} = \dfrac{13}{x} \Rightarrow 8x = 13 \Rightarrow x = \dfrac{13}{8} \Rightarrow x = 1.625$ inches

67. $y = kx,\ x = 6.2,\ y = 2800 \Rightarrow 2800 = k \cdot 6.2 \Rightarrow k \approx 451.613;$

    $y = 451.613x,\ x = 4.7 \Rightarrow y = 451.613(4.7) \Rightarrow y \approx 2123$ lb

69. $\dfrac{2\frac{2}{3}}{14} = \dfrac{x}{49} \Rightarrow \dfrac{8}{3} \cdot 49 = 14x \Rightarrow \dfrac{392}{3} = 14x \Rightarrow \dfrac{1}{14} \cdot \dfrac{392}{3} = \dfrac{1}{14} \cdot 14x \Rightarrow x = \dfrac{28}{3}$ or $9\frac{1}{3}$c

71. (a)    Direct, the ratios $\dfrac{R}{W}$ always equal 0.012

    (b)    $\dfrac{R}{W} = 0.012 \Rightarrow R = 0.012W$. See Figure 71.

    (c)    $W = 3200,\ R = 0.012(3200) \Rightarrow R = 38.4$ pounds.

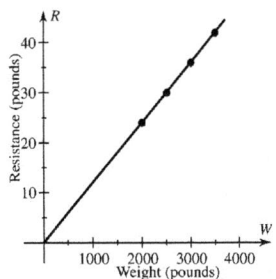   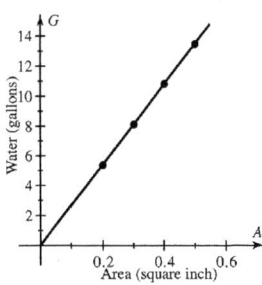

Figure 71                                                 Figure 73

73. (a)    Direct, the ratios $\dfrac{G}{A}$ always equal 27.

    (b)    $\dfrac{G}{A} = 27 \Rightarrow G = 27A$. See Figure 73.

    (c)    For each square-inch increase in the cross-sectional area of the hose, the flow increases by 27 gal/min.

75. (a)    $k = FL \Rightarrow k = 120 \cdot 10 \Rightarrow k = 1200 \Rightarrow F = \dfrac{k}{L} \Rightarrow F = \dfrac{1200}{L}$

    (b)    $L = 15 \Rightarrow F = \dfrac{1200}{15} \Rightarrow F = 80$ pounds

77. (a)    Direct

    (b)    $k = \dfrac{y}{x} \Rightarrow k = \dfrac{-95}{5} \Rightarrow k = -19 \Rightarrow y = kx \Rightarrow y = -19x$

    (c)    Negative, for each 1-mile increase in altitude the temperature decreases by 19°F.

    (d)    $y = kx,\ k = -19,\ x = 2.5,\ y = -19 \cdot 2.5 \Rightarrow y = -47.5 \Rightarrow 47.5°$F decrease

79.  $\dfrac{30}{3} = \dfrac{18}{x} \Rightarrow 30x = 54 \Rightarrow x = \dfrac{54}{30} \Rightarrow x = 1.8$ ohms

81.  $z = kx^2 y^3 \Rightarrow 31.9 = (2)^2 (2.5)^3 k \Rightarrow 31.9 = 62.5k \Rightarrow k = 0.5104$ and so $z = 0.5104 x^2 y^3$

83.  $S = kwt^2 \Rightarrow 300 = k(5)(3)^2 \Rightarrow 300 = 45k \Rightarrow k = \dfrac{20}{3}$ and so $S = \dfrac{20}{3} wt^2$

    When $w = 5$ and $t = 2,\ S = \dfrac{20}{3}(5)(2)^2 \approx 133$ lb.

85.  $W_m = kW_E \Rightarrow 28 = 175k \Rightarrow k = 0.16;$ so $W_m = 0.16W_E$

    A 220-pound person would weigh $W_m = 0.16(220) = 35.2$ pounds on the moon.

87. $V = kIR \Rightarrow 220 = (10)(22)k \Rightarrow k = 1$; so $V = IR$

When $I = 15$ and $R = 50$ the voltage is $V = (15)(50) = 750$.

## Checking Basic Concepts  Section 7.7

1. (a) $\dfrac{x}{9} = \dfrac{2}{5} \Rightarrow 5x = 18 \Rightarrow x = \dfrac{18}{5}$

   (b) $\dfrac{4}{3} = \dfrac{5}{b} \Rightarrow 4b = 15 \Rightarrow b = \dfrac{15}{4}$

2. (a) $\dfrac{4}{6} = \dfrac{8}{x} \Rightarrow 4x = 48 \Rightarrow x = \dfrac{48}{4} \Rightarrow x = 12$

   (b) $\dfrac{2}{148} = \dfrac{5}{x} \Rightarrow 2x = 740 \Rightarrow x = \dfrac{740}{2} \Rightarrow x = 370$ minutes

3. $k = xy,\ x = 15,\ y = 4 \Rightarrow k = 15 \cdot 4 \Rightarrow k = 60;\ y = \dfrac{k}{x},\ k = 60,\ x = 10 \Rightarrow y = \dfrac{60}{10} \Rightarrow y = 6$

4. (a) Direct, the ratios $\dfrac{y}{x}$ always equal $\dfrac{3}{2},\ \dfrac{3}{2};\ \dfrac{x}{y} = \dfrac{3}{2},\ \dfrac{6}{4} = \dfrac{3}{2},\ \dfrac{9}{6} = \dfrac{3}{2},\ \dfrac{12}{8} = \dfrac{3}{2}$

   (b) Inverse, the products $xy$ always equal 24,

   $xy = 24,\ 2 \cdot 12 = 24,\ 4 \cdot 6 = 24,\ 6 \cdot 4 = 24,\ 8 \cdot 3 = 24$

5. $\dfrac{272}{17} = \dfrac{x}{10} \Rightarrow 2720 = 17x \Rightarrow \dfrac{2720}{17} = x \Rightarrow x = 160 \Rightarrow x = \$160$

## Chapter 7 Review

1. $x = -2,\ \dfrac{3}{x-3} = \dfrac{3}{-2-3} = \dfrac{3}{-5} = -\dfrac{3}{5}$

2. $x = 3,\ \dfrac{4x}{5-x^2} = \dfrac{4(3)}{5-(3)^2} = \dfrac{12}{5-9} = \dfrac{12}{-4} = -3$

3. $x = 7,\ \dfrac{-x}{7-x} = \dfrac{-(7)}{7-(7)} = \dfrac{-7}{0} \Rightarrow$ undefined

4. $x = 2,\ \dfrac{4x}{x^2-3x+2} = \dfrac{4(2)}{(2)^2-3(2)+2} = \dfrac{8}{4-6+2} = \dfrac{8}{0} \Rightarrow$ undefined

5. $x = -2$, $\dfrac{3x}{x-1} = \dfrac{3(-2)}{-2-1} = \dfrac{-6}{-3} = 2$;  $x = -1$, $\dfrac{3x}{x-1} = \dfrac{3(-1)}{-1-1} = \dfrac{-3}{-2} = \dfrac{3}{2}$;

$x = 0$, $\dfrac{3x}{x-1} = \dfrac{3(0)}{0-1} = \dfrac{0}{-1} = 0$ ;  $x = 1$, $\dfrac{3x}{x-1} = \dfrac{3(1)}{1-1} = \dfrac{3}{0} \Rightarrow$ undefined ;

$x = 2$, $\dfrac{3x}{x-1} = \dfrac{3(2)}{2-1} = \dfrac{6}{1} = 6$ .  See Figure 5.

| $x$ | -2 | -1 | 0 | 1 | 2 |
|---|---|---|---|---|---|
| $\dfrac{3x}{x-1}$ | 2 | $\dfrac{3}{2}$ | 0 | — | 6 |

Figure 5

6. $\dfrac{8}{x^2-4} \Rightarrow x^2 - 4 \neq 0 \Rightarrow (x-2)(x+2) \neq 0 \Rightarrow x - 2 \neq 0 \Rightarrow x \neq 2$ or $x + 2 \neq 0 \Rightarrow x \neq -2$

Therefore $x = 2,\ -2$ make the expression undefined.

7. $\dfrac{25x^3 y^4}{15x^5 y} = \dfrac{5}{3} x^{3-5} y^{4-1} = \dfrac{5}{3} x^{-2} y^3 = \dfrac{5y^3}{3x^2}$

8. $\dfrac{x^2 - 36}{x+6} = \dfrac{(x+6)(x-6)}{x+6} = \dfrac{x+6}{x+6} \cdot \dfrac{x-6}{1} = 1 \cdot (x-6) = x - 6$

9. $\dfrac{x-9}{9-x} = \dfrac{x-9}{-1(-9+x)} = \dfrac{x-9}{-1(x-9)} = -1$

10. $\dfrac{x^2 - 5x}{5x} = \dfrac{x(x-5)}{x(5)} = \dfrac{x}{x} \cdot \dfrac{x-5}{5} = 1 \cdot \dfrac{x-5}{5} = \dfrac{x-5}{5}$

11. $\dfrac{2x^2 + 5x - 3}{2x^2 + x - 1} = \dfrac{(x+3)(2x-1)}{(x+1)(2x-1)} = \dfrac{x+3}{x+1} \cdot \dfrac{2x-1}{2x-1} = \dfrac{x+3}{x+1} \cdot 1 = \dfrac{x+3}{x+1}$

12. $\dfrac{3x^2 + 10x - 8}{3x^2 + x - 2} = \dfrac{(x+4)(3x-2)}{(x+1)(3x-2)} = \dfrac{x+4}{x+1} \cdot \dfrac{3x-2}{3x-2} = \dfrac{x+4}{x+1} \cdot 1 = \dfrac{x+4}{x+1}$

13. (a)    expression

    (b)    $\dfrac{x+1}{(x+1)(x-3)} = \dfrac{x+1}{x+1} \cdot \dfrac{1}{x-3} = 1 \cdot \dfrac{1}{x-3} = \dfrac{1}{x-3}$

14. (a)    equation

    (b)    $\dfrac{x}{3(4-1)} = 2 \Rightarrow \dfrac{x}{9} = 2 \Rightarrow x = 2 \cdot 9 \Rightarrow x = 18$

15. $\dfrac{x-3}{x+1} \cdot \dfrac{2x+2}{x-3} = \dfrac{x-3}{x+1} \cdot \dfrac{2(x+1)}{x-3} = \dfrac{x-3}{x-3} \cdot \dfrac{x+1}{x+1} \cdot 2 = 1 \cdot 1 \cdot 2 = 2$

16. $\dfrac{2x+5}{(x+5)(x-1)} \cdot \dfrac{x-1}{2x+5} = \dfrac{2x+5}{2x+5} \cdot \dfrac{x-1}{x-1} \cdot \dfrac{1}{x+5} = 1 \cdot 1 \cdot \dfrac{1}{x+5} = \dfrac{1}{x+5}$

17. $\dfrac{z+3}{z-4} \cdot \dfrac{z-4}{(z+3)^2} = \dfrac{z+3}{z+3} \cdot \dfrac{1}{z+3} \cdot \dfrac{z-4}{z-4} = 1 \cdot \dfrac{1}{z+3} \cdot 1 = \dfrac{1}{z+3}$

18. $\dfrac{x^2}{x^2-4}\cdot\dfrac{x+2}{x}=\dfrac{x(x)}{(x-2)(x+2)}\cdot\dfrac{x+2}{x}=\dfrac{x}{x}\cdot\dfrac{x+2}{x+2}\cdot\dfrac{x}{x-2}=1\cdot1\cdot\dfrac{x}{x-2}=\dfrac{x}{x-2}$

19. $\dfrac{x+1}{2x}\div\dfrac{3x+3}{5x}=\dfrac{x+1}{2x}\cdot\dfrac{5x}{3(x+1)}=\dfrac{5}{6}\cdot\dfrac{x+1}{x+1}\cdot\dfrac{x}{x}=\dfrac{5}{6}\cdot1\cdot1=\dfrac{5}{6}$

20. $\dfrac{4}{x^3}\div\dfrac{x+1}{2x^2}=\dfrac{4}{x^3}\cdot\dfrac{2x^2}{x+1}=\dfrac{8}{x}\cdot\dfrac{x^2}{x^2}\cdot\dfrac{1}{x+1}=\dfrac{8}{x}\cdot1\cdot\dfrac{1}{x+1}=\dfrac{8}{x(x+1)}$

21. $\dfrac{x-5}{x+2}\div\dfrac{2x-10}{x+2}=\dfrac{x-5}{x+2}\cdot\dfrac{x+2}{2(x-5)}=\dfrac{1}{2}\cdot\dfrac{x-5}{x-5}\cdot\dfrac{x+2}{x+2}=\dfrac{1}{2}\cdot1\cdot1=\dfrac{1}{2}$

22. $\dfrac{x^2-6x+5}{x^2-25}\div\dfrac{x-1}{x+5}=\dfrac{(x-1)(x-5)}{(x-5)(x+5)}\cdot\dfrac{x+5}{x-1}=\dfrac{x-1}{x-1}\cdot\dfrac{x+5}{x+5}\cdot\dfrac{x-5}{x-5}=1\cdot1\cdot1=1$

23. $\dfrac{x^2-y^2}{x+y}\div\dfrac{x-y}{x+y}=\dfrac{(x-y)(x+y)}{x+y}\cdot\dfrac{x+y}{x-y}=\dfrac{x-y}{x-y}\cdot\dfrac{x+y}{x+y}\cdot(x+y)=1\cdot1\cdot(x+y)=x+y$

24. $\dfrac{a^3-b^3}{a+b}\div\dfrac{a-b}{2a+2b}=\dfrac{(a-b)\left(a^2+ab+b^2\right)}{a+b}\cdot\dfrac{2(a+b)}{a-b}=\dfrac{a-b}{a-b}\cdot\dfrac{a+b}{a+b}\cdot\dfrac{2\left(a^2+ab+b^2\right)}{1}=$

$1\cdot1\cdot2\left(a^2+ab+b^2\right)=2\left(a^2+ab+b^2\right)$

25. $\dfrac{2}{x+10}+\dfrac{8}{x+10}=\dfrac{2+8}{x+10}=\dfrac{10}{x+10}$

26. $\dfrac{9}{x-1}-\dfrac{8}{x-1}=\dfrac{9-8}{x-1}=\dfrac{1}{x-1}$

27. $\dfrac{x+2y}{2x}+\dfrac{x-2y}{2x}=\dfrac{(x+2y)+(x-2y)}{2x}=\dfrac{2x}{2x}=1$

28. $\dfrac{x}{x+3}+\dfrac{3}{x+3}=\dfrac{x+3}{x+3}=1$

29. $\dfrac{x}{x^2-1}-\dfrac{1}{x^2-1}=\dfrac{x-1}{x^2-1}=\dfrac{x-1}{(x-1)(x+1)}=\dfrac{x-1}{x-1}\cdot\dfrac{1}{x+1}=1\cdot\dfrac{1}{x+1}=\dfrac{1}{x+1}$

30. $\dfrac{2x}{x^2-25}+\dfrac{10}{x^2-25}=\dfrac{2x+10}{x^2-25}=\dfrac{2(x+5)}{(x-5)(x+5)}=\dfrac{2}{x-5}\cdot\dfrac{x+5}{x+5}=\dfrac{2}{x-5}\cdot1=\dfrac{2}{x-5}$

31. $\dfrac{3}{xy}-\dfrac{1}{xy}=\dfrac{3-1}{xy}=\dfrac{2}{xy}$

32. $\dfrac{x+y}{2y}+\dfrac{x-y}{2y}=\dfrac{x+y+x-y}{2y}=\dfrac{2x}{2y}=\dfrac{2}{2}\cdot\dfrac{x}{y}=1\cdot\dfrac{x}{y}=\dfrac{x}{y}$

33. $3x=3\cdot x$ and $5x=5\cdot x\Rightarrow3\cdot5\cdot x=15x$

34. $5x^2=5\cdot x^2$ and $10x=2\cdot5\cdot x\Rightarrow2\cdot5\cdot x^2=10x^2$

35. $x$ and $x-5$ are both prime $\Rightarrow x(x-5)$

36. $10x^2 = 2 \cdot 5 \cdot x^2$ and $x^2 - x = x(x-1) \Rightarrow 2 \cdot 5 \cdot x^2 \cdot (x-1) = 10x^2(x-1)$

37. $x^2 - 1 = (x+1)(x-1)$ and $(x+1)^2 \Rightarrow (x+1)^2(x-1)$

38. $x^2 - 4x = x(x-4)$ and $x^2 - 16 = (x+4)(x-4) \Rightarrow x(x-4)(x+4)$

39. $\dfrac{3}{8}$, $D = 24 \Rightarrow \dfrac{3}{8} \cdot \dfrac{3}{3} = \dfrac{9}{24}$

40. $\dfrac{4}{3x}$, $D = 12x \Rightarrow \dfrac{4}{3x} \cdot \dfrac{4}{4} = \dfrac{16}{12x}$

41. $\dfrac{3x}{x-2}$, $D = x^2 - 4 \Rightarrow \dfrac{3x}{x-2} \cdot \dfrac{x+2}{x+2} = \dfrac{3x^2 + 6}{x^2 - 4}$

42. $\dfrac{2}{x+1}$, $D = x^2 + x \Rightarrow \dfrac{2}{x+1} \cdot \dfrac{x}{x} = \dfrac{2x}{x^2 + x}$

43. $\dfrac{3}{5x}$, $D = 5x^2 - 5x \Rightarrow \dfrac{3}{5x} \cdot \dfrac{x-1}{x-1} = \dfrac{3x-3}{5x^2 - 5x}$

44. $\dfrac{2x}{2x-3}$, $D = 2x^2 + x - 6 \Rightarrow \dfrac{2x}{2x-3} \cdot \dfrac{x+2}{x+2} = \dfrac{2x^2 + 4x}{2x^2 + x - 6}$

45. $\dfrac{5}{8} + \dfrac{1}{6} = \dfrac{5}{8} \cdot \dfrac{3}{3} + \dfrac{1}{6} \cdot \dfrac{4}{4} = \dfrac{15}{24} + \dfrac{4}{24} = \dfrac{19}{24}$

46. $\dfrac{3}{4x} + \dfrac{1}{x} = \dfrac{3}{4x} + \dfrac{1}{x} \cdot \dfrac{4}{4} = \dfrac{3}{4x} + \dfrac{4}{4x} = \dfrac{7}{4x}$

47. $\dfrac{5}{9x} - \dfrac{2}{3x} = \dfrac{5}{9x} - \dfrac{2}{3x} \cdot \dfrac{3}{3} = \dfrac{5}{9x} - \dfrac{6}{9x} = \dfrac{-1}{9x} = -\dfrac{1}{9x}$

48. $\dfrac{7}{x-1} - \dfrac{3}{x} = \dfrac{7}{x-1} \cdot \dfrac{x}{x} - \dfrac{3}{x} \cdot \dfrac{x-1}{x-1} = \dfrac{7x}{x(x-1)} - \dfrac{3x-3}{x(x-1)} = \dfrac{4x+3}{x(x-1)}$

49. $\dfrac{1}{x+1} + \dfrac{1}{x-1} = \dfrac{1}{x+1} \cdot \dfrac{x-1}{x-1} + \dfrac{1}{x-1} \cdot \dfrac{x+1}{x+1} = \dfrac{x-1}{(x+1)(x-1)} + \dfrac{x+1}{(x+1)(x-1)} = \dfrac{2x}{(x-1)(x+1)}$

50. $\dfrac{4}{3x^2} - \dfrac{3}{2x} = \dfrac{4}{3x^2} \cdot \dfrac{2}{2} - \dfrac{3}{2x} \cdot \dfrac{3x}{3x} = \dfrac{8}{6x^2} - \dfrac{9x}{6x^2} = \dfrac{8-9x}{6x^2}$

51. $\dfrac{1+x}{3x} - \dfrac{3}{2x} = \dfrac{1+x}{3x} \cdot \dfrac{2}{2} - \dfrac{3}{2x} \cdot \dfrac{3}{3} = \dfrac{2+2x}{6x} - \dfrac{9}{6x} = \dfrac{2x-7}{6x}$

52. $\dfrac{x}{x^2 - 1} - \dfrac{1}{x-1} = \dfrac{x}{(x+1)(x-1)} - \dfrac{1}{x-1} \cdot \dfrac{x+1}{x+1} = \dfrac{x}{(x+1)(x-1)} - \dfrac{x+1}{(x+1)(x-1)} = -\dfrac{1}{(x-1)(x+1)}$

53. $\dfrac{2}{x-y} - \dfrac{3}{x+y} = \dfrac{2}{x-y} \cdot \dfrac{x+y}{x+y} - \dfrac{3}{x+y} \cdot \dfrac{x-y}{x-y} = \dfrac{2x+2y}{(x-y)(x+y)} - \dfrac{3x-3y}{(x-y)(x+y)} = \dfrac{5y-x}{(x-y)(x+y)}$

54. $\dfrac{2}{x} - \dfrac{1}{2x} + \dfrac{2}{3x} = \dfrac{2}{x} \cdot \dfrac{6}{6} - \dfrac{1}{2x} \cdot \dfrac{3}{3} + \dfrac{2}{3x} \cdot \dfrac{2}{2} = \dfrac{12}{6x} - \dfrac{3}{6x} + \dfrac{4}{6x} = \dfrac{12-3+4}{6x} = \dfrac{12+1}{6x} = \dfrac{13}{6x}$

55. $\dfrac{3}{2y}+\dfrac{1}{2x}=\dfrac{3}{2y}\cdot\dfrac{x}{x}+\dfrac{1}{2x}\cdot\dfrac{y}{y}=\dfrac{3x}{2xy}+\dfrac{y}{2xy}=\dfrac{3x+y}{2xy}$

56. $\dfrac{x}{y-x}+\dfrac{y}{x-y}=\dfrac{x}{y-x}+\dfrac{y}{-1(y-x)}=\dfrac{x}{y-x}+\dfrac{-y}{y-x}=\dfrac{x-y}{y-x}=\dfrac{x-y}{-1(x-y)}=-1\cdot1=-1$

57. $\dfrac{\frac{3}{4}}{\frac{7}{11}}=\dfrac{3}{4}\div\dfrac{7}{11}=\dfrac{3}{4}\cdot\dfrac{11}{7}=\dfrac{33}{28}$

58. $\dfrac{\frac{x}{5}}{\frac{2x}{7}}=\dfrac{x}{5}\div\dfrac{2x}{7}=\dfrac{x}{5}\cdot\dfrac{7}{2x}=\dfrac{7x}{10x}=\dfrac{7}{10}\cdot\dfrac{x}{x}=\dfrac{7}{10}\cdot1=\dfrac{7}{10}$

59. $\dfrac{\frac{m}{n}}{\frac{2m}{n^2}}=\dfrac{m}{n}\div\dfrac{2m}{n^2}=\dfrac{m}{n}\cdot\dfrac{n^2}{2m}=\dfrac{m}{2m}\cdot\dfrac{n^2}{n}=\left(\dfrac{1}{2}\cdot\dfrac{m}{m}\right)\left(\dfrac{n}{n}\cdot n\right)=\left(\dfrac{1}{2}\cdot1\right)\cdot1\cdot n=\dfrac{n}{2}$

60. $\dfrac{\frac{3}{p-1}}{\frac{1}{p+1}}=\dfrac{3}{p-1}\div\dfrac{1}{p+1}=\dfrac{3}{p-1}\cdot\dfrac{p+1}{1}=\dfrac{3(p+1)}{p-1}$

61. $\dfrac{\frac{3}{m-1}}{\frac{2m-2}{m+1}}=\dfrac{3}{m-1}\div\dfrac{2m-2}{m+1}=\dfrac{3}{m-1}\cdot\dfrac{m+1}{2m-2}=\dfrac{3}{m-1}\cdot\dfrac{m+1}{2(m-1)}=\dfrac{3(m+1)}{2(m-1)^2}$

62. $\dfrac{\frac{2}{2n+1}}{\frac{8}{2n-1}}=\dfrac{2}{2n+1}\div\dfrac{8}{2n-1}=\dfrac{2}{2n+1}\cdot\dfrac{2n-1}{8}=\dfrac{2}{8}\cdot\dfrac{2n-1}{2n+1}=\dfrac{2n-1}{4(2n+1)}$

63. $\dfrac{\frac{1}{2x}-\frac{1}{3x}}{\frac{2}{3x}-\frac{1}{6x}}=\dfrac{\frac{3}{6x}-\frac{2}{6x}}{\frac{4}{6x}-\frac{1}{6x}}=\dfrac{\frac{1}{6x}}{\frac{3}{6x}}=\dfrac{1}{6x}\div\dfrac{3}{6x}=\dfrac{1}{6x}\cdot\dfrac{6x}{3}=\dfrac{1}{3}\cdot\dfrac{6x}{6x}=\dfrac{1}{3}\cdot1=\dfrac{1}{3}$

64. $\dfrac{\frac{2}{xy}-\frac{1}{y}}{\frac{2}{xy}+\frac{1}{y}}=\dfrac{\frac{2}{xy}-\frac{1}{y}}{\frac{2}{xy}+\frac{1}{y}}\cdot\dfrac{xy}{xy}=\dfrac{\frac{2xy}{xy}-\frac{xy}{y}}{\frac{2xy}{xy}+\frac{xy}{y}}=\dfrac{2\cdot\frac{xy}{xy}-x\cdot\frac{y}{y}}{2\cdot\frac{xy}{xy}+x\cdot\frac{y}{y}}=\dfrac{2\cdot1-x\cdot1}{2\cdot1+x\cdot1}=\dfrac{2-x}{2+x}$

65. $\dfrac{\frac{1}{x}-\frac{1}{x+1}}{\frac{x}{x+1}}=\dfrac{\frac{x+1}{x(x+1)}-\frac{x}{x(x+1)}}{\frac{x}{x+1}}=\dfrac{\frac{1}{x(x+1)}}{\frac{x}{x+1}}=\dfrac{1}{x(x+1)}\div\dfrac{x}{x+1}=\dfrac{1}{x(x+1)}\cdot\dfrac{x+1}{x}=\dfrac{1}{x\cdot x}\cdot\dfrac{x+1}{x+1}=\dfrac{1}{x^2}\cdot1=\dfrac{1}{x^2}$

66. $\dfrac{\frac{2}{x-1}-\frac{1}{x+1}}{\frac{1}{x^2-1}}=\dfrac{\frac{2x+2}{(x-1)(x+1)}-\frac{x-1}{(x-1)(x+1)}}{\frac{1}{x^2-1}}=\dfrac{\frac{x+3}{x^2-1}}{\frac{1}{x^2-1}}=\dfrac{x+3}{x^2-1}\div\dfrac{1}{x^2-1}=\dfrac{x+3}{x^2-1}\cdot\dfrac{x^2-1}{1}=$

   $\dfrac{x+3}{1}\cdot\dfrac{x^2-1}{x^2-1}=(x+3)\cdot1=x+3$

67. $\dfrac{x}{5}=\dfrac{4}{7}\Rightarrow7x=20\Rightarrow x=\dfrac{20}{7}$   Check: $\dfrac{\frac{20}{7}}{5}=\dfrac{4}{7}\Rightarrow7\left(\dfrac{20}{7}\right)=20\Rightarrow20=20$

68. $\dfrac{4}{x}=\dfrac{3}{2}\Rightarrow3x=8\Rightarrow x=\dfrac{8}{3}$   Check: $\dfrac{4}{\frac{8}{3}}=\dfrac{3}{2}\Rightarrow8=3\left(\dfrac{8}{3}\right)\Rightarrow8=8$

69. $\dfrac{3}{z+1} = \dfrac{1}{2z} \Rightarrow 6z = z+1 \Rightarrow 5z = 1 \Rightarrow z = \dfrac{1}{5}$

Check: $\dfrac{3}{\frac{1}{5}+1} = \dfrac{1}{2\left(\frac{1}{5}\right)} \Rightarrow \dfrac{3}{\frac{6}{5}} = \dfrac{1}{\frac{2}{5}} \Rightarrow \dfrac{3}{1} \cdot \dfrac{5}{6} = \dfrac{1}{1} \cdot \dfrac{5}{2} \Rightarrow \dfrac{15}{6} = \dfrac{5}{2} \Rightarrow \dfrac{5}{2} = \dfrac{5}{2}$

70. $\dfrac{x+2}{x} = \dfrac{3}{5} \Rightarrow 3x = 5x+10 \Rightarrow -2x = 10 \Rightarrow x = \dfrac{10}{-2} \Rightarrow x = -5$

Check: $\dfrac{-5+2}{-5} = \dfrac{3}{5} \Rightarrow \dfrac{-3}{-5} = \dfrac{3}{5} \Rightarrow \dfrac{3}{5} = \dfrac{3}{5}$

71. $\dfrac{1}{x+1} = \dfrac{2}{x-2} \Rightarrow x-2 = 2x+2 \Rightarrow -x = 4 \Rightarrow x = -4$

Check: $\dfrac{1}{-4+1} = \dfrac{2}{-4-2} \Rightarrow \dfrac{1}{-3} = \dfrac{2}{-6} \Rightarrow -\dfrac{1}{3} = -\dfrac{1}{3}$

72. $\dfrac{x}{3} = \dfrac{-1}{x+4} \Rightarrow x^2 + 4x = -3 \Rightarrow x^2 + 4x + 3 = 0 \Rightarrow (x+1)(x+3) = 0 \Rightarrow x+1 = 0 \Rightarrow x = -1,$

$x+3 = 0 \Rightarrow x = -3 \Rightarrow x = -1, \; -3$

Check: $\dfrac{-1}{3} = \dfrac{-1}{-1+4} \Rightarrow -\dfrac{1}{3} = -\dfrac{1}{3}, \; \dfrac{-3}{3} = \dfrac{-1}{-3+4} \Rightarrow -1 = \dfrac{-1}{1} \Rightarrow -1 = -1$

73. $\dfrac{1}{5x} + \dfrac{3}{5x} = \dfrac{1}{5} \Rightarrow \dfrac{4}{5x} = \dfrac{1}{5} \Rightarrow 5x = 20 \Rightarrow x = 4$   Check: $\dfrac{1}{5(4)} + \dfrac{3}{5(4)} = \dfrac{1}{5} \Rightarrow \dfrac{4}{20} = \dfrac{1}{5} \Rightarrow \dfrac{1}{5} = \dfrac{1}{5}$

74. $\dfrac{1}{x-1} + \dfrac{2x}{x-1} = 1 \Rightarrow \dfrac{2x+1}{x-1} = 1 \Rightarrow \dfrac{(2x+1)(x-1)}{x-1} = \dfrac{x-1}{1} \Rightarrow 2x+1 = x-1 \Rightarrow x = -2$

Check: $\dfrac{1}{-2-1} + \dfrac{2(-2)}{-2-1} = 1 \Rightarrow \dfrac{-3}{-3} = 1 \Rightarrow 1 = 1$

75. $\dfrac{1}{x} + \dfrac{2}{3x} = \dfrac{1}{3} \Rightarrow \dfrac{1(3)}{3(x)} + \dfrac{2}{3x} = \dfrac{1(x)}{3(x)} \Rightarrow \dfrac{3}{3x} + \dfrac{2}{3x} = \dfrac{x}{3x} \Rightarrow \dfrac{5}{3x} = \dfrac{x}{3x} \Rightarrow 15x = 3x^2 \Rightarrow \dfrac{15}{3} = \dfrac{x^2}{x} \Rightarrow$

$5 = x \Rightarrow x = 5$   Check: $\dfrac{1}{5} + \dfrac{2}{3(5)} = \dfrac{1}{3} \Rightarrow \dfrac{1}{5} + \dfrac{2}{15} = \dfrac{1}{3} \Rightarrow \dfrac{3}{15} + \dfrac{2}{15} = \dfrac{5}{15} \Rightarrow \dfrac{5}{15} = \dfrac{5}{15} \Rightarrow \dfrac{1}{3} = \dfrac{1}{3}$

76. $\dfrac{1}{x+3} + \dfrac{2x}{x+3} = \dfrac{3}{2} \Rightarrow \dfrac{2x+1}{x+3} = \dfrac{3}{2} \Rightarrow \dfrac{(2x+1)(x+3)(2)}{x+3} = \dfrac{3(x+3)(2)}{2} \Rightarrow$

$\dfrac{(4x+2)(x+3)}{x+3} = \dfrac{(3x+9)(2)}{2} \Rightarrow 4x+2 = 3x+9 \Rightarrow x = 7$

Check: $\dfrac{1}{7+3} + \dfrac{2(7)}{7+3} = \dfrac{3}{2} \Rightarrow \dfrac{15}{10} = \dfrac{3}{2} \Rightarrow \dfrac{3}{2} = \dfrac{3}{2}$

77. $\dfrac{5}{x} - \dfrac{3}{x+1} = \dfrac{1}{2} \Rightarrow \dfrac{5(x)(x+1)(2)}{x} - \dfrac{3(x)(x+1)(2)}{x+1} = \dfrac{1(x)(x+1)(2)}{2} \Rightarrow$

$\dfrac{(10x+10)(x)}{x} - \dfrac{6x(x+1)}{x+1} = \dfrac{\left(x^2+x\right)(2)}{2} \Rightarrow 10x+10-6x = x^2+x \Rightarrow 4x+10 = x^2+x \Rightarrow$

$x^2 - 3x - 10 = 0 \Rightarrow (x-5)(x+2) = 0,\ x-5 = 0 \Rightarrow x = 5,\ x+2 = 0 \Rightarrow x = -2 \Rightarrow x = 5,\ -2$

Check: $\dfrac{5}{-2} - \dfrac{3}{-2+1} = \dfrac{1}{2} \Rightarrow \dfrac{5}{-2} - \dfrac{3}{-1} = \dfrac{1}{2} \Rightarrow \dfrac{5}{-2} - \dfrac{6}{-2} = \dfrac{1}{2} \Rightarrow \dfrac{-1}{-2} = \dfrac{1}{2} \Rightarrow \dfrac{1}{2} = \dfrac{1}{2}$

Check: $\dfrac{5}{5} - \dfrac{3}{5+1} = \dfrac{1}{2} \Rightarrow \dfrac{5}{5} - \dfrac{3}{6} = \dfrac{1}{2} \Rightarrow 1 - \dfrac{1}{2} = \dfrac{1}{2} \Rightarrow \dfrac{1}{2} = \dfrac{1}{2}$

78. $\dfrac{1}{x-1} - \dfrac{1}{x+1} = \dfrac{1}{4} \Rightarrow \dfrac{1(x-1)(x+1)(4)}{x-1} - \dfrac{1(x-1)(x+1)(4)}{x+1} = \dfrac{1(x-1)(x+1)(4)}{4} \Rightarrow$

$\dfrac{(4x+4)(x-1)}{x-1} - \dfrac{(4x-4)(x+1)}{x+1} = \dfrac{\left(x^2-1\right)(4)}{4} \Rightarrow 4x+4-(4x-4) = x^2-1 \Rightarrow 8 = x^2-1 \Rightarrow$

$9 = x^2 \Rightarrow \pm 3 = x \Rightarrow x = -3,\ 3$

Check: $\dfrac{1}{-3-1} - \dfrac{1}{-3+1} = \dfrac{1}{4} \Rightarrow \dfrac{1}{-4} - \dfrac{1}{-2} = \dfrac{1}{4} \Rightarrow \dfrac{1}{-4} - \dfrac{2}{-4} = \dfrac{1}{4} \Rightarrow \dfrac{-1}{-4} = \dfrac{1}{4} \Rightarrow \dfrac{1}{4} = \dfrac{1}{4}$

Check: $\dfrac{1}{3-1} - \dfrac{1}{3+1} = \dfrac{1}{4} \Rightarrow \dfrac{1}{2} - \dfrac{1}{4} = \dfrac{1}{4} \Rightarrow \dfrac{2}{4} - \dfrac{1}{4} = \dfrac{1}{4} \Rightarrow \dfrac{1}{4} = \dfrac{1}{4}$

79. $\dfrac{4}{p} - \dfrac{5}{p+2} = 0 \Rightarrow \dfrac{4(p)(p+2)}{p} - \dfrac{5(p)(p+2)}{p+2} = 0(p)(p+2) \Rightarrow \dfrac{(4p+8)(p)}{p} - \dfrac{5p(p+2)}{p+2} = 0 \Rightarrow$

$4p+8-5p = 0 \Rightarrow -p+8 = 0 \Rightarrow -p = -8 \Rightarrow p = 8$

Check: $\dfrac{4}{8} - \dfrac{5}{8+2} = 0 \Rightarrow \dfrac{4}{8} - \dfrac{5}{10} = 0 \Rightarrow \dfrac{1}{2} - \dfrac{1}{2} = 0 \Rightarrow 0 = 0$

80. $\dfrac{1}{x-3} - \dfrac{1}{x+3} = \dfrac{1}{x^2-9} \Rightarrow \dfrac{1}{x-3} \cdot \dfrac{x+3}{x+3} - \dfrac{1}{x+3} \cdot \dfrac{x-3}{x-3} = \dfrac{1}{(x+3)(x-3)} \Rightarrow$

$\dfrac{x+3}{(x-3)(x+3)} - \dfrac{x-3}{(x-3)(x+3)} = \dfrac{1}{(x-3)(x+3)} \Rightarrow 6 = 1$, which is false, so no solution.

81. $\dfrac{1}{x+1} = \dfrac{-x}{x+1} \Rightarrow x+1 = -x(x+1) \Rightarrow x+1 = -x^2-x$

$\Rightarrow x^2 + 2x + 1 = 0 \Rightarrow (x+1)^2 = 0 \Rightarrow x+1 = 0,\ x = -1$

Check: $-1$ is extraneous because $\dfrac{1}{-1+1}$ is undefined. There are no solutions.

82. $\dfrac{2}{x} = \dfrac{2}{x^2+x} - 4 \Rightarrow \dfrac{2}{x} = \dfrac{2-4\left(x^2+x\right)}{x^2+x} \Rightarrow \dfrac{2}{x} = \dfrac{2-4x^2-4x}{x^2+x}$

$\Rightarrow 2\left(x^2+x\right) = x\left(2-4x^2-4x\right) \Rightarrow 2x^2+2x = 2x-4x^3-4x^2$

$\Rightarrow 4x^3+6x^2 = 0 \Rightarrow 2x^2\left(2x+3\right) = 0 \Rightarrow x = 0 \text{ and } 2x+3 = 0, \ x = -\dfrac{3}{2}$

Check: 0 is extraneous because $\dfrac{2}{0}$ is undefined.

$\dfrac{2}{-\frac{3}{2}} = \dfrac{2}{\left(-\frac{3}{2}\right)^2+\left(-\frac{3}{2}\right)} - 4 \Rightarrow 2 \div \left(-\dfrac{3}{2}\right) = \dfrac{2}{\frac{9}{4}-\frac{3}{2}} - 4 \Rightarrow 2 \cdot \left(-\dfrac{2}{3}\right) = \dfrac{2}{\frac{3}{4}} - 4$

$\Rightarrow -\dfrac{4}{3} = 2 \div \dfrac{3}{4} - 4 \Rightarrow -\dfrac{4}{3} = 2 \cdot \dfrac{4}{3} - 4 \Rightarrow -\dfrac{4}{3} = \dfrac{8}{3} - \dfrac{12}{3} \Rightarrow -\dfrac{4}{3} = -\dfrac{4}{3}$ The solution is $-\dfrac{3}{2}$.

83. $\dfrac{2}{x^2-2x} + \dfrac{1}{x^2-4} = \dfrac{1}{x^2+2x} \Rightarrow \dfrac{2(x)(x-2)(x+2)}{x(x-2)} + \dfrac{1(x)(x-2)(x+2)}{(x-2)(x+2)} = \dfrac{1(x)(x-2)(x+2)}{x(x+2)} \Rightarrow$

$\dfrac{(2x+4)(x)(x-2)}{x(x-2)} + \dfrac{x\left(x^2-4\right)}{x^2-4} = \dfrac{(x-2)(x)(x+2)}{x(x+2)} \Rightarrow 2x+4+x = x-2 \Rightarrow 3x+4 = x-2 \Rightarrow$

$2x = -6 \Rightarrow x = -3$ Check: $\dfrac{2}{(-3)^2-2(-3)} + \dfrac{1}{(-3)^2-4} = \dfrac{1}{(-3)^2+2(-3)} \Rightarrow \dfrac{2}{15} + \dfrac{1}{5} = \dfrac{1}{3} \Rightarrow$

$\dfrac{2}{15} + \dfrac{3}{15} = \dfrac{5}{15} \Rightarrow \dfrac{5}{15} = \dfrac{5}{15} \Rightarrow \dfrac{1}{3} = \dfrac{1}{3}$

84. $\dfrac{3}{x^2-3x} - \dfrac{1}{x^2-9} = \dfrac{1}{x^2+3x} \Rightarrow \dfrac{3(x)(x-3)(x+3)}{x(x-3)} - \dfrac{1(x)(x-3)(x+3)}{(x-3)(x+3)} = \dfrac{1(x)(x-3)(x+3)}{x(x+3)} \Rightarrow$

$\dfrac{(3x+9)(x)(x-3)}{x(x-3)} - \dfrac{x\left(x^2-9\right)}{x^2-9} = \dfrac{(x-3)(x)(x+3)}{x(x+3)} \Rightarrow 3x+9-x = x-3 \Rightarrow 2x+9 = x-3 \Rightarrow$

$x = -12$ Check: $\dfrac{3}{(-12)^2-3(-12)} - \dfrac{1}{(-12)^2-9} = \dfrac{1}{(-12)^2+3(-12)} \Rightarrow \dfrac{3}{180} - \dfrac{1}{135} = \dfrac{1}{108} \Rightarrow$

$\dfrac{9}{540} - \dfrac{4}{540} = \dfrac{5}{540} \Rightarrow \dfrac{5}{540} = \dfrac{5}{540} \Rightarrow \dfrac{1}{108} = \dfrac{1}{108}$

85. $\dfrac{1}{x^2} - \dfrac{5}{x^2+4x} = \dfrac{1}{x^2+4x} \Rightarrow \dfrac{1\left(x^2\right)\left(x^2+4x\right)}{x^2} - \dfrac{5\left(x^2\right)\left(x^2+4x\right)}{\left(x^2+4x\right)} = \dfrac{1\left(x^2\right)\left(x^2+4x\right)}{x^2+4x} \Rightarrow$

$\dfrac{\left(x^2+4x\right)\left(x^2\right)}{x^2} - \dfrac{5x^2\left(x^2+4x\right)}{x^2+4x} = \dfrac{x^2\left(x^2+4x\right)}{x^2+4x} \Rightarrow x^2+4x-5x^2 = x^2 \Rightarrow -4x^2+4x = x^2 \Rightarrow$

$-5x^2 = -4x \Rightarrow \dfrac{x^2}{x} = \dfrac{-4}{-5} \Rightarrow x = \dfrac{4}{5}$

Check: $\dfrac{1}{\left(\frac{4}{5}\right)^2} - \dfrac{5}{\left(\frac{4}{5}\right)^2 + 4\left(\frac{4}{5}\right)} = \dfrac{1}{\left(\frac{4}{5}\right)^2 + 4\left(\frac{4}{5}\right)} \Rightarrow \dfrac{1}{\frac{16}{25}} - \dfrac{5}{\frac{96}{25}} = \dfrac{1}{\frac{96}{25}} \Rightarrow \dfrac{1}{1}\cdot\dfrac{25}{16} - \dfrac{5}{1}\cdot\dfrac{25}{96} = \dfrac{1}{1}\cdot\dfrac{25}{96} \Rightarrow$

$\dfrac{25}{16} - \dfrac{125}{96} = \dfrac{25}{96} \Rightarrow \dfrac{150}{96} - \dfrac{125}{96} = \dfrac{25}{96} \Rightarrow \dfrac{25}{96} = \dfrac{25}{96}$

86. $\dfrac{5}{x^2-1} - \dfrac{1}{x^2+2x+1} = \dfrac{3}{x^2-1} \Rightarrow \dfrac{5(x+1)(x+1)(x-1)}{(x+1)(x-1)} - \dfrac{1(x+1)(x+1)(x-1)}{(x+1)(x+1)} =$

$\dfrac{3(x+1)(x+1)(x-1)}{(x+1)(x-1)} \Rightarrow \dfrac{(5x+5)(x^2-1)}{(x^2-1)} - \dfrac{(x-1)(x^2+2x+1)}{x^2+2x+1} = \dfrac{(3x+3)(x^2-1)}{x^2-1}$

$\Rightarrow 5x+5-(x-1) = 3x+3 \Rightarrow 4x+6 = 3x+3 \Rightarrow x=-3$

Check: $\dfrac{5}{(-3)^2-1} - \dfrac{1}{(-3)^2+2(-3)+1} = \dfrac{3}{(-3)^2-1} \Rightarrow \dfrac{5}{8} - \dfrac{1}{4} = \dfrac{3}{8} \Rightarrow \dfrac{5}{8} - \dfrac{2}{8} = \dfrac{3}{8} \Rightarrow \dfrac{3}{8} = \dfrac{3}{8}$

87. (a)  Equation

(b)  $\dfrac{4}{x} - x = 0 \Rightarrow 4-x^2 = 0 \Rightarrow (2-x)(2+x) = 0 \Rightarrow 2-x=0 \ \text{or}\ 2+x=0 \Rightarrow x=2,-2$

88. (a)  Expression

(b)  $\dfrac{x^2}{x-3} - \dfrac{9}{x-3} \Rightarrow \dfrac{x^2-9}{x-3} \Rightarrow \dfrac{(x+3)(x-3)}{x-3} \Rightarrow \dfrac{x+3}{1}\cdot\dfrac{x-3}{x-3} \Rightarrow x+3$

89. $\dfrac{1}{a} + \dfrac{2}{b} = \dfrac{3}{c}$ for $b \Rightarrow \dfrac{1}{a}\cdot\dfrac{bc}{bc} + \dfrac{2}{b}\cdot\dfrac{ac}{ac} = \dfrac{3}{c}\cdot\dfrac{ab}{ab} \Rightarrow bc+2ac = 3ab \Rightarrow 2ac = 3ab-bc \Rightarrow$

$2ac = b(3a-c) \Rightarrow b = \dfrac{2ac}{3a-c}$

90. $y = \dfrac{x}{x-1}$ for $x \Rightarrow y(x-1) = x\cdot1 \Rightarrow xy-y = x \Rightarrow xy-x = y \Rightarrow x(y-1) = y \Rightarrow x = \dfrac{y}{y-1}$

91. $\dfrac{x}{6} = \dfrac{1}{3} \Rightarrow 3x=6 \Rightarrow x = \dfrac{6}{3} = 2$

92. $\dfrac{5}{x} = \dfrac{7}{3} \Rightarrow 15 = 7x \Rightarrow \dfrac{15}{7} = x \Rightarrow x = \dfrac{15}{7}$

93. (a)  $\dfrac{6}{x} = \dfrac{13}{20}$

(b)  $\dfrac{6}{x} = \dfrac{13}{20} \Rightarrow 120 = 13x \Rightarrow \dfrac{120}{13} = x \Rightarrow x = \dfrac{120}{13}$

94. (a)  $\dfrac{341}{11} = \dfrac{x}{8}$

(b)  $\dfrac{341}{11} = \dfrac{x}{8} \Rightarrow 11x = 2728 \Rightarrow x = \dfrac{2728}{11} \Rightarrow x = \$248$

95. (a)  $k = \dfrac{y}{x},\ \ y = 8,\ x = 2 \Rightarrow k = \dfrac{8}{2} \Rightarrow k = 4$

    (b)  $y = kx,\ \ k = 4,\ x = 5 \Rightarrow y = 4 \cdot 5 \Rightarrow y = 20$

96. (a)  $k = \dfrac{y}{x},\ \ y = 21,\ x = 7 \Rightarrow k = \dfrac{21}{7} \Rightarrow k = 3$

    (b)  $y = kx,\ \ k = 3,\ x = 5 \Rightarrow y = 3 \cdot 5 \Rightarrow y = 15$

97. (a)  $k = xy,\ \ x = 4,\ y = 2.5 \Rightarrow k = 4 \cdot 2.5 \Rightarrow k = 10$

    (b)  $y = \dfrac{k}{x},\ \ k = 10,\ x = 5 \Rightarrow y = \dfrac{10}{5} \Rightarrow y = 2$

98. (a)  $k = xy,\ \ x = 3,\ y = 7 \Rightarrow k = 3 \cdot 7 \Rightarrow k = 21$

    (b)  $y = \dfrac{k}{x},\ \ k = 21,\ x = 5 \Rightarrow y = \dfrac{21}{5}$

99.  $z = kxy \Rightarrow 483 = k(23)(7) \Rightarrow 483 = 161k \Rightarrow k = \dfrac{483}{161} \Rightarrow k = 3$

100.  $z = kxy^2 \Rightarrow 891 = k(22)(3)^2 \Rightarrow 891 = 198k \Rightarrow k = \dfrac{891}{198} \Rightarrow k = \dfrac{9}{2} \Rightarrow z = \dfrac{9}{2}xy^2$

   So when $x = 10$ and $y = 4$, $z = \dfrac{9}{2}(10)(4)^2 = 720$

101. (a)  Inverse, because as $x$ increases $y$ decreases and $k = xy \Rightarrow 2 \cdot 30 = 3 \cdot 20 = 4 \cdot 15$, etc.

     (b)  $k = xy,\ \ x = 2,\ y = 30 \Rightarrow k = 2 \cdot 30 \Rightarrow k = 60;\ \ y = \dfrac{k}{x} \Rightarrow y = \dfrac{60}{x}$

     (c)  See Figure 101.

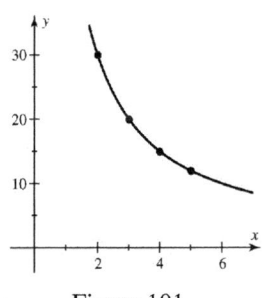   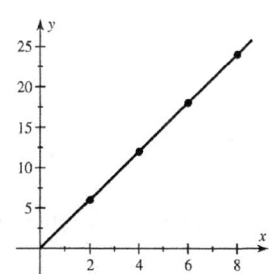

Figure 101                           Figure 102

102. (a)  Direct, because as $x$ increases $y$ also increases and $k = \dfrac{y}{x} \Rightarrow \dfrac{6}{2} = \dfrac{12}{4} = \dfrac{18}{6}$, etc.

     (b)  $k = \dfrac{y}{x},\ \ x = 2,\ y = 6 \Rightarrow k = \dfrac{6}{2} \Rightarrow k = 3;\ \ y = kx \Rightarrow y = 3x$

     (c)  See Figure 102.

103. Direct, $k = \dfrac{y}{x},\ \ y = 2,\ x = 4 \Rightarrow k = \dfrac{2}{4} \Rightarrow k = \dfrac{1}{2}$

104. Inverse, $k = xy,\ \ x = 2,\ y = 6 \Rightarrow k = 2 \cdot 6 \Rightarrow k = 12$

105. (a)   $x = 10,\ T = \dfrac{1}{15-x} \Rightarrow T = \dfrac{1}{15-10} = \dfrac{1}{5} = 0.2,$  when the rate of arrival is 10 cars/min., the wait

   is 0.2 minutes or 12 seconds.

   (b)   $x = 5,\ T = \dfrac{1}{15-x} \Rightarrow T = \dfrac{1}{15-5} = \dfrac{1}{10};\quad x = 10,\ T = \dfrac{1}{15-x} \Rightarrow T = \dfrac{1}{15-10} = \dfrac{1}{5};$

   $x = 13,\ T = \dfrac{1}{15-x} \Rightarrow T = \dfrac{1}{15-13} = \dfrac{1}{2};\quad x = 14,\ T = \dfrac{1}{15-x} \Rightarrow T = \dfrac{1}{15-14} = \dfrac{1}{1} = 1;$

   $x = 14.9,\ T = \dfrac{1}{15-x} \Rightarrow T = \dfrac{1}{15-14.9} = \dfrac{1}{0.1} = 10;$  See Figure 105.

   (c)   It increases dramatically.

| $x$ | 5 | 10 | 13 | 14 | 14.9 |
|---|---|---|---|---|---|
| $T$ | $\frac{1}{10}$ | $\frac{1}{5}$ | $\frac{1}{2}$ | 1 | 10 |

   Figure 105

106. If $r \cdot t = d$, then $50 \cdot t = 150 \Rightarrow t = 3$ and $75t = 150 \Rightarrow t = 2$. The combined $d = 300$ and the

   combined $t = 5$. Therefore, $r \cdot 5 = 300 \Rightarrow r = 60 \Rightarrow 60$ mph.

107. $\dfrac{t}{100} + \dfrac{t}{160} = 1 \Rightarrow \dfrac{800t}{100} + \dfrac{800t}{160} = 800 \Rightarrow 8t + 5t = 800 \Rightarrow 13t = 800 \Rightarrow t = \dfrac{800}{13} \Rightarrow t \approx 61.5$ hrs.

108. Let $x$ = slower jogger speed and $x + 2$ = faster jogger speed, then $\dfrac{10}{x} - \dfrac{10}{x+2} = \dfrac{1}{6} \Rightarrow$

   $60x + 120 - 60x = x^2 + 2x \Rightarrow x^2 + 2x - 120 = 0 \Rightarrow (x-10)(x+12) = 0 \Rightarrow$

   $x - 10 = 0 \Rightarrow x = 10,\ x + 12 = 0 \Rightarrow x = -12 \Rightarrow x = 10, -12.$  Therefore 10 mph and 12 mph.

109. $\dfrac{16}{x-4} = \dfrac{48}{x+4} \Rightarrow 16x + 64 = 48x - 192 \Rightarrow 256 = 32x \Rightarrow \dfrac{256}{32} = x \Rightarrow x = 8$ mph

110. $\dfrac{5}{6} = \dfrac{x}{40} \Rightarrow 6x = 200 \Rightarrow x = \dfrac{200}{6} \Rightarrow x \approx 33.3$ ft.

111. $k = xy,\ x = 0.25,\ y = 400 \Rightarrow k = 0.25 \cdot 400 \Rightarrow k = 100$

   $y = \dfrac{k}{x},\ x = 0.50,\ k = 100 \Rightarrow y = \dfrac{100}{0.50} \Rightarrow y = 200$ vehicles

112. $k = \dfrac{y}{x},\ x = 17,\ y = 612 \Rightarrow k = \dfrac{612}{17} \Rightarrow k = 36\quad y = kx,\ k = 36,\ x = 13 \Rightarrow y = 36 \cdot 13 \Rightarrow y = \$468$

113. $k = xy,\ x = 12,\ y = 30 \Rightarrow k = 12 \cdot 30 \Rightarrow k = 360\quad y = \dfrac{k}{x},\ x = 10,\ k = 360 \Rightarrow y = \dfrac{360}{10} \Rightarrow y = 36$ lbs.

114. $\dfrac{20}{1} = \dfrac{32}{x} \Rightarrow 20x = 32 \Rightarrow x = \dfrac{32}{20} \Rightarrow x = 1.6$ in.

115. $30 \cdot 25 = 750$ and $90 \cdot 25 = 2250$, therefore 750 to 2250 sec. or 12.5 to 37.5 min.

116. $y = \dfrac{k}{x},\ x = 18,\ y = 900 \Rightarrow 900 = \dfrac{k}{18} \Rightarrow 16,200 = k;\quad y = \dfrac{16,200}{x},\ x = 21 \Rightarrow y = \dfrac{16,200}{21} \Rightarrow y \approx 771$ lb

117. $P = k\,d^2 w^3 \Rightarrow 10{,}823 = (6)^2 (20)^3 k \Rightarrow 10{,}823 = 288{,}000\,k \Rightarrow k = \dfrac{10{,}823}{288{,}000}$ and so

$P = \dfrac{10{,}823}{288{,}000} d^2 w^3.$ When $d = 10$ and $w = 12$ $P = \dfrac{10{,}823}{288{,}000}(10)^2 (12)^3 = 6493.8$ watts.

118. $S = k\,w t^2 \Rightarrow 650 = k(8)(5)^2 \Rightarrow 650 = 200\,k \Rightarrow k = \dfrac{13}{4}$ and so $S = \dfrac{13}{4} w t^2.$

When $w = 6$ and $t = 6,$ $S = \dfrac{13}{4}(6)(6)^2 = 702$ pounds.

## Chapter 7 Test

1. $x = 3,$ $\dfrac{3x}{2x-1} \Rightarrow \dfrac{3(3)}{2(3)-1} = \dfrac{9}{6-1} = \dfrac{9}{5}$

2. $\dfrac{x-1}{x+2},$ the equation is undefined when $x + 2 = 0 \Rightarrow x = -2.$

3. $\dfrac{x^2 - 25}{x-5} = \dfrac{(x-5)(x+5)}{x-5} = \dfrac{x-5}{x-5} \cdot (x+5) = 1 \cdot (x+5) = x+5$

4. $\dfrac{3x^2 - 15x}{3x} = \dfrac{3x(x-5)}{3x} = \dfrac{3x}{3x} \cdot (x-5) = 1 \cdot (x-5) = x-5$

5. $\dfrac{x-2}{x+4} \cdot \dfrac{3x+12}{x-2} = \dfrac{x-2}{x-2} \cdot \dfrac{3(x+4)}{x+4} = 1 \cdot 1 \cdot 3 = 3$

6. $\dfrac{z+1}{z+3} \cdot \dfrac{2z+6}{z+1} = \dfrac{z+1}{z+1} \cdot \dfrac{2(z+3)}{z+3} = 1 \cdot 1 \cdot 2 = 2$

7. $\dfrac{x+1}{5x} \div \dfrac{2x+2}{x-1} = \dfrac{x+1}{5x} \cdot \dfrac{x-1}{2x+2} = \dfrac{x+1}{2(x+1)} \cdot \dfrac{x-1}{5x} = \dfrac{x-1}{2(5x)} \cdot 1 = \dfrac{x-1}{10x}$

8. $\dfrac{2}{x^2} \div \dfrac{x+3}{3x} = \dfrac{2}{x^2} \cdot \dfrac{3x}{x+3} = \dfrac{x}{x} \cdot \dfrac{2 \cdot 3}{x(x+3)} = 1 \cdot \dfrac{6}{x(x+3)} = \dfrac{6}{x(x+3)}$

9. $\dfrac{x}{x+4} + \dfrac{3x+1}{x+4} = \dfrac{4x+1}{x+4}$

10. $\dfrac{4t+1}{2t-3} - \dfrac{3t-6}{2t-3} = \dfrac{4t+1-(3t-6)}{2t-3} = \dfrac{t+7}{2t-3}$

11. $6x^2 = 2 \cdot 3 \cdot x \cdot x$ and $3x^2 - 3x = 3x(x-1)$ so the least common multiple is

$2 \cdot 3 \cdot x \cdot x \cdot (x-1) = 6x^2 (x-1)$

12. $7x^2 - 7x = 7x(x-1) \Rightarrow \dfrac{4}{7x} = \dfrac{4(x-1)}{7x(x-1)} = \dfrac{4x-4}{7x^2 - 7x}$

13. $\dfrac{1}{y^2+y}-\dfrac{y-1}{y^2-y}=\dfrac{1}{y(y+1)}\cdot\dfrac{y-1}{y-1}-\dfrac{y-1}{y(y-1)}\cdot\dfrac{y+1}{y+1}=\dfrac{y-1}{y(y+1)(y-1)}-\dfrac{y^2-1}{y(y+1)(y-1)}=$

$\dfrac{y-y^2}{y(y+1)(y-1)}=\dfrac{-y(y-1)}{y(y+1)(y-1)}=\dfrac{-1}{y+1}=-\dfrac{1}{y+1}$

14. $\dfrac{1}{xy}+\dfrac{x}{y}-\dfrac{1}{y^2}=\dfrac{y}{xy^2}+\dfrac{x^2y}{xy^2}-\dfrac{x}{xy^2}=\dfrac{x^2y-x+y}{xy^2}$

15. $\dfrac{\frac{a}{3b}}{\frac{5a}{b^2}}=\dfrac{a}{3b}\div\dfrac{5a}{b^2}=\dfrac{a}{3b}\cdot\dfrac{b^2}{5a}=\dfrac{a}{a}\cdot\dfrac{b}{b}\cdot\dfrac{b}{15}=1\cdot1\cdot\dfrac{b}{15}=\dfrac{b}{15}$

16. $\dfrac{1+\frac{1}{p-1}}{1-\frac{1}{p-1}}=\dfrac{\frac{p-1}{p-1}+\frac{1}{p-1}}{\frac{p-1}{p-1}-\frac{1}{p-1}}=\dfrac{\frac{p}{p-1}}{\frac{p-2}{p-1}}=\dfrac{p}{p-1}\div\dfrac{p-2}{p-1}=\dfrac{p}{p-1}\cdot\dfrac{p-1}{p-2}=\dfrac{p-1}{p-1}\cdot\dfrac{p}{p-2}=1\cdot\dfrac{p}{p-2}=\dfrac{p}{p-2}$

17. $\dfrac{2}{7}=\dfrac{5}{x}\Rightarrow 2x=35\Rightarrow x=\dfrac{35}{2}$

18. $\dfrac{x+3}{2x}=1\Rightarrow x+3=2x\Rightarrow 3=x\Rightarrow x=3$

19. $\dfrac{1}{2x}+\dfrac{2}{5x}=\dfrac{9}{10}=\dfrac{1\cdot10x}{2x}+\dfrac{2\cdot10x}{5x}=\dfrac{9\cdot10x}{10}\Rightarrow 5+4=9x\Rightarrow 9=9x\Rightarrow x=1$

20. $\dfrac{1}{x-1}+\dfrac{2}{x+2}=\dfrac{3}{2}\Rightarrow\dfrac{2(x-1)(x+2)}{x-1}+\dfrac{2(2)(x-1)(x+2)}{x+2}=\dfrac{3(2)(x-1)(x+2)}{2}\Rightarrow$

$2x+4+4x-4=3x^2+3x-6\Rightarrow 6x=3x^2+3x-6\Rightarrow 3x^2-3x-6=0\Rightarrow$

$3(x+1)(x-2)=0,\ x+1=0\Rightarrow x=-1,\ x-2=0\Rightarrow x=2\Rightarrow x=-1,\ 2$

21. $\dfrac{1}{x^2-1}-\dfrac{4}{x+1}=\dfrac{3}{x-1}\Rightarrow\dfrac{1}{(x+1)(x-1)}-\dfrac{4}{x+1}=\dfrac{3}{x-1}\Rightarrow$

$\dfrac{1(x+1)(x-1)}{(x+1)(x-1)}-\dfrac{4(x+1)(x-1)}{x+1}=\dfrac{3(x+1)(x-1)}{x-1}\Rightarrow 1-(4x-4)=3x+3\Rightarrow$

$-4x+5=3x+3\Rightarrow -7x=-2\Rightarrow x=\dfrac{-2}{-7}\Rightarrow x=\dfrac{2}{7}$

22. $\dfrac{1}{x^2-4x}+\dfrac{2}{x^2-16}=\dfrac{2}{x^2+4x}\Rightarrow\dfrac{1}{x(x-4)}+\dfrac{2}{(x-4)(x+4)}=\dfrac{2}{x(x+4)}\Rightarrow$

$\dfrac{1(x)(x+4)(x-4)}{x(x-4)}+\dfrac{2(x)(x+4)(x-4)}{(x-4)(x+4)}=\dfrac{2(x)(x+4)(x-4)}{x(x+4)}\Rightarrow x+4+2x=2x-8\Rightarrow$

$3x+4=2x-8\Rightarrow x=-12$

23. $\dfrac{x}{2x-1}=\dfrac{1-x}{2x-1}\Rightarrow x=1-x\Rightarrow x+x=1-x+x\Rightarrow 2x=1\ \Rightarrow\dfrac{2x}{2}=\dfrac{1}{2}\Rightarrow x=\dfrac{1}{2}$

However, $x=\dfrac{1}{2}$ is an extraneous solution because $\dfrac{x}{2x-1}=\dfrac{\frac{1}{2}}{2\left(\frac{1}{2}\right)-1}=\dfrac{\frac{1}{2}}{1-1}=\dfrac{\frac{1}{2}}{0}$ is undefined.

So there are no solutions.

24. $\dfrac{x}{x-5}+\dfrac{x}{x+5}=\dfrac{10x}{x^2-25}\Rightarrow\dfrac{x(x+5)}{(x-5)(x+5)}+\dfrac{x(x-5)}{(x-5)(x+5)}=\dfrac{10x}{(x-5)(x+5)}$

$\Rightarrow\dfrac{x(x+5)+x(x-5)}{(x-5)(x+5)}=\dfrac{10x}{(x-5)(x+5)}\Rightarrow x(x+5)+x(x-5)=10x$

$\Rightarrow x^2+5x+x^2-5x=10x\Rightarrow 2x^2=10x\Rightarrow 2x^2-10x=0$

$\Rightarrow 2x(x-5)=0\Rightarrow 2x=0,\ x-5=0\Rightarrow x=0,\ 5$   However, $x=5$ is an extraneous solution because

$\dfrac{x}{x-5}=\dfrac{5}{5-5}=\dfrac{5}{0}$ is undefined. So the solution is $x=0$.

25. $y=\dfrac{2}{3x-5}$ for $x$, $y(3x-5)=2\Rightarrow 3x-5=\dfrac{2}{y}\Rightarrow 3x=\dfrac{2}{y}+5\Rightarrow x=\left(\dfrac{2}{y}+5\right)\dfrac{1}{3}\Rightarrow$

$x=\dfrac{2}{3y}+\dfrac{5}{3}\Rightarrow x=\dfrac{2}{3y}+\dfrac{5y}{3y}\Rightarrow x=\dfrac{2+5y}{3y}$

26. $\dfrac{a+b}{ab}=1$ for $b$,  $ab=a+b\Rightarrow ab-b=a\Rightarrow b(a-1)=a\Rightarrow b=\dfrac{a}{a-1}$

27. (a)   $k=\dfrac{y}{x}\Rightarrow k=\dfrac{14}{4}\Rightarrow k=\dfrac{7}{2}$

(b)   $y=kx\Rightarrow y=\dfrac{7}{2}(6)\Rightarrow y=\dfrac{42}{2}\Rightarrow y=21$

28.  Inversely, as $x$ increases $y$ decreases and for all $k=xy$, $k=32$.

29. $\dfrac{t}{40}+\dfrac{t}{60}=1\Rightarrow\dfrac{120t}{40}+\dfrac{120t}{60}=120\Rightarrow 3t+2t=120\Rightarrow 5t=120\Rightarrow t=24$ hours

30. $\dfrac{5}{4}=\dfrac{x}{54}\Rightarrow 4x=270\Rightarrow x=67.5$ ft.

31. $N=\dfrac{x^2}{900-30x}$ for $x=24$, $N=\dfrac{24^2}{900-30(24)}\Rightarrow N=\dfrac{576}{900-720}\Rightarrow N=\dfrac{576}{180}\Rightarrow N=\dfrac{16}{5}$ or 3.2; when

the arrival rate is 24 people/hr., there are about 3 people in line.

## Chapter 7 Extended and Discovery Exercises

1. (a) $x = 3$, $N = \dfrac{3^2}{225 - 15(3)} = \dfrac{9}{225 - 45} = \dfrac{9}{180} = \dfrac{1}{20} \Rightarrow N = 0.05$;

   $x = 9$, $N = \dfrac{9^2}{225 - 15(9)} = \dfrac{81}{225 - 135} = \dfrac{81}{90} = \dfrac{9}{10} \Rightarrow N = 0.9$;

   $x = 12$, $N = \dfrac{12^2}{225 - 15(12)} = \dfrac{144}{225 - 180} = \dfrac{144}{45} = \dfrac{16}{5} \Rightarrow N = 3.2$;

   $x = 13$, $N = \dfrac{13^2}{225 - 15(13)} = \dfrac{169}{225 - 195} = \dfrac{169}{30} \Rightarrow N = 5.6\overline{3}$;

   $x = 14$, $N = \dfrac{14^2}{225 - 15(14)} = \dfrac{196}{225 - 210} = \dfrac{196}{15} \Rightarrow N = 13.0\overline{6}$   See Figure 1a.

   (b)   $x = 15$, because $\dfrac{15^2}{225 - 15(15)} = \dfrac{225}{0}$, which is undefined.

   (c)   See Figure 1c.

   (d)   See Figure 1d.

   (e)   As $x$ approaches 15, the wait increases dramatically.

   (f)   The formula is only valid for arrival rates under 15 cars per hour, but it can be inferred that the line of cars will continue to grow.

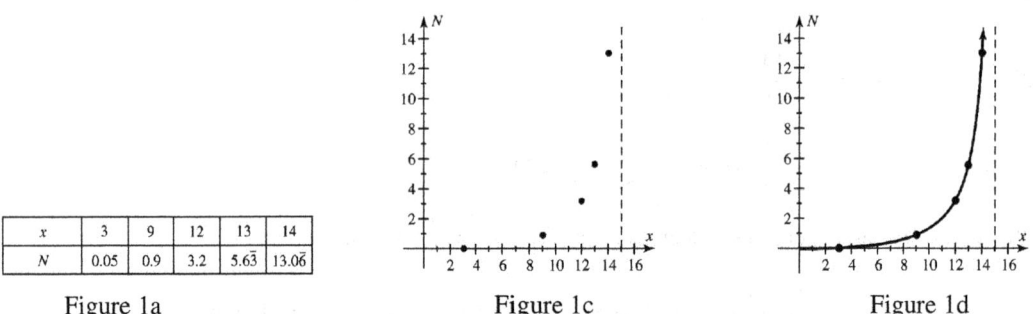

| $x$ | 3 | 9 | 12 | 13 | 14 |
|---|---|---|---|---|---|
| $N$ | 0.05 | 0.9 | 3.2 | $5.6\overline{3}$ | $13.0\overline{6}$ |

Figure 1a                                    Figure 1c                                    Figure 1d

2. (a)   $x = -4$, $y = \dfrac{1}{-4 - 1} \Rightarrow y = -\dfrac{1}{5}$;   $x = -3$, $y = \dfrac{1}{-3 - 1} \Rightarrow y = -\dfrac{1}{4}$;

   $x = -2$, $y = \dfrac{1}{-2 - 1} \Rightarrow y = -\dfrac{1}{3}$;   $x = -1$, $y = \dfrac{1}{-1 - 1} \Rightarrow y = -\dfrac{1}{2}$;

   $x = 0$, $y = \dfrac{1}{0 - 1} \Rightarrow y = -\dfrac{1}{1} = -1$;   $x = 1$, $y = \dfrac{1}{1 - 1} \Rightarrow y = \dfrac{1}{0} \Rightarrow$ undefined;

   $x = 2$, $y = \dfrac{1}{2 - 1} \Rightarrow y = \dfrac{1}{1} \Rightarrow y = 1$;   $x = 3$, $y = \dfrac{1}{3 - 1} \Rightarrow y = \dfrac{1}{2}$;

   $x = 4$, $y = \dfrac{1}{4 - 1} \Rightarrow y = \dfrac{1}{3}$;   See Figure 2a.

(b) $x = 1$ yields $\dfrac{1}{0}$ which is undefined.

(c) See Figure 2c-e.

(d) See Figure 2c-e.

(e) See Figure 2c-e.

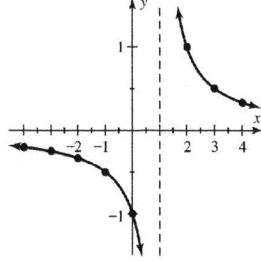

| $x$ | −4 | −3 | −2 | −1 | 0 | 1 | 2 | 3 | 4 |
|---|---|---|---|---|---|---|---|---|---|
| $y$ | −0.2 | −0.25 | $-0.\overline{3}$ | −0.5 | −1 | — | 1 | 0.5 | $0.\overline{3}$ |

Figure 2a                                    Figure 2c-e

3. (a) $x = -4,\ y = \dfrac{1}{-4+1} \Rightarrow y = -\dfrac{1}{3};\quad x = -3,\ y = \dfrac{1}{-3+1} \Rightarrow y = -\dfrac{1}{2};$

$x = -2,\ y = \dfrac{1}{-2+1} \Rightarrow y = -\dfrac{1}{1} = -1;\quad x = -1,\ y = \dfrac{1}{-1+1} \Rightarrow y = \dfrac{1}{0} \Rightarrow$ undefined;

$x = 0,\ y = \dfrac{1}{0+1} \Rightarrow y = \dfrac{1}{1} = 1;\quad x = 1,\ y = \dfrac{1}{1+1} \Rightarrow y = \dfrac{1}{2};$

$x = 2,\ y = \dfrac{1}{2+1} \Rightarrow y = \dfrac{1}{3};\quad x = 3,\ y = \dfrac{1}{3+1} \Rightarrow y = \dfrac{1}{4};$

$x = 4,\ y = \dfrac{1}{4+1} \Rightarrow y = \dfrac{1}{5};$ See Figure 3a.

(b) $x = -1$ yields $\dfrac{1}{0}$ which is undefined.

(c) See Figure 3c-e.

(d) See Figure 3c-e.

(e) See Figure 3c-e.

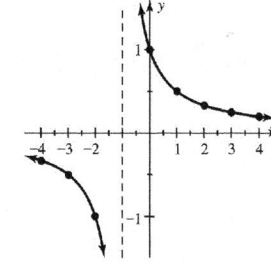

| $x$ | −4 | −3 | −2 | −1 | 0 | 1 | 2 | 3 | 4 |
|---|---|---|---|---|---|---|---|---|---|
| $y$ | $-0.\overline{3}$ | −0.5 | −1 | — | 1 | 0.5 | $0.\overline{3}$ | 0.25 | 0.2 |

Figure 3a                                    Figure 3c-e

4. (a)   $x = -4,\ y = \dfrac{4}{16+1} \Rightarrow y = \dfrac{4}{17};\quad x = -3,\ y = \dfrac{4}{9+1} \Rightarrow y = \dfrac{4}{10};$

$x = -2,\ y = \dfrac{4}{4+1} \Rightarrow y = \dfrac{4}{5};\quad x = -1,\ y = \dfrac{4}{1+1} \Rightarrow y = \dfrac{4}{2} \Rightarrow y = 2;$

$x = 0,\ y = \dfrac{4}{0+1} \Rightarrow y = \dfrac{4}{1} = 4;\quad x = 1,\ y = \dfrac{4}{1+1} \Rightarrow y = \dfrac{4}{2} = 2;$

$x = 2,\ y = \dfrac{4}{4+1} \Rightarrow y = \dfrac{4}{5};\quad x = 3,\ y = \dfrac{4}{9+1} \Rightarrow y = \dfrac{4}{10};$

$x = 4,\ y = \dfrac{4}{16+1} \Rightarrow y = \dfrac{4}{17};$   See Figure 4a.

(b)   No points are undefined.

(c)   See Figure 4c-e.

(d)   See Figure 4c-e.

(e)   See Figure 4c-e.

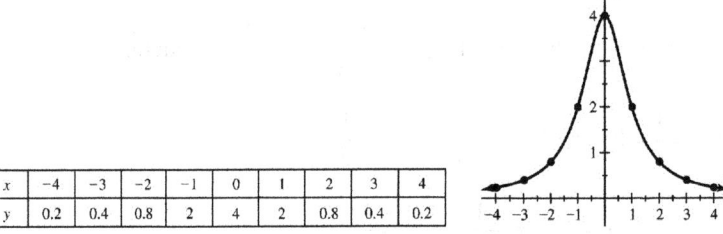

| x | -4 | -3 | -2 | -1 | 0 | 1 | 2 | 3 | 4 |
|---|----|----|----|----|---|---|---|---|---|
| y | 0.2 | 0.4 | 0.8 | 2 | 4 | 2 | 0.8 | 0.4 | 0.2 |

Figure 4a                                Figure 4c-e

5. (a)   $x = -4,\ y = \dfrac{-4}{-4+1} \Rightarrow y = \dfrac{4}{3};\quad x = -3,\ y = \dfrac{-3}{-3+1} \Rightarrow y = \dfrac{3}{2};$

$x = -2,\ y = \dfrac{-2}{-2+1} \Rightarrow y = \dfrac{2}{1} = 2;\quad x = -1,\ y = \dfrac{-1}{-1+1} \Rightarrow y = \dfrac{-1}{0} \Rightarrow \text{undefined};$

$x = 0,\ y = \dfrac{0}{0+1} \Rightarrow y = \dfrac{0}{1} = 0;\quad x = 1,\ y = \dfrac{1}{1+1} \Rightarrow y = \dfrac{1}{2};$

$x = 2,\ y = \dfrac{2}{2+1} \Rightarrow y = \dfrac{2}{3};\quad x = 3,\ y = \dfrac{3}{3+1} \Rightarrow y = \dfrac{3}{4};$

$x = 4,\ y = \dfrac{4}{4+1} \Rightarrow y = \dfrac{4}{5};$   See Figure 5a.

(b)   $x = -1$ yields $\dfrac{-1}{0}$ which is undefined.

(c)   See Figure 5c-e.

(d)   See Figure 5c-e.

(e)   See Figure 5c-e.

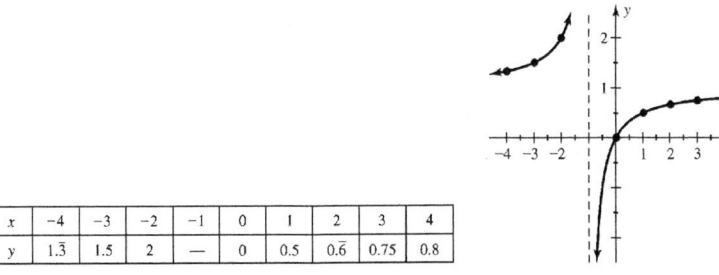

| x | −4 | −3 | −2 | −1 | 0 | 1 | 2 | 3 | 4 |
|---|----|----|----|----|---|---|---|---|---|
| y | $1.\overline{3}$ | 1.5 | 2 | — | 0 | 0.5 | $0.\overline{6}$ | 0.75 | 0.8 |

Figure 5a                                      Figure 5c-e

6. (a)   $x = -4, \ y = \dfrac{-4}{-4-1} \Rightarrow y = \dfrac{4}{5}; \quad x = -3, \ y = \dfrac{-3}{-3-1} \Rightarrow y = \dfrac{3}{4};$

   $x = -2, \ y = \dfrac{-2}{-2-1} \Rightarrow y = \dfrac{2}{3}; \quad x = -1, \ y = \dfrac{-1}{-1-1} \Rightarrow y = \dfrac{1}{2};$

   $x = 0, \ y = \dfrac{0}{0-1} \Rightarrow y = \dfrac{0}{-1} = 0; \quad x = 1, \ y = \dfrac{1}{1-1} \Rightarrow y = \dfrac{1}{0} \Rightarrow$ undefined;

   $x = 2, \ y = \dfrac{2}{2-1} \Rightarrow y = \dfrac{2}{1} = 2; \quad x = 3, \ y = \dfrac{3}{3-1} \Rightarrow y = \dfrac{3}{2};$

   $x = 4, \ y = \dfrac{4}{4-1} \Rightarrow y = \dfrac{4}{3};$   See Figure 6a.

   (b)   $x = 1$ yields $\dfrac{1}{0}$ which is undefined.

   (c)   See Figure 6c-e.

   (d)   See Figure 6c-e.

   (e)   See Figure 6c-e.

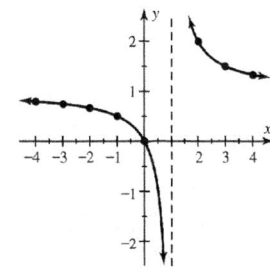

| x | −4 | −3 | −2 | −1 | 0 | 1 | 2 | 3 | 4 |
|---|----|----|----|----|---|---|---|---|---|
| y | 0.8 | 0.75 | $0.\overline{6}$ | 0.5 | 0 | — | 2 | 1.5 | $1.\overline{3}$ |

Figure 6a                                      Figure 6c-e

## Chapters 1-7  Cumulative Review Exercises

1.  $\pi r^2 h = \pi (2)^2 (6) = \pi (4)(6) = 24\pi \approx 75.4$

2.  $2x - 2$

3.  $\dfrac{1}{2} \div \dfrac{5}{4} = \dfrac{1}{2} \cdot \dfrac{4}{5} = \dfrac{4}{10} = \dfrac{2}{5}$

4. $\dfrac{5}{8}+\dfrac{1}{8}=\dfrac{6}{8}=\dfrac{3}{4}$

5. $-2+7x+4-5x=2x+2$

6. $-4(4-y)+(5-3y)=-16+4y+5-3y=y-11$

7. $-2x+11=13 \Rightarrow -2x=2 \Rightarrow x=-1$

    Check: $-2(-1)+11 \overset{?}{=} 13 \Rightarrow 2+11 \overset{?}{=} 13 \Rightarrow 13=13,$ so $x=-1$

8. $-3x+1 \ge x \Rightarrow -4x+1 \ge 0 \Rightarrow -4x \ge -1 \Rightarrow x \le \dfrac{1}{4}$

9. See Figure 9.

    $x$-intercept: $2x-3(0)=6 \Rightarrow 2x=6 \Rightarrow x=3$;  $y$-intercept: $2(0)-3y=6 \Rightarrow -3y=6 \Rightarrow y=-2$

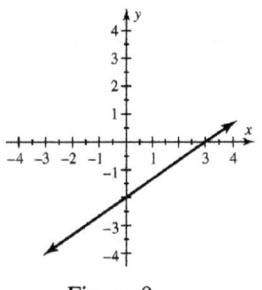
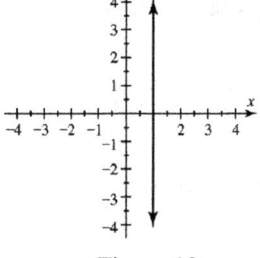
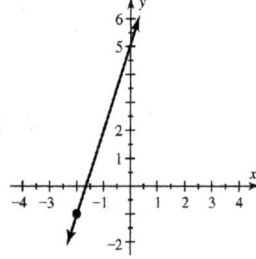

Figure 9                                      Figure 10                                      Figure 11

10. See Figure 10.  $x$-intercept: $x=1$    $y$-intercept: none

11. See Figure 11. Using $y=mx+b \Rightarrow y=3x+5$

12. Using $(-2,-5)$ and $(-1,-3)$, $m=\dfrac{-5-(-3)}{-2-(-1)}=\dfrac{-2}{-1}=2,$ the

    $y$-intercept is $(0,-1)$ or $b=-1$, so $y=2x-1$.

13. Parallel lines have equal slopes $\Rightarrow m=-\dfrac{2}{3},$ now using $(2,-1)$ and

    $y-y_1=m(x-x_1) \Rightarrow y-(-1)=-\dfrac{2}{3}(x-2) \Rightarrow y+1=-\dfrac{2}{3}x+\dfrac{4}{3} \Rightarrow y=-\dfrac{2}{3}x+\dfrac{1}{3}$

14. $m=\dfrac{4-2}{2-(-1)}=\dfrac{2}{3},$ using $(2,4)$ and

    $y-y_1=m(x-x_1) \Rightarrow y-4=\dfrac{2}{3}(x-2) \Rightarrow y-4=\dfrac{2}{3}x-\dfrac{4}{3} \Rightarrow y=\dfrac{2}{3}x+\dfrac{8}{3}$

15. Using $(2,-6)$ for $4x+y=2 \Rightarrow 4(2)+(-6)=2 \Rightarrow 8+(-6)=2,$ Yes; Using $(2,-6)$ for

    $x-4y=9 \Rightarrow 2-4(-6)=9 \Rightarrow 2+24=9,$ No; Using $(1,-2)$ for

    $4x+y=2 \Rightarrow 4(1)+(-2)=2 \Rightarrow 4+(-2)=2,$ Yes; Using $(1,-2)$ for

    $x-4y=9 \Rightarrow 1-4(-2)=9 \Rightarrow 1+8=9,$ Yes; $(1,-2)$ is true for both equations, therefore a solution

    to the system of equations.

16. Using substitution $2r + t = -4 \Rightarrow t = -2r - 4$, substituting into

$-3r - t = 2 \Rightarrow -3r - (-2r - 4) = 2 \Rightarrow -r + 4 = 2$

$\Rightarrow -r = -2 \Rightarrow r = 2$ and $t = -2(2) - 4 \Rightarrow t = -4 - 4 \Rightarrow t = -8 \Rightarrow (2, -8)$ is the solution.

17. $\begin{array}{r} 2x - y = 5 \\ -2x + y = -5 \\ \hline 0 = 0, \end{array}$ if $0 = 0$ then infinitely many solutions.

18. $\begin{array}{r} (4x - 6y = 12)3 = \quad 12x - 18y = 36 \\ (-6x + 9y = 18)2 = -12x + 18y = 36 \\ \hline 0 = 72, \end{array}$ if $0 = 72$ then no solutions.

19. $3z^2 \cdot 5z^6 = 3 \cdot 5 \cdot z^2 \cdot z^6 = 15z^8$

20. $(ab)^3 = a^3 b^3$

21. $(2y - 3)(5y + 2) = 10y^2 + 4y - 15y - 6 = 10y^2 - 11y - 6$

22. $\left(x^2 - y^2\right)^2 = \left(x^2 - y^2\right)\left(x^2 - y^2\right) = x^4 - x^2 y^2 - x^2 y^2 + y^4 = x^4 - 2x^2 y^2 + y^4$

23. $\left(3x^2\right)^{-3} = 3^{-3} x^{-6} = \dfrac{1}{3^3} \cdot \dfrac{1}{x^6} = \dfrac{1}{27x^6}$

24. $\dfrac{4x^2}{2x^4} = \dfrac{4}{2} \cdot \dfrac{x^2}{x^4} = \dfrac{2}{1} \cdot \dfrac{1}{x^2} = \dfrac{2}{x^2}$

25. $0.00123 = 1.23 \times 10^{-3}$

26. $\begin{array}{r} 2x + 1 + \dfrac{4}{x-1} \\ x - 1 \overline{\smash{\big)}\, 2x^2 - \phantom{0}x + 3} \\ \underline{2x^2 - 2x} \\ x + 3 \\ \underline{x - 1} \\ 4 \end{array}$

27. $6 + 13x - 5x^2 = (3 - x)(2 + 5x)$ by FOIL

28. $9z^2 - 4 = (3z + 2)(3z - 2)$, difference of squares

29. $t^2 + 16t + 64 = (t + 8)(t + 8) = (t + 8)^2$ by FOIL

30. $x^3 - 16x = x\left(x^2 - 16\right) \Rightarrow$ difference of squares, $x(x - 4)(x + 4)$

31. $y^2 + 5y - 14 = 0 \Rightarrow (y + 7)(y - 2) = 0$, then $y + 7 = 0 \Rightarrow y = -7$ or $y - 2 = 0 \Rightarrow y = 2$, so $y = -7, 2$

32. $x^3 = 4x \Rightarrow x^3 - 4x = 0 \Rightarrow x\left(x^2 - 4\right) = 0 \Rightarrow x(x - 2)(x + 2) = 0$

$\Rightarrow x = 0$ or $x - 2 = 0 \Rightarrow x = 2$ or $x + 2 = 0 \Rightarrow x = -2$, so $x = 0, -2, 2$

33. $\dfrac{3x}{4y} \cdot \dfrac{y}{9x^2} = \dfrac{3xy}{36x^2y} = \dfrac{3}{36} \cdot \dfrac{x}{x^2} \cdot \dfrac{y}{y} = \dfrac{1}{12} \cdot \dfrac{1}{x} \cdot 1 = \dfrac{1}{12x}$

34. $\dfrac{x}{x^2-4} \div \dfrac{2x}{x-2} = \dfrac{x}{(x-2)(x+2)} \cdot \dfrac{x-2}{2x} = \dfrac{x(x-2)}{2x(x-2)(x+2)}$

$= \dfrac{1}{2} \cdot \dfrac{x}{x} \cdot \dfrac{x-2}{x-2} \cdot \dfrac{1}{x+2} = \dfrac{1}{2} \cdot 1 \cdot 1 \cdot \dfrac{1}{x+2} = \dfrac{1}{2(x+2)}$

35. $\dfrac{1+\frac{2}{x}}{1-\frac{2}{x}} \cdot \dfrac{x}{x} = \dfrac{x+2}{x-2}$

36. $z = 3x - 2y \Rightarrow z + 2y = 3x \Rightarrow x = \dfrac{z+2y}{3}$

37. $\dfrac{4}{3x} - \dfrac{3}{4x} = 1 \Rightarrow \dfrac{4}{3x} \cdot \dfrac{4}{4} - \dfrac{3}{4x} \cdot \dfrac{3}{3} = 1 \Rightarrow \dfrac{16}{12x} - \dfrac{9}{12x} = 1 \Rightarrow \dfrac{7}{12x} = 1 \Rightarrow 7 = 12x \Rightarrow x = \dfrac{7}{12}$

38. $\dfrac{1}{x-1} + \dfrac{2}{x+2} = \dfrac{3}{2} \Rightarrow \dfrac{x+2}{(x-1)(x+2)} + \dfrac{2(x-1)}{(x-1)(x+2)} = \dfrac{3}{2} \Rightarrow \dfrac{x+2+2x-2}{(x-1)(x+2)} = \dfrac{3}{2} \Rightarrow \dfrac{3x}{(x-1)(x+2)} = \dfrac{3}{2}$

$\Rightarrow 6x = 3(x-1)(x+2) \Rightarrow 6x = 3(x^2+x-2) \Rightarrow 2x = x^2 + x - 2$

$\Rightarrow x^2 - x - 2 = 0 \Rightarrow (x-2)(x+1) = 0 \Rightarrow x-2 = 0, \ x = 2 \ \text{or} \ x+1 = 0, \ x = -1, \ \text{so} \ x = -1, 2$

39. $y = kx, \ x = 14, \ y = 7 \Rightarrow 7 = k \cdot 14 \Rightarrow k = \dfrac{7}{14} = \dfrac{1}{2}; \ y = \dfrac{1}{2}x, \ x = 11 \Rightarrow y = \dfrac{1}{2}(11) \Rightarrow y = 5.5$

40. Inverse variation, because as $x$ increases, $y$ decreases and $k = xy \Rightarrow k = 1 \cdot 20 = 2 \cdot 10 = 4 \cdot 5$, etc.

$k = xy \Rightarrow k = 1 \cdot 20 = 20 \Rightarrow y = \dfrac{k}{x} \Rightarrow y = \dfrac{20}{x}$

41. (a) $12x + 9x = 21x$

(b) $21x = 1890 \Rightarrow x = 90$ min

42. Let $x$ represent the minutes running and $60 - x$ represent the minutes walking.

$10x + 4(60 - x) = 450 \Rightarrow 10x + 240 - 4x = 450 \Rightarrow 6x = 210 \Rightarrow x = 35$, so 35 minutes running and 25 minutes walking.

43. (a) $x + x + y = 180 \Rightarrow 2x + y = 180$ and $2x = y + 32 \Rightarrow 2x - y = 32$

(b) Using elimination,

$2x + y = 180$

$\underline{2x - y = \phantom{0}32}$

$4x = 212 \Rightarrow x = 53 \Rightarrow 53°, \ 53°, \ 74°$

44. $y = \dfrac{k}{x}, \ x = 10, \ y = 1100 \Rightarrow 1100 = \dfrac{k}{10} \Rightarrow k = 11,000; \ y = \dfrac{11,000}{x}, \ x = 22 \Rightarrow y = \dfrac{11,000}{22} \Rightarrow y = 500$ lb

# Chapter 8: Introduction to Functions

## 8.1: Functions and Their Representations

1. function

3. symbolic

5. domain

7. one

9. True

11. $(a, b)$

13. 1

15. Yes, there is only one output for each input.

17. No, one exam can have many students who pass.

19. Yes, there is only one sales tax for each purchase.

21. $f(-1) = 4(-1) - 2 = -6$; $f(0) = 4(0) - 2 = -2$

23. $f(0) = \sqrt{0} = 0$; $f\left(\dfrac{9}{4}\right) = \sqrt{\dfrac{9}{4}} = \dfrac{3}{2}$

25. $f(-5) = (-5)^2 = 25$; $f\left(\dfrac{3}{2}\right) = \left(\dfrac{3}{2}\right)^2 = \dfrac{9}{4}$

27. $f(-8) = 3$; $f\left(\dfrac{7}{3}\right) = 3$

29. $f(-2) = 5 - (-2)^3 = 5 - (-8) = 13$; $f(3) = 5 - 3^3 = 5 - 27 = -22$

31. $f(-5) = \dfrac{2}{-5+1} = \dfrac{2}{-4} = -\dfrac{1}{2}$; $f(4) = \dfrac{2}{4+1} = \dfrac{2}{5}$

33. (a) Because there are 36 inches in 1 yard, the formula is $I(x) = 36x$.

    (b) $I(10) = 36(10) = 360$. There are 360 inches in 10 yards.

35. (a) Because there are 5280 feet in 1 mile, the formula is $M(x) = \dfrac{x}{5280}$.

    (b) $M(10) = \dfrac{10}{5280} \approx 0.0019$. Ten feet is equivalent to about 0.0019 mile.

37. (a) Because there are 43,560 square feet in 1 acre, the formula is $A(x) = 43,560x$.

    (b) $A(10) = 43,560(10) = 435,600$. There are 435,600 square feet in 10 acres.

39. $f = \{(1,3),(2,-4),(3,0)\}$; $D = \{1,2,3\}$, $R = \{-4,0,3\}$

41. $f = \{(a,b),(c,d),(e,a),(d,b)\}$; $D = \{a,c,d,e\}$, $R = \{a,b,d\}$

43. See Figure 43.

45. See Figure 45.

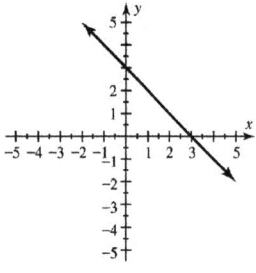

Figure 43

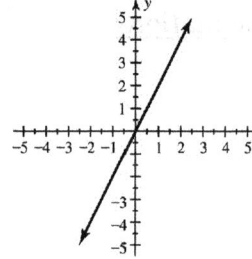

Figure 45

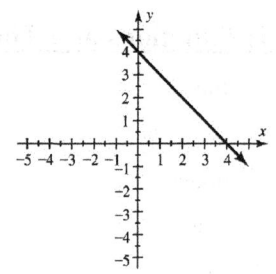

Figure 47

47. See Figure 47.

49. See Figure 49.

51. See Figure 51.

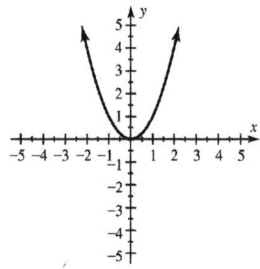

Figure 49

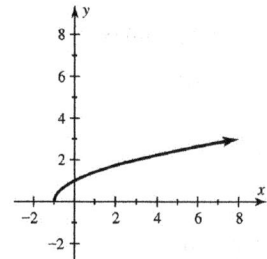

Figure 51

53. $f(0) = 3; f(2) = -1$

55. $f(-2) = 0; f(1) = 2$

57. $f(1) = -4; f(2) = -3$

59. $f(0) = 5.5; f(2) = 3.7$

61. $f(1990) = 26.9$ mpg ; In 1990 average fuel efficiency was 26.9 mpg.

63. Symbolic: $y = x + 5$. Numerical: See the table in Figure 63a.

    Graphical: See the graph in Figure 63b.

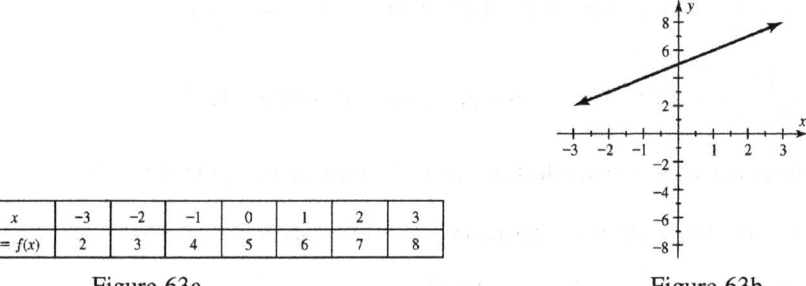

| $x$ | −3 | −2 | −1 | 0 | 1 | 2 | 3 |
|------|----|----|----|---|---|---|---|
| $y = f(x)$ | 2 | 3 | 4 | 5 | 6 | 7 | 8 |

Figure 63a                                          Figure 63b

65. Symbolic: $y = 5x - 2$. Numerical: See the table in Figure 65a.

    Graphical: See the graph in Figure 65b.

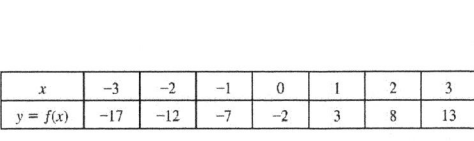

| $x$ | –3 | –2 | –1 | 0 | 1 | 2 | 3 |
|---|---|---|---|---|---|---|---|
| $y = f(x)$ | –17 | –12 | –7 | –2 | 3 | 8 | 13 |

Figure 65a

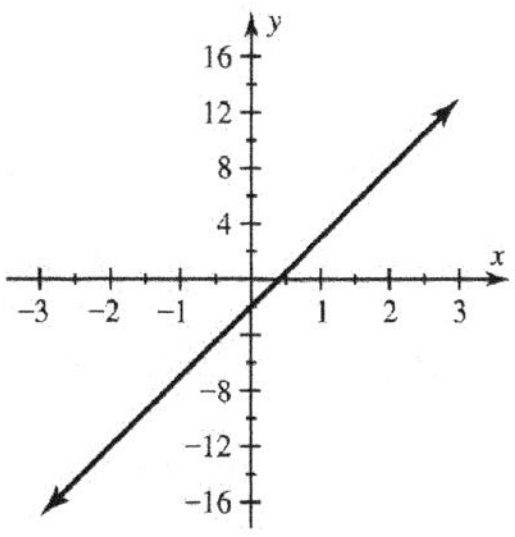

Figure 65b

67. Subtract $\frac{1}{2}$ from the input $x$ to obtain the output $y$.

69. Divide the input $x$ by 3 to obtain the output $y$.

71. Subtract 1 from the input $x$ and then take the square root to obtain the output $y$.

73. Symbolic: $f(x) = 0.50x$. Graphical: See the graph in Figure 73a. Numerical: See the table in

   Figure 73b.

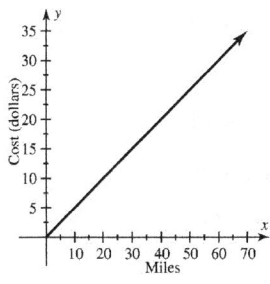

| Miles | 10 | 20 | 30 | 40 | 50 | 60 | 70 |
|---|---|---|---|---|---|---|---|
| Cost | $5 | $10 | $15 | $20 | $25 | $30 | $35 |

Figure 73a

Figure 73b

75. $S(2011) = 225(2011) - 450,650 = 452,475 - 450,650 = 1825$  In 2011 there were 1825 billion or

   1.825 trillion World Wide Web searches.

77. $D: -2 \le x \le 2$; $R: 0 \le y \le 2$

79. $D: -2 \le x \le 4$; $R: -2 \le y \le 2$

81. $D$: all real numbers; $R: y \ge -1$

83. $D: -3 \le x \le 3$; $R: -3 \le y \le 2$

85. $D = \{1, 2, 3, 4\}; R = \{5, 6, 7\}$

87. Any real number is a valid input for this function. $D$: all real numbers.

89. Any real number is a valid input for this function. $D$: all real numbers.

91. The denominator of this function cannot equal zero. $D: x \ne 5$.

93. The denominator of this function will never equal zero because the variable is squared and added to

   one. $D$: all real numbers.

95. The radicand must be greater than or equal to zero. $D: x \ge 1$.

97. Any real number is a valid input for this function. $D$: all real numbers.

99. The denominator of this function cannot equal zero. $D: x \ne 0$.

101. (a)    $W(2008) = 1726$;  In 2008 there were 1726 whales sighted in Maui.

    (b)    $D = \{2005, 2006, 2007, 2008, 2009\}$; $R = \{649, 1265, 959, 1726, 1010\}$

    (c)    The number of whale sightings increased every other year.

103. $D = \{1, 2, 3, ..., 20\}$; $R = \{200, 400, 600, ..., 4000\}$

105. No. The value 1 in the domain corresponds to more than one value in the range.

107. Yes. Each value in the domain corresponds to exactly one value in the range.

109. (a)    May is month number 5. The corresponding value for $P$ is 0.2.

    (b)    Yes. Each month has exactly one average precipitation.

    (c)    Months 2, 3, 7 and 11.

111. Yes. The graph passes the vertical line test. $D$ : all real numbers; $R$ : all real numbers

113. No. The graph does not pass the vertical line test.

115. Yes. The graph passes the vertical line test. $D : -4 \le x \le 4$; $R : 0 \le y \le 4$

117. Yes. The graph passes the vertical line test. $D$ : all real numbers; $R : y = 3$

119. No. The graph does not pass the vertical line test when $x = -2$.

121. Yes. The graph passes the vertical line test. $D = \{-6, -4, 2, 4\}$; $R = \{-4, 2\}$

123. Yes. Each value in the domain corresponds to exactly one value in the range.

125. No. The value 5 in the domain corresponds to more than one value in the range.

127. Walks away from home, then turns around and walks back a little slower.

129. See Figure 129.

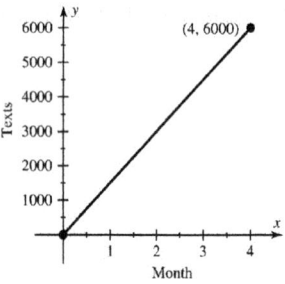

Figure 129

131. See Figure 131. There are 10 tick marks on the positive $x$-axis and 10 tick marks on the positive $y$-axis.

$[-10, 10, 1]$ by $[-10, 10, 1]$     $[0, 100, 10]$ by $[-50, 50, 10]$     $[1980, 1995, 1]$ by $[12{,}000, 16{,}000, 1000]$

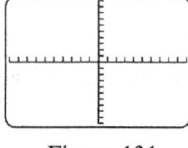

Figure 131

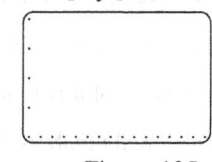

Figure 133          Figure 135

133. See Figure 133. There are 10 tick marks on the positive $x$-axis and 5 tick marks on the positive $y$-axis.

135. See Figure 135. There are 16 tick marks on the positive $x$-axis and 5 tick marks on the positive $y$-axis.

137. See Figure 137.

[-6, 6, 1] by [-6, 6, 1]   [-30, 30, 5] by [-50, 50, 5]   [-200, 200, 50] by [-250, 250, 50]

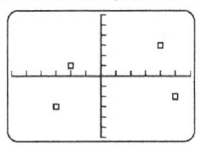
Figure 137

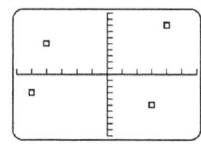
Figure 139

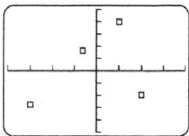
Figure 141

139. See Figure 139.

141. See Figure 141.

143. Table of Values                           Graphical

[-10, 10, 1] by [-10, 10 1]

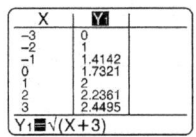

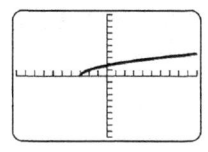

145. Table of Values                           Graphical

[-10, 10, 1] by [-10, 10, 1]

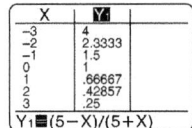

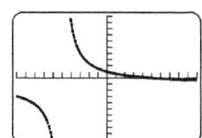

## Section 8.2 Linear Functions

1. $mx + b$

3. line

5. 7

7. True

9. Carpet costs \$2 per square foot. Ten square feet of carpet costs \$20.

11. Yes; $m = \dfrac{1}{2}$, $b = -6$

13. No, the variable is raised to the power of 2

15. Yes; $m = 0$, $b = -9$

17. Yes; $m = -9$, $b = 0$

19. Yes. The graph is a straight line.

21. No. The graph is not a straight line.

23. Yes. For each unit increase in $x$, the values of $f(x)$ increase by 3 units, so $m = \dfrac{3}{1} = 3$.

Because $f(x) = -6$ when $x = 0$, the $y$-intercept is $b = -6$. The function can be written

$f(x) = 3x - 6$.

25. Yes. For each 2-unit increase in $x$, the values of $f(x)$ decrease by 3 units, so $m = -\dfrac{3}{2}$.

Because $f(x) = 3$ when $x = 0$, the $y$-intercept is $b = 3$. The function can be written

$$f(x) = -\frac{3}{2}x + 3.$$

27. No. For each unit increase in $x$, the values of $f(x)$ do not increase by a constant amount.

29. Yes. For each unit increase in $x$, the values of $f(x)$ increase by 2 units, so $m = \dfrac{2}{1} = 2$.

Because $f(x) = 0$ when $x = 0$, the $y$-intercept is $b = 0$  The function can be written $f(x) = 2x$.

31. Yes. For each unit increase in $x$, the values of $f(x)$ increase by 0 units, so $m = 0$.

Because $f(x) = -4$ when $x = 0$, the $y$-intercept is $b = -4$  The function can be written $f(x) = -4$.

33. $f(-4) = 4(-4) = -16$ ; $f(5) = 4(5) = 20$

35. $f\left(-\dfrac{2}{3}\right) = 5 - \left(-\dfrac{2}{3}\right) = \dfrac{15}{3} + \dfrac{2}{3} = \dfrac{17}{3}$; $f(3) = 5 - 3 = 2$

37. $f\left(-\dfrac{3}{4}\right) = -22$ ; $f(13) = -22$

39. $f(-1) = -2$ ; $f(0) = 0$

41. $f(-2) = -1$ ; $f(4) = -4$

43. $f(-3) = 1$; $f(1) = 1$

45. $f(x) = 6x$; $f(3) = 6(3) = 18$

47. $f(x) = \dfrac{x}{6} - \dfrac{1}{2}$; $f(3) = \dfrac{(3)}{6} - \dfrac{1}{2} = \dfrac{1}{2} - \dfrac{1}{2} = 0$

49. The graph should have a positive slope and pass through (0, 0).  d

51. The graph should have a positive slope and pass through (0, –2).  b

53. See Figure 53.

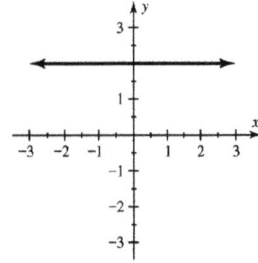
Figure 53

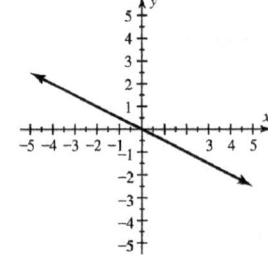
Figure 55

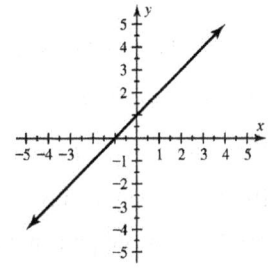
Figure 57

55. See Figure 55.

57. See Figure 57.

59. See Figure 59.

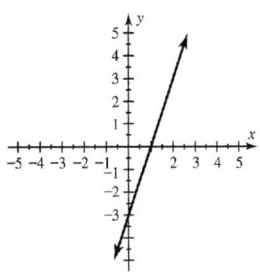

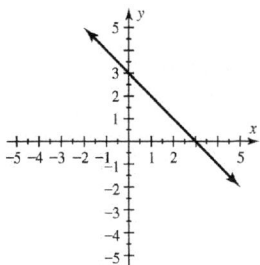

| Figure 59 | Figure 61 |

61. See Figure 61.

63. Since each pound is divided into 16 ounces: $f(x) = \dfrac{1}{16}x$

65. Since the car travels 65 miles each hour: $f(t) = 65t$

67. Since every day has 24 hours: $f(x) = 24$

69. *a*

71. (a)   *f* multiplies the input *x* by –2 and then adds 1 to obtain the output *y*.

    (b)   See Figure 71b.

    (c)   See Figure 71c.

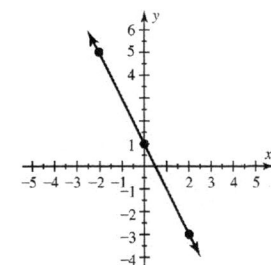

| $x$ | –2 | 0 | 2 |
|---|---|---|---|
| $y = f(x)$ | 5 | 1 | –3 |

| Figure 71b | Figure 71c |

73. (a)   *f* multiplies the input *x* by $\frac{1}{2}$ and then subtracts 1 to obtain the output *y*.

    (b)   See Figure 73b.

    (c)   See Figure 73c.

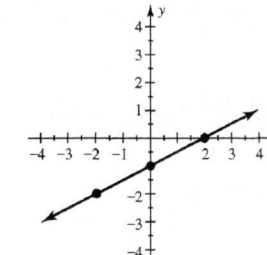

| $x$ | –2 | 0 | 2 |
|---|---|---|---|
| $y = f(x)$ | –2 | –1 | 0 |

| Figure 73b | Figure 73c |

75.        (a)                              (b)

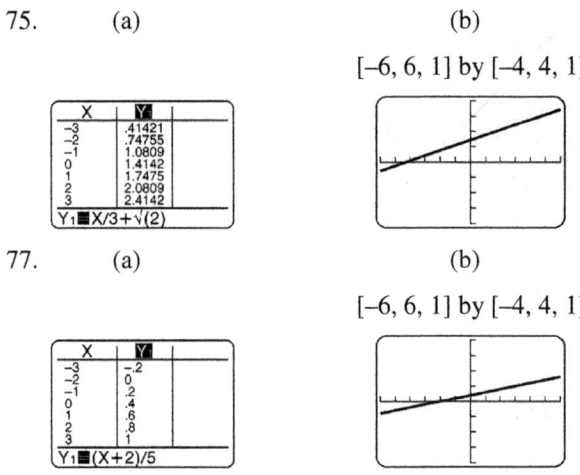

[–6, 6, 1] by [–4, 4, 1]

77.        (a)                              (b)

[–6, 6, 1] by [–4, 4, 1]

79. The graph should increase since the cost of tuition has been rising but it should not start at zero.  b

81. The graph should be a horizontal line since this distance has not changed over the past 10 years.  c

83. (a)    We divide the number of miles traveled by the miles per gallon.  $G(x) = \dfrac{x}{E}$

    (b)    We will multiply the function that was created in part (a) by 3.  $C(x) = \dfrac{3x}{E}$

85.  $M = \dfrac{-3+1}{2} = \dfrac{-2}{2} = -1$

87.  $M = \left(\dfrac{-3+1}{2}, \dfrac{-2+2}{2}\right) = \left(\dfrac{-2}{2}, \dfrac{0}{2}\right) = (-1, 0)$

89.  $M = \left(\dfrac{-4+3}{2}, \dfrac{3+(-2)}{2}\right) = \left(\dfrac{-1}{2}, \dfrac{1}{2}\right)$

91.  $M = \left(\dfrac{10+30}{2}, \dfrac{30+(-30)}{2}\right) = \left(\dfrac{40}{2}, \dfrac{0}{2}\right) = (20, 0)$

93.  $M = \left(\dfrac{-9+(-7)}{2}, \dfrac{-3+1}{2}\right) = \left(\dfrac{-16}{2}, \dfrac{-2}{2}\right) = (-8, -1)$

95.  $M = \left(\dfrac{\frac{1}{2}+\left(-\frac{5}{2}\right)}{2}, \dfrac{\frac{1}{3}+\left(-\frac{2}{3}\right)}{2}\right) = \left(\dfrac{\frac{-4}{2}}{2}, \dfrac{\frac{-1}{3}}{2}\right) = \left(-\dfrac{4}{4}, -\dfrac{1}{6}\right) = \left(-1, -\dfrac{1}{6}\right)$

97.  $M = \left(\dfrac{-0.3+0.7}{2}, \dfrac{0.1+0.4}{2}\right) = \left(\dfrac{0.4}{2}, \dfrac{0.5}{2}\right) = (0.2, 0.25)$

99.  $M = \left(\dfrac{2000+2010}{2}, \dfrac{5+13}{2}\right) = \left(\dfrac{4010}{2}, \dfrac{18}{2}\right) = (2005, 9)$

101.  $M = \left(\dfrac{a+3a}{2}, \dfrac{-b+5b}{2}\right) = \left(\dfrac{4a}{2}, \dfrac{4b}{2}\right) = (2a, 2b)$

103. (a)    $m = \dfrac{-3-5}{4-0} = \dfrac{-8}{4} = -2$; $y$−intercept: $5 \Rightarrow$ $f(x) = -2x+5$ and $f(2) = -2(2)+5 = -4+5 = 1$

(b)    $M = \left(\dfrac{0+4}{2}, \dfrac{5+(-3)}{2}\right) = \left(\dfrac{4}{2}, \dfrac{2}{2}\right) = (2, 1)$, so $f(2) = 1$

(c)    They are equal.

105. (a)    $m = \dfrac{3-(-1)}{7-(-3)} = \dfrac{4}{10} = \dfrac{2}{5}$; $y$−intercept: $3 = \dfrac{2}{5}(7)+b \Rightarrow 3 = \dfrac{14}{5}+b \Rightarrow \dfrac{15}{5}-\dfrac{14}{5} = b \Rightarrow \dfrac{1}{5} = b \Rightarrow$

$f(x) = \dfrac{2}{5}x+\dfrac{1}{5}$; $f(2) = \dfrac{2}{5}(2)+\dfrac{1}{5} = \dfrac{4}{5}+\dfrac{1}{5} = 1$

(b)    $M = \left(\dfrac{7+(-3)}{2}, \dfrac{3+(-1)}{2}\right) = \left(\dfrac{4}{2}, \dfrac{2}{2}\right) = (2, 1)$, so $f(2) = 1$

(c)    They are equal.

107. $M = \left(\dfrac{1990+2010}{2}, \dfrac{78.8+80.8}{2}\right) = \left(\dfrac{4000}{2}, \dfrac{159.6}{2}\right) = (2000, 79.8)$

In 2000, the life expectancy was about 79.8 years

109. $M = \left(\dfrac{1970+2010}{2}, \dfrac{205+308}{2}\right) = \left(\dfrac{3980}{2}, \dfrac{513}{2}\right) = (1990, 256.5)$

The population in 1990 was about 256.5 million.

111. $M = \left(\dfrac{1999+2009}{2}, \dfrac{40,700+49,800}{2}\right) = \left(\dfrac{4008}{2}, \dfrac{90,500}{2}\right) = (2004, 45,250)$

In 2004, the median income was $45,250.

113. (a)    Symbolic: $f(x) = 70$    Graphical: A graph of the function is shown in Figure 113a.

(b)    A table of the function is shown in Figure 113b.

(c)    The function $f$ is a constant function.

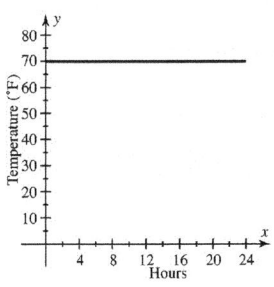

| Hours | 0 | 4 | 8 | 12 | 16 | 20 | 24 |
|---|---|---|---|---|---|---|---|
| Temp. (°F) | 70 | 70 | 70 | 70 | 70 | 70 | 70 |

Figure 113a                                                              Figure 113b

115. For each 2-hour increase in $t$, the distance increases by 120 miles, so $m = \dfrac{120}{2} = 60$. Because

$D = 50$ when $t = 0$, the $y$-intercept is $b = 50$. The function can be written $D(t) = 60t+50$.

117. (a)    Because the average kid under 18 sent 93 texts per day, the function is $K(x) = 93x$.

(b)    Because the average adult sent 1 text per day, the function is $A(x) = x$.

(c)   $K(365) = 93(365) = 33,945$, On average, someone under 18 sends 33,945 texts in 1 year, while someone over 65 sends 365 texts.

119. (a)   According to the table the sales in 2000 were 1.6 million.

(b)   For each unit increase in $x$, the values of $f(x)$ increase by 0.1 units, so the annual increase in sales was 0.1 million.

(c)   For each unit increase in $x$, the values of $f(x)$ increase by 0.1 units, so $m = \dfrac{0.1}{1} = 0.1$.

Because $f(x) = 1.6$ when $x = 0$, the $y$-intercept is $b = 1.6$ The function can be written

$f(x) = 0.1x + 1.6$

(d)   Let $x = 6 \Rightarrow f(6) = 0.1(6) + 1.6 = 0.6 + 1.6 = 2.2$. The result is 2.2 million.

121. (a)   Let $x = 4 \Rightarrow S(4) = 110(4) + 123 = 440 + 123 = 563$ million

(b)   In 2006 there were about 123 million people who used Skype.

(c)   The number of Skype users increased, on average, by 110 million per year.

123. (a)   Because each 1°C increase in temperature results in a 0.5 cubic centimeter increase in volume, $m = 0.5$. Because the volume is 137 cubic centimeters when the temperature is 0°C, the $y$-intercept is $b = 137$. The formula is $V(T) = 0.5T + 137$.

(b)   $V(50) = 0.5(50) + 137 = 25 + 137 = 162$ cubic centimeters.

125.   $f(x) = 40x$ In 30 days each additional pound of muscle will burn $40(30) = 1200$ calories. Then the amount of muscle necessary to lose 1 pound of fat is $3500 \div 1200 \approx 2.92$ lb.

127. (a)   According to the table 55% of those with a cell phone also subscribed to a data package.

(b)   The percentage changed by 4% per year.

(c)   For each year increase the percentage increases by 4, so $m = \dfrac{4}{1} = 4$. Because in 2007 (the starting year) the percentage is 55 the $y$-intercept is $b = 55$. The function can be written

$P(x) = 4x + 55$.

(d)   Let $x = 3 \Rightarrow P(3) = 4(3) + 55 = 12 + 55 = 67$ The result is 67% in 2010.

## Checking Basic Concepts  Sections 8.1 & 8.2

1. Symbolic: $f(x) = x^2 - 1$ Graphical: The graph is shown in Figure 1.

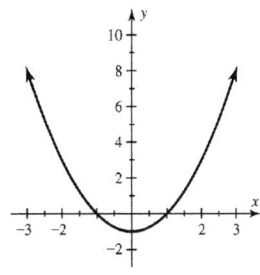

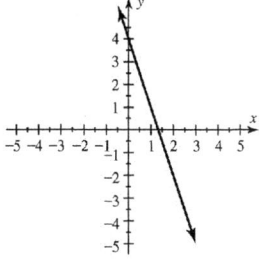

Figure 1                                    Figure 4

2. (a)   $D : -3 \le x \le 3; \ R : -4 \le y \le 4$

   (b)   $f(0) = 0; \ f(2) = 4$

   (c)   No. The graph is not a straight line.

3. (a)   Yes. The function is of the form $f(x) = mx + b$.

   (b)   No. The function can not be written in the form $f(x) = mx + b$.

   (c)   Yes. The function could be written $f(x) = 0x - 7$ which is of the form $f(x) = mx + b$.

   (d)   Yes. The function could be written $f(x) = 3x + 9$ which is of the form $f(x) = mx + b$.

4. The graph is shown in Figure 4.   $f(-2) = 4 - 3(-2) = 4 + 6 = 10$.

5. For each 2-unit increase in $x$, the values of $f(x)$ increase by 1 unit, so $m = \dfrac{1}{2}$.   Because

   $f(x) = -1$ when $x = 0$, the $y$-intercept is $b = -1$.   The function can be written $f(x) = \dfrac{1}{2}x - 1$.

6. (a)   $f(20) = 0.225(20) + 27.7 = 4.5 + 27.7 = 32.2$; In 1990 the median age was about 32 years.

   (b)   The number 0.225 means that the median age increased by 0.225 year each year.   The number 27.7 means that the initial median age in 1970 was 27.7 years.

7.   $M = \left( \dfrac{-3+5}{2}, \dfrac{4+(-6)}{2} \right) = \left( \dfrac{2}{2}, \dfrac{-2}{2} \right) = (1, -1)$

## Section 8.3 Compound Inequalities

1. An example of a compound inequality containing the word *and* is $x > 1$ and $x \le 7$; *Answers may vary.*

3. No, $1 \not\ge 3$.

5. Yes. The inequality can be written in either form.

7. Yes, $x = 2$ satisfies both inequalities. No, $x = 6$ does not satisfy $x - 1 < 5$.

9. No, $x = 0$ does not satisfy either inequality. Yes, $x = 3$ satisfies $2x \ge 3$.

11. No, $x = -3$ does not satisfy $2 - x \le 4$. Yes, $x = 0$ satisfies both inequalities.

13. $[2, 10]$

15. $(5, 8]$

17. $(-\infty, 4)$

19. $(-2, \infty)$

21. $[-2, 5)$

23. $(-8, 8]$

25. $(3, \infty)$

27. $(-\infty, -2] \cup [4, \infty)$

29. $(-\infty, 1) \cup [5, \infty)$

31. $(-3, 5]$

33. $(-\infty, -2)$

35. $(-\infty, 4)$

37. $(-\infty, 1) \cup (2, \infty)$

39. $x \le 3$ and $x \ge -1 \Rightarrow -1 \le x \le 3$; $\{x \mid -1 \le x \le 3\}$; See Figure 39.

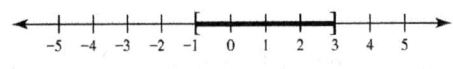
Figure 39

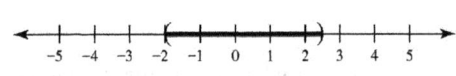
Figure 41

41. $2x < 5$ and $2x > -4 \Rightarrow x < \dfrac{5}{2}$ and $x > -2 \Rightarrow -2 < x < 2.5$; $\{x \mid -2 < x < 2.5\}$; See Figure 41.

43. $x + 2 > 5$ and $3 - x < 10 \Rightarrow x > 3$ and $-x < 7 \Rightarrow x > -7 \Rightarrow x > 3$; $\{x \mid x > 3\}$; See Figure 43.

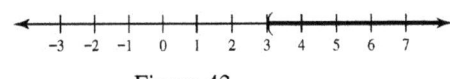
Figure 43

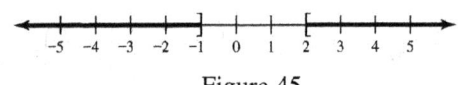
Figure 45

45. $x \le -1$ or $x \ge 2$; $\{x \mid x \le -1 \text{ or } x \ge 2\}$; See Figure 45.

47. $5 - x > 1$ or $x + 3 \ge -1 \Rightarrow -x > -4$ or $x \ge -4 \Rightarrow x < 4$ or $x \ge -4$; All real numbers; See Figure 47.

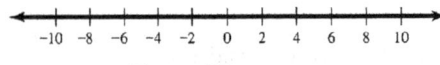
Figure 47

49. $x - 3 \le 4 \Rightarrow x \le 7$ and $x + 5 \ge -1 \Rightarrow x \ge -6$

The solutions must satisfy both of these inequalities. The interval is $[-6, 7]$.

51. $3t - 1 > -1 \Rightarrow 3t > 0 \Rightarrow t > 0$ and $2t - \dfrac{1}{2} > 6 \Rightarrow 2t > \dfrac{13}{2} \Rightarrow t > \dfrac{13}{4}$

The solutions must satisfy both of these inequalities. The interval is $\left( \dfrac{13}{4}, \infty \right)$.

53. $x - 4 \geq -3 \Rightarrow x \geq 1$ or $x - 4 \leq 3 \Rightarrow x \leq 7$

The solutions may satisfy either one or both of these inequalities. The interval is $(-\infty, \infty)$.

55. $-x < 1 \Rightarrow x > -1$ or $5x + 1 < -10 \Rightarrow 5x < -11 \Rightarrow x < -\dfrac{11}{5}$

The solutions may satisfy either one or both of these inequalities. The interval is

$\left(-\infty, -\dfrac{11}{5}\right) \cup (-1, \infty)$.

57. $1 - 7x < -48 \Rightarrow -7x < -49 \Rightarrow x > 7$ and $3x + 1 \leq -9 \Rightarrow 3x \leq -10 \Rightarrow x \leq -\dfrac{10}{3}$

The solutions must satisfy both of these inequalities. This is not possible. No solutions.

59. $-2 \leq t + 4 < 5 \Rightarrow -6 \leq t < 1; [-6, 1)$

61. $-\dfrac{5}{8} \leq y - \dfrac{3}{8} < 1 \Rightarrow -\dfrac{1}{4} \leq y < \dfrac{11}{8}; \left[-\dfrac{1}{4}, \dfrac{11}{8}\right)$

63. $-27 \leq 3x \leq 9 \Rightarrow -9 \leq x \leq 3; [-9, 3]$

65. $\dfrac{1}{2} < -2y \leq 8 \Rightarrow -\dfrac{1}{4} > y \geq -4 \Rightarrow -4 \leq y < -\dfrac{1}{4}; \left[-4, -\dfrac{1}{4}\right)$

67. $-4 < 5z + 1 \leq 6 \Rightarrow -5 < 5z \leq 5 \Rightarrow -1 < z \leq 1; (-1, 1]$

69. $3 \leq 4 - n \leq 6 \Rightarrow -1 \leq -n \leq 2 \Rightarrow 1 \geq n \geq -2 \Rightarrow -2 \leq n \leq 1; [-2, 1]$

71. $-1 < 2z - 1 < 3 \Rightarrow 0 < 2z < 4 \Rightarrow 0 < z < 2; (0, 2)$

73. $-2 \leq 5 - \dfrac{1}{3}m < 2 \Rightarrow -7 \leq -\dfrac{1}{3}m < -3 \Rightarrow 21 \geq m > 9 \Rightarrow 9 < m \leq 21; (9, 21]$

75. $100 \leq 10(5x - 2) \leq 200 \Rightarrow 10 \leq 5x - 2 \leq 20 \Rightarrow 12 \leq 5x \leq 22 \Rightarrow \dfrac{12}{5} \leq x \leq \dfrac{22}{5}; \left[\dfrac{12}{5}, \dfrac{22}{5}\right]$

77. $-3 < \dfrac{3z + 1}{4} < 1 \Rightarrow -12 < 3z + 1 < 4 \Rightarrow -13 < 3z < 3 \Rightarrow -\dfrac{13}{3} < z < 1; \left(-\dfrac{13}{3}, 1\right)$

79. $-\dfrac{5}{2} \leq \dfrac{2 - m}{4} \leq \dfrac{1}{2} \Rightarrow -10 \leq 2 - m \leq 2 \Rightarrow -12 \leq -m \leq 0 \Rightarrow 12 \geq m \geq 0 \Rightarrow 0 \leq m \leq 12; [0, 12]$

81. The values of $3x$ are between $-3$ and $6$ when $-1 \leq x \leq 2$. The interval is $[-1, 2]$.

83. The values of $1 - x$ are between $-1$ and $2$ when $-1 < x < 2$. The interval is $(-1, 2)$.

85. The values of $y_1$ are between the lines $y = -2$ and $y = 2$ when $-3 \leq x \leq 1$. The interval is $[-3, 1]$.

87. The values of $y_1$ are below the line $y = -2$ or above the line $y = 2$ when $x < -2$ or $x > 0$.

The interval is $(-\infty, -2) \cup (0, \infty)$.

89. (a)    The car is moving toward Omaha since the distance is decreasing.

    (b)    The car is 100 miles from Omaha when $x = 4$ hr. The car is 200 miles from Omaha when $x = 2$ hr.

    (c)    The car is 100 to 200 miles from Omaha when the elapsed time is between 2 and 4 hours.

    (d)    The car's distance from Omaha is greater than or equal to 200 miles during the first 2 hours.

91. (a)    Graphs $y_1$ and $y_2$ intersect when $x = 2$.

    (b)    Graphs $y_2$ and $y_3$ intersect when $x = 4$.

    (c)    The graph of $y_2$ is between the graphs of $y_1$ and $y_3$ when $x \leq 4$ and $x \geq 2$. The solution set is $\{x \mid 2 \leq x \leq 4\}$.

    d)    The graph of $y_2$ is below the graph of $y_1$ when $x < 2$ and $x \geq 0$. The solution set is $\{x \mid 0 \leq x < 2\}$.

93. Numerical: Table $Y_1 = 2X - 4$ with TblStart = 0 and $\Delta$Tbl = 1. See Figure 93a.

    Graphical: Graph $Y_1 = -2$, $Y_2 = 2X - 4$ and $Y_3 = 4$ in [–5, 5, 1] by [–5, 5, 1]. See Figure 93b.

    The solution is $[1, 4]$.

[–5, 5, 1] by [–5, 5, 1]          [–5, 5, 1] by [–5, 5, 1]

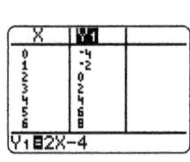     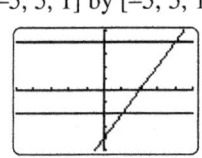     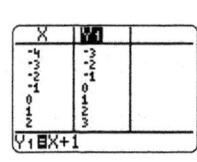     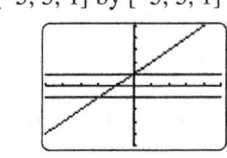

Figure 93a            Figure 93b            Figure 95a            Figure 95b

95. Numerical: Table $Y_1 = X + 1$ with TblStart = –4 and $\Delta$Tbl = 1. See Figure 95a.

    Graphical: Graph $Y_1 = -1$, $Y_2 = X + 1$ and $Y_3 = 1$ in [–5, 5, 1] by [–5, 5, 1]. See Figure 95b.

    The solution is $(-\infty, -2) \cup (0, \infty)$.

97. Numerical: Table $Y_1 = 25(X - 2000) + 45$ with TblStart = 2000 and $\Delta$Tbl = 1. See Figure 97a.

    Graphical: Graph $Y_1 = 95$, $Y_2 = 25(X - 2000) + 45$ and $Y_3 = 295$ in [2000, 2020, 1] by [0, 325, 25].

    See Figure 97b. The solution is $[2002, 2010]$.

[2000, 2020, 1] by [0, 325, 25]      [0, 10, 1] by [0, 20, 1]

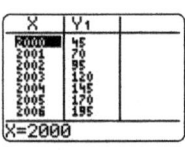     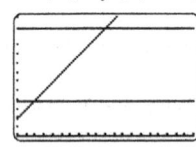     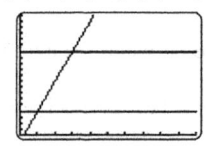     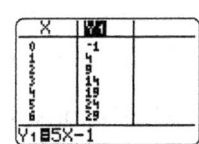

Figure 97a            Figure 97b            Figure 99a            Figure 99b

99. $4 \leq 5x - 1 \leq 14 \Rightarrow 5 \leq 5x \leq 15 \Rightarrow 1 \leq x \leq 3; [1, 3]$

    Graphical: Graph $Y_1 = 4$, $Y_2 = 5X - 1$ and $Y_3 = 14$ in [0, 10, 1] by [0, 20, 1]. See Figure 99a.

    Numerical: Table $Y_1 = 5X - 1$ with TblStart = 0 and $\Delta$Tbl = 1. See Figure 99b.

    The solution is $[1, 3]$.

101. $4 - x \geq 1$ or $4 - x < 3 \Rightarrow -x \geq -3$ or $-x < -1 \Rightarrow x \leq 3$ or $x > 1; (-\infty, \infty)$

   Graphical: Graph $Y_1 = 1$, $Y_2 = 4 - X$ and $Y_3 = 3$ in $[-10, 10, 1]$ by $[-10, 10, 1]$. See Figure 101a.

   Numerical: Table $Y_1 = 4 - X$ with TblStart $= -1$ and $\Delta$Tbl $= 1$. See Figure 101b.

   The solution is $(-\infty, \infty)$.

   $[-10, 10, 1]$ by $[-10, 10, 1]$  $[-10, 10, 1]$ by $[-10, 10, 1]$

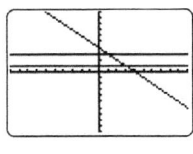   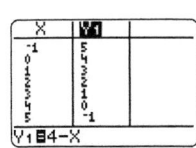   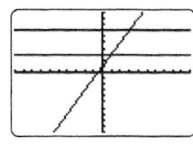   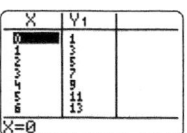

   Figure 101a           Figure 101b           Figure 103a           Figure 103b

103. $2x + 1 < 3$ or $2x + 1 \geq 7 \Rightarrow x < 1$ or $x \geq 3; (-\infty, 1) \cup [3, \infty)$

   Graphical: Graph $Y_1 = 3$, $Y_2 = 2X + 1$ and $Y_3 = 7$ in $[-10, 10, 1]$ by $[-10, 10, 1]$. See Figure 103a.

   Numerical: Table $Y_1 = 2X + 1$ with TblStart $= 0$ and $\Delta$Tbl $= 1$. See Figure 103b.

   The solution is $(-\infty, 1) \cup [3, \infty)$.

105. $c < x + b \leq d \Rightarrow c - b < x \leq d - b; (c - b, d - b]$

107. (a)   See Figure 107a. From 2006 to 2008

   (b)   See Figures 107b. From 2006 to 2008

   (c)   $15 \leq 2.5x - 5000 \leq 20 \Rightarrow 5015 \leq 2.5x \leq 5020 \Rightarrow 2006 \leq x \leq 2008$

   $[2000, 2010, 1]$ by $[10, 25, 1]$    $[2000, 2010, 1]$ by $[10, 25, 1]$

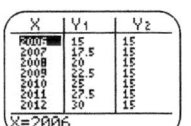   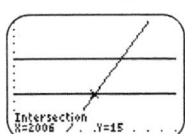   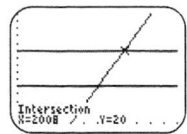

   Figure 107a                    Figure 107b

109. $51.3 \leq 60 - 5.8x \leq 57.1 \Rightarrow -8.7 \leq -5.8x \leq -2.9 \Rightarrow 1.5 \geq x \geq 0.5$; The dew point ranges from $57.1°F$ to $51.3°F$ from 0.5 to 1.5 miles high.

111. The perimeter of the rectangle is given by $2(x + 3) + 2(2x)$.

   $40 \leq 2(x + 3) + 2(2x) \leq 60 \Rightarrow 40 \leq 6x + 6 \leq 60 \Rightarrow 34 \leq 6x \leq 54 \Rightarrow 5.\overline{6} \leq x \leq 9$

113. $22.5 \leq 70 - 19x \leq 41.5 \Rightarrow -47.5 \leq -19x \leq -28.5 \Rightarrow 2.5 \geq x \geq 1.5 \Rightarrow 1.5 \leq x \leq 2.5$

   The air temperature is from $22.5°F$ to $41.5°F$ at altitudes from 1.5 to 2.5 miles.

115. (a)   $m = \dfrac{500 - 250}{2010 - 2000} = \dfrac{250}{10} = 25$. Since $x$ represents the number of years after 2000 the

   $y$- intercept is 250. The function is $M(x) = 25x + 250$.

   (b)   $300 \leq 25x + 250 \leq 400 \Rightarrow 50 \leq 25x \leq 150 \Rightarrow 2 \leq x \leq 6$. The result is 2002 to 2006.

## Section 8.4: Other Functions and Their Properties

1. domain

3. $(-\infty, \infty)$

5. absolute value

7. 2

9. rational

11. c

13. $D = (-\infty, \infty);\ R = (-\infty, \infty)$

15. $D = (-\infty, \infty);\ R = (-\infty, \infty)$

17. $D = (-\infty, \infty);\ R = [2, \infty)$

19. $D = (-\infty, \infty);\ R = (-\infty, 0]$

21. $x + 1 \geq 0 \Rightarrow x \geq -1;\ D = [-1, \infty);\ R = [0, \infty)$

23. $D = (-\infty, \infty);\ R = [0, \infty)$

25. $x - 1 = 0 \Rightarrow x = 1;\ D = (-\infty, 1) \cup (1, \infty)$

27. $6 - 3x = 0 \Rightarrow 6 = 3x \Rightarrow 2 = x;\ D = (-\infty, 2) \cup (2, \infty)$

29. $t^2 - 4 = 0 \Rightarrow (t + 2)(t - 2) = 0 \Rightarrow t = -2 \text{ or } t = 2;\ D = (-\infty, -2) \cup (-2, 2) \cup (2, \infty)$

31. $t^2 - 2t = 0 \Rightarrow t(t - 2) = 0 \Rightarrow t = 0 \text{ or } t = 2;\ D = (-\infty, 0) \cup (0, 2) \cup (2, \infty)$

33. $z^3 - 1 = 0 \Rightarrow z = 1;\ D = (-\infty, 1) \cup (1, \infty)$

35. $x^2 - 2x - 3 = 0 \Rightarrow (x - 3)(x + 1) = 0 \Rightarrow x = 3 \text{ or } x = -1;\ D = (-\infty, -1) \cup (-1, 3) \cup (3, \infty)$

37. $D = (-\infty, \infty);\ R = (-\infty, \infty)$

39. $D = [-2, 2];\ R = [-2, 2]$

41. $D = [-2, 3];\ R = [-2, 2]$

43. Yes $f(x) = 5x - 11$ represents a polynomial function. The degree is 1, and it is a linear function.

45. Yes, $f(x) = x^3$ represents a polynomial function. The degree is 3, and it is a cubic function.

47. No, $f(x) = \dfrac{6}{x + 5}$ is not a polynomial function. It is a rational function

49. Yes, $f(x) = 1 + 2x - x^2$ is a polynomial function. The degree is 2, and it is a quadratic function.

51. No, $f(x) = 5x^{-2}$ is not a polynomial function. It is a rational function, $f(x) = \dfrac{5}{x^2}$.

53. Yes, $f(x) = x^4 + 2x^2$ is a polynomial function. The degree is 4, and it is a fourth degree polynomial function.

55. $g(3) = |4 \cdot 3| = |12| = 12 \quad g(0) = |4 \cdot 0| = |0| = 0$

57. $g(1) = |1-2| = |-1| = 1$   $g\left(-\dfrac{3}{4}\right) = \left|-\dfrac{3}{4} - 2\right| = \left|-\dfrac{3}{4} - \dfrac{8}{4}\right| = \left|-\dfrac{11}{4}\right| = \dfrac{11}{4}$

59. $g(3) = 3^2 - 3 - 6 = 9 - 3 - 6 = 0$,   $g(-3) = (-3)^2 - (-3) - 6 = 9 + 3 - 6 = 6$

61. $g(2) = -2(2)^3 + 2 = -2(8) + 2 = -16 + 2 = -14$,   $g(-2) = -2(-2)^3 + (-2) = -2(-8) - 2 = 16 - 2 = 14$

63. $g(0) = 0^2 - 2(0) - 6 = 0 - 0 - 6 = -6$,   $g(-3) = (-3)^2 - 2(-3) - 6 = 9 + 6 - 6 = 9$

65. $g(11) = \dfrac{1}{11}$,   $g(-7) = -\dfrac{1}{7}$

67. $g(5) = -\dfrac{5}{5+1} = -\dfrac{5}{6}$,   $g(-1) = -\dfrac{-1}{-1+1} = \dfrac{1}{0}$, which is undefined.

69. $g(-5) = \dfrac{(-5)^2}{(-5)^2 - (-5)} = \dfrac{25}{25+5} = \dfrac{25}{30} = \dfrac{5}{6}$,   $g(1) = \dfrac{1^2}{1^2 - 1} = \dfrac{1}{1-1} = \dfrac{1}{0}$, which is undefined.

71. From the graph, $f(0) = 1$ and $f(1) = -1$.

   $f(0) = 1 - 2(0) = 1 - 0 = 1$,   $f(1) = 1 - 2(1) = 1 - 2 = -1$

73. From the graph, $f(-1) = -2$ and $f(2) = -2$.

   $f(-1) = 3(-1) - (-1)^3 = -3 - (-1) = -3 + 1 = -2$,   $f(2) = 3(2) - 2^3 = 6 - 8 = -2$

75. From the graph, $f(-2) = -4$ and $f(2) = 0$.

   $f(-2) = (-2) - \dfrac{1}{2}(-2)^2 = -2 - \dfrac{1}{2}(4) = -2 - 2 = -4$,   $f(2) = 2 - \dfrac{1}{2}(2)^2 = 2 - \dfrac{1}{2}(4) = 2 - 2 = 0$

77. From the graph, $f(-3) = -1$ and $f(-1)$ is undefined.

   $f(-3) = \dfrac{2}{-3+1} = \dfrac{2}{-2} = -1$,   $f(-1) = \dfrac{2}{-1+1} = \dfrac{2}{0}$, which is undefined.

79.

| $x$ | $-2$ | $-1$ | $0$ | $1$ | $2$ |
|-----|------|------|-----|-----|-----|
| $f(x) = \dfrac{1}{x-1}$ | $-\dfrac{1}{3}$ | $-\dfrac{1}{2}$ | $-1$ | $-$ | $1$ |

When $x = 1$, $\dfrac{1}{x-1} = \dfrac{1}{1-1} = \dfrac{1}{0}$, so $f(1)$ is undefined.

81. See Figure 81.

83. See Figure 83.

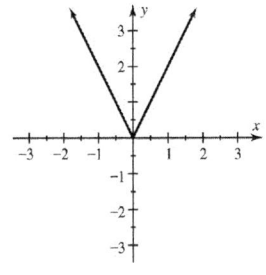

Figure 81

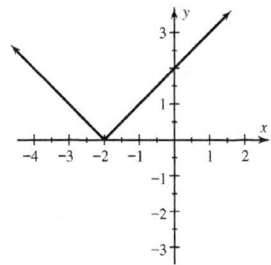

Figure 83

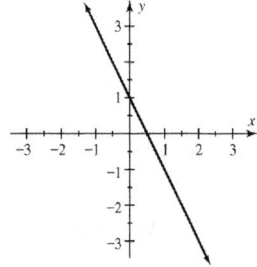

Figure 85

85. See Figure 85.

87. See Figure 87.

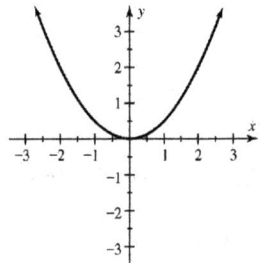

Figure 87

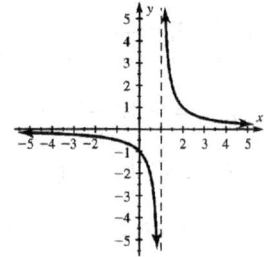

Figure 89

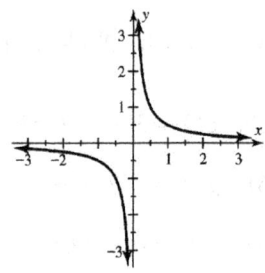

Figure 91

89. See Figure 89.

91. See Figure 91.

93. See Figure 93.

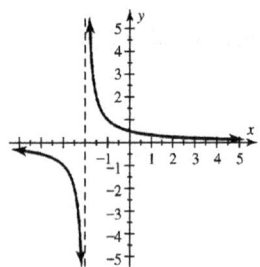

Figure 93

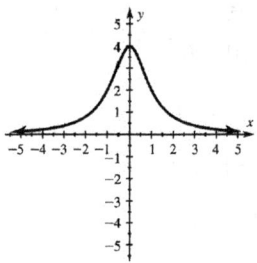

Figure 95

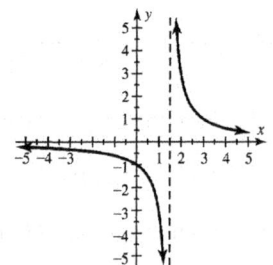

Figure 97

95. See Figure 95.

97. See Figure 97.

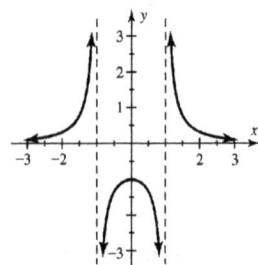

Figure 99

99. See Figure 99.

101. (a)    $(f+g)(3) = 5(3) + (3+1) = 15 + 4 = 19$

   (b)    $(f-g)(-2) = 5(-2) - (-2+1) = -10 - (-1) = -9$

   (c)    $(fg)(5) = (5 \cdot 5)(5+1) = 25(6) = 150$

   (d)    $(f/g)(0) = (5 \cdot 0)/(0+1) = \dfrac{0}{1} = 0$

103. (a)    $(f+g)(3) = (2 \cdot 3 - 1) + (4 \cdot 3^2) = 6 - 1 + 4 \cdot 9 = 5 + 36 = 41$

   (b)    $(f-g)(-2) = (2(-2) - 1) - (4(-2)^2) = -4 - 1 - 4(4) = -5 - 16 = -21$

   (c)    $(fg)(5) = (2 \cdot 5 - 1)(4 \cdot 5^2) = (9)(100) = 900$

(d)    $(f/g)(0) = (2 \cdot 0 - 1)/(4 \cdot 0^2) = \dfrac{-1}{0}$, which is undefined

105. (a)    $(f+g)(x) = (x+1)+(x+2) = 2x+3$

   (b)    $(f-g)(x) = (x+1)-(x+2) = -1$

   (c)    $(fg)(x) = (x+1)(x+2) = x^2 +3x+2$

   (d)    $(f/g)(x) = \dfrac{x+1}{x+2}$

107. (a)    $(f+g)(x) = (1-x)+(x^2) = x^2 -x+1$

   (b)    $(f-g)(x) = (1-x)-(x^2) = 1-x-x^2$

   (c)    $(fg)(x) = (1-x)(x^2) = x^2 -x^3$

   (d)    $(f/g)(x) = \dfrac{1-x}{x^2}$

109.  $f(a) = a^2 - 2a$

111. The graph should increase quickly at first and then continue to increase at a slower rate. The answer is graph c.

113. The graph should increase slowly at first and then continue to increase at a faster rate. The answer is graph d.

115. (a)    No; the values for $P(0)$ and $P(10)$ match the table but the other values do not.
       *Answers may vary.*

   (b)    Yes, $P$ does provide a reasonable model because the values of $P$ are very close to the values in the table.

   (c)    No, since the race is only 10 seconds long; $0 \le t \le 10$

117. (a)    $T(4) = \dfrac{1}{5-(4)} = \dfrac{1}{1} = 1$;  When cars leave the ramp at a rate of 4 vehicles per minute, the wait is 1 minute.

   (b)    As more cars try to exit, the waiting time increases. This agrees with intuition.

   (c)    $3 = \dfrac{1}{5-x} \Rightarrow 3(5-x) = 1 \Rightarrow 15-3x = 1 \Rightarrow -3x = -14 \Rightarrow x = 4.\overline{6}$ vehicles per minute.

119. (a)    $D(0.05) = \dfrac{900}{10.5+30(0.05)} = \dfrac{900}{12} = 75$;  The braking distance is 75 feet when the uphill grade is 0.05.

   (b)    $60 = \dfrac{900}{10.5+30x} \Rightarrow 60(10.5+30x) = 900 \Rightarrow 630+1800x = 900 \Rightarrow 1800x = 270 \Rightarrow x = 0.15$

121. (a)    $f(10) = 2.4(10)^2 -14(10)+23 = 240-140+23 = 123$ thousand, which is close to the actual value.

(b)    $f(17) = 2.4(17)^2 - 14(17) + 23 = 693.6 - 238 + 23 = 478.6$ thousand, which is too high. AIDS

deaths did not continue to rise as rapidly as the model predicts.

123.   $R(1) = \dfrac{100}{1.2(1)+1} \approx 45.45$, $R(3) = \dfrac{100}{1.2(3)+1} \approx 21.7$ After 1 day (3 days) students remember about

45% (22%) of what they learned.

125.   (a)    $C(100) = 0.3(100) + 100 = 130$. It costs $130 thousand to make 100 notebook computers.

(b)    The $y$-intercept for $C$ is 100. The company has $100 thousand in fixed costs even if they make

0 computers. The $y$-intercept for $R$ is 0. If the company sells 0 computers, their revenue is $0.

(c)    $P(x) = R(x) - C(x) = 0.75x - (0.3x + 100) = 0.45x - 100$

(d)    A profit will occur when revenue is greater than cost.

$0.75x > 0.3x + 100 \Rightarrow 0.45x > 100 \Rightarrow x > 222.22$. So 223 or more computers need to be sold.


## Checking Basic Concepts Sections 8.3 and 8.4

1.  (a)    Yes, because $2(3) - 1 = 6 - 1 = 5 \geq 3$ is true.

(b)    No, because $3 + 2 = 5 \not< 4$.

2.  (a)    $-5 \leq 2x + 1 \leq 3 \Rightarrow -6 \leq 2x \leq 2 \Rightarrow -3 \leq x \leq 1$

The solution is $[-3, 1]$.

(b)    $1 - x \leq -2$      or      $1 - x \geq 2$

$-x \leq -3$      or      $-x \geq 1$

$x \geq 3$      or      $x \leq -1$

The solution is $(-\infty, -1] \cup [3, \infty)$.

(c)    $-2 < \dfrac{4 - 3x}{2} \leq 6 \Rightarrow -4 < 4 - 3x \leq 12 \Rightarrow -8 < -3x \leq 8 \Rightarrow \dfrac{8}{3} > x \geq -\dfrac{8}{3}$

The solution is $\left[ -\dfrac{8}{3}, \dfrac{8}{3} \right)$.

3.  (a)    $D = (-\infty, \infty)$

(b)    $t - 1 = 0 \Rightarrow t = 1$;   $D = (-\infty, 1) \cup (1, \infty)$

(c)    $D = [0, \infty)$

4.  (a)    $D = [-2, 1]$; $R = [-3, 1]$

(b)    $f(0) = 1$; $f(-2) = -3$

5. See Figure 5.

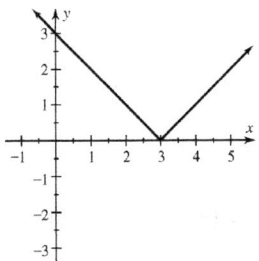

Figure 5

## Section 8.5 Absolute Value Equations and Inequalities

1. An example of an absolute value equation is $|3x+2| = 6$. *Answers may vary.*

3. Yes, since $|-3| = 3$.

5. Yes

7. 2 times

9. No, $|2(-3)-5| = |-11| = 11 \neq 1$.  Yes, $|2(3)-5| = |1| = 1$.

11. No, $|7-4(-2)| = |15| = 15 \nleq 5$.  Yes, $|7-4(2)| = |-1| = 1 \leq 5$.

13. Yes, $\left|7\left(-\dfrac{4}{7}\right)+4\right| = |-4+4| = |0| = 0 > -1$.  Yes, $|7(2)+4| = |18| = 18 > -1$.

15. $x = 0$ and $x = 4$

17. $x = -3$ and $x = 3$

19. $-3 < x < 3 \Rightarrow (-3,3)$

21. $x < -3$ or $x > 3 \Rightarrow (-\infty,-3) \cup (3,\infty)$

23. $x = -7$ and $x = 7$

25. An absolute value can not be negative.  No solution.

27. $4x = 9 \Rightarrow x = \dfrac{9}{4}$  or  $4x = -9 \Rightarrow x = -\dfrac{9}{4}$

29. Since $|-2x|-6 = 2 \Rightarrow |-2x| = 8$:  $-2x = 8 \Rightarrow x = -4$  or  $-2x = -8 \Rightarrow x = 4$

31. $2x+1 = 11 \Rightarrow 2x = 10 \Rightarrow x = 5$  or  $2x+1 = -11 \Rightarrow 2x = -12 \Rightarrow x = -6$

33. Since $|-2x+3|+3 = 4 \Rightarrow |-2x+3| = 1$:

$-2x+3 = 1 \Rightarrow -2x = -2 \Rightarrow x = 1$  or  $-2x+3 = -1 \Rightarrow -2x = -4 \Rightarrow x = 2$

35. $|3-4x| = 0 \Rightarrow 3-4x = 0 \Rightarrow 3 = 4x \Rightarrow \dfrac{3}{4} = x$

37. $\frac{1}{2}x - 1 = 5 \Rightarrow \frac{1}{2}x = 6 \Rightarrow x = 12$  or  $\frac{1}{2}x - 1 = -5 \Rightarrow \frac{1}{2}x = -4 \Rightarrow x = -8$

39. An absolute value can not be negative. No solution.

41. Since $\left|\frac{2}{3}z - 1\right| - 3 = 8 \Rightarrow \left|\frac{2}{3}z - 1\right| = 11$:

$\frac{2}{3}z - 1 = 11 \Rightarrow \frac{2}{3}z = 12 \Rightarrow z = 18$  or  $\frac{2}{3}z - 1 = -11 \Rightarrow \frac{2}{3}z = -10 \Rightarrow z = -15$

43. $z - 1 = 2z \Rightarrow -z = 1 \Rightarrow z = -1$  or  $z - 1 = -2z \Rightarrow 3z = 1 \Rightarrow z = \frac{1}{3}$

45. $3t + 1 = 2t - 4 \Rightarrow t = -5$  or  $3t + 1 = -2t + 4 \Rightarrow 5t = 3 \Rightarrow t = \frac{3}{5}$

47. $\frac{1}{4}x = 3 + \frac{1}{4}x \Rightarrow 0 = 3$ (no solution)  or  $\frac{1}{4}x = -3 - \frac{1}{4}x \Rightarrow \frac{1}{2}x = -3 \Rightarrow x = -6$

49. (a)   $2x = 8 \Rightarrow x = 4$  or  $2x = -8 \Rightarrow x = -4$

    (b)   $\{x \mid -4 < x < 4\}$

    (c)   $\{x \mid x < -4 \text{ or } x > 4\}$

51. (a)   $5 - 4x = 3 \Rightarrow -4x = -2 \Rightarrow x = \frac{1}{2}$  or  $5 - 4x = -3 \Rightarrow -4x = -8 \Rightarrow x = 2$

    (b)   $\left\{x \mid \frac{1}{2} \le x \le 2\right\}$

    (c)   $\left\{x \mid x \le \frac{1}{2} \text{ or } x \ge 2\right\}$

53. The solutions to $|x| \le 3$ satisfy $-3 \le x \le 3$ .

    The interval is $[-3, 3]$.

55. The solutions to $|k| > 4$ satisfy

    The interval is $(-\infty, -4) \cup (4, \infty)$.

57. The inequality $|t| \le -3$ has no solutions because absolute value is never negative.

59. The inequality $|z| > 0$ is true for any value of $z$ except $z = 0$. The interval is $(-\infty, 0) \cup (0, \infty)$.

61. The solutions to $|2x| > 7$ satisfy $x < c$ or $x > d$ where $c$ and $d$ are the solutions to $|2x| = 7$.

    $|2x| = 7$ is equivalent to $2x = -7 \Rightarrow x = -\frac{7}{2}$  or  $2x = 7 \Rightarrow x = \frac{7}{2}$.

    The interval is $\left(-\infty, -\frac{7}{2}\right) \cup \left(\frac{7}{2}, \infty\right)$.

63. The solutions to $|-4x+4| < 16$ satisfy $c < x < d$ where $c$ and $d$ are the solutions to $|-4x+4| = 16$.

    $|-4x+4| = 16$ is equivalent to $-4x+4 = -16 \Rightarrow x = 5$ or $-4x+4 = 16 \Rightarrow x = -3$.

    The interval is $(-3,\ 5)$.

65. First divide each side of $2|x+5| \geq 8$ by 2 to obtain $|x+5| \geq 4$.

    The solutions to $|x+5| \geq 4$ satisfy $x \leq c$ or $x \geq d$ where $c$ and $d$ are the solutions to $|x+5| = 4$.

    $|x+5| = 4$ is equivalent to $x+5 = -4 \Rightarrow x = -9$ or $x+5 = 4 \Rightarrow x = -1$.

    The interval is $(-\infty,\ -9] \cup [-1,\ \infty)$.

67. First add 1 to each side of $|8-6x| - 1 \leq 2$ to obtain $|8-6x| \leq 3$.

    The solutions to $|8-6x| \leq 3$ satisfy $c \leq x \leq d$ where $c$ and $d$ are the solutions to $|8-6x| = 3$.

    $|8-6x| = 3$ is equivalent to $8-6x = -3 \Rightarrow x = \dfrac{11}{6}$ or $8-6x = 3 \Rightarrow x = \dfrac{5}{6}$. The interval is $\left[\dfrac{5}{6},\ \dfrac{11}{6}\right]$.

69. First subtract 5 from each side of $5 + \left|\dfrac{2-x}{3}\right| \leq 9$ to obtain $\left|\dfrac{2-x}{3}\right| \leq 4$.

    The solutions to $\left|\dfrac{2-x}{3}\right| \leq 4$ satisfy $c \leq x \leq d$ where $c$ and $d$ are the solutions to $\left|\dfrac{2-x}{3}\right| = 4$.

    $\left|\dfrac{2-x}{3}\right| = 4$ is equivalent to $\dfrac{2-x}{3} = -4 \Rightarrow x = 14$ or $\dfrac{2-x}{3} = 4 \Rightarrow x = -10$.

    The interval is $[-10,\ 14]$.

71. The inequality $|2x-1| \leq -3$ has no solutions because absolute value is never negative.

73. First add 1 to each side of $|x+1| - 1 > -3$ to obtain $|x+1| > -2$.

    The inequality $|x+1| > -2$ is true for all values of $x$ because absolute value is never negative.

    The interval is $(-\infty,\ \infty)$.

75. The inequality $|2z-4| \leq -1$ has no solutions because absolute value is never negative.

77. The inequality $|3z-1| > -3$ is true for all values of $z$ because absolute value is never negative.

    The interval is $(-\infty,\ \infty)$.

79. The solutions to $\left|\dfrac{2-t}{3}\right| \geq 5$ satisfy $t \leq c$ or $t \geq d$ where $c$ and $d$ are the solutions to $\left|\dfrac{2-t}{3}\right| = 5$.

    $\left|\dfrac{2-t}{3}\right| = 5$ is equivalent to $\dfrac{2-t}{3} = -5 \Rightarrow t = 17$ or $\dfrac{2-t}{3} = 5 \Rightarrow x = -13$.

    The interval is $(-\infty,\ -13] \cup [17,\ \infty)$.

81. The solutions to $|t-1| \leq 0.1$ satisfy $c \leq t \leq d$ where $c$ and $d$ are the solutions to $|t-1| = 0.1$.

     $|t-1| = 0.1$ is equivalent to $t-1 = -0.1 \Rightarrow t = 0.9$ or $t-1 = 0.1 \Rightarrow t = 1.1$. The interval is $[0.9, \ 1.1]$.

83. The solutions to $|b-10| > 0.5$ satisfy $b < c$ or $b > d$ where $c$ and $d$ are the solutions to $|b-10| = 0.5$.

     $|b-10| = 0.5$ is equivalent to $b-10 = -0.5 \Rightarrow b = 9.5$ or $b-10 = 0.5 \Rightarrow b = 10.5$.

     The interval is $(-\infty, \ 9.5) \cup (10.5, \ \infty)$.

85. (a)    From the table, $y = 2$ when $x = -1$ or $x = 3$.

     (b)    $y < 2$ when $-1 < x < 3$. The interval is $(-1, \ 3)$.

     (c)    $y > 2$ when $x < -1$ or $x > 3$. The interval is $(-\infty, \ -1) \cup (3, \ \infty)$.

87. (a)    From the graph $y_1 = 1$ when $x = -1$ or $x = 0$.

     (b)    $y_1 \leq 1$ when $-1 \leq x \leq 0$. The interval is $[-1, \ 0]$.

     (c)    $y_1 \geq 1$ when $x \leq -1$ or $x \geq 0$. The interval is $(-\infty, \ -1] \cup [0, \ \infty)$.

89. Graph $Y_1 = \text{abs}(X)$ and $Y_2 = 1$ in $[-3, 3, 1]$ by $[-3, 3, 1]$. See Figures 89a and 89b.

     $(-\infty, \ -1] \cup [1, \ \infty)$

     $[-3, 3, 1]$ by $[-3, 3, 1]$           $[-3, 3, 1]$ by $[-3, 3, 1]$

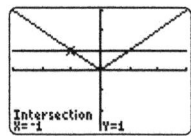         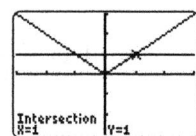

     Figure 89a                Figure 89b

91. Graph $Y_1 = \text{abs}(X-1)$ and $Y_2 = 3$ in $[-5, 5, 1]$ by $[-5, 5, 1]$. See Figures 91a and 91b. $[-2, \ 4]$

     $[-5, 5, 1]$ by $[-5, 5, 1]$           $[-5, 5, 1]$ by $[-5, 5, 1]$

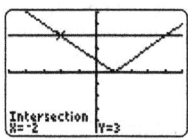         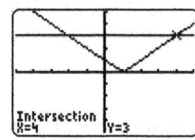

     Figure 91a                Figure 91b

93. Graph $Y_1 = \text{abs}(4-2X)$ and $Y_2 = 2$ in $[0, 5, 1]$ by $[0, 5, 1]$. See Figures 93a and 93b.

     $(-\infty, \ 1) \cup (3, \ \infty)$

     $[0, 5, 1]$ by $[0, 5, 1]$           $[0, 5, 1]$ by $[0, 5, 1]$

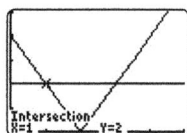         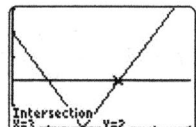

     Figure 93a                Figure 93b

95. Graph $Y_1 = abs(10 - 3X)$ and $Y_2 = 4$ in $[0, 6, 1]$ by $[0, 6, 1]$. See Figures 95a and 95b. $\left(2, 4.\overline{6}\right)$

$[0, 6, 1]$ by $[0, 6, 1]$          $[0, 6, 1]$ by $[0, 6, 1]$

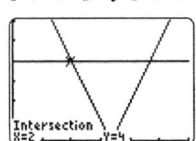      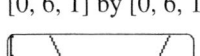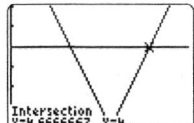

Figure 95a                        Figure 95b

97. Graph $Y_1 = abs(8.1 - X)$ and $Y_2 = -2$ in $[0, 15, 1]$ by $[-5, 5, 1]$. See Figure 97. $\left(-\infty, \infty\right)$

$[0, 15, 1]$ by $[-5, 5, 1]$

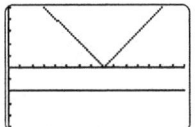

Figure 97

99. (a)  $3x = 9 \Rightarrow x = 3$  or  $3x = -9 \Rightarrow x = -3$; $\{x | -3 \le x \le 3\}$

(b)  Graph $Y_1 = abs(3X)$ and $Y_2 = 9$ in $[-5, 5, 1]$ by $[-5, 15, 1]$. See Figures 99a and 99b.

(c)  Table $Y_1 = abs(3X)$ with TblStart $= -7$ and $\Delta$Tbl $= 2$. See Figure 99c.

The solution set is $\{x | -3 \le x \le 3\}$.

$[-5, 5, 1]$ by $[-5, 15, 1]$          $[-5, 5, 1]$ by $[-5, 15, 1]$

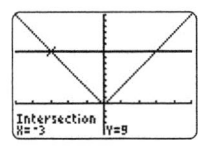      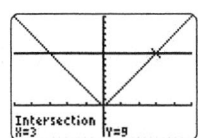

Figure 99a                        Figure 99b                        Figure 99c

101. (a)  $2x - 5 = 1 \Rightarrow 2x = 6 \Rightarrow x = 3$  or  $2x - 5 = -1 \Rightarrow 2x = 4 \Rightarrow x = 2$; $\left\{x | x < 2 \text{ or } x > 3\right\}$

(b)  Graph $Y_1 = abs(2X - 5)$ and $Y_2 = 1$ in $[0, 5, 1]$ by $[-1, 2, 1]$. See Figures 101a and 101b.

(c)  Table $Y_1 = abs(2X - 5)$ with TblStart $= 1$ and $\Delta$Tbl $= 0.5$. See Figure 101c.

The solution set is $\left\{x | x < 2 \text{ or } x > 3\right\}$.

$[0, 5, 1]$ by $[-1, 2, 1]$          $[0, 5, 1]$ by $[-1, 2, 1]$

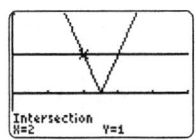      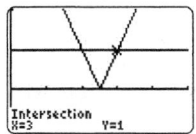

Figure 101a                        Figure 101b                        Figure 101c

103. $|x| \le 4$

105. $|y| > 2$

107. $|2x+1| \le 0.3$

109. $|\pi x| \ge 7$

111. two

113. (a)    $|T-43| \le 24 \Rightarrow -24 \le T-43 \le 24 \Rightarrow 19 \le T \le 67;\ \{T\,|\,19 \le T \le 67\}$

   (b)    The monthly average temperatures in Marquette, Michigan vary from 19°F to 67°F.

115. (a)    $|T-10| \le 36 \Rightarrow -36 \le T-10 \le 36 \Rightarrow -26 \le T \le 46;\ \{T\,|\,-26 \le T \le 46\}$

   (b)    The monthly average temperatures in Chesterfield, Canada vary from –26°F to 46°F.

117. (a)    $A = (29,028 + 22,834 + 20,320 + 19,340 + 18,510 + 16,066 + 7,310) \div 7 \approx 19,058$ feet

   (b)    Africa and Europe have elevations within 1000 feet of $A$.

   (c)    South America, North America, Africa, Europe and Antarctica have elevations within 5000 feet of $A$.

   (d)    $|E-A| \le 5000$

119. $|d-2.5| \le 0.002 \Rightarrow -0.002 \le d-2.5 \le 0.002 \Rightarrow 2.498 \le d \le 2.502;\ \{d\,|\,2.498 \le d \le 2.502\}$

   The diameter can vary from 2.498 inches to 2.502 inches.

121. $|d-3.8| \le 0.03$

123. The solutions to $\left|\dfrac{x-20}{20}\right| < 0.05$ satisfy $c < x < d$ where $c$ and $d$ are the solutions to $\left|\dfrac{x-20}{20}\right| = 0.05$.

   $\left|\dfrac{x-20}{20}\right| = 0.05$ is equivalent to $\dfrac{x-20}{20} = -0.05 \Rightarrow x = 19$ or $\dfrac{x-20}{20} = 0.05 \Rightarrow x = 21$.

   The interval is $(19,\ 21)$. The values must be between 19 and 21, exclusively.

## Checking Basic Concepts  Section 8.5

1. $\left|\dfrac{3}{4}x-1\right| - 3 = 5 \Rightarrow \left|\dfrac{3}{4}x-1\right| = 8 \Rightarrow \dfrac{3}{4}x-1 = -8 \Rightarrow \dfrac{3}{4}x = -7 \Rightarrow$

   $x = -\dfrac{28}{3}$ or $\dfrac{3}{4}x-1 = 8 \Rightarrow \dfrac{3}{4}x = 9 \Rightarrow x = 12$

2. (a)    $3x-6 = 8 \Rightarrow 3x = 14 \Rightarrow x = \dfrac{14}{3}$ or $3x-6 = -8 \Rightarrow 3x = -2 \Rightarrow x = -\dfrac{2}{3}$

   (b)    The solutions to $|3x-6| < 8$ satisfy $c < x < d$ where $c$ and $d$ are the solutions to $|3x-6| = 8$.

   From part (a), the interval is $\left(-\dfrac{2}{3},\ \dfrac{14}{3}\right)$.

   (c)    The solutions to $|3x-6| > 8$ satisfy $x < c$ or $x > d$ where $c$ and $d$ are the solutions

   to $|3x-6| = 8$. From part (a), the interval is $\left(-\infty,\ -\dfrac{2}{3}\right) \cup \left(\dfrac{14}{3},\ \infty\right)$.

3. The solutions to $\left|-2(3-x)\right| < 6$ satisfy $c < x < d$ where $c$ and $d$ are the solutions to $\left|-2(3-x)\right| = 6$.

$\left|-2(3-x)\right| = 6$ is equivalent to $-2(3-x) = -6 \Rightarrow x = 0$ or $-2(3-x) = 6 \Rightarrow x = 6$.

The interval is $(0, 6)$. Similarly, the solution to $\left|-2(3-x)\right| \geq 6$ is the interval $(-\infty, 0] \cup [6, \infty)$.

4. (a)  From the graph $y = 2$ when $x = 1$ or $x = 3$.

   (b)  $y \leq 2$ when $1 \leq x \leq 3$. The interval is $[1, 3]$.

   (c)  $y \geq 2$ when $x \leq 1$ or $x \geq 3$. The interval is $(-\infty, 1] \cup [3, \infty)$.

## Chapter 8 Review

1. $f(-2) = 3(-2) - 1 = -7$; $f\left(\dfrac{1}{3}\right) = 3\left(\dfrac{1}{3}\right) - 1 = 0$

2. $f(-3) = 5 - 3(-3)^2 = 5 - 27 = -22$; $f(1) = 5 - 3(1)^2 = 5 - 3 = 2$

3. $f(0) = \sqrt{0} - 2 = -2$; $f(9) = \sqrt{9} - 2 = 3 - 2 = 1$

4. $f(-5) = 5$; $f\left(\dfrac{7}{5}\right) = 5$

5. (a)  Since there are 2 pints in a quart, $P(q) = 2q$.

   (b)  $P(5) = 2(5) = 10$. There are 10 pints in 5 quarts.

6. (a)  Three less than four times a number is written $f(x) = 4x - 3$.

   (b)  $f(5) = 4(5) - 3 = 17$. Three less than four times five is 17.

7. $(3, -2)$

8. $f(4) = -6$; the answers are 4 and $-6$.

9. See Figure 9.

10. See Figure 10.

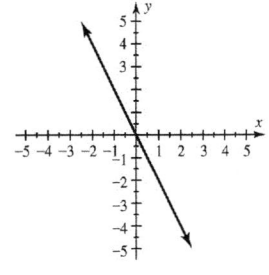

Figure 9

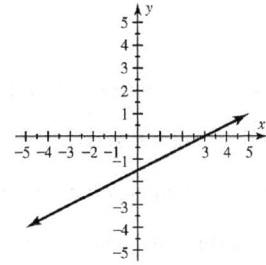

Figure 10

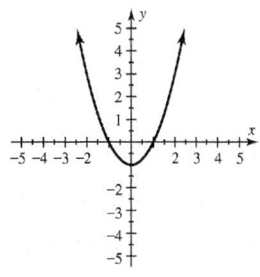

Figure 11

11. See Figure 11.

12. See Figure 12.

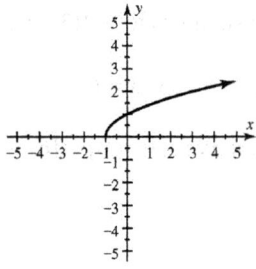

Figure 12

13. $f(0) = 1;\ f(-3) = 4$

14. $f(-2) = 1;\ f(1) = -2$

15. $f(-1) = 7;\ f(3) = -1$

16. Numerical: The table is shown in Figure 16a.

    Symbolic: $f(x) = 3x - 2$

    Graphical: The graph is shown in Figure 16b.

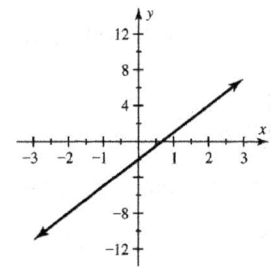

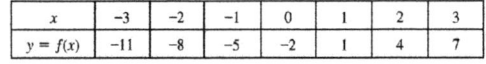

| $x$ | -3 | -2 | -1 | 0 | 1 | 2 | 3 |
|-----|----|----|----|----|----|----|----|
| $y = f(x)$ | -11 | -8 | -5 | -2 | 1 | 4 | 7 |

Figure 16a                                    Figure 16b

17. $D$: all real numbers; $R$: $y \le 4$

18. $D$: $-4 \le x \le 4$; $R$: $-4 \le y \le 0$

19. Yes. The graph passes the vertical line test.

20. No. The graph does not pass the vertical line test.

21. $D = \{-3, -1, 2, 4\}$; $R = \{-1, 3, 4\}$; Yes, $S$ is a function since each input has exactly one output.

22. $D = \{-1, 0, 1, 2\}$; $R = \{-2, 2, 3, 4, 5\}$; No, $S$ is not a function because the input $-1$ has more than one output.

23. Any real number is a valid input for this function. $D$: all real numbers.

24. The radicand must be greater than or equal to zero. $D$: $x \ge 0$.

25. The denominator of this function cannot equal zero. $D$: $x \ne 0$.

26. Any real number is a valid input for this function because the variable is squared. $D$: all real numbers.

27. The radicand must be greater than or equal to zero. $D$: $x \le 5$

28. The denominator of this function cannot equal zero. $D$: $x \ne -2$

29. Any real number is a valid input for this function. $D$: all real numbers

30. Any real number is a valid input for this function. $D$: all real numbers

31. No. The graph is not a straight line.

32. Yes. The graph is a straight line.

33. This function is linear because it is in the form $f(x) = mx + b$ with $m = -4$ and $b = 5$.

34. This function is linear because it can be written in the form $f(x) = mx + b$ with $m = -1$ and $b = 7$.

35. This function is not linear because it can not be written in the form $f(x) = mx + b$.

36. This function is linear because it can be written in the form $f(x) = mx + b$ with $m = 0$ and $b = 6$.

37. Yes. For each 2-unit increase in $x$, the values of $f(x)$ increase by 3 units, so $m = \dfrac{3}{2}$. Because

$f(x) = -3$ when $x = 0$, the $y$-intercept is $b = -3$. The function can be written $f(x) = \dfrac{3}{2}x - 3$.

38. No. For each unit increase in $x$, the values of $f(x)$ do not increase by a constant amount.

39. $f(-4) = \dfrac{1}{2}(-4) + 3 = -2 + 3 = 1$

40. $f(-2) = -3$ and $f(1) = 0$

41. See Figure 41.

42. See Figure 42.

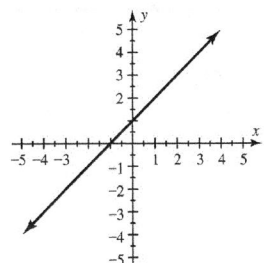
Figure 41

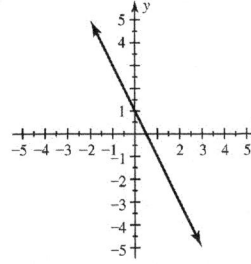
Figure 42

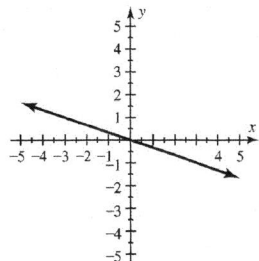
Figure 43

43. See Figure 43.

44. See Figure 44.

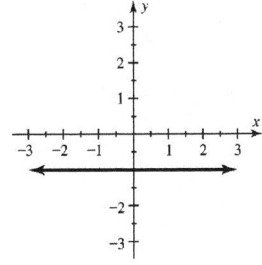
Figure 44

45. There are 24 hours in each day, so $H(x) = 24x$. $H(2) = 24(2) = 48$; there are 48 hours in 2 days.

46. (a)   See Figure 46a.                    (b)  See Figure 46b. Domain: $x \geq -2$

[–10, 10, 1] by [–10, 10, 1]

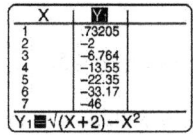

Figure 46a                                           Figure 46b

47.  $M = \left( \dfrac{-5+6}{2}, \dfrac{3+(-9)}{2} \right) = \left( \dfrac{1}{2}, \dfrac{-6}{2} \right) = (0.5, -3)$

48.  $M = \left( \dfrac{\frac{2}{3}+\frac{1}{6}}{2}, \dfrac{-\frac{3}{4}+\frac{3}{2}}{2} \right) = \left( \dfrac{\frac{4}{6}+\frac{1}{6}}{2}, \dfrac{-\frac{3}{4}+\frac{6}{4}}{2} \right) = \left( \dfrac{\frac{5}{6}}{2}, \dfrac{\frac{3}{4}}{2} \right) = \left( \dfrac{5}{12}, \dfrac{3}{8} \right)$

49.  $x+1 \leq 3 \Rightarrow x \leq 2$ and $x+1 \geq -1 \Rightarrow x \geq -2$;  $[-2, 2]$.  See Figure 49.

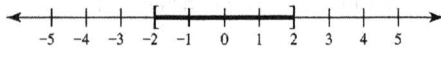

Figure 49                                              Figure 50

50.  $2x+7 < 5 \Rightarrow 2x < -2 \Rightarrow x < -1$ and $-2x \geq 6 \Rightarrow x \leq -3$;  $(-\infty, -3]$.  See Figure 50.

51.  $5x-1 \leq 3 \Rightarrow 5x \leq 4 \Rightarrow x \leq \dfrac{4}{5}$ or $1-x < -1 \Rightarrow 2 < x$;  $\left( -\infty, \dfrac{4}{5} \right] \cup (2, \infty)$.  See Figure 51.

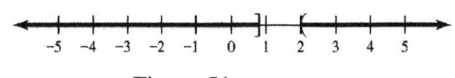

Figure 51                                              Figure 52

52.  $3x+1 > -1 \Rightarrow 3x > -2 \Rightarrow x > -\dfrac{2}{3}$ or $3x+1 < 10 \Rightarrow 3x < 9 \Rightarrow x < 3$;  $(-\infty, \infty)$.  See Figure 52.

53.  $2x+2$ is between $-2$ and $4$ when $-2 \leq x \leq 1$;  $[-2, 1]$

54.  (a)   The intersection point of $y_1$ and $y_2$ is $(-4, -100)$.  The solution is $x = -4$.

     (b)   The intersection point of $y_2$ and $y_3$ is $(2, 50)$.  The solution is $x = 2$.

     (c)   $y_2$ is between $y_1$ and $y_3$ when $-4 \leq x \leq 2$;  $[-4, 2]$

     (d)   $y_2$ is below $y_3$ when $x < 2$;  $(-\infty, 2)$

55.  (a)   The intersection point of $y_1$ and $y_2$ is $(2, 2)$.  The solution is $x = 2$.

     (b)   $y_1$ is below $y_2$ when $x > 2$;  $(2, \infty)$

     (c)   $y_1$ is above $y_2$ when $x < 2$;  $(-\infty, 2)$

56.  (a)   The intersection point of $f(x)$ and $g(x)$ is $(4, 20)$.  The solution is $x = 4$.

     (b)   The intersection point of $g(x)$ and $h(x)$ is $(2, 40)$.  The solution is $x = 2$.

     (c)   $g(x)$ is between $f(x)$ and $h(x)$ when $2 < x < 4$;  $(2, 4)$

57. $\left[-3, \dfrac{2}{3}\right]$

58. $(-6, 45]$

59. $\left(-\infty, \dfrac{7}{2}\right)$

60. $[1.8, \infty)$

61. $(-3, 4)$

62. $(-\infty, 4) \cup (10, \infty)$

63. $-4 < x + 1 < 6 \Rightarrow -5 < x < 5$; The solution set is $(-5, 5)$.

64. $20 \le 2x + 4 \le 60 \Rightarrow 16 \le 2x \le 56 \Rightarrow 8 \le x \le 28$; The solution set is $[8, 28]$.

65. $-3 < 4 - \dfrac{1}{3}x < 7 \Rightarrow -7 < -\dfrac{1}{3}x < 3 \Rightarrow 21 > x > -9 \Rightarrow -9 < x < 21$; The solution set is $(-9, 21)$.

66. $2 \le \dfrac{1}{2}x - 2 \le 12 \Rightarrow 4 \le \dfrac{1}{2}x \le 14 \Rightarrow 8 \le x \le 28$; The solution set is $[8, 28]$.

67. $-3 \le \dfrac{4 - 5x}{3} - 2 < 3 \Rightarrow -9 \le 4 - 5x - 6 < 9 \Rightarrow -9 \le -5x - 2 < 9 \Rightarrow -7 \le -5x < 11 \Rightarrow$

$\dfrac{7}{5} \ge x > -\dfrac{11}{5} \Rightarrow -\dfrac{11}{5} < x \le \dfrac{7}{5}; \left(-\dfrac{11}{5}, \dfrac{7}{5}\right]$

68. $30 \le \dfrac{2x - 6}{5} - 4 < 50 \Rightarrow 150 \le 2x - 6 - 20 < 250 \Rightarrow 150 \le 2x - 26 < 250 \Rightarrow$

$176 \le 2x < 276 \Rightarrow 88 \le x < 138$; $[88, 138)$

69. $D = (-\infty, \infty)$; $R = [0, \infty)$

70. $D = (-\infty, \infty)$; $R = [0, \infty)$

71. $2x - 8 = 0 \Rightarrow 2x = 8 \Rightarrow x = 4$; $D = (-\infty, 4) \cup (4, \infty)$

72. $D = [-3, 1]$; $R = [-3, 6]$

73. Yes, $f(x) = 1 + 2x - 3x^2$ represents a polynomial function. The degree is 2, and it is a quadratic function.

74. Yes, $f(x) = 5 + 7x$ represents a polynomial function. The degree is 1, and it is a linear function.

75. Yes, $f(x) = x^3 + 2x$ represents a polynomial function. The degree is 3, and it is a cubic function.

76. No, $f(x) = |2x - 1|$ is not a polynomial function. It is an absolute value function.

77. $g(3) = |1 - 4 \cdot 3| = |1 - 12| = |-11| = 11$, $g\left(-\dfrac{1}{4}\right) = \left|1 - 4\left(-\dfrac{1}{4}\right)\right| = |1 + 1| = |2| = 2$

78. $g(3) = \dfrac{4}{4 - 3^2} = \dfrac{4}{4 - 9} = \dfrac{4}{-5} = -\dfrac{4}{5}$  $g(-2) = \dfrac{4}{4 - (-2)^2} = \dfrac{4}{4 - 4} = \dfrac{4}{0}$, which is undefined.

79. See Figure 79.

80. See Figure 80.

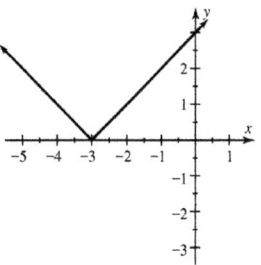

Figure 79                                              Figure 80

81. See Figure 81.

82. See Figure 82.

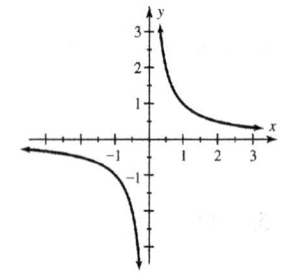                          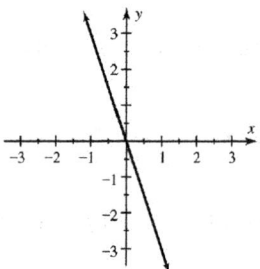

Figure 81                                              Figure 82

83. (a)   $(f+g)(3) = [2(3)^2 - 3(3)] + [2(3) - 3] = [2(9) - 9] + [6 - 3] = 18 - 9 + 3 = 12$

    (b)   $(fg)(3) = [2(3)^2 - 3(3)] \cdot [2(3) - 3] = (2 \cdot 9 - 9)(6 - 3) = (18 - 9)(3) = 9(3) = 27$

84. (a)   $(f-g)(x) = (x^2 - 1) - (x - 1) = x^2 - 1 - x + 1 = x^2 - x$

    (b)   $\left(\dfrac{f}{g}\right)(x) = \dfrac{x^2 - 1}{x - 1} = \dfrac{(x+1)(x-1)}{x-1} = x + 1$

85. No, $|12(-3) - 24| = |-60| = 60 \neq 24$.   No, $|12(2) - 24| = |0| = 0 \neq 24$.

86. No, $\left|5 - 3\left(\dfrac{4}{3}\right)\right| = |1| = 1 \not> 3$.   Yes, $|5 - 3(0)| = |5| = 5 > 3$.

87. No, $|3(-3) - 6| = |-15| = 15 \not\leq 6$.   Yes, $|3(4) - 6| = |6| = 6 \leq 6$.

88. No, $|2 + 3(-3)| + 4 = |-7| + 4 = 7 + 4 = 11 \not< 11$.   Yes, $\left|2 + 3\left(\dfrac{2}{3}\right)\right| + 4 = |4| + 4 = 4 + 4 = 8 < 11$.

89. (a)   From the table, $y_1 = 2$ when $x = 0$ or $x = 4$.

    (b)   From the table, $y_1 < 2$ when $0 < x < 4$; $(0,\ 4)$.

    (c)   From the table, $y_1 > 2$ when $x < 0$ or $x > 4$; $(-\infty,\ 0) \cup (4,\ \infty)$.

90. (a)   From the graph, $|2x+2| = 4$ when $x = -3$ or $x = 1$.

 (b)   From the graph, $|2x+2| \le 4$ when $-3 \le x \le 1$; $[-3, 1]$.

 (c)   From the graph, $|2x+2| \ge 4$ when $x \le -3$ or $x \ge 1$; $(-\infty, -3] \cup [1, \infty)$.

91.   $x = -22$ or $x = 22$

92.   $2x-9 = 7 \Rightarrow 2x = 16 \Rightarrow x = 8$   or   $2x-9 = -7 \Rightarrow 2x = 2 \Rightarrow x = 1$

93.   $4 - \dfrac{1}{2}x = 17 \Rightarrow -\dfrac{1}{2}x = 13 \Rightarrow x = -26$   or   $4 - \dfrac{1}{2}x = -17 \Rightarrow -\dfrac{1}{2}x = -21 \Rightarrow x = 42$

94.   First note that $\dfrac{1}{3}|3x-1| + 1 = 9 \Rightarrow \dfrac{1}{3}|3x-1| = 8 \Rightarrow |3x-1| = 24$.

 $3x-1 = 24 \Rightarrow 3x = 25 \Rightarrow x = \dfrac{25}{3}$   or   $3x-1 = -24 \Rightarrow 3x = -23 \Rightarrow x = -\dfrac{23}{3}$

95.   $2x-5 = 5-3x \Rightarrow 5x = 10 \Rightarrow x = 2$   or   $2x-5 = -5+3x \Rightarrow -x = 0 \Rightarrow x = 0$

96.   $-3+3x = -2x+6 \Rightarrow 5x = 9 \Rightarrow x = \dfrac{9}{5}$   or   $-3+3x = 2x-6 \Rightarrow x = -3$

97. (a)   $x+1 = 7 \Rightarrow x = 6$   or   $x+1 = -7 \Rightarrow x = -8$

 (b)   The solutions to $|x+1| \le 7$ satisfy $c \le x \le d$ where $c$ and $d$ are the solutions to $|x+1| = 7$.

   From part (a), the interval is $[-8, 6]$.

 (c)   The solutions to $|x+1| \ge 7$ satisfy $x \le c$ or $x \ge d$ where $c$ and $d$ are the solutions to $|x+1| = 7$.

   From part (a), the interval is $(-\infty, -8] \cup [6, \infty)$.

98. (a)   $1-2x = 6 \Rightarrow -2x = 5 \Rightarrow x = -\dfrac{5}{2}$   or   $1-2x = -6 \Rightarrow -2x = -7 \Rightarrow x = \dfrac{7}{2}$

 (b)   The solutions to $|1-2x| \le 6$ satisfy $c \le x \le d$ where $c$ and $d$ are the solutions to $|1-2x| = 6$.

   From part (a), the interval is $\left[-\dfrac{5}{2}, \dfrac{7}{2}\right]$.

 (c)   The solutions to $|1-2x| \ge 6$ satisfy $x \le c$ or $x \ge d$ where $c$ and $d$ are

   the solutions to $|1-2x| = 6$.  From part (a), the interval is $\left(-\infty, -\dfrac{5}{2}\right] \cup \left[\dfrac{7}{2}, \infty\right)$.

99.   The solutions to $|x| > 3$ satisfy $x < c$ or $x > d$ where $c$ and $d$ are the solutions to $|x| = 3$.

   $|x| = 3$ is equivalent to $x = -3$ or $x = 3$.  The interval is $(-\infty, -3) \cup (3, \infty)$.

100.   The solutions to $|-5x| < 20$ satisfy $c < x < d$ where $c$ and $d$ are the solutions to $|-5x| = 20$.

   $|-5x| = 20$ is equivalent to $-5x = -20 \Rightarrow x = 4$ or $-5x = 20 \Rightarrow x = -4$.  The interval is $(-4, 4)$.

101.   The solutions to $|4x-2| \le 14$ satisfy $c \le x \le d$ where $c$ and $d$ are the solutions to $|4x-2| = 14$.

   $|4x-2| = 14$ is equivalent to $4x-2 = -14 \Rightarrow x = -3$ or $4x-2 = 14 \Rightarrow x = 4$.

   The interval is $[-3, 4]$.

102. The solutions to $\left|1-\dfrac{4}{5}x\right| \geq 3$ satisfy $x \leq c$ or $x \geq d$ where $c$ and $d$ are the solutions to $\left|1-\dfrac{4}{5}x\right| = 3$.

$\left|1-\dfrac{4}{5}x\right| = 3$ is equivalent to $1-\dfrac{4}{5}x = -3 \Rightarrow x = 5$ or $1-\dfrac{4}{5}x = 3 \Rightarrow x = -\dfrac{5}{2}$.

The interval is $\left(-\infty,\ -\dfrac{5}{2}\right] \cup [5,\ \infty)$.

103. The solutions to $|t-4.5| \leq 0.1$ satisfy $c \leq t \leq d$ where $c$ and $d$ are the solutions to $|t-4.5| = 0.1$.

$|t-4.5| = 0.1$ is equivalent to $t-4.5 = -0.1 \Rightarrow t = 4.4$ or $t-4.5 = 0.1 \Rightarrow t = 4.6$.

The interval is $[4.4,\ 4.6]$.

104. First divide each side of $-2|13t-5| \geq -4$ by $-2$ to obtain $|13t-5| \leq 2$.

The solutions to $|13t-5| \leq 2$ satisfy $c \leq t \leq d$ where $c$ and $d$ are the solutions to $|13t-5| = 2$.

$|13t-5| = 2$ is equivalent to $13t-5 = -2 \Rightarrow t = \dfrac{3}{13}$ or $13t-5 = 2 \Rightarrow t = \dfrac{7}{13}$.

The interval is $\left[\dfrac{3}{13},\ \dfrac{7}{13}\right]$.

105. The inequality $|5-4x| > -5$ is true for all values of $x$ because absolute value is never negative.

The interval is $(-\infty,\ \infty)$.

106. Since absolute value can never be negative, $|2t-3| \leq 0$ is equivalent to $|2t-3| = 0$.

$|2t-3| = 0$ is equivalent to $2t-3 = 0 \Rightarrow t = \dfrac{3}{2}$. The only solution is $\dfrac{3}{2}$.

107. Graph $Y_1 = \text{abs}(2X)$ and $Y_2 = 3$ in $[-3, 3, 1]$ by $[0, 5, 1]$. See Figures 107a and 107b.

From the graph, $|2x| \geq 3$ in the interval $(-\infty,\ -1.5] \cup [1.5,\ \infty)$.

[-3, 3, 1] by [0, 5, 1]                    [-3, 3, 1] by [0, 5, 1]

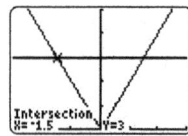          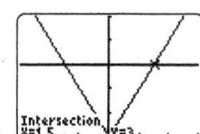

Figure 107a                              Figure 107b

108. Graph $Y_1 = \text{abs}\left((1/2)X-1\right)$ and $Y_2 = 2$ in $[-7, 7, 1]$ by $[0, 4, 1]$. See Figures 108a and 108b.

From the graph, $\left|\dfrac{1}{2}x-1\right| \leq 2$ in the interval $[-2,\ 6]$.

[-7, 7, 1] by [0, 4, 1]

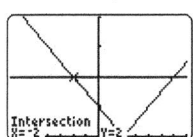

Figure 108a

[-7, 7, 1] by [0, 4, 1]

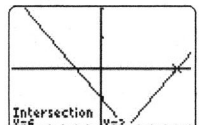

Figure 108b

109.  $|x| \le 0.05$

110.  $|5x - 1| > 4$

111. (a)  $f(1910) = -0.0492(1910) + 119.1 \approx 25.1$  years

(b) Graph $Y_1 = -0.0492X + 119.1$ in [1885, 1965, 10] by [22, 26, 1]. See Figure 111.

The median age at first marriage for males has decreased over this time period.

(c) The slope is –0.0492. The median age decreased by about 0.0492 year per year.

[1885, 1965, 10] by [22, 26, 1]

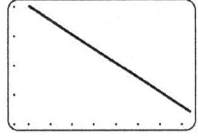

Figure 111

112. (a)  $f(2006) = 2.2$  million

(b) The number of marriages each year did not change over this time period.

113. (a)  $f(x) = 8x$

(b) The slope of the graph of $f$ is 8.

(c) The total fat changes at a rate of 8 grams per cup of milk.

114. (a) See Figure 114.

(b) Using the points (1950, 24.1) and (2010, 13.5) the slope is  $m = \dfrac{13.5 - 24.1}{2010 - 1950} = \dfrac{-10.6}{60} \approx 0.176.$

Then using the point (1950, 24.1) the equation is

$y = -0.176(x - 1950) + 24.1$ or $y = -0.176x + 367.3.$

(c)  $f(2000) = -0.176(2000 - 1950) + 24.1 = -0.176(50) + 24.1 = 15.3$

The birth rate is about 15 per 1000 people. *Answers may vary.*

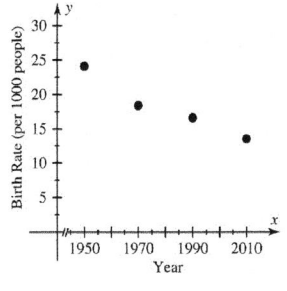

Figure 114

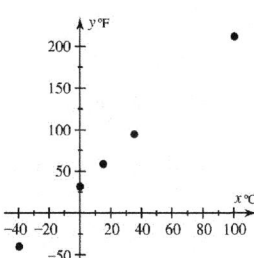

Figure 116

115. (a)    From the table $f(1995) = 113$. In 1995 there were 113 unhealthy days.

(b)    $D = \{1995, 1999, 2000, 2003, 2007\}$; $R = \{56, 87, 88, 100, 113\}$

(c)    The number of unhealthy days decreased and then increased over this time period.

116. (a)    See Figure 116. There is a linear relationship.

(b)    Using the points (0, 32) and (100, 212) the slope is $m = \dfrac{212 - 32}{100 - 0} = \dfrac{180}{100} = \dfrac{9}{5}$.

The $y$-intercept is 32. The function is given by $f(x) = \dfrac{9}{5}x + 32$. The slope of $\dfrac{9}{5}$ means that a

1°C change equals a $\dfrac{9}{5}$°F change.

(c)    $f(20) = \dfrac{9}{5}(20) + 32 = 36 + 32 = 68°$ F.

117. (a)    The distance between the riders is zero (they meet) when $x = 3$ hours.

(b)    The distance between the riders is 20 when $x = 2$ hours and $x = 4$ hours.

(c)    The riders are less than 20 miles apart between 2 and 4 hours exclusively.

(d)    The sum of the speeds of the two bicyclists is $\dfrac{60 \text{ miles}}{3 \text{ hours}} = 20$ mph.

118. (a)    $m = \dfrac{5 - 10}{2008 - 1992} = \dfrac{-5}{16} = -0.3125$  Since $x$ is the number of years after 1992 the $y$-intercept is

10 and the function is $f(x) = -0.3125x + 10$.

(b)    Let $x = 13 \Rightarrow f(13) = -0.3125(13) + 10 \approx 5.9$  The result is about 5.9 million.

119. (a)    $|A - 3.9| \le 1.7$

(b)    $A - 3.9 = 1.7 \Rightarrow A = 5.6$  or  $A - 3.9 = -1.7 \Rightarrow A = 2.2$; $\{A | 2.2 \le A \le 5.6\}$

120. The solutions to $\left|\dfrac{T - 35}{35}\right| < 0.08$ satisfy $-0.08 < \dfrac{T - 35}{35} < 0.08 \Rightarrow -2.8 < T - 35 < 2.8 \Rightarrow$

$32.2 < T < 37.8$. The interval is $(32.2, 37.8)$.

The values must be between 32.2 and 37.8, exclusively.

## Chapter 8 Test

1.  $f(4) = 3(4)^2 - \sqrt{4} = 3(16) - 2 = 48 - 2 = 46; (4, 46)$

2.  $C(x) = 4x$; $C(5) = 4(5) = 20$;  5 pounds of candy costs \$20.

3.  (a)    See Figure 3a.

(b)    See Figure 3b.

(c)    See Figure 3c.

(d)    See Figure 3d.

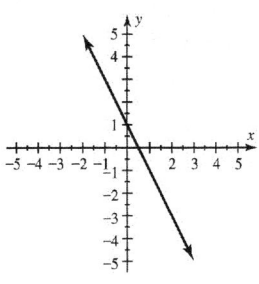

Figure 3a

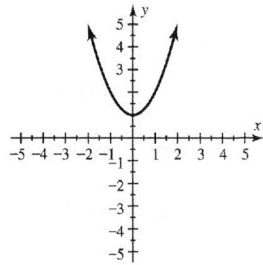

Figure 3b

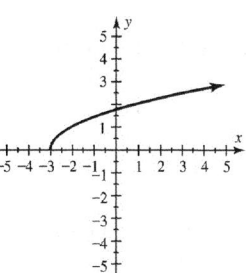

Figure 3c

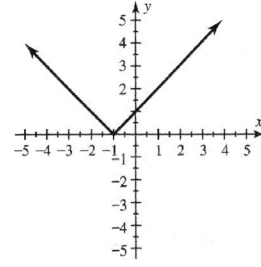

Figure 3d

4.  $f(-3) = 0;\ f(0) = -3;\ D:\{x|-3 \le x \le 3\}$ and $R:\{x|-3 \le y \le 0\}$

5.  Symbolic: $f(x) = x^2 - 5$                    Numerical: The table is shown in Figure 5a.

Graphical: The graph is shown in Figure 5b.

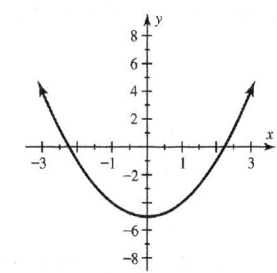

| $x$ | $-3$ | $-2$ | $-1$ | $0$ | $1$ | $2$ | $3$ |
|-----|------|------|------|-----|-----|-----|-----|
| $y = f(x)$ | 4 | $-1$ | $-4$ | $-5$ | $-4$ | $-1$ | 4 |

Figure 5a                                     Figure 5b

6.  No.  It does not pass the vertical line test.

7.  (a)    $D = \{-2, -1, 0, 5\}$

(b)    Any real number is a valid input for this function.  $D$: all real numbers

(c)    The radicand must be greater than or equal to zero.  $D: x \ge -4$

(d)    Any real number is a valid input for this function.  $D$: all real numbers

(e)    The denominator of this function cannot equal zero.  $D: x \ne 5$

8.  It is; $f(x) = -8x + 6$

9.  $2x + 6 < 2$ and $-3x \ge 3 \Rightarrow 2x < -4$ and $x \le -1 \Rightarrow x < -2$ and $x \le -1$;  See Figure 9.

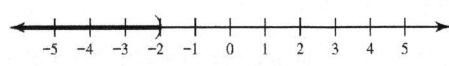

Figure 9

10. From the table, $-3x < -3$ when $x > 1$ and $-3x > 6$ when $x < -2$. The interval is $(-\infty, -2) \cup (1, \infty)$.

11. (a) The intersection point of $y_1$ and $y_2$ is $(-5, -300)$. The solution is $x = -5$.

    (b) The intersection point of $y_2$ and $y_3$ is $(5, 100)$. The solution is $x = 5$.

    (c) $y_2$ is between $y_1$ and $y_3$ when $-5 \le x \le 5$. The interval is $[-5, 5]$.

    (d) $y_2$ is below $y_3$ when $x < 5$. The interval is $(-\infty, 5)$.

12. $-2 < 2 + \frac{1}{2}x < 2 \Rightarrow -4 < \frac{1}{2}x < 0 \Rightarrow -8 < x < 0;\ (-8, 0)$

13. $2 - \frac{1}{3}x = 6 \Rightarrow -\frac{1}{3}x = 4 \Rightarrow x = -12$ or $2 - \frac{1}{3}x = -6 \Rightarrow -\frac{1}{3}x = -8 \Rightarrow x = 24$

14. (a) $-5 \le x \le 5; [-5, 5]$

    (b) The solution set is all real numbers except 0, or $(-\infty, 0) \cup (0, \infty)$.

15. Yes, $f(x) = 1 - 2x + x^3$ represents a polynomial function. The degree is 3, and it is a cubic function.

16. $h(-2) = -\frac{4(-2)}{5-(-2)} = -\frac{-8}{5+2} = \frac{8}{7};\ 5 - t = 0 \Rightarrow t = 5; D = (-\infty, 5) \cup (5, \infty)$

17. (a) $(f - g)(-2) = f(-2) - g(-2) = (-2)^2 + 1 - 2(-2) = 4 + 1 - (-4) = 5 + 4 = 9$

    (b) $(fg)(x) = (x^2 + 1)(2x) = 2x^3 + 2x$

18. (a) First divide the weight 150 by 2 to get 75. The function is $f(x) = 0.4x + 75$.

    (b) $0.4x + 75 = 89 \Rightarrow 0.4x = 14 \Rightarrow x = \frac{14}{0.4} \Rightarrow x = 35$ minutes

19. (a) $P(0) = -\frac{1}{300}(0)^2 + \frac{53}{30}(0) + 100 = 100;$

    $P(20) = -\frac{1}{300}(20)^2 + \frac{53}{30}(20) + 100 = -\frac{400}{300} + \frac{1060}{30} + 100$

    $= -\frac{400}{300} + \frac{10,600}{300} + \frac{30,000}{300} = \frac{40,200}{300} = 134;$

    $P(30) = -\frac{1}{300}(30)^2 + \frac{53}{30}(30) + 100 = -\frac{900}{300} + \frac{1590}{30} + 100$

    $= -\frac{900}{300} + \frac{15,900}{300} + \frac{30,000}{300} = \frac{45,000}{300} = 150;$

    $P(50) = -\frac{1}{300}(50)^2 + \frac{53}{30}(50) + 100 = -\frac{2500}{300} + \frac{2650}{30} + 100$

    $= -\frac{2500}{300} + \frac{26,500}{300} + \frac{30,000}{300} = \frac{54,000}{300} = 180;\ P(t)$ models the data in the table exactly.

    (b) Probably not because the race is over in 50 seconds; $D = [0, 50]$

20. (a) Graph $Y_1 = 1/(25 - X)$ in $[0, 25, 5]$ by $[0, 2, 0.5]$. Vertical asymptote: $x = 25$. See Figure 20.

(b)    $1 = \dfrac{1}{25-x} \Rightarrow 25 - x = 1 \Rightarrow -x = -24 \Rightarrow x = 24$ vehicles per minute

[0, 25, 5] by [0, 2, 0.5]

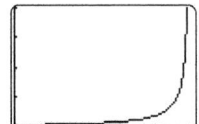

Figure 20

## Chapter 8 Extended and Discovery Exercises

1. (a)    The graph for tank A is linear. See Figure 1a. The graph for tank B is nonlinear.
          See Figure 1b.

   (b)    Tank B flows faster at first so it is the first to be half empty.

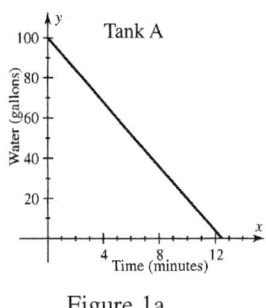

Figure 1a                          Figure 1b                          Figure 2

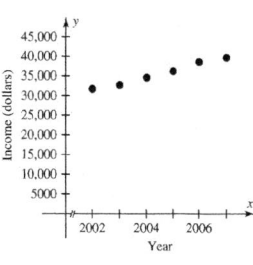

2. (a)    See Figure 2.

   (b)    Since the data appears to be nearly linear with an annual increase of about $1800 per year we
          may find a point-slope form of a linear equation using the point (2003, 32,271) and slope
          $m = 1800$  The equation is $f(x) = 1800(x - 2003) + 32,271$.  *Answers may vary.*

   (c)    $f(2000) = 1800(2000 - 2003) + 32,371 = \$26,971$

3. (a)    When the fish hatches it weighs about 7 mg.  It weighs about 105 mg at 6 weeks of age and
          about 158 mg at 12 weeks of age.  *Answers may vary.*

   (b)    From hatching to 6 weeks: $\dfrac{105 - 7}{6 - 0} = \dfrac{98}{6} \approx 16.3$ mg/week.

          From 6 weeks to 12 weeks: $\dfrac{158 - 105}{12 - 6} = \dfrac{53}{6} \approx 8.8$ mg/week.

   (c)    On average the fish gains about 16.3 mg per week during the first 6 weeks of its life and about
          8.8 mg per week during the second 6 weeks of its life.

   (d)    The fish gains weight the fastest during the first 6 weeks of its life.

4. (a)   The scatterplot is shown in Figure 4a.

   (b)   The relationship appears to be linear. This seems reasonable if one considers that when the number of megabytes doubles, the number of seconds should also double.

   (c)   Using the points $(0.129, 6.010)$ and $(1.260, 60.18)$

   $$m = \frac{60.18 - 6.010}{1.260 - 0.129} \approx 47.9, \; x_1 = 0.129 \text{ and } y_1 = 6.010.$$

   The equation is $y \approx 47.9(x - 0.129) + 6.010 \Rightarrow y \approx 47.9x - 0.1691.$

   Each additional megabyte of memory can record approximately 47.9 seconds of music.

   (d)   The scatterplot and the line are shown in Figure 4d.

   (e)   $47.9x - 0.1691 = 120$

   (f)   Symbolic: $47.9x - 0.1691 = 120 \Rightarrow 47.9x = 120.1691 \Rightarrow x \approx 2.5$ MB

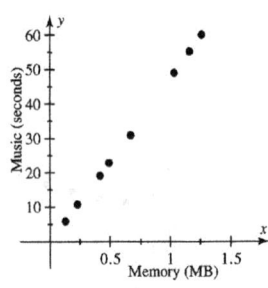

Figure 4a

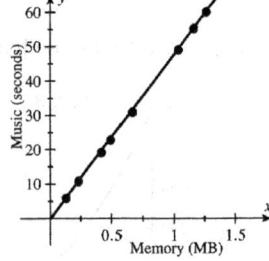

Figure 4d

## Chapters 1-8 Cumulative Review Exercises

1. $120 = 2 \cdot 2 \cdot 2 \cdot 3 \cdot 5 = 2^3 \cdot 3 \cdot 5$

2. $3n - 4 = n; \quad 3n - 4 = n \Rightarrow 2n - 4 = 0 \Rightarrow 2n = 4 \Rightarrow n = 2$

3. $\dfrac{1}{4} \div \dfrac{3}{4} - \dfrac{1}{2} = \dfrac{1}{4} \cdot \dfrac{4}{3} - \dfrac{1}{2} = \dfrac{4}{12} - \dfrac{6}{12} = -\dfrac{2}{12} = -\dfrac{1}{6}$

4. $12 - 3^2 \div 3 \cdot 2 = 12 - 9 \div 3 \cdot 2 = 12 - 3 \cdot 2 = 12 - 6 = 6$

5. $-3(3 - x) - 6 = 2x \Rightarrow -9 + 3x - 6 = 2x \Rightarrow -15 + 3x = 2x \Rightarrow -15 = -x \Rightarrow x = 15$

6. $0.075 = 0.075 \times 100\% = 7.5\%$

7. $A = \dfrac{1}{3}(2a - b) \Rightarrow 3A = 2a - b \Rightarrow 3A + b = 2a \Rightarrow a = \dfrac{3A + b}{2}$

8. $5 - 3t < 1 - t \Rightarrow 5 - 2t < 1 \Rightarrow -2t < -4 \Rightarrow t > 2$; the solution set is $\{t \mid t > 2\}$

9. $3x - 2y = -6 \Rightarrow -2y = -3x - 6 \Rightarrow y = \dfrac{3}{2}x + 3$, See Figure 9.

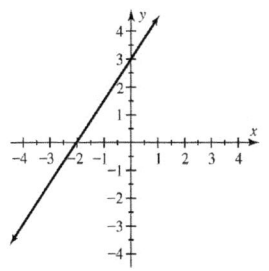

Figure 9

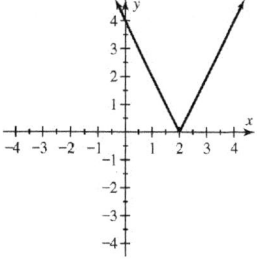

Figure 10

10. See Figure 10.

11. $m = \dfrac{1-4}{2-(-4)} = \dfrac{-3}{2+4} = \dfrac{-3}{6} = -\dfrac{1}{2};$

$y - 4 = -\dfrac{1}{2}(x-(-4)) \Rightarrow y - 4 = -\dfrac{1}{2}(x+4) \Rightarrow y = -\dfrac{1}{2}x - 2 + 4 \Rightarrow y = -\dfrac{1}{2}x + 2$

12. Multiply the first equation by 2 and the second equation by 3 and add the resulting equations.

$4x + 6y = 10$

$\dfrac{9x - 6y = 3}{13x \quad\quad = 13} \Rightarrow x = 1$   Substitute $x = 1$ into the original first equation and solve for $y$.

$2(1) + 3y = 5 \Rightarrow 3y = 3 \Rightarrow y = 1$. The solution is $(1, 1)$.

13. See Figure 13.

14. See Figure 14.

15. See Figure 15.

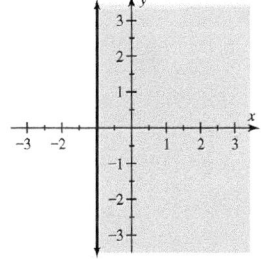

Figure 13

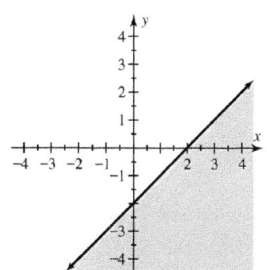

Figure 14

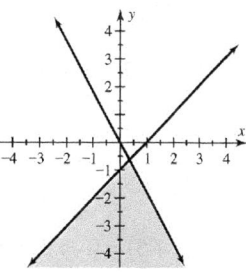

Figure 15

16. $(3x^2 + 2x - 4) - (4x^2 + 5) = 3x^2 - 4x^2 + 2x - 4 - 5 = -x^2 + 2x - 9$

17. $\dfrac{x^{-4}}{x^{-3}} = \dfrac{x^3}{x^4} = \dfrac{1}{x}$

18. $(3b^{-3})(2b^4) = 6b^{-3+4} = 6b$

19. $3(2t)^3 = 3(8t^3) = 24t^3$

20. $\left(\dfrac{2x^3}{x^2 y^{-1}}\right)^{-2} = \left(\dfrac{x^2 y^{-1}}{2x^3}\right)^2 = \dfrac{x^4 y^{-2}}{4x^6} = \dfrac{x^4}{4x^6 y^2} = \dfrac{1}{4x^2 y^2}$

21. $2x^2(x^3 - 4x^2 - 5) = 2x^5 - 8x^4 - 10x^2$

22. $(5x+1)(2x-7) = 10x^2 - 35x + 2x - 7 = 10x^2 - 33x - 7$

23. $(y-3)(y+3) = y^2 + 3y - 3y - 9 = y^2 - 9$

24. $(x-4y)^2 = x^2 - 2(4)xy + 16y^2 = x^2 - 8xy + 16y^2$

25. $2.5 \times 10^4 = 25,000$

26. $0.028 = 2.8 \times 10^{-2}$

27. $\dfrac{6x^3 - 4x^2 + 8x}{2x} = 3x^2 - 2x + 4$

28.
$$\begin{array}{r} 3x^2 + 5x + 5 \\ x-1\overline{\smash{\big)}\,3x^3 + 2x^2 + 1} \\ \underline{3x^3 - 3x^2} \\ 5x^2 + 1 \\ \underline{5x^2 - 5x} \\ 5x + 1 \\ \underline{5x - 5} \\ 6 \end{array}$$
   The quotient is $3x^2 + 5x + 5 + \dfrac{6}{x-1}$.

29. $10x^2y^3 - 15x^3y^2 = 5x^2y^2(2y - 3x)$

30. $x^3 + 3x^2 - x - 3 = x^2(x+3) - (x+3) = (x^2 - 1)(x+3) = (x-1)(x+1)(x+3)$

31. $2z^2 + z - 3 = (z-1)(2z+3)$

32. $16x^2 - 25 = (4x-5)(4x+5)$

33. $a^3 - 8 = (a-2)(a^2 + 2a + 4)$

34. $z^4 + 7z^2 + 6 = (z^2 + 1)(z^2 + 6)$

35. $x(x+5) = 0 \Rightarrow x = 0$ or $x = -5$

36. $4x^2 = 0 \Rightarrow x^2 = 0 \Rightarrow x = 0$

37. $2x^2 + 5x = 3 \Rightarrow 2x^2 + 5x - 3 = 0 \Rightarrow (2x-1)(x+3) = 0 \Rightarrow x = \dfrac{1}{2}$ or $x = -3$

38. $x^3 = x \Rightarrow x^3 - x = 0 \Rightarrow x(x^2 - 1) = 0 \Rightarrow x(x+1)(x-1) = 0 \Rightarrow x = 0$ or $x = -1$ or $x = 1$

39. $\dfrac{x^2 + 4x + 4}{x+2} = \dfrac{(x+2)(x+2)}{x+2} = x + 2$

40. $x = -2 \Rightarrow \dfrac{(-2)^2 + 1}{-2-1} = \dfrac{4+1}{-3} = \dfrac{5}{-3} = -\dfrac{5}{3}$   $x = 1 \Rightarrow \dfrac{1^2 + 1}{1-1} = \dfrac{1+1}{0}$, which is undefined.

41. $\dfrac{x^2 - 3x + 2}{x+2} \div \dfrac{x-1}{2x+4} = \dfrac{(x-2)(x-1)}{x+2} \cdot \dfrac{2(x+2)}{x-1} = 2(x-2) = 2x - 4$

42. $\dfrac{1}{x+3}+\dfrac{2}{x+1}=\dfrac{(x+1)}{(x+3)(x+1)}+\dfrac{2(x+3)}{(x+3)(x+1)}=\dfrac{x+1+2x+6}{(x+3)(x+1)}=\dfrac{3x+7}{(x+3)(x+1)}$

43. $\dfrac{x+2}{x-1}=\dfrac{2}{3}\Rightarrow 3(x+2)=2(x-1)\Rightarrow 3x+6=2x-2\Rightarrow x=-8$

44. $y=kx\Rightarrow 5=10k\Rightarrow k=\dfrac{1}{2}\Rightarrow y=\dfrac{1}{2}x;$ when $x=20,\ y=\dfrac{1}{2}(20)=10$

45. $f(-2)=(-2)^2-4(-2)=4+8=12$

46. $D=(-\infty,\infty);\ R=[-2,\infty)$  See Figure 46.

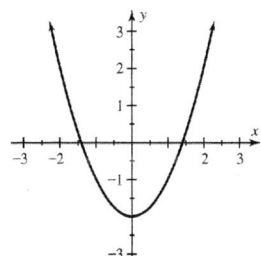

Figure 46

47. $x-2<3$ or $x-2>6\Rightarrow x<5$ or $x>8,$ so the solution set is $(-\infty,5)\cup(8,\infty).$

48. $|2x-4|=6\Rightarrow 2x-4=6$ or $2x-4=-6\Rightarrow 2x=10$ or $2x=-2\Rightarrow x=5$ or $x=-1$

49. $|2x-4|\le 6\Rightarrow 2x-4\le 6$ and $2x-4\ge -6\Rightarrow 2x\le 10$ and $2x\ge -2\Rightarrow x\le 5$ and $x\ge -1;$ the solution

is $[-1,5].$

50. $|x-4|>2\Rightarrow x-4>2$ or $x-4<-2\Rightarrow x>6$ or $x<2;$ the solution is $(-\infty,2)\cup(6,\infty).$

51. The data follow a linear pattern, and $m=\dfrac{216-144}{3-2}=\dfrac{72}{1}=72;\ d=72t$

52. (a)   $y=\dfrac{1}{4}x+2$

    (b)   $m=\dfrac{1}{4}$

    (c)   Rain is falling at a rate of $\dfrac{1}{4}$ inch per hour.

    (d)   $y=\dfrac{1}{4}(4)+2=1+2=3$ inches

53. Let $x$ represent the amount invested at 5% and $y$ represent the amount invested at 7%. The system
    required is $x+y=5000,\ 0.05x+0.07y=308$  Multiply the first equation by –0.07 and add the

    $$-0.07x-0.07y=-350$$

    result to the second equation. $\dfrac{0.05x+0.07y=\ \ \ 308}{-0.02x\ \ \ \ \ \ \ \ \ \ =-42}\Rightarrow x=2100$     Substitute $x=2100$ into the

    original first equation and solve for $y.$  $2100+y=5000\Rightarrow y=2900$

    $2100 is invested at 5% and $2900 is invested at 7%.

54. Let $x$ represent the height of the building. Then $\dfrac{8}{5} = \dfrac{x}{65} \Rightarrow 8(65) = 5x \Rightarrow 520 = 5x \Rightarrow x = 104$ feet

55. (a)   $9x + 11x = 20x$

    (b)   $20x = 1000 \Rightarrow x = 50$ minutes

56. Let $x$ represent the cost of the television. Then $0.05x = 82.50 \Rightarrow x = 1650$; the television costs $1650.

# Chapter 9: Systems of Linear Equations

## Section 9.1 Systems of Linear Equations in Three Variables

1. No, three planes cannot intersect at exactly 2 points.

3. Yes, since $1+2+3=6$.

5. Two

7. No

9. (1, 2, 3) satisfies all three equations.

11. (–1, 1, 2) satisfies all three equations.

13. Substitute $z=1$ into the second equation: $2y+(1)=-1 \Rightarrow 2y=-2 \Rightarrow y=-1$

   Substitute $z=1$ and $y=-1$ into the first equation: $x+(-1)-(1)=1 \Rightarrow x-2=1 \Rightarrow x=3$

   The solution is (3, –1, 1).

15. Substitute $z=2$ into the second equation: $2y+3(2)=3 \Rightarrow 2y=-3 \Rightarrow y=-\dfrac{3}{2}$

   Substitute $z=2$ and $y=-\dfrac{3}{2}$ into the first equation: $-x-3\left(-\dfrac{3}{2}\right)+(2)=-2 \Rightarrow -x=-\dfrac{17}{2} \Rightarrow x=\dfrac{17}{2}$

   The solution is $\left(\dfrac{17}{2},-\dfrac{3}{2},2\right)$.

17. Substitute $c=-2$ into the second equation: $-3b+(-2)=4 \Rightarrow -3b=6 \Rightarrow b=-2$

   Substitute $c=-2$ and $b=-2$ into the first equation: $a-(-2)+2(-2)=3 \Rightarrow a-2=3 \Rightarrow a=5$

   The solution is (5, –2, –2).

19. Add the first two equations together to eliminate the variable $x$.
$$\begin{array}{r} x \ +y-z = 11 \\ -x+2y+3z=-1 \\ \hline 3y+2z=10 \end{array}$$

   From the third equation, $2z=4 \Rightarrow z=2$.   And so $3y+2(2)=10 \Rightarrow 3y=6 \Rightarrow y=2$.

   Substitute $z=2$ and $y=2$ into the first equation: $x+(2)-(2)=11 \Rightarrow x=11$

   The solution is (11, 2, 2).

21. Add the first two equations together to eliminate both of the variables $x$ and $z$.
$$\begin{array}{r} x+y-z=-2 \\ -x \ +z=1 \\ \hline y=-1 \end{array}$$

   Substitute $y=-1$ into the third equation, $(-1)+2z=3 \Rightarrow 2z=4 \Rightarrow z=2$.

   Substitute $z=2$ and $y=-1$ into the first equation: $x+(-1)-(2)=-2 \Rightarrow x=1$

   The solution is (1, –1, 2).

23. Add the second and third equations together to eliminate the variable $y$.
$$\begin{array}{r} y + z = -1 \\ -y + 3z = 9 \\ \hline 4z = 8 \end{array}$$

And so $z = 2$. Substitute $z = 2$ into the second equation, $y + (2) = -1 \Rightarrow y = -3$.

Substitute $z = 2$ and $y = -3$ into the first equation: $x + (-3) - 2(2) = -7 \Rightarrow x = 0$

The solution is $(0, -3, 2)$.

25. Multiply the second equation by $-2$ and add the first and second equations to eliminate the variables $y$ and $z$.
$$\begin{array}{r} x + 2y + 2z = 1 \\ -2x - 2y - 2z = 0 \\ \hline -x = 1 \end{array}$$
And so $x = -1$. Add the first and third equations together to eliminate the variables $x$ and $y$.
$$\begin{array}{r} x + 2y + 2z = 1 \\ -x - 2y + 3z = -11 \\ \hline 5z = -10 \end{array}$$
And so $z = -2$. Substitute $x = -1$ and $z = -2$ into the second equation: $(-1) + y + (-2) = 0 \Rightarrow y = 3$ The solution is $(-1, 3, -2)$.

27. Multiply the second equation by $-1$ and add the first and second equations to eliminate the variables $y$ and $z$.
$$\begin{array}{r} x + y + z = 5 \\ -y - z = -6 \\ \hline x = -1 \end{array}$$
And so $x = -1$. Substitute $x = -1$ into the third equation:
$(-1) + z = 3 \Rightarrow z = 4$ Substitute $x = -1$ and $z = 4$ into the first equation $(-1) + y + (4) = 5 \Rightarrow y = 2$

The solution is $(-1, 2, 4)$.

29. Add the second and third equations to eliminate the variables $x$ and $z$.
$$\begin{array}{r} -x + y + 2z = 1 \\ x + y - 2z = 9 \\ \hline 2y = 10 \end{array}$$

And so, $y = 5$. Now add the first and second equations to eliminate the variable $x$.

$$\begin{array}{r} x + 2y + 3z = 24 \\ -x + y + 2z = 1 \\ \hline 3y + 5z = 25 \end{array}$$
Now substitute $y = 5$ in this new equation: $3(5) + 5z = 25 \Rightarrow 5z = 10 \Rightarrow z = 2$

Finally substitute $y = 5$ and $z = 2$ in the first equation: $x + 2(5) + 3(2) = 24 \Rightarrow x = 8$

The solution is $(8, 5, 2)$.

31. Add the first and second equations to eliminate the variable $y$.
$$\begin{array}{r} x + y + z = 2 \\ x - y + z = 1 \\ \hline 2x + 2z = 3 \end{array}$$

Multiply the third equation by $-2$ and add to this new equation to eliminate the variables $x$ and $z$.
$$\begin{array}{r} 2x + 2z = 3 \\ -2x + -2z = -6 \\ \hline 0 = -3 \end{array}$$
This is a contradiction, so there are no solutions.

33. Add the first two equations to eliminate the variable $y$.

$$\begin{array}{r} x+y+z=6 \\ x-y+z=2 \\ \hline 2x\quad+2z=8 \end{array}$$

$\Rightarrow x+z=4 \Rightarrow x=4-z$

Add the second and third equations to eliminate the variables $x$ and $z$.

$$\begin{array}{r} x-\ y+z=2 \\ -x+5y-z=6 \\ \hline 4y\quad=8 \end{array}$$

$\Rightarrow y=2.$

The system is dependent, and the solutions are all ordered triples of the form $(4-z,\,2,\,z)$.

35. Add the first and second equations to eliminate the variables $y$ and $z$.

$$\begin{array}{r} 2x+y+z=3 \\ 2x-y-z=9 \\ \hline 4x=12 \end{array}$$

And so, $x=3$. Add the second and third equations together to eliminate the variable $y$.

$$\begin{array}{r} 2x-y-z=9 \\ x+y-z=0 \\ \hline 3x-2z=9 \end{array}$$

Now substitute $x=3$ in this new equation: $3(3)-2z=9 \Rightarrow -2z=0 \Rightarrow z=0$

Finally substitute $x=3$ and $z=0$ in the third equation: $(3)+y-(0)=0 \Rightarrow y=-3$

The solution is $(3,-3,0)$.

37. Multiply the first equation by $-1$ and add the first and second equations to eliminate the variable $x$.

$$\begin{array}{r} -2x-6y+2z=-47 \\ 2x\ +y+3z=-28 \\ \hline -5y+5z=-75 \end{array}$$

Multiply the third equation by 2 and add the first and third equations to

eliminate the variables $x$ and $z$.

$$\begin{array}{r} 2x+6y-2z=47 \\ -2x+2y+2z=-7 \\ \hline 8y=40 \end{array}$$

And so $y=5$. Substitute $y=5$ into the first

*new* equation: $-5(5)+5z=-75 \Rightarrow 5z=-50 \Rightarrow z=-10$ Substitute $y=5$ and $z=-10$ into the

*original* third equation: $-x+(5)+(-10)=-\dfrac{7}{2} \Rightarrow x=-\dfrac{3}{2}$ The solution is $\left(-\dfrac{3}{2},5,-10\right)$.

39. Multiply the second equation by $-1$ and add the first and second equations to eliminate the variable $y$.

$$\begin{array}{r} x+3y-4z=\dfrac{13}{2} \\ 2x-3y\ +z=-\dfrac{1}{2} \\ \hline 3x-3z=6 \end{array}$$

Multiply this *new* equation by $-1$ and add it to the third equation to eliminate the

variable $x$.

$$\begin{array}{r} -3x+3z=-6 \\ 3x\ +z=4 \\ \hline 4z=-2 \end{array}$$

And so $z=-\dfrac{1}{2}$. Substitute $z=-\dfrac{1}{2}$ into the first *new* equation:

$3x-3\left(-\dfrac{1}{2}\right)=6 \Rightarrow 3x=\dfrac{9}{2} \Rightarrow x=\dfrac{3}{2}$ Substitute $x=\dfrac{3}{2}$ and $z=-\dfrac{1}{2}$ into the *original* first equation:

$\left(\dfrac{3}{2}\right)+3y-4\left(-\dfrac{1}{2}\right)=\dfrac{13}{2} \Rightarrow y=1$ The solution is $\left(\dfrac{3}{2},1,-\dfrac{1}{2}\right)$.

41. Multiply the second equation by –1 and add to the third equation to eliminate the variables $x$, $y$,

    $-x + y - z = -1$

    and $z$.  $\underline{\phantom{-}x - y + z = 3}$  This is a contradiction. There are no solutions.

    $\phantom{xxxxx}0 = 2$

43. Add the first two equations to eliminate the variable $y$.

    $\begin{array}{l} x + y + z = 5 \\ \underline{x - y + z = 3} \Rightarrow x + z = 4 \Rightarrow x = 4 - z \\ 2x \phantom{xx} + 2z = 8 \end{array}$

    Multiply the first equation by –1 and add to the second equation to eliminate the variables $x$ and $z$.

    $-x - y - z = -5$

    $\underline{\phantom{-}x - y + z = 3} \Rightarrow y = 1$  The system is dependent, and the solutions are all ordered pairs of the

    $\phantom{xx}-2y \phantom{xxx} = -2$

    form $(4 - z, 1, z)$.

45. Multiply the first equation by –1 and add to the second equation to eliminate the variables $x$, $y$, and $z$.

    $-x - y - z = -a$

    $\underline{\phantom{-}x + y + z = 2a}$  But $a \neq 0$, so this is a contradiction.  There are no solutions.

    $\phantom{xxx}0 = a$

47. (a)  $\begin{array}{l} x + 2y + 4z = 10 \\ x + 4y + 6z = 15 \\ 3y + 2z = 6 \end{array}$

    (b)  Using techniques similar to those used in exercises 19-44, the solution is $(2, 1, 1.5)$.

         A hamburger costs $2.00, fries cost $1.00 and a soft drink costs $1.50.

49. (a)  $\begin{array}{l} x + y + z = 180 \\ x \phantom{xx} - z = 55 \\ x - y - z = -10 \end{array}$

    (b)  Using techniques similar to those used in exercises 19-44, the solution is $(85, 65, 30)$.

         The angles are $x = 85°$, $y = 65°$, and $z = 30°$.

    (c)  These values check.  $\begin{array}{l} x + y + z = 180 \Rightarrow 85 + 65 + 30 = 180 \\ x \phantom{xx} - z = 55 \Rightarrow 85 - 30 = 55 \\ x - y - z = -10 \Rightarrow 85 - 65 - 30 = -10 \end{array}$

51. Add the first two equations to eliminate the variables $y$ and $z$.

    $\begin{array}{l} x + y + z = 180 \\ \underline{x - y - z = 40} \Rightarrow x = 110 \\ 2x \phantom{xxx} = 220 \end{array}$

    $x - z = 90 \Rightarrow z = x - 90 \Rightarrow z = 110 - 90 \Rightarrow z = 20$  Substitute the values for $x$ and $z$ into the first

    equation and solve for $y$.  $110 + y + 20 = 180 \Rightarrow y = 50$  The angle measures are $110°$, $50°$, and $20°$.

53. (a)  $\begin{array}{l} a + 600b + 4c = 525 \\ a + 400b + 2c = 365 \\ a + 900b + 5c = 805 \end{array}$

(b) Using techniques similar to those used in exercises 19-44, the solution is $(5, 1, -20)$.

That is, $a = 5$, $b = 1$, and $c = -20$ and so the equation is $F = 5 + A - 20W$.

(c) When $A = 500$ and $W = 3$, $F = 5 + (500) - 20(3) = 445$ fawns.

55. (a)
$$\begin{aligned} N + P + K &= 80 \\ N + P - K &= 8 \\ 9P - K &= 0 \end{aligned}$$

(b) Using techniques similar to those used in exercises 19-44, the solution is $(40, 4, 36)$.

The sample contains 40 pounds of nitrogen, 4 pounds of phosphorus and 36 pounds of potassium.

57. Let $x$, $y$ and $z$ represent the amounts invested at 4%, 5% and 7.5% respectively. The system needed

is: $\quad\begin{aligned} x \quad + y \quad + z &= 30{,}000 \\ 0.04x + 0.05y + 0.075z &= 1775 \\ x \quad + y \quad - z &= 2000 \end{aligned}\quad$ Using techniques similar to those used in exercises 19-44,

the solution is $(7500, 8500, 14{,}000)$. There was \$7500 invested at 4%, \$8500 invested at 5% and \$14,000 invested at 7.5%.

## Section 9.2 Matrix Solutions of Linear Systems

1. A rectangular array of numbers

3. $\begin{bmatrix} 1 & 3 & | & 10 \\ 2 & -6 & | & 4 \end{bmatrix}$; $2 \times 3$; *Answers may vary.*

5. $\begin{bmatrix} 1 & 0 & | & -3 \\ 0 & 1 & | & 5 \end{bmatrix}$; *Answers may vary.*

7. $3 \times 3$

9. $3 \times 2$

11. $\begin{bmatrix} 1 & -3 & | & 1 \\ -1 & 3 & | & -1 \end{bmatrix}$

13. $\begin{bmatrix} 2 & -1 & 2 & | & -4 \\ 1 & -2 & 0 & | & 2 \\ -1 & 1 & -2 & | & -6 \end{bmatrix}$

15. $\begin{aligned} x + 2y &= -6 \\ 5x - y &= 4 \end{aligned}$

17. $\begin{aligned} x - y + 2z &= 6 \\ 2x + y - 2z &= 1 \\ -x + 2y - z &= 3 \end{aligned}$

19. $\begin{aligned} x &= 4 \\ y &= -2 \\ z &= 7 \end{aligned}$

21. $\begin{bmatrix} 0 & 1 & 1 & 1 \\ 1 & 0 & 1 & 0 \\ 0 & 0 & 0 & 1 \\ 1 & 0 & 1 & 0 \end{bmatrix}$

23. $\begin{bmatrix} 1 & 1 & | & 4 \\ 1 & 3 & | & 10 \end{bmatrix} R_2 - R_1 \rightarrow \begin{bmatrix} 1 & 1 & | & 4 \\ 0 & 2 & | & 6 \end{bmatrix} (1/2)R_2 \rightarrow \begin{bmatrix} 1 & 1 & | & 4 \\ 0 & 1 & | & 3 \end{bmatrix} R_1 - R_2 \rightarrow \begin{bmatrix} 1 & 0 & | & 1 \\ 0 & 1 & | & 3 \end{bmatrix}$  The solution is $(1, 3)$.

25. $\begin{bmatrix} 2 & 3 & | & 3 \\ -2 & 2 & | & 7 \end{bmatrix} R_2 + R_1 \rightarrow \begin{bmatrix} 2 & 3 & | & 3 \\ 0 & 5 & | & 10 \end{bmatrix} (1/5)R_2 \rightarrow \begin{bmatrix} 2 & 3 & | & 3 \\ 0 & 1 & | & 2 \end{bmatrix} R_1 - 3R_2 \rightarrow \begin{bmatrix} 2 & 0 & | & -3 \\ 0 & 1 & | & 2 \end{bmatrix}$

$(1/2)R_2 \rightarrow \begin{bmatrix} 1 & 0 & | & -\frac{3}{2} \\ 0 & 1 & | & 2 \end{bmatrix}$  The solution is $\left(-\dfrac{3}{2}, 2\right)$.

27. $\begin{bmatrix} 1 & -1 & | & 5 \\ 1 & 3 & | & -1 \end{bmatrix} R_2 - R_1 \rightarrow \begin{bmatrix} 1 & -1 & | & 5 \\ 0 & 4 & | & -6 \end{bmatrix} (1/4)R_2 \rightarrow \begin{bmatrix} 1 & -1 & | & 5 \\ 0 & 1 & | & -\frac{3}{2} \end{bmatrix} R_1 + R_2 \rightarrow \begin{bmatrix} 1 & 0 & | & \frac{7}{2} \\ 0 & 1 & | & -\frac{3}{2} \end{bmatrix}$

The solution is $\left(\dfrac{7}{2}, -\dfrac{3}{2}\right)$.

29. $\begin{bmatrix} 4 & -8 & | & -10 \\ 1 & 1 & | & 2 \end{bmatrix} \begin{matrix} Exchange \\ R_2 \leftrightarrow R_1 \end{matrix} \begin{bmatrix} 1 & 1 & | & 2 \\ 4 & -8 & | & -10 \end{bmatrix} (-1/2)R_2 \rightarrow \begin{bmatrix} 1 & 1 & | & 2 \\ -2 & 4 & | & 5 \end{bmatrix} R_2 + 2R_1 \rightarrow \begin{bmatrix} 1 & 1 & | & 2 \\ 0 & 6 & | & 9 \end{bmatrix}$

$(1/6)R_2 \rightarrow \begin{bmatrix} 1 & 1 & | & 2 \\ 0 & 1 & | & \frac{3}{2} \end{bmatrix} R_1 - R_2 \rightarrow \begin{bmatrix} 1 & 0 & | & \frac{1}{2} \\ 0 & 1 & | & \frac{3}{2} \end{bmatrix}$  The solution is $\left(\dfrac{1}{2}, \dfrac{3}{2}\right)$.

31. $\begin{bmatrix} 1 & 1 & 1 & | & 6 \\ 0 & 2 & -1 & | & 1 \\ 0 & 1 & 1 & | & 5 \end{bmatrix} \begin{matrix} R_1 - R_3 \rightarrow \\ R_2 - 2R_3 \rightarrow \end{matrix} \begin{bmatrix} 1 & 0 & 0 & | & 1 \\ 0 & 0 & -3 & | & -9 \\ 0 & 1 & 1 & | & 5 \end{bmatrix} \begin{matrix} Exchange \\ R_2 \leftrightarrow R_3 \end{matrix} \begin{bmatrix} 1 & 0 & 0 & | & 1 \\ 0 & 1 & 1 & | & 5 \\ 0 & 0 & -3 & | & -9 \end{bmatrix}$

$(-1/3)R_3 \rightarrow \begin{bmatrix} 1 & 0 & 0 & | & 1 \\ 0 & 1 & 1 & | & 5 \\ 0 & 0 & 1 & | & 3 \end{bmatrix} R_2 - R_3 \rightarrow \begin{bmatrix} 1 & 0 & 0 & | & 1 \\ 0 & 1 & 0 & | & 2 \\ 0 & 0 & 1 & | & 3 \end{bmatrix}$  The solution is $(1, 2, 3)$.

33. $\begin{bmatrix} 1 & 2 & 3 & | & 6 \\ -1 & 3 & 4 & | & 0 \\ 1 & 1 & -2 & | & -6 \end{bmatrix} \begin{matrix} R_2 + R_1 \rightarrow \\ R_3 - R_1 \rightarrow \end{matrix} \begin{bmatrix} 1 & 2 & 3 & | & 6 \\ 0 & 5 & 7 & | & 6 \\ 0 & -1 & -5 & | & -12 \end{bmatrix} (-1)R_3 \rightarrow \begin{bmatrix} 1 & 2 & 3 & | & 6 \\ 0 & 5 & 7 & | & 6 \\ 0 & 1 & 5 & | & 12 \end{bmatrix}$

$\begin{matrix} R_1 - 2R_3 \rightarrow \\ R_2 - 5R_3 \rightarrow \end{matrix} \begin{bmatrix} 1 & 0 & -7 & | & -18 \\ 0 & 0 & -18 & | & -54 \\ 0 & 1 & 5 & | & 12 \end{bmatrix} \begin{matrix} Exchange \\ R_2 \leftrightarrow R_3 \end{matrix} \begin{bmatrix} 1 & 0 & -7 & | & -18 \\ 0 & 1 & 5 & | & 12 \\ 0 & 0 & -18 & | & -54 \end{bmatrix} (-1/18)R_3 \rightarrow \begin{bmatrix} 1 & 0 & -7 & | & -18 \\ 0 & 1 & 5 & | & 12 \\ 0 & 0 & 1 & | & 3 \end{bmatrix}$

$\begin{matrix} R_1 + 7R_3 \rightarrow \\ R_2 - 5R_3 \rightarrow \end{matrix} \begin{bmatrix} 1 & 0 & 0 & | & 3 \\ 0 & 1 & 0 & | & -3 \\ 0 & 0 & 1 & | & 3 \end{bmatrix}$  The solution is $(3, -3, 3)$.

35. $\begin{bmatrix} 1 & 1 & 1 & | & 0 \\ 2 & 1 & 2 & | & -1 \\ 1 & 1 & 0 & | & 0 \end{bmatrix} \begin{matrix} R_2 - 2R_1 \rightarrow \\ R_3 - R_1 \rightarrow \end{matrix} \begin{bmatrix} 1 & 1 & 1 & | & 0 \\ 0 & -1 & 0 & | & -1 \\ 0 & 0 & -1 & | & 0 \end{bmatrix} \begin{matrix} (-1)R_2 \rightarrow \\ (-1)R_3 \rightarrow \end{matrix} \begin{bmatrix} 1 & 1 & 1 & | & 0 \\ 0 & 1 & 0 & | & 1 \\ 0 & 0 & 1 & | & 0 \end{bmatrix} R_1 - R_2 \rightarrow \begin{bmatrix} 1 & 0 & 1 & | & -1 \\ 0 & 1 & 0 & | & 1 \\ 0 & 0 & 1 & | & 0 \end{bmatrix}$

$R_1 - R_3 \rightarrow \begin{bmatrix} 1 & 0 & 0 & | & -1 \\ 0 & 1 & 0 & | & 1 \\ 0 & 0 & 1 & | & 0 \end{bmatrix}$  The solution is $(-1, 1, 0)$.

37. $\begin{bmatrix} 1 & 1 & 1 & | & 3 \\ -1 & 0 & -1 & | & -2 \\ 1 & 1 & 2 & | & 4 \end{bmatrix} \begin{matrix} \\ R_2 + R_1 \to \\ R_3 - R_1 \to \end{matrix} \begin{bmatrix} 1 & 1 & 1 & | & 3 \\ 0 & 1 & 0 & | & 1 \\ 0 & 0 & 1 & | & 1 \end{bmatrix} \begin{matrix} R_1 - R_2 \to \\ \\ \\ \end{matrix} \begin{bmatrix} 1 & 0 & 1 & | & 2 \\ 0 & 1 & 0 & | & 1 \\ 0 & 0 & 1 & | & 1 \end{bmatrix} \begin{matrix} R_1 - R_3 \to \\ \\ \\ \end{matrix} \begin{bmatrix} 1 & 0 & 0 & | & 1 \\ 0 & 1 & 0 & | & 1 \\ 0 & 0 & 1 & | & 1 \end{bmatrix}$

The solution is $(1, 1, 1)$.

39. $\begin{bmatrix} 1 & 2 & 1 & | & 3 \\ 2 & 1 & -1 & | & -6 \\ -1 & -1 & 2 & | & 5 \end{bmatrix} \begin{matrix} \\ R_2 - 2R_1 \to \\ R_3 + R_1 \to \end{matrix} \begin{bmatrix} 1 & 2 & 1 & | & 3 \\ 0 & -3 & -3 & | & -12 \\ 0 & 1 & 3 & | & 8 \end{bmatrix} (-1/3)R_2 \to \begin{bmatrix} 1 & 2 & 1 & | & 3 \\ 0 & 1 & 1 & | & 4 \\ 0 & 1 & 3 & | & 8 \end{bmatrix}$

$\begin{matrix} \\ \\ R_3 - R_2 \to \end{matrix} \begin{bmatrix} 1 & 2 & 1 & | & 3 \\ 0 & 1 & 1 & | & 4 \\ 0 & 0 & 2 & | & 4 \end{bmatrix} \begin{matrix} R_1 - 2R_2 \to \\ \\ (1/2)R_3 \to \end{matrix} \begin{bmatrix} 1 & 0 & -1 & | & -5 \\ 0 & 1 & 1 & | & 4 \\ 0 & 0 & 1 & | & 2 \end{bmatrix} \begin{matrix} R_1 + R_3 \to \\ R_2 - R_3 \to \\ \\ \end{matrix} \begin{bmatrix} 1 & 0 & 0 & | & -3 \\ 0 & 1 & 0 & | & 2 \\ 0 & 0 & 1 & | & 2 \end{bmatrix}$

The solution is $(-3, 2, 2)$.

41. See example 7 in the text for graphing calculator instructions.

$[A] = \begin{bmatrix} 1 & 4 & | & 13 \\ 5 & -3 & | & -50 \end{bmatrix}$; rref$([A]) = \begin{bmatrix} 1 & 0 & | & -7 \\ 0 & 1 & | & 5 \end{bmatrix}$; The solution is $(-7, 5)$.

43. See example 7 in the text for graphing calculator instructions.

$[A] = \begin{bmatrix} 2 & -1 & 3 & | & 9 \\ -4 & 5 & 2 & | & 12 \\ 2 & 0 & 7 & | & 23 \end{bmatrix}$; rref$([A]) = \begin{bmatrix} 1 & 0 & 0 & | & 1 \\ 0 & 1 & 0 & | & 2 \\ 0 & 0 & 1 & | & 3 \end{bmatrix}$; The solution is $(1, 2, 3)$.

45. See example 7 in the text for graphing calculator instructions.

$[A] = \begin{bmatrix} 6 & 2 & 1 & | & 4 \\ -2 & 4 & 1 & | & -3 \\ 2 & -8 & 0 & | & -2 \end{bmatrix}$; rref$([A]) = \begin{bmatrix} 1 & 0 & 0 & | & 1 \\ 0 & 1 & 0 & | & 0.5 \\ 0 & 0 & 1 & | & -3 \end{bmatrix}$; The solution is $(1, 0.5, -3)$.

47. See example 7 in the text for graphing calculator instructions.

$[A] = \begin{bmatrix} 4 & 3 & 12 & | & -9.25 \\ -1 & 15 & 8 & | & -4.75 \\ 0 & 6 & 7 & | & -5.5 \end{bmatrix}$; rref$([A]) = \begin{bmatrix} 1 & 0 & 0 & | & 0.5 \\ 0 & 1 & 0 & | & 0.25 \\ 0 & 0 & 1 & | & -1 \end{bmatrix}$; The solution is $(0.5, 0.25, -1)$.

49. See example 7 in the text for graphing calculator instructions.

$[A] = \begin{bmatrix} 1.2 & -0.9 & 2.7 & | & 5.37 \\ 3.1 & -5.1 & 7.2 & | & 14.81 \\ 0.2 & 1.8 & -3.6 & | & -6.38 \end{bmatrix}$; rref$([A]) = \begin{bmatrix} 1 & 0 & 0 & | & 0.5 \\ 0 & 1 & 0 & | & -0.2 \\ 0 & 0 & 1 & | & 1.7 \end{bmatrix}$; The solution is $(0.5, -0.2, 1.7)$.

51. $\begin{bmatrix} 1 & 2 & | & 4 \\ -2 & -4 & | & -8 \end{bmatrix} R_2 + 2R_1 \to \begin{bmatrix} 1 & 2 & | & 4 \\ 0 & 0 & | & 0 \end{bmatrix}$

Row 2 represents the equation $0 = 0$. The system is dependent.

53. $\begin{bmatrix} 1 & 1 & 1 & | & 3 \\ 1 & 1 & -1 & | & 1 \\ 1 & 1 & 0 & | & 3 \end{bmatrix} \begin{matrix} \\ R_2 - R_1 \to \\ R_3 - R_1 \to \end{matrix} \begin{bmatrix} 1 & 1 & 1 & | & 3 \\ 0 & 0 & -2 & | & -2 \\ 0 & 0 & -1 & | & 0 \end{bmatrix} R_2 - 2R_3 \to \begin{bmatrix} 1 & 1 & 1 & | & 3 \\ 0 & 0 & 0 & | & -2 \\ 0 & 0 & -1 & | & 0 \end{bmatrix}$

Row 2 represents the equation $0 = -2$. The system is inconsistent.

55. $\begin{bmatrix} 1 & 2 & 3 & | & 14 \\ 2 & -3 & -2 & | & -10 \\ 3 & -1 & 1 & | & 4 \end{bmatrix} \begin{matrix} \\ R_2 - 2R_1 \to \\ R_3 - 3R_1 \to \end{matrix} \begin{bmatrix} 1 & 2 & 3 & | & 14 \\ 0 & -7 & -8 & | & -38 \\ 0 & -7 & -8 & | & -38 \end{bmatrix} R_3 - R_2 \to \begin{bmatrix} 1 & 2 & 3 & | & 14 \\ 0 & -7 & -8 & | & -38 \\ 0 & 0 & 0 & | & 0 \end{bmatrix}$

Row 3 represents the equation $0 = 0$. The system is dependent.

57. $\begin{bmatrix} a & 0 & 0 & | & 1 \\ 0 & b & 0 & | & 1 \\ 0 & 0 & ab & | & 2 \end{bmatrix} \begin{matrix} (1/a)R_1 \rightarrow \\ (1/b)R_2 \rightarrow \\ (1/(ab))R_3 \rightarrow \end{matrix} \begin{bmatrix} 1 & 0 & 0 & | & \frac{1}{a} \\ 0 & 1 & 0 & | & \frac{1}{b} \\ 0 & 0 & 1 & | & \frac{2}{ab} \end{bmatrix}$ The solution is $\left(\frac{1}{a}, \frac{1}{b}, \frac{2}{ab}\right)$.

59. The equation found in example 8 is $W = -374 + 19H + 6L$. When $H = 12$ and $L = 60$,

$W = -374 + 19(12) + 6(60) = 214$ lb.

61. (a)
$$\begin{aligned} a + 16b + 26c &= 80 \\ a + 28b + 45c &= 344 \\ a + 31b + 54c &= 416 \end{aligned}$$

(b)    Using the graphing calculator to solve the system, the solution is $a \approx -272.9$, $b \approx 19.8$, $c \approx 1.4$.

So the equation is $W \approx -272.9 + 19.8N + 1.4C$.

(c)    When $N = 22$ and $C = 38$, $W \approx -272.9 + 19.8(22) + 1.4(38) \approx 216$ lb.

63. Let $x$, $y$ and $z$ represent the time spent running at 5, 6 and 8 mph respectively. The system needed is:

$$\begin{aligned} x + y + z &= 2 \\ 5x + 6y + 8z &= 12.5 \\ x \quad - z &= 0 \end{aligned}$$ and so $[A] = \begin{bmatrix} 1 & 1 & 1 & | & 2 \\ 5 & 6 & 8 & | & 12.5 \\ 1 & 0 & -1 & | & 0 \end{bmatrix}$; $\text{rref}([A]) = \begin{bmatrix} 1 & 0 & 0 & | & 0.5 \\ 0 & 1 & 0 & | & 1 \\ 0 & 0 & 1 & | & 0.5 \end{bmatrix}$

The solution is (0.5, 1, 0.5). The runner ran 0.5 hr at 5 mph, 1 hr at 6 mph and 0.5 hr at 8 mph.

65. Let $x$, $y$ and $z$ represent the amount invested at 3%, 4% and 6% respectively. The system needed is:

$$\begin{aligned} x + y + z &= 3000 \\ 0.03x + 0.04y + 0.06z &= 145 \\ 3x \quad - z &= 0 \end{aligned}$$ and so $[A] = \begin{bmatrix} 1 & 1 & 1 & | & 3000 \\ 0.03 & 0.04 & 0.06 & | & 145 \\ 3 & 0 & -1 & | & 0 \end{bmatrix}$; $\text{rref}([A]) = \begin{bmatrix} 1 & 0 & 0 & | & 500 \\ 0 & 1 & 0 & | & 1000 \\ 0 & 0 & 1 & | & 1500 \end{bmatrix}$

The solution is (500, 1000, 1500). There was $500 invested at 3%, $1000 at 4% and $1500 at 6%.

67. (a)    See example below. Answers may vary.

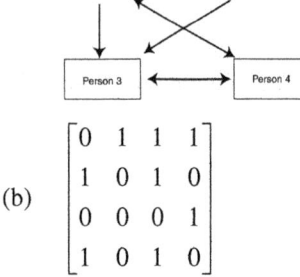

(b)    $\begin{bmatrix} 0 & 1 & 1 & 1 \\ 1 & 0 & 1 & 0 \\ 0 & 0 & 0 & 1 \\ 1 & 0 & 1 & 0 \end{bmatrix}$

## Checking Basic Concepts  Section 9.1 and 9.2

1. $(1, 3, -1)$ satisfies all three equations.

2. Multiply the first equation by –2 and add the first and second equations to eliminate the variable $x$.

$-2x+2y-2z=-4$

$\underline{2x-3y\ +z=-1}$  Add the first and third equations to eliminate the variables $x$ and $y$.

$-y-z=-5$

$x-y+z=2$

$\underline{-x+y+z=4}$  And so $z=3$. Substitute $z=3$ into the first *new* equation:

$2z=6$

$-y-(3)=-5 \Rightarrow -y=-2 \Rightarrow y=2$  Substitute $y=2$ and $z=3$ into the *original* first

equation: $x-(2)+(3)=2 \Rightarrow x=1$  The solution is (1, 2, 3).

3. (a) $\begin{bmatrix} 1 & 2 & 1 & | & 1 \\ 1 & 1 & 1 & | & -1 \\ 0 & 1 & 1 & | & 1 \end{bmatrix} \begin{matrix} \\ R_2-R_1 \to \\ \\ \end{matrix} \begin{bmatrix} 1 & 2 & 1 & | & 1 \\ 0 & -1 & 0 & | & -2 \\ 0 & 1 & 1 & | & 1 \end{bmatrix} \begin{matrix} R_1+2R_2 \to \\ -1R_2 \to \\ R_3+R_2 \to \end{matrix} \begin{bmatrix} 1 & 0 & 1 & | & -3 \\ 0 & 1 & 0 & | & 2 \\ 0 & 0 & 1 & | & -1 \end{bmatrix}$

$R_1-R_3 \to \begin{bmatrix} 1 & 0 & 0 & | & -2 \\ 0 & 1 & 0 & | & 2 \\ 0 & 0 & 1 & | & -1 \end{bmatrix}$;  The solution is (–2, 2, –1).

(b) $[A]=\begin{bmatrix} 1 & 2 & 1 & | & 1 \\ 1 & 1 & 1 & | & -1 \\ 0 & 1 & 1 & | & 1 \end{bmatrix}$; $\text{rref}([A])=\begin{bmatrix} 1 & 0 & 0 & | & -2 \\ 0 & 1 & 0 & | & 2 \\ 0 & 0 & 1 & | & -1 \end{bmatrix}$;  The solution is (–2, 2, –1).

4. Let $x$, $y$ and $z$ represent the amount invested at 1%, 2% and 4% respectively. The system needed is:

$x+\ \ y\ \ +z=1500$

$0.01x+0.02y+0.04z=46$  and so $[A]=\begin{bmatrix} 1 & 1 & 1 & | & 1500 \\ 0.01 & 0.02 & 0.04 & | & 46 \\ 2 & -1 & 0 & | & 0 \end{bmatrix}$; $\text{rref}([A])=\begin{bmatrix} 1 & 0 & 0 & | & 200 \\ 0 & 1 & 0 & | & 400 \\ 0 & 0 & 1 & | & 900 \end{bmatrix}$

$2x-y\ \ \ \ \ \ =0$

The solution is (200, 400, 900). There was $200 invested at 1%, $400 at 2% and $900 at 4%.

## Section 9.3 Determinants

1. square

3. system of linear equations

5. $\det A=1(-8)-3(-2)=-8+6=-2$

7. $\det A=-3(-1)-8(7)=3-56=-53$

9. $\det A=23(-13)-6(4)=-299-24=-323$

11. $\det A=1\left[(1)(7)-(-4)(-3)\right]-0\left[(-1)(7)-(-4)(2)\right]+0\left[(-1)(-3)-(1)(2)\right]=-5-0+0=-5$

13. $\det A=2\left[(-2)(8)-(1)(6)\right]-1\left[(-1)(8)-(1)(0)\right]+0\left[(-1)(6)-(-2)(0)\right]=-44+8+0=-36$

15. $\det A=(-1)\left[(-3)(7)-(-3)(5)\right]-3\left[(3)(7)-(-3)(5)\right]+2\left[(3)(5)-(-3)(5)\right]=6-108+60=-42$

17. $\det A=5\left[(-2)(5)-(0)(0)\right]-0\left[(0)(5)-(0)(0)\right]+0\left[(0)(0)-(-2)(0)\right]=-50-0+0=-50$

19. $\det A=0\left[(3)(9)-(5)(-9)\right]-0\left[(2)(9)-(5)(-3)\right]+0\left[(2)(-9)-(3)(-3)\right]=0-0+0=0$

21. Using the calculator we find $\det([A]) = -3555$.

23. Using the calculator we find $\det([A]) = -7466.5$.

25. $\det A = a(bc - 0) - 0(0 \cdot c - 0) + 0(0 - 0 \cdot b) = abc$

27. The triangle has vertices (3, 2), (5, 8) and (9, 5). The matrix needed is $A = \begin{bmatrix} 3 & 5 & 9 \\ 2 & 8 & 5 \\ 1 & 1 & 1 \end{bmatrix}$.

    The area is $|D| = \left| \dfrac{1}{2} \det([A]) \right| = 15 \text{ ft}^2$.

29. The triangle has vertices (–6, –4), (2, 6) and (6, –2). The matrix needed is $A = \begin{bmatrix} -6 & 2 & 6 \\ -4 & 6 & -2 \\ 1 & 1 & 1 \end{bmatrix}$.

    The area is $|D| = \left| \dfrac{1}{2} \det([A]) \right| = 52 \text{ ft}^2$.

31. Split the figure into two triangles with vertices (2, 1), (3, 6), (9, 3) and vertices (3, 6), (7, 7), (9, 3).

    The matrices needed are $A = \begin{bmatrix} 2 & 3 & 9 \\ 1 & 6 & 3 \\ 1 & 1 & 1 \end{bmatrix}$ and $B = \begin{bmatrix} 3 & 7 & 9 \\ 6 & 7 & 3 \\ 1 & 1 & 1 \end{bmatrix}$.

    The area is $\left| \dfrac{1}{2} \det([A]) \right| + \left| \dfrac{1}{2} \det([B]) \right| = 16.5 + 9 = 25.5 \text{ ft}^2$.

33. $E = \det \begin{bmatrix} 4 & 3 \\ 20 & -4 \end{bmatrix} = -16 - 60 = -76; \quad F = \det \begin{bmatrix} 5 & 4 \\ 6 & 20 \end{bmatrix} = 100 - 24 = 76$

    $D = \det \begin{bmatrix} 5 & 3 \\ 6 & -4 \end{bmatrix} = -20 - 18 = -38$; The solution is $x = \dfrac{E}{D} = \dfrac{-76}{-38} = 2$ and $y = \dfrac{F}{D} = \dfrac{76}{-38} = -2$.

35. $E = \det \begin{bmatrix} -3 & -5 \\ -8 & 6 \end{bmatrix} = -18 - 40 = -58; \quad F = \det \begin{bmatrix} 7 & -3 \\ -4 & -8 \end{bmatrix} = -56 - 12 = -68$

    $D = \det \begin{bmatrix} 7 & -5 \\ -4 & 6 \end{bmatrix} = 42 - 20 = 22$; The solution is $x = \dfrac{E}{D} = \dfrac{-58}{22} = -\dfrac{29}{11}$ and $y = \dfrac{F}{D} = \dfrac{-68}{22} = -\dfrac{34}{11}$.

37. $E = \det \begin{bmatrix} -61 & -3 \\ -23 & -4 \end{bmatrix} = 244 - 69 = 175; \quad F = \det \begin{bmatrix} 8 & -61 \\ -1 & -23 \end{bmatrix} = -184 - 61 = -245$

    $D = \det \begin{bmatrix} 8 & -3 \\ -1 & -4 \end{bmatrix} = -32 - 3 = -35$; The solution is $x = \dfrac{E}{D} = \dfrac{175}{-35} = -5$ and $y = \dfrac{F}{D} = \dfrac{-245}{-35} = 7$.

## Checking Basic Concepts Section 9.3

1. (a)  $\det A = -3(3) - (-2)(4) = -9 + 8 = -1$

   (b)  $\det A = 1[(1)(-1) - (2)(1)] - 5[(-2)(-1) - (2)(3)] + 0[(-2)(1) - (1)(3)] = -3 + 20 + 0 = 17$

2. $E = \det \begin{bmatrix} -14 & -1 \\ -36 & -4 \end{bmatrix} = 56 - 36 = 20;$ $F = \det \begin{bmatrix} 2 & -14 \\ 3 & -36 \end{bmatrix} = -72 - (-42) = -30$

$D = \det \begin{bmatrix} 2 & -1 \\ 3 & -4 \end{bmatrix} = -8 - (-3) = -5;$ The solution is $x = \dfrac{E}{D} = \dfrac{20}{-5} = -4$ and $y = \dfrac{F}{D} = \dfrac{-30}{-5} = 6.$

3. The triangle has vertices $(-1, 2)$, $(5, 6)$ and $(2, -3)$. The matrix needed is $A = \begin{bmatrix} -1 & 5 & 2 \\ 2 & 6 & -3 \\ 1 & 1 & 1 \end{bmatrix}.$

The area is $|D| = \left| \dfrac{1}{2} \det([A]) \right| = 21$ square units.

## Chapter 9 Review

1. Yes, since $3 + (-4) + 5 = 4.$

2. $(1, -1, 2)$ is a solution since it satisfies all three of the equations.

3. Add the first two equations together to eliminate the variable $x$.
$$\begin{array}{r} x - y - 2z = -11 \\ -x + 2y + 3z = 16 \\ \hline y + z = 5 \end{array}$$

From the third equation, $3z = 6 \Rightarrow z = 2.$ And so $y + (2) = 5 \Rightarrow y = 3.$

Substitute $z = 2$ and $y = 3$ into the first equation: $x - (3) - 2(2) = -11 \Rightarrow x = -4$

The solution is $(-4, 3, 2).$

4. Multiply the second equation by $-5$ and the third equation by $3$. Add these equations to eliminate the variable $z$.
$$\begin{array}{r} 10x - 5y - 15z = 10 \\ 3x - 6y + 15z = -78 \\ \hline 13x - 11y = -68 \end{array}$$
Multiply the first equation by $11$ and add it to this *new* equation to

eliminate the variable $y$.
$$\begin{array}{r} 11x + 11y = 44 \\ 13x - 11y = -68 \\ \hline 24x = -24 \end{array}$$
And so $x = -1.$ Substitute $x = -1$ into the first equation:

$(-1) + y = 4 \Rightarrow y = 5$ Substitute $x = -1$ and $y = 5$ into the second equation:

$-2(-1) + (5) + 3z = -2 \Rightarrow 3z = -9 \Rightarrow z = -3$ The solution is $(-1, 5, -3).$

5. Multiply the second equation by $-1$ and add it to the third equation to eliminate the variable $z$.
$$\begin{array}{r} -x - 2y - z = -7 \\ -2x + y + z = 7 \\ \hline -3x - y = 0 \end{array}$$
Multiply the first equation by $-1$ and add it to this *new* equation to eliminate the

variable $y$.
$$\begin{array}{r} -2x + y = 5 \\ -3x - y = 0 \\ \hline -5x = 5 \end{array}$$
And so $x = -1.$ Substitute $x = -1$ into the first equation:

$2(-1) - y = -5 \Rightarrow y = 3$ Substitute $x = -1$ and $y = 3$ into the third equation:

$-2(-1) + (3) + z = 7 \Rightarrow z = 2$ The solution is $(-1, 3, 2).$

6. Multiply the second equation by 2 and add the first and second equations to eliminate the variable $x$.

$2x+3y\ +z=6$

$\underline{-2x+4y+4z=6}$  Add the second and third equations together to eliminate the variable $x$.

$\quad 7y+5z=12$

$-x+2y+2z=3$

$\underline{\ x\ +y+2z=4}$  Multiply the first *new* equation by 4 and the second *new* equation by –5. Add

$\quad 3y+4z=7$

$\qquad\qquad\qquad\qquad\qquad\qquad 28y+20z=48$

these to eliminate the variable $z$.  $\underline{-15y-20z=-35}$  And so  $y=1$. Substitute  $y=1$ into the second

$\qquad\qquad\qquad\qquad\qquad\qquad\ \ 13y=13$

*new* equation: $3(1)+4z=7\Rightarrow 4z=4\Rightarrow z=1$  Substitute  $y=1$ and $z=1$ into the *original* third

equation: $x+(1)+2(1)=4\Rightarrow x=1$  The solution is $(1,\ 1,\ 1)$.

$\qquad\qquad\qquad\qquad\qquad\qquad\qquad\qquad x-y+3z=2$

7. Add the first two equations to eliminate the variable $y$.  $\underline{2x+y+4z=3}$  Multiply the second

$\qquad\qquad\qquad\qquad\qquad\qquad\qquad\qquad\ \ 3x\qquad +7z=5$

$\qquad\qquad\qquad\qquad\qquad\qquad\qquad\qquad\qquad\qquad -4x-2y-8z=-6$

equation by –2 and add to the third equation to eliminate the variable $y$.  $\underline{\ x+2y+\ z=5}$

$\qquad\qquad\qquad\qquad\qquad\qquad\qquad\qquad\qquad\qquad\ -3x\qquad -7z=-1$

$\qquad\qquad\qquad\qquad\qquad\qquad\qquad\qquad\qquad\qquad\ \ 3x+7z=5$

Add this new equation to the result of the first sum to eliminate the variables $x$ and $z$.  $\underline{-3x-7z=-1}$

$\qquad\qquad\qquad\qquad\qquad\qquad\qquad\qquad\qquad\qquad\qquad\qquad\ \ 0=4$

This is a contradiction. There are no solutions.

8. Solve the third equation for $x$: $x+z=2\Rightarrow x=2-z$. Substitute  $x=2-z$ into the second equation:

$2-z+y-z=1\Rightarrow y-2z=-1\Rightarrow y=2z-1$. The system is dependent, and all solutions are of the

form of the ordered triple $(2-z,\ 2z-1,\ z)$

9. $\begin{bmatrix} 1 & 1 & 1 & | & -6 \\ 1 & 2 & 1 & | & -8 \\ 0 & 1 & 1 & | & -5 \end{bmatrix} \begin{array}{c} \\ R_2-R_1 \rightarrow \\ \ \end{array} \begin{bmatrix} 1 & 1 & 1 & | & -6 \\ 0 & 1 & 0 & | & -2 \\ 0 & 1 & 1 & | & -5 \end{bmatrix} \begin{array}{c} R_1-R_3 \rightarrow \\ \\ R_3-R_2 \rightarrow \end{array} \begin{bmatrix} 1 & 0 & 0 & | & -1 \\ 0 & 1 & 0 & | & -2 \\ 0 & 0 & 1 & | & -3 \end{bmatrix};$   The solution is $(-1,\ -2,\ -3)$.

10. $\begin{bmatrix} 1 & 1 & 1 & | & -3 \\ -1 & 1 & 0 & | & 5 \\ 0 & 1 & 1 & | & -1 \end{bmatrix} \begin{array}{c} R_1-R_3 \rightarrow \\ R_2+R_1 \rightarrow \\ \ \end{array} \begin{bmatrix} 1 & 0 & 0 & | & -2 \\ 0 & 2 & 1 & | & 2 \\ 0 & 1 & 1 & | & -1 \end{bmatrix} \begin{array}{c} \\ R_2-R_3 \rightarrow \\ \ \end{array} \begin{bmatrix} 1 & 0 & 0 & | & -2 \\ 0 & 1 & 0 & | & 3 \\ 0 & 1 & 1 & | & -1 \end{bmatrix} \begin{array}{c} \\ \\ R_3-R_2 \rightarrow \end{array} \begin{bmatrix} 1 & 0 & 0 & | & -2 \\ 0 & 1 & 0 & | & 3 \\ 0 & 0 & 1 & | & -4 \end{bmatrix}$

The solution is $(-2,\ 3,\ -4)$.

11. $\begin{bmatrix} 1 & 2 & -1 & | & 1 \\ -1 & 1 & -2 & | & 5 \\ 0 & 2 & 1 & | & 10 \end{bmatrix} \begin{array}{c} R_1-R_3 \rightarrow \\ R_2+R_1 \rightarrow \\ \ \end{array} \begin{bmatrix} 1 & 0 & -2 & | & -9 \\ 0 & 3 & -3 & | & 6 \\ 0 & 2 & 1 & | & 10 \end{bmatrix} \begin{array}{c} \\ (1/3)R_2 \rightarrow \\ \ \end{array} \begin{bmatrix} 1 & 0 & -2 & | & -9 \\ 0 & 1 & -1 & | & 2 \\ 0 & 2 & 1 & | & 10 \end{bmatrix}$

$\begin{array}{c} \\ \\ R_3-2R_2 \rightarrow \end{array} \begin{bmatrix} 1 & 0 & -2 & | & -9 \\ 0 & 1 & -1 & | & 2 \\ 0 & 0 & 3 & | & 6 \end{bmatrix} \begin{array}{c} \\ \\ (1/3)R_3 \rightarrow \end{array} \begin{bmatrix} 1 & 0 & -2 & | & -9 \\ 0 & 1 & -1 & | & 2 \\ 0 & 0 & 1 & | & 2 \end{bmatrix} \begin{array}{c} R_1+2R_3 \rightarrow \\ R_2+R_3 \rightarrow \\ \ \end{array} \begin{bmatrix} 1 & 0 & 0 & | & -5 \\ 0 & 1 & 0 & | & 4 \\ 0 & 0 & 1 & | & 2 \end{bmatrix}$

The solution is $(-5,\ 4,\ 2)$.

12. $\begin{bmatrix} 2 & 2 & -2 & | & -14 \\ -2 & -3 & 2 & | & 12 \\ 1 & 1 & -4 & | & -22 \end{bmatrix} \begin{matrix} (1/2)R_1 \to \\ R_2 + R_1 \to \\ R_3 - (1/2)R_1 \to \end{matrix} \begin{bmatrix} 1 & 1 & -1 & | & -7 \\ 0 & -1 & 0 & | & -2 \\ 0 & 0 & -3 & | & -15 \end{bmatrix} \begin{matrix} R_1 + R_2 \to \\ -1R_2 \to \\ (-1/3)R_3 \to \end{matrix} \begin{bmatrix} 1 & 0 & -1 & | & -9 \\ 0 & 1 & 0 & | & 2 \\ 0 & 0 & 1 & | & 5 \end{bmatrix}$

$R_1 + R_3 \to \begin{bmatrix} 1 & 0 & 0 & | & -4 \\ 0 & 1 & 0 & | & 2 \\ 0 & 0 & 1 & | & 5 \end{bmatrix}$; The solution is $(-4, 2, 5)$.

13. See example 7 in section 9.2 in the text for graphing calculator instructions.

$[A] = \begin{bmatrix} 3 & -2 & 6 & | & -17 \\ -2 & -1 & 5 & | & 20 \\ 0 & 4 & 7 & | & 30 \end{bmatrix}$; $\mathrm{rref}([A]) = \begin{bmatrix} 1 & 0 & 0 & | & -7 \\ 0 & 1 & 0 & | & 4 \\ 0 & 0 & 1 & | & 2 \end{bmatrix}$; The solution is $(-7, 4, 2)$.

14. See example 7 in section 9.2 in the text for graphing calculator instructions.

$[A] = \begin{bmatrix} 19 & -13 & -7 & | & 7.4 \\ 22 & 33 & -8 & | & 110.5 \\ 10 & -56 & 9 & | & 23.7 \end{bmatrix}$; $\mathrm{rref}([A]) = \begin{bmatrix} 1 & 0 & 0 & | & 5.4 \\ 0 & 1 & 0 & | & 2.1 \\ 0 & 0 & 1 & | & 9.7 \end{bmatrix}$; The solution is $(5.4, 2.1, 9.7)$

15. $\det A = 6(2) - (-4)(-5) = 12 - 20 = -8$

16. $\det A = 0(9) - 5(-6) = 0 + 30 = 30$

17. $\det A = 3\big[(4)(1)-(-3)(7)\big] - 1\big[(-5)(1)-(-3)(-3)\big] + 0\big[(-5)(7)-(4)(-3)\big] = 75 - (-14) + 0 = 89$

18. $\det A = -2\big[(1)(8)-(-5)(-3)\big] - 2\big[(-1)(8)-(-5)(-7)\big] + 3\big[(-1)(-3)-(1)(-7)\big] =$

$14 - (-86) + 30 = 130$

19. Using the calculator we find $\det([A]) = 181,845$

20. Using the calculator we find $\det([A]) = 67.688$

21. The triangle has vertices $(-4, 6)$, $(-2, -4)$ and $(6, 2)$. The matrix needed is $A = \begin{bmatrix} -4 & -2 & 6 \\ 6 & -4 & 2 \\ 1 & 1 & 1 \end{bmatrix}$.

The area is $|D| = \left| \dfrac{1}{2} \det([A]) \right| = 46 \text{ ft}^2$.

22. The triangle has vertices $(-12, -8)$, $(4, 8)$ and $(8, -4)$. The matrix needed is $A = \begin{bmatrix} -12 & 4 & 8 \\ -8 & 8 & -4 \\ 1 & 1 & 1 \end{bmatrix}$.

The area is $D = \left| \dfrac{1}{2} \det([A]) \right| = 128 \text{ ft}^2$.

23. $E = \det \begin{bmatrix} 8 & 6 \\ 18 & -8 \end{bmatrix} = -64 - 108 = -172$; $F = \det \begin{bmatrix} 7 & 8 \\ 5 & 18 \end{bmatrix} = 126 - 40 = 86$

$D = \det \begin{bmatrix} 7 & 6 \\ 5 & -8 \end{bmatrix} = -56 - 30 = -86$; The solution is $x = \dfrac{E}{D} = \dfrac{-172}{-86} = 2$ and $y = \dfrac{F}{D} = \dfrac{86}{-86} = -1$.

24. $E = \det\begin{bmatrix} 25 & 5 \\ -3 & 4 \end{bmatrix} = 100 + 15 = 115;$  $F = \det\begin{bmatrix} -2 & 25 \\ 3 & -3 \end{bmatrix} = 6 - 75 = -69$

   $D = \det\begin{bmatrix} -2 & 5 \\ 3 & 4 \end{bmatrix} = -8 - 15 = -23;$  The solution is $x = \dfrac{E}{D} = \dfrac{115}{-23} = -5$ and $y = \dfrac{F}{D} = \dfrac{-69}{-23} = 3.$

25. $E = \det\begin{bmatrix} 1.5 & -6 \\ 8 & -5 \end{bmatrix} = -7.5 + 48 = 40.5;$  $F = \det\begin{bmatrix} 3 & 1.5 \\ 7 & 8 \end{bmatrix} = 24 - 10.5 = 13.5$

   $D = \det\begin{bmatrix} 3 & -6 \\ 7 & -5 \end{bmatrix} = -15 + 42 = 27;$  The solution is $x = \dfrac{E}{D} = \dfrac{40.5}{27} = \dfrac{3}{2}$ and $y = \dfrac{F}{D} = \dfrac{13.5}{27} = \dfrac{1}{2}.$

26. $E = \det\begin{bmatrix} -47 & 4 \\ 63 & -7 \end{bmatrix} = 329 - 252 = 77;$  $F = \det\begin{bmatrix} -5 & -47 \\ 6 & 63 \end{bmatrix} = -315 + 282 = -33$

   $D = \det\begin{bmatrix} -5 & 4 \\ 6 & -7 \end{bmatrix} = 35 - 24 = 11;$  The solution is $x = \dfrac{E}{D} = \dfrac{77}{11} = 7$ and $y = \dfrac{F}{D} = \dfrac{-33}{11} = -3.$

27. Let $x$ and $y$ represent pedestrian fatalities for 1994 and 2004 respectively. Then the system needed is

   $x + y = 10,130$ and $x - y = 848.$ Adding the two equations will eliminate the variable $y$.

   $\begin{array}{r} x + y = 10,130 \\ \underline{x - y = 848} \\ 2x = 10,978 \end{array}$  Thus, $x = 5489.$ And so $(5489) + y = 10,130 \Rightarrow y = 4641.$

   There were 5489 pedestrian fatalities in 1994 and 4641 in 2004.

28. Let $x$ and $y$ represent the number of \$8 and \$12 tickets respectively. Then the system needed is

   $x + y = 480$ and $8x + 12y = 4620.$ Multiplying the first equation by –8 and adding the two

   equations will eliminate the variable $x$.   $\begin{array}{r} -8x - 8y = -3840 \\ \underline{8x + 12y = 4620} \\ 4y = 780 \end{array}$  Thus, $y = 195.$ And so

   $x + (195) = 480 \Rightarrow x = 285.$ The solution is (285, 195).

   There were 285 tickets sold costing \$8 each and 195 tickets sold costing \$12 each.

29. (a)  $\begin{aligned} m + 3c + 5b &= 14 \\ m + 2c + 4b &= 11 \\ c + 3b &= 5 \end{aligned}$

   (b)  Using a graphing calculator to solve the system, the solution is (3, 2, 1).

   A malt costs \$3.00, a cone costs \$2.00 and an ice cream bar costs \$1.00.

30. Let $x$, $y$ and $z$ represent the measure of the largest, middle and smallest angle respectively. The

   system needed is: $\begin{aligned} x + y + z &= 180 \\ x - y - z &= 20 \\ x \quad\;\; - z &= 85 \end{aligned}$  Using a graphing calculator to solve the system, the solution

   is (100, 65, 15). The measures of the three angles are 100°, 65° and 15°.

31. Let $x$, $y$ and $z$ represent the amount of $1.50, $2.00 and $2.50 candy respectively. The system needed

$$\begin{array}{rrrl} x & +y & +z & =12 \\ 1.50x & +2.00y & +2.50z & =26 \\ -y & +z & & =2 \end{array}$$

is:  Using a graphing calculator to solve the system, the solution

is (2, 4, 6). There should be 2 lb of $1.50 candy, 4 lb of $2.00 candy and 6 lb of $2.50 candy.

32. (a)    $\begin{array}{l} a+202b+63c=40 \\ a+365b+70c=50 \\ a+446b+77c=55 \end{array}$

(b)    Using the graphing calculator to solve the system, the solution is

$a \approx 27.134$, $b \approx 0.061$, $c \approx 0.009$.  So the equation is $C \approx 27.134 + 0.061W + 0.009L$.

(c)    When $W = 300$ and $L = 68$,  $C \approx 27.134 + 0.061(300) + 0.009(68) = 46.046 \approx 46$ inches.

## Chapter 9 Test

1. No. Three planes cannot intersect at exactly three points.

2. Check (–4, 3, 3) in each equation:  $-4-3(3)+4(3)=-1 \Rightarrow -4-9+12=-1 \Rightarrow -1=-1$   True

   $2(-4)+3-3(3)=6 \Rightarrow -8+3-9=6 \Rightarrow -14=6$   False

   $-4-3+3=1 \Rightarrow -4=1$   False

   (–4, 3, 3) is not a solution to the system.

   Check (1, –2, –2) in each equation:  $1-3(-2)+4(-2)=-1 \Rightarrow 1+6-8=-1 \Rightarrow -1=-1$   True

   $2(1)+(-2)-3(-2)=6 \Rightarrow 2-2+6=6 \Rightarrow 6=6$   True

   $1-(-2)+(-2)=1 \Rightarrow 1+2-2=1 \Rightarrow 1=1$   True

   (1, –2, –2) is a solution to the system.

3. Multiply the third equation by –1 and add the second and third equations to eliminate the variables $y$

   and $z$.  $\begin{array}{r} -2x+y+z=5 \\ -y-z=3 \\ \hline -2x=8 \end{array}$  And so $x=-4$.  Substitute $x=-4$ into the first equation:

   $(-4)+3y=2 \Rightarrow 3y=6 \Rightarrow y=2$  Substitute $x=-4$ and $y=2$ into the second equation:

   $-2(-4)+(2)+z=5 \Rightarrow z=-5$  The solution is (–4, 2, –5).

4. Add the first and second equations to eliminate the variable $z$.
$$\begin{array}{r} x+\ y-z=1 \\ 2x-3y+z=0 \\ \hline 3x-2y=1 \end{array}$$

Multiply the first equation by 2 and add the first and third equations to eliminate the variable $z$.
$$2x+2y-2z=2$$
$$\underline{\ \ x-4y+2z=2\ \ }$$ So the two new equations are $3x-2y=1$ and $3x-2y=4$. This is a contradiction,
$$3x-2y=4$$

so there are no solutions.

5. Add the first and second equations to eliminate the variables $x$ and $y$.
$$\begin{array}{r} x-y+2z=\ \ 5 \\ -x+y-3z=-8 \\ \hline -z=-3 \Rightarrow z=3 \end{array}$$

Add the second and third equations to eliminate the variable $x$.
$$\begin{array}{r} -x+\ y-3z=-8 \\ x-2y+2z=\ \ 3 \\ \hline -y-\ z=-5 \end{array}.$$

Substitute $z=3$ and solve for $y$. $-y-3=-5 \Rightarrow -y=-2 \Rightarrow y=2$. Substitute $y=2$ and $z=3$ in the

first equation and solve for $x$. $x-2+2(3)=5 \Rightarrow x-2+6=5 \Rightarrow x=1$. The solution is $(1, 2, 3)$.

6. Add the first and second equation and eliminate both the $y$ and $z$ variable:
$$\begin{array}{r} x-y+z=1 \\ x+y-z=1 \Rightarrow x=1 \\ \hline 2x=2 \end{array}$$

Substituting $x=1$ into equation 2 we have $1+y-z=1 \Rightarrow y-z=0 \Rightarrow y=z$. The system is

dependent and the solution is $(1, z, z)$.

7. (a) $\begin{bmatrix} 2 & -4 & | & -10 \\ -3 & -2 & | & 7 \end{bmatrix}$

(b) $\begin{bmatrix} 2 & -4 & | & -10 \\ -3 & -2 & | & 7 \end{bmatrix} \begin{matrix} \frac{1}{2}R_1 \to \\ R_1+R_2 \to \end{matrix} \begin{bmatrix} 1 & -2 & | & -5 \\ -1 & -6 & | & -3 \end{bmatrix} R_1+R_2 \to$

$\begin{bmatrix} 1 & -2 & | & -5 \\ 0 & -8 & | & -8 \end{bmatrix} -\frac{1}{8}R_2 \to \begin{bmatrix} 1 & -2 & | & -5 \\ 0 & 1 & | & 1 \end{bmatrix} 2R_2+R_1 \to$

$\begin{bmatrix} 1 & 0 & | & -3 \\ 0 & 1 & | & 1 \end{bmatrix}$; The solution is $(-3, 1)$

8. (a) $\begin{bmatrix} 1 & 1 & 1 & | & 2 \\ 1 & -1 & -1 & | & 3 \\ 2 & 2 & 1 & | & 6 \end{bmatrix}$

(b) $\begin{bmatrix} 1 & 1 & 1 & | & 2 \\ 1 & -1 & -1 & | & 3 \\ 2 & 2 & 1 & | & 6 \end{bmatrix} \begin{matrix} \\ R_2-R_1 \to \\ R_3-2R_1 \to \end{matrix} \begin{bmatrix} 1 & 1 & 1 & | & 2 \\ 0 & -2 & -2 & | & 1 \\ 0 & 0 & -1 & | & 2 \end{bmatrix} \begin{matrix} R_1+(1/2)R_2 \to \\ (-1/2)R_2 \to \\ -1R_3 \to \end{matrix} \begin{bmatrix} 1 & 0 & 0 & | & \frac{5}{2} \\ 0 & 1 & 1 & | & -\frac{1}{2} \\ 0 & 0 & 1 & | & -2 \end{bmatrix}$

$R_2-R_3 \to \begin{bmatrix} 1 & 0 & 0 & | & \frac{5}{2} \\ 0 & 1 & 0 & | & \frac{3}{2} \\ 0 & 0 & 1 & | & -2 \end{bmatrix}$ The solution is $\left( \dfrac{5}{2}, \dfrac{3}{2}, -2 \right)$.

9. $\det\begin{bmatrix} -1 & 2 \\ -5 & 4 \end{bmatrix} = -4 - (-10) = 6;$

10. $\det A = 3\big[(2)(-3)-(8)(-6)\big]-6\big[(2)(-3)-(8)(-1)\big]+0\big[(2)(-6)-(2)(-1)\big]=126-12+0=114$

11. $E = \det\begin{bmatrix} 7 & -3 \\ 11 & 2 \end{bmatrix}=14-(-33)=47;\quad F=\det\begin{bmatrix} 5 & 7 \\ -4 & 11 \end{bmatrix}=55-(-28)=83$

$D=\det\begin{bmatrix} 5 & -3 \\ -4 & 2 \end{bmatrix}=10-12=-2;$  The solution is $x=\dfrac{E}{D}=\dfrac{47}{-2}=-\dfrac{47}{2}$ and $y=\dfrac{F}{D}=\dfrac{83}{-2}=-\dfrac{83}{2}.$

12. Let $x$, $y$, and $z$ represent the fractional parts of an hour the runner jogged at 6, 7, and 9 miles per hour,

$$6x+7y+9z=7.1$$

respectively. Then the system needed is $\quad x+\ y+\ z=1\quad$  Substitute $z+0.2$ for $x$ in the first and

$$x=\ z+0.2$$

second equations. $6(z+0.2)+7y+9z=7.1 \Rightarrow 6z+1.2+7y+9z=7.1 \Rightarrow 7y+15z=5.9$

$(z+0.2)+y+z=1 \Rightarrow y+2z=0.8$  Multiply the new second equation by $-7$ and add to the new first

$$7y+15z=\ 5.9$$

equation to eliminate the variable $y$.  $\underline{-7y-14z=-5.6}\quad$ Substitute $z=0.3$ into the new second

$$z=\ 0.3$$

equation and solve for $y$.  $y+2(0.3)=0.8 \Rightarrow y+0.6=0.8 \Rightarrow y=0.2$  Substitute $z=0.3$ in the third

equation and solve for $x$.  $x=0.3+0.2 \Rightarrow x=0.5$  The values $x=0.5$, $y=0.2$, and $z=0.3$ are in

hours, so converting to minutes, the runner jogged 6 miles per hour for 30 minutes, 7 miles per hour

for 12 minutes, and 9 miles per hour for 18 minutes.

13. Let $x$, $y$ and $z$ represent the measure of the largest, middle and smallest angle respectively.  The

$$x+y+z=180$$
system needed is:  $\begin{array}{l} x\quad\ -z=50 \\ -x+y+z=10 \end{array}$ and so $[A]=\begin{bmatrix} 1 & 1 & 1 & 180 \\ 1 & 0 & -1 & 50 \\ -1 & 1 & 1 & 10 \end{bmatrix}$; rref$([A])=\begin{bmatrix} 1 & 0 & 0 & 85 \\ 0 & 1 & 0 & 60 \\ 0 & 0 & 1 & 35 \end{bmatrix}$

The solution is $(85, 60, 35)$.  The angles are $85°$, $60°$, and $35°$.

14. (a)  $\begin{array}{l} a+20b+25c=168 \\ a+24b+40c=270 \\ a+30b+50c=405 \end{array}$

(b)    Write the corresponding augmented matrix and use Gauss-Jordan elimination to solve the

system.

$$\begin{bmatrix} 1 & 20 & 25 & | & 168 \\ 1 & 24 & 40 & | & 270 \\ 1 & 30 & 50 & | & 405 \end{bmatrix} \begin{matrix} \\ -R_1 + R_2 \rightarrow \\ -R_1 + R_3 \rightarrow \end{matrix} \begin{bmatrix} 1 & 20 & 25 & | & 168 \\ 0 & 4 & 15 & | & 102 \\ 0 & 10 & 25 & | & 237 \end{bmatrix}$$

$$\begin{matrix} -5 R_2 + R_1 \rightarrow \\ \\ -2.5 R_2 + R_3 \rightarrow \end{matrix} \begin{bmatrix} 1 & 0 & -50 & | & -342 \\ 0 & 4 & 15 & | & 102 \\ 0 & 0 & -12.5 & | & -18 \end{bmatrix} \begin{matrix} \frac{1}{4} R_2 \rightarrow \\ -\frac{1}{12.5} R_3 \rightarrow \end{matrix}$$

$$\begin{bmatrix} 1 & 0 & -50 & | & -342 \\ 0 & 1 & 3.75 & | & 25.5 \\ 0 & 0 & 1 & | & 1.44 \end{bmatrix} \begin{matrix} 50 R_3 + R_1 \rightarrow \\ -3.75 R_3 + R_2 \rightarrow \end{matrix} \begin{bmatrix} 1 & 0 & 0 & | & -270 \\ 0 & 1 & 0 & | & 20.1 \\ 0 & 0 & 1 & | & 1.44 \end{bmatrix}$$

Therefore, $a = -270, b = 20.1,$ and $c = 1.44$.

(c)    The model is $W = -270 + 20.1\,N + 1.44\,C,$ so $W = -270 + 20.1\,(26) + 1.44\,(44) = 315.96$ or

about 316 pounds.

## Chapter 9 Extended and Discovery Exercises

1.  $D = 1\big[(1)(3)-(1)(2)\big] - 2\big[(1)(3)-(1)(1)\big] + 0\big[(1)(2)-(1)(1)\big] = 1 - 4 + 0 = -3$

   $E = 6\big[(1)(3)-(1)(2)\big] - 9\big[(1)(3)-(1)(1)\big] + 9\big[(1)(2)-(1)(1)\big] = 6 - 18 + 9 = -3$

   $F = 1\big[(9)(3)-(9)(2)\big] - 2\big[(6)(3)-(9)(1)\big] + 0\big[(6)(2)-(9)(1)\big] = 9 - 18 + 0 = -9$

   $G = 1\big[(1)(9)-(1)(9)\big] - 2\big[(1)(9)-(1)(6)\big] + 0\big[(1)(9)-(1)(6)\big] = 0 - 6 + 0 = -6$

   $x = \dfrac{E}{D} = \dfrac{-3}{-3} = 1, \quad y = \dfrac{F}{D} = \dfrac{-9}{-3} = 3, \quad z = \dfrac{G}{D} = \dfrac{-6}{-3} = 2.$ The solution is $(1,\,3,\,2)$.

2.  $D = 0\big[(-1)(-1)-(1)(-1)\big] - 2\big[(1)(-1)-(1)(1)\big] + 1\big[(1)(-1)-(-1)(1)\big] = 0 + 4 + 0 = 4$

   $E = 1\big[(-1)(-1)-(1)(-1)\big] - (-1)\big[(1)(-1)-(1)(1)\big] + 3\big[(1)(-1)-(-1)(1)\big] = 2 - 2 + 0 = 0$

   $F = 0\big[(-1)(-1)-(3)(-1)\big] - 2\big[(1)(-1)-(3)(1)\big] + 1\big[(1)(-1)-(-1)(1)\big] = 0 + 8 + 0 = 8$

   $G = 0[(-1)(3)-(1)(-1)] - 2[(1)(3)-(1)(1)] + 1[(1)(-1)-(-1)(1)] = 0 - 4 + 0 = -4$

   $x = \dfrac{E}{D} = \dfrac{0}{4} = 0, \quad y = \dfrac{F}{D} = \dfrac{8}{4} = 2, \quad z = \dfrac{G}{D} = \dfrac{-4}{4} = -1.$ The solution is $(0,\,2,\,-1)$.

3.  $D = 1\big[(1)(2)-(1)(0)\big] - 1\big[(0)(2)-(1)(1)\big] + 0\big[(0)(0)-(1)(1)\big] = 2 + 1 + 0 = 3$

   $E = 2\big[(1)(2)-(1)(0)\big] - 0\big[(0)(2)-(1)(1)\big] + 1\big[(0)(0)-(1)(1)\big] = 4 + 0 - 1 = 3$

   $F = 1\big[(0)(2)-(1)(0)\big] - 1\big[(2)(2)-(1)(1)\big] + 0\big[(2)(0)-(0)(1)\big] = 0 - 3 + 0 = -3$

$G = 1\left[(1)(1)-(1)(0)\right]-1\left[(0)(1)-(1)(2)\right]+0\left[(0)(0)-(1)(2)\right]=1+2+0=3$

$x=\dfrac{E}{D}=\dfrac{3}{3}=1,\quad y=\dfrac{F}{D}=\dfrac{-3}{3}=-1,\quad z=\dfrac{G}{D}=\dfrac{3}{3}=1.$ The solution is $(1,-1,1)$.

4. $D=1\left[(-2)(-3)-(1)(-3)\right]-(-1)\left[(1)(-3)-(1)(2)\right]+0\left[(1)(-3)-(-2)(2)\right]=9-5+0=4$

$E=1\left[(-2)(-3)-(1)(-3)\right]-(-2)\left[(1)(-3)-(1)(2)\right]+5\left[(1)(-3)-(-2)(2)\right]=9-10+5=4$

$F=1\left[(-2)(-3)-(5)(-3)\right]-(-1)\left[(1)(-3)-(5)(2)\right]+0\left[(1)(-3)-(-2)(2)\right]=21-13+0=8$

$G=1[(-2)(5)-(1)(-2)]-(-1)[(1)(5)-(1)(1)]+0[(1)(-2)-(-2)(1)]=-8+4+0=-4$

$x=\dfrac{E}{D}=\dfrac{4}{4}=1,\quad y=\dfrac{F}{D}=\dfrac{8}{4}=2,\quad z=\dfrac{G}{D}=\dfrac{-4}{4}=-1.$ The solution is $(1,2,-1)$.

5. $D=1\left[(1)(2)-(-1)(1)\right]-(-1)\left[(0)(2)-(-1)(2)\right]+2\left[(0)(1)-(1)(2)\right]=3+2-4=1$

$E=7\left[(1)(2)-(-1)(1)\right]-5\left[(0)(2)-(-1)(2)\right]+6\left[(0)(1)-(1)(2)\right]=21-10-12=-1$

$F=1\left[(5)(2)-(6)(1)\right]-(-1)\left[(7)(2)-(6)(2)\right]+2\left[(7)(1)-(5)(2)\right]=4+2-6=0$

$G=1\left[(1)(6)-(-1)(5)\right]-(-1)\left[(0)(6)-(-1)(7)\right]+2\left[(0)(5)-(1)(7)\right]=11+7-14=4$

$x=\dfrac{E}{D}=\dfrac{-1}{1}=-1,\quad y=\dfrac{F}{D}=\dfrac{0}{1}=0,\quad z=\dfrac{G}{D}=\dfrac{4}{1}=4.$ The solution is $(-1,0,4)$.

6. $D=1\left[(-3)(-2)-(4)(-1)\right]-2\left[(2)(-2)-(4)(3)\right]+1\left[(2)(-1)-(-3)(3)\right]=10+32+7=49$

$E=-1\left[(-3)(-2)-(4)(-1)\right]-12\left[(2)(-2)-(4)(3)\right]+(-12)\left[(2)(-1)-(-3)(3)\right]=$

$-10+192-84=98$

$F=1\left[(12)(-2)-(-12)(-1)\right]-2\left[(-1)(-2)-(-12)(3)\right]+1\left[(-1)(-1)-(12)(3)\right]=$

$-36-76-35=-147$

$G=1\left[(-3)(-12)-(4)(12)\right]-2\left[(2)(-12)-(4)(-1)\right]+1\left[(2)(12)-(-3)(-1)\right]=-12+40+21=49$

$x=\dfrac{E}{D}=\dfrac{98}{49}=2,\quad y=\dfrac{F}{D}=\dfrac{-147}{49}=-3,\quad z=\dfrac{G}{D}=\dfrac{49}{49}=1.$ The solution is $(2,-3,1)$.

7. Since Denver is city 1 and Las Vegas is city 4, we look at either entry $a_{14}$ or $a_{41}$. The distance is 760 miles.

8. Add entry $a_{12}$ to entry $a_{23}$. The distance is 360 miles.

9. The dimension would be $20\times20$ and the matrix would contain 400 elements.

10. The elements on the main diagonal represent the distance from a city to itself, which is always zero.

11. $a_{14}+a_{41}=760+760=1520$ miles

12. $a_{11}+a_{44}=0+0=0$ miles

13. See Figure 13.

$$\begin{bmatrix} 0 & 130 & 95 & 75 \\ 130 & 0 & 186 & \star \\ 95 & 186 & 0 & 57 \\ 75 & \star & 57 & 0 \end{bmatrix}$$
$$\begin{bmatrix} 0 & 97 & \star & \star & 59 \\ 97 & 0 & 113 & 118 & \star \\ \star & 113 & 0 & 94 & \star \\ \star & 118 & 94 & 0 & 177 \\ 59 & \star & \star & 177 & 0 \end{bmatrix}$$

Figure 13                                    Figure 14

14. See Figure 14.

15. All maps for adjacency matrix $A$ must have the same distances between cities, but the location of each city may vary. The solution is not unique. One possible solution is shown in Figure 15.

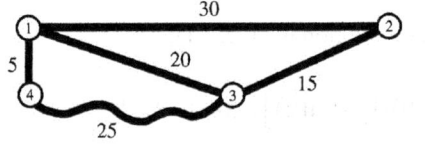
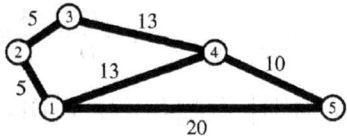

Figure 15                                    Figure 16

16. All maps for adjacency matrix $A$ must have the same distances between cities, but the location of each city may vary. The solution is not unique. One possible solution is shown in Figure 16.

17. (a) $$\begin{bmatrix} 1 & 19 & 57.5 & 32 & | & 125 \\ 1 & 26 & 65 & 42 & | & 316 \\ 1 & 30 & 72 & 48 & | & 436 \\ 1 & 30.5 & 75 & 54 & | & 514 \end{bmatrix}$$

(b) $a \approx -552.272$, $b \approx 8.733$, $c \approx 2.859$, $d \approx 10.843$

(c) $N = 24$, $L = 63$ and $C = 39$, $W \approx -552.272 + 8.733(24) + 2.859(63) + 10.843(39) \approx 260$

A bear with a 24-inch neck, 63-inch length and 39-inch chest weighs approximately 260 pounds.

## Chapters 1-9 Cumulative Review Exercises

1. $360 = 2 \cdot 2 \cdot 2 \cdot 3 \cdot 3 \cdot 5 = 2^3 \cdot 3^2 \cdot 5$

2. $2n + 7 = n - 2;$ $2n + 7 = n - 2 \Rightarrow n + 7 = -2 \Rightarrow n = -9$

3. $\dfrac{2}{3} + \dfrac{4}{7} \cdot \dfrac{21}{28} = \dfrac{2}{3} + \dfrac{4}{7} \cdot \dfrac{7 \cdot 3}{7 \cdot 4} = \dfrac{2}{3} + \dfrac{4}{7} \cdot \dfrac{7}{7} \cdot \dfrac{3}{4} = \dfrac{2}{3} + \dfrac{4 \cdot 3}{7 \cdot 4} = \dfrac{2}{3} + \dfrac{3}{7} \cdot \dfrac{4}{4} = \dfrac{2}{3} + \dfrac{3}{7} = \dfrac{14}{21} + \dfrac{9}{21} = \dfrac{23}{21}$

4. $\dfrac{3}{5} \div \dfrac{6}{5} - \dfrac{2}{3} = \dfrac{3}{5} \cdot \dfrac{5}{6} - \dfrac{2}{3} = \dfrac{3 \cdot 5}{5 \cdot 6} - \dfrac{2}{3} = \dfrac{3}{6} \cdot \dfrac{5}{5} - \dfrac{2}{3} = \dfrac{3}{6} - \dfrac{2}{3} = \dfrac{3}{6} - \dfrac{4}{6} = -\dfrac{1}{6}$

5. $30 - 4 \div 2 \cdot 6 = 30 - 2 \cdot 6 = 30 - 12 = 18$

6. $\dfrac{3^2 - 2^3}{20 - 5 \cdot 2} = \dfrac{9 - 8}{20 - 10} = \dfrac{1}{10}$

7. $2(x+1) - 6x = x - 4 \Rightarrow 2x + 2 - 6x = x - 4 \Rightarrow -4x + 2 = x - 4 \Rightarrow -5x = -6 \Rightarrow x = \dfrac{6}{5}$

8. Graph $Y_1 = 4 - 3X$ and $Y_2 = -Z$ in $[-10, 10, 1]$ by $[-10, 10, 1]$. The graphs intersect at $(2, -2)$, so the solution is $x = 2$. See Figure 8

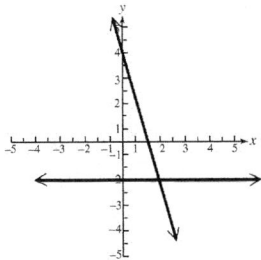

Figure 8

9. $124\% = 124\left(\dfrac{1}{100}\right) = \dfrac{124}{100} = \dfrac{31 \cdot 4}{25 \cdot 4} = \dfrac{31}{25}$; $124\% = 124(0.01) = 1.24$

10. $A = \dfrac{1}{2}bh \Rightarrow 30 = \dfrac{1}{2}b(10) \Rightarrow 30 = 5b \Rightarrow b = 6$ miles

11. $6t - 1 < 3 - t \Rightarrow 7t < 4 \Rightarrow t < \dfrac{4}{7}$

12. See Figure 12

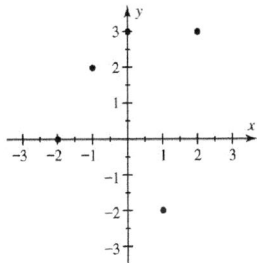

Figure 12

13. The $x$-intercept is $-4$; the $y$-intercept is 3. See Figure 13

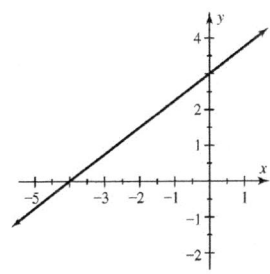

Figure 13

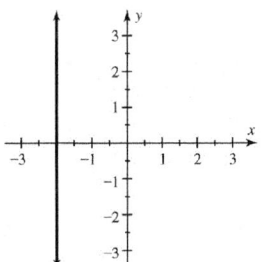

Figure 14

14. The $x$-intercept is $-2$; there is no $y$-intercept. See Figure 14

15. The $x$-intercept is 2; the $y$-intercept is 4. The slope is $-2$, so the equation is $y = -2x + 4$.

16. $y - 3 = -2(x - (-1)) \Rightarrow y = -2x - 2 + 3 \Rightarrow y = -2x + 1$

17. $m = \dfrac{8 - 5}{2 - (-3)} = \dfrac{3}{5}$; $y - 5 = \dfrac{3}{5}(x - (-3)) \Rightarrow y = \dfrac{3}{5}x + \dfrac{9}{5} + 5 \Rightarrow y = \dfrac{3}{5}x + \dfrac{34}{5}$

18. The given line can be written as $y = -\dfrac{1}{2}x + \dfrac{5}{2}$, so the slope of the line perpendicular to it is $m = 2$.

   $y - 1 = 2(x - (-1)) \Rightarrow y = 2x + 2 + 1 \Rightarrow y = 2x + 3$

19. The slope is 500 and the $P$-intercept is 4000; $P = 500x + 4000$

20. Multiply the first equation by 2 and add the result to the second equation.
$$\begin{array}{r} -4a + 2b = -10 \\ 4a - 3b = \phantom{-}0 \\ \hline -b = -10 \end{array} \Rightarrow b = 10$$

   Substitute $b = 10$ into the original first equation and solve for $a$.

   $-2a + 10 = -5 \Rightarrow -2a = -15 \Rightarrow a = 7.5$  The solution is $(7.5, 10)$.

21. See Figure 21

22. See Figure 22

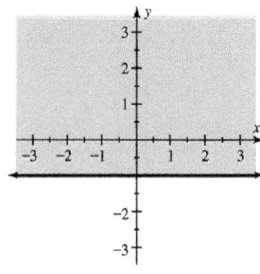    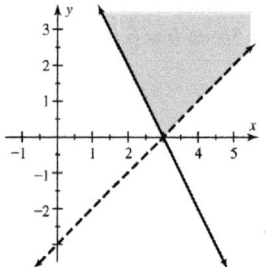

       Figure 21                                    Figure 22

23. $5 - 3^4 = 5 - 81 = -76$

24. $(8t^{-3})(3t^2)(t^5) = (8 \cdot 3)(t^{-3+2+5}) = 24t^4$

25. $\dfrac{2^{-4}}{4^{-2}} = \dfrac{4^2}{2^4} = \dfrac{16}{16} = 1$

26. $(2t^3)^{-2} = 2^{-2}t^{-6} = \dfrac{1}{2^2 t^6} = \dfrac{1}{4t^6}$

27. $(4a^2b^3)^2(2ab)^{-3} = \dfrac{(4a^2b^3)^2}{(2ab)^3} = \dfrac{16a^4b^6}{8a^3b^3} = 2ab^3$

28. $\left(\dfrac{2a^{-1}}{ab^{-2}}\right)^{-3} = \left(\dfrac{ab^{-2}}{2a^{-1}}\right)^3 = \dfrac{a^3 b^{-6}}{8a^{-3}} = \dfrac{a^6}{8b^6}$

29. $2a^2(a^2 - 2a + 3) = 2a^4 - 4a^3 + 6a^2$

30. $(5x + 1)(x - 7) = 5x^2 - 35x + x - 7 = 5x^2 - 34x - 7$

31. $(2x + 3y)^2 = 4x^2 + 2(2)(3)xy + 9y^2 = 4x^2 + 12xy + 9y^2$

32. $(a + b)(a - b) = a^2 - b^2$

33. $\dfrac{4x^3 - 8x^2 + 6x}{2x} = 2x^2 - 4x + 3$

34. The quotient is $x^3 - 4x^2 + 3x - 2 + \dfrac{1}{x-5}$.

$$\begin{array}{r} x^3 - 4x^2 + 3x - 2 \\ x-5{\overline{\smash{\big)}\,x^4 - 9x^3 + 23x^2 - 17x + 11}} \\ \underline{x^4 - 5x^3} \hphantom{xxxxxxxxxxxxxxxxxx} \\ -4x^3 + 23x^2 \hphantom{xxxxxxxxxx} \\ \underline{-4x^3 + 20x^2} \hphantom{xxxxxxxxx} \\ 3x^2 - 17x \hphantom{xxxx} \\ \underline{3x^2 - 15x} \hphantom{xxx} \\ -2x + 11 \\ \underline{-2x + 10} \\ 1 \end{array}$$

35. $10ab^2 - 25a^3b^5 = 5ab^2(2 - 5a^2b^3)$

36. $y^3 - 3y^2 + 2y - 6 = y^2(y-3) + 2(y-3) = (y^2 + 2)(y-3)$

37. $6z^2 + 7z - 3 = (2z+3)(3z-1)$

38. $4z^2 - 9 = (2z-3)(2z+3)$

39. $4y^2 - 20y + 25 = (2y-5)^2$

40. $a^3 - 27 = (a-3)(a^2 + 3a + 9)$

41. $4z^4 - 17z^2 + 15 = (z^2 - 3)(4z^2 - 5)$

42. $2a^3b + a^2b^2 - ab^3 = ab(2a^2 + ab - b^2) = ab(a+b)(2a-b)$

43. $(x-1)(x+2) = 0 \Rightarrow x = 1$ or $x = -2$

44. $6y^2 - 7y = 3 \Rightarrow 6y^2 - 7y - 3 = 0 \Rightarrow (3y+1)(2y-3) = 0 \Rightarrow y = -\dfrac{1}{3}$ or $y = \dfrac{3}{2}$

45. $\dfrac{x^2 - 16}{x+4} = \dfrac{(x+4)(x-4)}{x+4} = x - 4$

46. $\dfrac{2x^2 - 11x - 6}{6x^2 - 5x - 4} = \dfrac{(x-6)(2x+1)}{(3x-4)(2x+1)} = \dfrac{x-6}{3x-4}$

47. $\dfrac{x^2 - 3x + 2}{x+7} \div \dfrac{x-2}{2x+14} = \dfrac{(x-2)(x-1)}{x+7} \cdot \dfrac{2(x+7)}{x-2} = 2(x-1)$

48. $\dfrac{x}{2x+3} + \dfrac{x+3}{2x+3} = \dfrac{x+x+3}{2x+3} = \dfrac{2x+3}{2x+3} = 1$

49. $\dfrac{x+2}{5} = \dfrac{x}{4} \Rightarrow 4(x+2) = 5x \Rightarrow 4x + 8 = 5x \Rightarrow 8 = x$

50. $\dfrac{1}{3x}+\dfrac{5}{2x}=2\Rightarrow\dfrac{2}{6x}+\dfrac{15}{6x}=2\Rightarrow\dfrac{17}{6x}=2\Rightarrow17=12x\Rightarrow x=\dfrac{17}{12}$

51. $k=xy\Rightarrow k=25(4)\Rightarrow k=100$ so $100=xy$; when $x=10,\ 100=10y\Rightarrow y=10$.

52. $f(-3)=1-4(-3)=1+12=13$

53. $D=(-\infty,\infty);\ R=[-2,\infty)$  See Figure 53.

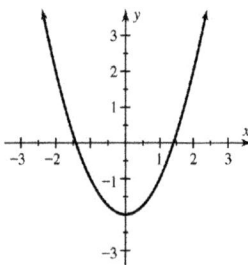

Figure 53

54. From the graph, $f(0)=2$ and $f(-2)=-2$.

55. $\dfrac{4x-9}{6}>\dfrac{1}{2}\Rightarrow\dfrac{4x-9}{6}>\dfrac{3}{6}\Rightarrow\dfrac{4x-9-3}{6}>0\Rightarrow\dfrac{4x-12}{6}>0\Rightarrow4x-12>0\Rightarrow4x>12\Rightarrow x>3$; the

solution set is $(3,\infty)$.

56. $\dfrac{2}{3}z-2\le\dfrac{1}{4}z-(2z+2)\Rightarrow\dfrac{8}{12}z-2\le\dfrac{3}{12}z-2z-2\Rightarrow\dfrac{5}{12}z+2z\le0\Rightarrow\dfrac{29}{12}z\le0\Rightarrow z\le0$; the solution

set is $(-\infty,0]$.

57. $x+2>1$ and $2x-1\le9\Rightarrow x>-1$ and $2x\le10\Rightarrow x>-1$ and $x\le5$; the solution set is $(-1,5]$.

See Figure 57.

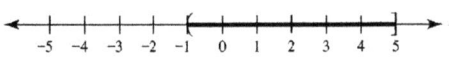                                                            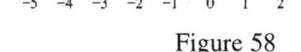

Figure 57                                                                    Figure 58

58. $4x+7<1$ or $3x+2\ge11\Rightarrow4x<-6$ or $3x\ge9\Rightarrow x<-\dfrac{3}{2}$ or $x\ge3$; the solution set is

$\left(-\infty,-\dfrac{3}{2}\right)\cup[3,\infty)$.  See Figure 58.

59. $-7\le2x-3\le5\Rightarrow-4\le2x\le8\Rightarrow-2\le x\le4$; the solution set is $[-2,4]$.

60. $-8\le-\dfrac{1}{2}x-3\le5\Rightarrow-5\le-\dfrac{1}{2}x\le8\Rightarrow10\ge x\ge-16$; the solution set is $[-16,10]$.

61. (a)   $y_1=2$ when $x=-3$ and $x=1$

    (b)   $y_1\le2$ when $-3\le x\le1;[-3,1]$

    (c)   $y_1\ge2$ when $x\le-3$ or $x\ge1;(-\infty,-3]\cup[1,\infty)$

62. $\left|\dfrac{2}{3}x-4\right|=8\Rightarrow\dfrac{2}{3}x-4=8$ or $\dfrac{2}{3}x-4=-8\Rightarrow\dfrac{2}{3}x=12$ or $\dfrac{2}{3}x=-4\Rightarrow x=18$ or $x=-6$

63. $|3x+5|>13 \Rightarrow 3x+5>13$ or $3x+5<-13 \Rightarrow 3x>8$ or $3x<-18 \Rightarrow x>\dfrac{8}{3}$ or $x<-6$; the solution

set is $\left(-\infty,-6\right)\cup\left(\dfrac{8}{3},\infty\right)$.

64. $-3|2t-11|\geq -9 \Rightarrow |2t-11|\leq 3 \Rightarrow 2t-11\leq 3$ and $2t-11\geq -3 \Rightarrow 2t\leq 14$ and $2t\geq 8 \Rightarrow t\leq 7$ and

$t\geq 4$; the solution set is [4, 7].

65. Add the second and third equations.
$$\begin{array}{r} 3x-y+4z=10 \\ 2x+y-2z=-1 \\ \hline 5x\phantom{+y}+2z=\phantom{-}9 \end{array}$$
Multiply the second equation by 3 and add to

the first equation.
$$\begin{array}{r} 2x+3y-\phantom{1}z=3 \\ 9x-3y+12z=30 \\ \hline 11x\phantom{+3y}+11z=33 \end{array} \Rightarrow x+z=3$$
substitute $3-z$ for $x$ in the result of the first sum

$\Rightarrow x=3-z$

of equations and solve for $z$. $5(3-z)+2z=9 \Rightarrow 15-5z+2z=9 \Rightarrow 15-3z=9 \Rightarrow -3z=-6 \Rightarrow z=2$

substitute $z=2$ into the result of the  second sum of equations and solve for $x$. $x=3-2 \Rightarrow x=1$

substitute $x=1$ and $z=2$ into the original third equation.

$2(1)+y-2(2)=-1 \Rightarrow 2+y-4=-1 \Rightarrow y=1$  The solution is (1, 1, 2).

66. $\begin{array}{r} x+\phantom{1}y-z=\phantom{-}4 \\ -x-\phantom{1}y-z=\phantom{-}0 \\ x-2y+z=-9 \end{array} \Rightarrow \left[\begin{array}{rrr|r} 1 & 1 & -1 & 4 \\ -1 & -1 & -1 & 0 \\ 1 & -2 & 1 & -9 \end{array}\right]$

$\left[\begin{array}{rrr|r} 1 & 1 & -1 & 4 \\ -1 & -1 & -1 & 0 \\ 1 & -2 & 1 & -9 \end{array}\right] \begin{array}{l} R_1+R_2 \to \\ -R_1+R_3 \to \end{array} \left[\begin{array}{rrr|r} 1 & 1 & -1 & 4 \\ 0 & 0 & -2 & 4 \\ 0 & -3 & 2 & -13 \end{array}\right] \begin{array}{l} -\frac{1}{3}R_3 \to \\ -\frac{1}{2}R_2 \to \end{array}$

$\left[\begin{array}{rrr|r} 1 & 1 & -1 & 4 \\ 0 & 1 & -\frac{2}{3} & \frac{13}{3} \\ 0 & 0 & 1 & -2 \end{array}\right] \begin{array}{l} R_3+R_1 \to \\ \frac{2}{3}R_3+R_2 \to \end{array} \left[\begin{array}{rrr|r} 1 & 1 & 0 & 2 \\ 0 & 1 & 0 & 3 \\ 0 & 0 & 1 & -2 \end{array}\right] -R_2+R_1 \to$

$\left[\begin{array}{rrr|r} 1 & 0 & 0 & -1 \\ 0 & 1 & 0 & 3 \\ 0 & 0 & 1 & -2 \end{array}\right]$

The solution is $(-1, 3, -2)$.

67. See Figure 67.

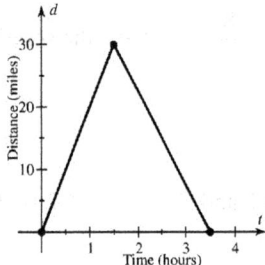

Figure 67

68. Let $a$ represent the number of adults and $c$ represent the number of children. Then the required

    system of equations is $\begin{array}{l} a+c=8 \\ 25a+15c=170 \end{array}$ solve the first equation for $a$ and substitute the result into

    the second equation and solve for $c$. $a+c=8 \Rightarrow a=8-c$; $25(8-c)+15c=170$

    $\Rightarrow 200-25c+15c=170 \Rightarrow -10c=-30 \Rightarrow c=3 \Rightarrow a+3=8 \Rightarrow a=5$.

    3 children and 5 adults went to the game.

69. Let $x$ represent the time it takes for both people to shovel the walk together. Then

    $\dfrac{1}{2}+\dfrac{1}{1.5}=\dfrac{1}{x} \Rightarrow \dfrac{3}{6}+\dfrac{4}{6}=\dfrac{1}{x} \Rightarrow \dfrac{7}{6}=\dfrac{1}{x} \Rightarrow 7x=6 \Rightarrow x=\dfrac{6}{7}$ hour

70. (a)   $h(2)=88(2)-16(2)^2=176-16(4)=176-64=112$ ft .

    (b)   $88t-16t^2=0 \Rightarrow t(88-16t)=0 \Rightarrow t=0$ or $t=5.5$; the golf ball strikes the ground after 5.5

          seconds.

71. Let $x$ represent the height of the building. Then $\dfrac{7}{4}=\dfrac{x}{35} \Rightarrow 4x=245 \Rightarrow x=61.25$ ft

72. (a)   $\begin{array}{l} 4b+3f+4m=23 \\ b+2f+\phantom{0}m=\phantom{0}7 \\ 3b+\phantom{0}f+2m=13 \end{array}$

    (b)   Multiply the second equation by –4 and add to the first equation.

          $\begin{array}{l} 4b+3f+4m=\phantom{-}23 \\ \underline{-4b-8f-4m=-28} \\ \phantom{-4b-}{-5f}\phantom{-4m}=-5 \end{array} \Rightarrow f=1$      Multiply the third equation by –2 and add to the first equation.

          $\begin{array}{l} 4b+3f+4m=\phantom{-}23 \\ \underline{-6b-2f-4m=-26} \\ -2b+\phantom{0}f\phantom{-4m}=-3 \end{array}$  substitute 1 for $f$ in this equation and solve for $b$.

          $-2b+1=-3 \Rightarrow -2b=-4 \Rightarrow b=2$  substitute $b=2$ and $f=1$ into the original second

          equation and solve for $m$. $2+2(1)+m=7 \Rightarrow 2+2+m=7 \Rightarrow m=3$

          A burger costs \$2, fries cost \$1, and a malt costs \$3.

# Chapter 10: Radical Expressions and Functions

## 10.1: Radical Expressions and Functions

1. $\pm 3$

3. 2

5. b

7. undefined

9. d

11. See Figure 11.

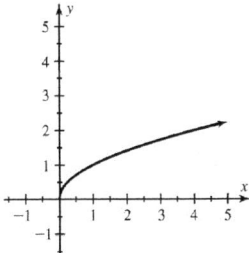

Figure 11

13. The domain is $\{x \mid x \geq 0\}$. See Figure 11 above.

15. $\sqrt{9} = 3$

17. $\sqrt{0.36} = 0.6$

19. $\sqrt{\dfrac{16}{25}} = \dfrac{4}{5}$

21. $\sqrt{x^2} = x$ since $x > 0$

23. $\sqrt[3]{27} = 3$

25. $\sqrt[3]{-64} = -4$

27. $\sqrt[3]{\dfrac{8}{27}} = \dfrac{2}{3}$

29. $-\sqrt[3]{x^9} = -\sqrt[3]{\left(x^3\right)^3} = -x^3$

31. $\sqrt[3]{(2x)^6} = \sqrt[3]{\left((2x)^2\right)^3} = (2x)^2 = 4x^2$

33. $\sqrt[4]{81} = 3$

35. $\sqrt[5]{-243} = \sqrt[5]{(-3)^5} = -3$

37. $\sqrt[4]{-16}$ is not possible to evaluate over the real numbers since the index is even and the radicand is negative.

39. $-\sqrt{5} \approx -2.24$

41. $\sqrt[3]{5} \approx 1.71$

43. $\sqrt[5]{-7} \approx -1.48$

45. $\sqrt{(-4)^2} = \sqrt{16} = 4$

47. $\sqrt{y^2} = |y|$

49. $\sqrt{(x-5)^2} = |x-5|$

51. $\sqrt{x^2 - 2x + 1} = \sqrt{(x-1)^2} = |x-1|$

53. $\sqrt[4]{y^4} = |y|$

55. $\sqrt[4]{x^{12}} = \sqrt[4]{(x^3)^4} = |x^3|$

57. $\sqrt[5]{x^5} = x$

59. $f(10) = \sqrt{10-1} = \sqrt{9} = 3;\ f(0)$ is not possible because $\sqrt{0-1} = \sqrt{-1}$

61. $f(-1) = \sqrt{3-3(-1)} = \sqrt{3+3} = \sqrt{6};\ f(5)$ is not possible because $\sqrt{3-3(5)} = \sqrt{3-15} = \sqrt{-12}$

63. $f(-4) = \sqrt{(-4)^2 - (-4)} = \sqrt{16+4} = \sqrt{20} = \sqrt{4\cdot5} = 2\sqrt{5};\ f(3) = \sqrt{3^2 - 3} = \sqrt{9-3} = \sqrt{6}$

65. $f(-3) = \sqrt[3]{(-3)^2 - 8} = \sqrt[3]{9-8} = \sqrt[3]{1} = 1;\ f(4) = \sqrt[3]{4^2 - 8} = \sqrt[3]{16-8} = \sqrt[3]{8} = 2$

67. $f(1) = \sqrt[3]{1-9} = \sqrt[3]{-8} = -2;\ f(10) = \sqrt[3]{10-9} = \sqrt[3]{1} = 1$

69. $f(-2) = \sqrt[3]{3-(-2)^2} = \sqrt[3]{3-4} = \sqrt[3]{-1} = -1;\ f(3) = \sqrt[3]{3-3^2} = \sqrt[3]{3-9} = \sqrt[3]{-6}$ or $-\sqrt[3]{6}$

71. $T(64) = \frac{1}{2}\sqrt{64} = \frac{1}{2}(8) = 4$

73. $f(4) = \sqrt{4+5} + \sqrt{4} = \sqrt{9} + \sqrt{4} = 3+2 = 5$

75. $x+2 \geq 0 \Rightarrow x \geq -2 \Rightarrow$ Domain: $[-2, \infty)$

77. $x-2 \geq 0 \Rightarrow x \geq 2 \Rightarrow$ Domain: $[2, \infty)$

79. $2x-4 \geq 0 \Rightarrow 2x \geq 4 \Rightarrow x \geq 2 \Rightarrow$ Domain: $[2, \infty)$

81. $1-x \geq 0 \Rightarrow -x \geq -1 \Rightarrow x \leq 1 \Rightarrow$ Domain: $(-\infty, 1]$

83. $8-5x \geq 0 \Rightarrow -5x \geq -8 \Rightarrow x \leq \frac{8}{5} \Rightarrow$ Domain: $\left(-\infty, \frac{5}{8}\right]$

85. $3x^2 + 4 \geq 0 \Rightarrow 3x^2 \geq -4 \Rightarrow x^2 \geq -\frac{4}{3} \Rightarrow$ Domain: $(-\infty, \infty)$

87. $2x+1 > 0 \Rightarrow 2x > -1 \Rightarrow x > -\frac{1}{2} \Rightarrow$ Domain: $\left(-\frac{1}{2}, \infty\right)$

89. See Figure 89.  This graph is shifted 2 units left.

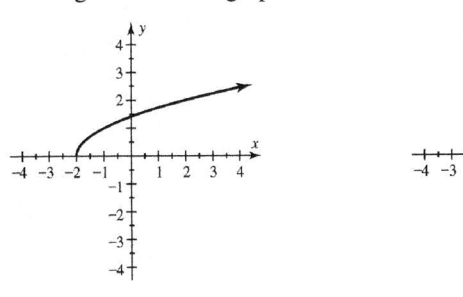

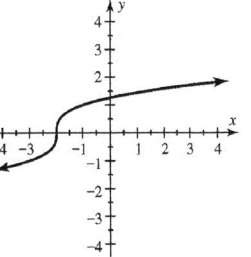

Figure 89                                Figure 91                                Figure 93

91. See Figure 91.  This graph is shifted 2 units upward.

93. See Figure 93.  This graph is shifted 2 units left.

95.  $f(x) = \sqrt{x} + 1$

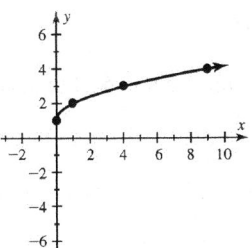

| $x$ | $\sqrt{x}+1$ |
|-----|------|
| $-1$ | — |
| 0 | 1 |
| 1 | 2 |
| 4 | 3 |
| 9 | 4 |

97.  $f(x) = \sqrt{3x}$

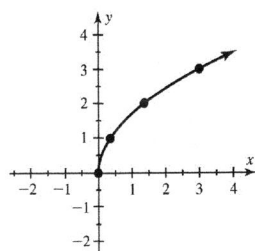

| $x$ | $\sqrt{3x}$ |
|-----|------|
| $-1$ | — |
| 0 | 0 |
| $\frac{1}{3}$ | 1 |
| $\frac{4}{3}$ | 2 |
| 3 | 3 |

99.  $f(x) = 2\sqrt[3]{x}$

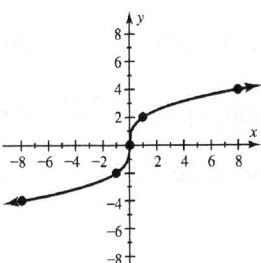

| $x$ | $2\sqrt[3]{x}$ |
|-----|------|
| $-8$ | $-4$ |
| $-1$ | $-2$ |
| 0 | 0 |
| 1 | 2 |
| 8 | 4 |

101.  $f(x) = \sqrt[3]{x-1}$

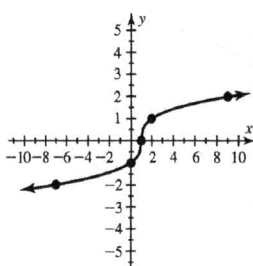

| $x$ | $\sqrt[3]{x-1}$ |
|-----|------|
| $-7$ | $-2$ |
| $0$ | $-1$ |
| $1$ | $0$ |
| $2$ | $1$ |
| $9$ | $2$ |

103.  $s = \dfrac{1}{2}(3+4+5) = 6 \;\Rightarrow\; A = \sqrt{6(6-3)(6-4)(6-5)} = \sqrt{6(3)(2)(1)} = \sqrt{36} = 6$

105.  $T(4) = \dfrac{\sqrt{4}}{2} = \dfrac{2}{2} = 1$ second

107.  $d = 1.22\sqrt{10,000} \Rightarrow d = 1.22(100) \Rightarrow d = 122$ miles

109.  $R(16) - R(15) = 108\sqrt{16} - 108\sqrt{15} \approx 13.72$. The result is about $14 thousand. Since the revenue is lower than the salary of the $16^{\text{th}}$ employee it is not a good decision to hire the additional employee.

111.  (a)    $P(25,000) = 400\sqrt{25,000} + 8000 \approx 71245.56$, If $25,000 is spent on equipment per worker, each worker will produce about $71,246 worth of goods.

(b)    See Figure 111.

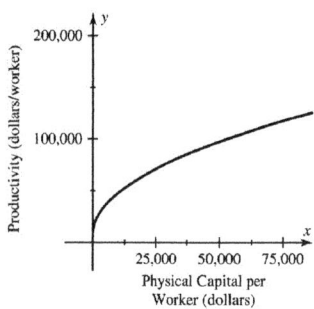

Figure 111

(c)    $P(50,000) - P(25,000) = \left(400\sqrt{50,000} + 8000\right) - \left(400\sqrt{25,000} + 8000\right) \approx 26,197$

$P(75,000) - P(50,000) = \left(400\sqrt{75,000} + 8000\right) - \left(400\sqrt{50,000} + 8000\right) \approx 20,102$

An additional $25,000 spent on equipment per worker. There is a point where the business starts to lose money.

## Section 10.2 Rational Exponents

1.  $4^{1/2} = \sqrt{4} = 2$

3.  $4^{-1/2} = \dfrac{1}{\sqrt{4}} = \dfrac{1}{2}$

5. $\sqrt{x} = x^{1/2}$

7. $\sqrt[n]{a}$

9. c, $\sqrt{x^3} = x^{3/2}$

11. g, $25^{-1/2} = \dfrac{1}{\sqrt{25}} = \dfrac{1}{5}$

13. d, $x^{1/5} = \sqrt[5]{x}$

15. a, $\sqrt{x} \cdot \sqrt[3]{x} = x^{1/2} \cdot x^{1/3} = x^{(1/2)+(1/3)} = x^{5/6} = \sqrt[6]{x^5}$

17. $7^{1/2} = \sqrt{7}$

19. $a^{1/3} = \sqrt[3]{a}$

21. $x^{5/6} = \sqrt[6]{x^5}$

23. $(x+5)^{1/2} = \sqrt{x+5}$

25. $b^{-2/3} = \dfrac{1}{b^{2/3}} = \dfrac{1}{\sqrt[3]{b^2}}$

27. $\sqrt{t} = t^{1/2}$

29. $\sqrt[3]{(x+1)} = (x+1)^{1/3}$

31. $\dfrac{1}{\sqrt{x+1}} = \dfrac{1}{(x+1)^{1/2}} = (x+1)^{-1/2}$

33. $\sqrt{a^2 - b^2} = \left(a^2 - b^2\right)^{1/2}$

35. $\dfrac{1}{\sqrt[3]{x^7}} = \dfrac{1}{(x)^{7/3}} = (x)^{-7/3}$

37. $16^{1/5} \approx 1.74$

39. $5^{1/3} \approx 1.71$

41. $9^{3/5} \approx 3.74$

43. $4^{-3/7} \approx 0.55$

45. $9^{1/2} = \sqrt{9} = 3$

47. $8^{1/3} = \sqrt[3]{8} = 2$

49. $\left(\dfrac{4}{9}\right)^{1/2} = \sqrt{\dfrac{4}{9}} = \dfrac{2}{3}$

51. $(-8)^{2/3} = \sqrt[3]{(-8)^2} = \sqrt[3]{64} = 4$

53. $\left(\dfrac{1}{8}\right)^{-1/3} = 8^{1/3} = \sqrt[3]{8} = 2$

55. $16^{-3/4} = \dfrac{1}{\sqrt[4]{16^3}} = \dfrac{1}{\left(\sqrt[4]{16}\right)^3} = \dfrac{1}{(2)^3} = \dfrac{1}{8}$

57. $\left(4^{1/2}\right)^{-3} = \dfrac{1}{\left(\sqrt{4}\right)^3} = \dfrac{1}{2^3} = \dfrac{1}{8}$

59. $z^{1/4} = \sqrt[4]{z}$

61. $y^{-2/5} = \dfrac{1}{\sqrt[5]{y^2}}$

63. $(3x)^{1/3} = \sqrt[3]{3x}$

65. $\sqrt{y} = y^{1/2}$

67. $\sqrt{x} \cdot \sqrt{x} = \left(x^{1/2}\right)^2 = x$

69. $\sqrt[3]{8x^2} = \sqrt[3]{8} \cdot \sqrt[3]{x^2} = 2x^{2/3}$

71. $\dfrac{\sqrt{49x}}{\sqrt[3]{x^2}} = \dfrac{\sqrt{49} \cdot \sqrt{x}}{x^{2/3}} = 7x^{1/2-2/3} = 7x^{-1/6} = \dfrac{7}{x^{1/6}}$

73. $\left(b^{1/a}\, b^{2/a}\right)^a = \left(b^{1/a+2/a}\right)^a = \left(b^{3/a}\right)^a = b^{\frac{3}{a} \cdot a} = b^3$

75. $\left(x^2\right)^{3/2} = x^{2 \cdot 3/2} = x^3$

77. $\sqrt[3]{x^3 y^6} = \left(x^3 y^6\right)^{1/3} = x^{3 \cdot 1/3} \cdot y^{6 \cdot 1/3} = xy^2$

79. $\sqrt{y^3} \cdot \sqrt[3]{y^2} = (y^3)^{1/2} \cdot (y^2)^{1/3} = y^{3 \cdot 1/2} \cdot y^{2 \cdot 1/3} = y^{3/2} \cdot y^{2/3} = y^{3/2+2/3} = y^{13/6}$

81. $\left(\dfrac{x^6}{27}\right)^{2/3} = \dfrac{x^{6 \cdot 2/3}}{27^{2/3}} = \dfrac{x^4}{\left(\sqrt[3]{27}\right)^2} = \dfrac{x^4}{3^2} = \dfrac{x^4}{9}$

83. $\left(\dfrac{x^2}{y^6}\right)^{-1/2} = \left(\dfrac{y^6}{x^2}\right)^{1/2} = \dfrac{y^{6 \cdot 1/2}}{x^{2 \cdot 1/2}} = \dfrac{y^3}{x}$

85. $\sqrt{\sqrt{y}} = \left(y^{1/2}\right)^{1/2} = y^{1/2 \cdot 1/2} = y^{1/4}$

87. $\left(a^{-1/2}\right)^{4/3} = a^{-1/2 \cdot 4/3} = a^{-2/3} = \dfrac{1}{a^{2/3}}$

89. $\dfrac{\left(k^{1/2}\right)^{-3}}{\left(k^2\right)^{1/4}} = \dfrac{k^{-3/2}}{k^{1/2}} = k^{-3/2-1/2} = k^{-4/2} = k^{-2} = \dfrac{1}{k^2}$

91. $\sqrt{b} \cdot \sqrt[4]{b} = b^{1/2} \cdot b^{1/4} = b^{1/2+1/4} = b^{3/4}$

93. $p^{1/2}\left(p^{3/2} + p^{1/2}\right) = p^{1/2+3/2} + p^{1/2+1/2} = p^2 + p$

95. $\sqrt[3]{x}\left(\sqrt{x} - \sqrt[3]{x^2}\right) = x^{1/3}\left(x^{1/2} - x^{2/3}\right) = x^{1/3+1/2} - x^{1/3+2/3} = x^{5/6} - x$

97. $\dfrac{\sqrt[3]{27x}}{\sqrt{x}} = \dfrac{\sqrt[3]{27} \cdot \sqrt[3]{x}}{\sqrt{x}} = \dfrac{3x^{1/3}}{x^{1/2}} = 3x^{(1/3)-(1/2)} = 3x^{-1/6} = \dfrac{3}{x^{1/6}}$

99. (a)

| $x$ | 0 | 20 | 40 | 60 | 80 |
|-----|---|----|----|----|----|
| $A(x)$ | 0 | 27% | 37% | 44% | 50% |

   (b)  The abandonment rates level off. The longer a person watches a video, the more likely they are to continue watching.

101. $N(h) = 1.6h^{-1/2} = 1.6(2.5)^{-1/2} = \dfrac{1.6}{\sqrt{2.5}} \approx 1.01$; The stepping frequency is about 1 step per second.

103. (a)  $A = 100\sqrt[3]{8^2} = 100\sqrt[3]{64} = 100 \cdot 4 = 400$ square inches

   (b)  $A = 100\sqrt[3]{W^2} \Rightarrow A = 100W^{2/3}$

105. (a)  $H(12) = 35.2(12)^{3/40} \approx 42.4$ cm,   $H(24) = 35.2(24)^{3/40} \approx 44.7$ cm

   (b)  According to our results above an infant's head size increases the most in the first year.

107. See exercise 101 for an example. *Answers may vary.*

## Checking Basic Concepts Sections 10.1 and 10.2

   1. (a)  $\pm 7$

      (b)  7

   2. (a)  $\sqrt[3]{-8} = -2$

      (b)  $-\sqrt[4]{81} = -3$

   3. (a)  $x^{3/2} = \sqrt{x^3}$ or $\left(\sqrt{x}\right)^3$

      (b)  $x^{2/3} = \sqrt[3]{x^2}$ or $\left(\sqrt[3]{x}\right)^2$

      (c)  $x^{-2/5} = \dfrac{1}{\sqrt[5]{x^2}}$ or $\dfrac{1}{\left(\sqrt[5]{x}\right)^2}$

   4. $\sqrt{(x-1)^2} = |x-1|$

   5. (a)  $f(9) = \sqrt{9} = \sqrt{3^2} = 3$

      (b)  $g(125) = \sqrt[3]{125} = \sqrt[3]{5^3} = 5$

(c)     $h\left(\dfrac{12}{7}\right) = \left(\dfrac{12}{7}\right)^{7/12} \approx 1.37$

## Section 10.3 Simplifying Radical Expressions

1. Yes

3. $\sqrt[3]{ab}$

5. $\dfrac{a}{b}$

7. No; $\sqrt{50} = \sqrt{25 \cdot 2} = \sqrt{25} \cdot \sqrt{2} = 5\sqrt{2}$ and $\sqrt{25} + \sqrt{25} = 5 + 5 = 2(5)$ or 10

9. No, because $1^3 \neq 3$.

11. $\sqrt{3} \cdot \sqrt{3} = \sqrt{3 \cdot 3} = \sqrt{9} = 3$

13. $\sqrt{2} \cdot \sqrt{50} = \sqrt{2 \cdot 50} = \sqrt{100} = 10$

15. $\sqrt[3]{4} \cdot \sqrt[3]{16} = \sqrt[3]{4 \cdot 16} = \sqrt[3]{64} = 4$

17. $\sqrt{\dfrac{9}{25}} = \dfrac{\sqrt{9}}{\sqrt{25}} = \dfrac{3}{5}$

19. $\sqrt{\dfrac{1}{2}} \cdot \sqrt{\dfrac{1}{8}} = \sqrt{\dfrac{1 \cdot 1}{2 \cdot 8}} = \sqrt{\dfrac{1}{16}} = \dfrac{\sqrt{1}}{\sqrt{16}} = \dfrac{1}{4}$

21. $\sqrt[3]{\dfrac{2}{3}} \cdot \sqrt[3]{\dfrac{4}{3}} \cdot \sqrt[3]{\dfrac{1}{3}} = \sqrt[3]{\dfrac{2}{3} \cdot \dfrac{4}{3} \cdot \dfrac{1}{3}} = \sqrt[3]{\dfrac{8}{27}} = \dfrac{2}{3}$

23. $\sqrt{x^3} \cdot \sqrt{x^3} = \sqrt{x^3 \cdot x^3} = \sqrt{x^6} = x^{6/2} = x^3$

25. $\sqrt[3]{\dfrac{7}{27}} = \dfrac{\sqrt[3]{7}}{\sqrt[3]{27}} = \dfrac{\sqrt[3]{7}}{3}$

27. $\sqrt[4]{\dfrac{x}{81}} = \dfrac{\sqrt[4]{x}}{\sqrt[4]{81}} = \dfrac{\sqrt[4]{x}}{3}$

29. $\sqrt{\dfrac{9}{z^2}} = \dfrac{\sqrt{9}}{\sqrt{z^2}} = \dfrac{3}{z}$

31. $\sqrt{\dfrac{x}{2}} \cdot \sqrt{\dfrac{x}{8}} = \sqrt{\dfrac{x \cdot x}{2 \cdot 8}} = \sqrt{\dfrac{x^2}{16}} = \dfrac{\sqrt{x^2}}{\sqrt{16}} = \dfrac{x}{4}$

33. $\dfrac{\sqrt{45}}{\sqrt{5}} = \sqrt{\dfrac{45}{5}} = \sqrt{9} = 3$

35. $\sqrt[3]{-4} \cdot \sqrt[3]{-16} = \sqrt[3]{-4 \cdot (-16)} = \sqrt[3]{64} = 4$

37. $\sqrt[4]{9} \cdot \sqrt[4]{9} = \sqrt[4]{9 \cdot 9} = \sqrt[4]{81} = 3$

39. $\dfrac{\sqrt[5]{64}}{\sqrt[5]{-2}} = \sqrt[5]{\dfrac{64}{-2}} = \sqrt[5]{-32} = -2$

41. $\dfrac{\sqrt{a^2 b}}{\sqrt{b}} = \sqrt{\dfrac{a^2 b}{b}} = \sqrt{a^2} = a$

43. $\dfrac{\sqrt[3]{54}}{\sqrt[3]{2}} = \sqrt[3]{\dfrac{54}{2}} = \sqrt[3]{27} = 3$

45. $\sqrt{4x^4} = \sqrt{4} \cdot \sqrt{\left(x^2\right)^2} = 2x^2$

47. $\sqrt[3]{-5a^6} = \sqrt[3]{-5} \cdot \sqrt[3]{\left(a^2\right)^3} = \sqrt[3]{-5} \cdot a^2 = -a^2 \sqrt[3]{5}$

49. $\sqrt[4]{16x^4 y} = \sqrt[4]{16} \cdot \sqrt[4]{x^4} \cdot \sqrt[4]{y} = 2x \sqrt[4]{y}$

51. $\sqrt{3x} \cdot \sqrt{12x} = \sqrt{3 \cdot 12 \cdot x \cdot x} = \sqrt{36x^2} = \sqrt{36} \cdot \sqrt{x^2} = 6x$

53. $\sqrt[3]{8x^6 y^3 z^9} = \sqrt[3]{8} \cdot \sqrt[3]{\left(x^2\right)^3} \cdot \sqrt[3]{y^3} \cdot \sqrt[3]{\left(z^3\right)^3} = 2x^2 y z^3$

55. $\sqrt[4]{\dfrac{3}{4}} \cdot \sqrt[4]{\dfrac{27}{4}} = \sqrt[4]{\dfrac{3}{4} \cdot \dfrac{27}{4}} = \sqrt[4]{\dfrac{81}{16}} = \dfrac{\sqrt[4]{81}}{\sqrt[4]{16}} = \dfrac{3}{2}$

57. $\sqrt[3]{12} \cdot \sqrt[3]{ab} = \sqrt[3]{12ab}$

59. $\sqrt[4]{25z} \cdot \sqrt[4]{25z} = \sqrt[4]{625z^2} = \sqrt[4]{625} \cdot \sqrt[4]{z^2} = 5z^{\frac{2}{4}} = 5z^{\frac{1}{2}} = 5\sqrt{z}$

61. $\sqrt[5]{\dfrac{7a}{b^2}} \cdot \sqrt[5]{\dfrac{b^2}{7a^6}} = \sqrt[5]{\dfrac{7ab^2}{7a^6 b^2}} = \sqrt[5]{\dfrac{1}{a^5}} = \dfrac{1}{a}$

63. $\sqrt{x+4} \cdot \sqrt{x-4} = \sqrt{(x+4)(x-4)} = \sqrt{x^2 - 16}$

65. $\sqrt[3]{a+1} \cdot \sqrt[3]{a^2 - a + 1} = \sqrt[3]{(a+1)\left(a^2 - a + 1\right)} = \sqrt[3]{a^3 + 1}$

67. $\dfrac{\sqrt{x^2 + 2x + 1}}{\sqrt{x+1}} = \sqrt{\dfrac{x^2 + 2x + 1}{x+1}} = \sqrt{\dfrac{(x+1)(x+1)}{x+1}} = \sqrt{x+1}$

69. $\sqrt{500} = \sqrt{100 \cdot 5} = \sqrt{100} \cdot \sqrt{5} = 10\sqrt{5}$; the answer is 10.

71. $\sqrt{8} = \sqrt{4 \cdot 2} = \sqrt{4} \cdot \sqrt{2} = 2\sqrt{2}$; the answer is 2.

73. $\sqrt{45} = \sqrt{9 \cdot 5} = \sqrt{9} \cdot \sqrt{5} = 3\sqrt{5}$; the answer is 3.

75. $\sqrt{200} = \sqrt{100 \cdot 2} = \sqrt{100} \cdot \sqrt{2} = 10\sqrt{2}$

77. $\sqrt[3]{81} = \sqrt[3]{27 \cdot 3} = \sqrt[3]{27} \cdot \sqrt[3]{3} = 3\sqrt[3]{3}$

79. $\sqrt[4]{64} = \sqrt[4]{16 \cdot 4} = \sqrt[4]{16} \cdot \sqrt[4]{4} = 2\sqrt[4]{4} = 2\sqrt[4]{2^2} = 2\sqrt{2}$

81. $\sqrt[5]{-64} = \sqrt[5]{-2^6} = \sqrt[5]{-2^5 \cdot 2} = \sqrt[5]{-2^5} \cdot \sqrt[5]{2} = -2\sqrt[5]{2}$

83. $\sqrt{b^5} = \sqrt{\left(b^2\right)^2 \cdot b} = \sqrt{\left(b^2\right)^2} \cdot \sqrt{b} = b^2\sqrt{b}$

85. $\sqrt{8n^3} = \sqrt{(2n)^2 \cdot 2n} = \sqrt{(2n)^2} \cdot \sqrt{2n} = 2n\sqrt{2n}$

87. $\sqrt{12a^2b^5} = \sqrt{\left(2ab^2\right)^2 \cdot 3b} = \sqrt{\left(2ab^2\right)^2} \cdot \sqrt{3b} = 2ab^2\sqrt{3b}$

89. $\sqrt[3]{-125x^4y^5} = \sqrt[3]{(-5xy)^3 \cdot xy^2} = \sqrt[3]{(-5xy)^3} \cdot \sqrt[3]{xy^2} = -5xy\sqrt[3]{xy^2}$

91. $\sqrt[3]{5t} \cdot \sqrt[3]{125t} = \sqrt[3]{625t^2} = \sqrt[3]{5^4t^2} = \sqrt[3]{5^3 \cdot 5t^2} = \sqrt[3]{5^3} \cdot \sqrt[3]{5t^2} = 5\sqrt[3]{5t^2}$

93. $\sqrt[4]{\dfrac{9t^5}{r^8}} \cdot \sqrt[4]{\dfrac{9r}{5t}} = \sqrt[4]{\dfrac{81rt^5}{5r^8t}} = \sqrt[4]{\dfrac{81t^4}{5r^7}} = \dfrac{\sqrt[4]{(3t)^4}}{\sqrt[4]{r^4 \cdot 5r^3}} = \dfrac{3t}{r\sqrt[4]{5r^3}}$

95. $\sqrt[3]{\dfrac{27x^2}{y^3}} = \dfrac{\sqrt[3]{27} \cdot \sqrt[3]{x^2}}{\sqrt[3]{y^3}} = \dfrac{3\sqrt[3]{x^2}}{y}$

97. $\sqrt{\dfrac{7a^2}{27}} \cdot \sqrt{\dfrac{7a}{3}} = \sqrt{\dfrac{49a^3}{81}} = \dfrac{\sqrt{49} \cdot \sqrt{a^3}}{\sqrt{81}} = \dfrac{7\sqrt{a^2} \cdot \sqrt{a}}{9} = \dfrac{7a\sqrt{a}}{9}$

99. $\left(\sqrt[mn]{a^mb^m}\right)^n = \left(\sqrt[m]{a^mb^m}\right)^{n/n} = \sqrt[m]{(ab)^m} = (ab)^{m/m} = ab$

101. $\sqrt{3} \cdot \sqrt[3]{3} = 3^{1/2} \cdot 3^{1/3} = 3^{1/2+1/3} = 3^{5/6} = \sqrt[6]{3^5}$

103. $\sqrt[4]{8} \cdot \sqrt[3]{4} = \sqrt[4]{2^3} \cdot \sqrt[3]{2^2} = 2^{3/4} \cdot 2^{2/3} = 2^{3/4+2/3} = 2^{17/12} = 2^{12/12+5/12} = 2 \cdot 2^{5/12} = 2\sqrt[12]{2^5}$

105. $\sqrt[4]{27} \cdot \sqrt[3]{9} \cdot \sqrt{3} = \sqrt[4]{3^3} \cdot \sqrt[3]{3^2} \cdot \sqrt{3} = 3^{3/4} \cdot 3^{2/3} \cdot 3^{1/2} = 3^{3/4+2/3+1/2} = 3^{23/12} = 3^{12/12} \cdot 3^{11/12} = 3\sqrt[12]{3^{11}}$

107. $\sqrt[4]{x^3} \cdot \sqrt[3]{x} = x^{3/4} \cdot x^{1/3} = x^{3/4+1/3} = x^{13/12} = x^{12/12} \cdot x^{1/12} = x\sqrt[12]{x}$

109. $\sqrt[4]{rt} \cdot \sqrt[3]{r^2t} = (rt)^{1/4} \cdot \left(r^2t\right)^{1/3} = r^{1/4}t^{1/4} \cdot r^{2/3}t^{1/3} = r^{1/4+2/3}t^{1/4+1/3} = r^{11/12}t^{7/12} = \sqrt[12]{r^{11}t^7}$

111. (a)    $S = \sqrt{25M} = \sqrt{5^2}\sqrt{M} = 5\sqrt{M}$

   (b)    $S = 5\sqrt{100} = 50$ mph

## Section 10.4 Operations on Radical Expressions

1. $2\sqrt{a}$

3. like

5. No; 6 and $3\sqrt{5}$ are not like radicals.

7. $\sqrt{t} + 5$

9. Not possible, since $\sqrt{12} = 2\sqrt{3}$ and $\sqrt{24} = 2\sqrt{6}$.

11. Since $\sqrt{28} = \sqrt{4 \cdot 7} = \sqrt{4} \cdot \sqrt{7} = 2\sqrt{7}$ and $\sqrt{63} = \sqrt{9 \cdot 7} = \sqrt{9} \cdot \sqrt{7} = 3\sqrt{7}$, the like radicals are

$\sqrt{7}, 2\sqrt{7}$, and $3\sqrt{7}$.

13. Since $\sqrt[3]{16} = \sqrt[3]{8 \cdot 2} = \sqrt[3]{8} \cdot \sqrt[3]{2} = 2\sqrt[3]{2}$ and $\sqrt[3]{-54} = \sqrt[3]{-27 \cdot 2} = \sqrt[3]{-27} \cdot \sqrt[3]{2} = -3\sqrt[3]{2}$, the like radicals are

$2\sqrt[3]{2}$ and $-3\sqrt[3]{2}$.

15. Not possible, since $\sqrt{x^2 y} = x\sqrt{y}$ and $\sqrt{4y^2} = 2y$.

17. Since $\sqrt[3]{8xy} = \sqrt[3]{8} \cdot \sqrt[3]{xy} = 2\sqrt[3]{xy}$ and $\sqrt[3]{x^4 y^4} = \sqrt[3]{(xy)^3 \cdot xy} = \sqrt[3]{(xy)^3} \cdot \sqrt[3]{xy} = xy\sqrt[3]{xy}$, the like radicals

are $2\sqrt[3]{xy}$ and $xy\sqrt[3]{xy}$.

19. $2\sqrt{3} + 7\sqrt{3} = 9\sqrt{3}$

21. $4\sqrt[3]{5} + 2\sqrt[3]{5} = 6\sqrt[3]{5}$

23. Not possible, since 7 and $4\sqrt{7}$ are not like radicals

25. Not possible, since $2\sqrt{3}$ and $3\sqrt{2}$ are not like radicals

27. Not possible, since $\sqrt{3}$ and $\sqrt[3]{3}$ are not like radicals

29. $\sqrt[3]{16} + 3\sqrt[3]{2} = \sqrt[3]{8 \cdot 2} + 3\sqrt[3]{2} = \sqrt[3]{8} \cdot \sqrt[3]{2} + 3\sqrt[3]{2} = 2\sqrt[3]{2} + 3\sqrt[3]{2} = 5\sqrt[3]{2}$

31. $\sqrt{2} + \sqrt{18} + \sqrt{32} = \sqrt{2} + \sqrt{9 \cdot 2} + \sqrt{16 \cdot 2} = \sqrt{2} + \sqrt{9} \cdot \sqrt{2} + \sqrt{16} \cdot \sqrt{2} = \sqrt{2} + 3\sqrt{2} + 4\sqrt{2} = 8\sqrt{2}$

33. $11\sqrt{11} - 5\sqrt{11} = 6\sqrt{11}$

35. $\sqrt{x} + \sqrt{x} - \sqrt{y} = 2\sqrt{x} - \sqrt{y}$

37. $\sqrt[3]{z} + \sqrt[3]{z} = 2\sqrt[3]{z}$

39. $2\sqrt[3]{6} - 7\sqrt[3]{6} = -5\sqrt[3]{6}$

41. $\sqrt[3]{y^6} - \sqrt[3]{y^3} = \sqrt[3]{(y^2)^3} - \sqrt[3]{y^3} = y^2 - y$

43. $3\sqrt{28} + 3\sqrt{7} = 3\sqrt{4 \cdot 7} + 3\sqrt{7} = 3 \cdot 2\sqrt{7} + 3\sqrt{7} = 9\sqrt{7}$

45. $\sqrt[4]{48} + 4\sqrt[4]{3} = \sqrt[4]{16 \cdot 3} + 4\sqrt[4]{3} = \sqrt[4]{16} \cdot \sqrt[4]{3} + 4\sqrt[4]{3} = 2\sqrt[4]{3} + 4\sqrt[4]{3} = 6\sqrt[4]{3}$

47. $\sqrt{9x} + \sqrt{16x} = \sqrt{9} \cdot \sqrt{x} + \sqrt{16} \cdot \sqrt{x} = 3\sqrt{x} + 4\sqrt{x} = 7\sqrt{x}$

49. $3\sqrt{2k} + \sqrt{8k} + \sqrt{18k} = 3\sqrt{2k} + \sqrt{4 \cdot 2k} + \sqrt{9 \cdot 2k} = 3\sqrt{2k} + \sqrt{4} \cdot \sqrt{2k} + \sqrt{9} \cdot \sqrt{2k}$

$= 3\sqrt{2k} + 2\sqrt{2k} + 3\sqrt{2k} = 8\sqrt{2k}$

51. $\sqrt{44} - 4\sqrt{11} = \sqrt{4 \cdot 11} - 4\sqrt{11} = 2\sqrt{11} - 4\sqrt{11} = -2\sqrt{11}$

53. $2\sqrt[3]{16} + \sqrt[3]{2} - \sqrt{2} = 2\sqrt[3]{8 \cdot 2} + \sqrt[3]{2} - \sqrt{2} = 2 \cdot 2\sqrt[3]{2} + \sqrt[3]{2} - \sqrt{2} = 5\sqrt[3]{2} - \sqrt{2}$

55. $\sqrt[3]{xy} - 2\sqrt[3]{xy} = -\sqrt[3]{xy}$

57. $\sqrt{4x+8} + \sqrt{x+2} = \sqrt{4(x+2)} + \sqrt{x+2} = 2\sqrt{x+2} + \sqrt{x+2} = 3\sqrt{x+2}$

59. $\sqrt{9x+18} - \sqrt{4x+8} = \sqrt{9(x+2)} - \sqrt{4(x+2)} = 3\sqrt{x+2} - 2\sqrt{x+2} = \sqrt{x+2}$

61. $\sqrt{x^3+x^2} - \sqrt{x+1} = \sqrt{x^2(x+1)} - \sqrt{x+1} = x\sqrt{x+1} - 1\sqrt{x+1} = (x-1)\sqrt{x+1}$

63. $\sqrt{25x^3} - \sqrt{x^3} = \sqrt{25} \cdot \sqrt{x^2} \cdot \sqrt{x} - \sqrt{x^2} \cdot \sqrt{x} = 5x\sqrt{x} - x\sqrt{x} = 4x\sqrt{x}$

65. $\sqrt[3]{\dfrac{7x}{8}} - \dfrac{\sqrt[3]{7x}}{3} = \dfrac{\sqrt[3]{7x}}{\sqrt[3]{8}} - \dfrac{\sqrt[3]{7x}}{3} = \dfrac{\sqrt[3]{7x}}{2} - \dfrac{\sqrt[3]{7x}}{3} = \dfrac{3\sqrt[3]{7x}}{6} - \dfrac{2\sqrt[3]{7x}}{6} = \dfrac{\sqrt[3]{7x}}{6}$

67. $\dfrac{4\sqrt{3}}{3} + \dfrac{\sqrt{3}}{6} = \dfrac{4\sqrt{3}}{3} \cdot \dfrac{2}{2} + \dfrac{\sqrt{3}}{6} = \dfrac{8\sqrt{3}}{6} + \dfrac{\sqrt{3}}{6} = \dfrac{8\sqrt{3}+\sqrt{3}}{6} = \dfrac{9\sqrt{3}}{6} = \dfrac{3\sqrt{3}}{2}$

69. $\dfrac{15\sqrt{8}}{4} - \dfrac{2\sqrt{2}}{5} = \dfrac{15 \cdot 2\sqrt{2}}{4} \cdot \dfrac{5}{5} - \dfrac{2\sqrt{2}}{5} \cdot \dfrac{4}{4} = \dfrac{150\sqrt{2}}{20} - \dfrac{8\sqrt{2}}{20} = \dfrac{150\sqrt{2}-8\sqrt{2}}{20} = \dfrac{142\sqrt{2}}{20} = \dfrac{71\sqrt{2}}{10}$

71. $2\sqrt[4]{64} - \sqrt[4]{324} + \sqrt[4]{4} = 2\sqrt[4]{16 \cdot 4} - \sqrt[4]{81 \cdot 4} + \sqrt[4]{4} = 4\sqrt[4]{4} - 3\sqrt[4]{4} + \sqrt[4]{4} = 2\sqrt[4]{4} = 2 \cdot 2^{\frac{2}{4}} = 2\sqrt{2}$

73. $5\sqrt[4]{x^5} - \sqrt[4]{x} = 5\sqrt[4]{x^4 \cdot x} - \sqrt[4]{x} = 5x\sqrt[4]{x} - \sqrt[4]{x} = (5x-1)\sqrt[4]{x}$

75. $\sqrt{64x^3} - \sqrt{x} + 3\sqrt{x} = \sqrt{(8x)^2 \cdot x} - \sqrt{x} + 3\sqrt{x} = 8x\sqrt{x} - \sqrt{x} + 3\sqrt{x} = 2\sqrt{x}(4x+1)$

77. $\sqrt[4]{81a^5b^5} - \sqrt[4]{ab} = \sqrt[4]{(3ab)^4 \cdot ab} - \sqrt[4]{ab} = 3ab\sqrt[4]{ab} - \sqrt[4]{ab} = (3ab-1)\sqrt[4]{ab}$

79. $5\sqrt[3]{\dfrac{n^4}{125}} - 2\sqrt[3]{n} = 5\sqrt[3]{\dfrac{n^3}{125} \cdot n} - 2\sqrt[3]{n} = 5 \cdot \dfrac{n}{5}\sqrt[3]{n} - 2\sqrt[3]{n} = n\sqrt[3]{n} - 2\sqrt[3]{n} = (n-2)\sqrt[3]{n}$

81. $(f+g)(x) = f(x) + g(x) = (5\sqrt{x}-2) + (-2\sqrt{x}+3) = 3\sqrt{x}+1$

    $(f-g)(x) = f(x) - g(x) = (5\sqrt{x}-2) - (-2\sqrt{x}+3) = 7\sqrt{x}-5$

83. $(f+g)(x) = f(x) + g(x) = (\sqrt[3]{8x}+1) + (2\sqrt[3]{x}-1) = 2\sqrt[3]{x}+1+2\sqrt[3]{x}-1 = 4\sqrt[3]{x}$

    $(f-g)(x) = f(x) - g(x) = (\sqrt[3]{8x}+1) - (2\sqrt[3]{x}-1) = 2\sqrt[3]{x}+1 - (2\sqrt[3]{x}-1) = 2$

85. $(\sqrt{x}-3)(\sqrt{x}+2) = (\sqrt{x})^2 + 2\sqrt{x} - 3\sqrt{x} - 6 = x - \sqrt{x} - 6$

87. $(3+\sqrt{7})(3-\sqrt{7}) = 3^2 - (\sqrt{7})^2 = 9 - 7 = 2$

89. $(11-\sqrt{2})(11+\sqrt{2}) = 11^2 - (\sqrt{2})^2 = 121 - 2 = 119$

91. $(\sqrt{x}+8)(\sqrt{x}-8) = (\sqrt{x})^2 - 8^2 = x - 64$

93. $(\sqrt{ab}-\sqrt{c})(\sqrt{ab}+\sqrt{c}) = (\sqrt{ab})^2 - (\sqrt{c})^2 = ab - c$

95. $(\sqrt{x}-7)(\sqrt{x}+8) = (\sqrt{x})^2 + 8\sqrt{x} - 7\sqrt{x} - 56 = x + \sqrt{x} - 56$

97. $\dfrac{1}{\sqrt{7}} = \dfrac{1}{\sqrt{7}} \cdot \dfrac{\sqrt{7}}{\sqrt{7}} = \dfrac{\sqrt{7}}{\left(\sqrt{7}\right)^2} = \dfrac{\sqrt{7}}{7}$

99. $\dfrac{4}{\sqrt{3}} = \dfrac{4}{\sqrt{3}} \cdot \dfrac{\sqrt{3}}{\sqrt{3}} = \dfrac{4\sqrt{3}}{\left(\sqrt{3}\right)^2} = \dfrac{4\sqrt{3}}{3}$

101. $\dfrac{5}{3\sqrt{5}} = \dfrac{5}{3\sqrt{5}} \cdot \dfrac{\sqrt{5}}{\sqrt{5}} = \dfrac{5\sqrt{5}}{3\left(\sqrt{5}\right)^2} = \dfrac{5\sqrt{5}}{3\cdot 5} = \dfrac{\sqrt{5}}{3}$

103. $\sqrt{\dfrac{b}{12}} = \dfrac{\sqrt{b}}{\sqrt{12}} \cdot \dfrac{\sqrt{12}}{\sqrt{12}} = \dfrac{\sqrt{12b}}{\left(\sqrt{12}\right)^2} = \dfrac{\sqrt{4\cdot 3b}}{12} = \dfrac{\sqrt{4}\cdot\sqrt{3b}}{12} = \dfrac{2\sqrt{3b}}{12} = \dfrac{\sqrt{3b}}{6}$

105. $\dfrac{rt}{2\sqrt{r^3}} = \dfrac{rt}{2\sqrt{r^2}\cdot\sqrt{r}} \cdot \dfrac{\sqrt{r}}{\sqrt{r}} = \dfrac{rt\sqrt{r}}{2r\left(\sqrt{r}\right)^2} = \dfrac{t\sqrt{r}}{2r}$

107. $\dfrac{1}{3-\sqrt{2}} = \dfrac{1}{3-\sqrt{2}} \cdot \dfrac{3+\sqrt{2}}{3+\sqrt{2}} = \dfrac{3+\sqrt{2}}{9-2} = \dfrac{3+\sqrt{2}}{7}$

109. $\dfrac{\sqrt{2}}{\sqrt{5}+2} = \dfrac{\sqrt{2}}{\sqrt{5}+2} \cdot \dfrac{\sqrt{5}-2}{\sqrt{5}-2} = \dfrac{\sqrt{10}-2\sqrt{2}}{5-4} = \dfrac{\sqrt{10}-2\sqrt{2}}{1} = \sqrt{10}-2\sqrt{2}$

111. $\dfrac{\sqrt{7}-2}{\sqrt{7}+2} = \dfrac{\sqrt{7}-2}{\sqrt{7}+2} \cdot \dfrac{\sqrt{7}-2}{\sqrt{7}-2} = \dfrac{7-4\sqrt{7}+4}{7-4} = \dfrac{11-4\sqrt{7}}{3}$

113. $\dfrac{1}{\sqrt{7}-\sqrt{6}} = \dfrac{1}{\sqrt{7}-\sqrt{6}} \cdot \dfrac{\sqrt{7}+\sqrt{6}}{\sqrt{7}+\sqrt{6}} = \dfrac{\sqrt{7}+\sqrt{6}}{7-6} = \dfrac{\sqrt{7}+\sqrt{6}}{1} = \sqrt{7}+\sqrt{6}$

115. $\dfrac{\sqrt{z}}{\sqrt{z}-3} = \dfrac{\sqrt{z}}{\sqrt{z}-3} \cdot \dfrac{\sqrt{z}+3}{\sqrt{z}+3} = \dfrac{z+3\sqrt{z}}{z-9}$

117. $\dfrac{\sqrt{a}+\sqrt{b}}{\sqrt{a}-\sqrt{b}} = \dfrac{\sqrt{a}+\sqrt{b}}{\sqrt{a}-\sqrt{b}} \cdot \dfrac{\sqrt{a}+\sqrt{b}}{\sqrt{a}+\sqrt{b}} = \dfrac{a+2\sqrt{ab}+b}{a-b}$

119. $\dfrac{1}{\sqrt{x+1}-\sqrt{x}} = \dfrac{1}{\sqrt{x+1}-\sqrt{x}} \cdot \dfrac{\sqrt{x+1}+\sqrt{x}}{\sqrt{x+1}+\sqrt{x}} = \dfrac{\sqrt{x+1}+\sqrt{x}}{x+1-x} = \dfrac{\sqrt{x+1}+\sqrt{x}}{1} = \sqrt{x+1}+\sqrt{x}$

121. $\dfrac{3}{\sqrt[3]{x}} = \dfrac{3}{x^{1/3}} \cdot \dfrac{x^{2/3}}{x^{2/3}} = \dfrac{3x^{2/3}}{x} = \dfrac{3\sqrt[3]{x^2}}{x}$

123. $\dfrac{1}{\sqrt[3]{x^2}} = \dfrac{1}{x^{2/3}} \cdot \dfrac{x^{1/3}}{x^{1/3}} = \dfrac{x^{1/3}}{x} = \dfrac{\sqrt[3]{x}}{x}$

125. $\sqrt{27}+\sqrt{48}+\sqrt{75} = \sqrt{9\cdot 3}+\sqrt{16\cdot 3}+\sqrt{25\cdot 3} = 3\sqrt{3}+4\sqrt{3}+5\sqrt{3} = 12\sqrt{3} \approx 20.8$ cm

127. A square with a diagonal of length $\sqrt{3}$ has sides of length $\dfrac{\sqrt{3}}{\sqrt{2}}$. The perimeter of this square is

$4\cdot\dfrac{\sqrt{3}}{\sqrt{2}} = 4\cdot\dfrac{\sqrt{3}}{\sqrt{2}} \cdot \dfrac{\sqrt{2}}{\sqrt{2}} = \dfrac{4\sqrt{6}}{2} = 2\sqrt{6}$.

129. A square with a diagonal of length 60 feet has sides of length $\dfrac{60}{\sqrt{2}}$ feet. The perimeter of this square

is $4 \cdot \dfrac{60}{\sqrt{2}} = \dfrac{240}{\sqrt{2}} \cdot \dfrac{\sqrt{2}}{\sqrt{2}} = \dfrac{240\sqrt{2}}{2} = 120\sqrt{2}$ feet.

131. A square with an area of $x$ square feet has sides of length $\sqrt{x}$ feet. Solve for $c$ in the Pythagorean

Theorem as follows: $a^2 + b^2 = c^2 \Rightarrow \left(\sqrt{x}\right)^2 + \left(\sqrt{x}\right)^2 = c^2 \Rightarrow 2x = c^2 \Rightarrow \sqrt{2x} = c$. The length of the

diagonal is $\sqrt{2x}$ feet.

## Checking Basic Concepts Sections 10.3 and 10.4

1. (a)   $\left(64^{-3/2}\right)^{1/3} = 64^{-3/2 \cdot 1/3} = 64^{-1/2} = \dfrac{1}{64^{1/2}} = \dfrac{1}{\sqrt{64}} = \dfrac{1}{8}$

   (b)   $\sqrt{5} \cdot \sqrt{20} = \sqrt{5 \cdot 20} = \sqrt{100} = 10$

   (c)   $\sqrt[3]{-8x^4 y} = \sqrt[3]{(-2x)^3 \cdot xy} = -2x\sqrt[3]{xy}$

   (d)   $\sqrt{\dfrac{4b}{5}} \cdot \sqrt{\dfrac{4b^3}{5}} = \sqrt{\dfrac{4b \cdot 4b^3}{5 \cdot 5}} = \dfrac{\sqrt{16b^4}}{\sqrt{25}} = \dfrac{4b^2}{5}$

2. $\sqrt[3]{7} \cdot \sqrt{7} = 7^{1/3} \cdot 7^{1/2} = 7^{1/3+1/2} = 7^{5/6} = \sqrt[6]{7^5}$

3. (a)   $\sqrt{3} \cdot \sqrt{12} = \sqrt{3 \cdot 12} = \sqrt{36} = 6$

   (b)   $\dfrac{\sqrt[3]{81}}{\sqrt[3]{3}} = \sqrt[3]{\dfrac{81}{3}} = \sqrt[3]{27} = 3$

   (c)   $\sqrt{36x^6} = \sqrt{36} \cdot \sqrt{\left(x^3\right)^2} = 6x^3$

4. (a)   $5\sqrt{6} + 2\sqrt{6} + \sqrt{7} = 7\sqrt{6} + \sqrt{7}$

   (b)   $8\sqrt[3]{x} - 3\sqrt[3]{x} = 5\sqrt[3]{x}$

   (c)   $\sqrt{9x} - \sqrt{4x} = \sqrt{9} \cdot \sqrt{x} - \sqrt{4} \cdot \sqrt{x} = 3\sqrt{x} - 2\sqrt{x} = \sqrt{x}$

5. (a)   $\sqrt[3]{xy^4} - \sqrt[3]{x^4 y} = \sqrt[3]{y^3 \cdot xy} - \sqrt[3]{x^3 \cdot xy} = y\sqrt[3]{xy} - x\sqrt[3]{xy} = (y - x)\sqrt[3]{xy}$

   (b)   $\left(4 - \sqrt{2}\right)\left(4 + \sqrt{2}\right) = 4^2 - \left(\sqrt{2}\right)^2 = 16 - 2 = 14$

6. $\dfrac{6}{2\sqrt{6}} = \dfrac{6}{2\sqrt{6}} \cdot \dfrac{\sqrt{6}}{\sqrt{6}} = \dfrac{6\sqrt{6}}{2 \cdot 6} = \dfrac{\sqrt{6}}{2}$

7. $\dfrac{2}{\sqrt{5} - 1} = \dfrac{2}{\sqrt{5} - 1} \cdot \dfrac{\sqrt{5} + 1}{\sqrt{5} + 1} = \dfrac{2\left(\sqrt{5} + 1\right)}{5 - 1} = \dfrac{2\left(\sqrt{5} + 1\right)}{4} = \dfrac{\sqrt{5} + 1}{2}$

## Section 10.5 More Radical Functions

1. See Figure 1.

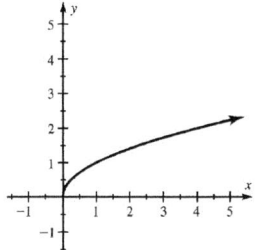

Figure 1

3. $\{x \mid x \geq 0\}$

5. $f(x) = x^p$, where $p$ is rational

7. The variable cannot be negative. The domain is $\{x \mid x \geq 0\}$.

9. $f(-2) = \sqrt{(-2)^2 - 1} = \sqrt{4 - 1} = \sqrt{3} \approx 1.73$, $f(-0) = \sqrt{(0)^2 - 1} = \sqrt{-1} \Rightarrow$ undefined

11. $f(-5) = \sqrt[4]{1 - (-5)} = \sqrt[4]{6} \approx 1.57$, $f(2) = \sqrt[4]{1 - (2)} = \sqrt[4]{-1} \Rightarrow$ undefined

13. $f(-3) = \sqrt[5]{4 - 3(-3)} = \sqrt[5]{4 + 9} = \sqrt[5]{13} \approx 1.67$, $f(1) = \sqrt[5]{4 - 3(1)} = \sqrt[5]{1} = 1$

15. $f(-5) = \sqrt[3]{1 - (-5)} = \sqrt[3]{1 + 5} = \sqrt[3]{6} \approx 1.82$, $f(2) = \sqrt[3]{1 - (2)} = \sqrt[3]{-1} = -1$

17. $f(x) = x^{1/2} = \sqrt{x}$

19. $f(x) = x^{2/3} = \sqrt[3]{x^2}$

21. $f(x) = x^{-1/5} = \dfrac{1}{x^{1/5}} = \dfrac{1}{\sqrt[5]{x}}$

23. $f(4) = 4^{5/2} = \left(\sqrt{4}\right)^5 = 2^5 = 32$; $f(5) = 5^{5/2} \approx 55.90$

25. $f(-32) = (-32)^{-7/5} = \dfrac{1}{(-32)^{7/5}} = \dfrac{1}{\left(\sqrt[5]{-32}\right)^7} = \dfrac{1}{(-2)^7} = -\dfrac{1}{128} \approx -0.01$; $f(10) = 10^{-7/5} = \dfrac{1}{10^{7/5}} \approx 0.04$

27. $f(256) = 256^{1/4} = \sqrt[4]{256} = 4$; $f(-10) = (-10)^{1/4} = \sqrt[4]{-10} \Rightarrow$ Not possible

29. $f(32) = 32^{2/5} = \left(\sqrt[5]{32}\right)^2 = 2^2 = 4$; $f(-32) = (-32)^{2/5} = \left(\sqrt[5]{-32}\right)^2 = (-2)^2 = 4$

31. See Figure 31. Domain: $[0, \infty)$

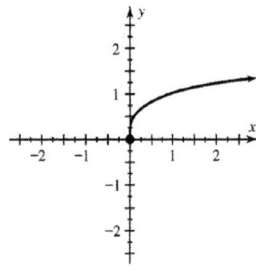

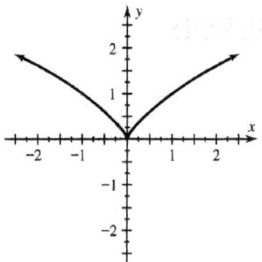

Figure 31                                                    Figure 33

33. See Figure 33. Domain: $(-\infty, \infty)$

35. Graph $Y_1 = X^{\wedge}(1/5)$ and $Y_2 = X^{\wedge}(1/3)$ in [0, 6, 1] by [0, 6, 1]. See Figure 35. Function $g(x)$

increases faster.

[0, 6, 1] by [0, 6, 1]                            [0, 6, 1] by [0, 6, 1]

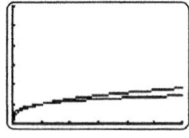

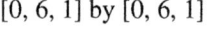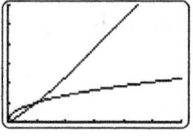

Figure 35                                                    Figure 37

37. Graph $Y_1 = X^{\wedge}1.2$ and $Y_2 = X^{\wedge}0.45$ in [0, 6, 1] by [0, 6, 1]. See Figure 37. Function $f(x)$ increases

faster.

39. $x^p > x^q$

41. (a)  $(f+g)(2) = \sqrt{8 \cdot 2} + \sqrt{2 \cdot 2} = \sqrt{16} + \sqrt{4} = 4 + 2 = 6$

    (b)  $(f-g)(x) = \sqrt{8x} - \sqrt{2x} = \sqrt{4 \cdot 2x} - \sqrt{2x} = 2\sqrt{2x} - \sqrt{2x} = \sqrt{2x}$

    (c)  $(fg)(x) = \left(\sqrt{8x}\right)\left(\sqrt{2x}\right) = \sqrt{8x \cdot 2x} = \sqrt{16x^2} = 4|x|$

    (d)  $(f/g)(x) = \dfrac{\sqrt{8x}}{\sqrt{2x}} = \sqrt{\dfrac{8x}{2x}} = \sqrt{4} = 2$

43. b

45. c

47. d

49. b

51. $S(150) = 342(150)^{0.425} \approx 2877$ square inches

53. According to the graph the result is about 35%.

55. No, for $x \geq 10$ the accuracy is less than double.

57. (a)  $T(0.8c) = 10\sqrt{1 - (0.8c/c)^2} = 10\sqrt{1 - 0.8^2} = 10\sqrt{1 - 0.64} = 10\sqrt{0.36} = 10 \cdot 0.6 = 6$ years

    (b)  The twin in the spaceship will be 4 years younger than the twin on Earth.

59. (a)  See Figure 59a.

(b)    $A(2) = k(2)^{2/3} = 0.254 \Rightarrow k = \dfrac{0.254}{2^{2/3}} \Rightarrow k \approx 0.16$

(c)    See Figure 59c.  Yes the graph does pass through the data points.

(d)    $A(2.5) = 0.16(2.5)^{2/3} \approx 0.295$ square meters

[0, 5, 1] by [0, 0.5, 0.1]                    [0, 5, 1] by [0, 0.5, 0.1]

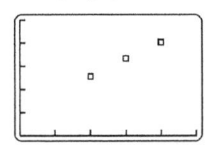                          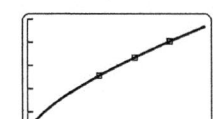

Figure 59a                                Figure 59c.

61. (a)    $L = kW^{\,1/3} \Rightarrow 0.422 = k \cdot 0.1^{1/3} \Rightarrow k = \dfrac{0.422}{0.1^{1/3}} \approx 0.91$

(b)    Plot the data and graph $Y_1 = 0.91X \wedge (1/3)$ in [0, 1.5, 0.1] by [0, 1, 0.1].  See Figure 61.

It increases.

(c)    $L = 0.91(0.7)^{1/3} \approx 0.808$ meters

(d)    $L = 0.91(0.65)^{1/3} \approx 0.788$;  A bird weighing 0.65 kg has a wing span of about 0.788 meters.

[0, 1600, 400] by [0, 220, 20]

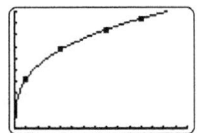

Figure 61


## Section 10.6 Equations Involving Radical Expressions

1. Square each side.

3. Yes

5. The Pythagorean theorem is used to find an unknown side of a right triangle. *Answers may vary.*

7. $d = \sqrt{(x_2 - x_1)^2 + (y_2 - y_1)^2}$

9. $\sqrt{2} \cdot \sqrt{2} = \sqrt{2 \cdot 2} = \sqrt{4} = 2$

11. $\sqrt{x} \cdot \sqrt{x} = \sqrt{x \cdot x} = \sqrt{x^2} = x$

13. $\left(\sqrt{2x+1}\right)^2 = 2x+1$

15. $\left(\sqrt[3]{5x^2}\right)^3 = 5x^2$

17. $\sqrt{x} = 8 \Rightarrow \left(\sqrt{x}\right)^2 = 8^2 \Rightarrow x = 64$

19. $\sqrt[4]{x} = 3 \Rightarrow \left(\sqrt[4]{x}\right)^4 = 3^4 \Rightarrow x = 81$

21. $\sqrt{2t+4} = 4 \Rightarrow \left(\sqrt{2t+4}\right)^2 = 4^2 \Rightarrow 2t+4 = 16 \Rightarrow 2t = 12 \Rightarrow t = 6$

23. $\sqrt{x+1} - 3 = 4 \Rightarrow \sqrt{x+1} = 7 \Rightarrow x+1 = 7^2 \Rightarrow x = 49 - 1 \Rightarrow x = 48$

25. $2\sqrt{x-2} + 1 = 5 \Rightarrow 2\sqrt{x-2} = 4 \Rightarrow \sqrt{x-2} = 2 \Rightarrow x-2 = 2^2 \Rightarrow x = 4+2 \Rightarrow x = 6$

27. $\sqrt{x+6} = x \Rightarrow \left(\sqrt{x+6}\right)^2 = x^2 \Rightarrow x+6 = x^2 \Rightarrow x^2 - x - 6 = 0 \Rightarrow (x+2)(x-3) = 0 \Rightarrow$

     $x = -2$ or $x = 3$. The solution $x = -2$ does not check. The solution is $x = 3$.

29. $\sqrt[3]{x} = 3 \Rightarrow \left(\sqrt[3]{x}\right)^3 = 3^3 \Rightarrow x = 27$

31. $\sqrt[3]{2z-4} = -2 \Rightarrow 2z-4 = (-2)^3 \Rightarrow 2z-4 = -8 \Rightarrow 2z = -4 \Rightarrow z = -2$

33. $\sqrt[4]{t+1} = 2 \Rightarrow t+1 = 2^4 \Rightarrow t+1 = 16 \Rightarrow t = 15$

35. $\sqrt{5z-1} = \sqrt{z+1} \Rightarrow \left(\sqrt{5z-1}\right)^2 = \left(\sqrt{z+1}\right)^2 \Rightarrow 5z-1 = z+1 \Rightarrow 4z = 2 \Rightarrow z = \frac{1}{2}$

37. $\sqrt{1-x} = 1-x \Rightarrow \left(\sqrt{1-x}\right)^2 = (1-x)^2 \Rightarrow 1-x = 1-2x+x^2 \Rightarrow x^2 - x = 0 \Rightarrow$

     $x(x-1) = 0 \Rightarrow x = 0$ or $x = 1$

39. $\sqrt{b^2-4} = b-2 \Rightarrow \left(\sqrt{b^2-4}\right)^2 = (b-2)^2 \Rightarrow b^2-4 = b^2-4b+4 \Rightarrow 4b = 8 \Rightarrow b = 2$

41. $\sqrt{1-2x} = x+7 \Rightarrow \left(\sqrt{1-2x}\right)^2 = (x+7)^2 \Rightarrow 1-2x = x^2+14x+49 \Rightarrow x^2+16x+48 = 0 \Rightarrow$

     $(x+12)(x+4) = 0 \Rightarrow x = -12$ or $x = -4$. The solution $x = -12$ does not check.

     The solution is $x = -4$.

43. $\sqrt{x} = \sqrt{x-5} + 1 \Rightarrow \left(\sqrt{x}\right)^2 = \left(\sqrt{x-5}+1\right)^2 \Rightarrow x = (x-5) + 2\sqrt{x-5} + 1 \Rightarrow$

     $2\sqrt{x-5} = 4 \Rightarrow \left(2\sqrt{x-5}\right)^2 = 4^2 \Rightarrow 4(x-5) = 16 \Rightarrow 4x-20 = 16 \Rightarrow 4x = 36 \Rightarrow x = 9$

45. $\sqrt{2t-2} + \sqrt{t} = 7 \Rightarrow \sqrt{2t-2} = 7 - \sqrt{t} \Rightarrow \left(\sqrt{2t-2}\right)^2 = \left(7-\sqrt{t}\right)^2 \Rightarrow 2t-2 = 49 - 14\sqrt{t} + t \Rightarrow$

     $14\sqrt{t} = 51 - t \Rightarrow \left(14\sqrt{t}\right)^2 = (51-t)^2 \Rightarrow 196t = 2601 - 102t + t^2 \Rightarrow t^2 - 298t + 2601 = 0 \Rightarrow$

     $(t-9)(t-289) = 0 \Rightarrow t = 9$ or $t = 289$. The solution $t = 289$ does not check.

     The solution is $t = 9$.

47. $x^2 = 49 \Rightarrow \sqrt{x^2} = \sqrt{49} \Rightarrow |x| = 7 \Rightarrow x = \pm 7$

49. $2z^2 = 200 \Rightarrow z^2 = 100 \Rightarrow \sqrt{z^2} = \sqrt{100} \Rightarrow |z| = 10 \Rightarrow z = \pm 10$

51. $(t+1)^2 = 16 \Rightarrow \sqrt{(t+1)^2} = \sqrt{16} \Rightarrow |t+1| = 4 \Rightarrow t+1 = \pm 4 \Rightarrow t = 3 \text{ or } t = -5$

53. $(4-2x)^2 = 100 \Rightarrow \sqrt{(4-2x)^2} = \sqrt{100} \Rightarrow |4-2x| = 10 \Rightarrow 4-2x = \pm 10$

   $\Rightarrow 4-2x = 10 \text{ or } 4-2x = -10 \Rightarrow -2x = 6 \text{ or } -2x = -14 \Rightarrow x = -3 \text{ or } x = 7$

55. $b^3 = 64 \Rightarrow \sqrt[3]{b^3} = \sqrt[3]{64} \Rightarrow b = 4$

57. $2t^3 = -128 \Rightarrow t^3 = -64 \Rightarrow \sqrt[3]{t^3} = \sqrt[3]{-64} \Rightarrow t = -4$

59. $(x+1)^3 = 8 \Rightarrow \sqrt[3]{(x+1)^3} = \sqrt[3]{8} \Rightarrow x+1 = 2 \Rightarrow x = 1$

61. $(2-5z)^3 = -125 \Rightarrow \sqrt[3]{(2-5z)^3} = \sqrt[3]{-125} \Rightarrow 2-5z = -5 \Rightarrow -5z = -7 \Rightarrow z = \frac{7}{5}$

63. $x^4 = 16 \Rightarrow \sqrt[4]{x^4} = \sqrt[4]{16} \Rightarrow |x| = 2 \Rightarrow x = \pm 2$

65. $x^5 = 12 \Rightarrow \sqrt[5]{x^5} = \sqrt[5]{12} \Rightarrow x = \sqrt[5]{12}$

67. $2(x+2)^4 = 162 \Rightarrow (x+2)^4 = 81 \Rightarrow \sqrt[4]{(x+2)^4} = \sqrt[4]{81} \Rightarrow |x+2| = 3 \Rightarrow x+2 = \pm 3 \Rightarrow x = 1 \text{ or } x = -5$

69. Graphical: Graph $Y_1 = \sqrt[3]{(X+5)}$ and $Y_2 = 2$ in $[-7, 7, 1]$ by $[0, 4, 1]$. See Figure 69.

   The solution is $x = 3$.

   $[-7, 7, 1]$ by $[0, 4, 1]$      $[0, 3, 1]$ by $[-2, 2, 1]$

   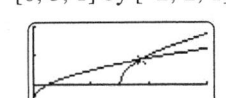

   Figure 69         Figure 71

71. Graphical: Graph $Y_1 = \sqrt{(2X-3)}$ and $Y_2 = \sqrt{(X)} - (1/2)$ in $[0, 3, 1]$ by $[-2, 2, 1]$. See Figure 71.

   The solution is $x \approx 1.88$.

73. Graphical: Graph $Y_1 = X^\wedge(5/3)$ and $Y_2 = 2 - 3X^2$ in $[-4, 4, 1]$ by $[-4, 4, 1]$. See Figures 73a & 73b.

   The solutions are $x = -1$ or $x \approx 0.70$.

   $[-4, 4, 1]$ by $[-4, 4, 1]$       $[-4, 4, 1]$ by $[-4, 4, 1]$       $[-3, 3, 1]$ by $[-3, 3, 1]$

   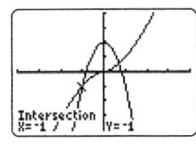       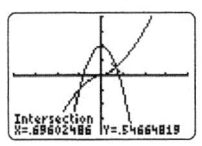      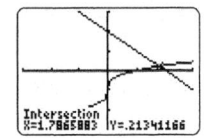

   Figure 73a          Figure 73b          Figure 75

75. Graphical: Graph $Y_1 = X^\wedge(1/3) - 1$ and $Y_2 = 2 - X$ in $[-3, 3, 1]$ by $[-3, 3, 1]$. See Figure 75.

   The solution is $z \approx 1.79$.

77. Graphical: Graph $Y_1 = \sqrt{(X+2)} + \sqrt{(3X+2)}$ and $Y_2 = 2$ in [−2, 2, 1] by [0, 5, 1]. See Figure 77.

The solution is $y \approx -0.47$.

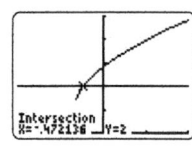          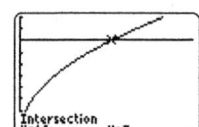          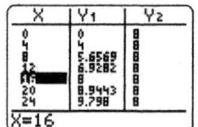

Figure 77                    Figure 79b                    Figure 79c

79. (a)    $2\sqrt{x} = 8 \Rightarrow \left(2\sqrt{x}\right)^2 = 8^2 \Rightarrow 4x = 64 \Rightarrow x = 16$

(b)    Graph $Y_1 = 2\sqrt{(X)}$ and $Y_2 = 8$ in [0, 30, 5] by [0, 10, 1]. See Figure 79b.

The solution is $x = 16$.

(c)    Table $Y_1 = 2\sqrt{(X)}$ and $Y_2 = 8$ with TblStart = 0 and $\Delta$Tbl = 4. See Figure 79c.

The solution is $x = 16$.

81. (a)    $\sqrt{6z-2} = 8 \Rightarrow \left(\sqrt{6z-2}\right)^2 = 8^2 \Rightarrow 6z - 2 = 64 \Rightarrow 6z = 66 \Rightarrow z = 11$

(b)    Graph $Y_1 = \sqrt{(6X-2)}$ and $Y_2 = 8$ in [0, 20, 2] by [0, 10, 1]. See Figure 81b.

The solution is $z = 11$.

(c)    Table $Y_1 = \sqrt{(6X-2)}$ and $Y_2 = 8$ with TblStart = 7 and $\Delta$Tbl = 1. See Figure 81c.

The solution is $z = 11$.

[0, 20, 2] by [0, 10, 1]

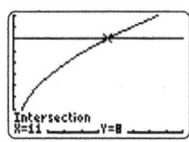          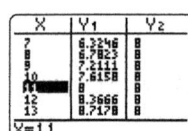

Figure 81b                    Figure 81c

83. $T = 2\pi\sqrt{\dfrac{L}{32}} \Rightarrow \dfrac{T}{2\pi} = \sqrt{\dfrac{L}{32}} \Rightarrow \left(\dfrac{T}{2\pi}\right)^2 = \dfrac{L}{32} \Rightarrow \dfrac{T^2}{4\pi^2} = \dfrac{L}{32} \Rightarrow 32 \cdot \dfrac{T^2}{4\pi^2} = L \Rightarrow L = \dfrac{8T^2}{\pi^2}$

85. $r = \sqrt{\dfrac{A}{\pi}} \Rightarrow r^2 = \dfrac{A}{\pi} \Rightarrow \pi r^2 = A \Rightarrow A = \pi r^2$

87. Yes, since $6^2 + 8^2 = 10^2$. That is $36 + 64 = 100$.

89. Yes, since $\left(\sqrt{5}\right)^2 + \left(\sqrt{9}\right)^2 = \left(\sqrt{14}\right)^2$. That is $5 + 9 = 14$.

91. Yes, since $7^2 + 24^2 = 25^2$. That is $49 + 576 = 625$.

93. No, since $8^2 + 8^2 \neq 16^2$. That is $64 + 64 \neq 256$.

95. $4^2 + 4^2 = c^2 \Rightarrow c^2 = 16 + 16 \Rightarrow c^2 = 32 \Rightarrow c = \sqrt{32} = 4\sqrt{2}$

97. $24^2 + b^2 = 25^2 \Rightarrow 576 + b^2 = 625 \Rightarrow b^2 = 49 \Rightarrow b = \sqrt{49} = 7$

99. $3^2 + 4^2 = c^2 \Rightarrow c^2 = 9 + 16 \Rightarrow c^2 = 25 \Rightarrow c = \sqrt{25} = 5$

101. $\left(\sqrt{3}\right)^2 + b^2 = 8^2 \Rightarrow 3 + b^2 = 64 \Rightarrow b^2 = 61 \Rightarrow b = \sqrt{61}$

103. $a^2 + 48^2 = 50^2 \Rightarrow a^2 + 2304 = 2500 \Rightarrow a^2 = 196 \Rightarrow a = \sqrt{196} = 14$

105. From $(-2, 1)$ to $(2, 3)$, $d = \sqrt{\left(2 - (-2)\right)^2 + \left(3 - 1\right)^2} = \sqrt{4^2 + 2^2} = \sqrt{16 + 4} = \sqrt{20} = 2\sqrt{5}.$

107. From $(10, 40)$ to $(30, -20)$,

$$d = \sqrt{\left(30 - 10\right)^2 + \left(-20 - 40\right)^2} = \sqrt{20^2 + \left(-60\right)^2} = \sqrt{400 + 3600} = \sqrt{4000} = 20\sqrt{10}.$$

109. $d = \sqrt{\left(4 - (-1)\right)^2 + \left(10 - 2\right)^2} = \sqrt{5^2 + 8^2} = \sqrt{25 + 64} = \sqrt{89}.$

111. $d = \sqrt{\left(4 - 0\right)^2 + \left(0 - (-3)\right)^2} = \sqrt{4^2 + 3^2} = \sqrt{16 + 9} = \sqrt{25} = 5.$

113. $\sqrt{\left(0 - x\right)^2 + \left(6 - 3\right)^2} = 5 \Rightarrow \sqrt{\left(-x\right)^2 + 3^2} = 5 \Rightarrow \sqrt{x^2 + 9} = 5 \Rightarrow \left(\sqrt{x^2 + 9}\right)^2 = 5^2 \Rightarrow$

$x^2 + 9 = 25 \Rightarrow x^2 = 16 \Rightarrow x = \sqrt{16} = \pm 4.$ Since $x$ is positive, $x = 4.$

115. $\sqrt{\left(62 - x\right)^2 + \left(6 - (-5)\right)^2} = 61 \Rightarrow \sqrt{\left(62 - x\right)^2 + 11^2} = 61 \Rightarrow \sqrt{\left(3844 - 124x + x^2\right) + 121} = 61 \Rightarrow$

$\sqrt{x^2 - 124x + 3965} = 61 \Rightarrow \left(\sqrt{x^2 - 124x + 3965}\right)^2 = 61^2 \Rightarrow x^2 - 124x + 3965 = 3721 \Rightarrow$

$x^2 - 124x + 244 = 0 \Rightarrow \left(x - 2\right)\left(x - 122\right) = 0 \Rightarrow x = 2 \text{ or } x = 122$

117. $400 = 100\sqrt[3]{W^2} \Rightarrow \dfrac{400}{100} = \sqrt[3]{W^2} \Rightarrow 4 = \sqrt[3]{W^2} \Rightarrow 4^3 = \left(\sqrt[3]{W^2}\right)^3 \Rightarrow 64 = W^2 \Rightarrow W = 8 \text{ lb}$

119. $50 = 7.3\sqrt[16]{x^7} \Rightarrow \dfrac{50}{7.3} = \sqrt[16]{x^7} \Rightarrow \dfrac{50}{7.3} = x^{7/16} \Rightarrow \left(\dfrac{50}{7.3}\right)^{16/7} = x \Rightarrow 81.3 \approx x,$ The result is about 81 sec.

121. $D(6) = 1.22\sqrt{6} \approx 2.988 \approx 3$ miles

123. $1.22\sqrt{h} = 20 \Rightarrow \sqrt{h} = \dfrac{20}{1.22} \Rightarrow \left(\sqrt{h}\right)^2 = \left(\dfrac{20}{1.22}\right)^2 \Rightarrow h \approx 268.745 \approx 269$ feet

125. $d^2 = 11.4^2 + 15.2^2 \Rightarrow d^2 = 129.96 + 231.04 \Rightarrow d^2 = 361 \Rightarrow d = \sqrt{361} = 19$ inches

127. The height can be found using proportions: $\dfrac{16}{9} = \dfrac{29}{x} \Rightarrow 16x = 261 \Rightarrow x = \dfrac{261}{16} \approx 16.3$ inches.

Then $d^2 = 29^2 + 16.3^2 \Rightarrow d^2 = 841 + 265.69 \Rightarrow d^2 = 1106.69 \Rightarrow d = \sqrt{1106.69} \approx 33.3$ inches.

129. (a) $\dfrac{60}{11}\sqrt{d} = 60 \Rightarrow \sqrt{d} = 60\left(\dfrac{11}{60}\right) \Rightarrow \sqrt{d} = 11 \Rightarrow \left(\sqrt{d}\right)^2 = 11^2 \Rightarrow d = 121$ feet

(b) $\dfrac{60}{11}\sqrt{d} = 100 \Rightarrow \sqrt{d} = 100\left(\dfrac{11}{60}\right) \Rightarrow \sqrt{d} = \dfrac{55}{3} \Rightarrow \left(\sqrt{d}\right)^2 = \left(\dfrac{55}{3}\right)^2 \Rightarrow d \approx 336$ feet

131. (a) $V = 30\sqrt{\dfrac{285}{178}} \approx 38$ mph. The accident vehicle was traveling about 38 mph.

(b) $45\sqrt{\dfrac{D}{255}} = 60 \Rightarrow \sqrt{\dfrac{D}{255}} = \dfrac{60}{45} \Rightarrow \left(\sqrt{\dfrac{D}{255}}\right)^2 = \left(\dfrac{4}{3}\right)^2 \Rightarrow \dfrac{D}{255} = \dfrac{16}{9} \Rightarrow D = 255\left(\dfrac{16}{9}\right) \approx 453$ feet

133. (a) $W(2v) = 3.8(2v)^3 = 3.8 \cdot 8 \cdot v^3 = 8 \cdot \left(3.8v^3\right);$ The wattage generated increases by a factor of 8.

(b) $W = 3.8v^3 \Rightarrow \dfrac{W}{3.8} = v^3 \Rightarrow \sqrt[3]{\dfrac{W}{3.8}} = \sqrt[3]{v^3} \Rightarrow v = \sqrt[3]{\dfrac{W}{3.8}}$

(c) $v = \sqrt[3]{\dfrac{30,400}{3.8}} = \sqrt[3]{8000} = 20$ mph

135. $c^2 = a^2 + a^2 \Rightarrow c^2 = 2a^2 \Rightarrow c = \sqrt{2a^2} = a\sqrt{2}$

## Checking Basic Concepts Sections 10.5 and 10.6

1. (a) See Figure 1a. $f(-1)$ is undefined

(b) See Figure 1b. $f(-1) = -1$

(c) See Figure 1c. $f(-1) = 1$

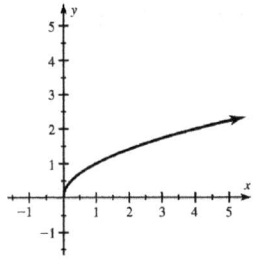 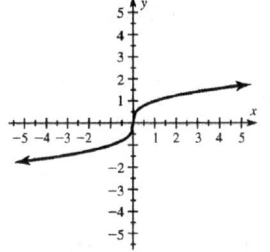 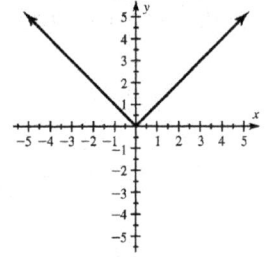

Figure 1a　　　　　　　　　Figure 1b　　　　　　　　　Figure 1c

2. $f(64) = 0.2(64)^{2/3} = 0.2\left(\sqrt[3]{64}\right)^2 = 0.2(4)^2 = 0.2 \cdot 16 = 3.2$

3. $x - 4 \geq 0 \Rightarrow x \geq 4 \Rightarrow$ Domain: $[4, \infty)$

4. (a) $\sqrt{2x-4} = 2 \Rightarrow \left(\sqrt{2x-4}\right)^2 = 2^2 \Rightarrow 2x - 4 = 4 \Rightarrow 2x = 8 \Rightarrow x = 4$

(b) $\sqrt[3]{x-1} = 3 \Rightarrow \left(\sqrt[3]{x-1}\right)^3 = 3^3 \Rightarrow x - 1 = 27 \Rightarrow x = 28$

(c) $\sqrt{3x} = 1 + \sqrt{x+1} \Rightarrow \left(\sqrt{3x}\right)^2 = \left(1 + \sqrt{x+1}\right)^2 \Rightarrow 3x = 1 + 2\sqrt{x+1} + x + 1 \Rightarrow$

$3x = x + 2 + 2\sqrt{x+1} \Rightarrow 2x - 2 = 2\sqrt{x+1} \Rightarrow x - 1 = \sqrt{x+1} \Rightarrow (x-1)^2 = \left(\sqrt{x+1}\right)^2 \Rightarrow$

$x^2 - 2x + 1 = x + 1 \Rightarrow x^2 - 3x = 0 \Rightarrow x(x-3) = 0 \Rightarrow x = 0$ or $x = 3$

The solution $x = 0$ does not check. The solution is $x = 3$.

5. $d = \sqrt{\left(2-(-3)\right)^2 + \left(-7-5\right)^2} = \sqrt{5^2 + \left(-12\right)^2} = \sqrt{25+144} = \sqrt{169} = 13$

6. $h^2 + 12.8^2 = 16^2 \Rightarrow h^2 + 163.84 = 256 \Rightarrow h^2 = 92.16 \Rightarrow h = \sqrt{92.16} = 9.6$ inches

7. $(x+1)^4 = 16 \Rightarrow \sqrt[4]{(x+1)^4} = \sqrt[4]{16} \Rightarrow |x+1| = 2 \Rightarrow x+1 = \pm 2 \Rightarrow x = -3$ or $x = 1$

## Section 10.7 Complex Numbers

1. $2 + 3i$;  *Answers may vary.*

3. $i$

5. $i\sqrt{a}$

7. $a + bi$

9. 4

11. 0

13. $\sqrt{-5} = i\sqrt{5}$

15. $\sqrt{-100} = i\sqrt{100} = i \cdot 10 = 10i$

17. $\sqrt{-144} = i\sqrt{144} = i \cdot 12 = 12i$

19. $\sqrt{-12} = i\sqrt{12} = i \cdot \sqrt{4 \cdot 3} = i \cdot \sqrt{4} \cdot \sqrt{3} = i \cdot 2 \cdot \sqrt{3} = 2i\sqrt{3}$

21. $\sqrt{-18} = i\sqrt{18} = i \cdot \sqrt{9 \cdot 2} = i \cdot \sqrt{9} \cdot \sqrt{2} = i \cdot 3 \cdot \sqrt{2} = 3i\sqrt{2}$

23. $(5+3i)+(-2-3i) = \left(5+(-2)\right)+\left(3+(-3)\right)i = 3+0i = 3$

25. $(2i)+(-8+5i) = \left(0+(-8)\right)+(2+5)i = -8+7i$

27. $(2-7i)-(1+2i) = (2-1)+(-7-2)i = 1-9i$

29. $(5i)-(10-2i) = (0-10)+\left(5-(-2)\right)i = -10+7i$

31. $(3+2i)(-1+5i) = -3+15i-2i+10i^2 = -3+13i+10(-1) = -3+13i-10 = -13+13i$

33. $4(5-3i) = 20-12i$

35. $(5+4i)(5-4i) = 25-16i^2 = 25-16(-1) = 25+16 = 41$

37. $(-4i)(5i) = -4 \cdot 5 \cdot i^2 = -20(-1) = 20$

39. $3i+(2-3i)-(1-5i) = 3i+2-3i-1+5i = 1+5i$

41. $(2+i)^2 = 4+4i+i^2 = 4+4i-1 = 3+4i$

43. $2i(-3+i) = -6i+2i^2 = -6i+2(-1) = -2-6i$

45. $i(1+i)^2 = i\left(1+2i+i^2\right) = i(1+2i-1) = i(2i) = 2i^2 = 2(-1) = -2$

47. $(a+3bi)(a-3bi) = a^2 - 9b^2 i^2 = a^2 - 9b^2(-1) = a^2 + 9b^2$

49. When 11 is divided by 4, the result is 2 with remainder 3. Thus $i^{11} = i^3 = -i$.

51. When 21 is divided by 4, the result is 5 with remainder 1. Thus $i^{21} = i^1 = i$.

53. When 58 is divided by 4, the result is 14 with remainder 2. Thus $i^{58} = i^2 = -1$.

55. When 64 is divided by 4, the result is 16 with remainder 0. Thus $i^{64} = i^0 = 1$.

57. $3 - 4i$

59. Since $-6i = 0 - 6i,$ the complex conjugate is $0 + 6i = 6i$.

61. $5 + 4i$

63. Since $-1 = -1 + 0i,$ the complex conjugate is $-1 - 0i = -1$.

65. $\dfrac{2}{1+i} = \dfrac{2}{1+i} \cdot \dfrac{1-i}{1-i} = \dfrac{2(1-i)}{1^2 - i^2} = \dfrac{2-2i}{1+1} = \dfrac{2-2i}{2} = \dfrac{2}{2} - \dfrac{2}{2}i = 1 - i$

67. $\dfrac{3i}{5-2i} = \dfrac{3i}{5-2i} \cdot \dfrac{5+2i}{5+2i} = \dfrac{3i(5+2i)}{5^2 - 4i^2} = \dfrac{15i + 6i^2}{25+4} = \dfrac{15i-6}{29} = -\dfrac{6}{29} + \dfrac{15}{29}i$

69. $\dfrac{8+9i}{5+2i} = \dfrac{8+9i}{5+2i} \cdot \dfrac{5-2i}{5-2i} = \dfrac{40 - 16i + 45i - 18i^2}{5^2 - 4i^2} = \dfrac{40 + 29i - 18(-1)}{25+4} = \dfrac{58 + 29i}{29} = 2 + i$

71. $\dfrac{5+7i}{1-i} = \dfrac{5+7i}{1-i} \cdot \dfrac{1+i}{1+i} = \dfrac{5 + 5i + 7i + 7i^2}{1^2 - i^2} = \dfrac{5 + 12i + 7(-1)}{1+1} = \dfrac{-2 + 12i}{2} = -1 + 6i$

73. $\dfrac{2-i}{i} = \dfrac{2-i}{i} \cdot \dfrac{-i}{-i} = \dfrac{-2i + i^2}{-i^2} = \dfrac{-2i + (-1)}{-(-1)} = \dfrac{-2i-1}{1} = -1 - 2i$

75. $\dfrac{1}{i} + \dfrac{1}{2i} = \dfrac{2}{2i} + \dfrac{1}{2i} = \dfrac{3}{2i} = \dfrac{3}{2i} \cdot \dfrac{-2i}{-2i} = \dfrac{-6i}{-4i^2} = \dfrac{-6i}{-4(-1)} = \dfrac{-6i}{4} = -\dfrac{3}{2}i$

77. $\dfrac{1}{-1+i} - \dfrac{2}{i} = \dfrac{i}{i(-1+i)} - \dfrac{2(-1+i)}{i(-1+i)} = \dfrac{i}{-1-i} - \dfrac{-2+2i}{-1-i} = \dfrac{i+2-2i}{-1-i} = \dfrac{2-i}{-1-i}$

$= \dfrac{2-i}{-1-i} \cdot \dfrac{-1+i}{-1+i} = \dfrac{-2 + 2i + i - i^2}{1 - i^2} = \dfrac{-2 + 3i - (-1)}{1 - (-1)} = \dfrac{-1 + 3i}{2} = -\dfrac{1}{2} + \dfrac{3}{2}i$

79. $Z = \dfrac{40 + 70i}{2 + 3i} = \dfrac{40 + 70i}{2 + 3i} \cdot \dfrac{2 - 3i}{2 - 3i} = \dfrac{80 - 120i + 140i - 210i^2}{2^2 - 9i^2} = \dfrac{290 + 20i}{13} = \dfrac{290}{13} + \dfrac{20}{13}i$

81. They are graphed using a real axis and an imaginary axis.

## Checking Basic Concepts Section 10.7

1. (a)  $\sqrt{-64} = i\sqrt{64} = i(8) = 8i$

    (b)  $\sqrt{-17} = i\sqrt{17}$

2. (a)  $(2-3i) + (1-i) = (2+1) + (-3 + (-1))i = 3 - 4i$

(b)  $4i - (2+i) = (0-2) + (4-1)i = -2 + 3i$

(c)  $(3-2i)(1+i) = 3 + 3i - 2i - 2i^2 = 3 + i - 2(-1) = 3 + i + 2 = 5 + i$

(d)  $\dfrac{3}{2-2i} = \dfrac{3}{2-2i} \cdot \dfrac{2+2i}{2+2i} = \dfrac{3(2+2i)}{2^2 - 4i^2} = \dfrac{6+6i}{4+4} = \dfrac{6+6i}{8} = \dfrac{6}{8} + \dfrac{6}{8}i = \dfrac{3}{4} + \dfrac{3}{4}i$

## **Chapter 10 Review**

1.  $\sqrt{4} = 2$

2.  $\sqrt{36} = 6$

3.  $\sqrt{9x^2} = \sqrt{9} \cdot \sqrt{x^2} = 3|x|$

4.  $\sqrt{(x-1)^2} = |x-1|$

5.  $\sqrt[3]{-64} = -4$

6.  $\sqrt[3]{-125} = -5$

7.  $\sqrt[3]{x^6} = \sqrt[3]{\left(x^2\right)^3} = x^2$

8.  $\sqrt[3]{27x^3} = \sqrt[3]{27} \cdot \sqrt[3]{x^3} = 3x$

9.  $\sqrt[4]{16} = 2$

10. $\sqrt[5]{-1} = -1$

11. $\sqrt[4]{x^8} = \sqrt[4]{\left(x^2\right)^4} = x^2$

12. $\sqrt[5]{(x+1)^5} = x+1$

13. $14^{1/2} = \sqrt{14}$

14. $(-5)^{1/3} = \sqrt[3]{-5}$

15. $\left(\dfrac{x}{y}\right)^{3/2} = \left(\sqrt{\dfrac{x}{y}}\right)^3$ or $\sqrt{\left(\dfrac{x}{y}\right)^3}$

16. $(xy)^{-2/3} = \dfrac{1}{(xy)^{2/3}} = \dfrac{1}{\sqrt[3]{(xy)^2}}$ or $\dfrac{1}{\left(\sqrt[3]{xy}\right)^2}$

17. $(-27)^{2/3} = \left(\sqrt[3]{-27}\right)^2 = (-3)^2 = 9$

18. $16^{1/4} = \sqrt[4]{16} = 2$

19. $16^{3/2} = \left(\sqrt{16}\right)^3 = 4^3 = 64$

20. $81^{3/4} = \left(\sqrt[4]{81}\right)^3 = 3^3 = 27$

21. $\left(z^3\right)^{2/3} = z^{3 \cdot 2/3} = z^2$

22. $\left(x^2 y^4\right)^{1/2} = x^{2 \cdot 1/2} \cdot y^{4 \cdot 1/2} = xy^2$

23. $\left(\dfrac{x^2}{y^6}\right)^{3/2} = \dfrac{x^{2 \cdot 3/2}}{y^{6 \cdot 3/2}} = \dfrac{x^3}{y^9}$

24. $\left(\dfrac{x^3}{y^6}\right)^{-1/3} = \dfrac{x^{3 \cdot (-1/3)}}{y^{6 \cdot (-1/3)}} = \dfrac{x^{-1}}{y^{-2}} = \dfrac{y^2}{x}$

25. $\sqrt{2} \cdot \sqrt{32} = \sqrt{64} = 8$

26. $\sqrt[3]{-4} \cdot \sqrt[3]{2} = \sqrt[3]{-8} = -2$

27. $\sqrt[3]{x^4} \cdot \sqrt[3]{x^2} = \sqrt[3]{x^6} = \sqrt[3]{\left(x^2\right)^3} = x^2$

28. $\dfrac{\sqrt{80}}{\sqrt{20}} = \sqrt{\dfrac{80}{20}} = \sqrt{4} = 2$

29. $\sqrt[3]{-\dfrac{x}{8}} = -\dfrac{\sqrt[3]{x}}{\sqrt[3]{8}} = -\dfrac{\sqrt[3]{x}}{2}$

30. $\sqrt{\dfrac{1}{3}} \cdot \sqrt{\dfrac{1}{3}} = \left(\sqrt{\dfrac{1}{3}}\right)^2 = \dfrac{1}{3}$

31. $\sqrt{48} = \sqrt{16 \cdot 3} = \sqrt{16} \cdot \sqrt{3} = 4\sqrt{3}$

32. $\sqrt{54} = \sqrt{9 \cdot 6} = \sqrt{9} \cdot \sqrt{6} = 3\sqrt{6}$

33. $\sqrt[3]{\dfrac{3}{x}} \cdot \sqrt[3]{\dfrac{9}{x^2}} = \sqrt[3]{\dfrac{3 \cdot 9}{x \cdot x^2}} = \sqrt[3]{\dfrac{27}{x^3}} = \dfrac{\sqrt[3]{27}}{\sqrt[3]{x^3}} = \dfrac{3}{x}$

34. $\sqrt{32a^3 b^2} = \sqrt{(4ab)^2 \cdot 2a} = \sqrt{(4ab)^2} \cdot \sqrt{2a} = 4ab\sqrt{2a}$

35. $\sqrt{3xy} \cdot \sqrt{27xy} = \sqrt{3 \cdot 27 \cdot xy \cdot xy} = \sqrt{81(xy)^2} = \sqrt{(9xy)^2} = 9xy$

36. $\sqrt[3]{-25z^2} \cdot \sqrt[3]{-5z^2} = \sqrt[3]{-25 \cdot (-5) \cdot z^2 \cdot z^2} = \sqrt[3]{125z^4} = \sqrt[3]{(5z)^3 \cdot z} = \sqrt{(5z)^3} \cdot \sqrt[3]{z} = 5z\sqrt[3]{z}$

37. $\sqrt{x^2 + 2x + 1} = \sqrt{(x+1)^2} = x+1$

38. $\sqrt[4]{\dfrac{2a^2}{b}} \cdot \sqrt[4]{\dfrac{8a^3}{b^3}} = \sqrt[4]{\dfrac{16a^5}{b^4}} = \sqrt[4]{\left(\dfrac{2a}{b}\right)^4 \cdot a} = \sqrt[4]{\left(\dfrac{2a}{b}\right)^4} \cdot \sqrt[4]{a} = \dfrac{2a\sqrt[4]{a}}{b}$

39. $2\sqrt{x} \cdot \sqrt[3]{x} = 2x^{1/2} \cdot x^{1/3} = 2x^{1/2+1/3} = 2x^{5/6} = 2\sqrt[6]{x^5}$

40. $\sqrt[3]{rt} \cdot \sqrt[4]{r^2 t^4} = (rt)^{1/3} \cdot \left(r^2 t^4\right)^{1/4} = r^{1/3} t^{1/3} \cdot r^{1/2} t = r^{1/3+1/2} t^{1/3+1} = r^{5/6} t^{4/3} = r^{5/6} t^{8/6} = \sqrt[6]{r^5 t^8}$ or $t\sqrt[6]{r^5 t^2}$

41. $3\sqrt{3} + \sqrt{3} = 4\sqrt{3}$

42. $\sqrt[3]{x} + 2\sqrt[3]{x} = 3\sqrt[3]{x}$

43. $3\sqrt[3]{5} - 6\sqrt[3]{5} = -3\sqrt[3]{5}$

44. $\sqrt[4]{y} - 2\sqrt[4]{y} = -\sqrt[4]{y}$

45. $2\sqrt{12} + 7\sqrt{3} = 2\sqrt{4 \cdot 3} + 7\sqrt{3} = 2\sqrt{4} \cdot \sqrt{3} + 7\sqrt{3} = 4\sqrt{3} + 7\sqrt{3} = 11\sqrt{3}$

46. $3\sqrt{18} - 2\sqrt{2} = 3\sqrt{9 \cdot 2} - 2\sqrt{2} = 3\sqrt{9} \cdot \sqrt{2} - 2\sqrt{2} = 9\sqrt{2} - 2\sqrt{2} = 7\sqrt{2}$

47. $7\sqrt[3]{16} - \sqrt[3]{2} = 7\sqrt[3]{8 \cdot 2} - \sqrt[3]{2} = 7\sqrt[3]{8} \cdot \sqrt[3]{2} - \sqrt[3]{2} = 14\sqrt[3]{2} - \sqrt[3]{2} = 13\sqrt[3]{2}$

48. $\sqrt{4x+4} + \sqrt{x+1} = \sqrt{4(x+1)} + \sqrt{x+1} = \sqrt{4} \cdot \sqrt{x+1} + \sqrt{x+1} = 3\sqrt{x+1}$

49. $\sqrt{4x^3} - \sqrt{x} = \sqrt{(2x)^2 \cdot x} - \sqrt{x} = \sqrt{(2x)^2} \cdot \sqrt{x} - \sqrt{x} = 2x\sqrt{x} - \sqrt{x} = (2x-1)\sqrt{x}$

50. $\sqrt[3]{ab^4} + 2\sqrt[3]{a^4 b} = \sqrt[3]{b^3 \cdot ab} + 2\sqrt[3]{a^3 \cdot ab} = \sqrt[3]{b^3} \cdot \sqrt[3]{ab} + 2\sqrt[3]{a^3} \cdot \sqrt[3]{ab} = (b+2a)\sqrt[3]{ab}$

51. $\left(1+\sqrt{2}\right)\left(3+\sqrt{2}\right) = 3 + \sqrt{2} + 3\sqrt{2} + \left(\sqrt{2}\right)^2 = 3 + 4\sqrt{2} + 2 = 5 + 4\sqrt{2}$

52. $\left(7-\sqrt{5}\right)\left(1+\sqrt{3}\right) = 7 + 7\sqrt{3} - \sqrt{5} - \sqrt{15}$

53. $\left(3+\sqrt{6}\right)\left(3-\sqrt{6}\right) = 3^2 - \left(\sqrt{6}\right)^2 = 9 - 6 = 3$

54. $\left(10-\sqrt{5}\right)\left(10+\sqrt{5}\right) = 10^2 - \left(\sqrt{5}\right)^2 = 100 - 5 = 95$

55. $\left(\sqrt{a}+\sqrt{2b}\right)\left(\sqrt{a}-\sqrt{2b}\right) = \left(\sqrt{a}\right)^2 - \left(\sqrt{2b}\right)^2 = a - 2b$

56. $\left(\sqrt{xy}-1\right)\left(\sqrt{xy}+2\right) = \left(\sqrt{xy}\right)^2 + 2\sqrt{xy} - \sqrt{xy} - 2 = xy + \sqrt{xy} - 2$

57. $\dfrac{4}{\sqrt{5}} = \dfrac{4}{\sqrt{5}} \cdot \dfrac{\sqrt{5}}{\sqrt{5}} = \dfrac{4\sqrt{5}}{5}$

58. $\dfrac{r}{2\sqrt{t}} = \dfrac{r}{2\sqrt{t}} \cdot \dfrac{\sqrt{t}}{\sqrt{t}} = \dfrac{r\sqrt{t}}{2t}$

59. $\dfrac{1}{\sqrt{2}+3} = \dfrac{1}{\sqrt{2}+3} \cdot \dfrac{\sqrt{2}-3}{\sqrt{2}-3} = \dfrac{\sqrt{2}-3}{2-9} = \dfrac{\sqrt{2}-3}{-7} = \dfrac{3-\sqrt{2}}{7}$

60. $\dfrac{2}{5-\sqrt{7}} = \dfrac{2}{5-\sqrt{7}} \cdot \dfrac{5+\sqrt{7}}{5+\sqrt{7}} = \dfrac{10+2\sqrt{7}}{25-7} = \dfrac{10+2\sqrt{7}}{18} = \dfrac{5+\sqrt{7}}{9}$

61. $\dfrac{1}{\sqrt{8}-\sqrt{7}} = \dfrac{1}{\sqrt{8}-\sqrt{7}} \cdot \dfrac{\sqrt{8}+\sqrt{7}}{\sqrt{8}+\sqrt{7}} = \dfrac{\sqrt{8}+\sqrt{7}}{8-7} = \dfrac{\sqrt{8}+\sqrt{7}}{1} = \sqrt{8}+\sqrt{7}$

62. $\dfrac{\sqrt{a}-\sqrt{b}}{\sqrt{a}+\sqrt{b}} = \dfrac{\sqrt{a}-\sqrt{b}}{\sqrt{a}+\sqrt{b}} \cdot \dfrac{\sqrt{a}-\sqrt{b}}{\sqrt{a}-\sqrt{b}} = \dfrac{a-2\sqrt{ab}+b}{a-b}$

63. See Figure 63.

64. See Figure 64.

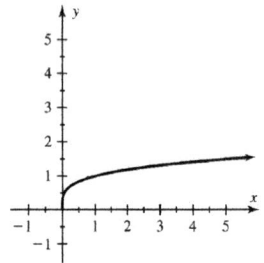    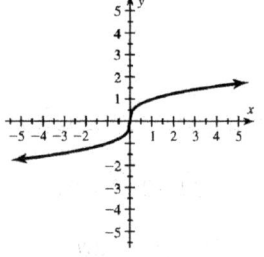    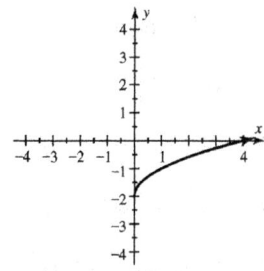

Figure 63                    Figure 64                    Figure 67

65. $f(x) = x^{1/2} = \sqrt{x}; f(4) = 4^{1/2} = \sqrt{4} = 2$

66. $f(x) = x^{2/7} = \sqrt[7]{x^2}; f(4) = 4^{2/7} = \sqrt[7]{4^2} = \sqrt[7]{16}$

67. See Figure 67. This graph is shifted 2 units downward.

68. See Figure 68. This graph is shifted 1 unit to the right.

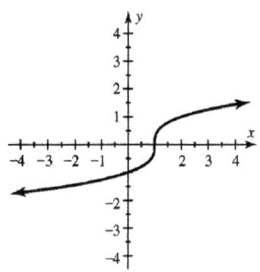

Figure 68

69. $x - 1 \geq 0 \Rightarrow x \geq 1 \Rightarrow$ Domain : $[1, \infty)$

70. $6 - 2x \geq 0 \Rightarrow -2x \geq -6 \Rightarrow x \leq 3 \Rightarrow$ Domain : $(-\infty, 3]$

71. $x^2 + 1 \geq 0 \Rightarrow x^2 \geq -1 \Rightarrow$ Domain : $(-\infty, \infty)$

72. $x + 2 > 0 \Rightarrow x > -2 \Rightarrow$ Domain : $(-2, \infty)$

73. $\sqrt{x+2} = x \Rightarrow \left(\sqrt{x+2}\right)^2 = x^2 \Rightarrow x+2 = x^2 \Rightarrow x^2 - x - 2 = 0 \Rightarrow (x-2)(x+1) = 0 \Rightarrow x = 2$ or $x = -1$.

The solution $x = -1$ does not check. The solution is $x = 2$.

74. $\sqrt{2x-1} = \sqrt{x+3} \Rightarrow \left(\sqrt{2x-1}\right)^2 = \left(\sqrt{x+3}\right)^2 \Rightarrow 2x-1 = x+3 \Rightarrow x = 4$

75. $\sqrt[3]{x-1} = 2 \Rightarrow \left(\sqrt[3]{x-1}\right)^3 = 2^3 \Rightarrow x-1 = 8 \Rightarrow x = 9$

76. $\sqrt[3]{3x} = 3 \Rightarrow \left(\sqrt[3]{3x}\right)^3 = 3^3 \Rightarrow 3x = 27 \Rightarrow x = 9$

77. $\sqrt{2x} = x-4 \Rightarrow \left(\sqrt{2x}\right)^2 = (x-4)^2 \Rightarrow 2x = x^2 - 8x + 16 \Rightarrow x^2 - 10x + 16 = 0 \Rightarrow$

$(x-2)(x-8) = 0 \Rightarrow x = 2$ or $x = 8$. The solution $x = 2$ does not check. The solution is $x = 8$.

78. $\sqrt{x} + 1 = \sqrt{x+2} \Rightarrow \left(\sqrt{x}+1\right)^2 = \left(\sqrt{x+2}\right)^2 \Rightarrow x + 2\sqrt{x} + 1 = x + 2 \Rightarrow 2\sqrt{x} = 1 \Rightarrow$

$\left(2\sqrt{x}\right)^2 = 1^2 \Rightarrow 4x = 1 \Rightarrow x = \dfrac{1}{4}$

79. Graph $Y_1 = (2X-1)^\wedge(1/3)$ and $Y_2 = 2$ in $[-4, 6, 1]$ by $[-3, 3, 1]$. See Figure 79.

The solution is $x = 4.5$.

[–4, 6, 1] by [–3, 3, 1]          [–5, 5, 1] by [–5, 5, 1]

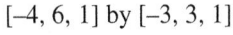

Figure 79                              Figure 80

80. Graph $Y_1 = X^\wedge(2/3)$ and $Y_2 = 3 - X$ in $[-5, 5, 1]$ by $[-5, 5, 1]$. See Figure 80.

The solution is $x \approx 1.62$.

81. $c^2 = 4^2 + 7^2 \Rightarrow c^2 = 16 + 49 \Rightarrow c^2 = 65 \Rightarrow c = \sqrt{65}$

82. $5^2 + b^2 = 8^2 \Rightarrow 25 + b^2 = 64 \Rightarrow b^2 = 39 \Rightarrow b = \sqrt{39}$

83. $d = \sqrt{\left(2-(-2)\right)^2 + \left(-2-3\right)^2} = \sqrt{4^2 + (-5)^2} = \sqrt{16+25} = \sqrt{41}$

84. $d = \sqrt{\left(-4-2\right)^2 + \left(1-(-3)\right)^2} = \sqrt{\left(-6\right)^2 + 4^2} = \sqrt{36+16} = \sqrt{52} = 2\sqrt{13}$

85. $x^2 = 121 \Rightarrow x = \pm\sqrt{121} \Rightarrow x = \pm 11$

86. $2z^2 = 32 \Rightarrow z^2 = 16 \Rightarrow z = \pm\sqrt{16} \Rightarrow z = \pm 4$

87. $(x-1)^2 = 16 \Rightarrow x - 1 = \pm\sqrt{16} \Rightarrow x - 1 = \pm 4 \Rightarrow x = 1 \pm 4 \Rightarrow x = -3$ or $x = 5$

88. $x^3 = 64 \Rightarrow x = \sqrt[3]{64} \Rightarrow x = 4$

89. $(x-1)^3 = 8 \Rightarrow x - 1 = \sqrt[3]{8} \Rightarrow x - 1 = 2 \Rightarrow x = 3$

90. $(2x-1)^3 = 27 \Rightarrow 2x - 1 = \sqrt[3]{27} \Rightarrow 2x - 1 = 3 \Rightarrow 2x = 4 \Rightarrow x = 2$

91. $x^4 = 256 \Rightarrow \sqrt[4]{x^4} = \sqrt[4]{256} \Rightarrow |x| = 4 \Rightarrow x = \pm 4$

92. $x^5 = -1 \Rightarrow \sqrt[5]{x^5} = \sqrt[5]{-1} \Rightarrow x = -1$

93. $(x-3)^5 = -32 \Rightarrow \sqrt[5]{(x-3)^5} = \sqrt[5]{-32} \Rightarrow x - 3 = -2 \Rightarrow x = 1$

94. $3(x+1)^4 = 3 \Rightarrow (x+1)^4 = 1 \Rightarrow \sqrt[4]{(x+1)^4} = \sqrt[4]{1} \Rightarrow |x+1| = 1 \Rightarrow x + 1 = \pm 1 \Rightarrow x = -2$ or $x = 0$

95. $(1-2i) + (-3+2i) = (1+(-3)) + (-2+2)i = -2 + 0i = -2$

96. $(1+3i)-(3-i)=(1-3)+(3-(-1))i=-2+4i$

97. $(1-i)(2+3i)=2+3i-2i-3i^2=2+i-3(-1)=2+i+3=5+i$

98. $\dfrac{3+i}{1-i}=\dfrac{3+i}{1-i}\cdot\dfrac{1+i}{1+i}=\dfrac{3+3i+i+i^2}{1^2-i^2}=\dfrac{3+4i-1}{1+1}=\dfrac{2+4i}{2}=\dfrac{2}{2}+\dfrac{4}{2}i=1+2i$

99. $\dfrac{i(4+i)}{2-3i}=\dfrac{4i+i^2}{2-3i}=\dfrac{-1+4i}{2-3i}=\dfrac{-1+4i}{2-3i}\cdot\dfrac{2+3i}{2+3i}=\dfrac{-2-3i+8i+12i^2}{4-9i^2}=\dfrac{-2+5i+12(-1)}{4-9(-1)}$

   $=\dfrac{-14+5i}{13}=-\dfrac{14}{13}+\dfrac{5}{13}i$

100. $(1-i)^2(1+i)=(1-2i+i^2)(1+i)=(1-2i-1)(1+i)=-2i(1+i)=-2i-2i^2=2-2i$

101. $\dfrac{\sqrt{h}}{2}=4.6\Rightarrow\sqrt{h}=9.2\Rightarrow\left(\sqrt{h}\right)^2=9.2^2\Rightarrow h=84.64\approx85$ feet

102. $d^2=90^2+90^2\Rightarrow d^2=8100+8100\Rightarrow d^2=16,200\Rightarrow d=\sqrt{16,200}\approx127.3$ feet

103. $T=\dfrac{1}{4}\sqrt{10}\approx0.79$ seconds

104. (a)   $A=s^2=\left(\sqrt{5}\right)^2=5$ square units

     (b)   $V=s^3=\left(\sqrt{5}\right)^3=5\sqrt{5}$ cubic units

     (c)   $d^2=\left(\sqrt{5}\right)^2+\left(\sqrt{5}\right)^2\Rightarrow d^2=5+5\Rightarrow d^2=10\Rightarrow d=\sqrt{10}$ units

     (d)   $d^2=\left(\sqrt{5}\right)^2+\left(\sqrt{10}\right)^2\Rightarrow d^2=5+10\Rightarrow d^2=15\Rightarrow d=\sqrt{15}$ units

105. $2\pi\sqrt{\dfrac{L}{32.2}}=1\Rightarrow\sqrt{\dfrac{L}{32.2}}=\dfrac{1}{2\pi}\Rightarrow\left(\sqrt{\dfrac{L}{32.2}}\right)^2=\left(\dfrac{1}{2\pi}\right)^2\Rightarrow\dfrac{L}{32.2}=\dfrac{1}{4\pi^2}\Rightarrow L=\dfrac{32.2}{4\pi^2}\approx0.82$ feet

106. $2\pi\sqrt{\dfrac{L}{5.1}}=1\Rightarrow\sqrt{\dfrac{L}{5.1}}=\dfrac{1}{2\pi}\Rightarrow\left(\sqrt{\dfrac{L}{5.1}}\right)^2=\left(\dfrac{1}{2\pi}\right)^2\Rightarrow\dfrac{L}{5.1}=\dfrac{1}{4\pi^2}\Rightarrow L=\dfrac{5.1}{4\pi^2}\approx0.13$ feet.  It is shorter.

107. $4(1+r)^{210}=281\Rightarrow(1+r)^{210}=\dfrac{281}{4}\Rightarrow\left((1+r)^{210}\right)^{1/210}=\left(\dfrac{281}{4}\right)^{1/210}\Rightarrow1+r=\left(\dfrac{281}{4}\right)^{1/210}\Rightarrow$

   $r=\left(\dfrac{281}{4}\right)^{1/210}-1\approx0.02.$  From 1790 to 2000 the annual percentage growth rate was about 2%.

108. (a)   $L=\sqrt{3.75(500)}=\sqrt{1875}\approx43$ mph

     (b)   $L=1.5\sqrt{500}\approx34$ mph.  A steeper bank allows for a higher speed limit.

          This agrees with intuition.

109. $x^2=7\Rightarrow x=\sqrt{7}\approx2.65$ feet

## Chapter 10 Test

1. $\sqrt[3]{-27} = \sqrt[3]{(-3)^3} = -3$

2. $\sqrt{(z+1)^2} = |z+1|$

3. $\sqrt{25x^4} = \sqrt{25} \cdot \sqrt{(x^2)^2} = 5x^2$

4. $\sqrt[3]{8z^6} = \sqrt[3]{8} \cdot \sqrt[3]{(z^2)^3} = 2z^2$

5. $\sqrt[4]{16x^4y^5} = \sqrt[4]{(2xy)^4 \cdot y} = \sqrt[4]{(2xy)^4} \cdot \sqrt[4]{y} = 2xy\sqrt[4]{y}$

6. $\left(\sqrt{3} - \sqrt{2}\right)\left(\sqrt{3} + \sqrt{2}\right) = \left(\sqrt{3}\right)^2 - \left(\sqrt{2}\right)^2 = 3 - 2 = 1$

7. $7^{2/5} = \sqrt[5]{7^2}$ or $\left(\sqrt[5]{7}\right)^2$

8. $\left(\dfrac{x}{y}\right)^{-2/3} = \left(\dfrac{y}{x}\right)^{2/3} = \sqrt[3]{\left(\dfrac{y}{x}\right)^2}$ or $\left(\sqrt[3]{\dfrac{y}{x}}\right)^2$

9. $(-8)^{4/3} = \left(\sqrt[3]{-8}\right)^4 = (-2)^4 = 16$

10. $36^{-3/2} = \dfrac{1}{36^{3/2}} = \dfrac{1}{\left(\sqrt{36}\right)^3} = \dfrac{1}{6^3} = \dfrac{1}{216}$

11. $\sqrt[3]{x^4} = x^{4/3}$

12. $\sqrt{x} \cdot \sqrt[5]{x} = x^{1/2} \cdot x^{1/5} = x^{(1/2+1/5)} = x^{7/10}$

13. $4 - x \geq 0 \Rightarrow -x \geq -4 \Rightarrow x \leq 4 \Rightarrow \text{Domain}: (-\infty, 4]$

14. See Figure 14.

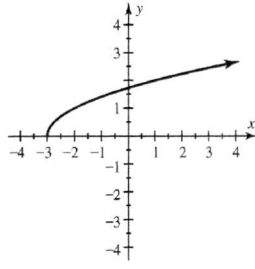

Figure 14

15. $\left(2z^{1/2}\right)^3 = 2^3 \cdot z^{1/2 \cdot 3} = 8z^{3/2}$

16. $\left(\dfrac{y^2}{z^3}\right)^{-1/3} = \left(\dfrac{z^3}{y^2}\right)^{1/3} = \dfrac{z^{3 \cdot 1/3}}{y^{2 \cdot 1/3}} = \dfrac{z}{y^{2/3}}$

17. $\sqrt{3} \cdot \sqrt{27} = \sqrt{81} = 9$

18. $\dfrac{\sqrt{y^3}}{\sqrt{4y}} = \sqrt{\dfrac{y^3}{4y}} = \sqrt{\dfrac{y^2}{4}} = \dfrac{\sqrt{y^2}}{\sqrt{4}} = \dfrac{y}{2}$

19. $7\sqrt{7} - 3\sqrt{7} + \sqrt{5} = 4\sqrt{7} + \sqrt{5}$

20. $7\sqrt[3]{x} - \sqrt[3]{x} = 6\sqrt[3]{x}$

21. $4\sqrt{18} + \sqrt{8} = 4(3)\sqrt{2} + 2\sqrt{2} = (12+2)\sqrt{2} = 14\sqrt{2}$

22. $\dfrac{\sqrt[3]{32}}{\sqrt[3]{4}} = \dfrac{\sqrt[3]{8 \cdot 4}}{\sqrt[3]{4}} = \dfrac{2\sqrt[3]{4}}{\sqrt[3]{4}} = 2$

23. (a) $\sqrt{x-2} = 5 \Rightarrow \left(\sqrt{x-2}\right)^2 = 5^2 \Rightarrow x - 2 = 25 \Rightarrow x = 27$

    (b) $\sqrt[3]{x+1} = 2 \Rightarrow \left(\sqrt[3]{x+1}\right)^3 = 2^3 \Rightarrow x + 1 = 8 \Rightarrow x = 7$

    (c) $(x-1)^3 = 8 \Rightarrow \sqrt[3]{(x-1)^3} = \sqrt[3]{8} \Rightarrow x - 1 = 2 \Rightarrow x = 3$

    (d) $\sqrt{2x+2} = x - 11 \Rightarrow \left(\sqrt{2x+2}\right)^2 = (x-11)^2 \Rightarrow 2x + 2 = x^2 - 22x + 121 \Rightarrow$

    $x^2 - 24x + 119 = 0 \Rightarrow (x-7)(x-17) = 0 \Rightarrow x = 7 \text{ or } x = 17.$

    The value $x = 7$ does not check. $x = 17$

24. (a) $\dfrac{2}{3\sqrt{7}} = \dfrac{2}{3\sqrt{7}} \cdot \dfrac{\sqrt{7}}{\sqrt{7}} = \dfrac{2\sqrt{7}}{21}$

    (b) $\dfrac{1}{1+\sqrt{5}} = \dfrac{1}{1+\sqrt{5}} \cdot \dfrac{1-\sqrt{5}}{1-\sqrt{5}} = \dfrac{1-\sqrt{5}}{1-5} = \dfrac{1-\sqrt{5}}{-4} = \dfrac{-1+\sqrt{5}}{4}$

25. Graph $Y_1 = \sqrt{3X} - X + 1$ and $Y_2 = (X-1)^\wedge(1/3)$. See Figure 25.

    The solution is approximately $x = 2.63$.

    $[-5,\ 5,\ 1]$ by $[-5,\ 5,\ 1]$

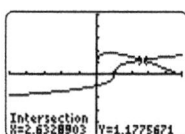

    Figure 25

26. $7^2 + b^2 = 13^2 \Rightarrow 49 + b^2 = 169 \Rightarrow b^2 = 120 \Rightarrow b = \sqrt{120} \approx 10.95$

27. $d = \sqrt{\left(-1-(-3)\right)^2 + (7-5)^2} = \sqrt{2^2 + 2^2} = \sqrt{4+4} = \sqrt{8} = 2\sqrt{2}$

28. $(-5+i) + (7-20i) = (-5+7) + (1+(-20))i = 2 - 19i$

29. $(3i) - (6-5i) = (0-6) + (3-(-5))i = -6 + 8i$

30. $\left(\dfrac{1}{2} - i\right)\left(\dfrac{1}{2} + i\right) = \dfrac{1}{4} + \dfrac{1}{2}i - \dfrac{1}{2}i - i^2 = \dfrac{1}{4} - (-1) = \dfrac{5}{4}$

31. $\dfrac{2i}{5+2i} = \dfrac{2i}{5+2i} \cdot \dfrac{5-2i}{5-2i} = \dfrac{2i(5-2i)}{5^2 - 4i^2} = \dfrac{10i - 4i^2}{25+4} = \dfrac{4+10i}{29} = \dfrac{4}{29} + \dfrac{10}{29}i$

32. (a)  $d = 1.22\sqrt{200} \approx 17.25$. The result is about 17.25 miles.

    (b)  $25 = 1.22\sqrt{x} \Rightarrow \dfrac{25}{1.22} = \sqrt{x} \Rightarrow \left(\dfrac{25}{1.22}\right)^2 = x \Rightarrow 419.9 \approx x$. The result is about 420 feet.

33.  $27.4W^{1/3} = 30 \Rightarrow W^{1/3} = \dfrac{30}{27.4} \Rightarrow \left(W^{1/3}\right)^3 = \left(\dfrac{30}{27.4}\right)^3 \Rightarrow W = \left(\dfrac{30}{27.4}\right)^3 \approx 1.31$ lb

## Chapter 10  Extended and Discovery Exercises

1. (a)  $k(47)^{1.12}(11)^{1.98} = 11.4 \Rightarrow k = \dfrac{11.4}{(47)^{1.12}(11)^{1.98}} \approx 0.001325$

    (b)  $V = 0.001325(105)^{1.12}(20)^{1.98} \approx 91.6$ ft$^3$

2. (a)  $S = 15.7(154)^{0.425}(65)^{0.725} \approx 2753.963261 \approx 2754$ in$^2$

    (b)  It increases by a factor of $2^{0.425} \approx 1.34$.

    (c)  It increases by a factor of $2^{0.725} \approx 1.65$.

3. (a)  Since the segment $AB$ is on land, the expression is $30x$.

    (b)  The legs of right triangle $BCD$ have lengths $CB = 1000 - x$ and $CD = 500$. Let $d = BD$.

    $d^2 = (1000 - x)^2 + 500^2 \Rightarrow d = \sqrt{(1000 - x)^2 + 500^2}$.

    (c)  Since the segment $BD$ is underwater, the expression is $50\sqrt{(1000 - x)^2 + 500^2}$.

    (d)  The expression is $30x + 50\sqrt{(1000 - x)^2 + 500^2}$.

    (e)  Graph $Y_1 = 30X + 50\sqrt{\left((1000 - X)^2 + 500^2\right)}$ in [0, 1000, 100] by [40,000, 60,000, 5000].

    See Figure 3.  The minimum cost is \$50,000 when $x = 625$ feet.

    [0, 1000, 100] by [40,000, 60,000, 5000]

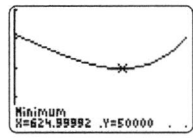

    Figure 3

## Chapters 1-10  Cumulative Review Exercises

1. $S = 4\pi(3)^2 = 4\pi(9) = 36\pi$

2. $D = \{-1, 0, 1\}; \ R = \{2, 4\}$

3. (a) $\left(\dfrac{ab^2}{b^{-1}}\right)^{-3} = \dfrac{a^{-3}b^{-6}}{b^3} = a^{-3}b^{-6-3} = a^{-3}b^{-9} = \dfrac{1}{a^3b^9}$

   (b) $\dfrac{\left(x^2 y\right)^3}{x^2\left(y^2\right)^{-3}} = \dfrac{x^6 y^3}{x^2 y^{-6}} = x^{6-2}y^{3-(-6)} = x^4 y^9$

   (c) $\left(rt\right)^2 \left(r^2 t\right)^3 = r^2 t^2 r^6 t^3 = r^{2+6} t^{2+3} = r^8 t^5$

4. $0.00043 = 4.3 \times 10^{-4}$

5. $f(3) = \dfrac{3}{3-2} = \dfrac{3}{1} = 3$; the denominator cannot equal 0, so $x \neq 2$.

6. All real numbers

7. See Figure 7.

8. See Figure 8.

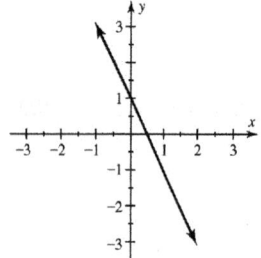

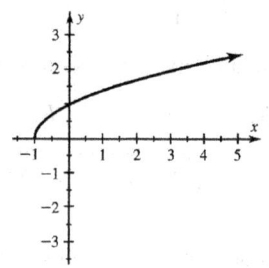

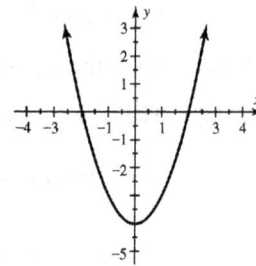

Figure 7                          Figure 8                          Figure 9

9. See Figure 9.

   (a) $D$: All real numbers; $R: y \geq -4$

   (b) $f(-2) = (-2)^2 - 4 = 4 - 4 = 0$

   (c) The $x$-intercepts are $-2$ and $2$.

   (d) $x^2 - 4 = 0 \Rightarrow (x-2)(x+2) = 0 \Rightarrow x = 2$ or $x = -2$

10. The line perpendicular to $y = -2x$ has slope $m = \dfrac{1}{2}$.

    $y - 2 = \dfrac{1}{2}\left(x - (-1)\right) \Rightarrow y = \dfrac{1}{2}x + \dfrac{1}{2} + 2 \Rightarrow y = \dfrac{1}{2}x + \dfrac{5}{2}$

11. $m = \dfrac{-5-7}{2-(-2)} = \dfrac{-12}{4} = -3; \ y - (-5) = -3(x-2) \Rightarrow y = -3x + 6 - 5 \Rightarrow y = -3x + 1; \ f(x) = 1 - 3x$

12. See Figure 12.

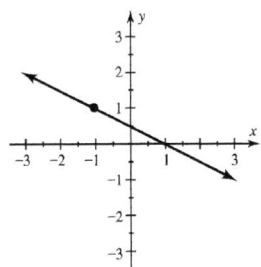

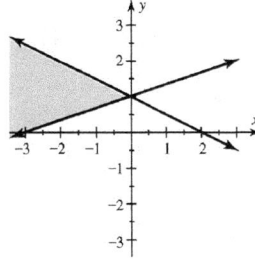

Figure 12                      Figure 18

13. $5x-(3-x)=\dfrac{1}{2}x \Rightarrow 5x-3+x-\dfrac{1}{2}x=0 \Rightarrow \dfrac{11}{2}x-3=0 \Rightarrow \dfrac{11}{2}x=3 \Rightarrow x=\dfrac{6}{11}$

14. $2x-5\le 4-x \Rightarrow 3x\le 9 \Rightarrow x\le 3; (-\infty, 3]$

15. $|x-2|\le 3 \Rightarrow x-2\le 3$ and $x-2\ge -3 \Rightarrow x\le 5$ and $x\ge -1;\ [-1, 5]$

16. $-1\le 1-2x\le 6 \Rightarrow -2\le -2x\le 5 \Rightarrow 1\ge x\ge -\dfrac{5}{2} \Rightarrow -\dfrac{5}{2}\le x\le 1;\ \left[-\dfrac{5}{2}, 1\right]$

17. (a)   Add the equations.   
$$\begin{array}{r} 2x-y=4 \\ x+y=8 \\ \hline 3x\quad =12 \end{array} \Rightarrow x=4 \qquad x+y=8 \Rightarrow 4+y=8 \Rightarrow y=4.$$

The solution is $(4,\ 4)$.

(b)   Multiply the second equation by $-3$ and add to the first equation.   
$$\begin{array}{r} 3x-4y=2 \\ -3x+4y=-3 \\ \hline 0=-1 \end{array}$$

The result is a contradiction; there are no solutions

18. Note that $x+2y\le 2 \Rightarrow y\le -\dfrac{1}{2}x+1$ and $-x+3y\ge 3 \Rightarrow y\ge \dfrac{1}{3}x+1$. See Figure 18.

19. Add the first two equations to eliminate the variable $z$.   
$$\begin{array}{r} x+2y-z=6 \\ x-3y+z=-2 \\ \hline 2x-y\quad =4 \end{array}$$
Add the first and third

equations to eliminate the variable $z$.   
$$\begin{array}{r} x+2y-z=6 \\ x+y+z=6 \\ \hline 2x+3y\quad =12 \end{array}$$
Multiply this new equation by $-1$ and add it

to the result of the first sum.   
$$\begin{array}{r} -2x-3y=-12 \\ 2x-y=\ \ 4 \\ \hline -4y=-8 \end{array} \Rightarrow y=2;\ 2x-2=4 \Rightarrow 2x=6 \Rightarrow x=3$$

Substitute $x=3$ and $y=2$ into the third equation to solve for $z$:

$3+2+z=6 \Rightarrow 5+z=6 \Rightarrow z=1.$ The solution is $(3,\ 2,\ 1)$.

20. $4(1 \cdot 1 - (-2) \cdot 0) - 2(2 \cdot 1 - 0 \cdot 0) + (-1)(2(-2) - 0 \cdot 1)$

    $= 4(1 - 0) - 2(2 - 0) - 1(-4 - 0) = 4(1) - 2(2) - 1(-4) = 4 - 4 + 4 = 4$

21. $4x(4 - x^3) = 16x - 4x^4 = -4x^4 + 16x$

22. $(x - 4)(x + 4) = x^2 - 16$

23. $(5x + 3)(x - 2) = 5x^2 - 10x + 3x - 6 = 5x^2 - 7x - 6$

24. $(4x + 9)^2 = (4x)^2 + 2(4x)(9) + (9)^2 = 16x^2 + 72x + 81$

25. $9x^2 - 16 = (3x - 4)(3x + 4)$

26. $x^2 - 4x + 4 = (x - 2)(x - 2) = (x - 2)^2$

27. $15x^3 - 9x^2 = 3x^2(5x - 3)$

28. $12x^2 - 5x - 3 = (4x - 3)(3x + 1)$

29. $r^3 - 1 = (r - 1)(r^2 + r + 1)$

30. $x^3 - 3x^2 + 5x - 15 = x^2(x - 3) + 5(x - 3) = (x - 3)(x^2 + 5)$

31. $x^2 - 3x + 2 = 0 \Rightarrow (x - 2)(x - 1) = 0 \Rightarrow x = 2$ or $x = 1$

32. $x^3 = 4x \Rightarrow x^3 - 4x = 0 \Rightarrow x(x^2 - 4) = 0 \Rightarrow x(x - 2)(x + 2) = 0 \Rightarrow x = 0$ or $x = 2$ or $x = -2$

33. $\dfrac{x^2 + 3x + 2}{x - 3} \div \dfrac{x + 1}{2x - 6} = \dfrac{(x + 2)(x + 1)}{x - 3} \cdot \dfrac{2(x - 3)}{x + 1} = 2(x + 2)$

34. $\dfrac{2}{x - 1} + \dfrac{5}{x} = \dfrac{2}{x - 1} \cdot \dfrac{x}{x} + \dfrac{5}{x} \cdot \dfrac{x - 1}{x - 1} = \dfrac{2x + 5(x - 1)}{x(x - 1)} = \dfrac{2x + 5x - 5}{x(x - 1)} = \dfrac{7x - 5}{x(x - 1)}$

35. $\sqrt{36x^2} = \sqrt{36} \cdot \sqrt{x^2} = 6x$

36. $\sqrt[3]{64} = 4$

37. $16^{-3/2} = \left(\sqrt{16}\right)^{-3} = 4^{-3} = \dfrac{1}{4^3} = \dfrac{1}{64}$

38. $\sqrt[4]{625} = \sqrt[4]{(5)^4} = 5$

39. $\sqrt{2x} \cdot \sqrt{8x} = \sqrt{(2x)(8x)} = \sqrt{16x^2} = 4x$

40. $\sqrt{x} \cdot \sqrt[4]{x} = x^{1/2} \cdot x^{1/4} = x^{3/4}$ or $\sqrt[4]{x^3}$

41. $\dfrac{\sqrt[3]{16x^4}}{\sqrt[3]{2x}} = \sqrt[3]{\dfrac{16x^4}{2x}} = \sqrt[3]{8x^3} = 2x$

42. $4\sqrt{12x} - 2\sqrt{3x} = 4\sqrt{4 \cdot 3x} - 2\sqrt{3x} = 4\sqrt{4} \cdot \sqrt{3x} - 2\sqrt{3x} = 4(2)\sqrt{3x} - 2\sqrt{3x} = 8\sqrt{3x} - 2\sqrt{3x} = 6\sqrt{3x}$

43. $\left(2x+\sqrt{3}\right)\left(x-\sqrt{3}\right) = 2x^2 - 2x\sqrt{3} + x\sqrt{3} - \left(\sqrt{3}\right)^2 = 2x^2 - x\sqrt{3} - 3$

44. The radicand must be greater than or equal to $0 \Rightarrow 1 - x \geq 0 \Rightarrow x \leq 1;\ (-\infty, 1]$

45. See Figure 45.

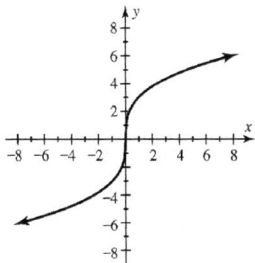

Figure 45

46. $d = \sqrt{\left(x_2 - x_1\right)^2 + \left(y_2 - y_1\right)^2} = \sqrt{\left(1-(-2)\right)^2 + (2-3)^2} = \sqrt{3^2 + (-1)^2} = \sqrt{9+1} = \sqrt{10}$

47. $(1-i)(2+3i) = 2+3i-2i-3i^2 = 2+i-3(-1) = 2+3+i = 5+i$

48. Use the Pythagorean Theorem: $a^2 + b^2 = c^2 \Rightarrow 5^2 + 12^2 = c^2 \Rightarrow 25+144 = c^2 \Rightarrow 169 = c^2 \Rightarrow c = 13$

49. $2\sqrt{x+3} = x \Rightarrow \sqrt{x+3} = \dfrac{x}{2} \Rightarrow x+3 = \left(\dfrac{x}{2}\right)^2 \Rightarrow x+3 = \dfrac{x^2}{4} \Rightarrow 4(x+3) = 4\left(\dfrac{x^2}{4}\right) \Rightarrow$

$4x+12 = x^2 \Rightarrow x^2 - 4x - 12 = 0 \Rightarrow (x-6)(x+2) = 0 \Rightarrow x = 6\ (x = -2\ \text{does not check}).$

50. $\sqrt[3]{x-1} = 3 \Rightarrow x-1 = 3^3 \Rightarrow x-1 = 27 \Rightarrow x = 28$

51. $\sqrt{x}+4 = 2\sqrt{x+5} \Rightarrow \left(\sqrt{x}+4\right)^2 = 4(x+5) \Rightarrow x+8\sqrt{x}+16 = 4x+20 \Rightarrow$

$8\sqrt{x} = 3x+4 \Rightarrow 64x = (3x+4)^2 \Rightarrow 64x = 9x^2 + 24x + 16 \Rightarrow 9x^2 - 40x + 16 = 0$

$\Rightarrow (9x-4)(x-4) = 0 \Rightarrow x = \dfrac{4}{9}\ \text{or}\ x = 4$

52. $\dfrac{1}{3}x^4 = 27 \Rightarrow x^4 = 81 \Rightarrow x = \sqrt[4]{81} \Rightarrow x = \pm 3$

53. Graph $Y_1 = (X^2-2)^{(1/3)} + X$ and $Y_2 = \sqrt{(X)}$. See Figure 53. The solution is $x \approx 1.41$.

$[-4.7,\ 4.7,\ 1]$ by $[-3.1,\ 3.1,\ 1]$          $[-6,\ 6,\ 1]$ by $[-4,\ 4,\ 1]$.

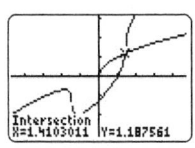          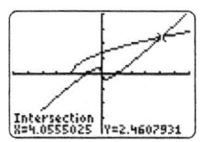

Figure 53                    Figure 54

54. Graph $Y_1 = X - X^{\left(1/3\right)}$ and $Y_2 = \sqrt{(X+2)}$. See Figure 54. The solution is $x \approx 4.06$.

55. The tank initially contains 300 gallons of water. Water is leaving the tank at 15 gallons per minute.

56. See Figure 56.

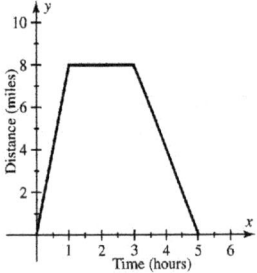

Figure 56

57. Let $x$ represent the amount invested at 5%. Then

$0.05x + 0.04(2000 - x) = 93 \Rightarrow 0.05x + 80 - 0.04x = 93 \Rightarrow 0.01x = 13 \Rightarrow x = 1300;$ $1300 is invested

at 5% and $700 is invested at 4%.

58. Let $x$ represent the length of a side of the base. Then $x^2(x-4) = 256 \Rightarrow x^3 - 4x^2 - 256 = 0$

Table $Y_1 = X \wedge 3 - 4X \wedge 2 - 256$ with TblStart $= 1$ and $\Delta$Tbl $= 1$ (not shown). $x = 8,$ so the

dimensions are 8 inches by 8 inches by 4 inches.

59. 5 feet 3 inches $= 5.25$ feet; let $h$ represent the height of the building; $\dfrac{5.25}{7.5} = \dfrac{h}{3.2}$

$\Rightarrow (5.25)(32) = 7.5h \Rightarrow 168 = 7.5h \Rightarrow h = 22.4$ feet

60. Let $x, y, z,$ represent the measures of the angles of the triangle from smallest to largest, respectively.

$$x + y + z = 180$$

The system needed is $z = x + y - 20$ . Substitute $z = x + y - 20$ into the third equation:

$$y + z = x + 90$$

$y + x + y - 20 = x + 90 \Rightarrow 2y = 110 \Rightarrow y = 55$. Substitute $y = 55$ and $z = x + y - 20$ into the first

equation: $x + 55 + x + 55 - 20 = 180 \Rightarrow 2x + 90 = 180 \Rightarrow 2x = 90 \Rightarrow x = 45$

Then $z = x + y - 20 \Rightarrow z = 45 + 55 - 20 \Rightarrow z = 80.$

The angle measures are $45°, 55°,$ and $80°.$

# Chapter 11: Quadratic Functions and Equations

## 11.1: Quadratic Functions and Their Graphs

1. parabola

3. axis of symmetry

5. See Figure 5. *Answers may vary.*

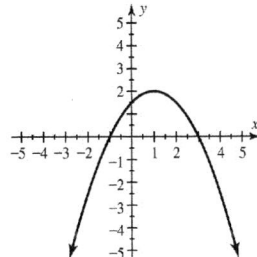

Figure 5

7. narrower

9. $ax^2 + bx + c$ with $a \neq 0$

11. True

13. False, The maximum $y$-value on the parabola is $b$.

15. $x = -\dfrac{b}{2a} = -\dfrac{(-4)}{2(1)} = 2$, $f(2) = (2)^2 - 4(2) - 2 = 4 - 8 - 2 = -6$; The vertex is $(2, -6)$.

17. $x = -\dfrac{b}{2a} = -\dfrac{(-2)}{2(-\frac{1}{3})} = -3$, $f(-3) = -\dfrac{1}{3}(-3)^2 - 2(-3) + 1 = -3 + 6 + 1 = 4$; The vertex is $(-3, 4)$.

19. $x = -\dfrac{b}{2a} = -\dfrac{(0)}{2(-2)} = 0$, $f(0) = 3 - 2(0)^2 = 3 - 0 = 3$; The vertex is $(0, 3)$.

21. $x = -\dfrac{b}{2a} = -\dfrac{(0.6)}{2(-0.3)} = 1$, $f(1) = -0.3(1)^2 + 0.6(1) + 1.1 = -0.3 + 0.6 + 1.1 = 1.4$; The vertex is $(1, 1.4)$.

23. $x = -\dfrac{b}{2a} = -\dfrac{(6)}{2(-1)} = \dfrac{6}{2} = 3$, $f(3) = 6(3) - (3)^2 = 18 - 9 = 9$; The vertex is $(3, 9)$.

25. $f(-2) = 0$ and $f(0) = -4$

27. $f(-3) = -2$ and $f(1) = -2$

29. The vertex is $(1, -2)$. The axis of symmetry is $x = 1$. The parabola opens upward.

31. The vertex is $(-2, 3)$. The axis of symmetry is $x = -2$. The parabola opens downward.

33. (a)　See Figure 33.

   (b)　The vertex is $(0, 0)$. The axis of symmetry is $x = 0$.

   (c)　$f(-2) = \frac{1}{2}(-2)^2 = \frac{1}{2}(4) = 2$ and $f(3) = \frac{1}{2}(3)^2 = \frac{1}{2}(9) = 4.5$

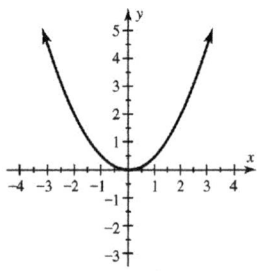

Figure 33

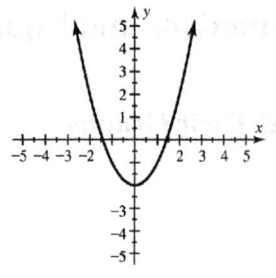

Figure 35

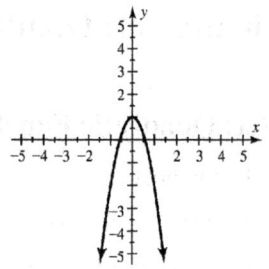

Figure 37

35. (a)    See Figure 35.

(b)    The vertex is $(0, -2)$. The axis of symmetry is $x = 0$.

(c)    $f(-2) = (-2)^2 - 2 = 4 - 2 = 2$ and $f(3) = (3)^2 - 2 = 9 - 2 = 7$

37. (a)    See Figure 37.

(b)    The vertex is $(0, 1)$. The axis of symmetry is $x = 0$.

(c)    $f(-2) = -3(-2)^2 + 1 = -12 + 1 = -11$ and $f(3) = -3(3)^2 + 1 = -27 + 1 = -26$

39. (a)    See Figure 39.

(b)    The vertex is $(1, 0)$. The axis of symmetry is $x = 1$.

(c)    $f(-2) = ((-2) - 1)^2 = (-3)^2 = 9$ and $f(3) = ((3) - 1)^2 = (2)^2 = 4$

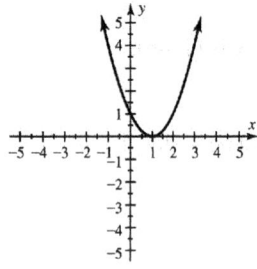

Figure 39

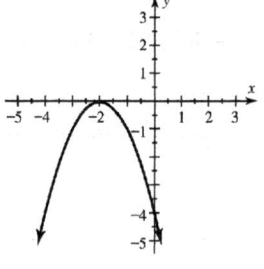

Figure 41

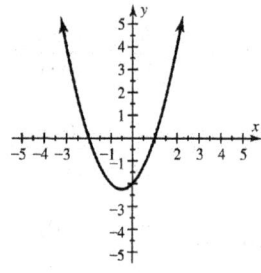

Figure 43

41. (a)    See Figure 41.

(b)    The vertex is $(-2, 0)$. The axis of symmetry is $x = -2$.

(c)    $f(-2) = -(-2 + 2)^2 = -(0)^2 = 0$ and $f(3) = -(3 + 2)^2 = -(5)^2 = -25$

43. (a)    See Figure 43.

(b)    The vertex is $(-0.5, -2.25)$. The axis of symmetry is $x = -0.5$.

(c)    $f(-2) = (-2)^2 + (-2) - 2 = 4 - 2 - 2 = 0$ and $f(3) = (3)^2 + (3) - 2 = 9 + 3 - 2 = 10$

45. (a)    See Figure 45.

(b)    The vertex is $(0, -3)$. The axis of symmetry is $x = 0$.

(c)    $f(-2) = 2(-2)^2 - 3 = 8 - 3 = 5$ and $f(3) = 2(3)^2 - 3 = 18 - 3 = 15$

47. (a)   See Figure 47.

    (b)   The vertex is $(1,\ 1)$. The axis of symmetry is $x = 1$.

    (c)   $f(-2) = 2(-2)-(-2)^2 = -4-4 = -8$ and $f(3) = 2(3)-(3)^2 = 6-9 = -3$

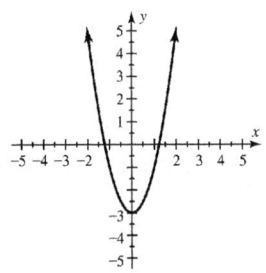

Figure 45

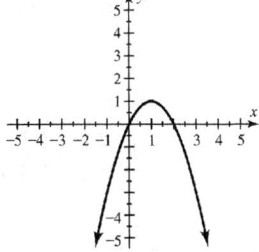

Figure 47

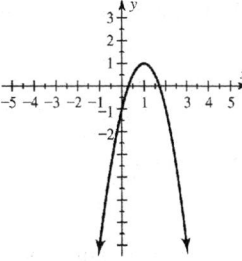

Figure 49

49. (a)   See Figure 49.

    (b)   The vertex is $(1,\ 1)$. The axis of symmetry is $x = 1$.

    (c)   $f(-2) = -2(-2)^2 + 4(-2)-1 = -8-8-1 = -17$ and $f(3) = -2(3)^2 + 4(3)-1 = -18+12-1 = -7$

51. (a)   See Figure 51.

    (b)   The vertex is $(2,\ 4)$. The axis of symmetry is $x = 2$.

    (c)   $f(-2) = \dfrac{1}{4}(-2)^2 -(-2)+5 = 1+2+5 = 8$ and $f(3) = \dfrac{1}{4}(3)^2 -(3)+5 = 2.25-3+5 = 4.25$

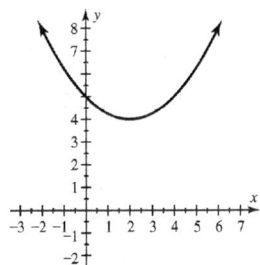

Figure 51

53. Because $-\dfrac{b}{2a} = -\dfrac{2}{2(1)} = -1$ and $f(-1) = (-1)^2 + 2(-1)-1 = -2$, the vertex is $(-1,\ -2)$.

    The minimum $y$-value on the graph is $-2$.

55. Because $-\dfrac{b}{2a} = -\dfrac{-5}{2(1)} = \dfrac{5}{2}$ and $f\left(\dfrac{5}{2}\right) = \left(\dfrac{5}{2}\right)^2 -5\left(\dfrac{5}{2}\right) = -\dfrac{25}{4}$, the vertex is $\left(\dfrac{5}{2},\ -\dfrac{25}{4}\right)$.

    The minimum $y$-value on the graph is $-\dfrac{25}{4}$.

57. Because $-\dfrac{b}{2a} = -\dfrac{2}{2(2)} = -\dfrac{1}{2}$ and $f\left(-\dfrac{1}{2}\right) = 2\left(-\dfrac{1}{2}\right)^2 +2\left(-\dfrac{1}{2}\right)-3 = -\dfrac{7}{2}$, the vertex is $\left(-\dfrac{1}{2},\ -\dfrac{7}{2}\right)$.

    The minimum $y$-value on the graph is $-\dfrac{7}{2}$.

59. Because $-\dfrac{b}{2a} = -\dfrac{2}{2(-1)} = 1$ and $f(1) = -(1)^2 + 2(1) + 5 = 6$, the vertex is $(1,\ 6)$.

    The maximum $y$-value on the graph is $6$.

61. Because $-\dfrac{b}{2a} = -\dfrac{4}{2(-1)} = 2$ and $f(2) = 4(2) - (2)^2 = 4$, the vertex is $(2,\ 4)$.

    The maximum $y$-value on the graph is $4$.

63. Because $-\dfrac{b}{2a} = -\dfrac{1}{2(-2)} = \dfrac{1}{4}$ and $f\left(\dfrac{1}{4}\right) = -2\left(\dfrac{1}{4}\right)^2 + \left(\dfrac{1}{4}\right) - 5 = -\dfrac{39}{8}$, the vertex is $\left(\dfrac{1}{4},\ -\dfrac{39}{8}\right)$.

    The maximum $y$-value on the graph is $-\dfrac{39}{8}$.

65. Let $x$ be one of the numbers, then $20 - x$ is the other number. $x(20 - x) = 20x - x^2$. The graph of

    $y = -x^2 + 20x$ is a parabola that opens downward, so its vertex is the maximum point on the graph.

    The $x$-value of the vertex is on the axis of symmetry, so $x = -\dfrac{b}{2a} = -\dfrac{20}{2(-1)} = -\dfrac{20}{-2} = 10$. So the

    numbers are 10 and 10.

67. See Figure 67. Compared to $y = x^2$, the graph is reflected across the $x$-axis.

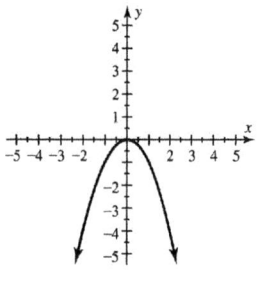
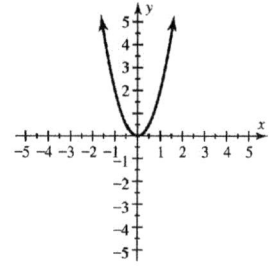
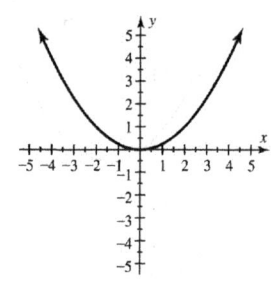

| Figure 67 | Figure 69 | Figure 71 |

69. See Figure 69. Compared to $y = x^2$, the graph is narrower.

71. See Figure 71. Compared to $y = x^2$, the graph is wider.

73. See Figure 73. Compared to $y = x^2$, the graph is reflected across the $x$-axis and is wider.

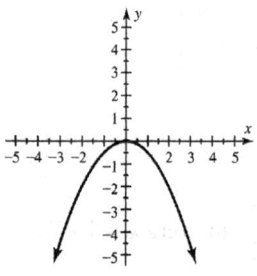

Figure 73

75. (a) $a = \dfrac{1}{2}$, so $a > 0$ and the graph opens upward. Because $0 < |a| < 1$, the graph is wider than the

graph of $y = x^2$.

(b) The axis of symmetry is $x = -\dfrac{b}{2a} = -\dfrac{1}{2(\frac{1}{2})} = -\dfrac{1}{1} = -1$.

$f(-1) = \dfrac{1}{2}(-1)^2 + (-1) - \dfrac{3}{2} = \dfrac{1}{2}(1) - \dfrac{2}{2} - \dfrac{3}{2} = \dfrac{1}{2} - \dfrac{2}{2} - \dfrac{3}{2} = -\dfrac{4}{2} = -2$ so the vertex is $(-1, -2)$.

(c) The $y$-intercept is $c = -\dfrac{3}{2}$. To find the $x$-intercepts, let $y = 0$ and solve for $x$:

$\dfrac{1}{2}x^2 + x - \dfrac{3}{2} = 0 \Rightarrow 2\left(\dfrac{1}{2}x^2 + x - \dfrac{3}{2}\right) = 2(0) \Rightarrow x^2 + 2x - 3 = 0 \Rightarrow$

$(x+3)(x-1) = 0 \Rightarrow x = -3$ or $x = 1$

(d) See Figure 75.

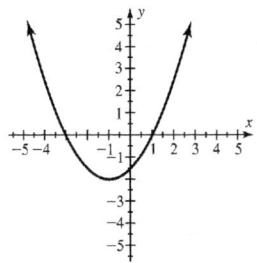

Figure 75

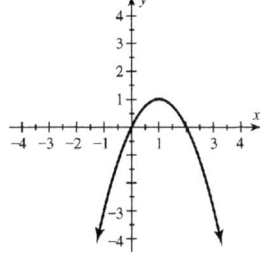

Figure 77

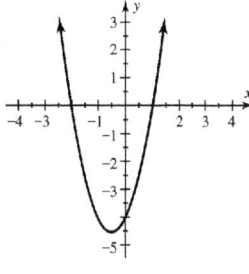

Figure 79

77. (a) $a = -1$, so $a < 0$ and the graph opens downward. Because $|a| = 1$, the graph has the same width

as the graph of $y = x^2$.

(b) The axis of symmetry is $x = -\dfrac{b}{2a} = -\dfrac{2}{2(-1)} = -\dfrac{2}{-2} = 1$. $f(1) = 2(1) - 1^2 = 2 - 1 = 1$, so the

vertex is $(1, 1)$.

(c) The $y$-intercept is $c = 0$. To find the $x$-intercepts, let $y = 0$ and solve for $x$:

$2x - x^2 = 0 \Rightarrow x(2 - x) = 0 \Rightarrow x = 0$ and $x = 2$.

(d) See Figure 77.

79. (a) $a = 2$, so $a > 0$ and the graph opens upward. Because $|a| > 1$, the graph is narrower than the

graph of $y = x^2$.

(b) The axis of symmetry is $x = -\dfrac{b}{2a} = -\dfrac{2}{2(2)} = -\dfrac{2}{4} = -\dfrac{1}{2}$.

$f\left(-\dfrac{1}{2}\right) = 2\left(-\dfrac{1}{2}\right)^2 + 2\left(-\dfrac{1}{2}\right) - 4 = 2\left(\dfrac{1}{4}\right) - 1 - 4 = \dfrac{1}{2} - \dfrac{2}{2} - \dfrac{8}{2} = -\dfrac{9}{2}$ so the vertex is $\left(-\dfrac{1}{2}, -\dfrac{9}{2}\right)$.

(c)    The $y$-intercept is $c = -4$. To find the $x$-intercepts, let $y = 0$ and solve for $x$:

$$2x^2 + 2x - 4 = 0 \Rightarrow 2(x^2 + x - 2) = 0 \Rightarrow 2(x + 2)(x - 1) = 0 \Rightarrow x = -2 \text{ or } x = 1.$$

(d)    See Figure 79.

81. Because $-\dfrac{b}{2a} = -\dfrac{\frac{200}{3}}{2\left(-\frac{100}{9}\right)} = 3$ and $f(3) = -\dfrac{100}{9}(3)^2 + \dfrac{200}{3}(3) = 100$, the vertex is $(3,\ 100)$.

The number of pieces that yields maximum satisfaction is 3.

83. d.    The stone's distance from the ground would increase and then decrease.

85. a.    The temperature would first decrease but after the repair, it would increase.

87. (a)    When the ball is hit, $t = 0$. Then $h(0) = -16(0)^2 + 64(0) + 2 = 2$ feet .

(b)    Find the $x$-coordinate of the vertex.    $-\dfrac{b}{2a} = -\dfrac{64}{2(-16)} = 2$ seconds

(c)    Find the $y$-coordinate of the vertex.    $h(2) = -16(2)^2 + 64(2) + 2 = 66$ feet .

89. The $x$-coordinate of the vertex represents the time when the ball reaches its maximum height. Here

we have $x = \dfrac{-b}{2a} = \dfrac{-66}{2(-16)} = \dfrac{66}{32} \approx 2$ seconds. The maximum height is

$$h(2) = -16(2)^2 + 66(2) + 6 \approx 74 \text{ feet.}$$

91. (a)    The revenue is increasing when $x \le 50$, and it is decreasing when $x \ge 50$.

(b)    From the graph, the maximum revenue is \$2500 when 50 tickets are sold.

(c)    If $x$ represents the number of tickets sold, then $100 - x$ represents the price of one ticket. The total revenue is given by $f(x) = x(100 - x)$ .

(d)    Since $f(x) = -x^2 + 100x$, the number of tickets that should be sold to maximize revenue is

$-\dfrac{b}{2a} = -\dfrac{100}{2(-1)} = 50$. The maximum revenue is $f(50) = 50(100 - 50) = 50(50) = \$2500$ .

93. (a)    $V(1) = 10.75(1)^2 - 24(1) + 35 = 10.75 - 24 + 35 = 21.75$

$V(2) = 10.75(2)^2 - 24(2) + 35 = 43 - 48 + 35 = 30$

$V(3) = 10.75(3)^2 - 24(3) + 35 = 96.75 - 72 + 35 = 59.75$

$V(4) = 10.75(4)^2 - 24(4) + 35 = 172 - 96 + 35 = 111$

In 2007 there were 21.75 million unique Facebook visitors in one month. The other values can be interpreted similarly.

(b)    The increases between consecutive years are 8.25 million, 29.75 million, and 51.25 million. A linear function does not model the data because these three values are not equal.

95. Because there are 1200 feet of fence and the width of the enclosure is $x$, the length is given by $1200 - 2x$. The area of the enclosure is $A(x) = x(1200 - 2x)$ or $A(x) = -2x^2 + 1200x$. The value of $x$ that will maximize the area is the $x$-coordinate of the vertex, $-\dfrac{b}{2a} = -\dfrac{1200}{2(-2)} = 300$. Thus the width is 300 feet and the length is $1200 - 2(300) = 1200 - 600 = 600$ feet. The enclosure measures 300 feet by 600 feet.

97. $S(8) = -0.227(8)^2 + 8.155(8) - 8.8 \approx 42$ inches

99. (a) $C(1990) = \dfrac{1}{300}(1990)^2 - \dfrac{199}{15}(1990) + \dfrac{39,619}{3} = 6$  In 1990, emissions were 6 billion metric tons.

    (b) $C(2020) = \dfrac{1}{300}(2020)^2 - \dfrac{199}{15}(2020) + \dfrac{39,619}{3} = 9$, $C(2020) - C(1990) = 9 - 6 = 3$

    There is an expected increase of 3 billion metric tons.

[20, 40, 5] by [0, 30, 5]          [20, 40, 5] by [0, 30, 5]

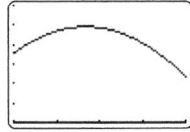          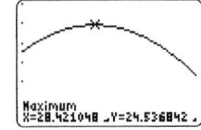

Figure 99a                Figure 99b

# Section 11.2  Parabolas and Modeling

1. $x^2 + 2$

3. $(1, 2)$

5. $f(x) = ax^2 + bx + c$ or $f(x) = a(x - h)^2 + k$

7. downward

9. a

11. The graph is shifted 3 units downward.

| $x$ | $-2$ | $-1$ | 0 | 1 | 2 |
|---|---|---|---|---|---|
| $y = x^2$ | 4 | 1 | 0 | 1 | 4 |
| $y = x^2 - 3$ | 1 | $-2$ | $-3$ | $-2$ | 1 |

13. The graph is shifted 3 units to the right.

| $x$ | $-2$ | $-1$ | 0 | 1 | 2 |
|---|---|---|---|---|---|
| $y = x^2$ | 4 | 1 | 0 | 1 | 4 |

| $x$ | 1 | 2 | 3 | 4 | 5 |
|---|---|---|---|---|---|
| $y = (x - 3)^2$ | 4 | 1 | 0 | 1 | 4 |

15. (a) The vertex form of the function is $f(x) = (x - 0)^2 + (-4)$. See Figure 15.

    (b) The vertex is $(0, -4)$.

    (c) Compared to the graph of $y = x^2$, the graph is shifted down 4 units.

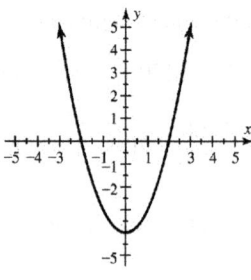

Figure 15

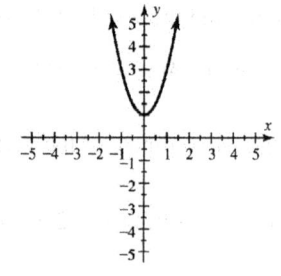

Figure 17

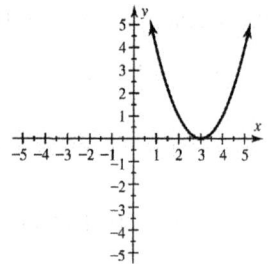

Figure 19

17. (a)    The vertex form of the function is $f(x) = 2(x-0)^2 + 1$. See Figure 17.

    (b)    The vertex is $(0,\ 1)$.

    (c)    Compared to the graph of $y = x^2$, the graph is narrower and is shifted up 1 unit.

19. (a)    The vertex form of the function is $f(x) = (x-3)^2 + 0$. See Figure 19.

    (b)    The vertex is $(3,\ 0)$.

    (c)    Compared to the graph of $y = x^2$, the graph is shifted right 3 units.

21. (a)    The vertex form of the function is $f(x) = -(x-0)^2 + 0$. See Figure 21.

    (b)    The vertex is $(0,\ 0)$.

    (c)    Compared to the graph of $y = x^2$, the graph is reflected across the $x$-axis.

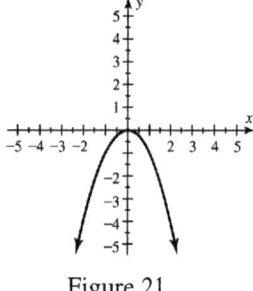

Figure 21

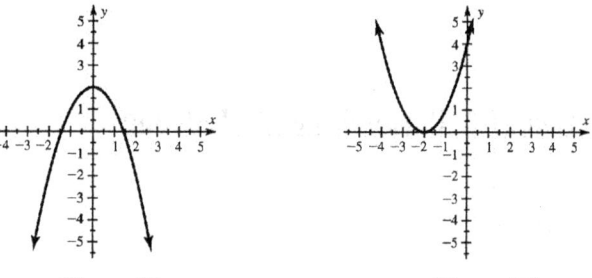

Figure 23                                    Figure 25

23. (a)    The vertex form of the function is $f(x) = -(x-0)^2 + 2$. See Figure 23.

    (b)    The vertex is $(0,\ 2)$.

    (c)    Compared to the graph of $y = x^2$, the graph is reflected across the $x$-axis and shifted up 2 units.

25. (a)    The vertex form of the function is $f(x) = (x-(-2))^2 + 0$. See Figure 25.

    (b)    The vertex is $(-2,\ 0)$.

    (c)    Compared to the graph of $y = x^2$, the graph is shifted left 2 units

27. (a)    The vertex form of the function is $f(x) = (x-(-1))^2 + (-2)$. See Figure 27.

    (b)    The vertex is $(-1,\ -2)$.

(c)    Compared to the graph of $y = x^2$, the graph is shifted left 1 unit and down 2 units.

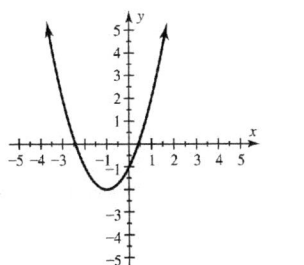

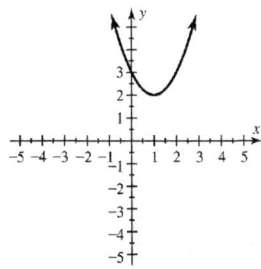

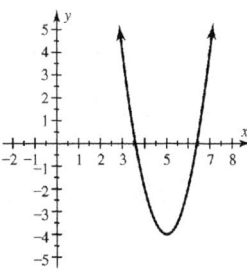

Figure 27                              Figure 29                              Figure 31

29. (a)    The vertex form of the function is $f(x) = (x-1)^2 + 2$. See Figure 29.

(b)    The vertex is $(1,\ 2)$.

(c)    Compared to the graph of $y = x^2$, the graph is shifted right 1 unit and up 2 units.

31. (a)    The vertex form of the function is $f(x) = 2(x-5)^2 + (-4)$. See Figure 31.

(b)    The vertex is $(5,\ -4)$.

(c)    Compared to the graph of $y = x^2$, the graph is narrower, shifted right 5 units and down 4 units.

33. (a)    The vertex form of the function is $f(x) = -\dfrac{1}{2}(x-(-3))^2 + 1$. See Figure 33.

(b)    The vertex is $(-3,\ 1)$.

(c)    Compared to the graph of $y = x^2$, the graph is wider, reflected across the x-axis, shifted left 3 units and up 1 unit.

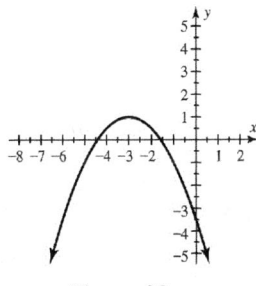

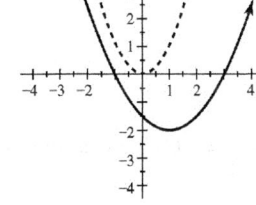

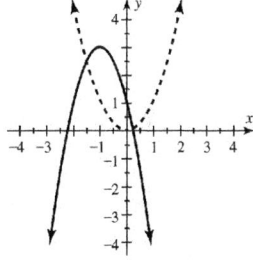

Figure 33                              Figure 35                              Figure 37

35. The graph of $f(x)$ is the graph of $y = x^2$ translated 1 unit right and 2 units downward. The graph of $f(x)$ is wider. See Figure 35.

37. The graph of $f(x)$ is the graph of $y = x^2$ translated 1 unit left and 3 units upward. The graph of $f(x)$ opens downward and is narrower. See Figure 37.

39. See Figure 39.

[–20, 20, 2] by [–20, 20, 2]

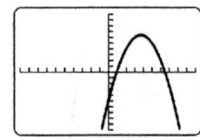

Figure 39

41. Since $h = 3$, $k = 4$ and $a = 3$, the equation is $y = 3(x-3)^2 + 4$. Expand to obtain the other form:

$$y = 3(x^2 - 6x + 9) + 4 \Rightarrow y = 3x^2 - 18x + 27 + 4 \Rightarrow y = 3x^2 - 18x + 31$$

43. Since $h = 5$, $k = -2$ and $a = -\dfrac{1}{2}$, the equation is $y = -\dfrac{1}{2}(x-5)^2 - 2$. Expand to obtain the other

   form: $y = -\dfrac{1}{2}(x^2 - 10x + 25) - 2 \Rightarrow y = -\dfrac{1}{2}x^2 + 5x - \dfrac{25}{2} - 2 \Rightarrow y = -\dfrac{1}{2}x^2 + 5x - \dfrac{29}{2}$

45. Since $h = 1$, $k = 2$ and $a = 1$, the equation is $y = (x-1)^2 + 2$.

47. Since $h = 0$, $k = -3$ and $a = -1$, the equation is $y = -(x-0)^2 - 3$ or $y = -x^2 - 3$.

49. Since $h = 0$, $k = -3$ and $a = 1$, the equation is $y = (x-0)^2 - 3$ or $y = x^2 - 3$.

51. Since $h = -1$, $k = 2$ and $a = -1$, the equation is $y = -(x+1)^2 + 2$.

53. (a)  We can use the vertex formula $x = -\dfrac{b}{2a}$, to find the x-coordinate of the vertex with $a = 4$

   and $b = -8$. $x = -\dfrac{(-8)}{2(4)} = \dfrac{8}{8} = 1$. To find the y-coordinate, let $x = 1$ in $y = 4x^2 - 8x + 5$,

   $y = 4(1)^2 - 8(1) + 5 = 1$  The vertex is $(1, 1)$.

   (b)  We can now find the vertex form with $a = 4$, $h = 1$, and $k = 1$.

   $y = a(x-h)^2 + k \Rightarrow y = 4(x-1)^2 + 1$

55. (a)  We can use the vertex formula $x = -\dfrac{b}{2a}$, to find the x-coordinate of the vertex with $a = -1$

   and $b = -2$. $x = -\dfrac{(-2)}{2(-1)} = \dfrac{2}{-2} = -1$. To find the y-coordinate, let $x = -1$ in $y = -x^2 - 2x - 3$,

   $y = -(-1)^2 - 2(-1) - 3 = -2$  The vertex is $(-1, -2)$.

   (b)  We can now find the vertex form with $a = -1$, $h = -1$, and $k = -2$.

   $y = a(x-h)^2 + k \Rightarrow y = -(x-(-1))^2 + (-2) \Rightarrow y = -(x+1)^2 - 2$

57. (a)  We can use the vertex formula $x = -\dfrac{b}{2a}$, to find the x-coordinate of the vertex with $a = -2$

   and $b = -4$. $x = -\dfrac{(-4)}{2(-2)} = \dfrac{4}{-4} = -1$. To find the y-coordinate, let $x = -1$ in

   $y = -2x^2 - 4x + 1$, $y = -2(-1)^2 - 4(-1) + 1 = 3$  The vertex is $(-1, 3)$.

(b)    We can now find the vertex form with $a = -2$, $h = -1$, and $k = 3$.

$$y = a(x-h)^2 + k \Rightarrow y = -2(x-(-1))^2 + 3 \Rightarrow y = -2(x+1)^2 + 3$$

59. The area of the rectangle in question is $1 \cdot 1 = 1$. Thus, to find the complete area of the square (complete the square) we must add the computed area of 1.

61. $y = x^2 + 2x \Rightarrow y = (x^2 + 2x + 1) - 1 \Rightarrow y = (x+1)^2 - 1$. The vertex is $(-1, -1)$.

63. $y = x^2 - 4x \Rightarrow y = (x^2 - 4x + 4) - 4 \Rightarrow y = (x-2)^2 - 4$. The vertex is $(2, -4)$.

65. $y = x^2 + 2x - 3 \Rightarrow y = (x^2 + 2x + 1) - 3 - 1 \Rightarrow y = (x+1)^2 - 4$. The vertex is $(-1, -4)$.

67. $y = x^2 - 4x + 5 \Rightarrow y = (x^2 - 4x + 4) + 5 - 4 \Rightarrow y = (x-2)^2 + 1$. The vertex is $(2, 1)$.

69. $y = x^2 + 3x - 2 \Rightarrow y = \left(x^2 + 3x + \dfrac{9}{4}\right) - 2 - \dfrac{9}{4} \Rightarrow y = \left(x + \dfrac{3}{2}\right)^2 - \dfrac{17}{4}$. The vertex is $\left(-\dfrac{3}{2}, -\dfrac{17}{4}\right)$.

71. $y = x^2 - 7x + 1 \Rightarrow y = \left(x^2 - 7x + \dfrac{49}{4}\right) + 1 - \dfrac{49}{4} \Rightarrow y = \left(x - \dfrac{7}{2}\right)^2 - \dfrac{45}{4}$. The vertex is $\left(\dfrac{7}{2}, -\dfrac{45}{4}\right)$.

73. $y = 3x^2 + 6x - 1 \Rightarrow y = 3(x^2 + 2x + 1) - 1 - 3 \Rightarrow y = 3(x+1)^2 - 4$. The vertex is $(-1, -4)$.

75. $y = 2x^2 - 3x \Rightarrow y = 2\left(x^2 - \dfrac{3}{2}x + \dfrac{9}{16}\right) - \dfrac{9}{8} \Rightarrow y = 2\left(x - \dfrac{3}{4}\right)^2 - \dfrac{9}{8}$. The vertex is $\left(\dfrac{3}{4}, -\dfrac{9}{8}\right)$.

77. $y = -2x^2 - 8x + 5 \Rightarrow y = -2(x^2 + 4x + 4) + 5 + 8 \Rightarrow y = -2(x+2)^2 + 13$. The vertex is $(-2, 13)$.

79. Since $y = 2$ when $x = 1$, $2 = a(1)^2 \Rightarrow 2 = 1a \Rightarrow a = 2$.

81. Since $y = 1.2$ when $x = 2$, $1.2 = a(2)^2 \Rightarrow 1.2 = 4a \Rightarrow a = 0.3$.

83. A scatterplot of the data (not shown) indicates that the data point $(1, -3)$ is the vertex of the parabola. The function has the form $f(x) = a(x-1)^2 - 3$. $f(2) = a(2-1)^2 - 3 = -1 \Rightarrow a - 3 = -1 \Rightarrow a = 2$.. Thus $f(x) = 2(x-1)^2 - 3$.

85. A scatterplot of the data (not shown) indicates that the data point $(1980, 6)$ is the vertex of the parabola. The function has the form $f(x) = a(x-1980)^2 + 6$.

$f(1990) = a(1990-1980)^2 + 6 = 55 \Rightarrow 100a + 6 = 55 \Rightarrow a = 0.5$. Thus $f(x) = 0.5(x-1980)^2 + 6$.

87. (a)    See Figure 87.

(b)    Since $D = 12$ when $x = 12$, we have $12 = a(12)^2 \Rightarrow 12 = 144a \Rightarrow a = \dfrac{12}{144} \Rightarrow a = \dfrac{1}{12}$.

The function is $D(x) = \dfrac{1}{12}x^2$.

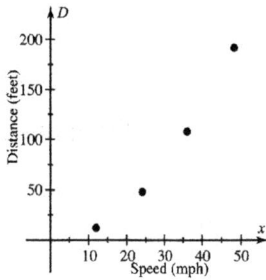

Figure 87

89. (a)    The GDP decreases and then increases.

    (b)    No, because the data decreases and then increases.

    (c)    Quadratic; This type of function can model data that decreases and then increases.

    (d)    (1995, 600). This point has the minimum $y$-value and the data is symmetric about this point.

    (e)    The scatterplot of the data indicates that the data point (1995, 600) is the vertex of the

           parabola. The function has the form $C(x) = a(x-1995)^2 + 600$.

           $C(1980) = a(1980-1995)^2 + 600 = 1000 \Rightarrow 224a = 400 \Rightarrow a \approx 1.8$.

           So $C(x) = 1.8(x-1995)^2 + 600$.

    (f)    See Figure 89.

    [1970, 2030, 10] by [0, 2500, 500]

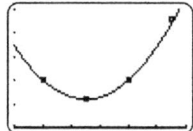

Figure 89

91. (a)    Graph the data (not shown) to see that the data point (1982, 1) is near the vertex of a parabola.

           Graph $Y_1 = (X-1982) \wedge 2 + 1$ to see if it is a reasonable model. A better model is

           $f(x) = 2(x-1982)^2 + 1$. *Answers may vary.*

    (b)    Using the model, $f(1992) = 2(1992-1982)^2 + 1 = 2(10)^2 + 1 = 2(100) + 1 = 201$. The model

           predicts 201 thousand deaths in 1992, which is close to the actual value of 202

           thousand. *Answers may vary.*

93. As we read the graph from left to right we can see that the function is increasing when $x > 2$ and

    decreasing when $x < 2$.

95. As we read the graph from left to right we can see that the function is increasing when $x < 1$ and

    decreasing when $x > 1$.

97. We can use the vertex formula $x = -\dfrac{b}{2a}$, to find the $x$-coordinate of the vertex with $a = 3$ and

$b = -4$. $x = -\dfrac{(-4)}{2(3)} = \dfrac{4}{6} = \dfrac{2}{3}$. Since the value of $a$ is positive, the graph is opening upward and

the function is increasing when $x > \dfrac{2}{3}$ and decreasing when $x < \dfrac{2}{3}$.

99. We can use the vertex formula $x = -\dfrac{b}{2a}$, to find the $x$-coordinate of the vertex with $a = -1$ and

$b = -3$. $x = -\dfrac{(-3)}{2(-1)} = -\dfrac{3}{2}$. Since the value of $a$ is negative, the graph is opening downward and

the function is increasing when $x < -\dfrac{3}{2}$ and decreasing when $x > -\dfrac{3}{2}$.

101. We can use the vertex formula $x = -\dfrac{b}{2a}$, to find the $x$-coordinate of the vertex with $a = -4$ and

$b = -1$. $x = -\dfrac{(-1)}{2(-4)} = -\dfrac{1}{8}$. Since the value of $a$ is negative, the graph is opening downward and

the function is increasing when $x < -\dfrac{1}{8}$ and decreasing when $x > -\dfrac{1}{8}$.

## Checking Basic Concepts Sections 11.1 and 11.2

1. (a) See Figure 1a. The vertex is $(0, -2)$, and the axis of symmetry is $x = 0$.

   (b) See Figure 1b. The vertex is $(1, -3)$, and the axis of symmetry is $x = 1$.

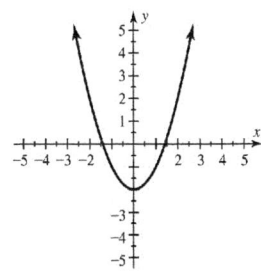
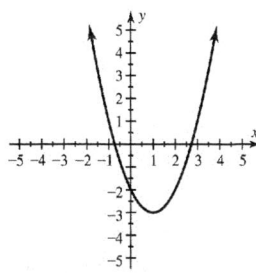

Figure 1a                         Figure 1b

2. The graph of $y_1$ opens upward whereas $y_2$ opens downward. Also, $y_1$ is narrower than $y_2$.

3. Because $-\dfrac{b}{2a} = -\dfrac{12}{2(-3)} = 2$ and $f(2) = -3(2)^2 + 12(2) - 5 = 7$, the vertex is $(2, 7)$.

   The maximum $y$-value on the graph is $7$.

4. (a) See Figure 4a. Compared to the graph of $y = x^2$, the graph is shifted right 1 unit and

   up 2 units.

   (b) See Figure 4b. Compared to the graph of $y = x^2$, the graph is reflected across the $x$-axis and

   shifted left 3 units.

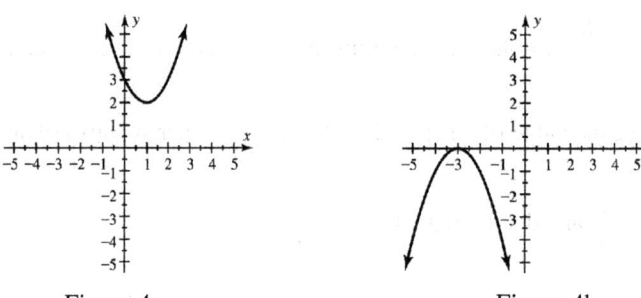

Figure 4a                    Figure 4b

5. (a)    $y = x^2 + 14x - 7 \Rightarrow y = (x^2 + 14x + 49) - 7 - 49 \Rightarrow y = (x + 7)^2 - 56$.

   (b)    $y = 4x^2 + 8x - 2 \Rightarrow y = 4(x^2 + 2x + 1) - 2 - 4 \Rightarrow y = 4(x + 1)^2 - 6$.

## Section 11.3 Quadratic Equations

1.  $x^2 + 3x - 2 = 0$; *Answers may vary.*  A quadratic equation can have 0, 1 or 2 solutions.

3.  Factoring, square root property, completing the square

5.  See Figure 5.  *Answers may vary*

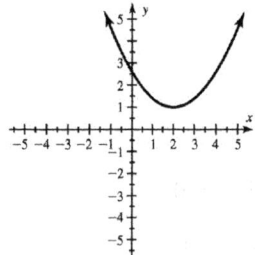

Figure 5

7.  $x = \pm 8$;  the square root property.

9.  Yes

11. No, there is no $x^2$ term.

13. Yes

15. No, the term $\sqrt{x}$ is not allowed in a quadratic equation.

17. (a)    $1 \pm \sqrt{7} \approx 1 \pm 2.65 = 3.65$ or $-1.65$

    (b)    $-2 \pm \sqrt{11} \approx -2 \pm 3.32 = 1.32$ or $-5.32$

19. (a)    $\dfrac{3 \pm \sqrt{13}}{5} \approx \dfrac{3 \pm 3.61}{5} = \dfrac{6.61}{5}$ or $-\dfrac{0.61}{5} \approx 1.32$ or $-0.12$

    (b)    $\dfrac{-5 \pm \sqrt{6}}{9} \approx \dfrac{-5 \pm 2.45}{9} = \dfrac{-2.55}{9}$ or $\dfrac{-7.45}{9} \approx -0.28$ or $-0.83$

21. $-2, 1$

23. No real solutions

25. –2, 3

27. –0.5

29. (a)  $x^2 - 4x - 5 = 0 \Rightarrow (x+1)(x-5) = 0 \Rightarrow x+1 = 0$ or $x-5 = 0 \Rightarrow x = -1$ or 5

   (b)  A graph of $y = x^2 - 4x - 5$ (not shown) intersects the $x$-axis at $-1$ and $5$.

   (c)  A table of $y = x^2 - 4x - 5$ (not shown) yields y-values of zero when $x = -1$ and $5$.

31. (a)  $x^2 + 2x = 3 \Rightarrow x^2 + 2x - 3 = 0 \Rightarrow (x+3)(x-1) = 0 \Rightarrow x+3 = 0$ or $x-1 = 0 \Rightarrow x = -3$ or 1

   (b)  A graph of $y = x^2 + 2x - 3$ (not shown) intersects the $x$-axis at $-3$ and $1$.

   (c)  A table of $y = x^2 + 2x - 3$ (not shown) yields y-values of zero when $x = -3$ and $1$.

33. (a)  $x^2 = 9 \Rightarrow x^2 - 9 = 0 \Rightarrow (x+3)(x-3) = 0 \Rightarrow x+3 = 0$ or $x-3 = 0 \Rightarrow x = -3$ or 3

   (b)  A graph of $y = x^2 - 9$ (not shown) intersects the $x$-axis at $-3$ and $3$.

   (c)  A table of $y = x^2 - 9$ (not shown) yields y-values of zero when $x = -3$ and $3$.

35. (a)  $4x^2 - 4x - 3 = 0 \Rightarrow (2x+1)(2x-3) = 0 \Rightarrow 2x+1 = 0$ or $2x-3 = 0 \Rightarrow x = -\dfrac{1}{2}$ or $\dfrac{3}{2}$

   (b)  A graph of $y = 4x^2 - 4x - 3$ (not shown) intersects the $x$-axis at $-\dfrac{1}{2}$ and $\dfrac{3}{2}$.

   (c)  A table of $y = 4x^2 - 4x - 3$ (not shown) yields y-values of zero when $x = -\dfrac{1}{2}$ and $\dfrac{3}{2}$.

37. (a)  $x^2 + 2x = -1 \Rightarrow x^2 + 2x + 1 = 0 \Rightarrow (x+1)(x+1) = 0 \Rightarrow x+1 = 0 \Rightarrow x = -1$.

   (b)  A graph of $y = x^2 + 2x + 1$ (not shown) intersects the $x$-axis at $-1$.

   (c)  A table of $y = x^2 + 2x + 1$ (not shown) yields y-values of zero when $x = -1$.

39. (a)  $x^2 + 2 = 0 \Rightarrow x^2 = -2 \Rightarrow$ no real solutions.

   (b)  A graph of $y = x^2 + 2$ (not shown) does not intersect the $x$-axis.

   (c)  A table of $y = x^2 + 2$ (not shown) does not yield y-values of zero.

41.  $x^2 + 2x - 35 = 0 \Rightarrow (x+7)(x-5) = 0 \Rightarrow$ Either $x+7 = 0 \Rightarrow x = -7$ or $x-5 = 0 \Rightarrow x = 5$

43.  $6x^2 - x - 1 = 0 \Rightarrow (3x+1)(2x-1) = 0 \Rightarrow$ Either $3x+1 = 0 \Rightarrow x = -\dfrac{1}{3}$ or $2x-1 = 0 \Rightarrow x = \dfrac{1}{2}$

45.  $4x^2 + 13x + 9 = x \Rightarrow 4x^2 + 12x + 9 = 0 \Rightarrow (2x+3)(2x+3) = 0 \Rightarrow 2x+3 = 0 \Rightarrow x = -\dfrac{3}{2}$

47.  $25x^2 - 350 = 125x \Rightarrow 25x^2 - 125x - 350 = 0 \Rightarrow 25(x^2 - 5x - 14) = 0 \Rightarrow 25(x-7)(x+2) = 0 \Rightarrow$

   Either $x-7 = 0 \Rightarrow x = 7$ or $x+2 = 0 \Rightarrow x = -2$

49.  $10x^2 - 27x + 18 = 0 \Rightarrow (5x-6)(2x-3) = 0 \Rightarrow$ Either $5x-6 = 0 \Rightarrow x = \dfrac{6}{5}$ or $2x-3 = 0 \Rightarrow x = \dfrac{3}{2}$

51. $x^2 = 144 \Rightarrow x = \pm\sqrt{144} \Rightarrow x = \pm 12$

53. $5x^2 - 64 = 0 \Rightarrow 5x^2 = 64 \Rightarrow x^2 = \dfrac{64}{5} \Rightarrow x = \pm\sqrt{\dfrac{64}{5}} \Rightarrow x = \pm\dfrac{8}{\sqrt{5}}$ or $\pm\dfrac{8\sqrt{5}}{5}$

55. $(x+1)^2 = 25 \Rightarrow x+1 = \pm\sqrt{25} \Rightarrow x+1 = \pm 5 \Rightarrow x = -1 \pm 5 \Rightarrow x = -6$ or $4$

57. $(x-1)^2 = 64 \Rightarrow x-1 = \pm\sqrt{64} \Rightarrow x-1 = \pm 8 \Rightarrow x = 1 \pm 8 \Rightarrow x = -7$ or $9$

59. $(2x-1)^2 = 5 \Rightarrow 2x-1 = \pm\sqrt{5} \Rightarrow 2x = 1 \pm\sqrt{5} \Rightarrow x = \dfrac{1 \pm\sqrt{5}}{2}$

61. $10(x-5)^2 = 50 \Rightarrow (x-5)^2 = 5 \Rightarrow x-5 = \pm\sqrt{5} \Rightarrow x = 5 \pm\sqrt{5}$

63. $\left(\dfrac{4}{2}\right)^2 = 2^2 = 4$

65. $\left(\dfrac{-5}{2}\right)^2 = \dfrac{25}{4}$

67. The term needed to complete the square is $\left(\dfrac{-8}{2}\right)^2 = (-4)^2 = 16$. The resulting perfect square is

$(x-4)^2$.

69. The term needed to complete the square is $\left(\dfrac{9}{2}\right)^2 = \dfrac{81}{4}$. The resulting perfect square is $\left(x+\dfrac{9}{2}\right)^2$.

71. $x^2 - 2x = 24 \Rightarrow x^2 - 2x + 1 = 24 + 1 \Rightarrow (x-1)^2 = 25 \Rightarrow x-1 = \pm\sqrt{25} \Rightarrow x-1 = \pm 5 \Rightarrow$

$x = 1 \pm 5 \Rightarrow x = -4$ or $6$

73. $x^2 + 6x - 2 = 0 \Rightarrow x^2 + 6x + 9 = 2 + 9 \Rightarrow (x+3)^2 = 11 \Rightarrow x+3 = \pm\sqrt{11} \Rightarrow x = -3 \pm\sqrt{11}$

75. $x^2 - 3x = 5 \Rightarrow x^2 - 3x + \dfrac{9}{4} = 5 + \dfrac{9}{4} \Rightarrow \left(x-\dfrac{3}{2}\right)^2 = \dfrac{29}{4} \Rightarrow x - \dfrac{3}{2} = \pm\sqrt{\dfrac{29}{4}} \Rightarrow x = \dfrac{3}{2} \pm\dfrac{\sqrt{29}}{2} = \dfrac{3 \pm\sqrt{29}}{2}$

77. $x^2 - 5x + 1 = 0 \Rightarrow x^2 - 5x = -1 \Rightarrow x^2 - 5x + \dfrac{25}{4} = -1 + \dfrac{25}{4} \Rightarrow \left(x-\dfrac{5}{2}\right)^2 = \dfrac{21}{4} \Rightarrow$

$x - \dfrac{5}{2} = \pm\sqrt{\dfrac{21}{4}} \Rightarrow x = \dfrac{5}{2} \pm\dfrac{\sqrt{21}}{2} = \dfrac{5 \pm\sqrt{21}}{2}$

79. $x^2 - 4 = 2x \Rightarrow x^2 - 2x = 4 \Rightarrow x^2 - 2x + 1 = 4 + 1 \Rightarrow (x-1)^2 = 5 \Rightarrow x-1 = \pm\sqrt{5} \Rightarrow x = 1 \pm\sqrt{5}$

81. $2x^2 - 3x = 4 \Rightarrow x^2 - \dfrac{3}{2}x = 2 \Rightarrow x^2 - \dfrac{3}{2}x + \dfrac{9}{16} = 2 + \dfrac{9}{16} \Rightarrow \left(x-\dfrac{3}{4}\right)^2 = \dfrac{41}{16} \Rightarrow x - \dfrac{3}{4} = \pm\sqrt{\dfrac{41}{16}} \Rightarrow$

$x = \dfrac{3}{4} \pm\dfrac{\sqrt{41}}{4} = \dfrac{3 \pm\sqrt{41}}{4}$

83. $4x^2 - 8x - 7 = 0 \Rightarrow x^2 - 2x - \dfrac{7}{4} = 0 \Rightarrow x^2 - 2x + 1 = \dfrac{7}{4} + 1 \Rightarrow (x-1)^2 = \dfrac{11}{4} \Rightarrow$

$x - 1 = \pm\sqrt{\dfrac{11}{4}} \Rightarrow x = 1 \pm \dfrac{\sqrt{11}}{2} \Rightarrow x = \dfrac{2 \pm \sqrt{11}}{2}$

85. $36x^2 + 18x + 1 = 0 \Rightarrow 36\left(x^2 + \dfrac{1}{2}x + \dfrac{1}{16}\right) = -1 + \dfrac{36}{16} \Rightarrow 36\left(x + \dfrac{1}{4}\right)^2 = \dfrac{20}{16} \Rightarrow$

$\left(x + \dfrac{1}{4}\right)^2 = \dfrac{20}{576} \Rightarrow x + \dfrac{1}{4} = \pm\sqrt{\dfrac{20}{576}} \Rightarrow x = -\dfrac{1}{4} \pm \dfrac{\sqrt{20}}{24} = \dfrac{-6 \pm 2\sqrt{5}}{24} = \dfrac{-3 \pm \sqrt{5}}{12}$

87. $3x^2 + 12x = 36 \Rightarrow 3x^2 + 12x - 36 = 0 \Rightarrow 3(x^2 + 4x - 12) = 0 \Rightarrow 3(x+6)(x-2) = 0 \Rightarrow$

Either $x + 6 = 0 \Rightarrow x = -6$ or $x - 2 = 0 \Rightarrow x = 2$

89. $x^2 + 4x = -2 \Rightarrow x^2 + 4x + 4 = -2 + 4 \Rightarrow (x+2)^2 = 2 \Rightarrow x + 2 = \pm\sqrt{2} \Rightarrow x = -2 \pm \sqrt{2}$

91. $3x^2 - 4 = 2 \Rightarrow 3x^2 = 6 \Rightarrow x^2 = 2 \Rightarrow x = \pm\sqrt{2}$

93. $-6x^2 + 70 = 16x \Rightarrow -6x^2 - 16x + 70 = 0 \Rightarrow -2(3x^2 + 8x - 35) = 0 \Rightarrow -2(3x-7)(x+5) = 0 \Rightarrow$

Either $3x - 7 = 0 \Rightarrow x = \dfrac{7}{3}$ or $x + 5 = 0 \Rightarrow x = -5$

95. $-3x(x-8) = 6 \Rightarrow -3x^2 + 24x = 6 \Rightarrow x^2 - 8x = -2 \Rightarrow x^2 - 8x + 16 = -2 + 16 \Rightarrow (x-4)^2 = 14$

$\Rightarrow x - 4 = \pm\sqrt{14} \Rightarrow x = 4 \pm \sqrt{14}$

97. $ax^2 - c = 0 \Rightarrow ax^2 = c \Rightarrow x^2 = \dfrac{c}{a} \Rightarrow x = \pm\sqrt{\dfrac{c}{a}}$

99. (a)    $x^2 - 3x - 18 = 0 \Rightarrow (x+3)(x-6) = 0 \Rightarrow$ Either $x + 3 = 0 \Rightarrow x = -3$ or $x - 6 = 0 \Rightarrow x = 6$

(b)    Graph $Y_1 = X^2 - 3X - 18$ in [–5, 8, 1] by [–25, 5, 5]. See Figures 99a & 99b.

The solutions are the $x$-intercepts, $x = -3$ or $x = 6$.

(c)    Table $Y_1 = X^2 - 3X - 18$ with TblStart = –6 and $\Delta$Tbl = 3. See Figure 99c.

Since $Y_1 = 0$ when $x = -3$ or when $x = 6$, the solutions are $x = -3$ or $x = 6$.

[–5, 8, 1] by [–25, 5, 5]          [–5, 8, 1] by [–25, 5, 5]

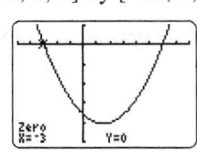          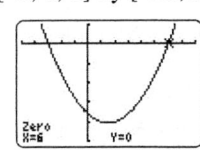          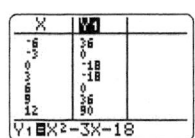

Figure 99a                    Figure 99b                    Figure 99c

101. (a)    $x^2 - 8x + 15 = 0 \Rightarrow (x-3)(x-5) = 0 \Rightarrow$ Either $x - 3 = 0 \Rightarrow x = 3$ or $x - 5 = 0 \Rightarrow x = 5$

(b)    Graph $Y_1 = X^2 - 8X + 15$ in [0, 10, 1] by [–5, 10, 1]. See Figures 101a & 101b.

The solutions are the $x$-intercepts, $x = 3$ or $x = 5$.

(c)   Table $Y_1 = X^2 - 8X + 15$ with TblStart = 1 and $\Delta$Tbl = 1.  See Figure 101c.

Since $Y_1 = 0$ when $x = 3$ or when $x = 5$, the solutions are $x = 3$ or $x = 5$.

[0, 10, 1] by [–5, 10, 1]          [0, 10, 1] by [–5, 10, 1]

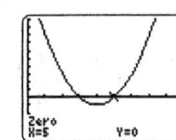          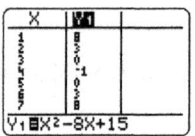

Figure 101a               Figure 101b               Figure 101c

103.  (a)   $4(x^2 + 35) = 48x \Rightarrow x^2 + 35 = 12x \Rightarrow x^2 - 12x + 35 = 0 \Rightarrow (x-5)(x-7) = 0 \Rightarrow$

Either $x - 5 = 0 \Rightarrow x = 5$ or $x - 7 = 0 \Rightarrow x = 7$

(b)   Graph $Y_1 = 4(X^2 + 35) - 48X$ in [–2, 10, 1] by [–6, 6, 1]. See Figures 103a & 103b.

The solutions are the $x$-intercepts, $x = 5$ or $x = 7$.

(c)   Table $Y_1 = 4(X^2 + 35) - 48X$ with TblStart = 3 and $\Delta$Tbl = 1.  See Figure 103c.

Since $Y_1 = 0$ when $x = 5$ or when $x = 7$, the solutions are $x = 5$ or $x = 7$.

[–2, 10, 1] by [–6, 6, 1]          [–2, 10, 1] by [–6, 6, 1]

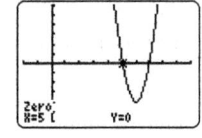          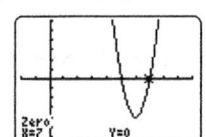          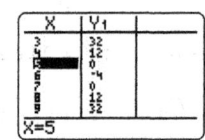

Figure 103a               Figure 103b               Figure 103c

105.  $x = y^2 - 1 \Rightarrow x + 1 = y^2 \Rightarrow \pm\sqrt{x+1} = y \Rightarrow y = \pm\sqrt{x+1}$

107.  $K = \dfrac{1}{2}mv^2 \Rightarrow 2K = mv^2 \Rightarrow \dfrac{2K}{m} = v^2 \Rightarrow \pm\sqrt{\dfrac{2K}{m}} = v \Rightarrow v = \pm\sqrt{\dfrac{2K}{m}}$

109.  $E = \dfrac{k}{r^2} \Rightarrow Er^2 = k \Rightarrow r^2 = \dfrac{k}{E} \Rightarrow r = \pm\sqrt{\dfrac{k}{E}}$

111.  $LC = \dfrac{1}{(2\pi f)^2} \Rightarrow 4\pi^2 f^2 LC = 1 \Rightarrow f^2 = \dfrac{1}{4\pi^2 LC} \Rightarrow f = \pm\sqrt{\dfrac{1}{4\pi^2 LC}} \Rightarrow f = \pm\dfrac{1}{2\pi\sqrt{LC}}$

113.  (a)   $\dfrac{1}{2}x^2 = 450 \Rightarrow x^2 = 900 \Rightarrow x = \pm\sqrt{900} \Rightarrow x = 30$ mph ($x = -30$ has no physical meaning)

(b)   $\dfrac{1}{2}x^2 = 800 \Rightarrow x^2 = 1600 \Rightarrow x = \pm\sqrt{1600} \Rightarrow x = 40$ mph ($x = -40$ has no physical meaning)

115.  $60 - 16t^2 = 0 \Rightarrow -16t^2 = -60 \Rightarrow t^2 = \dfrac{60}{16} \Rightarrow t = \pm\sqrt{\dfrac{60}{16}} \Rightarrow t = \dfrac{\sqrt{60}}{4} \approx 1.9$ seconds. The value $t \approx -1.9$

seconds has no physical meaning.  The toy takes about 1.9 seconds to hit the ground.  This is not

twice the time it takes to fall from a height of 30 feet.

117. $S = 42 \Rightarrow 42 = -0.227x^2 + 8.155x - 8.8 \Rightarrow -0.227x^2 + 8.155x = 50.8$

$\Rightarrow -0.227(x^2 - 35.925x) = 50.8 \Rightarrow x^2 - 35.925x = -223.789$

$\Rightarrow x^2 - 35.925x + 322.651 = -223.789 + 322.651 \Rightarrow (x - 17.9625)^2 = 98.862$

$\Rightarrow x - 17.9625 = \pm 9.943 \Rightarrow x = 17.9625 \pm 9.943 \Rightarrow x \approx 28$ or $8$

28 feet from the screen is not reasonable, so the answer is about 8 feet.

119. Let $x$ represent the number of hours the athletes run. The athletes are running along the legs of a right triangle. The distance between them is given by the length of the hypotenuse. One athlete runs $6x$ miles while the other runs $8x$ miles. By the Pythagorean Theorem we have the following

$(6x)^2 + (8x)^2 = 20^2 \Rightarrow 36x^2 + 64x^2 = 400 \Rightarrow 100x^2 = 400 \Rightarrow x^2 = 4 \Rightarrow x = \pm 2$.

The athletes are 20 miles apart after 2 hours.

121. (a) Let $x$ represent the width of the lot. Then $x + 6$ represents the length. So $x(x + 6) = 520$.

The quadratic equation is $x^2 + 6x - 520 = 0$.

(b) $x^2 + 6x - 520 = 0 \Rightarrow (x + 26)(x - 20) = 0 \Rightarrow x = -26$ or $20$

Since the value $-26$ has no physical meaning, the width is 20 feet.

123. The height of the seedling was about 22 centimeters when the temperature was about 23°C and 34°C.

125. $52 = -\dfrac{25}{144}x^2 + 70 \Rightarrow 18 = \dfrac{25}{144}x^2 \Rightarrow \dfrac{2592}{25} = x^2 \Rightarrow 10.2 \approx x$ Since $x$ represents the number of

months after January 2010, the result is October 2010.

## Section 11.4 The Quadratic Formula

1. We use the quadratic formula to solve equations of the form $ax^2 + bx + c = 0$.

3. $b^2 - 4ac$

5. Factoring, square root property, completing the square and the quadratic formula

7. $x^2 - k = 0 \Rightarrow x^2 = k \Rightarrow x = \pm\sqrt{k}$

9. $x = \dfrac{-11 \pm \sqrt{(11)^2 - 4(2)(-6)}}{2(2)} = \dfrac{-11 \pm \sqrt{169}}{4} = \dfrac{-11 \pm 13}{4} \Rightarrow x = -6$ or $x = \dfrac{1}{2}$

Graph $Y_1 = 2X^2 + 11X - 6$ in [–10, 5, 1] by [–25, 10, 5]. See Figures 9a & 9b. $x = -6$ or $x = \dfrac{1}{2}$

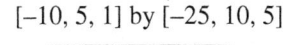

[–10, 5, 1] by [–25, 10, 5]        [–10, 5, 1] by [–25, 10, 5]

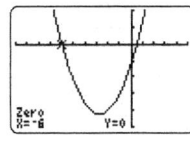       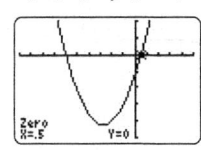

Figure 9a                    Figure 9b

11. $x = \dfrac{-2 \pm \sqrt{(2)^2 - 4(-1)(-1)}}{2(-1)} = \dfrac{-2 \pm \sqrt{0}}{-2} = \dfrac{-2}{-2} \Rightarrow x = 1$

Graph $Y_1 = -X^2 + 2X - 1$ in $[-5, 5, 1]$ by $[-5, 5, 1]$. See Figure 11. $x = 1$

$[-5, 5, 1]$ by $[-5, 5, 1]$                        $[-5, 5, 1]$ by $[-5, 5, 1]$

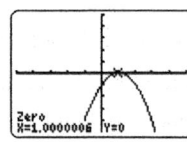                                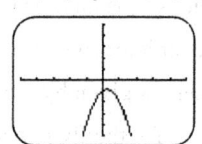

Figure 11                                    Figure 13

13. $x = \dfrac{-(1) \pm \sqrt{(1)^2 - 4(-2)(-1)}}{2(-2)} = \dfrac{-1 \pm \sqrt{-7}}{-4} \Rightarrow$ No real solutions.

Graph $Y_1 = -2X^2 + X - 1$ in $[-5, 5, 1]$ by $[-5, 5, 1]$. See Figure 13. No real solutions

15. $x = \dfrac{-(-6) \pm \sqrt{(-6)^2 - 4(1)(-16)}}{2(1)} = \dfrac{6 \pm \sqrt{100}}{2} = \dfrac{6 \pm 10}{2} \Rightarrow x = -2$ or $x = 8$

17. $x = \dfrac{-(-1) \pm \sqrt{(-1)^2 - 4(4)(-1)}}{2(4)} = \dfrac{1 \pm \sqrt{17}}{8}$

19. $x = \dfrac{-2 \pm \sqrt{(2)^2 - 4(-3)(-1)}}{2(-3)} = \dfrac{-2 \pm \sqrt{-8}}{-6} \Rightarrow$ No real solutions.

21. $x = \dfrac{-(-36) \pm \sqrt{(-36)^2 - 4(36)(9)}}{2(36)} = \dfrac{36 \pm \sqrt{0}}{72} = \dfrac{36}{72} = \dfrac{1}{2}$

23. $x = \dfrac{-(-6) \pm \sqrt{(-6)^2 - 4(2)(-2)}}{2(2)} = \dfrac{6 \pm \sqrt{52}}{4} = \dfrac{6 \pm 2\sqrt{13}}{4} = \dfrac{2(3 \pm \sqrt{13})}{2(2)} = \dfrac{3 \pm \sqrt{13}}{2}$

25. $x = \dfrac{-(-4) \pm \sqrt{(-4)^2 - 4(1)(1)}}{2(1)} = \dfrac{4 \pm \sqrt{12}}{2} = \dfrac{4 \pm 2\sqrt{3}}{2} = \dfrac{2(2 \pm \sqrt{3})}{2} = 2 \pm \sqrt{3}$

27. $x = \dfrac{-(-\frac{1}{2}) \pm \sqrt{(-\frac{1}{2})^2 - 4(\frac{3}{2})(-\frac{3}{2})}}{2(\frac{3}{2})} = \dfrac{\frac{1}{2} \pm \sqrt{\frac{37}{4}}}{3} = \dfrac{\frac{1}{2} \pm \frac{\sqrt{37}}{2}}{3} = \dfrac{1 \pm \sqrt{37}}{6}$

29. $x = \dfrac{-(-2) \pm \sqrt{(-2)^2 - 4(2)(-7)}}{2(2)} = \dfrac{2 \pm \sqrt{60}}{4} = \dfrac{2 \pm 2\sqrt{15}}{4} = \dfrac{2(1 \pm \sqrt{15})}{4} = \dfrac{1 \pm \sqrt{15}}{2}$

31. $x = \dfrac{-10 \pm \sqrt{(10)^2 - 4(-3)(-5)}}{2(-3)} = \dfrac{-10 \pm \sqrt{40}}{-6} = \dfrac{-10 \pm 2\sqrt{10}}{-6} = \dfrac{-2(5 \pm \sqrt{10})}{-6} = \dfrac{5 \pm \sqrt{10}}{3}$

33. (a)  Since the parabola opens upward, $a > 0$.

    (b)  The solutions are the $x$-intercepts, $x = -1$ or $x = 2$

    (c)  Since there are two real solutions, the discriminant is positive.

35. (a)  Since the parabola opens upward, $a > 0$.

    (b)  Since there are no $x$-intercepts, there are no real solutions.

(c)    Since there are no real solutions, the discriminant is negative.

37. (a)    Since the parabola opens downward, $a < 0$.

    (b)    The solution is the $x$-intercept, $x = 2$

    (c)    Since there is one real solution, the discriminant is zero.

39. (a)    $(1)^2 - 4(3)(-2) = 25$

    (b)    Since the discriminant is positive, there are two real solutions.

    (c)    Graph $Y_1 = 3X^2 + X - 2$ in [–3, 3, 1] by [–3, 3, 1]. See Figure 39. There are two $x$-intercepts.

[–3, 3, 1] by [–3, 3, 1]        [0, 4, 1] by [–3, 3, 1]

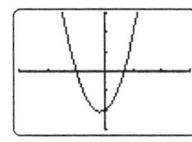

    Figure 39           Figure 41

41. (a)    $(-4)^2 - 4(1)(4) = 0$

    (b)    Since the discriminant is zero, there is only one real solution.

    (c)    Graph $Y_1 = X^2 - 4X + 4$ in [0, 4, 1] by [–3, 3, 1]. See Figure 41. There is one $x$-intercept.

43. (a)    $\left(\dfrac{3}{2}\right)^2 - 4\left(\dfrac{1}{2}\right)(2) = -\dfrac{7}{4}$

    (b)    Since the discriminant is negative, there are no real solutions.

    (c)    Graph $Y_1 = (1/2)X^2 + (3/2)X + 2$ in [–10, 10, 1] by [0, 10, 1]. See Figure 43. There are no $x$-intercepts.

[–10, 10, 1] by [0, 10, 1]     [–5, 5, 1] by [–8, 5, 1]

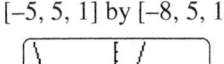
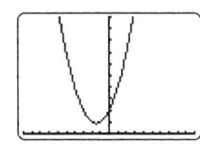
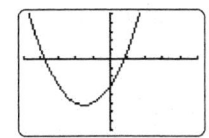

    Figure 43           Figure 45

45. (a)    $(3)^2 - 4(1)(-3) = 21$

    (b)    Since the discriminant is positive, there are two real solutions.

    (c)    Graph $Y_1 = X^2 + 3X - 3$ in [–5, 5, 1] by [–8, 5, 1]. See Figure 45. There are two $x$-intercepts.

47.    $x = \dfrac{-(-2) \pm \sqrt{(-2)^2 - 4(1)(-1)}}{2(1)} = \dfrac{2 \pm \sqrt{8}}{2} = \dfrac{2 \pm 2\sqrt{2}}{2} = \dfrac{2(1 \pm \sqrt{2})}{2} = 1 \pm \sqrt{2}$

49.    $x = \dfrac{-(-1) \pm \sqrt{(-1)^2 - 4(-2)(3)}}{2(-2)} = \dfrac{1 \pm \sqrt{25}}{-4} = \dfrac{1 \pm 5}{-4} = -\dfrac{3}{2}$ or 1

51.    $x = \dfrac{-1 \pm \sqrt{(1)^2 - 4(1)(5)}}{2(1)} = \dfrac{-1 \pm \sqrt{-19}}{2} \Rightarrow$ No real solutions. No $x$-intercepts.

53. $x = \dfrac{-(0) \pm \sqrt{(0)^2 - 4(1)(9)}}{2(1)} = \dfrac{0 \pm \sqrt{-36}}{2} \Rightarrow$   No real solutions. No $x$-intercepts.

55. $x = \dfrac{-4 \pm \sqrt{(4)^2 - 4(3)(-2)}}{2(3)} = \dfrac{-4 \pm \sqrt{40}}{6} = \dfrac{-4 \pm 2\sqrt{10}}{6} = \dfrac{2(-2 \pm \sqrt{10})}{6} = \dfrac{-2 \pm \sqrt{10}}{3}$

57. $x^2 + 9 = 0 \Rightarrow x^2 = -9 \Rightarrow x = \pm\sqrt{-9} \Rightarrow x = \pm 3i$

59. $x^2 + 80 = 0 \Rightarrow x^2 = -80 \Rightarrow x = \pm\sqrt{-80} \Rightarrow x = \pm\sqrt{-16 \cdot 5} \Rightarrow x = \pm 4i\sqrt{5}$

61. $x^2 + \dfrac{1}{4} = 0 \Rightarrow x^2 = -\dfrac{1}{4} \Rightarrow x = \pm\sqrt{-\dfrac{1}{4}} \Rightarrow x = \pm\dfrac{1}{2}i$

63. $16x^2 + 9 = 0 \Rightarrow 16x^2 = -9 \Rightarrow x^2 = -\dfrac{9}{16} \Rightarrow x = \pm\sqrt{-\dfrac{9}{16}} \Rightarrow x = \pm\dfrac{3}{4}i$

65. $x^2 = -6 \Rightarrow x = \pm\sqrt{-6} \Rightarrow x = \pm i\sqrt{6}$

67. $x^2 - 3 = 0 \Rightarrow x^2 = 3 \Rightarrow x = \pm\sqrt{3}$

69. $x^2 + 2 = 0 \Rightarrow x^2 = -2 \Rightarrow x = \pm\sqrt{-2} \Rightarrow x = \pm i\sqrt{2}$

71. $x = \dfrac{-(-1) \pm \sqrt{(-1)^2 - 4(1)(2)}}{2(1)} = \dfrac{1 \pm \sqrt{-7}}{2} = \dfrac{1 \pm i\sqrt{7}}{2} = \dfrac{1}{2} \pm i\dfrac{\sqrt{7}}{2}$

73. $x = \dfrac{-3 \pm \sqrt{3^2 - 4(2)(4)}}{2(2)} = \dfrac{-3 \pm \sqrt{-23}}{4} = \dfrac{-3 \pm i\sqrt{23}}{4} = -\dfrac{3}{4} \pm i\dfrac{\sqrt{23}}{4}$

75. $x = \dfrac{-(-4) \pm \sqrt{(-4)^2 - 4(1)(1)}}{2(1)} = \dfrac{4 \pm \sqrt{12}}{2} = \dfrac{4 \pm 2\sqrt{3}}{2} = 2 \pm \sqrt{3}$

77. $x = \dfrac{-1 \pm \sqrt{1^2 - 4(1)(2)}}{2(1)} = \dfrac{-1 \pm \sqrt{-7}}{2} = \dfrac{-1 \pm i\sqrt{7}}{2} = -\dfrac{1}{2} \pm i\dfrac{\sqrt{7}}{2}$

79. $x = \dfrac{-2 \pm \sqrt{(2)^2 - 4(5)(4)}}{2(5)} = \dfrac{-2 \pm \sqrt{-76}}{10} = \dfrac{-2 \pm 2i\sqrt{19}}{10} = -\dfrac{1}{5} \pm i\dfrac{\sqrt{19}}{5}$

81. $x = \dfrac{-\frac{3}{4} \pm \sqrt{(\frac{3}{4})^2 - 4(\frac{1}{2})(1)}}{2(\frac{1}{2})} = \dfrac{-\frac{3}{4} \pm \sqrt{-\frac{23}{16}}}{1} = -\dfrac{3}{4} \pm \sqrt{\dfrac{-23}{16}} = -\dfrac{3}{4} \pm \dfrac{\sqrt{-23}}{4} = -\dfrac{3}{4} \pm i\dfrac{\sqrt{23}}{4}$

83. $x = \dfrac{-1 \pm \sqrt{(1)^2 - 4(1)(4)}}{2(1)} = \dfrac{-1 \pm \sqrt{-15}}{2} = -\dfrac{1}{2} \pm i\dfrac{\sqrt{15}}{2}$

85. $x = \dfrac{-(-2) \pm \sqrt{(-2)^2 - 4(2)(-1)}}{2(2)} = \dfrac{2 \pm \sqrt{12}}{4} = \dfrac{2 \pm 2\sqrt{3}}{4} = \dfrac{2(1 \pm \sqrt{3})}{4} = \dfrac{1 \pm \sqrt{3}}{2}$

87. $x = \dfrac{-(-1) \pm \sqrt{(-1)^2 - 4(2)(2)}}{2(2)} = \dfrac{1 \pm \sqrt{-15}}{4} = \dfrac{1 \pm i\sqrt{15}}{4} = \dfrac{1}{4} \pm i\dfrac{\sqrt{15}}{4}$

89. $x^2 + 2x + 4 = 0 \Rightarrow x^2 + 2x + 1 = -4 + 1 \Rightarrow (x + 1)^2 = -3 \Rightarrow x + 1 = \pm\sqrt{-3} \Rightarrow$
$x = -1 \pm \sqrt{-3} \Rightarrow x = -1 \pm i\sqrt{3}$

91. $x(x+4) = -5 \Rightarrow x^2 + 4x + 5 = 0 \Rightarrow x^2 + 4x + 4 = -5 + 4 \Rightarrow (x+2)^2 = -1 \Rightarrow x + 2 = \pm\sqrt{-1} \Rightarrow$
    $x = -2 \pm \sqrt{-1} \Rightarrow x = -2 \pm i$

93. $2x^2 - 4x + 6 = 0 \Rightarrow x^2 - 2x + 3 = 0 \Rightarrow x^2 - 2x + 1 = -3 + 1 \Rightarrow (x-1)^2 = -2 \Rightarrow$
    $x - 1 = \pm\sqrt{-2} \Rightarrow x = 1 \pm \sqrt{-2} \Rightarrow x = 1 \pm i\sqrt{2}$

95. $x^2 - 3x + 2 = 0 \Rightarrow (x-1)(x-2) = 0 \Rightarrow x = 1 \text{ or } x = 2$

97. Multiply both sides of the equation by 4 to obtain the following result.

    $0.5x^2 - 1.75x - 1 = 0 \Rightarrow 2x^2 - 7x - 4 = 0 \Rightarrow (2x+1)(x-4) = 0 \Rightarrow x = -\dfrac{1}{2} \text{ or } x = 4$

99. $x = \dfrac{-(-5) \pm \sqrt{(-5)^2 - 4(1)(2)}}{2(1)} = \dfrac{5 \pm \sqrt{17}}{2}$

101. $x = \dfrac{-1 \pm \sqrt{(1)^2 - 4(2)(8)}}{2(2)} = \dfrac{-1 \pm \sqrt{-63}}{4} = \dfrac{-1 \pm 3i\sqrt{7}}{4} = -\dfrac{1}{4} \pm \dfrac{3}{4}i\sqrt{7}$

103. $4x^2 - 1 = 0 \Rightarrow (2x-1)(2x+1) = 0 \Rightarrow \text{Either } 2x - 1 = 0 \Rightarrow x = \dfrac{1}{2} \text{ or } 2x + 1 = 0 \Rightarrow x = -\dfrac{1}{2}$

105. $3x^2 + 6 = 0 \Rightarrow 3(x^2 + 2) = 0 \Rightarrow x^2 + 2 = 0 \Rightarrow x^2 = -2 \Rightarrow x = \pm\sqrt{-2} \Rightarrow x = \pm i\sqrt{2}$

107. $9x^2 + 1 = 6x \Rightarrow 9x^2 - 6x + 1 = 0 \Rightarrow (3x-1)(3x-1) = 0 \Rightarrow 3x - 1 = 0 \Rightarrow x = \dfrac{1}{3}$

109. $\dfrac{1}{9}x^2 + \dfrac{11}{3}x - 42 = 0 \Rightarrow x^2 + 33x - 378 = 0 \Rightarrow x = (x-9)(x+42) = 0 \Rightarrow x = 9 \text{ mph}$

    The value $x = -42$ has no physical meaning.

111. $\dfrac{1}{9}x^2 + \dfrac{11}{3}x - 390 = 0 \Rightarrow x^2 + 33x - 3510 = 0 \Rightarrow x = (x-45)(x+78) = 0 \Rightarrow x = 45 \text{ mph}$

    The value $x = -78$ has no physical meaning.

113. $-0.095x^2 + 1.85x + 6 = 12 \Rightarrow -0.095x^2 + 1.85x - 6 = 0 \Rightarrow$

    $x = \dfrac{-(1.85) \pm \sqrt{(1.85)^2 - 4(-0.095)(-6)}}{2(-0.095)} \Rightarrow x \approx 4.11 \text{ or } x \approx 15.36$

    Since $x$ represents the number of years after 1989 the result is about 1993 and 2004.

115. (a)  $G(0) = 0.4(0)^2 + 1.8(0) + 6 = 6$, In October 2010 Groupon's value was \$6 billion.

    (b)  $0.4x^2 + 1.8x + 6 = 15 \Rightarrow 0.4x^2 + 1.8x - 9 = 0 \Rightarrow$

    $x = \dfrac{-(1.8) \pm \sqrt{(1.8)^2 - 4(0.4)(-9)}}{2(0.4)} \Rightarrow x = 3 \text{ or } x = -7.5.$

    Since x represents the number of months since October 2010 the result is January 2011. The value $x = -7.5$ has no physical meaning.

117. $2.39x^2 + 5.04x + 5.1 = 200 \Rightarrow 2.39x^2 + 5.04x - 194.9 = 0$

$$x = \frac{-5.04 \pm \sqrt{(5.04)^2 - 4(2.39)(-194.9)}}{2(2.39)} = \frac{-5.04 \pm \sqrt{1888.6456}}{4.78} \Rightarrow x \approx 8.04 \text{ ; about 1992}$$

The value $x \approx -10.15$ has no meaning in this problem. Our answer agrees with the graph.

119. Let $x$ represent the speed of the airplane without wind. The first trip takes $\dfrac{500}{x-20}$ hours. The second

trip takes $\dfrac{500}{x+10}$ hours. The total is $\dfrac{500}{x-20} + \dfrac{500}{x+10} = 4$. Solve this equation for $x$.

$\dfrac{500}{x-20} + \dfrac{500}{x+10} = 4 \Rightarrow \dfrac{125}{x-20} + \dfrac{125}{x+10} = 1 \Rightarrow 125(x+10) + 125(x-20) = (x-20)(x+10) \Rightarrow$

$250x - 1250 = x^2 - 10x - 200 \Rightarrow x^2 - 260x + 1050 = 0 \Rightarrow$

$$x = \frac{-(-260) \pm \sqrt{(-260)^2 - 4(1)(1050)}}{2(1)} = \frac{260 \pm \sqrt{63,400}}{2} = 130 \pm 5\sqrt{634} \approx 256 \text{ mph}$$

121. Let x represent the height of the screen. Then $x+3$ represents the width. The equation is

$x(x+3) = 154$.

(a) Graph $Y_1 = X^2 + 3X - 154$ in [–20, 20, 2] by [–200, 100, 50]. See Figure 121a.

$x = 11, x+3 = 14$   The screen is 11 inches by 14 inches.

(b) Table $Y_1 = X^2 + 3X - 154$ with TblStart = 7 and $\Delta$Tbl = 1. See Figure 121b.

$x = 11, x+3 = 14$   The screen is 11 inches by 14 inches.

(c) $x^2 + 3x - 154 = 0 \Rightarrow (x-11)(x+14) = 0 \Rightarrow x = 11$; The value $x = -14$ has no physical

meaning. The screen is 11 inches by 14 inches.

[–20, 20, 2] by [–200, 100, 50]

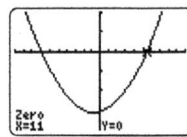
Figure 121a

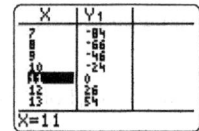
Figure 121b

123. (a) The rate of change is not constant.

(b) From the table, the height was 7 centimeters after about 75 seconds. *Answers may vary.*

(c) $x = \dfrac{-(-0.15) \pm \sqrt{(0.15)^2 - 4(0.0004)(9)}}{2(0.0004)} = \dfrac{0.15 \pm \sqrt{0.0081}}{0.0008} = \dfrac{0.15 \pm 0.09}{0.0008} \Rightarrow x = 75$ seconds

The value $x = 300$ is not possible for this problem.

## Checking Basic Concepts Sections 11.3 and 11.4

1.  Symbolic: $2x^2 - 7x + 3 = 0 \Rightarrow (2x - 1)(x - 3) = 0 \Rightarrow$ Either $2x - 1 = 0 \Rightarrow x = \dfrac{1}{2}$ or $x - 3 = 0 \Rightarrow x = 3$

    Graphical: Graph $Y_1 = 2X^2 - 7X + 3$ in $[-5, 5, 1]$ by $[-5, 5, 1]$.

    See Figures 1a & 1b. $x = \dfrac{1}{2}$ or $x = 3$

    $[-5, 5, 1]$ by $[-5, 5, 1]$

    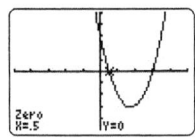

    | Figure 1a | Figure 1b |

2.  $x^2 = 5 \Rightarrow x = \pm\sqrt{5}$

3.  $x^2 - 4x + 1 = 0 \Rightarrow x^2 - 4x + 4 = -1 + 4 \Rightarrow (x - 2)^2 = 3 \Rightarrow x - 2 = \pm\sqrt{3} \Rightarrow x = 2 \pm \sqrt{3}$

4.  $x^2 + y^2 = 1 \Rightarrow y^2 = 1 - x^2 \Rightarrow y = \pm\sqrt{1 - x^2}$

5.  (a) $2x^2 = 3x + 1 \Rightarrow 2x^2 - 3x - 1 = 0 \Rightarrow x = \dfrac{-(-3) \pm \sqrt{(-3)^2 - 4(2)(-1)}}{2(2)} = \dfrac{3 \pm \sqrt{17}}{4}$

    (b) $9x^2 - 24x + 16 = 0 \Rightarrow x = \dfrac{-(-24) \pm \sqrt{(-24)^2 - 4(9)(16)}}{2(9)} = \dfrac{24 \pm \sqrt{0}}{18} = \dfrac{24}{18} = \dfrac{4}{3}$

    (c) $x = \dfrac{-1 \pm \sqrt{1^2 - 4(1)(2)}}{2(1)} = \dfrac{-1 \pm \sqrt{-7}}{2} = \dfrac{-1 \pm i\sqrt{7}}{2} = -\dfrac{1}{2} \pm i\dfrac{\sqrt{7}}{2}$

6.  (a) $(-5)^2 - 4(1)(5) = 5$; Since the discriminant is positive, there are two real solutions.

    (b) $(-5)^2 - 4(2)(4) = -7$; Since the discriminant is negative, there are no real solutions.

    (c) $(-56)^2 - 4(49)(16) = 0$; Since the discriminant is zero, there is one real solution

7.  (a) $x^2 + 5 = 0 \Rightarrow x^2 = -5 \Rightarrow x = \pm\sqrt{-5} \Rightarrow x = \pm i\sqrt{5}$

    (b) $x^2 + x + 3 = 0 \Rightarrow x = \dfrac{-1 \pm \sqrt{(1)^2 - 4(1)(3)}}{2(1)} = \dfrac{-1 \pm \sqrt{-11}}{2} = -\dfrac{1}{2} \pm i\dfrac{\sqrt{11}}{2}$

## Section 11.5 Quadratic Inequalities

1.  An inequality has an inequality sign rather than an equals sign.

3.  No, since $9 \nless 7$.

5.  $-2 < x < 4$

7.  Yes

9.  Yes

11. No, this inequality is linear.

13. Yes, since $2(3)^2 + (3) - 1 = 20$ and $20 > 0$.

15. No, since $(0)^2 + 2 = 2$ and $2 \nleq 0$.

17. No, since $(-3)^2 - 3(-3) = 18$ and $18 \nleq 1$.

19. (a)  The solutions are the $x$-intercepts, $x = -3$ or $x = 2$.

    (b)  $-3 < x < 2$

    (c)  $x < -3$ or $x > 2$

21. (a)  The solutions are the $x$-intercepts, $x = -2$ or $x = 2$.

    (b)  $-2 < x < 2$

    (c)  $x < -2$ or $x > 2$

23. (a)  The solutions are the $x$-intercepts, $x = -10$ or $x = 5$.

    (b)  $x < -10$ or $x > 5$

    (c)  $-10 < x < 5$

25. Graph $Y_1 = \pi X \wedge 2 - \sqrt{(3)} X$ and $Y_2 = 3/11$ in $[-4.7, 4.7, 1]$ by $[-3.1, 3.1, 1]$.

    See Figures 25a and 25b. The solution is $[-0.128, 0.679]$.

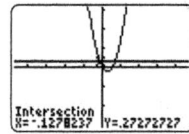

      Figure 25a          Figure 25b

27. (a)  The solutions are $x = -2$ or $x = 2$.

    (b)  $-2 < x < 2$

    (c)  $x < -2$ or $x > 2$

29. (a)  The solutions are $x = -4$ or $x = 0$.

    (b)  $-4 < x < 0$

    (c)  $x < -4$ or $x > 0$

31. $x^2 + 10x + 21 = 0 \Rightarrow (x + 7)(x + 3) = 0 \Rightarrow x = -7$ or $x = -3$

    Since the parabola opens upward, the solution is $[-7, -3]$.

33. $3x^2 - 9x + 6 = 0 \Rightarrow 3(x - 1)(x - 2) = 0 \Rightarrow x = 1$ or $x = 2$

    Since the parabola opens upward, the solution is $(-\infty, 1) \cup (2, \infty)$.

35. $x^2 = 10 \Rightarrow x = -\sqrt{10}$ or $x = \sqrt{10}$  Since the parabola opens upward, the solution is $(-\sqrt{10}, \sqrt{10})$.

37. $x(6 - x) = 0 \Rightarrow x = 0$ or $x = 6$  Since the parabola opens upward, the solution is $(-\infty, 0) \cup (6, \infty)$.

39. $x(4 - x) = 2 \Rightarrow -x^2 + 4x - 2 = 0 \Rightarrow x = \dfrac{-4 \pm \sqrt{(4)^2 - 4(-1)(-2)}}{2(-1)} = 2 \pm \sqrt{2}$

    Since the parabola opens downward, the solution is $(-\infty, 2 - \sqrt{2}] \cup [2 + \sqrt{2}, \infty)$.

41. (a) $x^2 - 4 = 0 \Rightarrow x^2 = 4 \Rightarrow x = -2$ or $x = 2$

   (b) Since the parabola opens upward, the solution is $-2 < x < 2$.

   (c) Since the parabola opens upward, the solution is $x < -2$ or $x > 2$.

43. (a) $x^2 + x - 1 = 0 \Rightarrow x = \dfrac{-1 \pm \sqrt{(1)^2 - 4(1)(-1)}}{2(1)} \Rightarrow x = \dfrac{-1 \pm \sqrt{5}}{2}$

   (b) Since the parabola opens upward, the solution is $\dfrac{-1 - \sqrt{5}}{2} < x < \dfrac{-1 + \sqrt{5}}{2}$.

   (c) Since the parabola opens upward, the solution is $x < \dfrac{-1 - \sqrt{5}}{2}$ or $x > \dfrac{-1 + \sqrt{5}}{2}$.

45. First replace the inequality symbol with an equals sign and solve the resulting equation.

   $x^2 + 4x + 3 = 0 \Rightarrow (x+3)(x+1) = 0 \Rightarrow x = -3$ or $-1$. The parabola given by $y = x^2 + 4x + 3$ lies below the x-axis when $-3 < x < -1$. The interval is $(-3, -1)$.

47. First replace the inequality symbol with an equals sign and solve the resulting equation.

   $2x^2 - x - 15 = 0 \Rightarrow (2x+5)(x-3) = 0 \Rightarrow x = -2.5$ or $3$. The parabola given by $y = 2x^2 - x - 15$ lies above the x-axis when $x \le -2.5$ or $x \ge 3$. The interval is $(-\infty, -2.5] \cup [3, \infty)$.

49. First replace the inequality symbol with an equals sign and solve the resulting equation.

   $2x^2 = 8 \Rightarrow x^2 = 4 \Rightarrow x^2 - 4 = 0 \Rightarrow (x+2)(x-2) = 0 \Rightarrow x = -2$ or $2$. The parabola given by $y = 2x^2 - 8$ lies below the x-axis when $-2 \le x \le 2$. The interval is $[-2, 2]$.

51. The value of $x^2$ is greater than $-5$ for all values of $x$. The parabola given by $y = x^2 + 5$ lies entirely above the x-axis. The interval is $(-\infty, \infty)$.

53. First replace the inequality symbol with an equals sign and solve the resulting equation.

   $-x^2 + 3x = 0 \Rightarrow -x(x-3) = 0 \Rightarrow x = 0$ or $3$. The parabola given by $y = -x^2 + 3x$ lies above the x-axis when $0 < x < 3$. The interval is $(0, 3)$.

55. $x^2 + 2 \le 0$ has no solutions; the left side of the inequality is always greater than or equal to 2.

57. $(x-2)^2 \le 0 \Rightarrow (x-2)^2 = 0$ because $(x-2)^2$ can never be less than 0.

   $(x-2)^2 = 0 \Rightarrow x - 2 = 0 \Rightarrow x = 2$.

59. First replace the inequality symbol with an equals sign and solve the resulting equation.

   $(x+1)^2 = 0 \Rightarrow x + 1 = 0 \Rightarrow x = -1$. The parabola given by $y = (x+1)^2$ lies above the x-axis when $x < -1$ or $x > -1$. The interval is $(-\infty, -1) \cup (-1, \infty)$.

61. $x(1-x) \ge -2 \Rightarrow x(1-x) + 2 \ge 0$. Replace the inequality symbol with an equals sign and solve the resulting equation. $x(1-x) + 2 = 0 \Rightarrow x - x^2 + 2 = 0 \Rightarrow -x^2 + x + 2 = 0 \Rightarrow (-x+2)(x+1) = 0 \Rightarrow x = 2$ or $x = -1$. The parabola given by $y = -x^2 + x + 2$ lies above or on the x-axis when $-1 \le x \le 2$.

The interval is $[-1, 2]$.

63. (a)  Graph $Y_1 = 0.0000375X^2 - 0.175X + 1000$ and $Y_2 = 850$ in [0, 4000, 1000] by

[500, 1200, 100]. See Figures 63a & 63b. The elevation is 850 feet or less from 1131 feet to

3535 feet (approximately).

(b)  The elevation is 850 feet or more before 1131 feet or after 3535 feet (approximately).

[0, 4000, 1000] by [500, 1200, 100]    [0, 4000, 1000] by [500, 1200, 100]

                    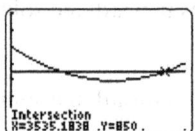

Figure 63a                          Figure 63b

65. (a)  From the formula, $f(1985) = -0.05107(1985)^2 + 194.74(1985) - 184,949 \approx 383$.

From the graph $f(1985) \approx 383$.

(b)  The death rate was 500 or less about 1969 and after.

(c)  $-0.05107x^2 + 194.74x - 184,949 = 500 \Rightarrow -0.05107x^2 + 194.74x - 185,449 = 0 \Rightarrow$

$$x = \frac{-194.74 \pm \sqrt{(194.74)^2 - 4(-0.05107)(-185,449)}}{2(-0.05107)} \approx 1845 \text{ or } 1969.$$

The only valid solution is 1969. The death rate was 500 or less about 1969 or after.

67.  Let $w$ represent the width of the pen. Then $w + 5$ represents the length and the area is given by

$w(w+5) = w^2 + 5w$. We want to find $w$ so that $w^2 + 5w \geq 176$ and $w^2 + 5w \leq 500$.

$w^2 + 5w - 176 = 0 \Rightarrow (w+16)(w-11) = 0 \Rightarrow w = -16$ or $w = 11$, so $w \geq 11$. The values $w \leq -16$ have

no meaning. $w^2 + 5w - 500 = 0 \Rightarrow (w+25)(w-20) = 0 \Rightarrow w = -25$ or $w = 20$, so $w \leq 20$. The values

$w \leq -25$ have no meaning. The width must be from 11 feet to 20 feet.

## Section 11.6 Equations in Quadratic Form

1.  $u^2 - 7u + 6 = 0 \Rightarrow (u-1)(u-6) = 0 \Rightarrow u = 1$ or $6$

When $u = 1$, $x^2 = 1 \Rightarrow x = \pm 1$. When $u = 6$, $x^2 = 6 \Rightarrow x = \pm\sqrt{6}$.

3.  $3u^2 + u - 10 = 0 \Rightarrow (u+2)(3u-5) = 0 \Rightarrow u = -2$ or $\dfrac{5}{3}$

When $u = -2$, $z^3 = -2 \Rightarrow z = -\sqrt[3]{2}$. When $u = \dfrac{5}{3}$, $z^3 = \dfrac{5}{3} \Rightarrow z = \sqrt[3]{\dfrac{5}{3}}$.

5.  $4u^2 + 17u + 15 = 0 \Rightarrow (u+3)(4u+5) = 0 \Rightarrow u = -3$ or $-\dfrac{5}{4}$

When $u = -3$, $n^{-1} = -3 \Rightarrow n = -\dfrac{1}{3}$. When $u = -\dfrac{5}{4}$, $n^{-1} = -\dfrac{5}{4} \Rightarrow n = -\dfrac{4}{5}$.

7. Let $u = x^2$. Then $u^2 = 8u + 9 \Rightarrow u^2 - 8u - 9 = 0 \Rightarrow (u+1)(u-9) = 0 \Rightarrow u = -1$ or $9$.

When $u = -1$, $x^2 = -1 \Rightarrow$ no solutions. When $u = 9$, $x^2 = 9 \Rightarrow x = \pm\sqrt{9} = -3$ or $3$.

9. Let $u = x^3$. Then $3u^2 - 5u - 2 = 0 \Rightarrow (3u+1)(u-2) = 0 \Rightarrow u = -\dfrac{1}{3}$ or $2$.

When $u = -\dfrac{1}{3}$, $x^3 = -\dfrac{1}{3} \Rightarrow x = -\sqrt[3]{\dfrac{1}{3}}$. When $u = 2$, $x^3 = 2 \Rightarrow x = \sqrt[3]{2}$.

11. Let $u = z^{-1}$. Then $2u^2 + 11u = 40 \Rightarrow 2u^2 + 11u - 40 = 0 \Rightarrow (u+8)(2u-5) = 0 \Rightarrow u = -8$ or $\dfrac{5}{2}$.

When $u = -8$, $z^{-1} = -8 \Rightarrow z = -\dfrac{1}{8}$. When $u = \dfrac{5}{2}$, $z^{-1} = \dfrac{5}{2} \Rightarrow z = \dfrac{2}{5}$.

13. Let $u = x^{1/3}$. Then $u^2 - 2u + 1 = 0 \Rightarrow (u-1)^2 = 0 \Rightarrow u = 1$. When $u = 1$, $x^{1/3} = 1 \Rightarrow x = 1^3 = 1$.

15. Let $u = x^{1/5}$. Then $u^2 - 33u + 32 = 0 \Rightarrow (u-1)(u-32) = 0 \Rightarrow u = 1$ or $32$.

When $u = 1$, $x^{1/5} = 1 \Rightarrow x = 1^5 = 1$. When $u = 32$, $x^{1/5} = 32 \Rightarrow x = 32^5 = 33,554,432$.

17. Let $u = x^{1/2}$. Then $u^2 - 13u + 36 = 0 \Rightarrow (u-4)(u-9) = 0 \Rightarrow u = 4$ or $9$.

When $u = 4$, $x^{1/2} = 4 \Rightarrow x = 4^2 = 16$. When $u = 9$, $x^{1/2} = 9 \Rightarrow x = 9^2 = 81$.

19. Let $u = z^{1/4}$. Then $u^2 - 2u + 1 = 0 \Rightarrow (u-1)^2 = 0 \Rightarrow u = 1$. When $u = 1$, $z^{1/4} = 1 \Rightarrow x = 1^4 = 1$.

21. Let $u = x+1$. Then $u^2 - 5u - 14 = 0 \Rightarrow (u+2)(u-7) = 0 \Rightarrow u = -2$ or $7$.

When $u = -2$, $x+1 = -2 \Rightarrow x = -3$. When $u = 7$, $x+1 = 7 \Rightarrow x = 6$.

23. Let $u = x^2 - 1$. Then $u^2 - 4 = 0 \Rightarrow (u+2)(u-2) = 0 \Rightarrow u = -2$ or $2$.

When $u = -2$, $x^2 - 1 = -2 \Rightarrow x^2 = -1 \Rightarrow$ no solutions.

When $u = 2$, $x^2 - 1 = 2 \Rightarrow x^2 = 3 \Rightarrow x = -\sqrt{3}$ or $\sqrt{3}$.

25. $x^4 - 16 = 0 \Rightarrow (x^2 - 4)(x^2 + 4) = 0 \Rightarrow (x-2)(x+2)(x^2+4) = 0 \Rightarrow x = 2$ or $x = -2$ or

$x^2 = -4 \Rightarrow x = \pm\sqrt{-4} \Rightarrow x = \pm 2i$

27. $x^3 + x = 0 \Rightarrow x(x^2+1) = 0 \Rightarrow x = 0$ or $x^2 = -1 \Rightarrow x = \pm\sqrt{-1} \Rightarrow x = \pm i$

29. $x^4 - 2 = x^2 \Rightarrow x^4 - x^2 - 2 = 0 \Rightarrow (x^2-2)(x^2+1) = 0 \Rightarrow x^2 = 2$ or $x^2 = -1 \Rightarrow x = \pm\sqrt{2}$ or $x = \pm i$

31. $\dfrac{1}{x} + \dfrac{1}{x^2} = -\dfrac{1}{2} \Rightarrow 2x^2\left(\dfrac{1}{x} + \dfrac{1}{x^2}\right) = 2x^2\left(-\dfrac{1}{2}\right) \Rightarrow 2x + 2 = -x^2 \Rightarrow x^2 + 2x + 2 = 0 \Rightarrow x^2 + 2x = -2$

$\Rightarrow x^2 + 2x + 1 = -2 + 1 \Rightarrow (x+1)^2 = -1 \Rightarrow x + 1 = \pm\sqrt{-1} \Rightarrow x = -1 \pm i$

33. $\dfrac{2}{x-2} - \dfrac{1}{x} = -\dfrac{1}{2} \Rightarrow 2x(x-2)\left(\dfrac{2}{x-2} - \dfrac{1}{x}\right) = 2x(x-2)\left(-\dfrac{1}{2}\right) \Rightarrow 4x - 2(x-2) = -x(x-2)$

$\Rightarrow 4x - 2x + 4 = -x^2 + 2x \Rightarrow x^2 + 4 = 0 \Rightarrow x^2 = -4 \Rightarrow x = \pm\sqrt{-4} \Rightarrow x = \pm 2i$

## Checking Basic Concepts Sections 11.5 and 11.6

1. $x^2 - x - 6 = 0 \Rightarrow (x+2)(x-3) = 0 \Rightarrow x = -2$ or $x = 3$

   Since the parabola opens upward, the solution is $(-\infty, \ -2) \cup (3, \ \infty)$

2. $3x^2 + 5x + 2 = 0 \Rightarrow (x+1)(3x+2) = 0 \Rightarrow x = -1$ or $x = -\dfrac{2}{3}$

   Since the parabola opens upward, the solution is $\left[ -1, \ -\dfrac{2}{3} \right]$.

3. Let $u = x^3$. Then $u^2 + 6u - 16 = 0 \Rightarrow (u+8)(u-2) = 0 \Rightarrow u = -8$ or $2$.

   When $u = -8$, $x^3 = -8 \Rightarrow x = \sqrt[3]{-8} = -2$. When $u = 2$, $x^3 = 2 \Rightarrow x = \sqrt[3]{2}$.

4. Let $u = x^{1/3}$. Then $u^2 - 7u - 8 = 0 \Rightarrow (u+1)(u-8) = 0 \Rightarrow u = -1$ or $8$.

   When $u = -1$, $x^{1/3} = -1 \Rightarrow x = (-1)^3 = -1$. When $u = 8$, $x^{1/3} = 8 \Rightarrow x = 8^3 = 512$.

5. $x^4 + 2x^2 + 1 = 0 \Rightarrow (x^2 + 1)^2 = 0 \Rightarrow x^2 + 1 = 0 \Rightarrow x^2 = -1 \Rightarrow x = \pm\sqrt{-1} \Rightarrow x = \pm i$

## Chapter 11 Review

1. Vertex: $(-3, 4)$; Axis of symmetry: $x = -3$; Opens downward;

2. Vertex: $(1, \ 0)$; Axis of symmetry: $x = 1$; Opens upward;

3. (a)   See Figure 3.

   (b)   The vertex is $(0, -2)$. The axis of symmetry is $x = 0$.

   (c)   $f(-1) = (-1)^2 - 2 = 1 - 2 = -1$

4. (a)   See Figure 4.

   (b)   The vertex is $(2, 1)$. The axis of symmetry is $x = 2$.

   (c)   $f(3) = -(3)^2 + 4(3) - 3 = -9 + 12 - 3 = 0$

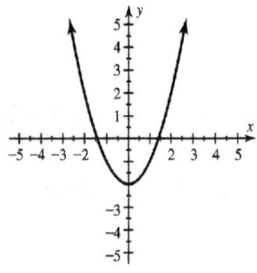

Figure 3

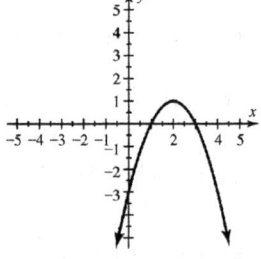

Figure 4

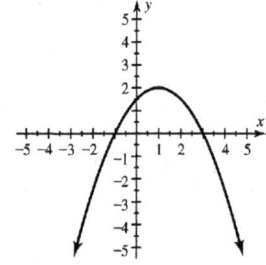

Figure 5

5. (a)   See Figure 5.

   (b)   The vertex is $(1, 2)$. The axis of symmetry is $x = 1$.

   (c)   $f(-2) = -\dfrac{1}{2}(-2)^2 + (-2) + \dfrac{3}{2} = -2 - 2 + \dfrac{3}{2} = -2.5$

6. (a)    See Figure 6.

   (b)    The vertex is (–2, –3). The axis of symmetry is $x = -2$.

   (c)    $f(-3) = 2(-3)^2 + 8(-3) + 5 = 18 - 24 + 5 = -1$

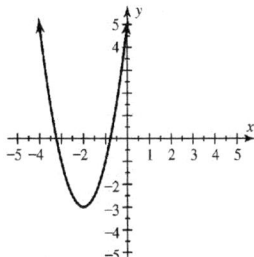

   Figure 6

7. Because $-\dfrac{b}{2a} = -\dfrac{-6}{2(2)} = \dfrac{3}{2}$ and $f\left(\dfrac{3}{2}\right) = 2\left(\dfrac{3}{2}\right)^2 - 6\left(\dfrac{3}{2}\right) + 1 = -\dfrac{7}{2}$, the vertex is $\left(\dfrac{3}{2},\ -\dfrac{7}{2}\right)$.

   The minimum $y$-value on the graph is $-\dfrac{7}{2}$.

8. Because $-\dfrac{b}{2a} = -\dfrac{2}{2(-3)} = \dfrac{1}{3}$ and $f\left(\dfrac{1}{3}\right) = -3\left(\dfrac{1}{3}\right)^2 + 2\left(\dfrac{1}{3}\right) - 5 = -\dfrac{14}{3}$, the vertex is $\left(\dfrac{1}{3},\ -\dfrac{14}{3}\right)$.

   The maximum $y$-value on the graph is $-\dfrac{14}{3}$.

9. $x = -\dfrac{b}{2a} = -\dfrac{(-4)}{2(1)} = 2$, $f(2) = (2)^2 - 4(2) - 2 = -6$; The vertex is (2, –6).

10. $x = -\dfrac{b}{2a} = -\dfrac{(0)}{2(-1)} = 0$, $f(0) = 5 - (0)^2 = 5$; The vertex is (0, 5).

11. $x = -\dfrac{b}{2a} = -\dfrac{(1)}{2(-\frac{1}{4})} = 2$, $f(2) = -\dfrac{1}{4}(2)^2 + (2) + 1 = 2$; The vertex is (2, 2).

12. $x = -\dfrac{b}{2a} = -\dfrac{(2)}{2(1)} = -1$, $f(-1) = 2 + 2(-1) + (-1)^2 = 1$; The vertex is (–1, 1).

13. (a)    See Figure 13.

    (b)    This graph is a shift of the graph of $y = x^2$ upward 2 units.

14. (a)    See Figure 14.

    (b)    This graph is more narrow than the graph of $y = x^2$.

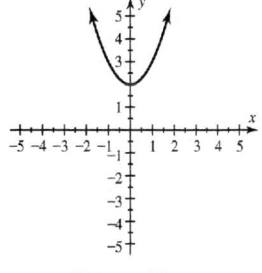

   Figure 13

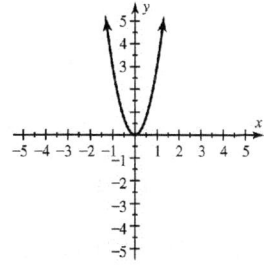

   Figure 14

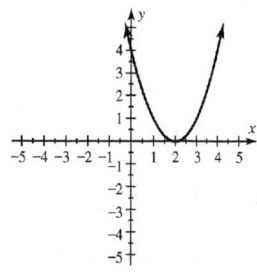

   Figure 15

15. (a)   See Figure 15.

    (b)   This graph is a shift of the graph of $y = x^2$ right 2 units.

16. (a)   See Figure 16.

    (b)   This graph is a shift of the graph of $y = x^2$ left 1 unit and downward 3 units.

17. (a)   See Figure 17.

    (b)   This graph is wider than the graph of $y = x^2$ and is shifted left 1 unit and upward 2 units.

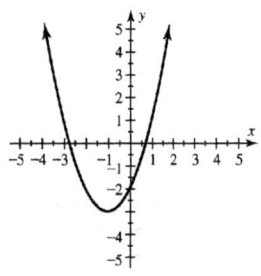 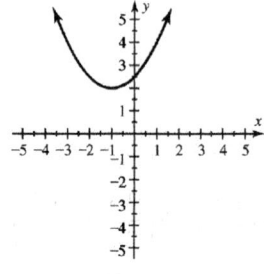 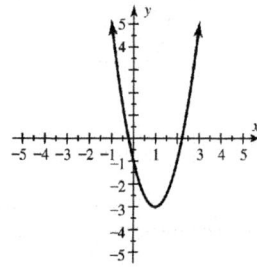

   Figure 16                      Figure 17                      Figure 18

18. (a)   See Figure 18.

    (b)   This graph is more narrow than the graph of $y = x^2$ and is shifted right 1 unit and downward 3 units.

19.   $y = a(x-h)^2 + k$ where $(h, k)$ is the vertex , gives $y = -4(x-2)^2 - 5$.

20.   Opening downward means $a = -1$. $y = a(x-h)^2 + k$, gives $y = -1(x+4)^2 + 6$.

21.   $y = x^2 + 4x - 7 \Rightarrow y = (x^2 + 4x + 4) - 7 - 4 \Rightarrow y = (x+2)^2 - 11$. The vertex is $(-2, -11)$.

22.   $y = x^2 - 7x + 1 \Rightarrow y = \left( x^2 - 7x + \dfrac{49}{4} \right) + 1 - \dfrac{49}{4} \Rightarrow y = \left( x - \dfrac{7}{2} \right)^2 - \dfrac{45}{4}$. The vertex is $\left( \dfrac{7}{2}, -\dfrac{45}{4} \right)$.

23.   $y = 2x^2 - 3x - 8 \Rightarrow y = 2\left( x^2 - \dfrac{3}{2}x + \dfrac{9}{16} \right) - 8 - \dfrac{9}{8} \Rightarrow y = 2\left( x - \dfrac{3}{4} \right)^2 - \dfrac{73}{8}$.

      The vertex is $\left( \dfrac{3}{4}, -\dfrac{73}{8} \right)$.

24.   $y = 3x^2 + 6x - 2 \Rightarrow y = 3(x^2 + 2x + 1) - 2 - 3 \Rightarrow y = 3(x+1)^2 - 5$. The vertex is $(-1, -5)$.

25.   $a(1)^2 - 1 = 2 \Rightarrow a = 3$

26.   $a(-1)^2 - 1 = -\dfrac{3}{4} \Rightarrow a = \dfrac{1}{4}$

27.   $f(x) = -5(x-3)^2 + 4 \Rightarrow f(x) = -5(x^2 - 6x + 9) + 4$

      $\Rightarrow f(x) = -5x^2 + 30x - 45 + 4 \Rightarrow f(x) = -5x^2 + 30x - 41$; the $y$-intercept is $c = -41$.

28.   $f(x) = 3(x+2)^2 - 4 \Rightarrow f(x) = 3(x^2 + 4x + 4) - 4$

      $\Rightarrow f(x) = 3x^2 + 12x + 12 - 4 \Rightarrow f(x) = 3x^2 + 12x + 8$; the $y$-intercept is $c = 8$.

29. –2, 3

30. –1

31. No real solutions.

32. –4, 6

33. –10, 5

34. –0.5, 0.25

35. (a)    Graph $Y_1 = X^2 - 5X - 50$ in [–10, 20, 5] by [–100, 20, 10]. See Figures 35a & 35b.

    The solutions are the $x$-intercepts, $x = -5$ and $x = 10$.

   (b)    Table $Y_1 = X^2 - 5X - 50$ with TblStart = –10 and $\Delta$Tbl = 5. See Figure 35c.

    Since $Y_1 = 0$ when $x = -5$ and when $x = 10$, the solutions are $x = -5$ or $x = 10$.

[–10, 20, 5] by [–100, 20, 10]        [–10, 20, 5] by [–100, 20, 10]

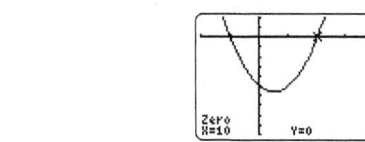

Figure 35a                 Figure 35b                 Figure 35c

36. (a)    Graph $Y_1 = (1/2)X^2 + X - (3/2)$ in [–5, 5, 1] by [–5, 5, 1]. See Figures 36a & 36b.

    The solutions are the $x$-intercepts, $x = -3$ or $x = 1$.

   (b)    Table $Y_1 = (1/2)X^2 + X - (3/2)$ with TblStart = –4 and $\Delta$Tbl = 1. See Figure 36c.

    Since $Y_1 = 0$ when $x = -3$ and when $x = 1$, the solutions are $x = -3$ and $x = 1$.

[–5, 5, 1] by [–5, 5, 1]        [–5, 5, 1] by [–5, 5, 1]

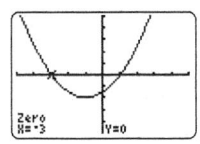        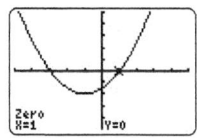

Figure 36a                 Figure 36b                 Figure 36c

37. (a)    Graph $Y_1 = (1/4)X^2 + (1/2)X - 2$ in [–5, 5, 1] by [–3, 3, 1]. See Figures 37a & 37b.

    The solutions are the $x$-intercepts, $x = -4$ and $x = 2$.

   (b)    Table $Y_1 = (1/4)X^2 + (1/2)X - 2$ with TblStart = –8 and $\Delta$Tbl = 2. See Figure 37c.

    Since $Y_1 = 0$ when $x = -4$ and when $x = 2$, the solutions are $x = -4$, and $x = 2$.

[–5, 5, 1] by [–3, 3, 1]        [–5, 5, 1] by [–3, 3, 1]

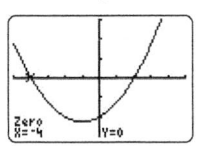        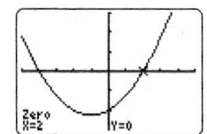

Figure 37a                 Figure 37b                 Figure 37c

38. (a)    Graph $Y_1 = (1/4)X^2 - (1/2)X - (3/4)$ in $[-5, 5, 1]$ by $[-3, 3, 1]$.  See Figures 38a & 38b.

The solutions are the $x$-intercepts,  $x = -1$ and $x = 3$ .

(b)    Table $Y_1 = (1/4)X^2 - (1/2)X - (3/4)$ with TblStart = $-2$ and $\Delta$Tbl = 1.  See Figure 38c.

Since $Y_1 = 0$ when $x = -1$ and when $x = 3$, the solutions are  $x = -1$ and $x = 3$ .

$[-5, 5, 1]$ by $[-3, 3, 1]$            $[-5, 5, 1]$ by $[-3, 3, 1]$

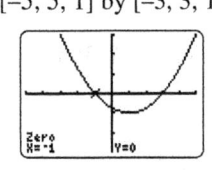          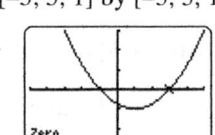          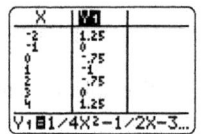

Figure 38a                    Figure 38b                    Figure 38c

39.  $x^2 + x - 20 = 0 \Rightarrow (x+5)(x-4) = 0 \Rightarrow x = -5$ or $x = 4$

40.  $x^2 + 11x + 24 = 0 \Rightarrow (x+8)(x+3) = 0 \Rightarrow x = -8$ or $x = -3$

41.  $15x^2 - 4x - 4 = 0 \Rightarrow (5x+2)(3x-2) = 0 \Rightarrow x = -\dfrac{2}{5}$ or $x = \dfrac{2}{3}$

42.  $7x^2 - 25x + 12 = 0 \Rightarrow (7x-4)(x-3) = 0 \Rightarrow x = \dfrac{4}{7}$ or $x = 3$

43.  $x^2 = 100 \Rightarrow x = \pm\sqrt{100} \Rightarrow x = \pm 10$

44.  $3x^2 = \dfrac{1}{3} \Rightarrow x^2 = \dfrac{1}{9} \Rightarrow x = \pm\sqrt{\dfrac{1}{9}} \Rightarrow x = \pm\dfrac{1}{3}$

45.  $4x^2 - 6 = 0 \Rightarrow x^2 = \dfrac{6}{4} \Rightarrow x = \pm\sqrt{\dfrac{6}{4}} \Rightarrow x = \pm\dfrac{\sqrt{6}}{2}$

46.  $5x^2 = x^2 - 4 \Rightarrow 4x^2 = -4 \Rightarrow x^2 = -1 \Rightarrow x = \pm\sqrt{-1} \Rightarrow$ No real solutions.

47.  $x^2 + 6x = -2 \Rightarrow x^2 + 6x + 9 = -2 + 9 \Rightarrow (x+3)^2 = 7 \Rightarrow x + 3 = \pm\sqrt{7} \Rightarrow x = -3 \pm\sqrt{7}$

48.  $x^2 - 4x = 6 \Rightarrow x^2 - 4x + 4 = 6 + 4 \Rightarrow (x-2)^2 = 10 \Rightarrow x - 2 = \pm\sqrt{10} \Rightarrow x = 2 \pm\sqrt{10}$

49.  $x^2 - 2x - 5 = 0 \Rightarrow x^2 - 2x + 1 = 5 + 1 \Rightarrow (x-1)^2 = 6 \Rightarrow x - 1 = \pm\sqrt{6} \Rightarrow x = 1 \pm\sqrt{6}$

50.  $2x^2 + 6x - 1 = 0 \Rightarrow x^2 + 3x + \dfrac{9}{4} = \dfrac{1}{2} + \dfrac{9}{4} \Rightarrow \left(x + \dfrac{3}{2}\right)^2 = \dfrac{11}{4} \Rightarrow x + \dfrac{3}{2} = \pm\sqrt{\dfrac{11}{4}} \Rightarrow x = \dfrac{-3 \pm\sqrt{11}}{2}$

51.  $F = \dfrac{k}{(R+r)^2} \Rightarrow F(R+r)^2 = k \Rightarrow (R+r)^2 = \dfrac{k}{F} \Rightarrow R + r = \pm\sqrt{\dfrac{k}{F}} \Rightarrow R = -r \pm\sqrt{\dfrac{k}{F}}$

52.  $2x^2 + 3y^2 = 12 \Rightarrow 3y^2 = 12 - 2x^2 \Rightarrow y^2 = \dfrac{12 - 2x^2}{3} \Rightarrow y = \pm\sqrt{\dfrac{12 - 2x^2}{3}}$

53.  $x = \dfrac{-(-9) \pm\sqrt{(-9)^2 - 4(1)(18)}}{2(1)} = \dfrac{9 \pm\sqrt{9}}{2} = \dfrac{9 \pm 3}{2} \Rightarrow x = 3$ or $x = 6$

54. $x = \dfrac{-(-24) \pm \sqrt{(-24)^2 - 4(1)(143)}}{2(1)} = \dfrac{24 \pm \sqrt{4}}{2} = \dfrac{24 \pm 2}{2} \Rightarrow x = 11 \text{ or } x = 13$

55. $x = \dfrac{-1 \pm \sqrt{(1)^2 - 4(6)(-1)}}{2(6)} = \dfrac{-1 \pm \sqrt{25}}{12} = \dfrac{-1 \pm 5}{12} \Rightarrow x = -\dfrac{1}{2} \text{ or } x = \dfrac{1}{3}$

56. $x = \dfrac{-(-5) \pm \sqrt{(-5)^2 - 4(5)(1)}}{2(5)} = \dfrac{5 \pm \sqrt{5}}{10}$

57. $x = \dfrac{-(-8) \pm \sqrt{(-8)^2 - 4(1)(-5)}}{2(1)} = \dfrac{8 \pm \sqrt{84}}{2} = \dfrac{8 \pm 2\sqrt{21}}{2} = \dfrac{2(4 \pm \sqrt{21})}{2} = 4 \pm \sqrt{21}$

58. $x = \dfrac{-(-6) \pm \sqrt{(-6)^2 - 4(2)(3)}}{2(2)} = \dfrac{6 \pm \sqrt{12}}{4} = \dfrac{6 \pm 2\sqrt{3}}{4} = \dfrac{2(3 \pm \sqrt{3})}{2(2)} = \dfrac{3 \pm \sqrt{3}}{2}$

59. $x^2 - 4 = 0 \Rightarrow (x-2)(x+2) = 0 \Rightarrow x = 2 \text{ or } x = -2$

60. $4x^2 - 1 = 0 \Rightarrow (2x-1)(2x+1) = 0 \Rightarrow x = \dfrac{1}{2} \text{ or } x = -\dfrac{1}{2}$

61. $2x^2 + 15 = 11x \Rightarrow 2x^2 - 11x + 15 = 0 \Rightarrow (2x-5)(x-3) = 0 \Rightarrow x = \dfrac{5}{2} \text{ or } x = 3$

62. $2x^2 + 15 = 13x \Rightarrow 2x^2 - 13x + 15 = 0 \Rightarrow (2x-3)(x-5) = 0 \Rightarrow x = \dfrac{3}{2} \text{ or } x = 5$

63. $x(5 - x) = 2x + 1 \Rightarrow 5x - x^2 - 2x - 1 = 0 \Rightarrow -x^2 + 3x - 1 = 0;\ a = -1, b = 3, c = -1$

$x = \dfrac{-b \pm \sqrt{b^2 - 4(ac)}}{2a} \Rightarrow x = \dfrac{-3 \pm \sqrt{3^2 - 4(-1)(-1)}}{2(-1)} \Rightarrow x = \dfrac{-3 \pm \sqrt{5}}{-2} \Rightarrow x = \dfrac{3 \pm \sqrt{5}}{2}$

64. $-2x(x-1) = x - \dfrac{1}{2} \Rightarrow 2(-2x)(x-1) = 2\left(x - \dfrac{1}{2}\right) \Rightarrow -4x^2 + 4x = 2x - 1 \Rightarrow -4x^2 + 2x + 1 = 0;$

$a = -4, b = 2, c = 1\quad x = \dfrac{-b \pm \sqrt{b^2 - 4ac}}{2a} \Rightarrow x = \dfrac{-2 \pm \sqrt{2^2 - 4(-4)(1)}}{2(-4)}$

$\Rightarrow x = \dfrac{-2 \pm \sqrt{20}}{-8} \Rightarrow x = \dfrac{-2 \pm 2\sqrt{5}}{-8} \Rightarrow x = \dfrac{1 \pm \sqrt{5}}{4}$

65. (a)   Since the parabola opens upward, $a > 0$.

    (b)   The solutions are the $x$-intercepts, $x = -2$ or $x = 3$.

    (c)   Since there are two unique solutions, the discriminant is positive.

66. (a)   Since the parabola opens upward, $a > 0$.

    (b)   The solution is the $x$-intercept, $x = 2$.

    (c)   Since there is one solution, the discriminant is zero.

67. (a)   Since the parabola opens downward, $a < 0$.

    (b)   There are no $x$-intercepts.  No real solutions.

    (c)   Since there are no real solutions, the discriminant is negative.

68. (a)    Since the parabola opens downward, $a < 0$.

    (b)    The solutions are the $x$-intercepts, $x = -4$ or $x = 2$.

    (c)    Since there are two unique solutions, the discriminant is positive.

69. (a)    $(-3)^2 - 4(2)(1) = 1$

    (b)    Since the discriminant is positive, there are two real solutions.

    (c)    Graph $Y_1 = 2X^2 - 3X + 1$ in $[0, 2, 1]$ by $[-1, 1, 1]$. See Figure 69. There are two $x$-intercepts.

70. (a)    $(2)^2 - 4(7)(-5) = 144$

    (b)    Since the discriminant is positive, there are two real solutions.

    (c)    Graph $Y_1 = 7X^2 + 2X - 5$ in $[-2, 2, 1]$ by $[-10, 5, 1]$. See Figure 70. There are two $x$-intercepts.

$[0, 2, 1]$ by $[-1, 1, 1]$          $[-2, 2, 1]$ by $[-10, 5, 1]$

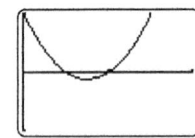                 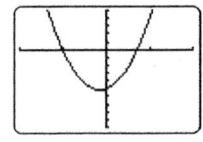

   Figure 69                          Figure 70

71. (a)    $(1)^2 - 4(3)(2) = -23$

    (b)    Since the discriminant is negative, there are no real solutions.

    (c)    Graph $Y_1 = 3X^2 + X + 2$ in $[-3, 3, 1]$ by $[-5, 10, 1]$. See Figure 71. There are no $x$-intercepts.

$[-3, 3, 1]$ by $[-5, 10, 1]$        $[0, 3, 1]$ by $[-5, 10, 1]$

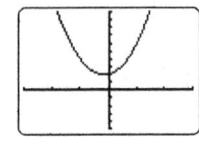                 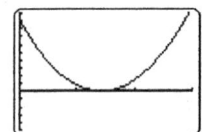

   Figure 71                          Figure 72

72. (a)    $(-12.6)^2 - 4(4.41)(9) = 0$

    (b)    Since the discriminant is zero, there is one real solution.

    (c)    Graph $4.41X^2 - 12.6X + 9$ in $[0, 3, 1]$ by $[-5, 10, 1]$. See Figure 72.

        There is one $x$-intercept.

73.  $x = \dfrac{-1 \pm \sqrt{(1)^2 - 4(1)(5)}}{2(1)} = \dfrac{-1 \pm \sqrt{-19}}{2} = -\dfrac{1}{2} \pm i\dfrac{\sqrt{19}}{2}$

74.  $2x^2 + 8 = 0 \Rightarrow 2x^2 = -8 \Rightarrow x^2 = -4 \Rightarrow x = \pm\sqrt{-4} \Rightarrow x = \pm 2i$

75.  $x = \dfrac{-(-1) \pm \sqrt{(-1)^2 - 4(2)(1)}}{2(2)} = \dfrac{1 \pm \sqrt{-7}}{4} = \dfrac{1}{4} \pm i\dfrac{\sqrt{7}}{4}$

76.  $x = \dfrac{-(-2) \pm \sqrt{(-2)^2 - 4(7)(5)}}{2(7)} = \dfrac{2 \pm \sqrt{-136}}{14} = \dfrac{2 \pm 2\sqrt{-34}}{14} = \dfrac{2(1 \pm \sqrt{-34})}{14} = \dfrac{1}{7} \pm i\dfrac{\sqrt{34}}{7}$

77. (a)   The solutions are the *x*-intercepts, $x = -2$ or $x = 6$.

    (b)   $-2 < x < 6$

    (c)   $x < -2$ or $x > 6$

78. (a)   The solutions are the *x*-intercepts, $x = -2$ or $x = 0$.

    (b)   $x < -2$ or $x > 0$

    (c)   $-2 < x < 0$

79. (a)   The solutions are $x = -4$ or $x = 4$.

    (b)   $-4 < x < 4$

    (c)   $x < -4$ or $x > 4$

80. (a)   The solutions are $x = -2$ or $x = 1$.

    (b)   $-2 < x < 1$

    (c)   $x < -2$ or $x > 1$

81. (a)   $x^2 - 2x - 3 = 0 \Rightarrow (x+1)(x-3) = 0 \Rightarrow x = -1$ or $x = 3$

    (b)   Since the parabola opens upward, the solution is $-1 < x < 3$.

    (c)   Since the parabola opens upward, the solution is $x < -1$ or $x > 3$.

82. (a)   $2x^2 - 7x - 15 = 0 \Rightarrow (2x+3)(x-5) = 0 \Rightarrow x = -\dfrac{3}{2}$ or $x = 5$

    (b)   Since the parabola opens upward, the solution is $-\dfrac{3}{2} \le x \le 5$.

    (c)   Since the parabola opens upward, the solution is $x \le -\dfrac{3}{2}$ or $x \ge 5$.

83.  $x^2 + 4x + 3 = 0 \Rightarrow (x+3)(x+1) = 0 \Rightarrow x = -3$ or $x = -1$

     Since the parabola opens upward, the solution is $[-3, \ -1]$.

84.  $5x^2 - 16x + 3 = 0 \Rightarrow (5x-1)(x-3) = 0 \Rightarrow x = \dfrac{1}{5}$ or $x = 3$

     Since the parabola opens upward, the solution is $\left( \dfrac{1}{5}, \ 3 \right)$.

85.  $6x^2 - 13x + 2 = 0 \Rightarrow (6x-1)(x-2) = 0 \Rightarrow x = \dfrac{1}{6}$ or $x = 2$

     Since the parabola opens upward, the solution is $\left( -\infty, \ \dfrac{1}{6} \right) \cup (2, \ \infty)$.

86.  $x^2 = 5 \Rightarrow \Rightarrow x = -\sqrt{5}$ or $x = \sqrt{5}$

     Since the parabola opens upward, the solution is $(-\infty, \ \sqrt{5}] \cup [\sqrt{5}, \ \infty)$.

87.  The graph of the parabola $y = (x-1)^2$ lies on or above the *x*-axis for all real numbers *x*. The interval

     is $(-\infty, \infty)$.

88. $x^2 + 3 < 2 \Rightarrow x^2 < -1$, which is not true for any real number $x$. The graph of the parabola $y = x^2 + 1$

lies above the $x$-axis for all real numbers, so there are no solutions.

89. Let $u = x^2$. Then $u^2 - 14u + 45 = 0 \Rightarrow (u - 5)(u - 9) = 0 \Rightarrow u = 5$ or $9$.

When $u = 5$, $x^2 = 5 \Rightarrow x = \pm\sqrt{5}$. When $u = 9$, $x^2 = 9 \Rightarrow x = \pm\sqrt{9} = \pm 3$.

90. Let $u = z^{-1}$. Then $2u^2 + u - 28 = 0 \Rightarrow (u + 4)(2u - 7) = 0 \Rightarrow u = -4$ or $\dfrac{7}{2}$.

When $u = -4$, $z^{-1} = -4 \Rightarrow z = -\dfrac{1}{4}$. When $u = \dfrac{7}{2}$, $z^{-1} = \dfrac{7}{2} \Rightarrow z = \dfrac{2}{7}$.

91. Let $u = x^{1/3}$. Then $u^2 - 9u + 8 = 0 \Rightarrow (u - 1)(u - 8) = 0 \Rightarrow u = 1$ or $8$.

When $u = 1$, $x^{1/3} = 1 \Rightarrow x = 1^3 = 1$. When $u = 8$, $x^{1/3} = 8 \Rightarrow x = 8^3 = 512$.

92. Let $u = x - 1$. Then $u^2 + 2u + 1 = 0 \Rightarrow (u + 1)^2 = 0 \Rightarrow u = -1$. When $u = -1$, $x - 1 = -1 \Rightarrow x = 0$.

93. $4x^4 + 4x^2 + 1 = 0 \Rightarrow (2x^2 + 1)^2 = 0 \Rightarrow 2x^2 + 1 = 0 \Rightarrow x^2 = -\dfrac{1}{2} \Rightarrow x = \pm\sqrt{-\dfrac{1}{2}} \Rightarrow x = \pm i\dfrac{\sqrt{1}}{\sqrt{2}}$

$\Rightarrow x = \pm i\dfrac{1}{\sqrt{2}} \Rightarrow x = \pm i\dfrac{\sqrt{2}}{2}$

94. $\dfrac{1}{x - 2} - \dfrac{3}{x} = -1 \Rightarrow x(x - 2)\left(\dfrac{1}{x - 2} - \dfrac{3}{x}\right) = x(x - 2)(-1) \Rightarrow x - 3(x - 2) = -x^2 + 2x \Rightarrow x^2 - 4x + 6 = 0$

$\Rightarrow x^2 - 4x = -6 \Rightarrow x^2 - 4x + 4 = -6 + 4 \Rightarrow (x - 2)^2 = -2 \Rightarrow x - 2 = \pm\sqrt{-2} \Rightarrow x = 2 \pm i\sqrt{2}$

95. (a)  $f(x) = x(12 - 2x)$

　　(b)  Note that $f(x) = x(12 - 2x) \Rightarrow f(x) = -2x^2 + 12x + 0$. Then $-\dfrac{b}{2a} = -\dfrac{12}{2(-2)} = 3$.

　　　　The maximum area occurs when $x = 3$. The dimensions should be 6 inches by 3 inches.

96. (a)  $-16t^2 + 44t + 4 = 32 \Rightarrow -16t^2 + 44t - 28 = 0 \Rightarrow 4t^2 - 11t + 7 = 0 \Rightarrow (t - 1)(4t - 7) = 0 \Rightarrow$

　　　　$t = 1$ or $t = \dfrac{7}{4}$. The height of the stone is 32 feet after 1 second and 1.75 seconds.

　　(b)  Find the vertex. $-\dfrac{b}{2a} = -\dfrac{44}{2(-16)} = 1.375$; $f(1.375) = -16(1.375)^2 + 44(1.375) + 4 = 34.25$

　　　　After 1.375 seconds, the stone is at a height of 34.25 feet.

97. (a)  $f(x) = x(90 - 3x)$

　　(b)  Graph $Y_1 = X(90 - 3X)$ in $[0, 30, 5]$ by $[0, 800, 100]$. See Figure 97a.

　　(c)  Graph $Y_1 = X(90 - 3X)$ and $Y_2 = 600$ in $[0, 30, 5]$ by $[0, 800, 100]$.

　　　　See Figures 97b &97c. 10 or 20 rooms.

　　(d)  Graph $Y_1 = X(90 - 3X)$ in $[0, 30, 5]$ by $[0, 800, 100]$. See Figure 97d. Rent 15 rooms.

[0, 30, 5] by [0, 800, 100]          [0, 30, 5] by [0, 800, 100]

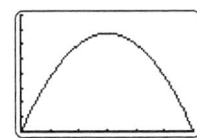                 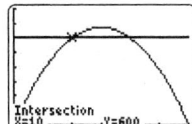

Figure 97a                           Figure 97b

[0, 30, 5] by [0, 800, 100]          [0, 30, 5] by [0, 800, 100]

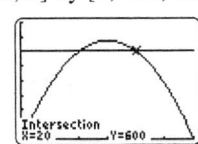                 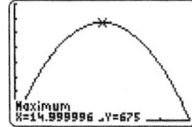

Figure 97c                           Figure 97d

98. (a)   $x(x+2) = 143$ or $x^2 + 2x - 143 = 0$

(b)   $x^2 + 2x - 143 = 0 \Rightarrow (x+13)(x-11) = 0 \Rightarrow x = -13$ or $11$. There are two possible number

pairs. Either $x = -13$, and the other number is $-11$, or $x = 11$, and the other number is $13$.

99. (a)   $\dfrac{x^2}{12} = 144 \Rightarrow x^2 = 1728 \Rightarrow x = \sqrt{1728} \approx 41.6$ mph ($-\sqrt{1728}$ mph has no meaning).

(b)   $\dfrac{x^2}{12} = 300 \Rightarrow x^2 = 3600 \Rightarrow x = \sqrt{3600} = 60$ mph ($-60$ mph has no meaning).

100. (a)   The vertex is $(1950, 220)$. In 1950 the per-capita energy consumption was at a low of 220

million Btu.

(b)   Graph $Y_1 = (1/4)(X - 1950)^2 + 220$ in $[1950, 1970, 5]$ by $[200, 350, 25]$. See Figure 100. It

increased.

(c)   $f(2010) = \dfrac{1}{4}(2010 - 1950)^2 + 220 = 1120$. The function is not a good model for 2010 because

the trend represented by the model did not continue after 1970.

[1950, 1970, 5] by [200, 350, 25].

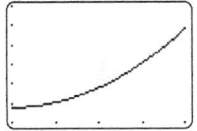

Figure 100

101. $\sqrt{123} \approx 11.1$; the screen is about 11.1 inches by 11.1 inches.

102. $x^2 + (x+70)^2 = 130^2 \Rightarrow x^2 + x^2 + 140x + 4900 = 16,900 \Rightarrow 2x^2 + 140x - 12,000 = 0 \Rightarrow$

$2(x-50)(x+120) = 0 \Rightarrow x = 50$ or $x = -120$. The solution is $x = 50$ feet

($x = -120$ has no meaning).

103.  $(30 + 2x)(50 + 2x) - 30(50) = 250 \Rightarrow 1500 + 160x + 4x^2 - 1500 = 250 \Rightarrow 4x^2 + 160x - 250 = 0 \Rightarrow$

$$x = \frac{-160 \pm \sqrt{(160)^2 - 4(4)(-250)}}{2(4)} = \frac{-160 \pm \sqrt{29,600}}{8} \Rightarrow x \approx -41.5 \text{ or } x \approx 1.5$$

The width of the strip of grass is about 1.5 feet. The value $x \approx -41.5$ has no physical meaning.

104.  We must find $r$ so that $750 \le \frac{1}{3}\pi r^2 (20) \le 1700$.

$$2250 \le \pi r^2 (20) \le 5100 \Rightarrow \frac{2250}{20\pi} \le r^2 \le \frac{5100}{20\pi} \Rightarrow \sqrt{\frac{2250}{20\pi}} \le r \le \sqrt{\frac{5100}{20\pi}} \Rightarrow 5.98 \le r \le 9.01$$

The values or $r$ can range from about 6 inches to about 9 inches.

## Chapter 11 Test

1.  $x = -\frac{b}{2a} = -\frac{1}{2(-\frac{1}{2})} = \frac{1}{1} = 1, f(1) = -\frac{1}{2}(1)^2 + (1) + 1 = \frac{3}{2}$ ; Vertex: $\left(1, \frac{3}{2}\right)$; Axis of symmetry:

$x = 1$. $f(-2) = -\frac{1}{2}(-2)^2 + (-2) + 1 = -\frac{1}{2}(4) + (-2) + 1 = -3$

2.  Because $-\frac{b}{2a} = -\frac{3}{2(1)} = -\frac{3}{2}$ and $f\left(-\frac{3}{2}\right) = \left(-\frac{3}{2}\right)^2 + 3\left(-\frac{3}{2}\right) - 5 = -\frac{29}{4}$, the vertex is $\left(-\frac{3}{2}, -\frac{29}{4}\right)$.

The minimum $y$-value on the graph is $-\frac{29}{4}$.

3.  Since $f(x) = 0$ when $x = -2$, $a(-2)^2 + 2 = 0 \Rightarrow 4a = -2 \Rightarrow a = -\frac{1}{2}$.

4.  (a)   Same as $y = x^2$ except shifted 1 unit right. See Figure 4a.

    (b)   Same as $y = x^2$ except shifted 2 units downward. See Figure 4b.

    (c)   Same as $y = x^2$ except shifted 3 units right, 2 units upward, and wider. See Figure 4c.

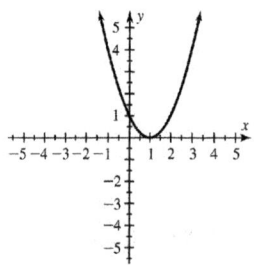     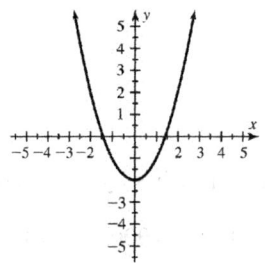     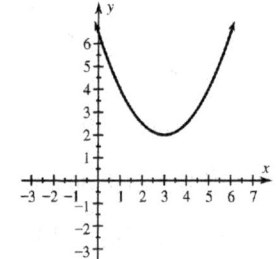

      Figure 4a                    Figure 4b                    Figure 4c

5.  $y = x^2 - 6x + 2 \Rightarrow y = (x^2 - 6x + 9) + 2 - 9 \Rightarrow y = (x - 3)^2 - 7$. The vertex is $(3, -7)$. The axis of

symmetry is $x = 3$.

6.  The solutions are the $x$-intercepts, $x = -1$ or $x = 2$. $f(1) = 2$.

7. $3x^2 + 11x - 4 = 0 \Rightarrow (x+4)(3x-1) = 0 \Rightarrow x = -4 \text{ or } x = \dfrac{1}{3}$

8. $2x^2 = 2 - 6x^2 \Rightarrow 8x^2 = 2 \Rightarrow x^2 = \dfrac{1}{4} \Rightarrow x = \pm\sqrt{\dfrac{1}{4}} \Rightarrow x = -\dfrac{1}{2} \text{ or } x = \dfrac{1}{2}$

9. $x^2 - 8x = 1 \Rightarrow x^2 - 8x + 16 = 1 + 16 \Rightarrow (x-4)^2 = 17 \Rightarrow x - 4 = \pm\sqrt{17} \Rightarrow x = 4 \pm \sqrt{17}$

10. $x = \dfrac{-3 \pm \sqrt{(3)^2 - 4(-2)(1)}}{2(-2)} = \dfrac{-3 \pm \sqrt{17}}{-4} = \dfrac{3 \pm \sqrt{17}}{4}$

11. $9x^2 - 16 = 0 \Rightarrow (3x+4)(3x-4) = 0 \Rightarrow x = \pm\dfrac{4}{3}$

12. $F = \dfrac{Gm^2}{r^2} \Rightarrow Fr^2 = Gm^2 \Rightarrow m^2 = \dfrac{Fr^2}{G} \Rightarrow m = \pm\sqrt{\dfrac{Fr^2}{G}}$

13. (a)   Since the parabola opens downward, $a < 0$.

    (b)   The solutions are the $x$-intercepts, $x = -3$ or $x = 1$.

    (c)   Since there are two real solutions, the discriminant is positive.

14. (a)   $(4)^2 - 4(-3)(-5) = -44$

    (b)   Since the discriminant is negative, there are no real solutions.

    (c)   Graph $Y_1 = -3X^2 + 4X - 5$ in $[-5, 5, 1]$ by $[-20, 10, 5]$. See Figure 14. It does not intersect

    the $x$-axis.

    $[-5, 5, 1]$ by $[-20, 10, 5]$

    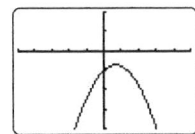

    Figure 14

15. (a)   The solutions are the $x$-intercepts, $x = -1$ or $x = 1$.

    (b)   $-1 < x < 1$

    (c)   $x < -1 \text{ or } x > 1$

16. (a)   The solutions are the $x$-intercepts, $x = -10$ or $x = 20$.

    (b)   $x < -10 \text{ or } x > 20$

    (c)   $-10 < x < 20$

17. (a)   $8x^2 - 2x - 3 = 0 \Rightarrow (2x+1)(4x-3) = 0 \Rightarrow x = -\dfrac{1}{2} \text{ or } x = \dfrac{3}{4}$

    (b)   Since the parabola opens upward, the solution is $\left[ -\dfrac{1}{2}, \dfrac{3}{4} \right]$.

    (c)   Since the parabola opens upward, the solution is $\left( -\infty, -\dfrac{1}{2} \right] \cup \left[ \dfrac{3}{4}, \infty \right)$.

18. $x^2 + 2x = 0 \Rightarrow x(x+2) = 0 \Rightarrow x = -2$ or $x = 0$. Solutions to $x^2 + 2x \le 0$ lie between and include these two values. The solution set is $[-2, 0]$.

19. Let $u = x^3$. Then $u^2 - 3u + 2 = 0 \Rightarrow (u-1)(u-2) = 0 \Rightarrow u = 1$ or $2$.

When $u = 1$, $x^3 = 1 \Rightarrow x = \sqrt[3]{1} = 1$. When $u = 2$, $x^3 = 2 \Rightarrow x = \sqrt[3]{2}$.

20. $2x^2 + 4x + 3 = 0 \Rightarrow x = \dfrac{-4 \pm \sqrt{(4)^2 - 4(2)(3)}}{2(2)} = \dfrac{-4 \pm \sqrt{-8}}{4} = -1 \pm i\dfrac{\sqrt{2}}{2}$

21. Graph $Y_1 = \sqrt{2} - \pi X^2$ and $Y_2 = 2.12X - 0.5\pi$ in $[-5, 5, 1]$ by $[-7, 3, 1]$. See Figures 21a and 21b. The solutions are approximately $x = -1.37$ and $x = 0.69$.

$[-5, 5, 1]$ by $[-7, 3, 1]$.          $[-5, 5, 1]$ by $[-7, 3, 1]$          $[0, 6, 1]$ by $[0, 150, 50]$

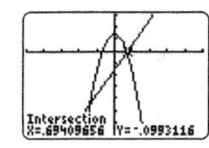

Figure 21a                    Figure 21b                    Figure 24

22. $\dfrac{x^2}{9} = 250 \Rightarrow x^2 = 2250 \Rightarrow x = \sqrt{2250} \approx 47.4$ mph (the value $x \approx -47.4$ mph has no

physical meaning).

23. (a)   $2z + 2x + 20 = 200 \Rightarrow 2x + 2z = 180 \Rightarrow z = 90 - x$, so the formula is $f(x) = (x+20)(90-x)$.

    (b)   Note that $f(x) = (x+20)(90-x) \Rightarrow f(x) = -x^2 + 70x + 1800$. The area is greatest for the

          value of $x$ at the vertex of the parabola, $x = -\dfrac{b}{2a} = -\dfrac{70}{2(-1)} = 35$. The enclosed area is greatest

          when $x = 35$.

24. (a)   Graph $Y_1 = -16X^2 + 88X + 8$ in $[0, 6, 1]$ by $[0, 150, 50]$. See Figure 24.

    (b)   $t = \dfrac{-88 \pm \sqrt{(88)^2 - 4(-16)(8)}}{2(-16)} = \dfrac{-88 \pm \sqrt{8256}}{-32} \approx 5.6$ seconds (the value $t \approx -0.089$ has

          no meaning).

    (c)   The stone reaches a maximum height of 129 feet after 2.75 seconds. See Figure 24.

## Chapter 11 Extended and Discovery Exercises

1. (a)   For the first 3 years of life, the likelihood of survival increases with age. After 3 years of life, it decreases with age.

   (b)   Plot the data in $[0, 10, 1]$ by $[0, 75, 5]$. See Figure 1b. A quadratic function could model these data since the data points form a parabola.

   (c)   Graph $Y_1 = -3.57X + 71.1$ in $[0, 10, 1]$ by $[0, 75, 5]$. See Figure 1c.

         Graph $Y_1 = -2.07X^2 + 17.1X + 33$ in $[0, 10, 1]$ by $[0, 75, 5]$. See Figure 1d.

         The function $f_2$ models the data better.

(d) Evaluate *f* at *x* = 6.5 to find the likelihood of a 5.5-year-old sparrowhawk living 1 more year.

$f_2(6.5) = -2.07(6.5)^2 + 17.1(6.5) + 33 \approx 56.7\%$

[0, 10, 1] by [0, 75, 5]    [0, 10, 1] by [0, 75, 5]    [0, 10, 1] by [0, 75, 5]

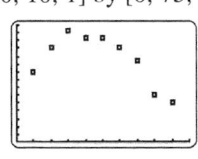    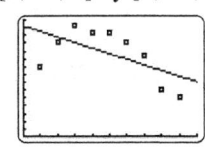    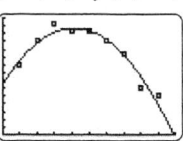

Figure 1b            Figure 1c            Figure 1d

2. (a) Plot the data in [–5, 35, 5] by [0, 100, 10]. See Figure 2.

   (b) A quadratic function could model these data since the data points form a parabola.

   (c) Using the data point (12, 95) as the vertex, the function has the form $f(x) = a(x-12)^2 + 95$.

   Using the data point (17, 89), $a(17-12)^2 + 95 = 89 \Rightarrow 25a = -6 \Rightarrow a = -0.24$ The function is

   given by $f(x) = -0.24(x-12)^2 + 95$. *Answers may vary.*

   (d) The *x*-coordinate of the vertex, 12°C, represents the the temperature that photosynthesis is most efficient.

[–5, 35, 5] by [0, 100, 10]    [–4, 4, 1] by [0, 6, 1]    [–4, 4, 1] by [0, 6, 1]

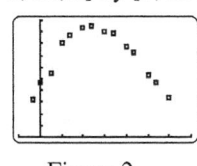    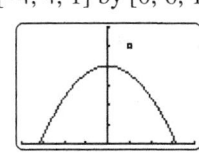    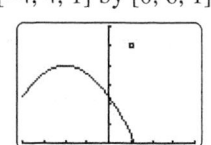

Figure 2            Figure 3a            Figure 3b

3. (a) Plot (1, 5) and graph $Y_1 = -0.4X^2 + 4$ in [–4, 4, 1] by [0, 6, 1]. See Figure 3a.

   (b) After 10 seconds the plane would have moved 2 kilometers. We will shift the parabola 2 units left. Plot (1, 5) and graph $Y_1 = -0.4(X+2)^2 + 4$ in [–4, 4, 1] by [0, 6, 1]. See Figure 3b.

4. We would need to translate the mountain both to the right and downward so that the airplane would appear to move to the left and upward. One example could be $f(x) = -0.4(x-2)^2 + 4 - 2$.

5. The discriminant is $(-1)^2 - 4(10)(-3) = 121 = 11^2$. The trinomial factors as $(2x+1)(5x-3)$.

6. The discriminant is $(-3)^2 - 4(4)(-6) = 105$. The trinomial will not factor.

7. The discriminant is $(2)^2 - 4(3)(-2) = 28$. The trinomial will not factor.

8. The discriminant is $(1)^2 - 4(2)(3) = -23$. The trinomial will not factor.

9. $x^3 - x^2 - 6x = 0 \Rightarrow x(x^2 - x - 6) = 0 \Rightarrow x(x-3)(x+2) = 0 \Rightarrow x = 0, 3,$ or $-2$

   In the interval $(-\infty, -2)$, we test $x = -3$. $f(-3) = (-3)^3 - (-3)^2 - 6(-3) = -18 < 0$

   In the interval $(-2, 0)$, we test $x = -1$. $f(-1) = (-1)^3 - (-1)^2 - 6(-1) = 4 > 0$

   In the interval $(0, 3)$, we test $x = 1$. $f(1) = (1)^3 - (1)^2 - 6(1) = -6 < 0$

   In the interval $(3, \infty)$, we test $x = 4$. $f(4) = (4)^3 - (4)^2 - 6(4) = 24 > 0$

   The solution to the given inequality is $(-2, 0) \cup (3, \infty)$.

10. $x^3 - 3x^2 + 2x = 0 \Rightarrow x(x^2 - 3x + 2) = 0 \Rightarrow x(x-1)(x-2) = 0 \Rightarrow x = 0,\ 1,\ \text{or } 2$

In the interval $(-\infty,\ 0)$, we test $x = -1$. $f(-1) = (-1)^3 - 3(-1)^2 + 2(-1) = -6 < 0$

In the interval $(0,\ 1)$, we test $x = 0.5$. $f(0.5) = (0.5)^3 - 3(0.5)^2 + 2(0.5) = 0.375 > 0$

In the interval $(1,\ 2)$, we test $x = 1.5$. $f(1.5) = (1.5)^3 - 3(1.5)^2 + 2(1.5) = -0.375 < 0$

In the interval $(2,\ \infty)$, we test $x = 3$. $f(3) = (3)^3 - 3(3)^2 + 2(3) = 6 > 0$

The solution to the given inequality is $(-\infty,\ 0) \cup (1,\ 2)$.

11. $x^3 - 7x^2 + 14x - 8 = 0 \Rightarrow (x^3 - 8) - 7x^2 + 14x = 0 \Rightarrow (x-2)(x^2 + 2x + 4) - 7x(x-2) = 0 \Rightarrow$

$(x-2)((x^2 + 2x + 4) - 7x) = 0 \Rightarrow (x-2)(x^2 - 5x + 4) = 0 \Rightarrow (x-2)(x-1)(x-4) = 0 \Rightarrow x = 1,\ 2,\ \text{or } 4$

In the interval $(-\infty,\ 1)$, we test $x = 0$. $f(0) = (0)^3 - 7(0)^2 + 14(0) - 8 = -8 \le 0$

In the interval $(1,\ 2)$, we test $x = 1.5$. $f(1.5) = (1.5)^3 - 7(1.5)^2 + 14(1.5) - 8 = 0.625 \ge 0$

In the interval $(2,\ 4)$, we test $x = 3$. $f(3) = (3)^3 - 7(3)^2 + 14(3) - 8 = -2 \le 0$

In the interval $(4,\ \infty)$, we test $x = 5$. $f(5) = (5)^3 - 7(5)^2 + 14(5) - 8 = 12 \ge 0$

The solution to the given inequality is $(-\infty,\ 1] \cup [2,\ 4]$.

12. $9x - x^3 = 0 \Rightarrow x(9 - x^2) = 0 \Rightarrow x(3-x)(3+x) = 0 \Rightarrow x = 0,\ 3,\ \text{or } -3$

In the interval $(-\infty,\ -3)$, we test $x = -4$. $f(-4) = 9(-4) - (-4)^3 = 28 \ge 0$

In the interval $(-3,\ 0)$, we test $x = -1$. $f(-1) = 9(-1) - (-1)^3 = -8 \le 0$

In the interval $(0,\ 3)$, we test $x = 1$. $f(1) = 9(1) - (1)^3 = 8 \ge 0$

In the interval $(3,\ \infty)$, we test $x = 4$. $f(4) = 9(4) - (4)^3 = -28 \le 0$

The solution to the given inequality is $(-\infty,\ -3] \cup [0,\ 3]$.

13. $x^4 - 5x^2 + 4 = 0 \Rightarrow (x^2 - 1)(x^2 - 4) = 0 \Rightarrow (x-1)(x+1)(x-2)(x+2) = 0 \Rightarrow x = -2,\ -1,\ 1,\ \text{or } 2$

In the interval $(-\infty,\ -2)$, we test $x = -3$. $f(-3) = (-3)^4 - 5(-3)^2 + 4 = 40 > 0$

In the interval $(-2,\ -1)$, we test $x = -1.5$. $f(-1.5) = (-1.5)^4 - 5(-1.5)^2 + 4 = -2.1875 < 0$

In the interval $(-1,\ 1)$, we test $x = 0$. $f(0) = (0)^4 - 5(0)^2 + 4 = 4 > 0$

In the interval $(1,\ 2)$, we test $x = 1.5$. $f(1.5) = (1.5)^4 - 5(1.5)^2 + 4 = -2.1875 < 0$

In the interval $(2,\ \infty)$, we test $x = 3$. $f(3) = (3)^4 - 5(3)^2 + 4 = 40 > 0$

The solution to the given inequality is $(-\infty,\ -2) \cup (-1,\ 1) \cup (2,\ \infty)$.

14. $1 - x^4 = 0 \Rightarrow (1 - x^2)(1 + x^2) = 0 \Rightarrow (1-x)(1+x)(1+x^2) = 0 \Rightarrow x = -1 \text{ or } 1$

In the interval $(-\infty,\ -1)$, we test $x = -2$. $f(-2) = 1 - (-2)^4 = -15 < 0$

In the interval $(-1,\ 1)$, we test $x = 0$. $f(0) = 1 - (0)^4 = 1 > 0$

In the interval $(1, \infty)$, we test $x = 2$. $f(2) = 1 - (2)^4 = -15 < 0$

The solution to the given inequality is $(-\infty, -1) \cup (1, \infty)$.

15. $\dfrac{3-x}{3x} = 0 \Rightarrow 3 - x = 0 \Rightarrow x = 3$. The expression is undefined when $3x = 0 \Rightarrow x = 0$.

In the interval $(-\infty, 0)$, we test $x = -1$. $f(-1) = \dfrac{3-(-1)}{3(-1)} = -\dfrac{4}{3} \le 0$

In the interval $(0, 3)$, we test $x = 1$. $f(1) = \dfrac{3-(1)}{3(1)} = \dfrac{2}{3} \ge 0$

In the interval $(3, \infty)$, we test $x = 4$. $f(4) = \dfrac{3-(4)}{3(4)} = -\dfrac{1}{12} \le 0$

The solution to the given inequality is $(0, 3]$.

16. $\dfrac{x-2}{x+2} = 0 \Rightarrow x - 2 = 0 \Rightarrow x = 2$. The expression is undefined when $x + 2 = 0 \Rightarrow x = -2$.

In the interval $(-\infty, -2)$, we test $x = -3$. $f(-3) = \dfrac{(-3)-2}{(-3)+2} = 5 > 0$

In the interval $(-2, 2)$, we test $x = 0$. $f(0) = \dfrac{(0)-2}{(0)+2} = -1 < 0$

In the interval $(2, \infty)$, we test $x = 3$. $f(3) = \dfrac{(3)-2}{(3)+2} = \dfrac{1}{5} > 0$

The solution to the given inequality is $(-\infty, -2) \cup (2, \infty)$.

17. $\dfrac{3-2x}{1+x} = 3 \Rightarrow 3 - 2x = 3 + 3x \Rightarrow 5x = 0 \Rightarrow x = 0$.

The expression is undefined when $1 + x = 0 \Rightarrow x = -1$.

In the interval $(-\infty, -1)$, we test $x = -2$. $f(-2) = \dfrac{3-2(-2)}{1+(-2)} - 3 = -10 < 0$

In the interval $(-1, 0)$, we test $x = -0.5$. $f(-0.5) = \dfrac{3-2(-0.5)}{1+(-0.5)} - 3 = 5 > 0$

In the interval $(0, \infty)$, we test $x = 1$. $f(1) = \dfrac{3-2(1)}{1+(1)} - 3 = -\dfrac{5}{2} < 0$

The solution to the given inequality is $(-\infty, -1) \cup (0, \infty)$.

18. $\dfrac{x+1}{4-2x} = 1 \Rightarrow x + 1 = 4 - 2x \Rightarrow 3x = 3 \Rightarrow x = 1$.

The expression is undefined when $4 - 2x = 0 \Rightarrow 2x = 4 \Rightarrow x = 2$.

In the interval $(-\infty, 1)$, we test $x = 0$. $f(0) = \dfrac{(0)+1}{4-2(0)} - 1 = -\dfrac{3}{4} \le 0$

In the interval $(1, 2)$, we test $x = 1.5$.   $f(1.5) = \dfrac{(1.5)+1}{4-2(1.5)} - 1 = \dfrac{3}{2} \geq 0$

In the interval $(2, \infty)$, we test $x = 3$.   $f(3) = \dfrac{(3)+1}{4-2(3)} - 1 = -3 \leq 0$

The solution to the given inequality is $[1, 2)$.

19. The expression is undefined when $x^2 - 4 = 0 \Rightarrow (x+2)(x-2) = 0 \Rightarrow x = -2$ or $2$.

In the interval $(-\infty, -2)$, we test $x = -3$.   $f(-3) = \dfrac{5}{(-3)^2 - 4} = 1 > 0$

In the interval $(-2, 2)$, we test $x = 0$.   $f(0) = \dfrac{5}{(0)^2 - 4} = -\dfrac{5}{4} < 0$

In the interval $(2, \infty)$, we test $x = 3$.   $f(3) = \dfrac{5}{(3)^2 - 4} = 1 > 0$

The solution to the given inequality is $(-2, 2)$.

20. $\dfrac{x}{x^2-1} = 0 \Rightarrow x = 0$.  The expression is undefined when $x^2 - 1 = 0 \Rightarrow (x+1)(x-1) = 0 \Rightarrow x = -1$ or $1$.

In the interval $(-\infty, -1)$, we test $x = -2$.   $f(-2) = \dfrac{(-2)}{(-2)^2 - 1} = -\dfrac{2}{3} \leq 0$

In the interval $(-1, 0)$, we test $x = -0.5$.   $f(-0.5) = \dfrac{(-0.5)}{(-0.5)^2 - 1} = \dfrac{2}{3} \geq 0$

In the interval $(0, 1)$, we test $x = 0.5$.   $f(0.5) = \dfrac{(0.5)}{(0.5)^2 - 1} = -\dfrac{2}{3} \leq 0$

In the interval $(1, \infty)$, we test $x = 2$.   $f(2) = \dfrac{(2)}{(2)^2 - 1} = \dfrac{2}{3} \geq 0$

The solution to the given inequality is $(-1, 0] \cup (1, \infty)$.

## Chapters 1-11 Cumulative Review Exercises

1. $F = \dfrac{5}{(-2)^2 + 1} = \dfrac{5}{4+1} = \dfrac{5}{5} = 1$

2. Natural number: $\sqrt[3]{8} = 2$; whole number: $0, \sqrt[3]{8}$; integer: $0, -5, \sqrt[3]{8}$;

   rational number: $0.\overline{4}, 0, -5, \sqrt[3]{8}, -\dfrac{4}{3}$; irrational number: $\sqrt{7}$

3. (a)  $\left(\dfrac{x^2 y^6}{x^{-3}}\right)^2 = \dfrac{x^4 y^{12}}{x^{-6}} = x^{4-(-6)} y^{12} = x^{10} y^{12}$

   (b)  $\dfrac{(xy^{-3})^2}{x(y^{-2})^{-1}} = \dfrac{x^2 y^{-6}}{xy^2} = x^{2-1} y^{-6-2} = x^1 y^{-8} = \dfrac{x}{y^8}$

(c)  $(a^2b)^2(ab^3)^{-4} = a^4b^2a^{-4}b^{-12} = a^{4-4}b^{2-12} = a^0b^{-10} = \dfrac{1}{b^{10}}$

4. $9{,}290{,}000 = 9.29 \times 10^6$

5. $f(-2) = \sqrt{2-(-2)} = \sqrt{4} = 2$ ; the radicand must be greater than or equal to zero, so

   $2 - x \geq 0 \Rightarrow x \leq 2$.

6. $(2, 5)$ lies on the graph of $f$.

7. See Figure 7.

8. See Figure 8.

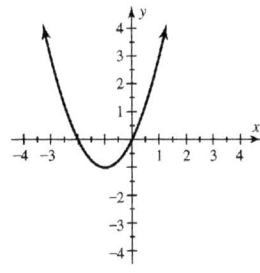

Figure 7

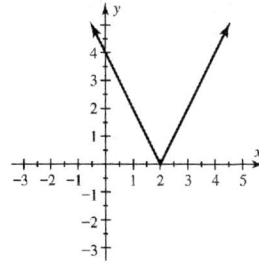

Figure 8

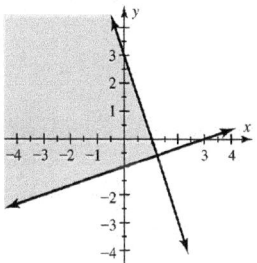

Figure 16

9. Parallel lines have the same slope, so $m = \dfrac{4-1}{-2-0} = \dfrac{3}{-2} = -\dfrac{3}{2}$.

   $y - (-1) = -\dfrac{3}{2}(x - 4) \Rightarrow y = -\dfrac{3}{2}x + 6 - 1 \Rightarrow y = -\dfrac{3}{2}x + 5$

10. $x = -3$

11. $2x - 3(x+2) = 6 \Rightarrow 2x - 3x - 6 = 6 \Rightarrow -x = 12 \Rightarrow x = -12$

12. $7 - x > 3x \Rightarrow 7 > 4x \Rightarrow \dfrac{7}{4} > x$ ; $\left(-\infty, \dfrac{7}{4}\right)$

13. $|3x - 2| \leq 1 \Rightarrow 3x - 2 \leq 1$ and $3x - 2 \geq -1 \Rightarrow 3x \leq 3$ and $3x \geq 1 \Rightarrow x \leq 1$ and $x \geq \dfrac{1}{3}$; $\left[\dfrac{1}{3}, 1\right]$

14. $-4 \leq 1 - x < 2 \Rightarrow -5 \leq -x < 1 \Rightarrow 5 \geq x > -1 \Rightarrow -1 < x \leq 5$; $(-1, 5]$

15. Multiply the first equation by 5 and add to the second equation.
    $$\begin{array}{r} -5x - 20y = -15 \\ 5x + \ \ y = -4 \\ \hline -19y = -19 \end{array} \Rightarrow y = 1$$

    Substitute $y = 1$ into the first equation: $-x - 4(1) = -3 \Rightarrow -x = -3 + 4 \Rightarrow -x = 1 \Rightarrow x = -1$

    The solution is $(-1, 1)$.

16. Note that $3x + y \leq 3 \Rightarrow y \leq -3x + 3$ and $x - 3y \leq 3 \Rightarrow -3y \leq -x + 3 \Rightarrow y \geq \dfrac{1}{3}x - 1$. See Figure 16.

17. Add the first two equations.
$$\begin{array}{r} x+y-z=3 \\ x-y+z=1 \\ \hline 2x\qquad\;\;=4 \end{array} \Rightarrow x=2$$

Add the second and third equations.

$$\begin{array}{r} x-\;\;y+z=1 \\ 2x-\;y-z=1 \\ \hline 3x-2y\qquad=2 \end{array} \Rightarrow 3(2)-2y=2 \Rightarrow 6-2y=2 \Rightarrow -2y=-4 \Rightarrow y=2$$

Substitute $x=2$ and $y=2$ into the first equation. $2+2-z=3 \Rightarrow 4-z=3 \Rightarrow -z=-1 \Rightarrow z=1$

The solution is $(2, 2, 1)$.

18. $(3x-2)(2x+7)=6x^2+21x-4x-14=6x^2+17x-14$

19. $3xy(x^2+y^2)=3x^3y+3xy^3$

20. $(\sqrt{x}+3)(\sqrt{x}-3)=(\sqrt{x})^2-(3)^2=x-9$

21. $x^3-x^2-2x=x(x^2-x-2)=x(x-2)(x+1)$

22. $4x^2-25=(2x-5)(2x+5)$

23. $x^2-3=0 \Rightarrow x^2=3 \Rightarrow x=\pm\sqrt{3}$

24. $x^2+1=2x \Rightarrow x^2-2x+1=0 \Rightarrow (x-1)^2=0 \Rightarrow x-1=0 \Rightarrow x=1$

25. $\dfrac{(x+3)^2}{x+2}\cdot\dfrac{x+2}{2x+6}=\dfrac{(x+3)(x+3)(x+2)}{(x+2)(2)(x+3)}=\dfrac{x+3}{2}$ or $\dfrac{1}{2}(x+3)$

26. $\dfrac{1}{x+2}-\dfrac{1}{x}=\dfrac{1}{x+2}\cdot\dfrac{x}{x}-\dfrac{1}{x}\cdot\dfrac{x+2}{x+2}=\dfrac{x}{x(x+2)}-\dfrac{x+2}{x(x+2)}=\dfrac{x-x-2}{x(x+2)}=\dfrac{-2}{x(x+2)}$

27. $\sqrt{16x^6}=\sqrt{16}\cdot\sqrt{x^6}=4x^3$

28. $16^{-3/2}=(16^{1/2})^{-3}=4^{-3}=\dfrac{1}{4^3}=\dfrac{1}{64}$

29. $\dfrac{\sqrt[3]{81x}}{\sqrt[3]{3x}}=\sqrt[3]{\dfrac{81x}{3x}}=\sqrt[3]{27}=3$

30. $\sqrt{8x}+\sqrt{2x}=\sqrt{4\cdot 2x}+\sqrt{2x}=2\sqrt{2x}+\sqrt{2x}=3\sqrt{2x}$

31. See Figure 31.

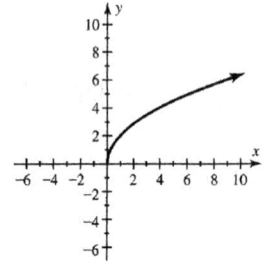

Figure 31

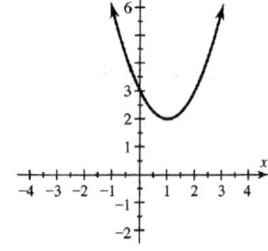

Figure 36

32. $d=\sqrt{(x_2-x_1)^2+(y_2-y_1)^2}=\sqrt{(4-(-1))^2+(3-2)^2}=\sqrt{(5)^2+(1)^2}=\sqrt{25+1}=\sqrt{26}$

33. $\dfrac{3-i}{2+i} = \dfrac{3-i}{2+i} \cdot \dfrac{2-i}{2-i} = \dfrac{6-3i-2i+i^2}{4-i^2} = \dfrac{6-5i-1}{4-(-1)} = \dfrac{5-5i}{5} = 1-i$

34. $3\sqrt{x+1} = 2x \Rightarrow 9(x+1) = 4x^2 \Rightarrow 9x+9 = 4x^2 \Rightarrow 4x^2 - 9x - 9 = 0 \Rightarrow (x-3)(4x+3) = 0 \Rightarrow x = 3$

$\left( x = -\dfrac{3}{4} \text{ does not check} \right)$

35. Graph $Y_1 = 2X$ and $Y_2 = \sqrt{\ }((2.1-X)+(0.1X) \wedge (1/3))$ in $[-4.7, 4.7, 1]$ by $[-3.1, 3.1, 1]$.

See Figure 35. The solution is $x \approx 0.79$.

$[-4.7, 4.7, 1]$ by $[-3.1, 3.1, 1]$

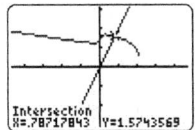

Figure 35

36. See Figure 36.

(a) $(1, 2)$

(b) $f(-1) = (-1)^2 - 2(-1) + 3 = 1 + 2 + 3 = 6$

(c) $x = 1$

(d) $f$ is increasing when $x \geq 1$

37. $f(x) = 2x^2 - 4x - 1 \Rightarrow f(x) = 2(x^2 - 2x) - 1 \Rightarrow f(x) = 2(x^2 - 2x + 1) - 1 - 2 \Rightarrow f(x) = 2(x-1)^2 - 3$

38. The graph is shifted 1 unit left and 2 units downward. The graph of $f(x)$ is narrower.

39. $x^2 + 6x = 2 \Rightarrow x^2 + 6x + 9 = 2 + 9 \Rightarrow (x+3)^2 = 11 \Rightarrow x + 3 = \pm\sqrt{11} \Rightarrow x = -3 \pm \sqrt{11}$

40. $2x^2 - 3x = 1 \Rightarrow 2x^2 - 3x - 1 = 0$; $a = 2, b = -3, c = -1$

$x = \dfrac{-b \pm \sqrt{b^2 - 4ac}}{2a} = \dfrac{-(-3) \pm \sqrt{(-3)^2 - 4(2)(-1)}}{2(2)} = \dfrac{3 \pm \sqrt{17}}{4}$

41. $x(4-x) = 3 \Rightarrow 4x - x^2 = 3 \Rightarrow x^2 - 4x + 3 = 0 \Rightarrow (x-3)(x-1) = 0 \Rightarrow x = 3$ or $x = 1$

42. (a) $x = -2$ or $x = 1$

(b) $-2 \leq x \leq 1$

43. $x^2 - 3x + 2 = 0 \Rightarrow (x-2)(x-1) = 0 \Rightarrow x = 1$ or $x = 2$. Since the parabola opens up, the solution is

$x < 1$ or $x > 2$.

44. $x^4 - 256 = 0 \Rightarrow (x^2 - 16)(x^2 + 16) = 0 \Rightarrow (x-4)(x+4)(x^2+16) = 0 \Rightarrow x = 4$ or $x = -4$ or

$x^2 = -16 \Rightarrow x = \pm\sqrt{-16} \Rightarrow x = \pm 4i$

45. c; $y = ax - b$ is a linear equation and $a > 0$.

46. f; $y = b$ is a horizontal line.

47. g; $y = -ax^2 + c$ is a quadratic equation whose graph is a parabola opening downward (since $a > 0$).

48.  e; $y = \dfrac{a}{x}$ is a rational equation, undefined when $x = 0$ and $y = 0$.

49.  d; $y = ax^3$ is a cubic equation.        50.        a; $y = |ax + b|$ has only nonnegative values.

51.  b; $y = a\sqrt{x}$ is a radical function defined only for $x \geq 0$.

52.  h; $y = a\sqrt[3]{x}$ is a cube root equation.

53.  (a)    $G(0) = 300$ gallons ; initially, the tank holds 300 gallons.

     (b)    The $t$-intercept is 6; after 6 minutes the tank is empty.

     (c)    $(0, 300)$ and $(6, 0)$ are on the graph so $m = \dfrac{0 - 300}{6 - 0} = \dfrac{-300}{6} = -50$ ; water is pumped out at 50

            gallons per minute.

     (d)    Since $(0, 300)$ is on the graph and $m = -50$, $G(t) = -50t + 300$ or $G(t) = 300 - 50t$

54.  Let $x$, $y$, and $z$ be the amounts invested at 4% , 5%, and 6% interest, respectively. The system needed

$$x + y + z = 4000 \qquad\qquad\qquad x + y + \ z = 4000$$
is $z = y + 1000 \qquad\qquad \Rightarrow \qquad\qquad y - \ z = -1000$
$$0.04x + 0.05y + 0.06z = 216 \qquad 4x + 5y + 6z = 21,600$$

Multiply the first equation by $-4$ and add to the third equation

$$\begin{array}{r} -4x - 4y - 4z = -16,000 \\ \underline{4x + 5y + 6z = \ \ 21,600} \\ y + 2z = \ \ \ \ 5600 \end{array}$$

Multiply the second equation by 2 and add to this new equation.

$$\begin{array}{r} 2y - 2z = -2000 \\ \underline{y + 2z = \ \ 5600} \\ 3y \ \ \ \ \ \ = \ \ 3600 \end{array} \qquad \Rightarrow y = 1200$$

Substitute $y = 1200$ into the second equation.

$$1200 - z = -1000 \Rightarrow -z = -2200 \Rightarrow z = 2200$$

Substitute $y = 1200$ and $z = 2200$ into the first equation.

$$x + 1200 + 2200 = 4000 \Rightarrow x = 600$$

$600 is invested at 4%, $1200 is invested at 5%, and $2200 is invested at 6%.

55.  Let $x$ represent the length of one of the two equal sides of the garden without the gate. Then the other

     sides of the garden have lengths $\dfrac{1}{2}(490 - 2x + 10) = \dfrac{1}{2}(500 - 2x) = 250 - x$. $(250 - x)x = 250x - x^2$ ;

     The graph of $y = -x^2 + 250x$ is a parabola opening downward; its vertex is its maximum point. The

     $x$-value of the vertex is $x = -\dfrac{b}{2a} = -\dfrac{250}{2(-1)} = 125$. The dimensions of the garden of largest area are

     125 feet by 125 feet.

56.  Let $h$ represent the height of the tree. $\dfrac{6}{10} = \dfrac{h}{55} \quad \Rightarrow \quad 10h = 6(55) \Rightarrow 10h = 330 \Rightarrow h = 33$ feet

# Chapter 12: Exponential and Logarithmic Functions

## 12.1: Composite and Inverse Functions

1. $g(f(7))$

3. No

5. No

7. adding 10

9. 8; 6

11. one-to-one

13. (a)  $f(-2)=(-2)^2=4$, then $(g\circ f)(-2)=g(f(-2))=g(4)=4+3=7$

(b)  $g(4)=4+3=7$, then $(f\circ g)(4)=f(g(4))=f(7)=7^2=49$

(c)  $(g\circ f)(x)=g(f(x))=g(x^2)=x^2+3$

(d)  $(f\circ g)(x)=f(g(x))=f(x+3)=(x+3)^2$

15. (a)  $f(-2)=2(-2)=-4$, then $(g\circ f)(-2)=g(f(-2))=g(-4)=(-4)^3-1=-65$

(b)  $g(4)=4^3-1=63$, then $(f\circ g)(4)=f(g(4))=f(63)=2(63)=126$

(c)  $(g\circ f)(x)=g(f(x))=g(2x)=(2x)^3-1=8x^3-1$

(d)  $(f\circ g)(x)=f(g(x))=f(x^3-1)=2(x^3-1)=2x^3-2$

17. (a)  $f(-2)=\dfrac{1}{2}(-2)=-1$, then $(g\circ f)(-2)=g(f(-2))=g(-1)=|(-1)-2|=3$

(b)  $g(4)=|4-2|=2$, then $(f\circ g)(4)=f(g(4))=f(2)=\dfrac{1}{2}(2)=1$

(c)  $(g\circ f)(x)=g(f(x))=g\left(\dfrac{1}{2}x\right)=\left|\dfrac{1}{2}x-2\right|$

(d)  $(f\circ g)(x)=f(g(x))=f(|x-2|)=\dfrac{1}{2}|x-2|$

19. (a)  $f(-2)=\dfrac{1}{-2}=-\dfrac{1}{2}$, then $(g\circ f)(-2)=g(f(-2))=g\left(-\dfrac{1}{2}\right)=3-5\left(-\dfrac{1}{2}\right)=\dfrac{11}{2}$

(b)  $g(4)=3-5(4)=-17$, then $(f\circ g)(4)=f(g(4))=f(-17)=\dfrac{1}{-17}=-\dfrac{1}{17}$

(c)  $(g\circ f)(x)=g(f(x))=g\left(\dfrac{1}{x}\right)=3-5\left(\dfrac{1}{x}\right)=3-\dfrac{5}{x}$

(d)  $(f\circ g)(x)=f(g(x))=f(3-5x)=\dfrac{1}{3-5x}$

21. (a)  $f(-2)=2(-2)=-4$, then $(g \circ f)(-2)=g(f(-2))=g(-4)=4(-4)^2-2(-4)+5=77$

    (b)  $g(4)=4(4)^2-2(4)+5=61$, then $(f \circ g)(4)=f(g(4))=f(61)=2(61)=122$

    (c)  $(g \circ f)(x)=g(f(x))=g(2x)=4(2x)^2-2(2x)+5=16x^2-4x+5$

    (d)  $(f \circ g)(x)=f(g(x))=f(4x^2-2x+5)=2(4x^2-2x+5)=8x^2-4x+10$

23. (a)  $(f \circ g)(0)=f(g(0))=f(-1)=1$

    (b)  $(g \circ f)(-1)=g(f(-1))=g(1)=2$

25. (a)  $(f \circ f)(-1)=f(f(-1))=f(1)=-1$

    (b)  $(g \circ g)(0)=g(g(0))=g(-1)=1$

27. (a)  $\left(f^{-1} \circ g\right)(-2)=f^{-1}(g(-2))=f^{-1}(0)=0$

    (b)  $\left(g^{-1} \circ f\right)(2)=g^{-1}(f(2))=g^{-1}(-2)=2$

29. (a)  $(f \circ g)(0)=f(g(0))=f(-1)=2$

    (b)  $(g \circ f)(1)=g(f(1))=g(2)=-3$

    (c)  $(f \circ f)(-1)=f(f(-1))=f(2)=-1$

31.  $f(1)=f(-1)=5$;  *Answers may vary.*

33.  $f(1)=f(-1)=101$;  *Answers may vary.*

35.  $f(2)=f(-2)=4$;  *Answers may vary.*

37.  This graph passes the horizontal line test.  The function is one-to-one.

39.  This graph does not pass the horizontal line test.  The function is not one-to-one.

41.  This graph passes the horizontal line test.  The function is one-to-one.

43.  Divide $x$ by 7.  $f(x)=7x$; $g(x)=\dfrac{x}{7}$

45.  Multiply $x$ by 2 and then subtract 5.  $f(x)=\dfrac{x+5}{2}$; $g(x)=2x-5$

47.  Add 3 to $x$ and then multiply the result by 2.  $f(x)=\dfrac{1}{2}x-3$; $g(x)=2(x+3)$

49.  Take the cube root of $x$ and then subtract 5.  $f(x)=(x+5)^3$; $g(x)=\sqrt[3]{x}-5$

51.  $\left(f \circ f^{-1}\right)(x)=f\left(f^{-1}(x)\right)=f\left(\dfrac{x}{4}\right)=4\left(\dfrac{x}{4}\right)=x$  and  $\left(f^{-1} \circ f\right)(x)=f^{-1}(f(x))=f^{-1}(4x)=\dfrac{4x}{4}=x$

53. $\left(f \circ f^{-1}\right)(x) = f\left(f^{-1}(x)\right) = f\left(\dfrac{x-5}{3}\right) = 3\left(\dfrac{x-5}{3}\right) + 5 = x - 5 + 5 = x$

$\left(f^{-1} \circ f\right)(x) = f^{-1}\left(f(x)\right) = f^{-1}(3x+5) = \dfrac{(3x+5)-5}{3} = \dfrac{3x}{3} = x$

55. $\left(f \circ f^{-1}\right)(x) = f\left(f^{-1}(x)\right) = f\left(\sqrt[3]{x}\right) = \left(\sqrt[3]{x}\right)^3 = x$

$\left(f^{-1} \circ f\right)(x) = f^{-1}\left(f(x)\right) = f^{-1}\left(x^3\right) = \sqrt[3]{x^3} = x$

57. $\left(f \circ f^{-1}\right)(x) = f\left(f^{-1}(x)\right) = f\left(\dfrac{1}{x}\right) = \dfrac{1}{\frac{1}{x}} = x$ and $\left(f^{-1} \circ f\right)(x) = f^{-1}\left(f(x)\right) = f^{-1}\left(\dfrac{1}{x}\right) = \dfrac{1}{\frac{1}{x}} = x$

59. $f(x) = 12x \Rightarrow y = 12x$, interchange $x$ and $y$ and solve for $y$. $x = 12y \Rightarrow y = \dfrac{x}{12} \Rightarrow f^{-1}(x) = \dfrac{x}{12}$

61. $f(x) = x + 8 \Rightarrow y = x + 8$, interchange $x$ and $y$ and solve for $y$.

$x = y + 8 \Rightarrow y = x - 8 \Rightarrow f^{-1}(x) = x - 8$

63. $f(x) = 5x - 2 \Rightarrow y = 5x - 2$, interchange $x$ and $y$ and solve for $y$.

$x = 5y - 2 \Rightarrow 5y = x + 2 \Rightarrow y = \dfrac{x+2}{5} \Rightarrow f^{-1}(x) = \dfrac{x+2}{5}$

65. $f(x) = -\dfrac{1}{2}x + 1 \Rightarrow y = -\dfrac{1}{2}x + 1$, interchange $x$ and $y$ and solve for $y$.

$x = -\dfrac{1}{2}y + 1 \Rightarrow -\dfrac{1}{2}y = x - 1 \Rightarrow y = -2(x-1) \Rightarrow f^{-1}(x) = -2(x-1)$

67. $f(x) = 8 - x \Rightarrow y = 8 - x$, interchange $x$ and $y$ and solve for $y$.

$x = 8 - y \Rightarrow y = 8 - x \Rightarrow f^{-1}(x) = 8 - x$

69. $f(x) = \dfrac{x+1}{2} \Rightarrow y = \dfrac{x+1}{2}$, interchange $x$ and $y$ and solve for $y$.

$x = \dfrac{y+1}{2} \Rightarrow y + 1 = 2x \Rightarrow y = 2x - 1 \Rightarrow f^{-1}(x) = 2x - 1$

71. $f(x) = \sqrt[3]{2x} \Rightarrow y = \sqrt[3]{2x}$, interchange $x$ and $y$ and solve for $y$.

$x = \sqrt[3]{2y} \Rightarrow 2y = x^3 \Rightarrow y = \dfrac{x^3}{2} \Rightarrow f^{-1}(x) = \dfrac{x^3}{2}$

73. $f(x) = x^3 - 8 \Rightarrow y = x^3 - 8$, interchange $x$ and $y$ and solve for $y$.

$x = y^3 - 8 \Rightarrow y^3 = x + 8 \Rightarrow y = \sqrt[3]{x+8} \Rightarrow f^{-1}(x) = \sqrt[3]{x+8}$

75. See Figure 75. The domain of $f$ = the range of $f^{-1} = \{0, 1, 2, 3, 4\}$.

The range of $f$ = the domain of $f^{-1} = \{0, 5, 10, 15, 20\}$.

| $x$ | 0 | 5 | 10 | 15 | 20 |
|------|---|---|----|----|----|
| $f^{-1}(x)$ | 0 | 1 | 2 | 3 | 4 |

Figure 75

| $x$ | 4 | 2 | 0 | $-2$ | $-4$ |
|------|---|---|---|------|------|
| $f^{-1}(x)$ | $-5$ | 0 | 5 | 10 | 15 |

Figure 77

77. See Figure 77. The domain of $f$ = the range of $f^{-1}$ = $\{-5, 0, 5, 10, 15\}$.

    The range of $f$ = the domain of $f^{-1}$ = $\{-4, -2, 0, 2, 4\}$.

79. See Figure 79.

81. See Figure 81.

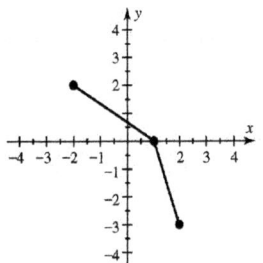

Figure 79

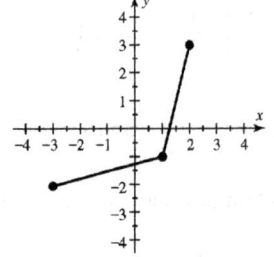

Figure 81

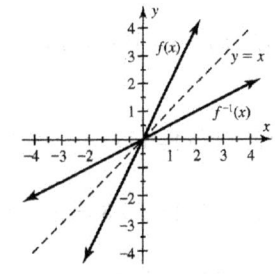

Figure 83

83. The graph of $f^{-1}$ is a reflection of the graph of $f$ across the line $y = x$. See Figure 83.

85. The graph of $f^{-1}$ is a reflection of the graph of $f$ across the line $y = x$. See Figure 85.

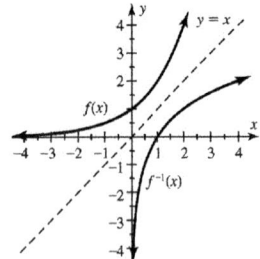

Figure 85

87. (a)   $(fg)(2) = (2^2 - 2)(2^2 + 2) = (4 - 2)(4 + 2) = (2)(6) = 12$

    (b)   $(f - g)(x) = (x^2 - 2) - (x^2 + 2) = x^2 - x^2 - 2 - 2 = -4$

    (c)   $(f \circ g)(x) = f(g(x)) = f(x^2 + 2) = (x^2 + 2)^2 - 2 = x^4 + 4x^2 + 4 - 2 = x^4 + 4x^2 + 2$

89. (a)   $(fg)(2) = \left(\dfrac{1}{2}\right)\left(\dfrac{2}{2}\right) = \dfrac{2}{4} = \dfrac{1}{2}$

    (b)   $(f - g)(x) = \dfrac{1}{x} - \dfrac{2}{x} = -\dfrac{1}{x}$

    (c)   $(f \circ g)(x) = f(g(x)) = f\left(\dfrac{2}{x}\right) = \dfrac{1}{\dfrac{2}{x}} = \dfrac{x}{2}$

91. (a)   $(C \circ r)(5) = C(r(5)) = C(2 \cdot 5) = C(10) = 2\pi \cdot 10 = 20\pi$

    After 5 seconds, the wave has a circumference of $20\pi \approx 62.8$ feet.

(b)  $(C \circ r)(t) = C\left(r(t)\right) = C(2t) = 2\pi \cdot 2t = 4\pi t$

93. (a)  $P(1980) = 16$; In 1980, 16% of people 25 or older completed 4 or more years of college.

(b)  See Figure 93.

(c)  $P^{-1}(16) = 1980$

| $x$ | 8 | 16 | 27 | 29 |
|---|---|---|---|---|
| $P^{-1}(x)$ | 1960 | 1980 | 2000 | 2010 |

Figure 93

95. (a)  $T(1) = 75°$ and $M(75) = 150$

(b)  $(M \circ T)(1) = M\left(T(1)\right) = M(75) = 150$

One hour after midnight there are 150 mosquitoes per 100 square feet.

(c)  $(M \circ T)(h)$ calculates the number of mosquitoes per 100 square feet, $h$ hours after midnight.

(d)  For $T(h)$, $m = \dfrac{50-80}{6-0} = \dfrac{-30}{6} = -5$.  Since the $y$-intercept is 80, the equation is

$T(h) = -5h + 80$.  For $M(T)$, $m = \dfrac{150-100}{75-50} = \dfrac{50}{25} = 2$.  Since the line passes through

(50, 100), the equation is $M(T) = 2(T-50) + 100 = 2T - 100 + 100 = 2T$.

(e)  $(M \circ T)(h) = M\left(T(h)\right) = M(-5h+80) = 2(-5h+80) = -10h + 160$

97. (a)  Yes, this is a one-to-one function because different inputs result in different outputs.

(b)  $f(x) = \dfrac{5}{9}x + 32 \Rightarrow y = \dfrac{5}{9}x + 32$, interchange $x$ and $y$ and solve for $y$.

$x = \dfrac{5}{9}y + 32 \Rightarrow \dfrac{5}{9}y = x - 32 \Rightarrow y = \dfrac{9}{5}(x-32) \Rightarrow f^{-1}(x) = \dfrac{9}{5}(x-32)$

This formula converts $x$ degrees Fahrenheit to an equivalent temperature in degrees Celsius.

99. Since there are 4 quarts in 1 gallon, the function $f(x) = 4x$ converts $x$ gallons to quarts.

$f(x) = 4x \Rightarrow y = 4x$, interchange $x$ and $y$ and solve for $y$.  $x = 4y \Rightarrow y = \dfrac{x}{4} \Rightarrow f^{-1}(x) = \dfrac{x}{4}$

This formula converts $x$ quarts to gallons.

101. (a)  There are 16 cups in a gallon.  The function is $C(x) = 16x$.

(b)  There are 48 teaspoons in a cup.  The function is $T(x) = 48x$.

(c)  $(T \circ C)(x) = T(C(x)) = T(16x) = 48(16x) = 768x$

(d)  $(T \circ C)(3) = T(C(3)) = T(48) = 48(48) = 2304$,  There are 2304 teaspoons in 3 gallons.

## Section 12.2 Exponential Functions

1.  $f(x) = Ca^x$

3. Domain: All real numbers; Range: All positive real numbers

5. $e \approx 2.718$

7. factor

9. $\dfrac{B-A}{A} \times 100$

11. $f(-2) = 3^{-2} = \dfrac{1}{3^2} = \dfrac{1}{9}$ and $f(2) = 3^2 = 9$

13. $f(0) = 5\left(2^0\right) = 5(1) = 5$ and $f(5) = 5\left(2^5\right) = 5(32) = 160$

15. $f(-2) = \left(\dfrac{1}{2}\right)^{-2} = 2^2 = 4$ and $f(3) = \left(\dfrac{1}{2}\right)^3 = \dfrac{1}{2^3} = \dfrac{1}{8}$

17. $f(-1) = 5(3)^{-(-1)} = 5(3)^1 = 15$ and $f(2) = 5(3)^{-2} = 5\left(\dfrac{1}{3^2}\right) = \dfrac{5}{9}$

19. $f(-3) = 1.8^{-3} \approx 0.17$ and $f(1.5) = 1.8^{1.5} \approx 2.41$

21. $f(-1) = 3(0.6)^{-1} = 5$ and $f(2) = 3(0.6)^2 = 1.08$

23. $f(0) = a^0 = 1;\ f(-1) = a^{-1} = \dfrac{1}{a}$

25. c. This function models exponential growth and passes through the point (0, 1).

27. d. This function models exponential decay and passes through the point (2, 1).

29. Since $y = 1$ when $x = 0, 1 = Ca^0 \Rightarrow C = 1$. Since $y = 2$ when $x = 1, 2 = 1(a)^1 \Rightarrow a = 2$.

31. Since $y = 4$ when $x = 0, 4 = Ca^0 \Rightarrow C = 4$. Since $y = 1$ when $x = 1, 1 = 4(a)^1 \Rightarrow a = \dfrac{1}{4}$.

33. See Figure 33. The graph illustrates exponential growth.

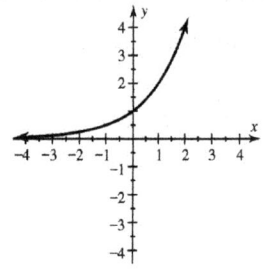

Figure 33

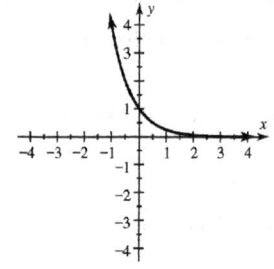

Figure 35

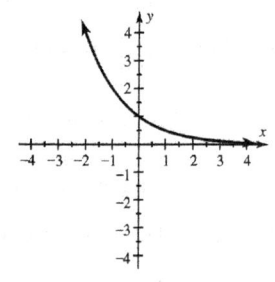

Figure 37

35. See Figure 35. The graph illustrates exponential decay.

37. See Figure 37. The graph illustrates exponential decay.

39. See Figure 39. The graph illustrates exponential growth.

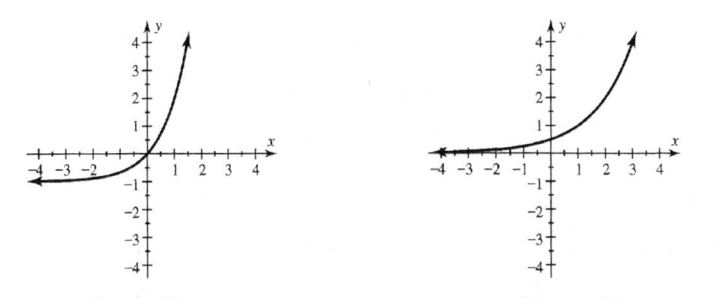

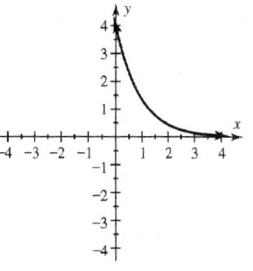

Figure 39                           Figure 41                           Figure 43

41. See Figure 41. The graph illustrates exponential growth.

43. See Figure 43. The graph illustrates exponential decay.

45. (a)   Exponential decay. For each unit increase in $x$, $f(x)$ decreases by a factor of $\frac{1}{4}$.

    (b)   Since $f(x) = 64$ when $x = 0$, $f(x) = 64\left(\frac{1}{4}\right)^x$

47. (a)   Linear growth. For each unit increase in $x$, $f(x)$ increases by 3 units.

    (b)   Since $f(x) = 8$ when $x = 0$, $f(x) = 3x + 8$

49. (a)   Exponential growth. For each unit increase in $x$, $f(x)$ increases by a factor of 1.25.

    (b)   Since $f(x) = 4$ when $x = 0$, $f(x) = 4(1.25)^x$

51. (a)   $\frac{B-A}{A} \times 100 \Rightarrow \frac{400-200}{200} \times 100 = \frac{200}{200} \times 100 = 100\%$

    (b)   $\frac{A-B}{B} \times 100 \Rightarrow \frac{200-400}{400} \times 100 = \frac{-200}{400} \times 100 = -50\%$

53. (a)   $\frac{B-A}{A} \times 100 \Rightarrow \frac{30-150}{150} \times 100 = \frac{-120}{150} \times 100 = -80\%$

    (b)   $\frac{A-B}{B} \times 100 \Rightarrow \frac{150-30}{30} \times 100 = \frac{120}{30} \times 100 = 400\%$

55. (a)   $rA = 1.2(1000) = 1200$   The account increased by $1200.

    (b)   The new value of the account is $A + rA = \$1000 + \$1200 = \$2200$.

    (c)   The account increased by a factor of $a = 1 + r = 1 + 1.2 = 2.2$.

57. (a)   $rA = 0.2(650) = 130$   The account increased by $130.

    (b)   The new value of the account is $A + rA = \$650 + \$130 = \$780$.

    (c)   The account increased by a factor of $a = 1 + r = 1 + 0.2 = 1.2$.

59. (a)   $rA = -0.1(800) = -80$   The account decreased by $80.

    (b)   The new value of the account is $A + rA = \$800 + \$(-80) = \$720$.

    (c)   The account decreased by a factor of $a = 1 + r = 1 + (-0.1) = 0.9$.

61. The general form of an exponential function is $f(x) = Ca^x$, where $C$ is the initial value, $a$ is the growth factor, and $x$ is the variable as an exponent. The initial value is $C = 9$, since $a > 1$ we have a growth factor $a = 1.07$. $a = 1 + r \Rightarrow 1.07 = 1 + r \Rightarrow r = 0.7 \Rightarrow$ percent change $R = 7\%$

63. The general form of an exponential function is $f(x) = Ca^x$, where $C$ is the initial value, $a$ is the growth factor, and $x$ is the variable as an exponent. The initial value is $C = 1.5$, since $0 < a < 1$ we have a decay factor $a = 0.45$. $a = 1 + r \Rightarrow 0.45 = 1 + r \Rightarrow r = -0.55 \Rightarrow$ percent change $R = -55\%$

65. $1500(1 + 0.09)^{10} = \$3551.05$

67. $200(1 + 0.20)^{50} = \$1,820,087.63$

69. $560(1 + 0.014)^{25} = \$792.75$

71. Yes. This is equivalent to having two accounts, each containing $1000 initially.

73. $A = P\left(1 + \dfrac{r}{n}\right)^{nt} \Rightarrow A = 700\left(1 + \dfrac{0.04}{4}\right)^{4 \cdot 3} = 700(1.01)^{12} \approx 788.78$. The result is $788.78.

75. $A = P\left(1 + \dfrac{r}{n}\right)^{nt} \Rightarrow A = 1200\left(1 + \dfrac{0.025}{12}\right)^{12 \cdot 7} \approx 1200(1.00208)^{84} \approx 1429.24$. The result is $1429.24.

77. $f(1.2) = e^{1.2} \approx 3.32$

79. $f(-2) = 1 - e^{-2} \approx 0.86$

81. Graph $Y_1 = e \char`\^ (0.5X)$ in $[-4, 4, 1]$ by $[0, 8, 1]$. See Figure 81.

    The graph illustrates exponential growth.

    [-4, 4, 1] by [0, 8, 1]          [-4, 4, 1] by [0, 8, 1]

    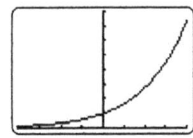          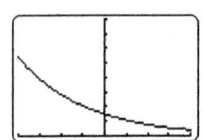

         Figure 81                        Figure 83

83. Graph $Y_1 = 1.5e \char`\^ (-0.32X)$ in $[-4, 4, 1]$ by $[0, 8, 1]$. See Figure 83.

    The graph illustrates exponential decay.

85. The initial value is $C = 5000$, $a = 1 + r \Rightarrow a = 1 + (-0.25) \Rightarrow a = 0.75 \Rightarrow f(x) = 5000(0.75)^x$

    $f(4) = 5000(0.75)^4 \approx 1582$

87. The initial value is $C = 50$, $a = 1 + r \Rightarrow a = 1 + (0.1) \Rightarrow a = 1.1 \Rightarrow f(x) = 50(1.1)^x$

    $f(4) = 50(1.1)^4 \approx 73$

89. A 20% raise is calculated as $A = 50,000(1 + 0.2)^1 = 60,000$  A \$20 raise is calculated as $A = 50,000 + 20 = 50,020$. A 20% raise per year is much better.

91. (a) $a = 1 + r \Rightarrow a = 1 + (-0.4) \Rightarrow a = 0.6$

    (b) The initial value is $C = 0.07$ and from part (a) we know $a = 0.6$. $B(x) = 0.07(0.6)^x$

    (c) $B(2) = 0.07(0.6)^2 = 0.0252$. After 2 hours the blood alcohol is 0.0252.

93. (a) The monthly growth factor is $a = 1.242$.

    (b) In July 2008 there were about 0.5 million tweets per month.

    (c) $T(24) = 0.5(1.242)^{24} \approx 90.8$. After 24 months there were 90.8 million tweets per month.

95. (a) Since $f(t) = 500$ when $t = 0$, $500 = Ca^0 \Rightarrow C = 500$. Since $f(t) = 1000$ when $t = 50$,

    $$1000 = 500a^{50/50} \Rightarrow a = \frac{1000}{500} = 2$$

    (b) $f(170) = 500(2)^{170/50} \approx 5278$ thousand bacteria per milliliter or 5.278 million bacteria per milliliter

    (c) The growth in the number of bacteria is exponential and doubles every 50 minutes.

97. Table $Y_1 = (0.905)^\wedge X$ with TblStart $= 0$ and $\Delta$Tbl $= 10$. See Figure 97.

    (a) $f(0) = (0.905)^0 = 1$. The probability that no vehicle will enter the intersection during a period of zero seconds is 1.

    (b) Since $f(30) = (0.905)^{30} = 0.05006$ and $f(31) = (0.905)^{31} = 0.04530$, this occurs after about 30 seconds.

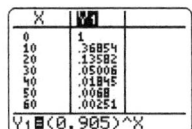

Figure 97

99. (a) $f(x) = 2.7e^{0.014x}$

    (b) $f(10) = 2.7e^{0.014(10)} \approx 3.1$; In 2020. the population of Nevada will be about 3.1 million.

101. (a) $f(x) = 38e^{0.0102x}$

     (b) $f(10) = 38e^{0.0102(10)} \approx 42$; In 2020, the population of California will be about 42 million.

103. (a) Example: Using a 3% rate of interest. $I = 1200(0.03)(1) = 36$. The result is $36.

     (b) Example: $I = 1200(0.03)(x)$. In this case we are assuming simple interest.

## Checking Basic Concepts Sections 12.1 and 12.2

1. (a) $f(1) = 2(1)^2 + 5(1) - 1 = 6$, and so $(g \circ f)(1) = g(f(1)) = g(6) = 6 + 1 = 7$

(b)   $(f \circ g)(x) = f(g(x)) = f(x+1) = 2(x+1)^2 + 5(x+1) - 1 = 2x^2 + 9x + 6$

2. See Figure 2.

(a)   No, this is not a one-to-one function because it does not pass the horizontal line test.

(b)   No, this function does not have an inverse because it is not one-to-one.

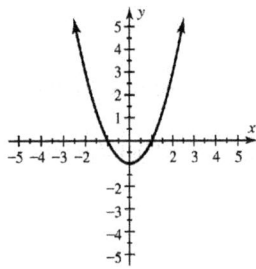

Figure 2                    Figure 5

3.  $f(x) = 4x - 3 \Rightarrow y = 4x - 3$, interchange $x$ and $y$ and solve for $y$.

$$x = 4y - 3 \Rightarrow 4y = x + 3 \Rightarrow y = \frac{x+3}{4} \Rightarrow f^{-1}(x) = \frac{x+3}{4}$$

4.  $f(-2) = 3(2^{-2}) = 3 \cdot \frac{1}{2^2} = 3 \cdot \frac{1}{4} = \frac{3}{4}$

5. See Figure 5.

6. Since $y = 2$ when $x = 0$, $2 = Ca^0 \Rightarrow C = 2$.  Since $y = 1$ when $x = 1$, $1 = 2(a)^1 \Rightarrow a = \frac{1}{2}$.

## Section 12.3 Logarithmic Functions

1. 10

3. D: $\{x \mid x > 0\}$; R: all real numbers

5. $k$

7. $x$

9. log 5

11. $\log 1 = 0$ because $10^0 = 1$.

13.

| $x$ | $10^{-5}$ | $10^0$ | $10^{0.5}$ | $10^{2.2}$ |
|---|---|---|---|---|
| $\log x$ | $-5$ | $0$ | $0.5$ | $2.2$ |

15.

| $x$ | $e^{-6}$ | $e^{-1}$ | $e^{5/7}$ | $e^{\pi}$ |
|---|---|---|---|---|
| $\ln x$ | $-6$ | $-1$ | $5/7$ | $\pi$ |

17.  $\log 10^5 = 5$

19.  $\log 10^{-4} = -4$

21.  $\log 1 = \log 10^0 = 0$

23. $\log \dfrac{1}{100} = \log 10^{-2} = -2$

25. $\log 10^{4.7} = 4.7$

27. $\log 10,000 = \log 10^4 = 4$

29. $\log \sqrt{10} = \log \sqrt{10} = \log(10)^{1/2} = \dfrac{1}{2}$

31. $\log(-23) \Rightarrow$ undefined, the domain of logarithmic functions are all positive real numbers.

33. $\log 0.001 = \log 10^{-3} = -3$

35. $10^{\log 2} = 2$

37. $10^{\log x^2} = x^2$

39. $10^{\log 5} = 5$

41. $\log 10^{(2x-7)} = 2x - 7$

43. $\log 25 \approx 1.398$

45. $\log 1.45 \approx 0.161$

47. See Figure 47. Compared to the graph of $y = \log x$, this graph is shifted 1 unit downward.

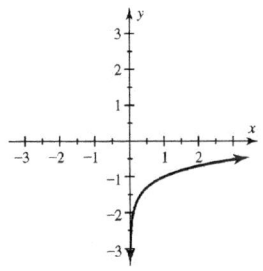 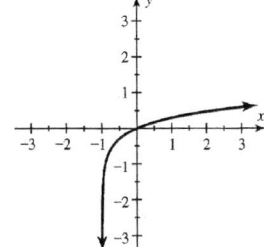 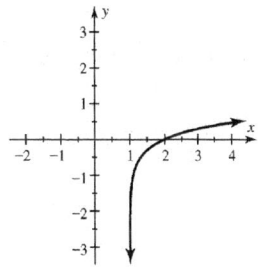

Figure 47                          Figure 49                          Figure 51

49. See Figure 49. Compared to the graph of $y = \log x$, this graph is shifted 1 unit to the left.

51. See Figure 51. Compared to the graph of $y = \log x$, this graph is shifted 1 unit to the right.

53. See Figure 53. Compared to the graph of $y = \log x$, this graph increases faster.

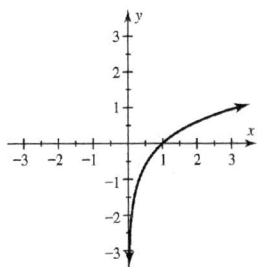

Figure 53

55. $\ln 1 = \ln e^0 = 0$

57. $\ln e^{-5x} = -5x$

59. $e^{\ln x^2} = x^2, x > 0$

61. $\ln 7 \approx 1.946$

63. $\ln \dfrac{4}{7} \approx -0.560$

65. Graph $Y_1 = \ln\left(\text{abs}(X)\right)$ in $[-4, 4, 1]$ by $[-4, 4, 1]$. See Figure 65. The graph is a reflection across

   the $y$-axis together with the graph of $y = \ln x$. The domain is $\{x \mid x \neq 0\}$.

   $[-4, 4, 1]$ by $[-4, 4, 1]$          $[-4, 4, 1]$ by $[-4, 4, 1]$

   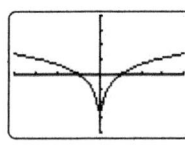          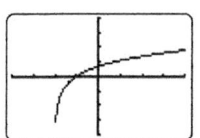

   Figure 65                    Figure 67

67. Graph $Y_1 = \ln(X + 2)$ in $[-4, 4, 1]$ by $[-4, 4, 1]$. See Figure 67. The graph is shifted 2 units to the

   left. The domain is $\{x \mid x > -2\}$.

69. $\log_5 5^{6x} = 6x$

71. $\log_2 \sqrt{\dfrac{1}{8}} = \log_2 \sqrt{2^{-3}} = \log_2 2^{-3/2} = -\dfrac{3}{2}$

73. $\log_2 2^8 = 8$

75. $\log_2 \sqrt{8} = \log_2 \left(2^3\right)^{1/2} = \log_2 2^{3/2} = \dfrac{3}{2}$

77. $\log_2 \sqrt[3]{\dfrac{1}{4}} = \log_2 \left(2^{-2}\right)^{1/3} = \log_2 2^{-2/3} = -\dfrac{2}{3}$

79. $\log_2 -8$ is undefined.

81. $\log_2 4 = \log_2 2^2 = 2$

83. $\log_2 \dfrac{1}{16} = \log_2 2^{-4} = -4$

85. $\log_3 \dfrac{1}{9} = \log_3 3^{-2} = -2$

87. $\log_5 \dfrac{1}{25} = \log_5 5^{-2} = -2$

89. $5^{\log_5 17} = 17$

91. $4^{\log_4 (2x)^2} = (2x)^2, x \neq 0$

93. $5^{\log_5 0.6z} = 0.6z, z > 0$

95. See Figure 95.

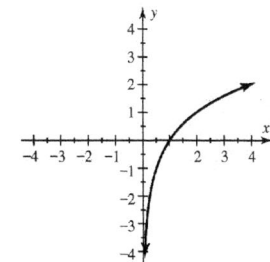

| $x$ | $\frac{1}{4}$ | $\frac{1}{2}$ | 1 | $\sqrt{2}$ | 64 |
|---|---|---|---|---|---|
| $\log_2 x$ | $-2$ | $-1$ | 0 | $\frac{1}{2}$ | 6 |

Figure 95                                    Figure 97

97. See Figure 97.

99. See Figure 99.

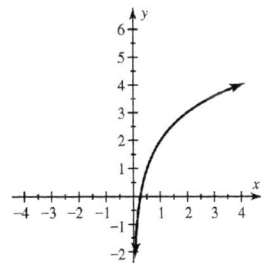

Figure 99

101. d. The graph of this function passes through the point $(1, 0)$ but does not pass through $(3, 1)$.

103. a. The graph of this function is shifted 2 units upward.

105. $f\left(10^{-4}\right) = 160 + 10\log\left(10^{-4}\right) = 160 + 10(-4) = 160 + (-40) = 120$ db.   Yes, this could cause pain.

107. (a)   The runway length increases but it does not double when the weight of the plane doubles.

   (b)   $L(50) = 1.3\ln(50) \approx 5.086$;  A 50-thousand pound plane needs a runway of at least 5086 feet

109. (a)   $M = 6 - 2.5\log\dfrac{10}{1} = 6 - 2.5\log(10) = 6 - 2.5(1) = 3.5$

   (b)   $M = 6 - 2.5\log\dfrac{100}{1} = 6 - 2.5\log(100) = 6 - 2.5(2) = 1$

   (c)   The intensity of the star decreases by 2.5.

111. (a)   $\log\dfrac{x}{1} = 6.0 \Rightarrow \log x = 6.0 \Rightarrow x = 10^6$ and $\log\dfrac{x}{1} = 8.0 \Rightarrow \log x = 8.0 \Rightarrow x = 10^8$

   (b)   $10^8 \div 10^6 = 100$ times

## Section 12.4 Properties of Logarithms

1. 4

3. 3

5. $\log m - \log n$

7. No

9. No; $\log(xy) = \log x + \log y$

11. $\log_a x = \dfrac{\log x}{\log a}$ or $\log_a x = \dfrac{\ln x}{\ln a}$

13. $\ln(15) = \ln(3 \cdot 5) = \ln 3 + \ln 5$

15. $\log xy = \log x + \log y$

17. $\log y^2 = \log(y \cdot y) = \log y + \log y$

19. $\log \dfrac{7}{3} = \log 7 - \log 3$

21. $\ln \dfrac{x}{y} = \ln x - \ln y$

23. $\log_2 \dfrac{45}{x} = \log_2 45 - \log_2 x$

25. $\log 45 + \log 5 = \log(45 \cdot 5) = \log 225$

27. $\ln x + \ln y = \ln xy$

29. $\ln 7x^2 + \ln 2x = \ln(7x^2 \cdot 2x) = \ln 14x^3$

31. $\ln x + \ln y^2 - \ln y = \ln xy^2 - \ln y = \ln \dfrac{xy^2}{y} = \ln xy$

33. $\log 20 - \log 4 = \log\left(\dfrac{20}{4}\right) = \log 5$

35. $\ln x^4 - \ln x^2 = \ln\left(\dfrac{x^4}{x^2}\right) = \ln x^2$

37. $\log 300x - \log 3x = \log\left(\dfrac{300x}{3x}\right) = \log 100 = 2$

39. $\log 3^6 = 6\log 3$

41. $\ln 2^x = x \ln 2$

43. $\log_2 5^{1/4} = \dfrac{1}{4}\log_2 5$

45. $\log_4 \sqrt[3]{z} = \log_4 z^{1/3} = \dfrac{1}{3}\log_4 z$

47. $\log x^{y-1} = (y-1)\log x$

49. $4\log z - \log z^3 = \log z^4 - \log z^3 = \log \dfrac{z^4}{z^3} = \log z$

51. $\log x + 2 \log x + 2 \log y = \log x + \log x^2 + \log y^2 = \log\left(x \cdot x^2 \cdot y^2\right) = \log x^3 y^2$

53. $\log x - 2 \log \sqrt{x} = \log x - \log\left(\sqrt{x}\right)^2 = \log x - \log x = 0$

55. $\ln 2^{x+1} - \ln 2 = \ln \dfrac{2^{x+1}}{2} = \ln 2^x$

57. $\ln \sqrt[3]{x} + \ln \sqrt{x} = \ln x^{1/3} + \ln x^{1/2} = \ln\left(x^{1/3} \cdot x^{1/2}\right) = \ln x^{5/6}$

59. $2 \log_a (x+1) - \log_a\left(x^2 - 1\right) = \log_a (x+1)^2 - \log_a\left(x^2 - 1\right) = \log_a \dfrac{(x+1)(x+1)}{(x+1)(x-1)} = \log_a \dfrac{x+1}{x-1}$

61. $\log xy^2 = \log x + \log y^2 = \log x + 2 \log y$

63. $\ln \dfrac{x^4 y}{z} = \ln x^4 y - \ln z = \ln x^4 + \ln y - \ln z = 4 \ln x + \ln y - \ln z$

65. $\log \dfrac{\sqrt[3]{z}}{\sqrt{y}} = \log \dfrac{z^{1/3}}{y^{1/2}} = \log z^{1/3} - \log y^{1/2} = \dfrac{1}{3} \log z - \dfrac{1}{2} \log y$

67. $\log\left(x^4 y^3\right) = \log x^4 + \log y^3 = 4 \log x + 3 \log y$

69. $\ln \dfrac{1}{y} - \ln \dfrac{1}{x} = \ln y^{-1} - \ln x^{-1} = -1 \ln y - (-1) \ln x = \ln x - \ln y$

71. $\log_4 \sqrt{\dfrac{x^3 y}{z^2}} = \log_4 \left(\dfrac{x^3 y}{z^2}\right)^{1/2} = \dfrac{1}{2} \log_4 \dfrac{x^3 y}{z^2} = \dfrac{1}{2}\left(\log_4 x^3 + \log_4 y - \log_4 z^2\right) =$

   $= \dfrac{1}{2}\left(3 \log_4 x + \log_4 y - 2 \log_4 z\right) = \dfrac{3}{2} \log_4 x + \dfrac{1}{2} \log_4 y - \log_4 z$

73. Graph $Y_1 = \log\left(X^\wedge 3\right)$ and $Y_2 = 3 \log(X)$ in [–6, 6, 1] by [–4, 4, 1]. See Figures 73a & 73b.

   By the power rule $\log x^3 = 3 \log x$.

   [–6, 6, 1] by [–4, 4, 1]    [–6, 6, 1] by [–4, 4, 1]    [–6, 6, 1] by [–4, 4, 1]    [–6, 6, 1] by [–4, 4, 1]

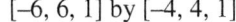
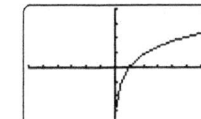
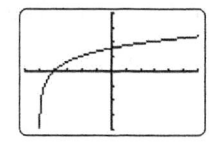
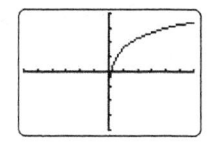

   Figure 73a            Figure 73b            Figure 75a            Figure 75b

75. Graph $Y_1 = \ln(X+5)$ and $Y_2 = \ln(X) + \ln(5)$ in [–6, 6, 1] by [–4, 4, 1]. See Figure 75a & 75b.

   Not the same.

77. $\log 16 = \log 2^4 = 4 \log 2 = 4(0.3) = 1.2$

79. $\log 65 = \log(5 \cdot 13) = \log 5 + \log 13 = 0.7 + 1.1 = 1.8$

81. $\log 130 = \log(2 \cdot 5 \cdot 13) = \log 2 + \log 5 + \log 13 = 0.3 + 0.7 + 1.1 = 2.1$

83. $\log \dfrac{5}{2} = \log 5 - \log 2 = 0.7 - 0.3 = 0.4$

85.  $\log \dfrac{1}{13} = \log 13^{-1} = -\log 13 = -1.1$

87.  $\log_3 5 = \dfrac{\log 5}{\log 3} \approx 1.46$

89.  $\log_2 25 = \dfrac{\log 25}{\log 2} \approx 4.64$

91.  $\log_9 102 = \dfrac{\log 102}{\log 9} \approx 2.10$

93.  $f(x) = 10 \log\left(10^{16} x\right) = 10\left(\log 10^{16} + \log x\right) = 10\left(16 + \log x\right) = 160 + 10 \log x$


## Checking Basic Concepts  Sections 12.3 and 12.4

1.  (a)   $\log 10^4 = 4$

   (b)   $\ln e^x = x$

   (c)   $\log_2 \dfrac{1}{8} = \log_2 2^{-3} = -3$

   (d)   $\log_5 \sqrt{5} = \log_5 5^{1/2} = \dfrac{1}{2}$

2.  See Figure 2.

   (a)   $D : \{x \mid x > 0\};\ R :$ all real numbers

   (b)   $f(1) = 0$

   (c)   Yes, for example $\log \dfrac{1}{10} = -1.$

   (d)   No, since negative numbers are not in the domain of $f(x) = \log x.$

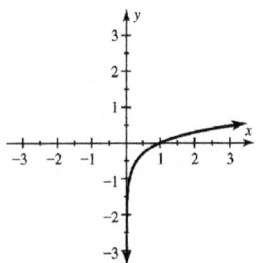

Figure 2

3.  (a)   $\log xy = \log x + \log y$

   (b)   $\ln \dfrac{x}{yz} = \ln x - \ln yz = \ln x - \left(\ln y + \ln z\right) = \ln x - \ln y - \ln z$

(c)   $\ln x^2 = 2 \ln x$

(d)   $\log \dfrac{x^2 y^3}{\sqrt{z}} = \log x^2 y^3 - \log z^{1/2} = \log x^2 + \log y^3 - \log z^{1/2} = 2 \log x + 3 \log y - \dfrac{1}{2} \log z$

4. (a)   $\log x + \log y = \log xy$

(b)   $\ln 2x - 3 \ln y = \ln 2x - \ln y^3 = \ln \dfrac{2x}{y^3}$

(c)   $2 \log_2 x + 3 \log_2 y - \log_2 z = \log_2 x^2 + \log_2 y^3 - \log_2 z = \log_2 x^2 y^3 - \log_2 z = \log_2 \dfrac{x^2 y^3}{z}$

## Section 12.5 Exponential and Logarithmic Equations

1. Add 5 to both sides.

3. Take the common logarithm of both sides.

5. $x$

7. $2x$

9. No, $\log \dfrac{5}{4} = \log 5 - \log 4$

11. One

13. $10^x = 1000 \Rightarrow 10^x = 10^3 \Rightarrow x = 3$

15. $2^x = 64 \Rightarrow 2^x = 2^6 \Rightarrow x = 6$

17. $2^{x-3} = 8 \Rightarrow 2^{x-3} = 2^3 \Rightarrow x - 3 = 3 \Rightarrow x = 6$

19. $4^x + 3 = 259 \Rightarrow 4^x = 256 \Rightarrow 4^x = 4^4 \Rightarrow x = 4$

21. $10^{0.4x} = 124 \Rightarrow \log 10^{0.4x} = \log 124 \Rightarrow 0.4x = \log 124 \Rightarrow x = \dfrac{\log 124}{0.4} \approx 5.23$

23. $e^{-x} = 1 \Rightarrow e^{-x} = e^0 \Rightarrow -x = 0 \Rightarrow x = 0$

25. $e^x = 25 \Rightarrow \ln e^x = \ln 25 \Rightarrow x = \ln 25 \Rightarrow x \approx 3.22$

27. $0.4^x = 2 \Rightarrow \ln 0.4^x = \ln 2 \Rightarrow x \ln 0.4 = \ln 2 \Rightarrow x = \dfrac{\ln 2}{\ln 0.4} \Rightarrow x \approx -0.76$

29. $e^x - 1 = 6 \Rightarrow e^x = 7 \Rightarrow \ln e^x = \ln 7 \Rightarrow x = \ln 7 \approx 1.95$

31. $2(10)^{x+2} = 35 \Rightarrow 10^{x+2} = \dfrac{35}{2} \Rightarrow \log 10^{x+2} = \log \dfrac{35}{2} \Rightarrow x + 2 = \log \dfrac{35}{2} \Rightarrow x = \log \dfrac{35}{2} - 2 \approx -0.76$

33. $3.1^{2x} - 4 = 16 \Rightarrow 3.1^{2x} = 20 \Rightarrow \log_{3.1} 3.1^{2x} = \log_{3.1} 20 \Rightarrow 2x = \log_{3.1} 20 \Rightarrow x = \dfrac{\log 20}{2 \log 3.1} \approx 1.32$

35. $e^{3x} = e^{2x-1} \Rightarrow 3x = 2x - 1 \Rightarrow x = -1$

37. $5^{4x} = 5^{x^2-5} \Rightarrow 4x = x^2 - 5 \Rightarrow x^2 - 4x - 5 = 0 \Rightarrow (x+1)(x-5) = 0 \Rightarrow x = -1 \text{ or } 5$

39. $e^{2x} \cdot e^x = 10 \Rightarrow e^{2x+x} = 10 \Rightarrow e^{3x} = 10 \Rightarrow \ln e^{3x} = \ln 10 \Rightarrow 3x = \ln 10 \Rightarrow x = \dfrac{\ln 10}{3} \approx 0.77$

41. $e^x = 2^{x+2} \Rightarrow \ln e^x = \ln 2^{x+2} \Rightarrow x = (x+2)\ln 2 \Rightarrow x = x\ln 2 + 2\ln 2 \Rightarrow x - x\ln 2 = 2\ln 2 \Rightarrow$

   $x(1-\ln 2) = 2\ln 2 \Rightarrow x = \dfrac{2\ln 2}{1-\ln 2} \approx 4.52$

43. $4^{0.5x} = 5^{x+2} \Rightarrow \log 4^{0.5x} = \log 5^{x+2} \Rightarrow 0.5x\log 4 = (x+2)\log 5 \Rightarrow 0.5x\log 4 = x\log 5 + 2\log 5 \Rightarrow$

   $0.5x\log 4 - x\log 5 = 2\log 5 \Rightarrow x(0.5\log 4 - \log 5) = 2\log 5 \Rightarrow x = \dfrac{2\log 5}{0.5\log 4 - \log 5} \approx -3.51$

45. (a)   The solution is the $x$-coordinate of the intersection point, $x = 1$.

   (b)   $0.2(10^x) = 2 \Rightarrow 10^x = 10 \Rightarrow x = 1$

47. (a)   The solution is the $x$-coordinate of the intersection point, $x = -2$.

   (b)   $2^{-x} = 4 \Rightarrow 2^{-x} = 2^2 \Rightarrow -x = 2 \Rightarrow x = -2$

49. $10^x = 0.1 \Rightarrow 10^x = 10^{-1} \Rightarrow x = -1$  For numerical support, table $Y_1 = 10 \wedge X$ and $Y_2 = 0.1$ with

   TblStart = $-3$ and $\Delta$Tbl = 1.  See Figure 49.

   $[-1, 1, 1]$ by $[0, 3, 1]$

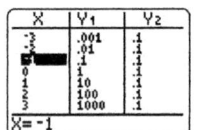   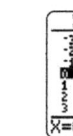   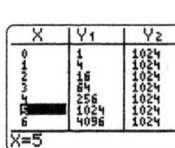   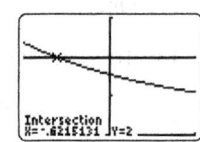

   Figure 49                Figure 51                Figure 53                Figure 55

51. $4e^x + 5 = 9 \Rightarrow 4e^x = 4 \Rightarrow e^x = 1 \Rightarrow e^x = e^0 \Rightarrow x = 0$  For numerical support, table

   $Y_1 = 4e \wedge X + 5$ and $Y_2 = 9$ with TblStart = $-3$ and $\Delta$Tbl = 1.  See Figure 51.

53. $4^x = 1024 \Rightarrow 4^x = 4^5 \Rightarrow x = 5$  For numerical support, table $Y_1 = 4 \wedge X$ and $Y_2 = 1024$ with

   TblStart = 0 and $\Delta$Tbl = 1.  See Figure 53.

55. $(0.55)^x + 0.55 = 2 \Rightarrow 0.55^x = 1.45 \Rightarrow \log_{0.55} 0.55^x = \log_{0.55} 1.45 \Rightarrow x = \dfrac{\log 1.45}{\log 0.55} \approx -0.62$

   For graphical support, graph $Y_1 = 0.55 \wedge X + 0.55$ and $Y_2 = 2$ in $[-1, 1, 1]$ by $[0, 3, 1]$.

   See Figure 55.

57. Graph $Y_1 = e \wedge X - X$ and $Y_2 = 2$ in $[-5, 5, 1]$ by $[-5, 5, 1]$.  See Figures 57a & 57b.

   The solutions are the $x$-coordinates of the intersection points, $x \approx -1.84$ and $x \approx 1.15$.

$[-5, 5, 1]$ by $[-5, 5, 1]$        $[-5, 5, 1]$ by $[-5, 5, 1]$

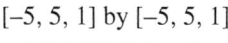

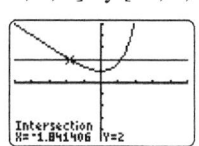

Figure 57a        Figure 57b

59. Graph $Y_1 = \ln(X)$ and $Y_2 = e^\wedge(-X)$ in $[-5, 5, 1]$ by $[-5, 5, 1]$. See Figure 59.

The solution is the $x$-coordinate of the intersection point, $x \approx 1.31$.

$[-5, 5, 1]$ by $[-5, 5, 1]$

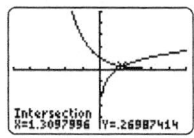

Figure 59

61. $\log x = 2 \Rightarrow 10^{\log x} = 10^2 \Rightarrow x = 100$

63. $\ln x = 5 \Rightarrow e^{\ln x} = e^5 \Rightarrow x = e^5 \approx 148.41$

65. $\log 2x = 7 \Rightarrow 10^{\log 2x} = 10^7 \Rightarrow 2x = 10,000,000 \Rightarrow x = 5,000,000$

67. $\log_2 x = 4 \Rightarrow 2^{\log_2 x} = 2^4 \Rightarrow x = 16$

69. $\log_2 5x = 2.3 \Rightarrow 2^{\log_2 5x} = 2^{2.3} \Rightarrow 5x = 2^{2.3} \Rightarrow x = \dfrac{2^{2.3}}{5} \approx 0.98$

71. $2 \log x + 5 = 7.8 \Rightarrow 2 \log x = 2.8 \Rightarrow \log x = 1.4 \Rightarrow 10^{\log x} = 10^{1.4} \Rightarrow x = 10^{1.4} \approx 25.12$

73. $5 \ln (2x+1) = 55 \Rightarrow \ln (2x+1) = 11 \Rightarrow e^{\ln (2x+1)} = e^{11} \Rightarrow 2x+1 = e^{11} \Rightarrow x = \dfrac{e^{11}-1}{2} \approx 29,936.57$

75. $\log x^2 = \log x \Rightarrow x^2 = x \Rightarrow x^2 - x = 0 \Rightarrow x(x-1) = 0 \Rightarrow x = 0$ or $1$

The solution $x = 0$ causes an undefined expression in the original equation. The only solution is 1.

77. $\ln x + \ln (x+1) = \ln 30 \Rightarrow \ln x(x+1) = \ln 30 \Rightarrow x(x+1) = 30 \Rightarrow x^2 + x - 30 = 0 \Rightarrow$

$(x+6)(x-5) = 0 \Rightarrow x = -6$ or $5$  The solution $x = -6$ causes an undefined expression in the original

equation. The only solution is 5.

79. $\log_3 3x - \log_3 (x+2) = \log_3 2 \Rightarrow \log_3 \dfrac{3x}{x+2} = \log_3 2 \Rightarrow \dfrac{3x}{x+2} = 2 \Rightarrow 3x = 2x + 4 \Rightarrow x = 4$

81. $\log_2 (x-1) + \log_2 (x+1) = 3 \Rightarrow \log_2 (x^2 - 1) = 3 \Rightarrow 2^{\log_2 (x^2-1)} = 2^3 \Rightarrow x^2 - 1 = 8 \Rightarrow$

$x^2 - 9 = 0 \Rightarrow (x+3)(x-3) = 0 \Rightarrow x = -3$ or $3$  The solution $x = -3$ causes an undefined expression

in the original equation. The only solution is 3.

83. $\log x = 1.6 \Rightarrow 10^{\log x} = 10^{1.6} \Rightarrow x = 10^{1.6} \approx 39.81$  For graphical support, graph

$Y_1 = \log(X)$ and $Y_2 = 1.6$ in $[0, 50, 10]$ by $[-2, 2, 1]$. See Figure 83.

[0, 50, 10] by [–2, 2, 1]          [–1, 2, 1] by [–2, 2, 1]

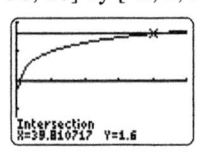          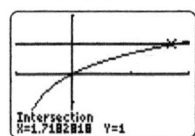

       Figure 83                    Figure 85

85.  $\ln(x+1)=1 \Rightarrow e^{\ln(x+1)} = e^1 \Rightarrow x+1 = e \Rightarrow x = e-1 \approx 1.72$  For graphical support, graph

     $Y_1 = \ln(X+1)$ and $Y_2 = 1$ in [–1, 2, 1] by [–2, 2, 1]. See Figure 85.

87.  $17-6\log_3 x = 5 \Rightarrow 6\log_3 x = 12 \Rightarrow \log_3 x = 2 \Rightarrow 3^{\log_3 x} = 3^2 \Rightarrow x = 3^2 = 9$  For graphical support,

     graph $Y_1 = 17-6\left(\ln(X)/\ln(3)\right)$ and $Y_2 = 5$ in [0, 12, 1] by [0, 12, 1]. See Figure 87.

     [0, 12, 1] by [0, 12, 1]     [

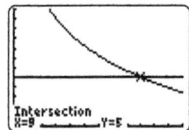

       Figure 87

89.  (a)    The solution is the $x$-coordinate of the intersection point, $x = 2$.

     (b)    $\ln x = 0.7 \Rightarrow e^{\ln x} = e^{0.7} \Rightarrow x = e^{0.7} \approx 2.01$

91.  (a)    The solution is the $x$-coordinate of the intersection point, $x = 2$.

     (b)    $5\log 2x = 3 \Rightarrow \log 2x = 0.6 \Rightarrow 10^{\log 2x} = 10^{0.6} \Rightarrow 2x = 10^{0.6} \Rightarrow x = \dfrac{10^{0.6}}{2} \approx 1.99$

93.  (a)    $W(5) = 1.73\left(10^{0.276(5)}\right) = 1.73\left(10^{1.38}\right) \approx 41.5$,  In 2010 China generated about 41.5 gigawatts.

     (b)    $22 = 1.73\left(10^{0.276(x)}\right) \Rightarrow \dfrac{22}{1.73} = \left(10^{0.276x}\right) \Rightarrow \log\left(\dfrac{22}{1.73}\right) = 0.276x \Rightarrow \dfrac{\log\left(\dfrac{22}{1.73}\right)}{0.276} = x \Rightarrow$

            $x \approx 4$  Since $x$ is the number of years after 2005 the result is 2009.

95.  $20 = 5(1.15)^x \Rightarrow 4 = (1.15)^x \Rightarrow \ln 4 = x\ln(1.15) \Rightarrow \dfrac{\ln 4}{\ln(1.15)} = x \Rightarrow x \approx 10$ hours

97.  $2000(1+0.15)^t = 6000 \Rightarrow 1.15^t = 3 \Rightarrow \log_{1.15} 1.15^t = \log_{1.15} 3 \Rightarrow t = \dfrac{\log 3}{\log 1.15} \approx 7.86 \approx 8$  years

99.  (a)    $f(1994) = 2339(1.24)^{(1994-1988)} \approx 8503$;  In 1994 there were about 8503 people waiting for

            liver transplants.

     (b)    $2339(1.24)^{(x-1988)} = 20,000 \Rightarrow (1.24)^{(x-1988)} = \dfrac{20,000}{2339} \Rightarrow$

            $\log_{1.24}(1.24)^{(x-1988)} = \log_{1.24}\left(\dfrac{20,000}{2339}\right) \Rightarrow x-1988 = \log_{1.24}\left(\dfrac{20,000}{2339}\right) \Rightarrow$

$$x = \log_{1.24}\left(\frac{20{,}000}{2339}\right) + 1988 = \frac{\log\left(\frac{20{,}000}{2339}\right)}{\log 1.24} + 1988 \approx 1998$$

101. $3 \log x = 3.960 \Rightarrow \log x = 1.320 \Rightarrow 10^{\log x} = 10^{1.320} \Rightarrow x = 10^{1.320} \approx 20.893 = 20{,}893$ lb

103. (a) $f(1) = 230\left(10^{-0.055 \cdot 1}\right) \approx 203$; In 1975 there were about 203 thousand bluefin tuna.

   (b) About 1979.

   (c) $230\left(10^{-0.055x}\right) = 115 \Rightarrow 10^{-0.055x} = 0.5 \Rightarrow \log 10^{-0.055x} = \log 0.5 \Rightarrow -0.055x = \log 0.5 \Rightarrow$

   $x = \dfrac{\log 0.5}{-0.055} \approx 5.47$ or about 1979.

105. From the data point $(1, 25)$, $25 = a + b \log 1 \Rightarrow 25 = a + b(0) \Rightarrow a = 25$.

   Using the data point $(10, 28)$ and the fact that $a = 25$, $28 = 25 + b \log 10 \Rightarrow 28 = 25 + b \Rightarrow b = 3$

107. (a) Graph $Y_1 = 645 \log(X+1) + 1925$ and $Y_2 = 2200$ in $[0, 4, 1]$ by $[0, 3000, 1000]$.

   See Figure 107. A person consuming 2200 calories would typically own about 1.67 acres.

   (b) $645 \log(x+1) + 1925 = 2200 \Rightarrow 645 \log(x+1) = 275 \Rightarrow \log(x+1) = \dfrac{275}{645} \Rightarrow$

   $10^{\log(x+1)} = 10^{275/645} \Rightarrow x+1 = 10^{275/645} \Rightarrow x = 10^{275/645} - 1 \approx 1.67$ acres

   $[0, 4, 1]$ by $[0, 3000, 1000]$

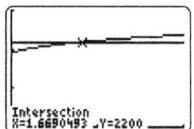

   Figure 107

109. (a) The data is nonlinear. It does not increase at a constant rate.

   (b) Each year the amount of fertilizer increases by a factor of 1.06 or 6%.

   (c) $5(1.06)^{(x-1950)} = 15 \Rightarrow 1.06^{(x-1950)} = 3 \Rightarrow \log_{1.06} 1.06^{(x-1950)} = \log_{1.06} 3 \Rightarrow$

   $x - 1950 = \log_{1.06} 3 \Rightarrow x = \log_{1.06} 3 + 1950 = \dfrac{\log 3}{\log 1.06} + 1950 \approx 1968.85$ or in 1968

111. $160 + 10 \log x = 100 \Rightarrow 10 \log x = -60 \Rightarrow \log x = -6 \Rightarrow 10^{\log x} = 10^{-6} \Rightarrow x = 10^{-6}$ w/cm$^2$

113. $0.48 \ln(x+1) + 27 = 28 \Rightarrow 0.48 \ln(x+1) = 1 \Rightarrow \ln(x+1) = \dfrac{1}{0.48} \Rightarrow e^{\ln(x+1)} = e^{1/0.48} \Rightarrow$

   $x + 1 = e^{1/0.48} \Rightarrow x = e^{1/0.48} - 1 \approx 7.03$ or about 7 miles

## Checking Basic Concepts Section 12.5

1. (a) $2(10^x) = 40 \Rightarrow 10^x = 20 \Rightarrow \log 10^x = \log 20 \Rightarrow x = \log 20 \approx 1.30$

(b)    $2^{3x}+3=150 \Rightarrow 2^{3x}=147 \Rightarrow \log_2 2^{3x}=\log_2 147 \Rightarrow 3x=\dfrac{\log 147}{\log 2} \Rightarrow x=\dfrac{\log 147}{3\log 2} \approx 2.40$

(c)    $\ln x = 4.1 \Rightarrow e^{\ln x}=e^{4.1} \Rightarrow x=e^{4.1} \approx 60.34$

(d)    $4\log 2x = 12 \Rightarrow \log 2x = 3 \Rightarrow 10^{\log 2x}=10^3 \Rightarrow 2x=1000 \Rightarrow x=500$

2.  $\log(x+4)+\log(x-4)=\log 48 \Rightarrow \log\left(x^2-16\right)=\log 48 \Rightarrow x^2-16=48 \Rightarrow$

   $x^2-64=0 \Rightarrow (x+8)(x-8)=0 \Rightarrow x=-8$ or $8$   The solution $x=-8$ causes an undefined expression

   in the original equation.  The only solution is 8.

3.  $500(1.03)^x=900 \Rightarrow 1.03^x=\dfrac{9}{5} \Rightarrow \log_{1.03} 1.03^x=\log_{1.03}\left(\dfrac{9}{5}\right) \Rightarrow x=\dfrac{\log\left(\frac{9}{5}\right)}{\log 1.03} \approx 20$ years

## Chapter 12 Review

1.  (a)    $f(-2)=2(-2)^2-4(-2)=16$, then $(g\circ f)(-2)=g\left(f(-2)\right)=g(16)=5(16)+1=81$

   (b)    $(f\circ g)(x)=f\left(g(x)\right)=f(5x+1)=2(5x+1)^2-4(5x+1)=50x^2-2$

2.  (a)    $f(-2)=\sqrt[3]{-2-6}=-2$, then $(g\circ f)(-2)=g\left(f(-2)\right)=g(-2)=4(-2)^3=-32$

   (b)    $(f\circ g)(x)=f\left(g(x)\right)=f\left(4x^3\right)=\sqrt[3]{4x^3-6}$

3.  (a)    $(f\circ g)(2)=f\left(g(2)\right)=f(3)=0$

   (b)    $(g\circ f)(1)=g\left(f(1)\right)=g(2)=3$

4.  (a)    $(f\circ g)(-1)=f\left(g(-1)\right)=f(2)=3$

   (b)    $(g\circ f)(2)=g\left(f(2)\right)=g(3)=-2$

   (c)    $(f\circ f)(1)=f\left(f(1)\right)=f(0)=-1$

5.  $f(1)=f(-1)=2$

6.  $f(0)=f(2)=1$

7.  This graph does not pass the horizontal line test.  The function is not one-to-one.

8.  This graph passes the horizontal line test.  The function is one-to-one.

9.  $\left(f\circ f^{-1}\right)(x)=f\left(f^{-1}(x)\right)=f\left(\dfrac{x+9}{2}\right)=2\left(\dfrac{x+9}{2}\right)-9=x+9-9=x$

   $\left(f^{-1}\circ f\right)(x)=f^{-1}\left(f(x)\right)=f^{-1}(2x-9)=\dfrac{(2x-9)+9}{2}=\dfrac{2x}{2}=x$

10. $\left(f \circ f^{-1}\right)(x) = f\left(f^{-1}(x)\right) = f\left(\sqrt[3]{x-1}\right) = \left(\sqrt[3]{x-1}\right)^3 + 1 = x - 1 + 1 = x$

$\left(f^{-1} \circ f\right)(x) = f^{-1}(f(x)) = f^{-1}\left(x^3 + 1\right) = \sqrt[3]{\left(x^3 + 1\right) - 1} = \sqrt[3]{x^3} = x$

11. $f(x) = 5x \Rightarrow y = 5x$, interchange $x$ and $y$ and solve for $y$. $x = 5y \Rightarrow y = \dfrac{x}{5} \Rightarrow f^{-1}(x) = \dfrac{x}{5}$

12. $f(x) = x - 11 \Rightarrow y = x - 11$, interchange $x$ and $y$ and solve for $y$.

$x = y - 11 \Rightarrow y = x + 11 \Rightarrow f^{-1}(x) = x + 11$

13. $f(x) = 2x + 7 \Rightarrow y = 2x + 7$, interchange $x$ and $y$ and solve for $y$.

$x = 2y + 7 \Rightarrow 2y = x - 7 \Rightarrow y = \dfrac{x-7}{2} \Rightarrow f^{-1}(x) = \dfrac{x-7}{2}$

14. $f(x) = \dfrac{4}{x} \Rightarrow y = \dfrac{4}{x}$, interchange $x$ and $y$ and solve for $y$. $x = \dfrac{4}{y} \Rightarrow xy = 4 \Rightarrow y = \dfrac{4}{x} \Rightarrow f^{-1}(x) = \dfrac{4}{x}$

15. See Figure 15. $D = \{3, 7, 8, 10\}; \ R = \{0, 1, 2, 3\}$

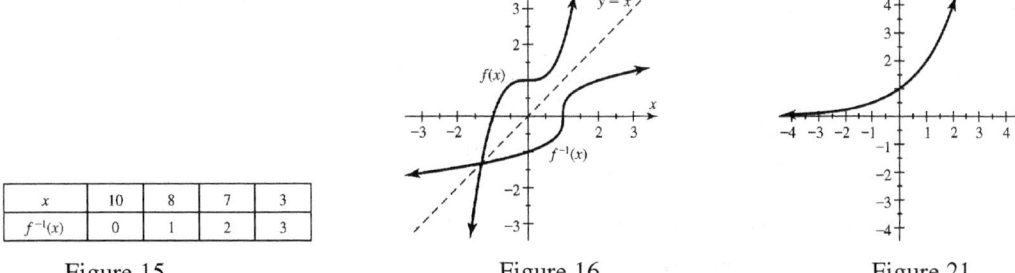

| $x$ | 10 | 8 | 7 | 3 |
|---|---|---|---|---|
| $f^{-1}(x)$ | 0 | 1 | 2 | 3 |

Figure 15                          Figure 16                          Figure 21

16. The graph of $f^{-1}$ is a reflection of the graph of $f$ across the line $y = x$. See Figure 16.

17. $f(-1) = 6^{-1} = \dfrac{1}{6}$ and $f(2) = 6^2 = 36$

18. $f(0) = 5\left(2^0\right) = 5(1) = 5$ and $f(3) = 5\left(2^{-3}\right) = 5\left(\dfrac{1}{8}\right) = \dfrac{5}{8}$

19. $f(-1) = \left(\dfrac{1}{3}\right)^{-1} = 3$ and $f(4) = \left(\dfrac{1}{3}\right)^4 = \dfrac{1}{3^4} = \dfrac{1}{81}$

20. $f(0) = 3\left(\dfrac{1}{6}\right)^0 = 3(1) = 3$ and $f(1) = 3\left(\dfrac{1}{6}\right)^1 = 3\left(\dfrac{1}{6}\right) = \dfrac{3}{6} = \dfrac{1}{2}$

21. See Figure 21. The graph illustrates exponential growth.

22. See Figure 22. The graph illustrates exponential decay.

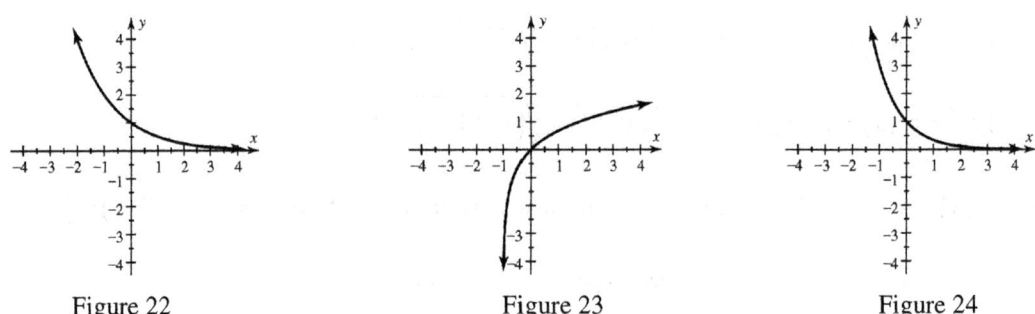

Figure 22                     Figure 23                     Figure 24

23. See Figure 23. The graph illustrates logarithmic growth.

24. See Figure 24. The graph illustrates exponential decay.

25. (a)  Exponential growth. For each unit increase in $x$, $f(x)$ increases by a factor of 2.

    (b)  Since $f(x) = 5$ when $x = 0$, $f(x) = 5(2)^x$

26. (a)  Linear growth. For each unit increase in $x$, $f(x)$ increases by 5 units.

    (b)  Since $f(x) = 5$ when $x = 0$, $f(x) = 5x + 5$

27. Since $y = \dfrac{1}{2}$ when $x = 0$, $\dfrac{1}{2} = Ca^0 \Rightarrow C = \dfrac{1}{2}$. Since $y = 1$ when $x = 1$, $1 = \dfrac{1}{2}(a)^1 \Rightarrow a = 2$.

28. Since $y = 2$ when $x = 2$, $2 = k\log_2 2 \Rightarrow 2 = k(1) \Rightarrow k = 2$.

29. $\dfrac{A-B}{B} \times 100 \Rightarrow \dfrac{120-150}{150} \times 100 = \dfrac{-30}{150} \times 100 = -20\%$

30. The account increased by a factor of $a = 1 + r = 1 + 0.07 = 1.07$.

31. (a)  $\dfrac{B-A}{A} \times 100 \Rightarrow \dfrac{1200-600}{600} \times 100 = \dfrac{600}{600} \times 100 = 100\%$

    (b)  $\dfrac{A-B}{B} \times 100 \Rightarrow \dfrac{600-1200}{1200} \times 100 = \dfrac{-600}{1200} \times 100 = -50\%$

32. (a)  $\dfrac{B-A}{A} \times 100 \Rightarrow \dfrac{1.00-2.20}{2.20} \times 100 = \dfrac{-1.20}{2.20} \times 100 = 54.\overline{54}\%$

    (b)  $\dfrac{A-B}{B} \times 100 \Rightarrow \dfrac{2.20-1.00}{1.00} \times 100 = \dfrac{1.20}{1.00} \times 100 = 120\%$

33. (a)  $rA = 2.10(500) = 1050$  The account increased by $1050.

    (b)  The new value of the account is $A + rA = \$500 + \$1050 = \$1550$.

    (c)  The account increased by a factor of $a = 1 + r = 1 + 2.1 = 3.1$.

34. (a)  $rA = -0.25(700) = -175$  The account decreased by $175.

    (b)  The new value of the account is $A + rA = \$700 + \$(-175) = \$525$.

    (c)  The account decreased by a factor of $a = 1 + r = 1 + (-0.25) = 0.75$.

35. The initial value is $C = 20,000$, $a = 1 + r \Rightarrow a = 1 + (-0.05) \Rightarrow a = 0.95 \Rightarrow f(x) = 20,000(0.95)^x$

    $f(2) = 20,000(0.95)^2 = 18,050$

36. The initial value is $C = 1500$, $a = 1 + r \Rightarrow a = 1 + (3.00) \Rightarrow a = 4 \Rightarrow f(x) = 1500(4)^x$

$$f(2) = 1500(4)^2 = 24,000$$

37. $1200(1 + 0.10)^9 = \$2829.54$

38. $900(1 + 0.18)^{40} = \$675,340.51$

39. $f(5.3) = 2e^{5.3} - 1 \approx 399.67$

40. $f(2.1) = 0.85^{2.1} \approx 0.71$

41. $f(55) = 2 \log 55 \approx 3.48$

42. $f(23) = \ln(2 \cdot 23 + 3) \approx 3.89$

43. $\log 0.001 = \log 10^{-3} = -3$

44. $\log \sqrt{10,000} = \log 100 = 2$

45. $\ln e^{-4} = -4$

46. $\log_4 16 = \log_4 4^2 = 2$

47. $\log 65 \approx 1.813$

48. $\ln 0.85 \approx -0.163$

49. $\ln 120 \approx 4.787$

50. $\log \dfrac{2}{5} \approx -0.398$

51. $10^{\log 7} = 7$

52. $\log_2 2^{5/9} = \dfrac{5}{9}$

53. $\ln e^{6-x} = 6 - x$

54. $e^{2 \ln x} = e^{\ln x^2} = x^2, \ x > 0$

55. $\ln xy = \ln x + \ln y$

56. $\log \dfrac{x}{y} = \log x - \log y$

57. $\ln x^2 y^3 = \ln x^2 + \ln y^3 = 2 \ln x + 3 \ln y$

58. $\log \dfrac{\sqrt{x}}{z^3} = \log \dfrac{x^{1/2}}{z^3} = \log x^{1/2} - \log z^3 = \dfrac{1}{2} \log x - 3 \log z$

59. $\log_2 \dfrac{x^2 y}{z} = \log_2 x^2 y - \log_2 z = \log_2 x^2 + \log_2 y - \log_2 z = 2 \log_2 x + \log_2 y - \log_2 z$

60. $\log_3 \sqrt[3]{\dfrac{x}{y}} = \log_3 \left(\dfrac{x}{y}\right)^{1/3} = \dfrac{1}{3}\log_3\left(\dfrac{x}{y}\right) = \dfrac{1}{3}\left(\log_3 x - \log_3 y\right) = \dfrac{1}{3}\log_3 x - \dfrac{1}{3}\log_3 y$

61. $\log 45 + \log 5 - \log 3 = \log(45 \cdot 5) - \log 3 = \log 225 - \log 3 = \log \dfrac{225}{3} = \log 75$

62. $\log_4 2x + \log_4 5x = \log_4(2x \cdot 5x) = \log_4\left(10x^2\right)$

63. $2\ln x - 3\ln y = \ln x^2 - \ln y^3 = \ln \dfrac{x^2}{y^3}$

64. $\log x^4 - \log x^3 + \log y = \log \dfrac{x^4}{x^3} + \log y = \log x + \log y = \log xy$

65. $\log 6^3 = 3\log 6$

66. $\ln x^2 = 2\ln x$

67. $\log_2 5^{2x} = (2x)\log_2 5$

68. $\log_4 (0.6)^{x+1} = (x+1)\log_4 0.6$

69. $10^x = 100 \Rightarrow 10^x = 10^2 \Rightarrow x = 2$

70. $2^{2x} = 256 \Rightarrow 2^{2x} = 2^8 \Rightarrow 2x = 8 \Rightarrow x = 4$

71. $3e^x + 1 = 28 \Rightarrow 3e^x = 27 \Rightarrow e^x = 9 \Rightarrow \ln e^x = \ln 9 \Rightarrow x = \ln 9 \approx 2.20$

72. $0.85^x = 0.2 \Rightarrow \log_{0.85} 0.85^x = \log_{0.85} 0.2 \Rightarrow x = \log_{0.85} 0.2 = \dfrac{\log 0.2}{\log 0.85} \approx 9.90$

73. $5\ln x = 4 \Rightarrow \ln x = 0.8 \Rightarrow e^{\ln x} = e^{0.8} \Rightarrow x = e^{0.8} \approx 2.23$

74. $\ln 2x = 5 \Rightarrow e^{\ln 2x} = e^5 \Rightarrow 2x = e^5 \Rightarrow x = \dfrac{e^5}{2} \approx 74.21$

75. $2\log x = 80 \Rightarrow \log x = 40 \Rightarrow 10^{\log x} = 10^{40} \Rightarrow x = 10^{40}$

76. $3\log x - 5 = 1 \Rightarrow 3\log x = 6 \Rightarrow \log x = 2 \Rightarrow 10^{\log x} = 10^2 \Rightarrow x = 100$

77. $2^{x+4} = 3^x \Rightarrow \log 2^{x+4} = \log 3^x \Rightarrow (x+4)\log 2 = x\log 3 \Rightarrow x\log 2 + 4\log 2 = x\log 3 \Rightarrow$

    $x\log 3 - x\log 2 = 4\log 2 \Rightarrow x(\log 3 - \log 2) = 4\log 2 \Rightarrow x = \dfrac{4\log 2}{\log 3 - \log 2} \approx 6.84$

78. $\ln(2x+1) + \ln(x-5) = \ln 13 \Rightarrow \ln(2x+1)(x-5) = \ln 13 \Rightarrow (2x+1)(x-5) = 13 \Rightarrow$

    $2x^2 - 9x - 5 = 13 \Rightarrow 2x^2 - 9x - 18 = 0 \Rightarrow (2x+3)(x-6) = 0 \Rightarrow x = -\dfrac{3}{2}$ or $6$  The solution $x = -\dfrac{3}{2}$

    causes an undefined expression in the original equation. The only solution is 6.

79. (a)    The solution is the $x$-coordinate of the intersection point, $x = 3$.

    (b)    $\dfrac{1}{2}\left(2^x\right) = 4 \Rightarrow 2^x = 8 \Rightarrow 2^x = 2^3 \Rightarrow x = 3$

80. (a)  The solution is the $x$-coordinate of the intersection point, $x = 4$.

   (b)  $\log_2 2x = 3 \Rightarrow 2^{\log_2 2x} = 2^3 \Rightarrow 2x = 8 \Rightarrow x = 4$

81. (a)  $(S \circ r)(8) = S(r(8)) = S(\sqrt{2 \cdot 8}) = S(4) = 4\pi(4)^2 = 64\pi$

   After 8 seconds, the balloon has a surface area of $64\pi \approx 201$ square inches.

   (b)  $(S \circ r)(t) = S(r(t)) = S(\sqrt{2t}) = 4\pi(\sqrt{2t})^2 = 4\pi \cdot 2t = 8\pi t$

82. (a)  Yes, this is a one-to-one function because different inputs result in different outputs.

   (b)  $f(x) = 0.08x \Rightarrow y = 0.08x$, interchange $x$ and $y$ and solve for $y$.

   $x = 0.08y \Rightarrow y = \dfrac{x}{0.08} \Rightarrow f^{-1}(x) = \dfrac{x}{0.08}$

   This formula calculates the cost of an item whose sales tax is $x$ dollars.

83. $1500(1+0.11)^t = 3000 \Rightarrow 1.11^t = 2 \Rightarrow \log_{1.11} 1.11^t = \log_{1.11} 2 \Rightarrow t = \dfrac{\log 2}{\log 1.11} \approx 6.64 \approx 7$ years

84. $100 = a + b \log 1 \Rightarrow 100 = a + b(0) \Rightarrow a = 100$

   $150 = 100 + b \log 10 \Rightarrow 150 = 100 + b(1) \Rightarrow b = 50$

85. $3 = Ca^0 \Rightarrow C = 3$ and $6 = 3a^1 \Rightarrow a = 2$

86. $\log \dfrac{x}{1} = 7 \Rightarrow 10^{\log x} = 10^7 \Rightarrow x = 10^7$

87. (a)  Graph $Y_1 = 2e^{\wedge}(0.051X)$ in $[0, 10, 2]$ by $[0, 4, 1]$. The function represents exponential growth.

   $[0, 10, 2]$ by $[0, 4, 1]$

   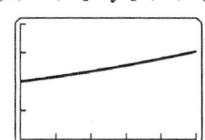

   (b)  In 2010, $x = 10$ thus $f(10) = 2e^{0.051(10)} \approx 3.3$ million

   (c)  $2e^{0.051x} = 3 \Rightarrow e^{0.051x} = \dfrac{3}{2} \Rightarrow \ln e^{0.051x} = \ln\left(\dfrac{3}{2}\right) \Rightarrow 0.051x = \ln(1.5) \Rightarrow x = \dfrac{\ln(1.5)}{0.051} \approx 7.95$

   That is about 8 years after 2000, which is 2008.

88. (a)  $N(0) = 1000e^{0.0014(0)} = 1000$; There were initially 1000 bacteria.

   (b)  $1000e^{0.0014x} = 2000 \Rightarrow e^{0.0014x} = 2 \Rightarrow \ln e^{0.0014x} = \ln 2 \Rightarrow 0.0014x = \ln 2 \Rightarrow$

   $x = \dfrac{\ln 2}{0.0014} \approx 495.11$ min.

89. (a)  $f(5) = 1.2 \ln 5 + 5 \approx 6.93$ m/sec

   (b)  $1.2 \ln x + 5 = 8 \Rightarrow 1.2 \ln x = 3 \Rightarrow \ln x = 2.5 \Rightarrow e^{\ln x} = e^{2.5} \Rightarrow x = e^{2.5} \approx 12.18$ meters

## Chapter 12 Test

1. $f(1) = 4(1)^3 - 5(1) = -1$, then $(g \circ f)(1) = g(f(1)) = g(-1) = (-1) + 7 = 6$

   $(f \circ g)(x) = f(g(x)) = f(x+7) = 4(x+7)^3 - 5(x+7)$

2. (a)  $(f \circ g)(-1) = f(g(-1)) = f(-1) = 3$

   (b)  $(g \circ f)(1) = g(f(1)) = g(1) = 3$

3. Two different inputs result in the same output. For example $-5 \neq 5$, but $f(-5) = f(5) = 0$.

4. $f(x) = 5 - 2x \Rightarrow y = 5 - 2x$, interchange $x$ and $y$ and solve for $y$.

   $x = 5 - 2y \Rightarrow 2y = 5 - x \Rightarrow y = \dfrac{5-x}{2} \Rightarrow f^{-1}(x) = \dfrac{5-x}{2}$

5. The graph of $f^{-1}$ is a reflection of the graph of $f$ across the line $y = x$. See Figure 5.

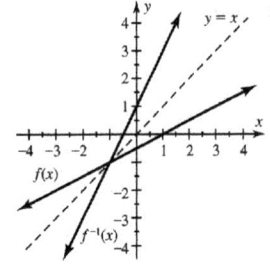
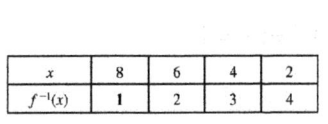
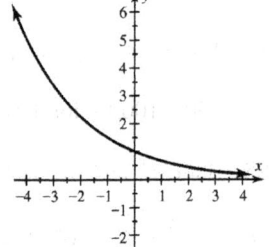

Figure 5                    Figure 6                    Figure 8

6. See Figure 6.  $D = \{2, 4, 6, 8\}$;  $R = \{1, 2, 3, 4\}$

7. $f(2) = 3\left(\dfrac{1}{4}\right)^2 = 3\left(\dfrac{1}{16}\right) = \dfrac{3}{16}$

8. See Figure 8. This graph represents exponential decay.

9. (a)  Exponential growth. For each unit increase in $x$, $f(x)$ increases by a factor of 2.

   (b)  Since $f(x) = 3$ when $x = 0$, $f(x) = 3(2)^x$

10. (a)  Linear growth. For each unit increase in $x$, $f(x)$ increases by 1.5 units.

    (b)  Since $f(x) = -1$ when $x = 0$, $f(x) = 1.5x - 1$

11. $1 = Ca^0 \Rightarrow C = 1$ and $2 = 1a^{-1} \Rightarrow a = \dfrac{1}{2}$

12. $\dfrac{A-B}{B} \times 100 \Rightarrow \dfrac{900 - 600}{600} \times 100 = \dfrac{300}{600} \times 100 = 50\%$

13. The account increased by a factor of $a = 1 + r = 1 + (0.05) = 1.05$.

14. $750(1 + 0.07)^5 = \$1051.91$

15. $f(21) = 1.5 \ln(21 - 5) = 1.5 \ln 16 \approx 4.16$

16. $\log \sqrt{10} = \log 10^{1/2} = \dfrac{1}{2}$

17. $\log_2 43 = \dfrac{\log 43}{\log 2} \approx 5.426$

18. See Figure 18. The graph is shifted 2 units to the right.

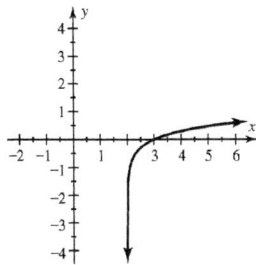

Figure 18

19. $\log \dfrac{x^3 y^2}{\sqrt{x}} = \log \dfrac{x^3 y^2}{x^{1/2}} = \log x^3 + \log y^2 - \log x^{1/2} = 3 \log x + 2 \log y - \dfrac{1}{2} \log z$

20. $4 \ln x - 5 \ln y + \ln z = \ln x^4 - \ln y^5 + \ln z = \ln \dfrac{x^4}{y^5} + \ln z = \ln \dfrac{x^4 z}{y^5}$

21. $\log 7^{2x} = 2x \log 7$

22. $\ln e^{1-3x} = 1 - 3x$

23. $2e^x = 50 \Rightarrow e^x = 25 \Rightarrow \ln e^x = \ln 25 \Rightarrow x = \ln 25 \approx 3.22$

24. $3(10)^x - 7 = 143 \Rightarrow 3(10)^x = 150 \Rightarrow 10^x = 50 \Rightarrow \log 10^x = \log 50 \Rightarrow x = \log 50 \approx 1.70$

25. $5 \log x = 9 \Rightarrow \log x = 1.8 \Rightarrow 10^{\log x} = 10^{1.8} \Rightarrow x = 10^{1.8} \approx 63.10$

26. $3 \ln 5x = 27 \Rightarrow \ln 5x = 9 \Rightarrow e^{\ln 5x} = e^9 \Rightarrow 5x = e^9 \Rightarrow x = \dfrac{e^9}{5} \approx 1620.62$

27. $5 = a + b \log 1 \Rightarrow 5 = a + b(0) \Rightarrow a = 5$ and $8 = 5 + b \log 10 \Rightarrow 8 = 5 + b(1) \Rightarrow b = 3$

28. (a)    $f(0) = 4(1.09)^0 = 4(1) = 4$ million

    (b)    $f(5) = 4(1.09)^{(5)} \approx 6.15$; After 5 hours there were about 6.15 million bacteria.

    (c)    This represents exponential growth.

    (d)    $4(1.09)^x = 8 \Rightarrow (1.09)^x = 2 \Rightarrow \ln (1.09)^x = \ln 2 \Rightarrow \ln(1.09)x = \ln 2 \Rightarrow x = \dfrac{\ln 2}{\ln(1.09)} \approx 8$

        There were 6 million bacteria after about 8 hours.

29. (a)    The initial value C is 5000. The account decreased by a factor of $a = 1 + r = 1 + (-0.2) = 0.98$.

        The function is $A(x) = 5000(0.98)^x$.

    (b)    $A(3) = 5000(0.98)^3 = 4705.96$   The result is $4705.96.

(c)    $4500 = 5000(0.98)^x \Rightarrow 0.9 = 0.98^x \Rightarrow \ln(0.9) = \ln(0.98)x \Rightarrow \dfrac{\ln(0.9)}{\ln(0.98)} = x \Rightarrow x \approx 5.2$

## Chapter 12 Extended and Discovery Exercises

1. $a^{5730} = 0.5 \Rightarrow \left(a^{5730}\right)^{1/5730} = 0.5^{1/5730} \Rightarrow a \approx 0.9998790392$

2. $P(10,000) = 0.9998790392^{10,000} \approx 0.298$ or $29.8\%$

3. $0.9998790392^x = 0.9 \Rightarrow \log_{0.9998790392} 0.9998790392^x = \log_{0.9998790392} 0.9 \Rightarrow$

   $x = \dfrac{\log 0.9}{\log 0.9998790392} \Rightarrow x \approx 871$ years

4. $0.9998790392^x = 0.01 \Rightarrow \log_{0.9998790392} 0.999879^x = \log_{0.9998790392} 0.01 \Rightarrow$

   $x = \dfrac{\log 0.01}{\log 0.9998790392} \Rightarrow x \approx 38,069$ years   (38,100 rounded to the nearest 100 years)

5. Plot the data and graph $Y_1 = 0.133\left(0.878(0.73 \wedge X) + 0.122(0.92 \wedge X)\right)$ in [0, 25, 5] by

   [0, 0.11, 0.01]. See Figure 5. The fit is quite good.

   [0, 25, 5] by [0, 0.11, 0.01]     [0, 25, 5] by [0, 0.11, 0.01]

   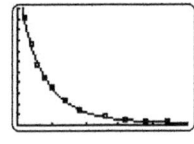        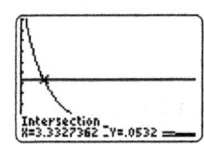

   Figure 5                    Figure 8

6. $f(0) = 0.133\left(0.878\left(0.73^0\right) + 0.122\left(0.92^0\right)\right) = 0.133(0.878 + 0.122) = 0.133(1) = 0.133$

   The initial concentration is $0.133\,\text{mg/mL}$.

7. The concentration decreases to 0 as the body eliminates the dye from the blood stream.

8. Note that 40% of 0.133 is 0.0532. Graph $Y_1 = 0.133\left(0.878(0.73 \wedge X) + 0.122(0.92 \wedge X)\right)$

   and $Y_2 = 0.0532$ in [0, 25, 5] by [0, 0.11, 0.01]. See Figure 8. This happens after about 3.33

   minutes. Solving this problem symbolically would be very difficult.

9. $f\left(10^{-4.7}\right) = -\log 10^{-4.7} = -(-4.7) = 4.7;$   This rain could cause the pH to drop below 5.6.

10. The ion concentration in seawater is $-\log x = 8.2 \Rightarrow \log x = -8.2 \Rightarrow 10^{\log x} = 10^{-8.2} \Rightarrow x = 10^{-8.2}.$

    This is $\dfrac{10^{-4.7}}{10^{-8.2}} \approx 3162$ times greater.

11. $A = 100\left[\dfrac{\left(1 + \frac{0.09}{26}\right)^{260} - 1}{\frac{0.09}{26}}\right] \approx \$42,055.97$

12. A 19-year-old student would have 46 years to deposit money before age 65. *Answers may vary.*

$$x\left[\frac{\left(1+\frac{0.12}{26}\right)^{1196}-1}{\frac{0.12}{26}}\right]=1,000,000 \Rightarrow x=1,000,000\left[\frac{\frac{0.12}{26}}{\left(1+\frac{0.12}{26}\right)^{1196}-1}\right]=\$18.80$$

## Chapters 1-12 Cumulative Review Exercises

1. Move the decimal point 4 places to the right: $0.000429 = 4.29 \times 10^{-4}$.

2. Natural: none; Whole: 0; Integer: $-3, 0$; Rational: $-\frac{11}{7}, -3, 0, 5.\overline{18}$; Irrational: $\sqrt{6}, \pi$

3. By substitution of values, the formula that best fits the data is (c).

4. This equation illustrates the commutative property for addition.

5. $\left(\dfrac{1}{d^2}\right)^{-2} = \left(d^2\right)^2 = d^4$

6. $\left(\dfrac{8a^2}{2b^3}\right)^{-3} = \left(\dfrac{4a^2}{b^3}\right)^{-3} = \left(\dfrac{b^3}{4a^2}\right)^3 = \dfrac{\left(b^3\right)^3}{\left(4a^2\right)^3} = \dfrac{b^9}{4^3\left(a^2\right)^3} = \dfrac{b^9}{64a^6}$

7. $\dfrac{\left(2x^{-2}y^3\right)^2}{xy^{-2}} = \dfrac{2^2\left(x^{-2}\right)^2\left(y^3\right)^2}{xy^{-2}} = \dfrac{4x^{-4}y^6}{xy^{-2}} = 4x^{-4-1}y^{6-(-2)} = 4x^{-5}y^8 = \dfrac{4y^8}{x^5}$

8. $\dfrac{x^{-3}y}{4x^2y^{-3}} = \dfrac{1}{4}x^{-3-2}y^{1-(-3)} = \dfrac{1}{4}x^{-5}y^4 = \dfrac{y^4}{4x^5}$

9. The graph rises 5 units for each 4 units of run. The slope is $\dfrac{5}{4}$. The $y$-intercept is 1. So $y = \dfrac{5}{4}x + 1$.

10. The function is defined for all values of the variable except $-3$. The domain is $\{x \mid x \neq -3\}$.

11. The slope is $m = \dfrac{5-1}{2-1} = \dfrac{4}{1} = 4$. Since $f(x) = -3$ when $x = 0$ the $y$-intercept is $-3$.

    Here $f(x) = 4x - 3$.

12. Vertical lines have equations of the form $x = k$. The equation of the vertical line passing through
    $(4, 7)$ is $x = 4$.

13. $m = \dfrac{-3-(-1)}{2-4} = \dfrac{-2}{-2} = 1$

14. See Figure 14.

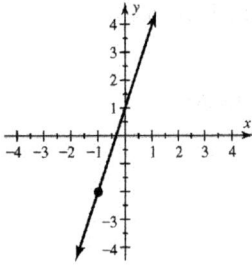

Figure 14

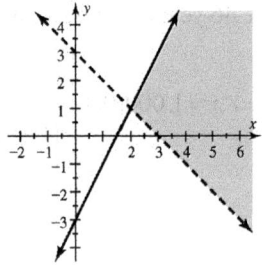

Figure 25

15. Since the line is perpendicular to $y = -\dfrac{1}{7}x - 8$, the slope is $m = 7$. Using the point-slope form gives

$$y = 7(x-1) + 1 \Rightarrow y = 7x - 7 + 1 \Rightarrow y = 7x - 6$$

16. Since the line is parallel to $y = 3x - 1$, the slope is $m = 3$. The $y$-intercept is given, so $b = 5$.

The equation is $y = 3x + 5$.

17. The graphs intersect at the point $(-1, 1)$, so $y_1 = y_2$ when $x = -1$.

18. Here $y < -4$ when $x < 0$.

19. $\dfrac{2}{3}(x-3) + 8 = -6 \Rightarrow \dfrac{2}{3}(x-3) = -14 \Rightarrow x - 3 = -21 \Rightarrow x = -18$

20. $\dfrac{1}{3}z + 6 < \dfrac{1}{4}z - (5z - 6) \Rightarrow 4z + 72 < 3z - 12(5z - 6) \Rightarrow 4z + 72 < 3z - 60z + 72 \Rightarrow$

$4z + 72 < -57z + 72 \Rightarrow 61z < 0 \Rightarrow z < 0$. The interval is $(-\infty, 0)$.

21. $\left(\dfrac{t+2}{3}\right) - 10 = \dfrac{1}{3}t - (5t + 8) \Rightarrow t + 2 - 30 = t - 3(5t + 8) \Rightarrow t - 28 = t - 15t - 24 \Rightarrow 15t = 4 \Rightarrow t = \dfrac{4}{15}$

22. $-10 \le -\dfrac{3}{5}x - 4 < -1 \Rightarrow -6 \le -\dfrac{3}{5}x < 3 \Rightarrow 10 \ge x > -5 \Rightarrow -5 < x \le 10$. The interval is $(-5, 10]$.

23. First divide each side of $-2|t - 4| \ge -12$ by $-2$ to obtain $|t - 4| \le 6$. The solutions to $|t - 4| \le 6$

satisfy $c \le t \le d$ where $c$ and $d$ are the solutions to $|t - 4| = 6$. $|t - 4| = 6$ is equivalent to $t - 4 = -6 \Rightarrow$

$t = -2$ and $t - 4 = 6 \Rightarrow t = 10$ The interval is $[-2, 10]$.

24. $\left|\dfrac{1}{2}x - 5\right| = 3 \Rightarrow \dfrac{1}{2}x - 5 = -3 \Rightarrow \dfrac{1}{2}x = 2 \Rightarrow x = 4$ or $\dfrac{1}{2}x - 5 = 3 \Rightarrow \dfrac{1}{2}x = 8 \Rightarrow x = 16$

25. See Figure 25.

26. $\det A = -1(4) - 3(-2) = -4 + 6 = 2$

27. Multiply the first equation by 2 and the second equation by 3. Add the equations to eliminate the

variable $y$. $\quad\dfrac{\begin{aligned}8x - 6y &= 2 \\ 15x + 6y &= 21\end{aligned}}{23x = 23}$ Thus, $x = 1$. And so $4(1) - 3y = 1 \Rightarrow y = 1$. The solution is $(1, 1)$.

28. Multiplying the first equation by 3 and adding the equations will eliminate both variables.

$6x - 9y = -6$
$\underline{-6x + 9y = \phantom{-}5}$    Thus, the system has no solutions.
$0 = -1$

29. Multiply the first equation by 5 and add the first and second equations to eliminate the variable $y$.

$10x - 5y + 15z = -10$
$\underline{\phantom{10}x + 5y \phantom{1}- 2z = \phantom{-}-8}$    Multiply the third equation by 5 and add the second and third equations to
$11x + 13z = -18$

eliminate the variable $y$.  
$x + 5y - 2z = -8$
$\underline{-15x - 5y - 15z = 30}$    Multiply the first *new* equation by 14 and the second
$-14x - 17z = 22$

*new* equation by 11. Add the equations to eliminate $x$.  
$154x + 182z = -252$
$\underline{-154x - 187z = \phantom{-}242}$    And so $z = 2$.
$-5z = -10$

Substitute $z = 2$ into the first *new* equation: $11x + 13(2) = -18 \Rightarrow x = -4$   Substitute

$x = -4$ and $z = 2$ into the *original* first equation: $2(-4) - y + 3(2) = -2 \Rightarrow y = 0$

The solution is $(-4, 0, 2)$.

30. $\begin{bmatrix} 1 & 1 & -1 & | & -1 \\ -1 & -1 & -1 & | & -1 \\ 1 & -2 & 1 & | & 1 \end{bmatrix} \begin{matrix} \\ R_2 + R_1 \rightarrow \\ R_3 - R_1 \rightarrow \end{matrix} \begin{bmatrix} 1 & 1 & -1 & | & -1 \\ 0 & 0 & -2 & | & -2 \\ 0 & -3 & 2 & | & 2 \end{bmatrix} \begin{matrix} \\ Exchange \\ R_2 \leftrightarrow R_3 \end{matrix} \begin{bmatrix} 1 & 1 & -1 & | & -1 \\ 0 & -3 & 2 & | & 2 \\ 0 & 0 & -2 & | & -2 \end{bmatrix}$

$\begin{matrix} \\ \\ (-1/2)R_3 \rightarrow \end{matrix} \begin{bmatrix} 1 & 1 & -1 & | & -1 \\ 0 & -3 & 2 & | & 2 \\ 0 & 0 & 1 & | & 1 \end{bmatrix} \begin{matrix} R_1 + R_3 \rightarrow \\ R_2 - 2R_3 \rightarrow \\ \end{matrix} \begin{bmatrix} 1 & 1 & 0 & | & 0 \\ 0 & -3 & 0 & | & 0 \\ 0 & 0 & 1 & | & 1 \end{bmatrix} \begin{matrix} R_1 + (1/3)R_2 \rightarrow \\ (-1/3)R_2 \rightarrow \\ \end{matrix} \begin{bmatrix} 1 & 0 & 0 & | & 0 \\ 0 & 1 & 0 & | & 0 \\ 0 & 0 & 1 & | & 1 \end{bmatrix}$

The solution is $(0, 0, 1)$.

31. From the graph of the region of feasible solutions (not shown), the vertices are (0, 0), (0, 4), (3, 3),

and (4, 0). The maximum value of $R$ occurs at one of the vertices. For $(0, 0)$, $R = 2(0) + 5(0) = 0$.

For $(0, 4)$, $R = 2(0) + 5(4) = 20$. For $(3, 3)$, $R = 2(3) + 5(3) = 21$. For $(4, 0)$, $R = 2(4) + 5(0) = 8$.

The maximum value is $R = 21$.

32. The triangle has vertices $(-2, -3), (-1, 2)$ and $(2, 1)$. The matrix needed is $A = \begin{bmatrix} -2 & -1 & 2 \\ -3 & 2 & 1 \\ 1 & 1 & 1 \end{bmatrix}$.

The area is $D = \left| \frac{1}{2} \det([A]) \right| = 8 \text{ in}^2$.

33. $2x^3 - 4x^2 + 2x = 2x(x^2 - 2x + 1) = 2x(x-1)^2$

34. $4a^2 - 25b^2 = (2a)^2 - (5b)^2 = (2a - 5b)(2a + 5b)$

35. $8t^3 - 27 = (2t)^3 - 3^3 = (2t - 3)(4t^2 + 6t + 9)$

36. $4a^3 - 2a^2 + 10a - 5 = 2a^2(2a - 1) + 5(2a - 1) = (2a^2 + 5)(2a - 1)$

37. $6x^2 - 7x - 10 = 0 \Rightarrow (6x + 5)(x - 2) = 0 \Rightarrow x = -\dfrac{5}{6}$ or $x = 2$

38. $9x^2 = 4 \Rightarrow 9x^2 - 4 = 0 \Rightarrow (3x + 2)(3x - 2) = 0 \Rightarrow x = -\dfrac{2}{3}$ or $x = \dfrac{2}{3}$

39. $x^4 - 2x^3 = 15x^2 \Rightarrow x^4 - 2x^3 - 15x^2 = 0 \Rightarrow x^2(x + 3)(x - 5) = 0 \Rightarrow x = -3, 0,$ or $5$

40. $5x - 10x^2 = 0 \Rightarrow 5x(1 - 2x) = 0 \Rightarrow x = 0$ or $x = \dfrac{1}{2}$

41. $\dfrac{x^2 + 5x + 6}{x^2 - 9} \cdot \dfrac{x - 3}{x + 2} = \dfrac{(x + 2)(x + 3)(x - 3)}{(x - 3)(x + 3)(x + 2)} = 1$

42. $\dfrac{x^2 - 2x - 8}{x^2 + x - 12} \div \dfrac{(x - 4)^2}{x^2 - 16} = \dfrac{x^2 - 2x - 8}{x^2 + x - 12} \cdot \dfrac{x^2 - 16}{(x - 4)^2} = \dfrac{(x + 2)(x - 4)(x - 4)(x + 4)}{(x + 4)(x - 3)(x - 4)(x - 4)} = \dfrac{x + 2}{x - 3}$

43. $\dfrac{2}{x + 2} - \dfrac{1}{x - 2} = \dfrac{-3}{x^2 - 4} \Rightarrow 2(x - 2) - 1(x + 2) = -3 \Rightarrow x - 6 = -3 \Rightarrow x = 3$

44. $\dfrac{3y}{y^2 + y - 2} = \dfrac{1}{y - 1} - 2 \Rightarrow 3y = 1(y + 2) - 2(y + 2)(y - 1) \Rightarrow 3y = y + 2 - 2y^2 - 2y + 4 \Rightarrow$

    $2y^2 + 4y - 6 = 0 \Rightarrow 2(y + 3)(y - 1) = 0 \Rightarrow y = -3$ or $y = 1$. The value $y = 1$ is not valid, thus $y = -3$.

45. $P = \dfrac{J + 2z}{J} \Rightarrow JP = J + 2z \Rightarrow JP - J = 2z \Rightarrow J(P - 1) = 2z \Rightarrow J = \dfrac{2z}{P - 1}$

46. $\dfrac{\dfrac{3}{x^2} + x}{x - \dfrac{3}{x^2}} = \dfrac{\dfrac{3}{x^2} + x}{x - \dfrac{3}{x^2}} \cdot \dfrac{x^2}{x^2} = \dfrac{3 + x^3}{x^3 - 3} = \dfrac{x^3 + 3}{x^3 - 3}$

47. $y = kx \Rightarrow 15 = 3k \Rightarrow k = 5 \Rightarrow y = 5x$, so when $x = 8$, $y = 5(8) = 40$

48.

$$\begin{array}{r} 3x^2 + 6x + 10 \\ x - 2 \overline{\smash{)}\ 3x^3 + 0x^2 - 2x - 15} \\ \underline{3x^3 - 6x^2} \\ 6x^2 - 2x \\ \underline{6x^2 - 12x} \\ 10x - 15 \\ \underline{10x - 20} \\ 5 \end{array}$$

The solution is: $3x^2 + 6x + 10 + \dfrac{5}{x - 2}$

49. $\left(\dfrac{x^6}{y^9}\right)^{2/3} = \dfrac{\left(x^6\right)^{2/3}}{\left(y^9\right)^{2/3}} = \dfrac{x^{6 \cdot (2/3)}}{y^{9 \cdot (2/3)}} = \dfrac{x^4}{y^6}$

50. $\sqrt[3]{-x^4} \cdot \sqrt[3]{-x^5} = \sqrt[3]{\left(-x^4\right)\left(-x^5\right)} = \sqrt[3]{x^9} = x^{9/3} = x^3$

51. $\sqrt{5ab} \cdot \sqrt{20ab} = \sqrt{5 \cdot 20 \cdot ab \cdot ab} = \sqrt{100\left(ab\right)^2} = \sqrt{\left(10ab\right)^2} = 10ab$

52. $2\sqrt{24} - \sqrt{54} = 2\sqrt{4 \cdot 6} - \sqrt{9 \cdot 6} = 2\sqrt{4} \cdot \sqrt{6} - \sqrt{9} \cdot \sqrt{6} = 4\sqrt{6} - 3\sqrt{6} = \sqrt{6}$

53. $\sqrt[3]{a^5 b^4} + 3\sqrt[3]{a^5 b} = \sqrt[3]{\left(ab\right)^3 \cdot a^2 b} + 3\sqrt[3]{a^3 \cdot a^2 b} = ab\sqrt[3]{a^2 b} + 3a\sqrt[3]{a^2 b} = \left(b + 3\right)a\sqrt[3]{a^2 b}$

54. $\left(5 + \sqrt{5}\right)\left(5 - \sqrt{5}\right) = 5^2 - \left(\sqrt{5}\right)^2 = 25 - 5 = 20$

55. $\dfrac{2}{5 - \sqrt{3}} = \dfrac{2}{5 - \sqrt{3}} \cdot \dfrac{5 + \sqrt{3}}{5 + \sqrt{3}} = \dfrac{10 + 2\sqrt{3}}{25 - 3} = \dfrac{10 + 2\sqrt{3}}{22} = \dfrac{2\left(5 + \sqrt{3}\right)}{2 \cdot 11} = \dfrac{5 + \sqrt{3}}{11}$

56. For the function to be defined, $x - 4 > 0 \Rightarrow x > 4$. The interval is $(4, \infty)$.

57. $2\left(x + 1\right)^2 = 50 \Rightarrow \left(x + 1\right)^2 = 25 \Rightarrow x + 1 = \pm\sqrt{25} \Rightarrow x + 1 = \pm 5 \Rightarrow x = -6$ or $4$

58. $\sqrt{x + 6} = x \Rightarrow x + 6 = x^2 \Rightarrow x^2 - x - 6 = 0 \Rightarrow \left(x + 2\right)\left(x - 3\right) = 0 \Rightarrow x = -2$ or $3$

   The value $x = -2$ does not check. The only solution is 3.

59. $\left(-2 + 3i\right) - \left(-5 - 2i\right) = -2 - \left(-5\right) + 3i + 2i = 3 + 5i$

60. $\dfrac{3 - i}{1 + 3i} = \dfrac{3 - i}{1 + 3i} \cdot \dfrac{1 - 3i}{1 - 3i} = \dfrac{3 - 9i - i + 3i^2}{1 - 9i^2} = \dfrac{3 - 10i - 3}{1 + 9} = \dfrac{-10i}{10} = -i$

61. $-\dfrac{b}{2a} = -\dfrac{-12}{2(3)} = \dfrac{12}{6} = 2;\ f(2) = 3(2)^2 - 12(2) + 13 = 1.$ The vertex is $(2, 1)$.

62. $-\dfrac{b}{2a} = -\dfrac{6}{2(-2)} = \dfrac{6}{4} = \dfrac{3}{2};\ f\left(\dfrac{3}{2}\right) = -2\left(\dfrac{3}{2}\right)^2 + 6\left(\dfrac{3}{2}\right) - 1 = \dfrac{7}{2}.$ The vertex is $\left(\dfrac{3}{2}, \dfrac{7}{2}\right)$. The maximum

   value is $\dfrac{7}{2}$.

63. Compared to $y = x^2$, the graph of $f(x)$ is shifted right 3 units and up 2 units.

64. $y = x^2 + 6x - 2 \Rightarrow y = \left(x^2 + 6x + 9\right) - 2 - 9 \Rightarrow y = \left(x + 3\right)^2 - 11.$ The vertex is $(-3, -11)$.

65. $x^2 - 13x + 40 = 0 \Rightarrow \left(x - 5\right)\left(x - 8\right) = 0 \Rightarrow x = 5$ or $x = 8$

66. $2d^2 - 5 = d \Rightarrow 2d^2 - d - 5 = 0.$ Let $a = 2, b = -1$ and $c = -5$ in the quadratic formula.

   $d = \dfrac{-(-1) \pm \sqrt{(-1)^2 - 4(2)(-5)}}{2(2)} = \dfrac{1 \pm \sqrt{41}}{4}$

67. $z^2 - 4z = -2 \Rightarrow z^2 - 4z + 4 = -2 + 4 \Rightarrow (z-2)^2 = 2 \Rightarrow z - 2 = \pm\sqrt{2} \Rightarrow z = 2 \pm \sqrt{2}$

68. $x^4 - 10x^2 + 24 = 0 \Rightarrow (x^2 - 6)(x^2 - 4) = 0 \Rightarrow x^2 - 6 = 0$ or $x^2 - 4 = 0 \Rightarrow x = \pm\sqrt{6}$ or $x = \pm 2$

69. (a)　The graph intersects the $x$-axis at $-1$ and $3$.

　　(b)　Because the parabola opens downward, $a < 0$.

　　(c)　Because there are two real solutions, the discriminant is positive.

70. $x^2 + 5x - 14 = 0 \Rightarrow (x+7)(x-2) = 0 \Rightarrow x = -7$ or $x = 2$

　　Since the parabola opens upward, the solution is $(-\infty, -7] \cup [2, \infty)$.

71. (a)　$g(1) = 2(1) + 1 = 3$, then $(f \circ g)(1) = f(g(1)) = f(3) = (3)^2 - 2 = 7$

　　(b)　$(g \circ f)(x) = g(f(x)) = g(x^2 - 2) = 2(x^2 - 2) + 1 = 2x^2 - 4 + 1 = 2x^2 - 3$　　72.

　　　$f(-4) = f(3) = 6$

73. $f(x) = \dfrac{3}{x} \Rightarrow y = \dfrac{3}{x}$, interchange $x$ and $y$ and solve for $y$.　$x = \dfrac{3}{y} \Rightarrow xy = 3 \Rightarrow y = \dfrac{3}{x} \Rightarrow f^{-1}(x) = \dfrac{3}{x}$

74. $A = 800(1 + 0.075)^{15} \approx \$2367.10$

75. $\log_3 81 = \log_3 3^4 = 4$

76. $e^{\ln(2x)} = 2x,\ x > 0$

77. $\log \dfrac{\sqrt{x}}{y^2} = \log \dfrac{x^{1/2}}{y^2} = \log x^{1/2} - \log y^2 = \dfrac{1}{2}\log x - 2\log y$

78. $2\ln x + \ln 5x = \ln x^2 + \ln 5x = \ln(x^2 \cdot 5x) = \ln(5x^3)$

79. $6\log x - 2 = 9 \Rightarrow 6\log x = 11 \Rightarrow \log x = \dfrac{11}{6} \Rightarrow 10^{\log x} = 10^{11/6} \Rightarrow x \approx 68.13$

80. $2^{3x} = 17 \Rightarrow \log_2 2^{3x} = \log_2 17 \Rightarrow 3x = \dfrac{\log 17}{\log 2} \Rightarrow x = \dfrac{\log 17}{3 \log 2} \approx 1.36$

81. $12,000(1+r)^5 = 14,600 \Rightarrow (1+r)^5 = \dfrac{73}{60} \Rightarrow 1 + r = \sqrt[5]{\dfrac{73}{60}} \Rightarrow r = \sqrt[5]{\dfrac{73}{60}} - 1 \approx 0.04$ or $4\%$

82. $27.4\sqrt[3]{W} = 36 \Rightarrow \sqrt[3]{W} = \dfrac{36}{27.4} \Rightarrow W = \left(\dfrac{36}{27.4}\right)^3 \approx 2.27$ pounds

83. (a)　$-\dfrac{b}{2a} = -\dfrac{-975}{2(0.25)} = 1950$

　　(b)　$f(1950) = 0.25(1950)^2 - 975(1950) + 950,845 = 220$ million Btu

84. $\dfrac{x^2}{12} = 350 \Rightarrow x^2 = 4200 \Rightarrow x = \sqrt{4200} \approx 64.8$ miles per hour

85.  $8000(1+r)^{45} = 1,000,000 \Rightarrow (1+r)^{45} = 125 \Rightarrow 1+r = 125^{1/45} \Rightarrow r = 125^{1/45} - 1 \approx 0.113$ or $11.3\%$

86.  (a)    $f(8) = 1.4 \ln 8 + 7 \approx 9.91$ meters per second

    (b)    $1.4 \ln x + 7 = 10 \Rightarrow 1.4 \ln x = 3 \Rightarrow \ln x = \dfrac{3}{1.4} \Rightarrow e^{\ln x} = e^{3/1.4} \Rightarrow x \approx 8.52$ meters

## Chapter 13: Conic Sections

### 13.1: Parabolas and Circles

1. Parabola, ellipse and hyperbola

3. No, it does not pass the vertical line test for functions.

5. No, it does not pass the vertical line test for functions.

7. left

9. circle; $(h, k)$

11. d; The equation fits the form $x = a(y - k)^2 + h$ where $(h, k)$ is the vertex.

13. c; The equation fits the form $(x - h)^2 + (y - k)^2 = r^2$ where $(h, k)$ is the center and $r$ is the radius.

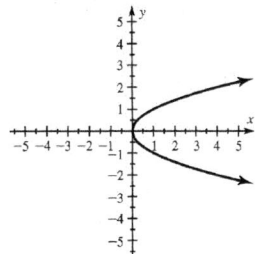
Figure 15

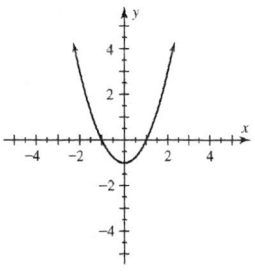
Figure 17

Figure 19

15. Since $x = (y - 0)^2 + 0$, the vertex is (0, 0) and the axis of symmetry is $y = 0$.  See Figure 15.

17. Since $x = (y - 0)^2 + 1$, the vertex is (1, 0) and the axis of symmetry is $y = 0$.  See Figure 17.

19. Since $y = x^2 - 1 = (x - 0)^2 - 1$, the vertex is (0, -1) and the axis of symmetry is $x = 0$.

    See Figure 19.

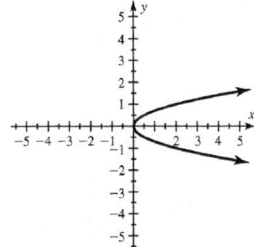
Figure 21

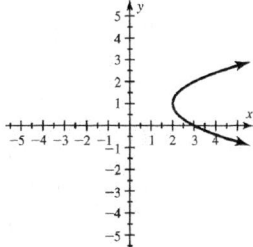
Figure 23

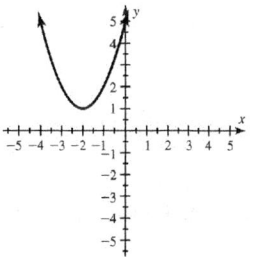
Figure 25

21. Since $x = 2(y - 0)^2 + 0$, the vertex is (0, 0) and the axis of symmetry is $y = 0$.  See Figure 21.

23. Since $x = (y - 1)^2 + 2$, the vertex is (2, 1) and the axis of symmetry is $y = 1$.  See Figure 23.

25. Since $y = (x + 2)^2 + 1$, the vertex is (-2, 1) and the axis of symmetry is $x = -2$.  See Figure 25.

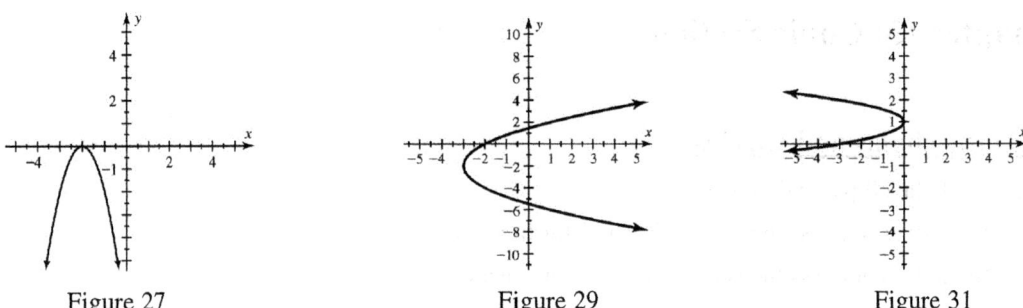

Figure 27                                          Figure 29                                          Figure 31

27. Since $y = -2(x+2)^2$, the vertex is (–2, 0) and the axis of symmetry is $x = -2$. See Figure 27.

29. Since $x = \frac{1}{2}(y+1)^2 - 3$, the vertex is (–3, –1) and the axis of symmetry is $y = -1$. See Figure 29.

31. Since $x = -3(y-1)^2 + 0$, the vertex is (0, 1) and the axis of symmetry is $y = 1$. See Figure 31.

33. See Figure 33. $x = -\frac{b}{2a} = -\frac{-1}{2(2)} = \frac{1}{4}$ and $y = 2\left(\frac{1}{4}\right)^2 - \left(\frac{1}{4}\right) + 1 = \frac{7}{8}$. Vertex: $\left(\frac{1}{4}, \frac{7}{8}\right)$.

Axis of symmetry: $x = \frac{1}{4}$.

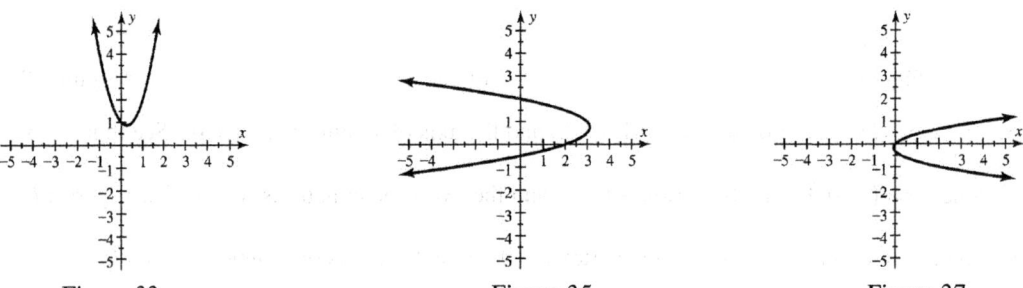

Figure 33                                          Figure 35                                          Figure 37

35. See Figure 35. $y = -\frac{b}{2a} = -\frac{3}{2(-2)} = \frac{3}{4}$ and $x = -2\left(\frac{3}{4}\right)^2 + 3\left(\frac{3}{4}\right) + 2 = \frac{25}{8}$. Vertex: $\left(\frac{25}{8}, \frac{3}{4}\right)$.

Axis of symmetry: $y = \frac{3}{4}$.

37. See Figure 37. $y = -\frac{b}{2a} = -\frac{1}{2(3)} = -\frac{1}{6}$ and $x = 3\left(-\frac{1}{6}\right)^2 + \left(-\frac{1}{6}\right) = -\frac{1}{12}$. Vertex: $\left(-\frac{1}{12}, -\frac{1}{6}\right)$.

Axis of symmetry: $y = -\frac{1}{6}$.

39. See Figure 39. $y = -\frac{b}{2a} = -\frac{2}{2(1)} = -\frac{2}{2} = -1$ and $x = (-1)^2 + 2(-1) + 1 = 0$.

Vertex: $(0, -1)$. Axis of symmetry: $y = -1$

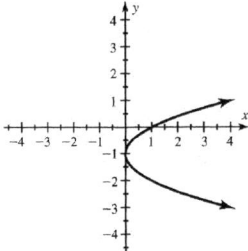

Figure 39

41. Since the parabola opens upward and the vertex is (0, 0), the equation has the form $y = a(x-0)^2 + 0.$

Since the parabola passes through (1, 1), $1 = a(1-0)^2 + 0 \Rightarrow a = 1.$ The equation is $y = x^2.$

43. Since the parabola opens to the right and the vertex is (–2, –1), the equation has the form

$x = a(y+1)^2 - 2.$ Since the parabola passes through (–1, 0), $-1 = a(0+1)^2 - 2 \Rightarrow a = 1.$

The equation is $x = (y+1)^2 - 2.$

45. By plotting the points by hand, we see that the parabola must open upward.

47. By plotting the points and axis by hand, we see that the parabola must open downward.

49. Since the parabola opens to the right and the vertex is (0, 0), the possible $x$-values are $x \geq 0.$

51. Since the parabola opens to the right and the vertex is to the left of the $y$-axis, the parabola has two $y$-intercepts.

53. $x = 3(0)^2 - (0) + 1 \Rightarrow x = 1$

55. $(x-0)^2 + (y-0)^2 = 1^2 \Rightarrow x^2 + y^2 = 1$

57. $(x-(-1))^2 + (y-5)^2 = 3^2 \Rightarrow (x+1)^2 + (y-5)^2 = 9$

59. $(x-(-4))^2 + (y-(-6))^2 = (\sqrt{2})^2 \Rightarrow (x+4)^2 + (y+6)^2 = 2$

61. Since the center is (0, 0) and the radius is 4, the equation is $x^2 + y^2 = 16.$

63. Since the center is (–3, 2) and the radius is 1, the equation is $(x+3)^2 + (y-2)^2 = 1.$

65. The radius is 2 and the center is (0, 0). Solving the equation for $y$ results in $y = \pm\sqrt{4 - x^2}.$

See Figure 65.

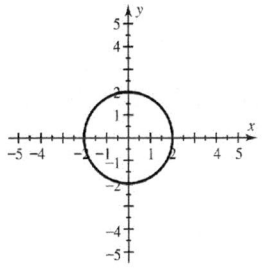
Figure 65

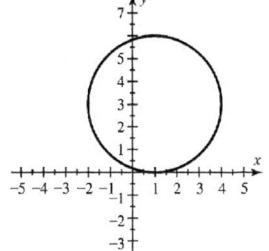
Figure 67

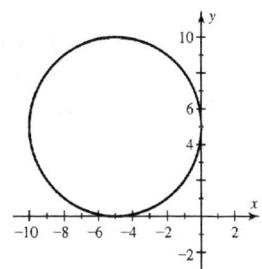
Figure 69

67. The radius is 3 and the center is (1, 3). Solving the equation for $y$ results in $y = 3 \pm \sqrt{9 - (x-1)^2}$.

   See Figure 67.

69. The radius is 5 and the center is (–5, 5). Solving the equation for $y$ results in $y = 5 \pm \sqrt{25 - (x+5)^2}$.

   See Figure 69.

71. $x^2 + 6x + y^2 - 2y = -1 \Rightarrow x^2 + 6x + 9 + y^2 - 2y + 1 = -1 + 9 + 1 \Rightarrow (x+3)^2 + (y-1)^2 = 9$

   The radius is 3 and the center is (–3, 1). Solving the equation for $y$ results in $y = 1 \pm \sqrt{9 - (x+3)^2}$.

   See Figure 71.

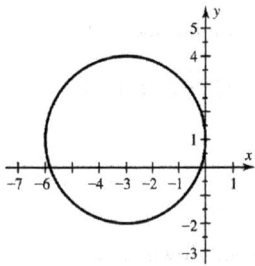     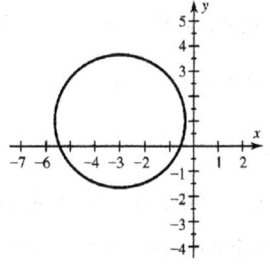

Figure 71                                Figure 73

73. $x^2 + 6x + y^2 - 2y + 3 = 0 \Rightarrow x^2 + 6x + 9 + y^2 - 2y + 1 = -3 + 9 + 1 \Rightarrow (x+3)^2 + (y-1)^2 = 7$

   The radius is $\sqrt{7}$ and the center is (–3, 1). Solving the equation for $y$ results in $y = 1 \pm \sqrt{7 - (x+3)^2}$.

   See Figure 73.

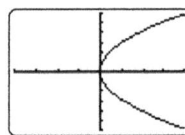     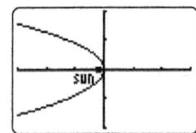

Figure 75                    Figure 77

75. (a)   Graph $Y_1 = (32/11025)X^2$ using the DrawInv feature in [–40, 40, 10] by [–120, 120, 20].

       See Figure 75.

   (b)   When $y = 105$, $x = \dfrac{32}{11,025}(105)^2 = 32$ feet.

77. (a)   Plot (–0.1, 0) and graph $Y_1 = -2.5X^2$ using the DrawInv feature in [–1.5, 1.5, 0.5] by

       [–1, 1, 0.5]. See Figure 77.

   (b)   $d = \sqrt{(-2.5 - (-0.1))^2 + (1-0)^2} = \sqrt{(-2.4)^2 + 1^2} = \sqrt{6.76} = 2.6$ A.U. or 241,800,000 miles.

## Section 13.2 Ellipses and Hyperbolas

1. See Figure 1.

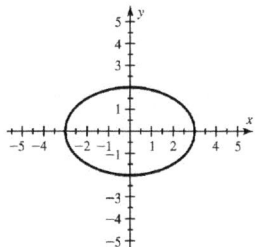

Figure 1

3. horizontal

5. 2

7. left and right

9. They are the diagonals extended.

11. a; Since the given equation is in the form $\dfrac{x^2}{b^2} + \dfrac{y^2}{a^2} = 1$ where the vertices are $(0, a)$ and $(0, -a)$.

13. The ellipse has a vertical major axis with vertices $(0, \pm 5)$ and minor axis endpoints $(\pm 3, 0)$.

See Figure 13.

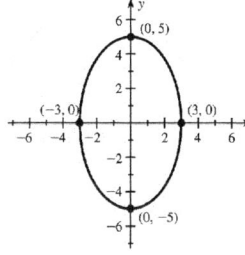

Figure 13

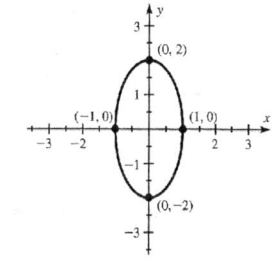

Figure 15                        Figure 17

15. The ellipse has a horizontal major axis with vertices $(\pm 3, 0)$ and minor axis endpoints $(0, \pm 2)$.

See Figure 15.

17. The ellipse has a vertical major axis with vertices $(0, \pm 2)$ and minor axis endpoints $(\pm 1, 0)$.

See Figure 17.

19. The ellipse has a horizontal major axis with vertices $\left(\pm\sqrt{7}, 0\right)$ and minor axis endpoints $\left(0, \pm\sqrt{5}\right)$.

See Figure 19.

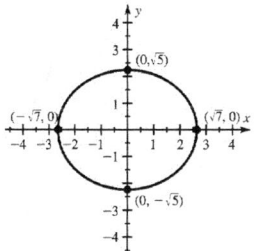

Figure 19

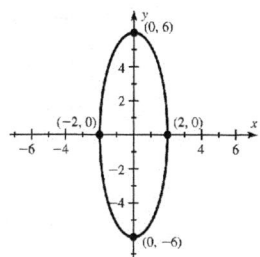

Figure 21

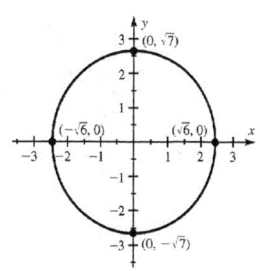

Figure 23

21. $36x^2 + 4y^2 = 144 \Rightarrow \dfrac{36x^2}{144} + \dfrac{4y^2}{144} = 1 \Rightarrow \dfrac{x^2}{4} + \dfrac{y^2}{36} = 1$  The ellipse has a vertical major axis with

vertices $(0, \pm 6)$ and minor axis endpoints $(\pm 2, 0)$.  See Figure 21.

23. $6y^2 + 7x^2 = 42 \Rightarrow \dfrac{6y^2}{42} + \dfrac{7x^2}{42} = 1 \Rightarrow \dfrac{y^2}{7} + \dfrac{x^2}{6} = 1$  The ellipse has a vertical major axis with vertices

$\left(0, \pm\sqrt{7}\right)$ and minor axis endpoints $\left(\pm\sqrt{6}, 0\right)$.  See Figure 23.

25. Horizontal major axis with vertices $(\pm 3, 0)$ and minor axis endpoints $(0, \pm 2)$  $\Rightarrow \dfrac{x^2}{9} + \dfrac{y^2}{4} = 1$

27. Vertical major axis with vertices $(0, \pm 5)$ and minor axis endpoints $(\pm 4, 0)$  $\Rightarrow \dfrac{y^2}{25} + \dfrac{x^2}{16} = 1$

29. The hyperbola has a horizontal transverse axis with vertices $(\pm 2, 0)$ and asymptotes $y = \pm\dfrac{3}{2}x.$

See Figure 29.

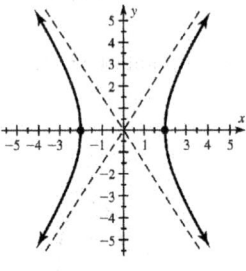

Figure 29

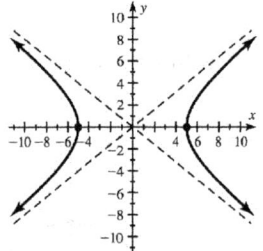

Figure 31

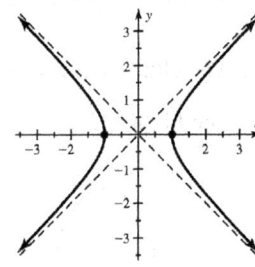

Figure 33

31. The hyperbola has a horizontal transverse axis with vertices $(\pm 5, 0)$ and asymptotes $y = \pm\dfrac{4}{5}x.$

See Figure 31.

33. The hyperbola has a horizontal transverse axis with vertices $(\pm 1, 0)$ and asymptotes $y = \pm x.$

See Figure 33.

35. The hyperbola has a horizontal transverse axis with vertices $\left(\pm\sqrt{3}, 0\right)$ and asymptotes $y = \pm\dfrac{2}{\sqrt{3}}x.$

See Figure 35.

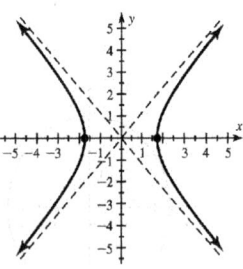

Figure 35

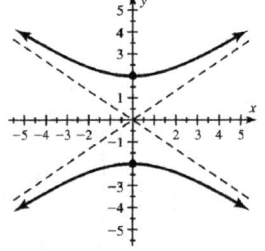

Figure 37

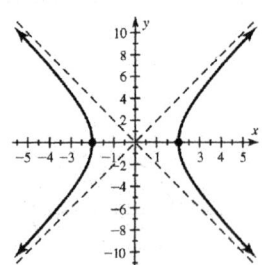

Figure 39

37. $9y^2 - 4x^2 = 36 \Rightarrow \dfrac{9y^2}{36} - \dfrac{4x^2}{36} = 1 \Rightarrow \dfrac{y^2}{4} - \dfrac{x^2}{9} = 1$  The hyperbola has a vertical transverse axis with

    vertices $(0, \pm 2)$ and asymptotes $y = \pm \dfrac{2}{3}x.$  See Figure 37.

39. $16x^2 - 4y^2 = 64 \Rightarrow \dfrac{16x^2}{64} - \dfrac{4y^2}{64} = 1 \Rightarrow \dfrac{x^2}{4} - \dfrac{y^2}{16} = 1$  The hyperbola has a horizontal transverse axis

    with vertices $(\pm 2, 0)$ and asymptotes $y = \pm 2x.$  See Figure 39.

41. Horizontal transverse axis with vertices $(\pm 1, 0)$ and asymptotes $y = \pm \dfrac{1}{1}x \Rightarrow x^2 - y^2 = 1$

43. Vertical transverse axis with vertices $(0, \pm 2)$ and asymptotes $y = \pm \dfrac{2}{3}x \Rightarrow \dfrac{y^2}{4} - \dfrac{x^2}{9} = 1$

45. (a)  $A = \pi(5)(4) \approx 62.83;\ P = 2\pi\sqrt{\dfrac{5^2 + 4^2}{2}} \approx 28.45$

    (b)  $A = \pi\left(\sqrt{7}\right)\left(\sqrt{2}\right) \approx 11.75;\ P = 2\pi\sqrt{\dfrac{\left(\sqrt{7}\right)^2 + \left(\sqrt{2}\right)^2}{2}} \approx 13.33$

47. (a)  $\dfrac{x^2}{39.44^2} + \dfrac{y^2}{38.20^2} = 1 \Rightarrow \dfrac{y^2}{38.20^2} = 1 - \dfrac{x^2}{39.44^2} \Rightarrow y = \pm\sqrt{38.20^2\left(1 - \dfrac{x^2}{39.44^2}\right)}$

    Graph $Y_1 = \sqrt{\left(38.20^2\left(1 - \left(X^2/39.44^2\right)\right)\right)}$ and $Y_2 = -\sqrt{\left(38.20^2\left(1 - \left(X^2/39.44^2\right)\right)\right)}$ and plot the

    point $(9.81, 0)$ in $[-60, 60, 10]$ by $[-40, 40, 10]$.  See Figure 47.

    (b)  $P = 2\pi\sqrt{\dfrac{39.44^2 + 38.20^2}{2}} \approx 243.9$ A.U. or about $2.27 \times 10^{10}$ miles

        $A = \pi(39.44)(38.20) \approx 4733$ A.U. or about $4.09 \times 10^{19}$ square miles

$[-60, 60, 10]$ by $[-40, 40, 10]$        $[-21, 21, 5]$ by $[-14, 14, 5]$

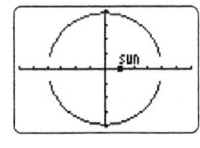

Figure 47

49. The maximum and minimum heights occur when $y = 0$.  These values are calculated below.

    Earth: $\dfrac{(x - 164)^2}{3960^2} + 0 = 1 \Rightarrow (x - 164)^2 = 3960^2 \Rightarrow x - 164 = \pm 3960 \Rightarrow x = 4124$ or $x = -3796$.

    Explorer VII: $\dfrac{x^2}{4464^2} + 0 = 1 \Rightarrow x^2 = 4464^2 \Rightarrow x = \pm 4464 \Rightarrow x = 4464$ or $x = -4464$.

    The maximum height is $-3796 - (-4464) = 668$ miles.  The minimum height is

    $4464 - 4124 = 340$ miles.

51. The height is half the length of the minor axis and the width is the full length of the major axis.

$$400x^2 + 10,000y^2 = 4,000,000 \Rightarrow \frac{400x^2}{4,000,000} + \frac{10,000y^2}{4,000,000} = 1 \Rightarrow \frac{x^2}{10,000} + \frac{y^2}{400} = 1$$

The height is $\sqrt{400} = 20$ feet and the width is $2\sqrt{10,000} = 2(100) = 200$ feet.

53. Example: Satellite and Planet orbits, *Answers may vary.*

## Checking Basic Concepts Sections 13.1 and 13.2

1. See Figure 1. Vertex: (1, 2). Axis of symmetry: $y = 2$.

2. The equation is $(x-1)^2 + (y+2)^2 = 4$.

$$(x-1)^2 + (y+2)^2 = 4 \Rightarrow (y+2)^2 = 4 - (x-1)^2 \Rightarrow y = -2 \pm \sqrt{4 - (x-1)^2}. \text{ See Figure 2}$$

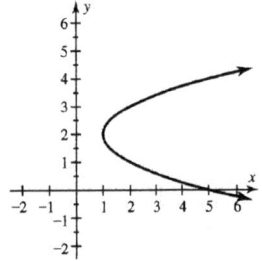     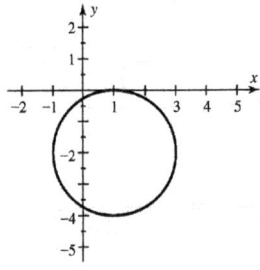     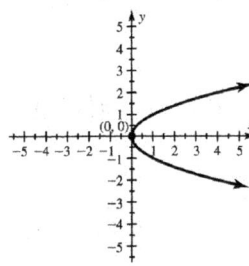

Figure 1                          Figure 2                          Figure 4a

3. x-intercepts: $\frac{x^2}{4} + \frac{0^2}{9} = 1 \Rightarrow \frac{x^2}{4} = 1 \Rightarrow x^2 = 4 \Rightarrow x = \pm 2$

   y-intercepts: $\frac{0^2}{4} + \frac{y^2}{9} = 1 \Rightarrow \frac{y^2}{9} = 1 \Rightarrow y^2 = 9 \Rightarrow y = \pm 3$

4. (a)   This parabola has vertex (0, 0) and axis of symmetry $y = 0$. See Figure 4a.

   (b)   This ellipse has a vertical major axis with vertices $(0, \pm 5)$ and minor axis endpoints $(\pm 4, 0)$.

        See Figure 4b.

   (c)   This hyperbola has a horizontal transverse axis with vertices $(\pm 2, 0)$ and asymptotes

        $y = \pm \frac{3}{2}x.$ See Figure 4c.

   (d)   This circle is centered at $(1, -2)$ and has radius 3. See Figure 4d.

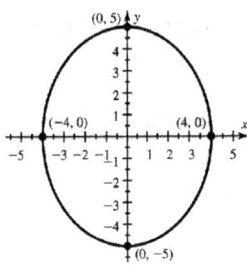     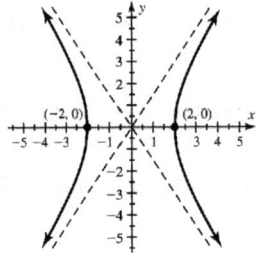     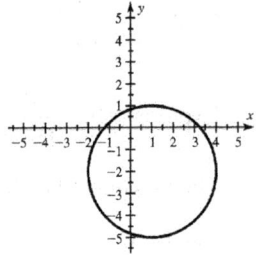

Figure 4b                         Figure 4c                         Figure 4d

## Section 13.3 Nonlinear Systems of Equations and Inequalities

1. Any number.

3. Two, the line intersects the circle twice.

5. No. $5(-2)^2 - 2(-1)^2 = 5(4) - 2(1) = 18 \not> 18$

7. See Figure 7. *Answers may vary.*

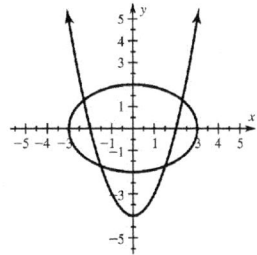

Figure 7

9. The solutions are the intersection points $(1, 3)$ and $(-1, -3)$. Both solutions check.

11. The solutions are the intersection points $(0, -1)$ and $(0, 1)$. Both solutions check.

13. Substitute $y = 2x$ into the second equation.

$$x^2 + (2x)^2 = 45 \Rightarrow x^2 + 4x^2 = 45 \Rightarrow 5x^2 = 45 \Rightarrow x^2 = 9 \Rightarrow x = \pm 3$$

When $x = -3$, $y = 2(-3) = -6$. When $x = 3$, $y = 2(3) = 6$. The solutions are $(-3, -6)$ and $(3, 6)$.

15. From the first equation, $y = 1 - x$. Substitute $y = 1 - x$ into the second equation.

$$x^2 - (1 - x)^2 = 3 \Rightarrow x^2 - 1 + 2x - x^2 = 3 \Rightarrow 2x - 1 = 3 \Rightarrow 2x = 4 \Rightarrow x = 2$$

When $x = 2$, $y = 1 - (2) = -1$ The solution is $(2, -1)$.

17. From the first equation, $x^2 = y$. Substitute $x^2 = y$ into the second equation.

$$y + y^2 = 6 \Rightarrow y^2 + y - 6 = 0 \Rightarrow (y + 3)(y - 2) = 0 \Rightarrow \text{Either } y = -3 \text{ or } y = 2$$

When $y = -3$, $x = \pm\sqrt{-3}$, (not real numbers). When $y = 2$, $x = \pm\sqrt{2}$.

The solutions are $\left(-\sqrt{2}, 2\right)$ and $\left(\sqrt{2}, 2\right)$.

19. Solve the second equation for $x : x = y - 2$. Substitute $x = y - 2$ into the first equation:

$$3(y - 2)^2 + 2y^2 = 5 \Rightarrow 3(y^2 - 4y + 4) + 2y^2 = 5$$

$$\Rightarrow 3y^2 - 12y + 12 + 2y^2 = 5 \Rightarrow 5y^2 - 12y + 7 = 0 \Rightarrow (5y - 7)(y - 1) = 0$$

$$\Rightarrow y = \frac{7}{5} \text{ or } y = 1 \Rightarrow x = -\frac{3}{5} \text{ or } x = -1. \text{ The solutions are } \left(-\frac{3}{5}, \frac{7}{5}\right) \text{ and } (-1, 1).$$

21. Multiply the second equation by $-1$ and add to the first equation.

$$\begin{array}{l} x^2 + y^2 = 4 \\ \underline{-x^2 + 9y^2 = -9} \\ \phantom{xxxx}10y^2 = -5 \end{array}$$

$y^2 = -\dfrac{1}{2} \Rightarrow y = \pm\sqrt{-\dfrac{1}{2}}$, which is not a real number. There are no real solutions.

23. Add the equations.

$$\begin{array}{l} x^2 + y^2 = 10 \\ \underline{2x^2 - y^2 = 17} \Rightarrow x^2 = 9 \Rightarrow x = \pm 3 \\ 3x^2 \phantom{xx} = 27 \end{array}$$

$(3)^2 + y^2 = 10 \Rightarrow y^2 = 1 \Rightarrow y = \pm 1$

$(-3)^2 + y^2 = 10 \Rightarrow y^2 = 1 \Rightarrow y = \pm 1$  The solutions are $(\pm 3, \pm 1)$

25. Solve the second equation for $y$.  $2x^2 - y = 1 - 3x \Rightarrow y = 2x^2 + 3x - 1$

Graph $Y_1 = X^2 - 3$ and $Y_2 = 2X^2 + 3X - 1$ in $[-5, 5, 1]$ by $[-5, 5, 1]$.  See Figures 25a & 25b.

The solutions are the intersection points $(-2, 1)$ and $(-1, -2)$.

$[-5, 5, 1]$ by $[-5, 5, 1]$ \qquad\qquad $[-5, 5, 1]$ by $[-5, 5, 1]$

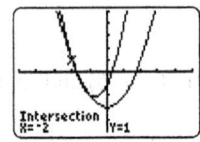  \qquad\qquad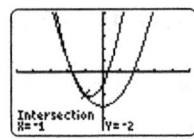

Figure 25a \qquad\qquad\qquad\qquad Figure 25b

27. Solve both equations for $y$.  $y - x = -4 \Rightarrow y = x - 4$ and $x - y^2 = -2 \Rightarrow y^2 = x + 2 \Rightarrow y = \pm\sqrt{x+2}$

Graph $Y_1 = X - 4$, $Y_2 = \sqrt{(X+2)}$ and $Y_3 = -\sqrt{(X+2)}$ in $[-9.4, 9.4, 1]$ by $[-6.2, 6.2, 1]$.

See Figures 27a & 27b.  The solutions are the intersection points $(7, 3)$ and $(2, -2)$.

$[-9.4, 9.4, 1]$ by $[-6.2, 6.2, 1]$ \qquad $[-9.4, 9.4, 1]$ by $[-6.2, 6.2, 1]$

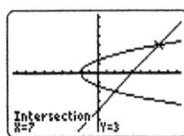  \qquad\qquad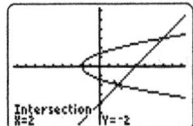

Figure 27a \qquad\qquad\qquad\qquad Figure 27b

29. (a) Substitute $y = -2x$ into the second equation.

$x^2 + (-2x) = 3 \Rightarrow x^2 - 2x - 3 = 0 \Rightarrow (x+1)(x-3) = 0 \Rightarrow$ Either $x = -1$ or $x = 3$

When $x = -1$, $y = -2(-1) = 2$.  When $x = 3$, $y = -2(3) = -6$.

The solutions are $(-1, 2)$ and $(3, -6)$.

(b) Solve the second equation for $y$.  $x^2 + y = 3 \Rightarrow y = 3 - x^2$  Graph $Y_1 = -2X$ and $Y_2 = 3 - X^2$ in

$[-10, 10, 1]$ by $[-10, 10, 1]$.  See Figures 29a & 29b.

(c)　Table $Y_1 = -2X$ and $Y_2 = 3 - X^2$ with TblStart $= -3$ and $\Delta$Tbl $= 1$.  See Figure 29c.

[–10, 10, 1] by [–10, 10, 1]　　[–10, 10, 1] by [–10, 10, 1]

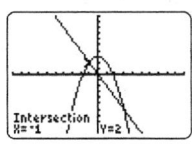　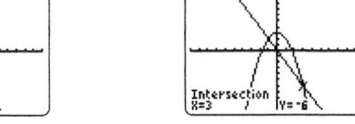　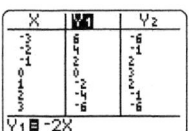

Figure 29a　　　　　　　　　Figure 29b　　　　　　　　　Figure 29c

31.　(a)　From the second equation, $y = x$.  Substitute $y = x$ into the first equation.

$x \cdot x = 1 \Rightarrow x^2 = 1 \Rightarrow x = \pm 1$　When $x = -1$, $y = -1$.  When $x = 1$, $y = 1$.

The solutions are (–1, –1) and (1, 1).

(b)　Solve both equations for $y$.　$xy = 1 \Rightarrow y = \dfrac{1}{x}$ and $x - y = 0 \Rightarrow y = x$

Graph $Y_1 = 1/X$ and $Y_2 = X$ in [–4.7, 4.7, 1] by [–3.1, 3.1, 1].  See Figures 31a & 31b.

(c)　Table $Y_1 = 1/X$ and $Y_2 = X$ with TblStart $= -3$ and $\Delta$Tbl $= 1$.  See Figure 31c.

[–4.7, 4.7, 1] by [–3.1, 3.1, 1]　　　　　[–4.7, 4.7, 1] by [–3.1, 3.1, 1]

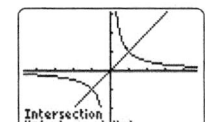　　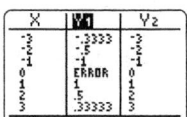

Figure 31a　　　　　　　　　Figure 31b　　　　　　　　　Figure 31c

33.　Sketch the parabola given by $y = x^2$ using a solid line.  Try test point (0, 2) to determine shading.

Since $2 \geq 0^2$, we shade the portion of the $xy$-plane containing the point (0, 2).  See Figure 33.

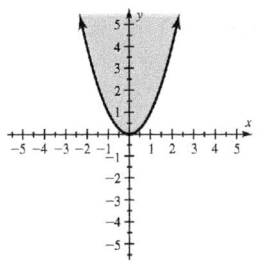　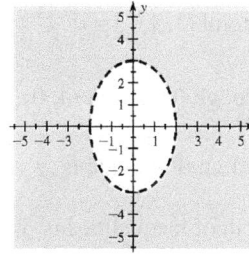　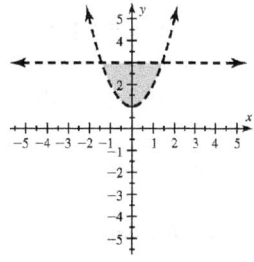

Figure 33　　　　　　　　　Figure 35　　　　　　　　　Figure 37

35.　Sketch the ellipse given by $\dfrac{x^2}{4} + \dfrac{y^2}{9} = 1$ using a dashed line.  Try test point (4, 0) to determine

shading.  Since $\dfrac{4^2}{4} + \dfrac{0^2}{9} = 4 > 1$, we shade the portion of the $xy$-plane containing the point (4, 0).

See Figure 35.

37.　Sketch the parabola given by $y = x^2 + 1$ using a dashed line and the line given by $y = 3$ using a

dashed line.  Since the point (0, 2) satisfies both inequalities, we shade the portion of the $xy$-plane

containing the point (0, 2).  See Figure 37.  One solution is (0, 2).  *Answers may vary.*

39. Sketch the circle given by $x^2 + y^2 = 1$ using a solid line and the line given by $y = x$ using a dashed

line. Since the point $\left(\dfrac{1}{2}, -\dfrac{1}{2}\right)$ satisfies both inequalities, we shade the portion of the $xy$-plane

containing the point $\left(\dfrac{1}{2}, -\dfrac{1}{2}\right)$. See Figure 39. One solution is (0.5, –0.5). *Answers may vary.*

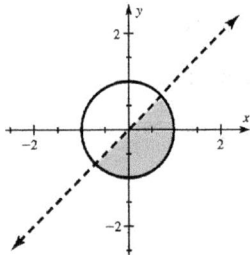    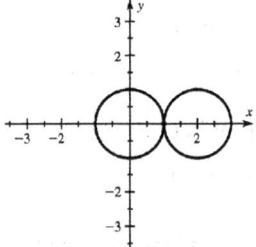    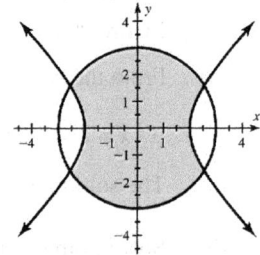

Figure 39                              Figure 41                              Figure 43

41. Sketch the circle given by $x^2 + y^2 = 1$ using a solid line and the circle given by $(x - 2)^2 + y^2 = 1$

using a solid line. Since the point (1, 0) is the only point that satisfies both inequalities, we mark

only the point (1, 0) on the graph. See Figure 41. The only solution is (1, 0).

43. Sketch the hyperbola given by $x^2 - y^2 = 4$ using a solid line and the circle given by $x^2 + y^2 = 9$

using a solid line. Since the point (0, 0) satisfies both inequalities, we shade the portion of the $xy$

plane containing the point (0, 0). See Figure 43. One solution is (0, 0). *Answers may vary.*

45. a

47. A parabola with vertex (0, 0) has an equation of the form $y = a(x - 0)^2 + 0$ or $y = ax^2$. Since the

parabola also passes through the point (1, 1), $1 = a \cdot 1^2 \Rightarrow a = 1$. The equation of the parabola is

$y = x^2$. Since the line passes through (0, 4) and (4, 0), its slope is $m = \dfrac{0 - 4}{4 - 0} = \dfrac{-4}{4} = -1$.

The $y$-intercept is (0, 4). The equation of the line is $y = -x + 4$ or $y = 4 - x$. Since the line is dashed

and the parabola is solid, the system of inequalities is $y \geq x^2$ and $y < 4 - x$.

49. See Figure 49.

[–10, 10, 1] by [–10, 10, 1]          [0, 5, 1] by [0, 10, 1]

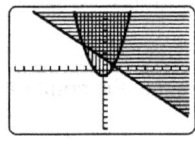          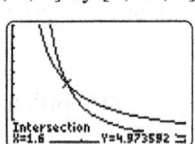

Figure 49                              Figure 51

51. The necessary equations are $\pi r^2 h = 40$ and $2\pi rh = 50$. Start by solving each of these equations

for $h$. $\pi r^2 h = 40 \Rightarrow h = \dfrac{40}{\pi r^2}$ and $2\pi rh = 50 \Rightarrow h = \dfrac{50}{2\pi r}$

(a) Graph $Y_1 = 40/(\pi X^2)$ and $Y_2 = 50/(2\pi X)$ in [0, 5, 1] by [0, 10, 1]. See Figure 51.

The graphs intersect near the point (1.6, 4.97). The dimensions are

$r = 1.6$ inches and $h \approx 4.97$ inches.

(b) $\dfrac{40}{\pi r^2} = \dfrac{50}{2\pi r} \Rightarrow 80\pi r = 50\pi r^2 \Rightarrow 8r = 5r^2 \Rightarrow 5r^2 - 8r = 0 \Rightarrow r(5r - 8) = 0 \Rightarrow r = 0$ or $r = \dfrac{8}{5}$

Since $r = 0$ has no physical meaning, $r = \dfrac{8}{5} = 1.6$. And so $h = \dfrac{50}{2\pi\left(\frac{8}{5}\right)} = \dfrac{125}{8\pi} \approx 4.97$.

The dimensions are $r = 1.6$ inches and $h \approx 4.97$ inches.

53. (a) $xy = 143$ and $2x + 2y = 48$

(b) Graph $Y_1 = 143/X$ and $Y_2 = 24 - X$ in [0, 20, 4] by [0, 20, 4]. See Figure 53.

The dimensions of the room are $x = 11$ ft and $y = 13$ ft.

[0, 20, 4] by [0, 20, 4]

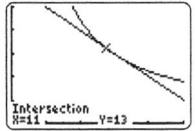

Figure 53

## Checking Basic Concepts Section 13.3

1. Symbolically: From the second equation $y = 2x - 3$. Substitute $y = 2x - 3$ into the first equation.

$x^2 - (2x - 3) = 2x \Rightarrow x^2 - 4x + 3 = 0 \Rightarrow (x - 1)(x - 3) = 0 \Rightarrow x = 1$ or $x = 3$

When $x = 1$, $y = 2(1) - 3 = -1$. When $x = 3$, $y = 2(3) - 3 = 3$. The solutions are $(1, -1)$ and $(3, 3)$.

Graphically: Solve each equation for $y$. $x^2 - y = 2x \Rightarrow y = x^2 - 2x$ and $2x - y = 3 \Rightarrow y = 2x - 3$

Graph $Y_1 = X^2 - 2X$ and $Y_2 = 2X - 3$ in [-5, 5, 1] by [-5, 5, 1]. See Figures 1a & 1b.

[-5, 5, 1] by [-5, 5, 1]    [-5, 5, 1] by [-5, 5, 1]

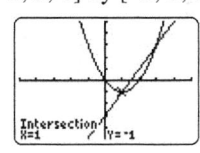   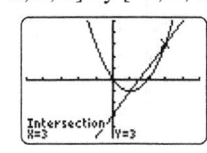

Figure 1a               Figure 1b

2. Since graphing these equations would result in a parabola intersected twice by a line, there are two solutions.

3. (a) The point (0, 3) is in the shaded region and is a solution. The point (4, 4) is not in the shaded region and is not a solution. *Answers may vary*

(b)   A parabola with vertex (0, 4) has an equation of the form

$y = a(x-0)^2 + 4$ or $y = ax^2 + 4$.  Since the parabola also passes through the point (2, 0),

$0 = a \cdot 2^2 + 4 \Rightarrow 4a = -4 \Rightarrow a = -1$.  The equation of the parabola is

$y = -x^2 + 4$ or $y = 4 - x^2$.

Since the line passes through (0, 2) and (2, 0), its slope is   $m = \dfrac{0-2}{2-0} = \dfrac{-2}{2} = -1$.  The $y$-

intercept is (0, 2).  The equation of the line is  $y = -x + 2$ or $y = 2 - x$.  Since both the line and

the parabola are solid, the system of inequalities is  $y \le 4 - x^2$ and $y \ge 2 - x$.

4.  Sketch the circle given by $x^2 + y^2 = 4$ using a solid line and the line given by  $y = 1$ using a dashed
line.  The point (0, 0) satisfies both inequalities.  Shade the portion of the $xy$-plane containing
(0, 0).  See Figure 4.

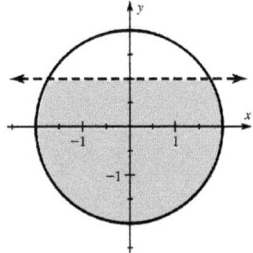

Figure 4

## Chapter 13 Review

1.  Since  $x = 2(y+0)^2 + 0$, the vertex is (0, 0) and the axis of symmetry is $y = 0$.  See Figure 1.

2.  Since  $x = -(y+1)^2 + 0$, the vertex is (0, –1) and the axis of symmetry is $y = -1$.  See Figure 2.

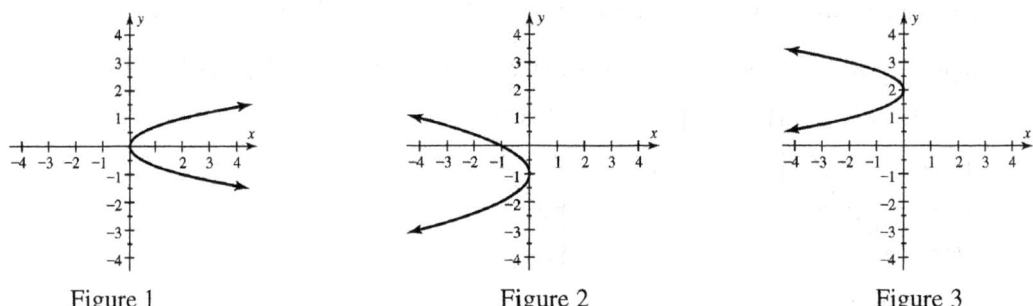

Figure 1                           Figure 2                           Figure 3

3.  Since  $x = -2(y-2)^2 + 0$, the vertex is (0, 2) and the axis of symmetry is $y = 2$.  See Figure 3.

4.  Since  $x = (y+2)^2 - 1$, the vertex is (–1, –2) and the axis of symmetry is $y = -2$.  See Figure 4.

5.  Since  $x = -3(y-0)^2 + 1$, the vertex is (1, 0) and the axis of symmetry is $y = 0$.  See Figure 5.

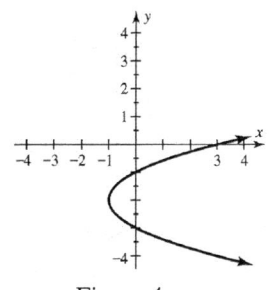

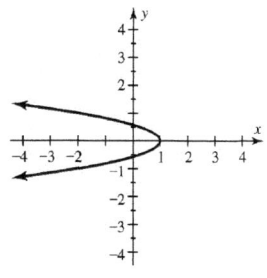

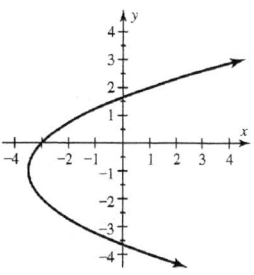

Figure 4                                    Figure 5                                    Figure 6

6. Since $x = \frac{1}{2}(y+1)^2 - \frac{7}{2}$, the vertex is $\left(-\frac{7}{2}, -1\right)$ and the axis of symmetry is $y = -1$. See Figure 6.

7. Since the vertex is (0, 0), the parabola has an equation of the form $x = a(x-0)^2 + 0$ or $x = ay^2$.

   Since the parabola passes through the point (1, 1), $1 = a(1) \Rightarrow a = 1$. The equation is $x = y^2$.

8. This is a circle of radius 4 centered at (–2, 2). The equation is $(x+2)^2 + (y-2)^2 = 16$.

9. $(x-0)^2 + (y-0)^2 = 1^2 \Rightarrow x^2 + y^2 = 1$

10. $(x-2)^2 + (y-(-3))^2 = 4^2 \Rightarrow (x-2)^2 + (y+3)^2 = 16$

11. The radius is 5 and the center is (0, 0). Solving the equation for $y$ results in $y = \pm\sqrt{25 - x^2}$.

    See Figure 11.

12. The radius is 3 and the center is (2, 0). Solving the equation for $y$ results in $y = \pm\sqrt{9 - (x-2)^2}$.

    See Figure 12.

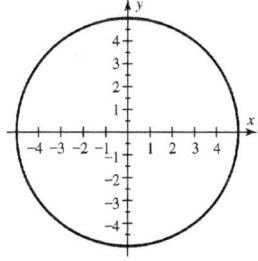

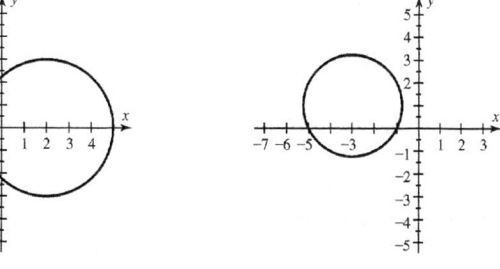

Figure 11                                   Figure 12                                   Figure 13

13. The radius is $\sqrt{5}$ and the center is (–3, 1). Solving the equation for $y$ results in $y = 1 \pm \sqrt{5 - (x+3)^2}$.

    See Figure 13.

14. $x^2 - 2x + y^2 + 2y = 7 \Rightarrow x^2 - 2x + 1 + y^2 + 2y + 1 = 7 + 1 + 1 \Rightarrow (x-1)^2 + (y+1)^2 = 9$ The radius is 3

    and the center is (1, –1). Solving the equation for $y$ results in $y = -1 \pm \sqrt{9 - (x-1)^2}$. See Figure 14.

15. The ellipse has a vertical major axis with vertices $(0, \pm 5)$ and minor axis endpoints $(\pm 2, 0)$.

    See Figure 15.

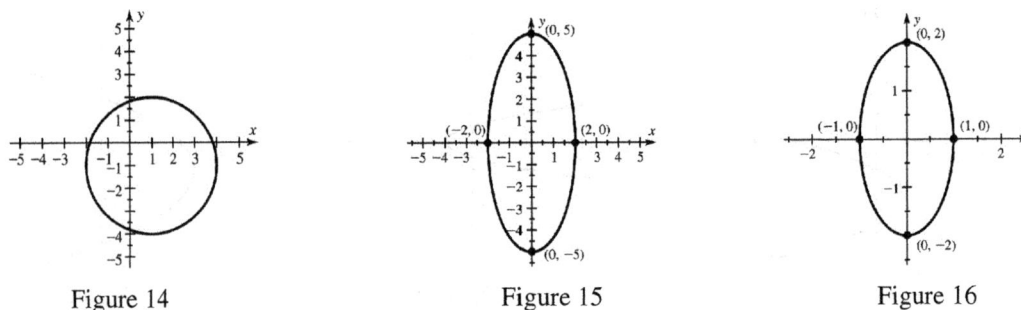

Figure 14                    Figure 15                    Figure 16

16. The ellipse has a vertical major axis with vertices $(0, \pm 2)$ and minor axis endpoints $(\pm 1, 0)$.

    See Figure 16.

17. $25x^2 + 20y^2 = 500 \Rightarrow \dfrac{25x^2}{500} + \dfrac{20y^2}{500} = 1 \Rightarrow \dfrac{x^2}{20} + \dfrac{y^2}{25} = 1$  The ellipse has a vertical major axis with

    vertices $(0, \pm 5)$ and minor axis endpoints $\left(\pm\sqrt{20}, 0\right)$.  See Figure 17.

18. $4x^2 + 9y^2 = 36 \Rightarrow \dfrac{4x^2}{36} + \dfrac{9y^2}{36} = 1 \Rightarrow \dfrac{x^2}{9} + \dfrac{y^2}{4} = 1$  The ellipse has a horizontal major axis with

    vertices $(\pm 3, 0)$ and minor axis endpoints $(0, \pm 2)$.  See Figure 18.

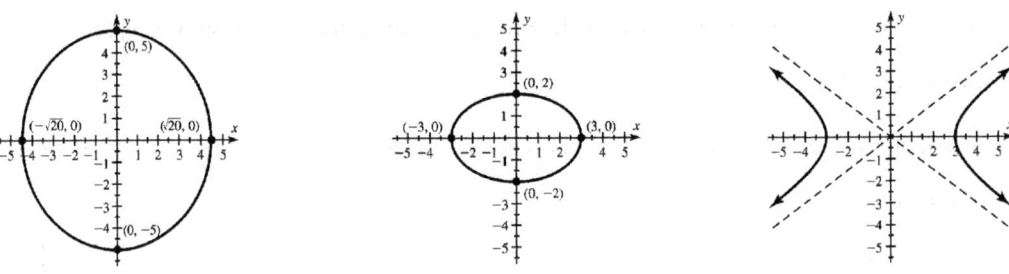

Figure 17                    Figure 18                    Figure 21

19. Vertical major axis with vertices $(0, \pm 4)$ and minor axis endpoints $(\pm 2, 0)$  $\Rightarrow \dfrac{y^2}{16} + \dfrac{x^2}{4} = 1$

20. Horizontal transverse axis with vertices $(\pm 1, 0)$ and asymptotes $y = \pm 2x$  $\Rightarrow x^2 - \dfrac{y^2}{4} = 1$

21. The hyperbola has a horizontal transverse axis with vertices $(\pm 3, 0)$ and asymptotes $y = \pm\dfrac{2}{3}x$.

    See Figure 21.

22. The hyperbola has a vertical transverse axis with vertices $(0, \pm 5)$ and asymptotes $y = \pm\dfrac{5}{4}x$.

    See Figure 22.

23. The hyperbola has a vertical transverse axis with vertices $(0, \pm 1)$ and asymptotes $y = \pm x$.

    See Figure 23.

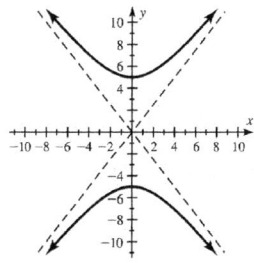

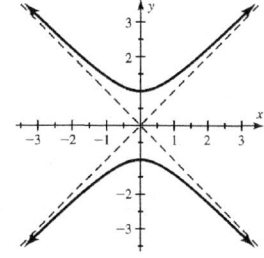

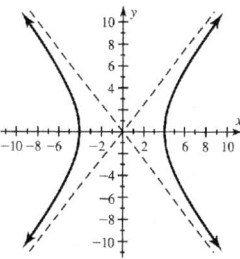

Figure 22                                   Figure 23                                   Figure 24

24. $25x^2 - 16y^2 = 400 \Rightarrow \dfrac{25x^2}{400} - \dfrac{16y^2}{400} = 1 \Rightarrow \dfrac{x^2}{16} - \dfrac{y^2}{25} = 1$  The hyperbola has a horizontal transverse

   axis with vertices $(\pm 4, 0)$ and asymptotes $y = \pm\dfrac{5}{4}x.$  See Figure 24.

25. The solutions are the intersection points $(0, 3)$ and $(3, 0)$.  Both solutions check.

26. The solutions are the intersection points $(-1, -2)$ and $(1, 2)$.  Both solutions check.

27. The solutions are the intersection points $(0, 0)$ and $(2, 2)$.  Both solutions check.

28. The solutions are the intersection points $(-2, -1)$, $(-2, 1)$, $(2, -1)$ and $(2, 1)$.  All four solutions check.

29. Substitute $y = x$ into the second equation.  $x^2 + (x)^2 = 32 \Rightarrow 2x^2 = 32 \Rightarrow x^2 = 16 \Rightarrow x = \pm 4$

   When $x = -4$, $y = -4$.  When $x = 4$, $y = 4$.  The solutions are $(-4, -4)$ and $(4, 4)$.

30. From the first equation, $y = x - 4$.  Substitute $y = x - 4$ into the second equation.

   $x^2 + (x - 4)^2 = 16 \Rightarrow x^2 + x^2 - 8x + 16 = 16 \Rightarrow 2x^2 - 8x = 0 \Rightarrow 2x(x - 4) = 0 \Rightarrow x = 0$ or $x = 4$

   When $x = 0$, $y = (0) - 4 = -4$.  When $x = 4$, $y = (4) - 4 = 0$.  The solutions are $(0, -4)$ and $(4, 0)$.

31. Substitute $y = x^2$ into the second equation.  $2x^2 + (x^2) = 3 \Rightarrow 3x^2 = 3 \Rightarrow x^2 = 1 \Rightarrow x = \pm 1$

   When $x = -1$, $y = (-1)^2 = 1$.  When $x = 1$, $y = (1)^2 = 1$.  The solutions are $(-1, 1)$ and $(1, 1)$.

32. Substitute $y = x^2 + 1$ into the second equation.

   $2x^2 - (x^2 + 1) = 3x - 3 \Rightarrow x^2 - 3x + 2 = 0 \Rightarrow (x - 1)(x - 2) = 0 \Rightarrow x = 1$ or $x = 2$

   When $x = 1$, $y = (1)^2 + 1 = 2$.  When $x = 2$, $y = (2)^2 + 1 = 5$.  The solutions are $(1, 2)$ and $(2, 5)$.

33. Solve both equations for $y$.  $2x - y = 4 \Rightarrow y = 2x - 4$ and $x^2 + y = 4 \Rightarrow y = 4 - x^2$

   Graph $Y_1 = 2X - 4$ and $Y_2 = 4 - X^2$ in $[-8, 8, 1]$ by $[-25, 10, 5]$.  See Figures 33a & 33b.

   The solutions are the intersection points $(-4, -12)$ and $(2, 0)$.

   $[-8, 8, 1]$ by $[-25, 10, 5]$        $[-8, 8, 1]$ by $[-25, 10, 5]$

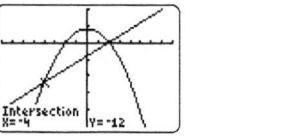

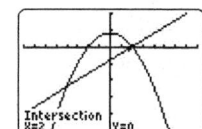

   Figure 33a                    Figure 33b

34. Solve both equations for $y$.  $x^2 + y = 0 \Rightarrow y = -x^2$ and $x^2 + y^2 = 2 \Rightarrow y = \pm\sqrt{2 - x^2}$

    Graph $Y_1 = -X^2$, $Y_2 = \sqrt{(2 - X^2)}$ and $Y_3 = -\sqrt{(2 - X^2)}$ in $[-4.7, 4.7, 1]$ by $[-4.1, 2.1, 1]$.

    See Figures 34a and 34b. The solutions are the intersection points $(-1, -1)$ and $(1, -1)$.

    $[-4.7, 4.7, 1]$ by $[-4.1, 2.1, 1]$          $[-4.7, 4.7, 1]$ by $[-4.1, 2.1, 1]$

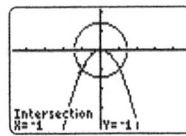          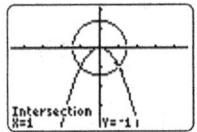

         Figure 34a                        Figure 34b

35. (a)    Substitute $y = x$ into the second equation.

    $$x^2 + 2(x) = 8 \Rightarrow x^2 + 2x - 8 = 0 \Rightarrow (x + 4)(x - 2) = 0 \Rightarrow x = -4 \text{ or } x = 2$$

    When $x = -4$, $y = -4$.  When $x = 2$, $y = 2$.  The solutions are $(-4, -4)$ and $(2, 2)$.

    (b)    Solve the second equation for $y$.  $x^2 + 2y = 8 \Rightarrow y = 4 - \dfrac{x^2}{2}$

    Graph $Y_1 = X$ and $Y_2 = 4 - (X^2/2)$ in $[-10, 10, 1]$ by $[-10, 10, 1]$.  See Figures 35a & 35b.

    (c)    Table $Y_1 = X$ and $Y_2 = 4 - (X^2/2)$ with TblStart $= -2$ and $\Delta$Tbl $= 1$.  See Figure 35c.

    $[-10, 10, 1]$ by $[-10, 10, 1]$          $[-10, 10, 1]$ by $[-10, 10, 1]$

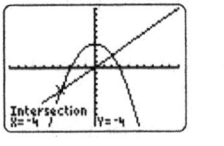          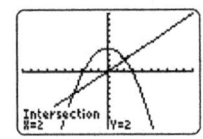          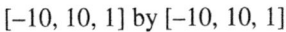

         Figure 35a                        Figure 35b                        Figure 35c

36. (a)    Substitute $y = x^3$ into the second equation.  $x^2 - x^3 = 0 \Rightarrow x^2(1 - x) = 0 \Rightarrow x = 0 \text{ or } x = 1$

    When $x = 0$, $y = 0^3 = 0$.  When $x = 1$, $y = 1^3 = 1$.  The solutions are $(0, 0)$ and $(1, 1)$.

    (b)    Solve the second equation for $y$.  $x^2 - y = 0 \Rightarrow y = x^2$

    Graph $Y_1 = X \wedge 3$ and $Y_2 = X^2$ in $[-1.5, 1.5, 1]$ by $[-1.5, 1.5, 1]$.  See Figures 36a & 36b.

    (c)    Table $Y_1 = X \wedge 3$ and $Y_2 = X^2$ with TblStart $= -1$ and $\Delta$Tbl $= 0.5$.  See Figure 36c.

    $[-1.5, 1.5, 1]$ by $[-1.5, 1.5, 1]$          $[-1.5, 1.5, 1]$ by $[-1.5, 1.5, 1]$

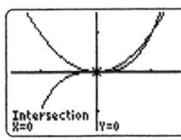          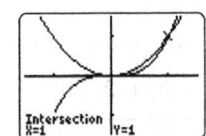

         Figure 36a                        Figure 36b                        Figure 36c

37. Sketch the parabola given by $y = 2x^2$ using a solid line. Try test point (0, 2) to determine shading.

   Since $2 \geq 2(0^2)$, we shade the portion of the $xy$-plane containing the point (0, 2). See Figure 37.

38. Sketch the line given by $y = 2x - 3$ using a dashed line. Try test point (3, 0) to determine shading.

   Since $0 < 2(3) - 3$, we shade the portion of the $xy$-plane containing the point (3, 0). See Figure 38.

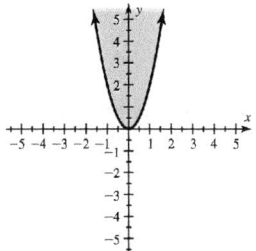

Figure 37

Figure 38

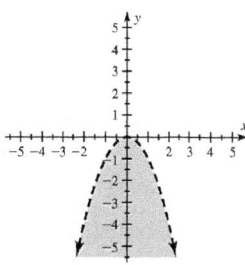

Figure 39

39. Sketch the parabola given by $y = -x^2$ using a dashed line. Try test point (0, –1) to determine

   shading. Since $-1 < -(0)^2$, we shade the portion of the $xy$-plane containing the point (0, –1).

   See Figure 39.

40. Sketch the ellipse given by $\dfrac{x^2}{9} + \dfrac{y^2}{16} = 1$ using a solid line. Try test point (0, 0) to determine shading.

   Since $0 + 0 \leq 1$, we shade the portion of the $xy$-plane containing the point (0, 0). See Figure 40.

41. Sketch the parabola given by $y = x^2 + 1$ using a solid line and the line given by $y = 2$ using a solid

   line. Since the point (0, 1.5) satisfies both inequalities, we shade the portion of the $xy$-plane

   containing the point (0, 1.5). See Figure 41.

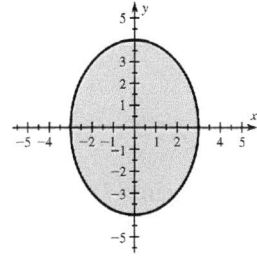

Figure 40

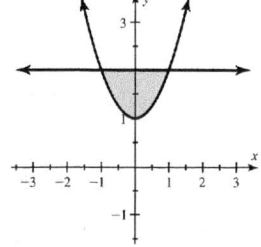

Figure 41

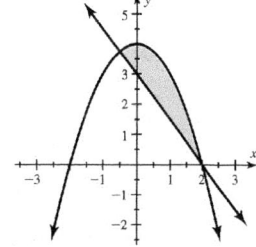

Figure 42

42. Sketch the parabola given by $y = 4 - x^2$ using a solid line and the line given by $3x + 2y = 6$ using a

   solid line. Since the point (1, 2) satisfies both inequalities, we shade the portion of the $xy$-plane

   containing the point (1, 2). See Figure 42.

43. Sketch the parabola given by $y = x^2$ using a dashed line and the parabola given by $y = 4 - x^2$ using

   a dashed line. Since the point (0, 2) satisfies both inequalities, we shade the portion of the $xy$-plane

   containing the point (0, 2). See Figure 43.

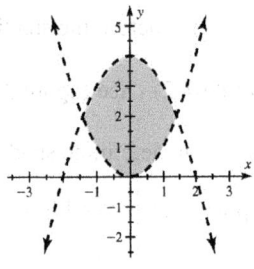

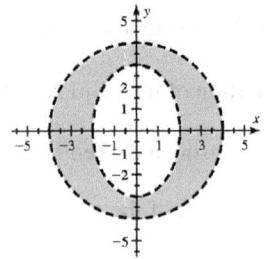

Figure 43                                   Figure 44

44. Sketch the ellipse given by $\dfrac{x^2}{4} + \dfrac{y^2}{9} = 1$ using a dashed line and the circle given by $x^2 + y^2 = 16$

   using a dashed line. Since the point (3, 0) satisfies both inequalities, we shade the portion of the $xy$

   plane containing the point (3, 0). See Figure 44.

45. A parabola with vertex (0, –2) has an equation of the form $y = a(x-0)^2 - 2$ or $y = ax^2 - 2$. Since

   the parabola also passes through the point (1, –1), $-1 = a \cdot 1^2 - 2 \Rightarrow a = 1$. The equation of the

   parabola is $y = x^2 - 2$. Since the line passes through (0, 2) and (2, 0), its slope is

   $m = \dfrac{0-2}{2-0} = \dfrac{-2}{2} = -1$. The $y$-intercept is (0, 2). The equation of the line is $y = -x + 2$ or $y = 2 - x$.

   Since both the line and the parabola are solid, the system of inequalities is $y \geq x^2 - 2$ and $y \leq 2 - x$.

46. A circle of radius 2 centered at (0, 0) has the equation $x^2 + y^2 = 4$. Since the line passes through the

   points (0, 0) and (1, 1), its slope is $m = \dfrac{1-0}{1-0} = \dfrac{1}{1} = 1$. The $y$-intercept is (0, 0). The equation of the

   line is $y = x$. Since both the circle and the line are solid, the system of inequalities is

   $y \geq x$ and $x^2 + y^2 \leq 4$.

47. (a)   $xy = 1000$ and $2x + 2y = 130$

   (b)   Solve each equation for $y$.   $xy = 1000 \Rightarrow y = \dfrac{1000}{x}$ and $2x + 2y = 130 \Rightarrow y = 65 - x$

         Graph $Y_1 = 1000/X$ and $Y_2 = 65 - X$ in [0, 50, 10] by [0, 50, 10]. See Figure 47.

         The table has dimensions $x = 25$ inches and $y = 40$ inches.

   (c)   From the second equation, $y = 65 - x$. Substitute $y = 65 - x$ into the first equation.

         $x(65 - x) = 1000 \Rightarrow x^2 - 65x + 1000 = 0 \Rightarrow (x - 25)(x - 40) = 0 \Rightarrow x = 25$ or $x = 40$

         When $x = 25$, $y = 65 - 25 = 40$. When $x = 40$, $y = 65 - 40 = 25$.

         These answers are equivalent.

[0, 50, 10] by [0, 50, 10]          [0, 16, 4] by [0, 16, 4]          [0, 5, 1] by [0, 50, 10]

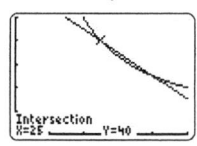          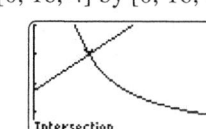          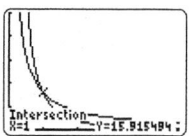

Figure 47                          Figure 48                          Figure 49

48. (a)    $xy = 60$ and $y - x = 7$

(b)    Solve each equation for $y$.  $xy = 60 \Rightarrow y = \dfrac{60}{x}$ and $y - x = 7 \Rightarrow y = x + 7$

Graph $Y_1 = 60/X$ and $Y_2 = X + 7$ in [0, 16, 4] by [0, 16, 4].  See Figure 48.

The numbers are $x = 5$ and $y = 12$.

(c)    From the second equation, $y = x + 7$.  Substitute $y = x + 7$ into the first equation.

$x(x + 7) = 60 \Rightarrow x^2 + 7x - 60 = 0 \Rightarrow (x + 12)(x - 5) = 0 \Rightarrow x = -12$ or $x = 5$

The only positive value is .  $x = 5$.  When $x = 5$, $y = 5 + 7 = 12$.

49. Solve each equation for $h$.  $V = \pi r^2 h \Rightarrow h = \dfrac{V}{\pi r^2}$ and $A = 2\pi rh \Rightarrow h = \dfrac{A}{2\pi r}$

Graph $Y_1 = 50/\left(\pi X^2\right)$ and $Y_2 = 100/(2\pi X)$ in [0, 5, 1] by [0, 50, 10].  See Figure 49.

The unique answer is $r = 1$ foot and $h \approx 15.92$ feet.

50. Solve each equation for $h$.  $V = \pi r^2 h \Rightarrow h = \dfrac{V}{\pi r^2}$ and $A = 2\pi rh + 2\pi r^2 \Rightarrow h = \dfrac{A - 2\pi r^2}{2\pi r}$

Graph $Y_1 = 35/\left(\pi X^2\right)$ and $Y_2 = \left(80 - 2\pi X^2\right)/(2\pi X)$ in [0, 4, 1] by [-5, 25, 5].

See Figures 50a &50b.  Two solutions are possible.  The answer is not unique.

Either $r \approx 0.94$ inches and $h \approx 12.60$ inches or $r \approx 3.00$ inches and $h \approx 1.23$ inches.

[0, 4, 1] by [-5, 25, 5]    [0, 4, 1] by [-5, 25, 5]    [-7.5, 7.5, 1] by [-5, 5, 1]    [-3, 3, 1] by [-2, 2, 1]

    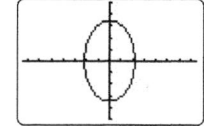    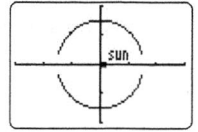

Figure 50a                  Figure 50b                  Figure 51                  Figure 52

51. (a)    $\dfrac{x^2}{5} + \dfrac{y^2}{12} = 1 \Rightarrow \dfrac{y^2}{12} = 1 - \dfrac{x^2}{5} \Rightarrow y^2 = 12\left(1 - \dfrac{x^2}{5}\right) \Rightarrow y = \pm\sqrt{12\left(1 - \dfrac{x^2}{5}\right)}$

Graph $Y_1 = \sqrt{\left(12\left(1 - X^2/5\right)\right)}$ and $Y_2 = -\sqrt{\left(12\left(1 - X^2/5\right)\right)}$ in [-7.5, 7.5, 1] by [-5, 5, 1].

See Figure 51.

(b)    $A = \pi\left(\sqrt{12}\right)\left(\sqrt{5}\right) \approx 24.33$ square units and $P = 2\pi\sqrt{\dfrac{5 + 12}{2}} \approx 18.32$ units

52. (a) $\dfrac{x^2}{1.524^2}+\dfrac{y^2}{1.517^2}=1 \Rightarrow \dfrac{y^2}{1.517^2}=1-\dfrac{x^2}{1.524^2} \Rightarrow y=\pm\sqrt{1.517^2\left(1-\dfrac{x^2}{1.524^2}\right)}$

Plot the point (0.15, 0) and Graph

$Y_1 = \sqrt{\left(1.517^2\left(1-X^2/1.524^2\right)\right)}$ and $Y_2 = -\sqrt{\left(1.517^2\left(1-X^2/1.524^2\right)\right)}$ in [–3, 3, 1] by

[–2, 2, 1]. See Figure 52.

(b)   $P = 2\pi\sqrt{\dfrac{1.524^2+1.517^2}{2}} \approx 9.55$ A.U. or about $8.9\times10^8$ miles

$A = \pi(1.524)(1.517) \approx 7.26$ square A.U. or about $6.3\times10^{16}$ square miles

## Chapter 13 Test

1. Since $y=-(x-1)^2+2$, the vertex is (1, 2) and the axis of symmetry is $x=1$. See Figure 1.

2. Since $x=(y-4)^2-2$, the vertex is (–2, 4) and the axis of symmetry is $y=4$. See Figure 2.

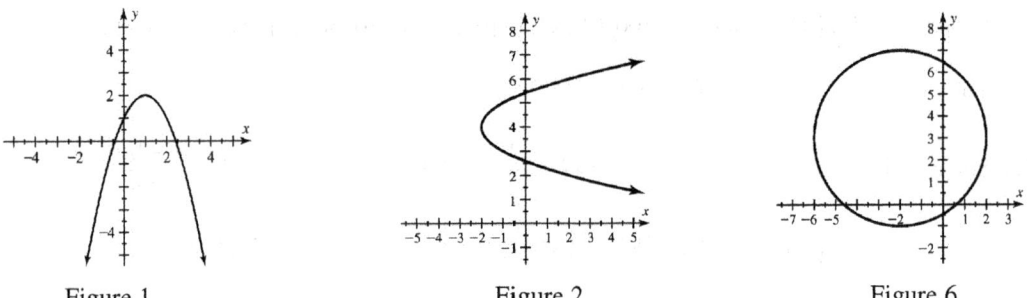

Figure 1                                        Figure 2                                        Figure 6

3. Since the parabola opens to the left and the vertex is (1, 0), the equation has the form

   $x=a(y-0)^2+1$. Since the parabola passes through (0, 1), $0=a(1-0)^2+1 \Rightarrow a=-1$.

   The equation is $x=-y^2+1$.

4. Since the center is (2, –4) and the radius is 2, the equation is $(x-2)^2+(y+4)^2=4$.

5. $\left(x-(-5)\right)^2+(y-2)^2=10^2 \Rightarrow (x+5)^2+(y-2)^2=100$

6. $x^2+4x+y^2-6y=3 \Rightarrow x^2+4x+4+y^2-6y+9=3+4+9 \Rightarrow (x+2)^2+(y-3)^2=16$

   The radius is 4 and the center is (–2, 3). Solving the equation for $y$ results in $y=3\pm\sqrt{16-(x+2)^2}$.

   See Figure 6.

7. The ellipse has a vertical major axis with vertices $(0,\pm7)$ and minor axis endpoints $(\pm4,0)$.

   See Figure 7.

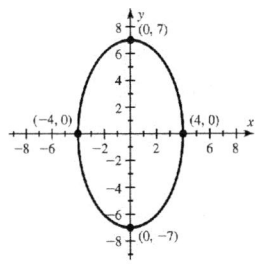

Figure 7

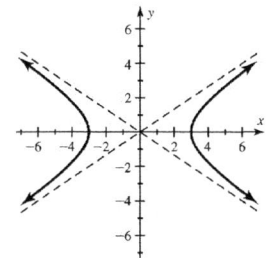

Figure 9

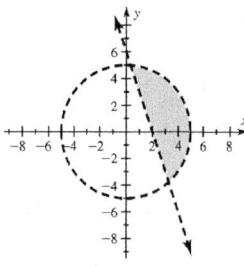

Figure 13

8. Horizontal major axis with vertices $(\pm 10, 0)$ and minor axis endpoints $(0, \pm 8)$ $\Rightarrow \dfrac{x^2}{100} + \dfrac{y^2}{64} = 1$

9. $4x^2 - 9y^2 = 36 \Rightarrow \dfrac{4x^2}{36} - \dfrac{9y^2}{36} = 1 \Rightarrow \dfrac{x^2}{9} - \dfrac{y^2}{4} = 1$ The hyperbola has a horizontal transverse axis with

   vertices $(\pm 3, 0)$ and asymptotes $y = \pm \dfrac{2}{3} x.$ See Figure 9

10. The solutions are the intersection points $(0, -4)$ and $(4, 0)$. Both solutions check.

11. From the first equation, $y = x - 3$. Substitute $y = x - 3$ into the second equation.

    $x^2 + (x-3)^2 = 17 \Rightarrow x^2 + x^2 - 6x + 9 = 17 \Rightarrow 2(x-4)(x+1) = 0 \Rightarrow x = -1 \text{ or } x = 4$

    When $x = -1$, $y = (-1) - 3 = -4$. When $x = 4$, $y = (4) - 3 = 1$. The solutions are $(-1, -4)$ and $(4, 1)$.

12. Solve both equations for $y$. $2x^2 - y = 4 \Rightarrow y = 2x^2 - 4$ and $x^2 + y = 8 \Rightarrow y = 8 - x^2$

    Graph $Y_1 = 2X^2 - 4$, $Y_2 = 8 - X^2$ in $[-10, 10, 1]$ by $[-10, 10, 1]$. See Figures 12a & 12b.

    The solutions are the intersection points $(-2, 4)$ and $(2, 4)$.

    $[-10, 10, 1]$ by $[-10, 10, 1]$

    $[-10, 10, 1]$ by $[-10, 10, 1]$

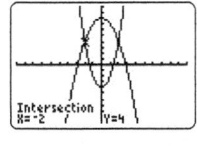

Figure 12a

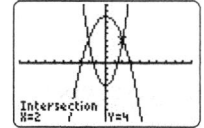

Figure 12b

13. Sketch the line given by $3x + y = 6$ using a dashed line and the circle given by $x^2 + y^2 = 25$ using a

    dashed line. Since the point $(4, 0)$ satisfies both inequalities, we shade the portion of the $xy$-plane

    containing the point $(4, 0)$. See Figure 13.

14. A parabola with vertex $(0, -4)$ has an equation of the form $y = a(x - 0)^2 - 4$ or $y = ax^2 - 4$. Since

    the parabola also passes through the point $(2, 0)$, $0 = a \cdot 2^2 - 4 \Rightarrow a = 1$. The equation of the

    parabola is $y = x^2 - 4$. The other parabola is a reflection of this parabola across the $x$-axis. Its

    equation is $y = 4 - x^2$. Since both the parabolas are solid, the system of inequalities is

    $y \le 4 - x^2$ and $y \ge x^2 - 4$.

15. (a)    $xy = 5000$ and $2x + 2y = 300$

    (b)    $\dfrac{5000}{x} = 150 - x \Rightarrow 5000 = 150x - x^2 \Rightarrow x^2 - 150x + 5000 = 0 \Rightarrow (x - 50)(x - 100) = 0 \Rightarrow$

    Either $x = 50$ or $x = 100$.   When $x = 50$, $y = 150 - 50 = 100$.

    When $x = 100$, $y = 150 - 100 = 50$.   Since the width is shorter than the length,

    $x = 50$ and $y = 100$.   The solution is 50 by 100 feet.

16. From the hint, the two equations to graph are $y = \dfrac{1183}{x^2}$ and $y = \dfrac{702 - x^2}{4x}$.   Figures not shown.   There

    are two possible solutions: Either $x \approx 22.08$ and $y \approx 2.43$ or $x \approx 7.29$ and $y \approx 22.24$.   The answer is

    not unique.

17. (a)    $\dfrac{x^2}{19.18^2} + \dfrac{y^2}{19.16^2} = 1 \Rightarrow \dfrac{y^2}{19.16^2} = 1 - \dfrac{x^2}{19.18^2} \Rightarrow y = \pm \sqrt{19.16^2 \left( 1 - \dfrac{x^2}{19.18^2} \right)}$

    Plot the point (0.9, 0) and graph

    $Y_1 = \sqrt{\left( 19.16^2 \left( 1 - X^2 / 19.18^2 \right) \right)}$ and $Y_2 = -\sqrt{\left( 19.16^2 \left( 1 - X^2 / 19.18^2 \right) \right)}$ in [–30, 30, 10] by

    [–20, 20, 10].   See Figure 17.

    (b)    The minimum distance occurs when $x = 19.18$ A.U.   The distance is $19.18 - 0.9 = 18.28$ A.U.

    This is about 1,700,040,000 miles.

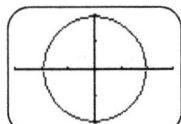

Figure 17

## Chapter 13 Extended and Discovery Exercises

1. (a)    $x^2 = 4y \Rightarrow x^2 = 4(1)y$.   Since $p = 1$, the focus is (0, 1).   See Figure 1a.

    (b)    $y^2 = -8x \Rightarrow y^2 = 4(-2)x$.   Since $p = -2$, the focus is (–2, 0).   See Figure 1b.

    (c)    $x = 2y^2 \Rightarrow y^2 = \dfrac{1}{2}x \Rightarrow y^2 = 4 \left( \dfrac{1}{8} \right)x$.   Since $p = \dfrac{1}{8}$, the focus is $\left( \dfrac{1}{8}, 0 \right)$.   See Figure 1c.

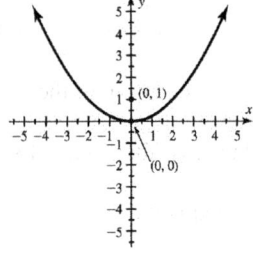

Figure 1a

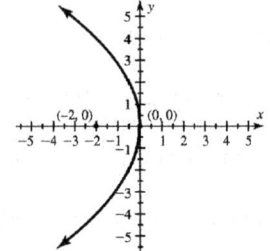

Figure 1b

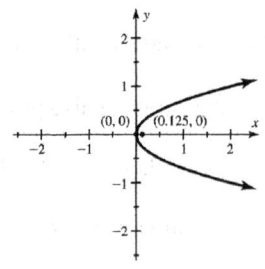

Figure 1c

2. (a)   Since the vertex of the parabola is $(0, 0)$, the cross section of the dish has an equation of the

form $y = ax^2$. Since the parabola passes through $(150, 44)$,

$$44 = a(150)^2 \Rightarrow 44 = 22,500a \Rightarrow a = \frac{44}{22,500} = \frac{11}{5625}. \text{ The equation is } y = \frac{11}{5625}x^2.$$

(b)   Writing this equation in the form $x^2 = 4py$ yields $x^2 = 4\left(\dfrac{5625}{44}\right)y.$ The focus is located at

$\left(0, \dfrac{5625}{44}\right).$ That is, the focus is $\dfrac{5625}{44} \approx 127.8$ feet from the vertex.

3. (a)   This is an ellipse with horizontal major axis, centered at $(3, 1)$. See Figure 3a.

(b)   This is an ellipse with vertical major axis, centered at $(-1, -2)$. See Figure 3b.

(c)   This is a hyperbola centered at $(-1, 3)$. The asymptotes are $y = \pm\dfrac{3}{2}(x+1)+3.$ See Figure 3c.

(d)   This is a hyperbola centered at $(-1, 4)$. The asymptotes are $y = \pm2(x+1)+4.$ See Figure 3d.

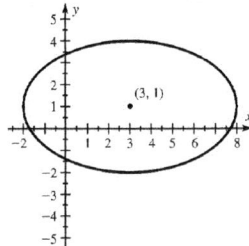
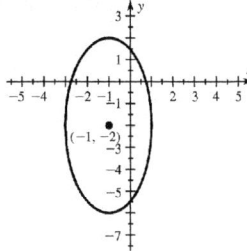
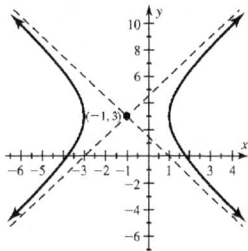

Figure 3a                    Figure 3b                    Figure 3c

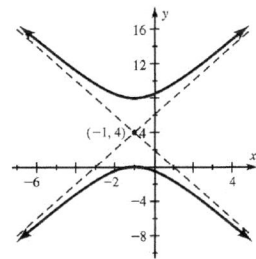

Figure 3d

4. (a)   $\dfrac{(x+3)^2}{16} + \dfrac{(y-5)^2}{4} = 1$

(b)   $\dfrac{(y+3)^2}{25} + \dfrac{(x-2)^2}{9} = 1$

5. (a)   $9x^2 - 18x + 4y^2 + 24y + 9 = 0 \Rightarrow 9\left(x^2 - 2x + 1\right) + 4\left(y^2 + 6y + 9\right) = -9 + 9 + 36 \Rightarrow$

$9(x-1)^2 + 4(y+3)^2 = 36 \Rightarrow \dfrac{(x-1)^2}{4} + \dfrac{(y+3)^2}{9} = 1 \Rightarrow \text{Center: } (1, -3)$

(b)   $25x^2 + 150x - 16y^2 + 32y - 191 = 0 \Rightarrow 25(x^2 + 6x + 9) - 16(y^2 - 2y + 1) = 191 + 225 - 16 \Rightarrow$

$25(x+3)^2 - 16(y-1)^2 = 400 \Rightarrow \dfrac{(x+3)^2}{16} - \dfrac{(y-1)^2}{25} = 1 \Rightarrow$ Center: $(-3, 1)$

## Chapters 1-13 Cumulative Review Exercises

1.  $K = (4)^2 + (-3)^2 = 16 + 9 = 25$

2.  $\dfrac{\left(a^{-2}b\right)^2}{a^{-1}\left(b^3\right)^{-2}} = \dfrac{a^{-4}b^2}{a^{-1}b^{-6}} = a^{-4-(-1)}b^{2-(-6)} = a^{-3}b^8 = \dfrac{b^8}{a^3}$

3.  $7.345 \times 10^{-3} = 0.007345$

4.  $f(-4) = \dfrac{-4}{-4-4} = \dfrac{-4}{-8} = \dfrac{1}{2}$; the denominator cannot equal zero, so $x - 4 \neq 0 \Rightarrow x \neq 4$.

5.  $f(3) = 4$

6.  See Figure 6.

7.  See Figure 7.

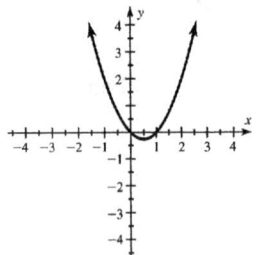       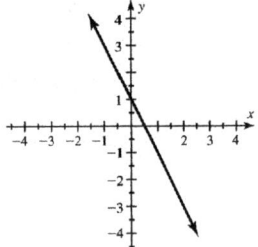

Figure 6                        Figure 7

8.  A line perpendicular to $y = -\dfrac{2}{3}x + 1$ has slope $m = \dfrac{3}{2}$.

$y - (-2) = \dfrac{3}{2}(x - 2) \Rightarrow y = \dfrac{3}{2}x - 3 - 2 \Rightarrow y = \dfrac{3}{2}x - 5$

9.  $2(1 - x) - 4x = x \Rightarrow 2 - 2x - 4x - x = 0 \Rightarrow -7x = -2 \Rightarrow x = \dfrac{2}{7}$

10. $-5 \leq 1 - 2x < 3 \Rightarrow -6 \leq -2x < 2 \Rightarrow 3 \geq x > -1 \Rightarrow -1 < x \leq 3; (-1, 3]$

11. $x^2 - 4 \leq 0$; replace the inequality symbol with an equals sign and solve the resulting equation.

$x^2 - 4 = 0 \Rightarrow (x-2)(x+2) = 0 \Rightarrow x = 2$ or $x = -2$. The graph of $y = x^2 - 4$ is a parabola that lies on

or below the $x$-axis when $x \geq -2$ and $x \leq 2$. The solution is $[-2, 2]$.

12. $|1 - x| \geq 2 \Rightarrow 1 - x \geq 2$ or $1 - x \leq -2 \Rightarrow -x \geq 1$ or $-x \leq -3 \Rightarrow x \leq -1$ or $x \geq 3$;

The solution is $(-\infty, -1] \cup [3, \infty)$.

13. Add the equations. $\begin{array}{r} -2x+y=1 \\ 5x-y=2 \\ \hline 3x \quad =3 \end{array} \Rightarrow x=1.$ Substitute $x=1$ into the first equation.

$-2(1)+y=1 \Rightarrow -2+y=1 \Rightarrow y=3.$ The solution is $(1, 3)$.

14. Note that $x+y \le 4 \Rightarrow y \le -x+4$ and $x-y \ge 2 \Rightarrow y \le x-2.$ See Figure 14.

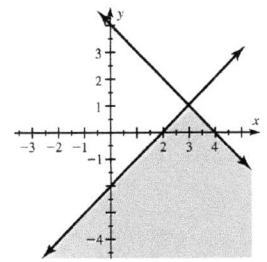

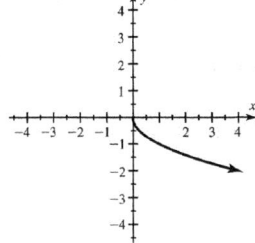

Figure 14          Figure 27

15. $(2x-1)(x+5)=2x^2+10x-x-5=2x^2+9x-5$

16. $xy(2x-3y^2+1)=2x^2y-3xy^3+xy$

17. $6x^2-13x-5=(3x+1)(2x-5)$

18. $x^3-4x=x(x^2-4)=x(x-2)(x+2)$

19. $x^2+3x+2=0 \Rightarrow (x+2)(x+1)=0 \Rightarrow x=-2$ or $x=-1$

20. $x^2+1=-3x \Rightarrow x^2+3x+1=0; a=1, b=3, c=1$   $x=\dfrac{-b\pm\sqrt{b^2-4ac}}{2a}=\dfrac{-3\pm\sqrt{3^2-4(1)(1)}}{2(1)}=\dfrac{-3\pm\sqrt{5}}{2}$

21. $\dfrac{x-2}{x+2}\div\dfrac{2x-4}{3x+6}=\dfrac{x-2}{x+2}\cdot\dfrac{3(x+2)}{2(x-2)}=\dfrac{3}{2}$

22. $\dfrac{1}{x+1}+\dfrac{1}{x-1}=\dfrac{1}{x+1}\cdot\dfrac{x-1}{x-1}+\dfrac{1}{x-1}\cdot\dfrac{x+1}{x+1}=\dfrac{x-1}{x^2-1}+\dfrac{x+1}{x^2-1}=\dfrac{x-1+x+1}{x^2-1}=\dfrac{2x}{x^2-1}$

23. $\sqrt{8x^2}=\sqrt{4x^2\cdot 2}=2x\sqrt{2}$

24. $8^{2/3}=\left(\sqrt[3]{8}\right)^2=2^2=4$

25. $\sqrt[3]{2x}\cdot\sqrt[3]{32x^2}=\sqrt[3]{2x\cdot 32x^2}=\sqrt[3]{64x^3}=4x$

26. $3\sqrt{3x}+\sqrt{12x}=3\sqrt{3x}+2\sqrt{3x}=5\sqrt{3x}$

27. See Figure 27.

28. $d=\sqrt{(x_2-x_1)^2+(y_2-y_1)^2}=\sqrt{(-2-2)^2+(0-(-3))^2}=\sqrt{(-4)^2+(3)^2}=\sqrt{16+9}=\sqrt{25}=5$

29. $(2+3i)(2-3i)=(2)^2-(3i)^2=4-9i^2=4-9(-1)=4+9=13$

30. $\sqrt{x+2}=x \Rightarrow x+2=x^2 \Rightarrow x^2-x-2=0 \Rightarrow (x-2)(x+1)=0 \Rightarrow x=2$ ($x=-1$ does not check.)

31. $x = -\dfrac{b}{2a} = -\dfrac{(-6)}{2(1)} = \dfrac{-6}{-2} = 3;\ y = (3)^2 - 6(3) + 3\ \Rightarrow y = 9 - 18 + 3 \Rightarrow y = -6.$ The vertex is $(3, -6)$.

32. $f(x) = x^2 - 2x + 3 \Rightarrow f(x) = \left(x^2 - 2x + 1\right) + 3 - 1 \Rightarrow f(x) = (x-1)^2 + 2$

33. The graph of $f(x)$ is shifted 4 units right.

34. $x(3-x) = 2 \Rightarrow 3x - x^2 - 2 = 0 \Rightarrow x^2 - 3x + 2 = 0 \Rightarrow (x-2)(x-1) = 0 \Rightarrow x = 2$ or $x = 1$

35. $x^3 + x = 0 \Rightarrow x\left(x^2 + 1\right) = 0 \Rightarrow x = 0$ or $x^2 = -1 \Rightarrow x = \pm\sqrt{-1} \Rightarrow x = \pm i$

36. (a)   $\log 10,000 = 4$ because $10^4 = 10,000$

    (b)   $\log_2 8 = 3$

    (c)   $\log_3 3^x = x$

    (d)   $e^{\ln 6} = 6$

    (e)   $\log 2 + \log 50 = \log(2 \cdot 50) = \log(100) = 2$ because $10^2 = 100$

    (f)   $\log_2 24 - \log_2(3) = \log_2\left(\dfrac{24}{3}\right) = \log_2(8) = 3$

37. (a)   $(f \circ g)(2) = f(g(2)) = f(2 \cdot 2) = f(4) = 4^2 + 1 = 16 + 1 = 17$

    (b)   $(g \circ f)(x) = g(f(x)) = g\left(x^2 + 1\right) = 2\left(x^2 + 1\right) = 2x^2 + 2$

38. Let $x = 2 - 3y$. Then $x - 2 = -3y \Rightarrow y = \dfrac{x-2}{-3}$ or $f^{-1}(x) = -\dfrac{1}{3}x + \dfrac{2}{3}$

39. $A = 1000(1 + 0.05)^6 = 1000(1.05)^6 \approx \$1340.10$

40. $\log \dfrac{x^2 \sqrt{y}}{z^3} = \log\left(x^2 \sqrt{y}\right) - \log\left(z^3\right) = \log\left(x^2\right) + \log\left(y^{1/2}\right) - \log\left(z^3\right) = 2\log x + \dfrac{1}{2}\log y - 3\log z$

41. $2e^x - 1 = 17 \Rightarrow 2e^x = 18 \Rightarrow e^x = 9 \Rightarrow \ln e^x = \ln 9 \Rightarrow x = \ln 9$

42. $3 + \log 4x = 5 \Rightarrow \log 4x = 2 \Rightarrow 10^{\log 4x} = 10^2 \Rightarrow 4x = 100 \Rightarrow x = 25$

43. See Figures 43a, 43b, 43c, 43d

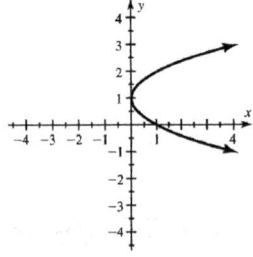

Figure 43a

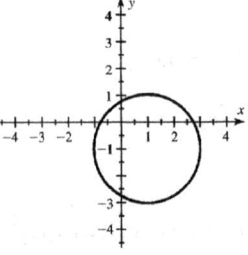

Figure 43b

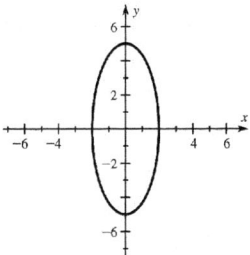

Figure 43c

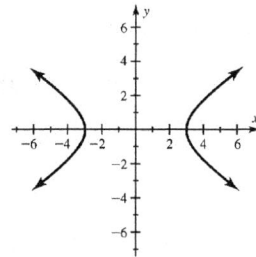

Figure 43d

44. Multiply the first equation by –1 and add to the second equation.

$$\begin{array}{r} -x^2 - y^2 = -1 \\ x^2 + 9y^2 = 9 \\ \hline 8y^2 = 8 \end{array} \Rightarrow y^2 = 1 \Rightarrow y = \pm 1$$

$x^2 + (\pm 1)^2 = 1 \Rightarrow x^2 + 1 = 1 \Rightarrow x^2 = 0 \Rightarrow x = 0$  The solutions are (0, 1) and (0, –1).

45. See Figure 45.

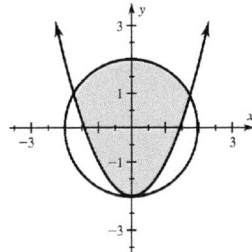

Figure 45

46. (a)  $D(0) = 400 - 50(0) = 400$;  initially, the driver is 400 miles from home.

  (b)  $400 - 50x = 0 \Rightarrow 400 = 50x \Rightarrow x = 8$;  after 8 hours the driver arrives at home.

  (c)  $-50$;  the driver is traveling 50 miles per hour toward home.

47. Let $x, y, z$ represent the amounts invested at 5%, 6%, and 7%, respectively. The system needed is

$$\begin{array}{lll} x + y + z = 2000 & & x + y + z = 2000 \\ y = x + 500 & \Rightarrow & -x + y \quad\quad = 500 \\ 0.05x + 0.06y + 0.07z = 120 & & 5x + 6y + 7z = 12,000 \end{array}$$

Multiply the first equation by –7 and add to the third equation.

$$\begin{array}{r} -7x - 7y - 7z = -14,000 \\ 5x + 6y + 7z = 12,000 \\ \hline -2x - y \quad\quad = -2000 \end{array}$$  Add this new equation to the second equation.

$$\begin{array}{r} -2x - y = -2000 \\ -x + y = 500 \\ \hline -3x \quad\quad = -1500 \end{array} \Rightarrow x = 500 \Rightarrow y = 500 + 500 = 1000$$

Substitute $x = 500$ and $y = 1000$ into the first equation. $500 + 1000 + z = 2000 \Rightarrow z = 500$,

$500 is invested at 5%, $1000 is invested at 6%, and $500 is invested at 7%.

48. Let $x$ represent the length of the garden. Then the area is $x(600-x) = 600x - x^2$. Find the $x$-value of

the vertex of $y = -x^2 + 600x$.   $x = -\dfrac{b}{2a} = -\dfrac{600}{2(-1)} = \dfrac{-600}{-2} = 300$

The dimensions should be 300 feet by 300 feet.

49. $2e^{0.02x} = 4 \Rightarrow e^{0.02x} = 2 \Rightarrow \ln e^{0.02x} = \ln 2 \Rightarrow 0.02x = \ln 2 \Rightarrow x = \dfrac{\ln 2}{0.02} \Rightarrow 50 \ln 2 \approx 34.7$ years

50. $V = \pi r^2 h = 60$ and $S = 2\pi rh = 50$, solve for $h \Rightarrow h = \dfrac{50}{2\pi r}$, substitute into $\pi r^2 h = 60$

$\Rightarrow \pi r^2 \left( \dfrac{50}{2\pi r} \right) = 60 \Rightarrow 25r = 60 \Rightarrow r = \dfrac{60}{25} = 2.4$ inches

# Chapter 14: Sequences and Series

## Section 14.1: Sequences

1.  1, 2, 3, 4; *Answers may vary.*

3.  function; natural numbers

5.  6

7.  $f(2)$

9.  $f(1) = 1^2 = 1,\ f(2) = 2^2 = 4,\ f(3) = 3^2 = 9,\ f(4) = 4^2 = 16 \Rightarrow 1,\ 4,\ 9,\ 16$

11. $f(1) = \dfrac{1}{1+5} = \dfrac{1}{6},\ f(2) = \dfrac{1}{2+5} = \dfrac{1}{7},\ f(3) = \dfrac{1}{3+5} = \dfrac{1}{8},\ f(4) = \dfrac{1}{4+5} = \dfrac{1}{9} \Rightarrow \dfrac{1}{6},\ \dfrac{1}{7},\ \dfrac{1}{8},\ \dfrac{1}{9}$

13. $f(1) = 5\left(\dfrac{1}{2}\right)^1 = \dfrac{5}{2},\ f(2) = 5\left(\dfrac{1}{2}\right)^2 = \dfrac{5}{4},\ f(3) = 5\left(\dfrac{1}{2}\right)^3 = \dfrac{5}{8},\ f(4) = 5\left(\dfrac{1}{2}\right)^4 = \dfrac{5}{16} \Rightarrow \dfrac{5}{2},\ \dfrac{5}{4},\ \dfrac{5}{8},\ \dfrac{5}{16}$

15. $f(1) = 9,\ f(2) = 9,\ f(3) = 9,\ f(4) = 9 \Rightarrow 9,\ 9,\ 9,\ 9$

17. $a_1 = 1^3 = 1,\ a_2 = 2^3 = 8,\ a_3 = 3^3 = 27 \Rightarrow 1,\ 8,\ 27$

19. $a_1 = \dfrac{4(1)}{3+1} = 1,\ a_2 = \dfrac{4(2)}{3+2} = \dfrac{8}{5},\ a_3 = \dfrac{4(3)}{3+3} = 2 \Rightarrow 1,\ \dfrac{8}{5},\ 2$

21. $a_1 = 2(1)^2 + 1 - 1 = 2,\ a_2 = 2(2)^2 + 2 - 1 = 9,\ a_3 = 2(3)^2 + 3 - 1 = 20 \Rightarrow 2,\ 9,\ 20$

23. $a_1 = -2,\ a_2 = -2,\ a_3 = -2 \Rightarrow -2,\ -2,\ -2$

25. $a_1 = b(1) + c = b + c;\ a_2 = b(2) + c = 2b + c$

27. $\dfrac{1}{2}(a_1 + a_4) = \dfrac{1}{2}(10 + 4) = \dfrac{1}{2}(14) = 7$

29. The points shown are (1, 3), (2, 4), (3, 5), (4, 3) and (5, 1).  The sequence is 3, 4, 5, 3, 1.

31. The points shown are (1, 6), (2, 5), (3, 4), (4, 3), (5, 2) and (6, 1).  The sequence is 6, 5, 4, 3, 2, 1.

33. Numerical: See Figure 33a.  Graphical: See Figure 33b.

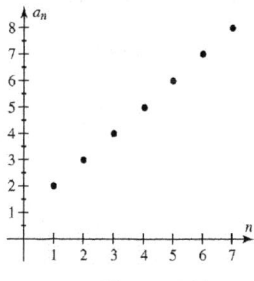

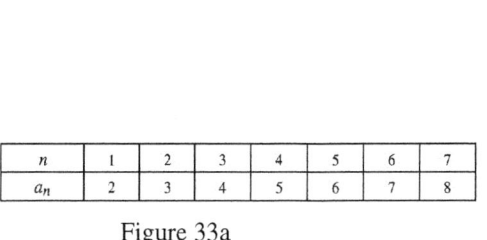

| $n$ | 1 | 2 | 3 | 4 | 5 | 6 | 7 |
|-----|---|---|---|---|---|---|---|
| $a_n$ | 2 | 3 | 4 | 5 | 6 | 7 | 8 |

Figure 33a                                              Figure 33b

35. Numerical: See Figure 35a.  Graphical: See Figure 35b.

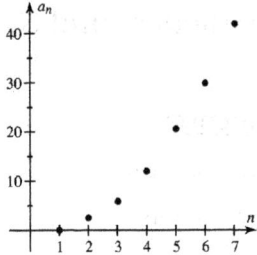

| $n$ | 1 | 2 | 3 | 4 | 5 | 6 | 7 |
|---|---|---|---|---|---|---|---|
| $a_n$ | 0 | 2 | 6 | 12 | 20 | 30 | 42 |

Figure 35a                                    Figure 35b

37. Numerical: See Figure 37a.  Graphical: See Figure 37b.

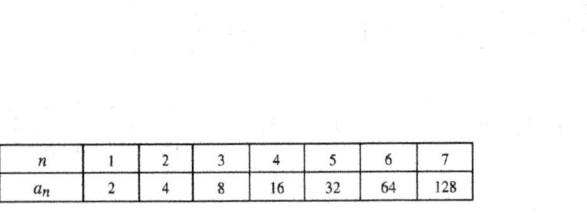

| $n$ | 1 | 2 | 3 | 4 | 5 | 6 | 7 |
|---|---|---|---|---|---|---|---|
| $a_n$ | 2 | 4 | 8 | 16 | 32 | 64 | 128 |

Figure 37a                                    Figure 37b

39. Symbolic: $a_n = 30n$ for $n = 1, 2, 3, \ldots, 7$  Numerical: See Figure 39a.  Graphical: See Figure 39b.

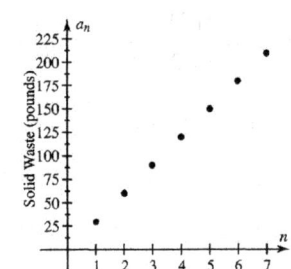

| $n$ | 1 | 2 | 3 | 4 | 5 | 6 | 7 |
|---|---|---|---|---|---|---|---|
| $a_n$ | 30 | 60 | 90 | 120 | 150 | 180 | 210 |

Figure 39a                                    Figure 39b

41. (a)   $a_1 = 1^2 = 1, a_2 = 2^2 = 4, a_3 = 3^2 = 9, a_4 = 4^2 = 16 \Rightarrow 1, 4, 9, 16$

    (b)   $a_1 = 4(1) = 4, a_2 = 4(2) = 8, a_3 = 4(3) = 12, a_4 = 4(4) = 16 \Rightarrow 4, 8, 12, 16$

43. (a)   After 1 year it is worth $25,000(0.80) = \$20,000$.  After 2 years it is worth

          $20,000(0.80) = \$16,000$.

    (b)   $a_n = 25,000(0.8)^n$

    (c)   See Figure 43.

| $n$ | 1 | 2 | 3 | 4 | 5 | 6 | 7 |
|---|---|---|---|---|---|---|---|
| $a_n$ | 20,000 | 16,000 | 12,800 | 10,240 | 8192 | 6553.6 | 5242.9 |

Figure 43

45. (a)   $a_n = 2048(0.5)^{n-1}$ for $n = 1, 2, 3, \ldots, 7$

    (b)   See Figure 45b.

    (c)   See Figure 45c

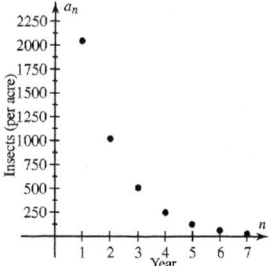

| $n$ | 1 | 2 | 3 | 4 | 5 | 6 | 7 |
|---|---|---|---|---|---|---|---|
| $a_n$ | 2048 | 1024 | 512 | 256 | 128 | 64 | 32 |

Figure 45b                                          Figure 45c

## Section 14.2 Arithmetic and Geometric Sequences

1. linear

3. $a_n = 3n + 1$; the common difference is 3. *Answers may vary.*

5. add; previous

7. 19; 4

9. $a_n = a_1 (r)^{n-1}$

11. Yes, the common difference is 10.

13. Yes, the common difference is –1.

15. No, there is no common difference.

17. Yes, the common difference is 3.

19. Yes, the common difference is –3.

21. No, there is no common difference.

23. Yes, the common difference is 1.

25. No, there is no common difference.

27. Yes, the common difference is 2.

29. $a_n = 7 + (n-1)(-2) \Rightarrow a_n = 7 - 2n + 2 \Rightarrow a_n = -2n + 9$

31. Note: $d = \dfrac{6-(-2)}{2} = 4$, thus $a_n = -2 + (n-1)(4) \Rightarrow a_n = -2 + 4n - 4 \Rightarrow a_n = 4n - 6$

33. Note: $d = \dfrac{8-16}{4} = -2$ and $a_1 = 16 - 7(-2) = 30$, thus

$a_n = 30 + (n-1)(-2) \Rightarrow a_n = 30 - 2n + 2 \Rightarrow a_n = -2n + 32$

35. $a_{32} = -3 + (32-1)(2) \Rightarrow -3 + (31)(2) = -3 + 62 = 59$

37. Note: $d = 0 - (-3) = 3$, thus $a_9 = -3 + (9-1)(3) \Rightarrow -3 + (8)(3) = -3 + 24 = 21$

39. Yes, the common ratio is 3.

41. Yes, the common ratio is 0.8.

43. No, there is no common ratio.

45. Yes, the common ratio is 2.

47. No, there is no common ratio.

49. Yes, the common ratio is 4.

51. Yes, the common ratio is 2.

53. No, there is no common ratio.

55. $a_n = 1.5(4)^{n-1}$

57. Note: $r = \dfrac{6}{-3} = -2$, thus $a_n = -3(-2)^{n-1}$

59. Note: $16 = 1 \cdot r^2 \Rightarrow r^2 = 16 \Rightarrow r = 4 \,(\text{since } r > 0)$, thus $a_n = 1(4)^{n-1}$

61. $a_8 = 2(3)^{8-1} = 2(3)^7 = 4374$

63. Note: $r = \dfrac{3}{-1} = -3$, thus $a_6 = -1(-3)^{6-1} = -1(-3)^5 = 243$

65. (a)    3000, 6000, 9000, 12,000, 15,000; This sequence is arithmetic.

    (b)    $a_n = 3000n$

    (c)    $a_{20} = 3000(20) = 60,000;$ When there are 20 people, the ventilation should be 60,000 cubic

           feet per hour.

    (d)    See Figure 65.  The points are collinear.

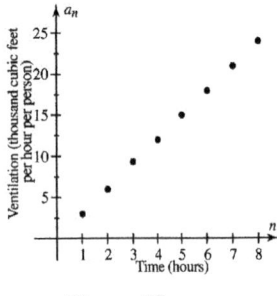

Figure 65                                      Figure 67

67. (a)    Since 20% of the chlorine dissipates, 80% will remain.  The function is $a_n = 3(0.8)^{n-1}$.

    (b)    See Figure 67.  The points are not collinear.  The sequence is geometric.

69. (a)    $a_n = 5(0.85)^{n-1}$

    (b)    The sequence is geometric and the common ratio is 0.85.

    (c)    $a_8 = 5(0.85)^{8-1} = 5(0.85)^7 \approx 1.6;$ On the 8th bounce, the ball reaches a maximum height of

           about 1.6 feet.

71. (a)    The number of seats can be modeled by an arithmetic sequence whose common difference is 2.

    (b)    Since $a_1 = 40$ and $d = 2$, $a_n = 40 + 2(n-1)$ or $a_n = 2n + 38$

    (c)    $a_{20} = 40 + 2(20 - 1) = 40 + 2(19) = 40 + 38 = 78$ seats

## Checking Basic Concepts Sections 14.1 and 14.2

1. $a_1 = \dfrac{1}{1+4} = \dfrac{1}{5}$, $a_2 = \dfrac{2}{2+4} = \dfrac{1}{3}$, $a_3 = \dfrac{3}{3+4} = \dfrac{3}{7}$, $a_4 = \dfrac{4}{4+4} = \dfrac{1}{2}$ $\Rightarrow \dfrac{1}{5}, \dfrac{1}{3}, \dfrac{3}{7}, \dfrac{1}{2}$

2. Graphical: See Figure 2a.  Numerical: See Figure 2b.

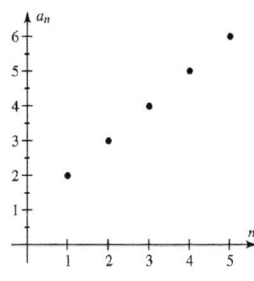

| $n$ | 1 | 2 | 3 | 4 | 5 |
|---|---|---|---|---|---|
| $a_n$ | 2 | 3 | 4 | 5 | 6 |

Figure 2a                            Figure 2b

3. (a)   Arithmetic.  Here $d = 1-(-2) = 3$ and $a_1 = -2$, thus $a_n = -2+(n-1)(3) \Rightarrow a_n = 3n-5$

    (b)   Geometric.  Here $r = \dfrac{-6}{3} = -2$ and $a_1 = 3$, thus $a_n = 3(-2)^{n-1}$

4. $a_n = 5+(n-1)(2) \Rightarrow a_n = 5+2n-2 \Rightarrow a_n = 2n+3$

5. $a_n = 5(2)^{n-1}$

## Section 14.3 Series

1. series

3. arithmetic (the common difference is 2)

5. $n\left(\dfrac{a_1 + a_n}{2}\right)$ or $\dfrac{n}{2}\left(2a_1 + (n-1)d\right)$

7. sum

9. arithmetic

11. $6\left(\dfrac{3+13}{2}\right) = 6(8) = 48$

13. $40\left(\dfrac{1+40}{2}\right) = 40(20.5) = 820$

15. $5\left(\dfrac{-7+5}{2}\right) = 5(-1) = -5$

17. Here $r = 3$ and $n = 7$ thus $S_7 = 3\left(\dfrac{1-3^7}{1-3}\right) = 3\left(\dfrac{-2186}{-2}\right) = 3(1093) = 3279$

19. Here $r = -2$ and $n = 8$ thus $S_8 = 1\left(\dfrac{1-(-2)^8}{1-(-2)}\right) = 1\left(\dfrac{-255}{3}\right) = -85$

21.  Here $r = 3$ and $n = 6$ thus $S_6 = 0.5\left(\dfrac{1-3^6}{1-3}\right) = 0.5\left(\dfrac{-728}{-2}\right) = 0.5(364) = 182$

23.  $S_{20} = 2000\left(\dfrac{(1+0.08)^{20} - 1}{0.08}\right) \approx \$91,523.93$

25.  $S_5 = 10,000\left(\dfrac{(1+0.11)^5 - 1}{0.11}\right) \approx \$62,278.01$

27.  $2(1) + 2(2) + 2(3) + 2(4) \Rightarrow 2 + 4 + 6 + 8 = 20$

29.  $4 + 4 + 4 + 4 + 4 + 4 + 4 + 4 = 32$

31.  $1^2 + 2^2 + 3^2 + 4^2 + 5^2 + 6^2 + 7^2 \Rightarrow 1 + 4 + 9 + 16 + 25 + 36 + 49 = 140$

33.  $\left(4^2 - 4\right) + \left(5^2 - 5\right) \Rightarrow 12 + 20 = 32$

35.  $\displaystyle\sum_{k=1}^{6} k^4$

37.  $\displaystyle\sum_{k=1}^{5} \dfrac{1}{k^2}$

39.  $\displaystyle\sum_{k=1}^{n} k = n\left(\dfrac{a_1 + a_n}{2}\right) = n\left(\dfrac{1+n}{2}\right) = \dfrac{n(n+1)}{2}$

41.  (a)    $8518 + 9921 + 10,706 + 14,035 + 14,307 + 12,249$

     (b)    $8518 + 9921 + 10,706 + 14,035 + 14,307 + 12,249 = 69,736$

43   (a)    Each air filter removes 80% or 0.8 of the impurities, so 20% or 0.2 passes through it

            100% or 1 represent the amount of impurities entering the first air filter, the amount removed

            by $n$ filters equals $(0.8)(1) + (0.8)(0.2) + (0.8)(0.04) + (0.8)(0.008) + \cdots + (0.8)(0.2)^{n-1}$. In

            summation notation, $\displaystyle\sum_{k=1}^{n} 0.8(0.2)^{k-1}$

     (b)    To remove 96% or 0.96 of the impurities requires 2 filters because

            $\displaystyle\sum_{k=1}^{2} 0.8(0.2)^{k-1} = (0.8)(1) + (0.8)(0.2) = 0.8 + 0.16 = 0.96$

45.  (a)    Each successive square has half the area of the square before it.  $1, \dfrac{1}{2}, \dfrac{1}{4}, \dfrac{1}{8}, \dfrac{1}{16}$

     (b)    Here $r = \dfrac{1}{2}$ and $n = 10$ thus $S_5 = 1\left(\dfrac{1 - \left(\frac{1}{2}\right)^{10}}{1 - \left(\frac{1}{2}\right)}\right) = \left(\dfrac{\frac{1023}{1024}}{\frac{1}{2}}\right) = \dfrac{1023}{512}.$

47. The sum is $14+13+12+11+10+9+8+7+6$. This is an arithmetic series with $a_1 = 14$ and $a_9 = 6$.

   The sum is $S_9 = 9\left(\dfrac{14+6}{2}\right) = 9(10) = 90$ logs.

49. This is an arithmetic series with $a_1 = 35,000$, $n = 20$ and $d = 2000$

   The sum is $S_{20} = \dfrac{20}{2}\left(2(35,000)+(20-1)(2000)\right) = 10(70,000+38,000) =$

   $10(108,000) = \$1,080,000$

51. This is a geometric sequence given by $a_n = 10(0.75)^n$. The distance it falls is

   $a_4 = 10(0.75)^4 \approx 3.16$ feet.

53. *Answers may vary.*


## Section 14.4 The Binomial Theorem

1. 5

3. Row 4

5. $4! = 1 \cdot 2 \cdot 3 \cdot 4 = 24$

7. $\dfrac{n!}{(n-r)!r!}$

9. $\dfrac{n!}{(n-1)!} = \dfrac{n \cdot (n-1)!}{(n-1)!} = n$

11. Row 4 of Pascal's triangle is 1, 3, 3, 1. $(x+y)^3 = x^3 + 3x^2 y + 3xy^2 + y^3$

13. Row 5 of Pascal's triangle is 1, 4, 6, 4, 1.

   $(2x+1)^4 = (2x)^4 + 4(2x)^3 (1) + 6(2x)^2 (1)^2 + 4(2x)(1)^3 + (1)^4 \Rightarrow$

   $(2x+1)^4 = 16x^4 + 32x^3 + 24x^2 + 8x + 1$

15. Row 6 of Pascal's triangle is 1, 5, 10, 10, 5, 1. $(a-b)^5 = a^5 - 5a^4 b + 10a^3 b^2 - 10a^2 b^3 + 5ab^4 - b^5$

17. Row 4 of Pascal's triangle is 1, 3, 3, 1. $(x^2+1)^3 = (x^2)^3 + 3(x^2)^2 (1) + 3(x^2)(1)^2 + (1)^3 \Rightarrow$

   $(x^2+1)^3 = x^6 + 3x^4 + 3x^2 + 1$

19. $3! = 1 \cdot 2 \cdot 3 = 6$

21. $\dfrac{4!}{3!} = \dfrac{1 \cdot 2 \cdot 3 \cdot 4}{1 \cdot 2 \cdot 3} = 4$

23. $\dfrac{2!}{0!} = \dfrac{1 \cdot 2}{1} = 2$

25. $\dfrac{5!}{2!3!} = \dfrac{1 \cdot 2 \cdot 3 \cdot 4 \cdot 5}{(1 \cdot 2)(1 \cdot 2 \cdot 3)} = 2 \cdot 5 = 10$

27. $_5C_4 = \dfrac{5!}{4!1!} = \dfrac{1 \cdot 2 \cdot 3 \cdot 4 \cdot 5}{(1 \cdot 2 \cdot 3 \cdot 4)(1)} = 5$

29. $_6C_5 = \dfrac{6!}{5!1!} = \dfrac{1 \cdot 2 \cdot 3 \cdot 4 \cdot 5 \cdot 6}{(1 \cdot 2 \cdot 3 \cdot 4 \cdot 5)(1)} = 6$

31. $_4C_0 = \dfrac{4!}{0!4!} = \dfrac{1 \cdot 2 \cdot 3 \cdot 4}{(1)(1 \cdot 2 \cdot 3 \cdot 4)} = 1$

33. $_{12}C_7 = 792$

35. $_9C_5 = 126$

37. $_{19}C_{11} = 75,582$

39. $(m+n)^3 = (_3C_0)m^3 + (_3C_1)m^2n + (_3C_2)mn^2 + (_3C_3)n^3 \Rightarrow (m+n)^3 = m^3 + 3m^2n + 3mn^2 + n^3$

41. $(x-y)^4 = (_4C_0)x^4 - (_4C_1)x^3y + (_4C_2)x^2y^2 - (_4C_3)xy^3 + (_4C_4)y^4 \Rightarrow$

    $(x-y)^4 = x^4 - 4x^3y + 6x^2y^2 - 4xy^3 + y^4$

43. $(2a+1)^3 = (_3C_0)(2a)^3 + (_3C_1)(2a)^2(1) + (_3C_2)(2a)(1)^2 + (_3C_3)(1)^3 \Rightarrow$

    $(2a+1)^3 = 8a^3 + 12a^2 + 6a + 1$

45. $(x+2)^5 = (_5C_0)x^5 + (_5C_1)x^4(2) + (_5C_2)x^3(2)^2 + (_5C_3)x^2(2)^3 + (_5C_4)x(2)^4 + (_5C_5)(2)^5 \Rightarrow$

    $(x+2)^5 = x^5 + 10x^4 + 40x^3 + 80x^2 + 80x + 32$

47. $(3+2m)^4 = (_4C_0)(3)^4 + (_4C_1)(3)^3(2m) + (_4C_2)(3)^2(2m)^2 + (_4C_3)(3)(2m)^3 + (_4C_4)(2m)^4 \Rightarrow$

    $(3+2m)^4 = 81 + 216m + 216m^2 + 96m^3 + 16m^4$

49. $(2x-y)^3 = (_3C_0)(2x)^3 - (_3C_1)(2x)^2y + (_3C_2)(2x)y^2 - (_3C_3)y^3 \Rightarrow$

    $(2x-y)^3 = 8x^3 - 12x^2y + 6xy^2 - y^3$

51. Here $r = 0$ and $n = 8$. The first term is $(_8C_0)a^{8-0}b^0 = a^8$.

53. Here $r = 3$ and $n = 7$. The fourth term is $(_7C_3)x^{7-3}y^3 = 35x^4y^3$.

55. Here $r = 0$ and $n = 9$. The first term is $(_9C_0)(2m)^{9-0}n^0 = 512m^9$.

## Checking Basic Concepts  Sections 14.3 and 14.4

1. (a)  Geometric. The common ratio is $\dfrac{1}{2}$.

(b)  Arithmetic.  The common difference is 2.

2.  $12\left(\dfrac{4+48}{2}\right)=12(26)=312$

3.  Here  $r=-2$  and  $n=10$  thus  $S_{10}=1\left(\dfrac{1-(-2)^{10}}{1-(-2)}\right)=1\left(\dfrac{-1023}{3}\right)=-341$

4.  Row 5 of Pascal's triangle is 1, 4, 6, 4, 1.  $(x-y)^4=x^4-4x^3y+6x^2y^2-4xy^3+y^4$

5.  $(x+2)^3=\left({}_3C_0\right)x^3+\left({}_3C_1\right)x^2(2)+\left({}_3C_2\right)x(2)^2+\left({}_3C_3\right)(2)^3\Rightarrow(x+2)^3=x^3+6x^2+12x+8$

## Chapter 14 Review

1.  $f(1)=1^3=1,\ f(2)=2^3=8,\ f(3)=3^3=27,\ f(4)=4^3=64\Rightarrow1,8,27,64$

2.  $f(1)=5-2(1)=3,\ f(2)=5-2(2)=1,\ f(3)=5-2(3)=-1,\ f(4)=5-2(4)=-3\Rightarrow3,1,-1,-3$

3.  $f(1)=\dfrac{2(1)}{1^2+1}=1,\ f(2)=\dfrac{2(2)}{2^2+1}=\dfrac{4}{5},\ f(3)=\dfrac{2(3)}{3^2+1}=\dfrac{3}{5},\ f(4)=\dfrac{2(4)}{4^2+1}=\dfrac{8}{17}\Rightarrow1,\dfrac{4}{5},\dfrac{3}{5},\dfrac{8}{17}$

4.  $f(1)=(-2)^1=-2,\ f(2)=(-2)^2=4,\ f(3)=(-2)^3=-8,\ f(4)=(-2)^4=16\Rightarrow-2,4,-8,16$

5.  The points shown are $(1,-2),\ (2,0),\ (3,4),$ and $(4,2)$.  The sequence is $-2,0,4,2$.

6.  The points shown are $(1,5),\ (2,3),\ (3,2),$ and $(4,1)$.  The sequence is $5,3,2,1$.

7.  Numerical: See Figure 7a.  Graphical: See Figure 7b.

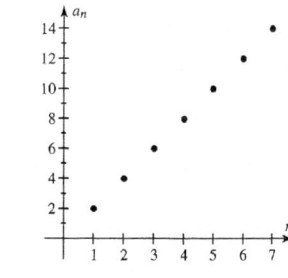

| $n$ | 1 | 2 | 3 | 4 | 5 | 6 | 7 |
|-----|---|---|---|---|---|----|----|
| $a_n$ | 2 | 4 | 6 | 8 | 10 | 12 | 14 |

Figure 7a                                                                  Figure 7b

8.  Numerical: See Figure 8a.  Graphical: See Figure 8b.

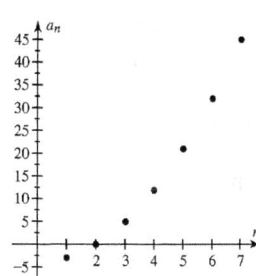

| $n$ | 1 | 2 | 3 | 4 | 5 | 6 | 7 |
|-----|----|---|---|----|----|----|----|
| $a_n$ | -3 | 0 | 5 | 12 | 21 | 32 | 45 |

Figure 8a                                                                  Figure 8b

9.  Numerical: See Figure 9a.  Graphical: See Figure 9b.

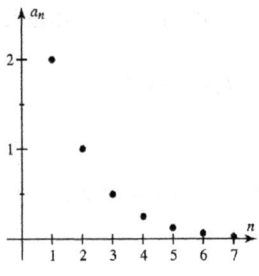

Figure 9b

| $n$ | 1 | 2 | 3 | 4 | 5 | 6 | 7 |
|---|---|---|---|---|---|---|---|
| $a_n$ | 2 | 1 | 0.5 | 0.25 | 0.125 | 0.0625 | 0.0313 |

Figure 9a

10. Numerical: See Figure 10a.  Graphical: See Figure 10b.

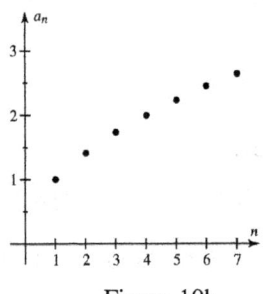

Figure 10b

| $n$ | 1 | 2 | 3 | 4 | 5 | 6 | 7 |
|---|---|---|---|---|---|---|---|
| $a_n$ | 1 | 1.4142 | 1.7321 | 2 | 2.2361 | 2.4495 | 2.6458 |

Figure 10a

11. Yes, the common difference is 5.

12. No, there is no common difference.

13. No, there is no common difference.

14. Yes, the common difference is $-\dfrac{1}{3}$.

15. Yes, the common difference is $-3$.

16. No, there is no common difference.

17. Yes, the common difference is $-1$.

18. No, there is no common difference.

19. $a_n = -3 + (n-1)(4) \Rightarrow a_n = -3 + 4n - 4 \Rightarrow a_n = 4n - 7$

20. Note: $d = -3 - 2 = -5,$ thus $a_n = 2 + (n-1)(-5) \Rightarrow a_n = 2 - 5n + 5 \Rightarrow a_n = -5n + 7$

21. Yes, the common ratio is 4.

22. No, there is no common ratio.

23. No, there is no common ratio.

24. Yes, the common ratio is 0.7.

25. No, there is no common ratio.

26. Yes, the common ratio is $-\dfrac{1}{3}$.

27. No, there is no common ratio.

28. Yes, the common ratio is 2.

29. $a_n = 5(0.9)^{n-1}$

30. Note: $r = \dfrac{8}{2} = 4$, thus $a_n = 2(4)^{n-1}$

31. $9\left(\dfrac{4+44}{2}\right) = 9(24) = 216$

32. $5\left(\dfrac{4.5+(-1.5)}{2}\right) = 5(1.5) = 7.5$

33. Here $r = -4$ and $n = 7$ thus $S_7 = 1\left(\dfrac{1-(-4)^7}{1-(-4)}\right) = \left(\dfrac{16,385}{5}\right) = 3277$

34. Here $r = \dfrac{1}{2}$ and $n = 9$ thus $S_9 = 1\left(\dfrac{1-\left(\frac{1}{2}\right)^9}{1-\left(\frac{1}{2}\right)}\right) = \left(\dfrac{\frac{511}{512}}{\frac{1}{2}}\right) = \dfrac{511}{256}$

35. $(2(1)+1)+(2(2)+1)+(2(3)+1)+(2(4)+1)+(2(5)+1) \Rightarrow 3+5+7+9+11$

36. $\dfrac{1}{1+1}+\dfrac{1}{2+1}+\dfrac{1}{3+1}+\dfrac{1}{4+1} \Rightarrow \dfrac{1}{2}+\dfrac{1}{3}+\dfrac{1}{4}+\dfrac{1}{5}$

37. $1^3 + 2^3 + 3^3 + 4^3 \Rightarrow 1+8+27+64$

38. $(1-2)+(1-3)+(1-4)+(1-5)+(1-6)+(1-7) \Rightarrow -1+(-2)+(-3)+(-4)+(-5)+(-6)$

39. $\displaystyle\sum_{k=1}^{20} k$

40. $\displaystyle\sum_{k=1}^{20} \dfrac{1}{k}$

41. $\displaystyle\sum_{k=1}^{9} \dfrac{k}{k+1}$

42. $\displaystyle\sum_{k=1}^{7} k^2$

43. Row 4 of Pascal's triangle is 1, 3, 3, 1. $(x+4)^3 = x^3 + 3x^2(4) + 3x(4)^2 + (4)^3 \Rightarrow$

$(x+4)^3 = x^3 + 12x^2 + 48x + 64$

44. Row 5 of Pascal's triangle is 1, 4, 6, 4, 1.

$(2x+1)^4 = (2x)^4 + 4(2x)^3(1) + 6(2x)^2(1)^2 + 4(2x)(1)^3 + (1)^4 \Rightarrow$

$(2x+1)^4 = 16x^4 + 32x^3 + 24x^2 + 8x + 1$

45. Row 6 of Pascal's triangle is 1, 5, 10, 10, 5, 1. $(x-y)^5 = x^5 - 5x^4 y + 10x^3 y^2 - 10x^2 y^3 + 5xy^4 - y^5$

46. Row 7 of Pascal's triangle is 1, 6, 15, 20, 15, 6, 1.

$(a-1)^6 = a^6 - 6a^5(1) + 15a^4(1)^2 - 20a^3(1)^3 + 15a^2(1)^4 - 6a(1)^5 + (1)^6 \Rightarrow$

$(a-1)^6 = a^6 - 6a^5 + 15a^4 - 20a^3 + 15a^2 - 6a + 1$

47. $3! = 1 \cdot 2 \cdot 3 = 6$

48. $\dfrac{5!}{3!2!} = \dfrac{1 \cdot 2 \cdot 3 \cdot 4 \cdot 5}{(1 \cdot 2 \cdot 3)(1 \cdot 2)} = 2 \cdot 5 = 10$

49. $_6C_3 = \dfrac{6!}{3!3!} = \dfrac{1 \cdot 2 \cdot 3 \cdot 4 \cdot 5 \cdot 6}{(1 \cdot 2 \cdot 3)(1 \cdot 2 \cdot 3)} = 2 \cdot 5 \cdot 2 = 20$

50. $_4C_3 = \dfrac{4!}{3!1!} = \dfrac{1 \cdot 2 \cdot 3 \cdot 4}{(1 \cdot 2 \cdot 3)(1)} = 4$

51. $(m+2)^4 = \left(_4C_0\right)m^4 + \left(_4C_1\right)m^3(2) + \left(_4C_2\right)m^2(2)^2 + \left(_4C_3\right)m(2)^3 + \left(_4C_4\right)(2)^4 \Rightarrow$

$(m+2)^4 = m^4 + 8m^3 + 24m^2 + 32m + 16$

52. $(a+b)^5 = \left(_5C_0\right)a^5 + \left(_5C_1\right)a^4b + \left(_5C_2\right)a^3b^2 + \left(_5C_3\right)a^2b^3 + \left(_5C_4\right)ab^4 + \left(_5C_5\right)b^5 \Rightarrow$

$(a+b)^5 = a^5 + 5a^4b + 10a^3b^2 + 10a^2b^3 + 5ab^4 + b^5$

53. $(x-3y)^4 = \left(_4C_0\right)x^4 - \left(_4C_1\right)x^3(3y) + \left(_4C_2\right)x^2(3y)^2 - \left(_4C_3\right)x(3y)^3 + \left(_4C_4\right)(3y)^4 \Rightarrow$

$(x-3y)^4 = x^4 - 12x^3y + 54x^2y^2 - 108xy^3 + 81y^4$

54  $(3x-2)^3 = \left(_3C_0\right)(3x)^3 - \left(_3C_1\right)(3x)^2(2) + \left(_3C_2\right)(3x)(2)^2 - \left(_3C_3\right)(2)^3 \Rightarrow$

$(3x-2)^3 = 27x^3 - 54x^2 + 36x - 8$

55. Symbolic: $a_n = 45{,}000(1.10)^{\wedge}(n-1)$ for $n = 1, 2, 3, \ldots, 7$. This is a geometric sequence.

Numerical: See Figure 55a. Graphical: See Figure 55b.

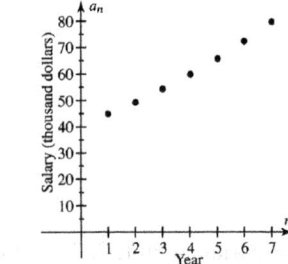

| $n$   | 1      | 2      | 3      | 4      | 5      | 6      | 7      |
|-------|--------|--------|--------|--------|--------|--------|--------|
| $a_n$ | 45,000 | 49,500 | 54,450 | 59,895 | 65,885 | 72,473 | 79,720 |

Figure 55a                                                                          Figure 55b

56. Symbolic: $a_n = 45{,}000 + 5000(n-1)$ for $n = 1, 2, 3, \ldots, 7$. This is an arithmetic sequence.

Numerical: See Figure 56a. Graphical: See Figure 56b.

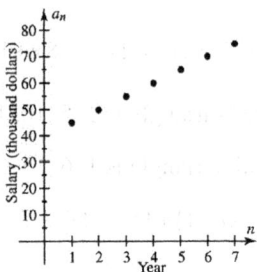

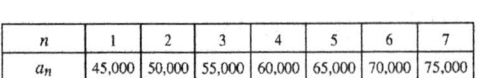

| $n$   | 1      | 2      | 3      | 4      | 5      | 6      | 7      |
|-------|--------|--------|--------|--------|--------|--------|--------|
| $a_n$ | 45,000 | 50,000 | 55,000 | 60,000 | 65,000 | 70,000 | 75,000 |

Figure 56a                                                                          Figure 56b

57. Symbolic: $a_n = 49n$ for $n = 1, 2, 3, \ldots, 7$

Graphical: See Figure 57a.  Numerical: See Figure 57b.

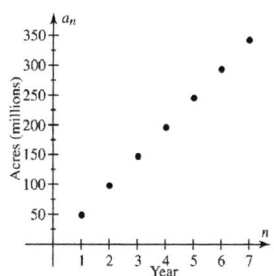

Figure 57a

| $n$ | 1 | 2 | 3 | 4 | 5 | 6 | 7 |
|---|---|---|---|---|---|---|---|
| $a_n$ | 49 | 98 | 147 | 196 | 245 | 294 | 343 |

Figure 57b

58. (a)  $a_n = 1087(1.025)^{n-1}$

(b)  The sequence is geometric.  The common ratio is 1.025.

(c)  $a_5 = 1087(1.025)^{5-1} \approx 1200$; The average mortgage payment in 2000 was about \$1200.

(d)  See Figure 58.

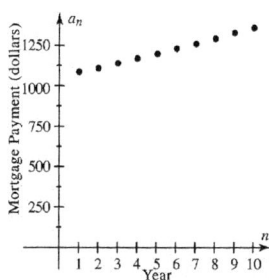

Figure 58

| $n$ | 1 | 2 | 3 | 4 | 5 | 6 | 7 |
|---|---|---|---|---|---|---|---|
| $a_n$ | 0 | 2 | 6 | 12 | 20 | 30 | 42 |

Figure 3

## Chapter 14 Test

1.  $f(1) = \dfrac{1^2}{1+1} = \dfrac{1}{2}, f(2) = \dfrac{2^2}{2+1} = \dfrac{4}{3}, f(3) = \dfrac{3^2}{3+1} = \dfrac{9}{4}, f(4) = \dfrac{4^2}{4+1} = \dfrac{16}{5} \Rightarrow \dfrac{1}{2}, \dfrac{4}{3}, \dfrac{9}{4}, \dfrac{16}{5}$

2.  The points shown are (1, –3), (2, 2), (3, 1), (4, –2) and (5, 3).  The sequence is –3, 2, 1, –2, 3.

3.  See Figure 3.

4.  Row 5 of Pascal's triangle is 1, 4, 6, 4, 1.

$(2x-1)^4 = (2x)^4 - 4(2x)^3(1) + 6(2x)^2(1)^2 - 4(2x)(1)^3 + (1)^4 \Rightarrow$

$(2x-1)^4 = 16x^4 - 32x^3 + 24x^2 - 8x + 1$

5.  The sequence is arithmetic.  The common difference is –3.

6.  The sequence is geometric.  The common ratio is –2.

7.  $a_n = 2 + (n-1)(-3) \Rightarrow a_n = 2 - 3n + 3 \Rightarrow a_n = -3n + 5$

8. Note: $2 \cdot r^2 = 4.5 \Rightarrow r^2 = \dfrac{4.5}{2} \Rightarrow r = 1.5$, thus $a_n = 2(1.5)^{n-1}$

9. Yes, the common ratio is 2.5.

10. No, there is no common ratio.

11. $9\left(\dfrac{-1+23}{2}\right) = 9(11) = 99$

12. Here $r = -\dfrac{2}{3}$ and $n = 7$ thus $S_7 = 1\left(\dfrac{1-\left(-\frac{2}{3}\right)^7}{1-\left(-\frac{2}{3}\right)}\right) = \left(\dfrac{\frac{2315}{2187}}{\frac{5}{3}}\right) = \dfrac{463}{729}$

13. $3(2)+3(3)+3(4)+3(5)+3(6)+3(7) \Rightarrow 6+9+12+15+18+21$

14. $\displaystyle\sum_{k=1}^{60} k^3$

15. $\dfrac{7!}{4!3!} = \dfrac{1\cdot2\cdot3\cdot4\cdot5\cdot6\cdot7}{(1\cdot2\cdot3\cdot4)(1\cdot2\cdot3)} = 5\cdot7 = 35$

16. $_5C_3 = \dfrac{5!}{3!2!} = \dfrac{1\cdot2\cdot3\cdot4\cdot5}{(1\cdot2\cdot3)(1\cdot2)} = 2\cdot5 = 10$

17. This is an arithmetic series with $a_1 = 50$ and $d = 7$. The total number of seats is

$$S_{45} = \dfrac{45}{2}\left(2(50)+(45-1)(7)\right) = 22.5(408) = 9180 \text{ seats.}$$

18. Symbolic: $a_n = 180,000(0.96)^\wedge(n-1)$ for $n = 1,2,3,4,5$. This is a geometric sequence.

Numerical: See Figure 18a. Graphical: See Figure 18b.

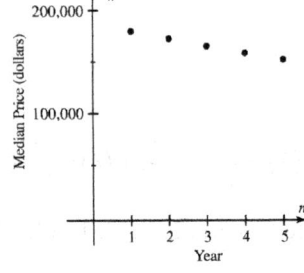

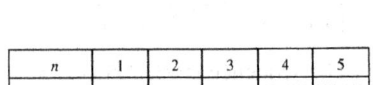

| $n$ | 1 | 2 | 3 | 4 | 5 |
|---|---|---|---|---|---|
| $a_n$ | 180,000 | 172,800 | 165,888 | 159,252 | 152,882 |

Figure 18a                                    Figure 18b

19. (a)    $a_n = 2000(2)^{n-1}$

(b)    The sequence is geometric. The common ratio is 2.

(c)    $a_6 = 2000(2)^{6-1} \approx 64,000$; After 30 days, there are 64,000 caterpillars.

(d)    See Figure 19.

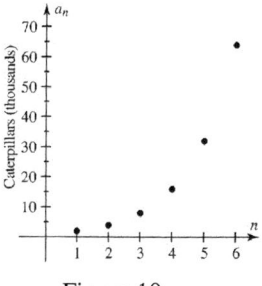

Figure 19

## Chapter 14 Extended and Discovery Exercises

1. Each term of the sequence is obtained by adding the two previous terms.  1, 1, 2, 3, 5, 8, 13, 21, 34, 55, 89, 144

2. (a)   Table $u(n) = 2.85u(n-1) - 0.19(u(n-1))^2$ with TblStart = 1 and $\Delta$Tbl = 1. See Figure 2a.

   Note: be sure to set $n$Min = 1 and u($n$Min) = {1}.

   (b)   Graph $u(n) = 2.85u(n-1) - 0.19(u(n-1))^2$ in [0, 22, 2] by [0, 11, 1]. See Figure 2b.

   Note: be sure to set $n$Min = 1 and $n$Max = 20 in the WINDOW settings.

   The moth population increases and then oscillates until it settles to a constant number (about 9.737 thousand).

[0, 22, 2] by [0, 11, 1]

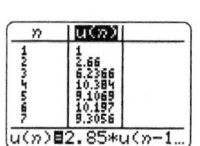      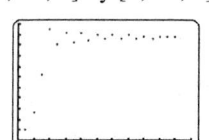

            Figure 2a                    Figure 2b

3. (a)   $\pi \approx \left[ 90 \left( \dfrac{1}{1^4} + \dfrac{1}{2^4} + \dfrac{1}{3^4} + \dfrac{1}{4^4} \right) \right]^{1/4} \approx 3.138997889$

   (b)   $\pi \approx 3.141590776$; This is correct to 5 decimal places.

4. (a)   $S = \dfrac{2}{1 - \left( -\frac{1}{2} \right)} = \dfrac{2}{\frac{3}{2}} = \dfrac{4}{3}$

   (b)   $S = \dfrac{1}{1 - \left( \frac{1}{3} \right)} = \dfrac{1}{\frac{2}{3}} = \dfrac{3}{2}$

   (c)   $S = \dfrac{0.1}{1 - (0.1)} = \dfrac{0.1}{0.9} = \dfrac{1}{9} = 0.\overline{1}$

   (d)   $S = \dfrac{0.12}{1 - (0.01)} = \dfrac{0.12}{0.99} = \dfrac{4}{33} = 0.\overline{12}$

## Chapters 1-14 Cumulative Review Exercises

1. This equation illustrates a distributive property.

2. The domain corresponds to the $x$-coordinates and the range corresponds to the $y$-coordinates.

   $$D = \{-6, -2, 0, 2\}; \; R = \{0, 1, 3, 5\}$$

3. $\dfrac{x^{-2}y^3}{\left(3xy^{-2}\right)^3} = \dfrac{x^{-2}y^3}{3^3 x^3 \left(y^{-2}\right)^3} = \dfrac{x^{-2}y^3}{27x^3 y^{-6}} = \dfrac{1}{27}x^{-2-3}y^{3-(-6)} = \dfrac{1}{27}x^{-5}y^9 = \dfrac{y^9}{27x^5}$

4. $\left(\dfrac{3b}{6a^2}\right)^{-4} = \left(\dfrac{b}{2a^2}\right)^{-4} = \left(\dfrac{2a^2}{b}\right)^4 = \dfrac{\left(2a^2\right)^4}{b^4} = \dfrac{2^4\left(a^2\right)^4}{b^4} = \dfrac{16a^8}{b^4}$

5. $\left(\dfrac{1}{z^2}\right)^{-5} = \left(z^2\right)^5 = z^{10}$

6. $\dfrac{8x^{-3}y^2}{4x^3 y^{-1}} = 2x^{-3-3}y^{2-(-1)} = 2x^{-6}y^3 = \dfrac{2y^3}{x^6}$

7. The function is defined for all values of the variable except 8. The domain is $\{x \mid x \neq 8\}$.

8. The slope is $m = \dfrac{1-5}{0-(-2)} = \dfrac{-4}{2} = -2$. Since $f(x) = 1$ when $x = 0$ the $y$-intercept is 1.

   Here $f(x) = -2x + 1$.

9. Horizontal lines have equations of the form $y = k$. The equation of the horizontal line passing

   through $(2, 3)$ is $y = 3$.

10. The equation is in the form $f(x) = mx + b$. The slope is $-3$ and the $y$-intercept is 5.

11. Since the line is perpendicular to $y = -\dfrac{2}{3}x - 4$, the slope is $m = \dfrac{3}{2}$. Using the point-slope form

   gives $y = \dfrac{3}{2}(x-1) + 4 \Rightarrow y = \dfrac{3}{2}x - \dfrac{3}{2} + 4 \Rightarrow y = \dfrac{3}{2}x + \dfrac{5}{2}$

12. Since the line is parallel to $y = 2x - 7$, the slope is $m = 2$. Using the point-slope form gives

   $y = 2(x-5) + 2 \Rightarrow y = 2x - 10 + 2 \Rightarrow y = 2x - 8$

13. $\dfrac{2}{5}(x-4) = -12 \Rightarrow x - 4 = -30 \Rightarrow x = -26$

14. $\dfrac{2}{5}z + \dfrac{1}{4}z > 2 - (z-1) \Rightarrow 8z + 5z > 40 - 20(z-1) \Rightarrow 13z > 40 - 20z + 20 \Rightarrow$

   $13z > 60 - 20z \Rightarrow 33z > 60 \Rightarrow z > \dfrac{60}{33} \Rightarrow z > \dfrac{20}{11}$. The interval is $\left(\dfrac{20}{11}, \infty\right)$.

15. First divide each side of $-3|t-5| \le -18$ by $-3$ to obtain $|t-5| \ge 6$.

The solutions to $|t-5| \ge 6$ satisfy $t \le c$ or $t \ge d$ where $c$ and $d$ are the solutions to $|t-5| = 6$.

$|t-5| = 6$ is equivalent to $t-5 = -6 \Rightarrow t = -1$ and $t-5 = 6 \Rightarrow t = 11$.

The interval is $(-\infty, -1] \cup [11, \infty)$.

16. $\left|4 + \dfrac{2}{3}x\right| = 6 \Rightarrow 4 + \dfrac{2}{3}x = -6 \Rightarrow \dfrac{2}{3}x = -10 \Rightarrow x = -15$ or $4 + \dfrac{2}{3}x = 6 \Rightarrow \dfrac{2}{3}x = 2 \Rightarrow x = 3$

17. $\dfrac{1}{4}t - (2t+5) + 6 = \dfrac{t+3}{4} \Rightarrow \dfrac{1}{4}t - 2t + 1 = \dfrac{t+3}{4} \Rightarrow t - 8t + 4 = t + 3 \Rightarrow -8t = -1 \Rightarrow t = \dfrac{1}{8}$

18. $-3 \le \dfrac{2}{3}x + 5 < 11 \Rightarrow -8 \le \dfrac{2}{3}x < 6 \Rightarrow -12 \le x < 9$. The interval is $[-12, 9)$.

19. By substitution, $(3, -2)$ is a solution to the given system of equations.

20. See Figure 20.

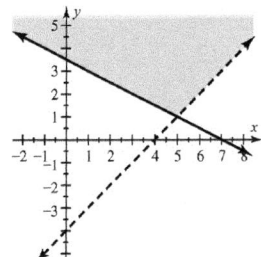

Figure 20

21. Multiply the first equation by 2 and add the equations to eliminate the variable $x$.

$\begin{array}{l} 2x - 4y = 2 \\ -2x + 7y = 4 \\ \hline 3y = 6 \end{array}$ Thus, $y = 2$. And so $x - 2(2) = 1 \Rightarrow x = 5$. The solution is $(5, 2)$.

22. $\begin{bmatrix} 1 & 1 & 1 & 5 \\ -2 & -1 & 1 & -10 \\ 1 & 2 & 8 & 1 \end{bmatrix} \begin{array}{l} \\ R_2 + 2R_1 \rightarrow \\ R_3 - R_1 \rightarrow \end{array} \begin{bmatrix} 1 & 1 & 1 & 5 \\ 0 & 1 & 3 & 0 \\ 0 & 1 & 7 & -4 \end{bmatrix} \begin{array}{l} R_1 - R_2 \rightarrow \\ \\ R_3 - R_2 \rightarrow \end{array} \begin{bmatrix} 1 & 0 & -2 & 5 \\ 0 & 1 & 3 & 0 \\ 0 & 0 & 4 & -4 \end{bmatrix}$

$\begin{array}{l} \\ \\ (1/4)R_3 \rightarrow \end{array} \begin{bmatrix} 1 & 0 & -2 & 5 \\ 0 & 1 & 3 & 0 \\ 0 & 0 & 1 & -1 \end{bmatrix} \begin{array}{l} R_1 + 2R_3 \rightarrow \\ R_2 - 3R_3 \rightarrow \\ \end{array} \begin{bmatrix} 1 & 0 & 0 & 3 \\ 0 & 1 & 0 & 3 \\ 0 & 0 & 1 & -1 \end{bmatrix}$ The solution is $(3, 3, -1)$.

23. From the graph of the region of feasible solutions (not shown), the vertices are $(0, 0)$, $(0, 2.5)$, $(2, 2)$, and $(2.5, 0)$. The maximum value of $R$ occurs at one of the vertices. For $(0, 0)$, $R = 3(0) + 8(0) = 0$.

For $(0, 2.5)$, $R = 3(0) + 8(2.5) = 20$. For $(2, 2)$, $R = 3(2) + 8(2) = 22$.

For $(2.5, 0)$, $R = 3(2.5) + 8(0) = 7.5$. The maximum value is $R = 22$.

24. $\det A = 4(2) - 3(-3) = 8 + 9 = 17$

25. $2x^3\left(4x^4 - 3x^3 + 5\right) = 8x^7 - 6x^6 + 10x^3$

26. $(2z - 7)(3z + 4) = 6z^2 + 8z - 21z - 28 = 6z^2 - 13z - 28$

27. $4x^2 - 9y^2 = (2x)^2 - (3y)^2 = (2x - 3y)(2x + 3y)$

28. $2a^3 - a^2 + 8a - 4 = a^2(2a - 1) + 4(2a - 1) = \left(a^2 + 4\right)(2a - 1)$

29. $4x^2 - x - 3 = 0 \Rightarrow (4x + 3)(x - 1) = 0 \Rightarrow x = -\dfrac{3}{4}$ or $x = 1$

30. $x^4 - 10x^3 = -24x^2 \Rightarrow x^4 - 10x^3 + 24x^2 = 0 \Rightarrow x^2(x - 4)(x - 6) = 0 \Rightarrow x = 0, 4,$ or $6$

31. $\dfrac{x^2 - 7x + 10}{x^2 - 25} \cdot \dfrac{x + 5}{x + 1} = \dfrac{(x - 2)(x - 5)(x + 5)}{(x - 5)(x + 5)(x + 1)} = \dfrac{x - 2}{x + 1}$

32. $\dfrac{x^2 + 7x + 12}{x^2 - 9} \div \dfrac{x^2 - 5x + 6}{(x - 3)^2} = \dfrac{x^2 + 7x + 12}{x^2 - 9} \cdot \dfrac{(x - 3)^2}{x^2 - 5x + 6} = \dfrac{(x + 4)(x + 3)(x - 3)(x - 3)}{(x - 3)(x + 3)(x - 3)(x - 2)} = \dfrac{x + 4}{x - 2}$

33. $\dfrac{2}{x + 5} = \dfrac{-3}{x^2 - 25} + \dfrac{1}{x - 5} \Rightarrow 2(x - 5) = -3 + 1(x + 5) \Rightarrow 2x - 10 = x + 2 \Rightarrow x = 12$

34. $\dfrac{2y}{y^2 - 3y + 2} = \dfrac{1}{y - 2} + 2 \Rightarrow 2y = 1(y - 1) + 2(y - 2)(y - 1) \Rightarrow 2y = y - 1 + 2y^2 - 6y + 4 \Rightarrow$

$2y^2 - 7y + 3 = 0 \Rightarrow (2y - 1)(y - 3) = 0 \Rightarrow y = \dfrac{1}{2}$ or $y = 3$.

35. $R = \dfrac{3C - 2W}{5} \Rightarrow 5R = 3C - 2W \Rightarrow 5R - 3C = -2W \Rightarrow \dfrac{5R - 3C}{-2} = W \Rightarrow W = \dfrac{3C - 5R}{2}$

36. $\dfrac{\dfrac{1}{x^2} + \dfrac{2}{x}}{\dfrac{1}{x^2} - \dfrac{4}{x}} = \dfrac{\dfrac{1}{x^2} + \dfrac{2}{x}}{\dfrac{1}{x^2} - \dfrac{4}{x}} \cdot \dfrac{x^2}{x^2} = \dfrac{1 + 2x}{1 - 4x}$

37. $\sqrt[3]{x^4 y^4} - 2\sqrt[3]{xy} = \sqrt[3]{(xy)^3 \cdot xy} - 2\sqrt[3]{xy} = xy\sqrt[3]{xy} - 2\sqrt[3]{xy} = (xy - 2)\sqrt[3]{xy}$

38. $\left(4 + \sqrt{2}\right)\left(4 - \sqrt{2}\right) = 4^2 - \left(\sqrt{2}\right)^2 = 16 - 2 = 14$

39. $8(x - 3)^2 = 200 \Rightarrow (x - 3)^2 = 25 \Rightarrow x - 3 = \pm\sqrt{25} \Rightarrow x - 3 = \pm 5 \Rightarrow x = -2$ or $8$

40. $3\sqrt{2x + 6} = 6x \Rightarrow \sqrt{2x + 6} = 2x \Rightarrow 2x + 6 = 4x^2 \Rightarrow 4x^2 - 2x - 6 = 0 \Rightarrow 2(2x - 3)(x + 1) \Rightarrow$

$x = \dfrac{3}{2}$ or $x = -1$. The value $x = -1$ does not check. The only solution is $\dfrac{3}{2}$.

41. $(-3 + i)(-4 - 2i) = 12 + 6i - 4i - 2i^2 = 12 + 2i + 2 = 14 + 2i$

42. $\dfrac{2 - 6i}{1 + 2i} = \dfrac{2 - 6i}{1 + 2i} \cdot \dfrac{1 - 2i}{1 - 2i} = \dfrac{2 - 4i - 6i + 12i^2}{1 - 4i^2} = \dfrac{2 - 10i - 12}{1 + 4} = \dfrac{-10 - 10i}{5} = -2 - 2i$

43. $-\dfrac{b}{2a} = -\dfrac{8}{2(3)} = -\dfrac{8}{6} = -\dfrac{4}{3}$; $f\left(-\dfrac{4}{3}\right) = 3\left(-\dfrac{4}{3}\right)^2 + 8\left(-\dfrac{4}{3}\right) + 5 = -\dfrac{1}{3}$. The vertex is $\left(-\dfrac{4}{3}, -\dfrac{1}{3}\right)$.

The minimum value is $-\dfrac{1}{3}$.

44. $y = 2x^2 + 8x + 17 \Rightarrow y = 2\left(x^2 + 4x + 4\right) + 17 - 8 \Rightarrow y = 2\left(x+2\right)^2 + 9$. The vertex is $(-2, 9)$.

45. $x^2 - 4x + 13 = 0 \Rightarrow x^2 - 4x + 4 = -13 + 4 \Rightarrow (x-2)^2 = -9 \Rightarrow x - 2 = \pm\sqrt{-9} \Rightarrow x = 2 \pm 3i$

46. $z^2 - 4z = 32 \Rightarrow z^2 - 4z - 32 = 0 \Rightarrow (z+4)(z-8) = 0 \Rightarrow z = -4$ or $z = 8$

47. (a)  The graph intersects the $x$-axis at $-3$ and $1$.

   (b)  Because the parabola opens upward, $a > 0$.

   (c)  Because there are two real solutions, the discriminant is positive.

48. $x^2 + 2x - 3 = 0 \Rightarrow (x+3)(x-1) = 0 \Rightarrow x = -3, 1$.  Because the parabola opens up the

interval is $(-3, 1)$.

49. (a)  $g(-2) = 3(-2) - 2 = -8$, then $(f \circ g)(-2) = f\left(g(-2)\right) = f(-8) = (-8)^2 + 1 = 65$

   (b)  $(g \circ f)(x) = g\left(f(x)\right) = g\left(x^2 + 1\right) = 3\left(x^2 + 1\right) - 2 = 3x^2 + 3 - 2 = 3x^2 + 1$

50. $f(x) = \dfrac{3x+1}{2} \Rightarrow y = \dfrac{3x+1}{2}$, interchange $x$ and $y$ and solve for $y$.

$x = \dfrac{3y+1}{2} \Rightarrow 2x = 3y + 1 \Rightarrow 3y = 2x - 1 \Rightarrow y = \dfrac{2x-1}{3} \Rightarrow f^{-1}(x) = \dfrac{2x-1}{3}$

51. $\ln\left(x^3 \sqrt{y}\right) = \ln\left(x^3 y^{1/2}\right) = \ln x^3 + \ln y^{1/2} = 3\ln x + \dfrac{1}{2}\ln y$

52. $2\log x - \log 4xy = \log x^2 - \log 4xy = \log \dfrac{x^2}{4xy} = \log \dfrac{x}{4y}$

53. $8\log x + 3 = 17 \Rightarrow 8\log x = 14 \Rightarrow \log x = \dfrac{7}{4} \Rightarrow 10^{\log x} = 10^{7/4} \Rightarrow x \approx 56.23$

54. $4^{2x} = 5 \Rightarrow \log_4 4^{2x} = \log_4 5 \Rightarrow 2x = \log_4 5 \Rightarrow 2x = \dfrac{\log 5}{\log 4} \Rightarrow x = \dfrac{\log 5}{2\log 4} \approx 0.58$

55. See Figure 55.  The vertex is $(1, 3)$, and the axis of symmetry is $y = 3$.

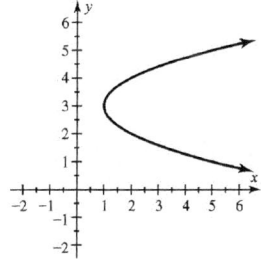

Figure 55

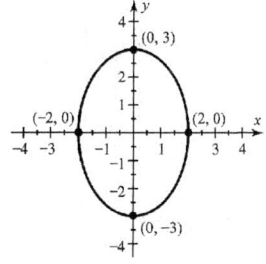

Figure 57

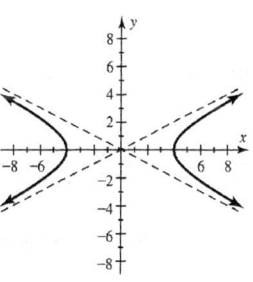

Figure 58

56. $x^2 - 6x + y^2 + 2y = -6 \Rightarrow \left(x^2 - 6x + 9\right) + \left(y^2 + 2y + 1\right) = -6 + 9 + 1 \Rightarrow (x-3)^2 + (y+1)^2 = 4$

   The center is $(3, -1)$, and the radius is 2.

57. See Figure 57.

58. See Figure 58.

59. Vertical transverse axis with vertices $(0, \pm 2)$ and asymptotes $y = \pm \dfrac{2}{4} x \Rightarrow \dfrac{y^2}{4} - \dfrac{x^2}{16} = 1$

60. Horizontal major axis with vertices $(\pm 4, 0)$ and minor axis endpoints $(0, \pm 2) \Rightarrow \dfrac{x^2}{16} + \dfrac{y^2}{4} = 1$

61. Substitute $y = x^2 + 1$ in the second equation and solve for $x$.

   $x^2 + 2\left(x^2 + 1\right) = 5 \Rightarrow 3x^2 - 3 = 0 \Rightarrow 3\left(x^2 - 1\right) = 0 \Rightarrow 3(x+1)(x-1) = 0 \Rightarrow x = -1 \text{ or } x = 1$

   When $x = -1$, $y = (-1)^2 + 1 = 2$. When $x = 1$, $y = (1)^2 + 1 = 2$. The solutions are $(-1, 2)$ and $(1, 2)$.

62. See Figure 62.

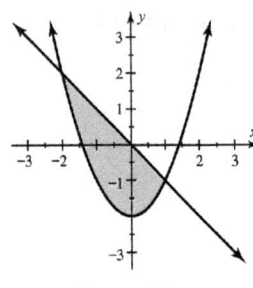
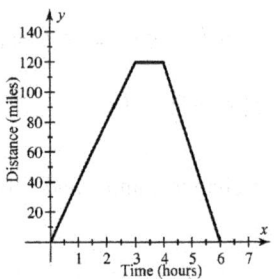

Figure 62                               Figure 74

63. This sequence is arithmetic, the common difference is $-2$.

64. This sequence is geometric, the common ratio is $0.2$.

65. This sequence is geometric, the common ratio is $4$.

66. This sequence is arithmetic, the common difference is $6$.

67. Note: $d = 5 - 2 = 3$, thus $a_n = 2 + (n-1)(3) \Rightarrow a_n = 2 + 3n - 3 \Rightarrow a_n = 3n - 1$

68. Note: $r = \dfrac{12}{4} = 3$, thus $a_n = 4(3)^{n-1}$

69. $9\left(\dfrac{3+35}{2}\right) = 9(19) = 171$

70. Here $r = -2$ and $n = 11$ thus $S_{11} = 1\left(\dfrac{1-(-2)^{11}}{1-(-2)}\right) = 1\left(\dfrac{2049}{3}\right) = 683$

71. $(2x+3)^4 = \left(_4C_0\right)(2x)^4 + \left(_4C_1\right)(2x)^3(3) + \left(_4C_2\right)(2x)^2\left(3^2\right) + \left(_4C_3\right)(2x)\left(3^3\right) + \left(_4C_4\right)\left(3^4\right) \Rightarrow$

   $(2x+3)^4 = 16x^4 + 96x^3 + 216x^2 + 216x + 81$

72. $(2a-5b)^3 = (_3C_0)(2a)^3 - (_3C_1)(2a)^2(5b) + (_3C_2)(2a)(5b)^2 - (_3C_3)(5b)^3 \Rightarrow$

$(2a-5b)^3 = 8a^3 - 60a^2b + 150ab^2 - 125b^3$

73. $r = \sqrt{\dfrac{14}{\pi}} \approx 2.11$ inches

74. See Figure 74.

75. (a) $f(x) = \dfrac{170}{2} + 0.4x \Rightarrow f(x) = 0.4x + 85$

   (b) $0.4(90) + \dfrac{W}{2} = 130 \Rightarrow 36 + \dfrac{W}{2} = 130 \Rightarrow \dfrac{W}{2} = 94 \Rightarrow W = 188$ pounds

76. Let $x$ and $y$ represent the speed of the airplane and the speed of the wind respectively. Then the system needed is $x - y = 360$ and $x + y = 400$. Adding the two equations will eliminate $y$.

   $x - y = 360$
   $\underline{x + y = 400}$ Thus, $x = 380$. And so $(380) + y = 400 \Rightarrow y = 20$.
   $2x = 760$

   The speed of the airplane is 380 mph and the speed of the wind is 20 mph.

77. Let $x$ represent the width of the tent floor. Then $2x - 6$ represents the length.

   $x(2x-6) = 108 \Rightarrow 2x^2 - 6x - 108 = 0 \Rightarrow x^2 - 3x - 54 = 0 \Rightarrow (x+6)(x-9) = 0 \Rightarrow x = -6$ or $9$

   Since the value $x = -6$ has no physical meaning, the solution is 9. The dimensions are 9 feet by 12 feet.

78. Let $x$ represent time required to weed the garden if they worked together. Then $\dfrac{x}{60} + \dfrac{x}{90} = 1$.

   $180 \cdot \left(\dfrac{x}{60} + \dfrac{x}{90}\right) = 1 \cdot 180 \Rightarrow 3x + 2x = 180 \Rightarrow 5x = 180 \Rightarrow x = \dfrac{180}{5} = 36$ minutes

79. (a) $xy = 96$ and $3x - y = 12$

   (b) Solve the second equation for $y$ and substitute the result in the first equation. $3x - y = 12 \Rightarrow y = 3x - 12$

   $x(3x-12) = 96 \Rightarrow 3x^2 - 12x - 96 = 0 \Rightarrow x^2 - 4x - 32 = 0 \Rightarrow (x+4)(x-8) = 0 \Rightarrow$

   $x = -4$ or $x = 8$. Since the numbers must be positive, the solution is $x = 8$. The numbers are 8 and 12.

80. Find the sum of the series $1 + 3 + 5 + \cdots + 23$. The sum is $12\left(\dfrac{1+23}{2}\right) = 12(12) = 144$ musicians.